Lecture Notes in Computer Science   10328

Commenced Publication in 1973
Founding and Former Series Editors:
Gerhard Goos, Juris Hartmanis, and Jan van Leeuwen

Editorial Board

More information about this series at http://www.springer.com/series/7407

Friedrich Eisenbrand · Jochen Koenemann (Eds.)

Integer Programming and Combinatorial Optimization

19th International Conference, IPCO 2017
Waterloo, ON, Canada, June 26–28, 2017
Proceedings

 Springer

Editors
Friedrich Eisenbrand
Ecole Polytechnique Federale
 de Lausanne - EPFL
Lausanne
Switzerland

Jochen Koenemann
University of Waterloo
Waterloo, ON
Canada

ISSN 0302-9743   ISSN 1611-3349 (electronic)
Lecture Notes in Computer Science
ISBN 978-3-319-59249-7   ISBN 978-3-319-59250-3   (eBook)
DOI 10.1007/978-3-319-59250-3

Library of Congress Control Number: 2017941547

LNCS Sublibrary: SL1 – Theoretical Computer Science and General Issues

Printed on acid-free paper

This Springer imprint is published by Springer Nature
The registered company is Springer International Publishing AG
The registered company address is: Gewerbestrasse 11, 6330 Cham, Switzerland

Preface

This volume contains the 36 extended abstracts presented at IPCO 2017, the 19th Conference on Integer Programming and Combinatorial Optimization, held during June 26–28, 2017, in Waterloo, Canada. IPCO is under the auspices of the Mathematical Optimization Society. The first IPCO conference took place at the University of Waterloo in May 1990 and it returned to Waterloo for the first time this year. IPCO is held every year, except for those in which the International Symposium on Mathematical Programming is held.

The conference had a Program Committee consisting of 14 members. In response to the call for papers, we received 125 submissions, of which three were withdrawn prior to the decision progress. The Program Committee met in Leysin, Switzerland, in January 2017. Each submission was reviewed by at least three Program Committee members. There were many high-quality submissions, of which the committee selected 36 to appear in the conference proceedings. We expect the full versions of the extended abstracts appearing here to be submitted for publication in refereed journals.

This year, IPCO was preceded by a Summer School during June 24–25, 2017, with lectures by Sanjeeb Dash, Anupam Gupta, and Aleksander Madry. We would like to thank:

- The authors who submitted their research to IPCO
- The members of the Program Committee, who spent much time and energy reviewing the submissions
- The expert additional reviewers whose opinion was crucial in the paper selection
- The members of the local Organizing Committee, who made this conference possible

April 2017

Friedrich Eisenbrand
Jochen Koenemann

Organization

Program Committee

Nikhil Bansal	Eindhoven University of Technology, The Netherlands
Gérard Cornuéjols	Carnegie Mellon University, USA
Daniel Dadush	Centrum Wiskunde and Informatica, The Netherlands
Santanu S. Dey	Georgia Institute of Technology, USA
Friedrich Eisenbrand	EPFL, Switzerland
Samuel Fiorini	Université Libre de Bruxelles, Belgium
Anupam Gupta	Carnegie Mellon University, USA
Satoru Iwata	University of Tokyo, Japan
Jochen Koenemann	University of Waterloo, Canada
Kurt Mehlhorn	MPI Informatik, Germany
Seffi Naor	Technion, Haifa, Israel
Britta Peis	RWTH Aachen, Germany
Laura Sanità	University of Waterloo, Canada
Laurence Wolsey	CORE, Université catholique de Louvain, Belgium
Rico Zenklusen	ETH Zurich, Switzerland

Additional Reviewers

Abdi, Ahmad
Aboulker, Pierre
Adjiashvili, David
Ahmed, Shabbir
Althaus, Ernst
Antoniadis, Antonios
Azar, Yossi
Bar-Noy, Amotz
Barvinkok, Alexander
Basu, Amitabh
Bei, Xiaohui
Benchetrit, Yohann
Blekherman, Greg
Bonifaci, Vincenzo
Bonomo, Flavia
Boros, Endre
Buchbinder, Niv
Byrka, Jaroslaw
Bérczi, Kristóf

Cevallos, Alfonso
Chakrabarty, Deeparnab
Chalermsook, Parinya
Cheriyan, Joe
Chestnut, Stephen
Chubanov, Sergei
Conforti, Michele
Cornuejols, Gerard
Cseh, Ágnes
Cygan, Marek
Dash, Sanjeeb
Di Summa, Marco
Ene, Alina
Even, Guy
Faenza, Yuri
Fairstein, Yaron
Farczadi, Linda
Feldman, Moran
Filmus, Yuval

Fleiner, Tamas
Fukasawa, Ricardo
Fukunaga, Takuro
Gairing, Martin
Garg, Naveen
Gaubert, Stephane
Goldner, Kira
Gottschalk, Corinna
Grandoni, Fabrizio
Gribling, Sander
Groß, Martin
Gunluk, Oktay
Gupta, Swati
Gupte, Akshay
Guzman, Cristobal
Hansen, Thomas Dueholm
Harks, Tobias
Hartvigsen, David
Hassin, Refael

Hildebrand, Robert
Hirai, Hiroshi
Hoefer, Martin
Hosten, Serkan
Huang, Chien-Chung
Huchette, Joey
Huynh, Tony
Im, Sungjin
Jansen, Klaus
Jerrum, Mark
Joret, Gwenaël
Kaibel, Volker
Kakimura, Naonori
Kalaitzis, Christos
Kamiyama, Naoyuki
Kaplan, Haim
Kapralov, Michael
Kesselheim, Thomas
Kimura, Kei
Klimm, Max
Koeppe, Matthias
Krishnaswamy,
  Ravishankar
Kuhn, Daniel
Kumar, Amit
Kutiel, Gilad
Lau, Lap Chi
Lee, Yin Tat
Letchford, Adam
Li, Shi
Linderoth, Jeff
Lokshtanov, Daniel
Marchetti-Spaccamela,
  Alberto
Matuschke, Jannik
McCormick, Tom

Mccormick, Tom
Megow, Nicole
Mehta, Aranyak
Mestre, Julian
Miyazaki, Shuichi
Mnich, Matthias
Mohar, Bojan
Molinaro, Marco
Murota, Kazuo
Mömke, Tobias
Nagano, Kiyohito
Nagarajan, Viswanath
Newman, Alantha
Nikolov, Aleksandar
Onak, Krzysztof
Oriolo, Gianpaolo
Panigrahi, Debmalya
Pashkovich, Kanstantsin
Pfetsch, Marc
Puleo, Gregory
Rabani, Yuval
Ralphs, Ted
Ravi, R.
Richard, Jean Philippe
Roytman, Alan
Röglin, Heiko
Salvagnin, Domenico
Schaudt, Oliver
Schwartz, Roy
Schweitzer, Pascal
Sebo, Andras
Seddighin, Saeed
Shigeno, Maiko
Singh, Mohit
Singla, Sahil
Sinnl, Markus

Skopalik, Alexander
Skutella, Martin
Soma, Tasuku
Stephens-Davidowitz,
  Noah
Straszak, Damian
Strehler, Martin
Svensson, Ola
Swamy, Chaitanya
Takazawa, Kenjiro
Talmon, Ohad
Tanigawa, Shin-Ichi
Toriello, Alejandro
Trick, Michael
Tönnis, Andreas
Uetz, Marc
Umboh, Seeun William
van Stee, Rob
Van Vyve, Mathieu
van Zuylen, Anke
Vegh, Laszlo
Ventura, Paolo
Verschae, José
Vielma, Juan Pablo
Vishnoi, Nisheeth
von Heymann, Frederik
Wajc, David
Weismantel, Robert
Weltge, Stefan
Wierz, Andreas
Wiese, Andreas
Wollan, Paul
Woods, Kevin
Xie, Weijun
Yu, Josephine
Zhou, Hang

Contents

X   Contents

The Two-Point Fano and Ideal Binary Clutters

Ahmad Abdi$^{(\boxtimes)}$ and Bertrand Guenin

Department of Combinatorics and Optimization,
University of Waterloo, Waterloo, Canada
{a3abdi,bguenin}@uwaterloo.ca

Abstract. Let $\mathbb{F}$ be a binary clutter. We prove that if $\mathbb{F}$ is non-ideal, then either $\mathbb{F}$ or its blocker $b(\mathbb{F})$ has one of $\mathbb{L}_7, \mathbb{O}_5, \mathbb{LC}_7$ as a minor. $\mathbb{L}_7$ is the non-ideal clutter of the lines of the Fano plane, $\mathbb{O}_5$ is the non-ideal clutter of odd circuits of the complete graph K_5, and the *two-point Fano* $\mathbb{LC}_7$ is the ideal clutter whose sets are the lines, and their complements, of the Fano plane that contain exactly one of two fixed points. In fact, we prove the following stronger statement: if $\mathbb{F}$ is a minimally non-ideal binary clutter different from $\mathbb{L}_7, \mathbb{O}_5, b(\mathbb{O}_5)$, then through every element, either $\mathbb{F}$ or $b(\mathbb{F})$ has a two-point Fano minor.

1 Introduction

Let E be a finite set. A *clutter* $\mathbb{F}$ over *ground set* $E(\mathbb{F}) := E$ is a family of subsets of E, where no subset is contained in another. We say that $\mathbb{F}$ is *binary* if the symmetric difference of any odd number of sets in $\mathbb{F}$ contains a set of $\mathbb{F}$. We say that $\mathbb{F}$ is *ideal* if the polyhedron

$$Q(\mathbb{F}) := \left\{ x \in \mathbb{R}_+^E : \sum(x_e : e \in C) \geq 1 \quad C \in \mathbb{F} \right\}$$

has only integral extreme points; otherwise it is *non-ideal*. When is a binary clutter ideal? We will be studying this question.

Let us describe some examples of ideal and non-ideal binary clutters. Given a graph G and distinct vertices s, t, the clutter of st-paths of G over the edge-set is binary. An immediate consequence of Menger's theorem [12], as well as Ford and Fulkerson's theorem [6], is that this binary clutter is ideal [3]. The clutter of *lines of the Fano plane*

$$\mathbb{L}_7 := \left\{ \{1,2,6\}, \{1,4,7\}, \{1,3,5\}, \{2,5,7\}, \{2,3,4\}, \{3,6,7\}, \{4,5,6\} \right\}$$

is binary, and it is non-ideal as $\left(\frac{1}{3}, \frac{1}{3}, \ldots, \frac{1}{3} \right)$ is an extreme point of $Q(\mathbb{L}_7)$. (See Fig. 1.) The clutter of odd circuits of K_5 over its ten edges, denoted $\mathbb{O}_5$, is also binary, and it is non-ideal as $\left(\frac{1}{3}, \frac{1}{3}, \ldots, \frac{1}{3} \right)$ is an extreme point of $Q(\mathbb{O}_5)$.

We say that two clutters are *isomorphic* if relabeling the ground set of one yields the other. There are two fundamental clutter operations that preserve being binary and ideal, let us describe them. The *blocker of* $\mathbb{F}$, denoted $b(\mathbb{F})$, is another clutter over the same ground set whose sets are the (inclusionwise)

© Springer International Publishing AG 2017
F. Eisenbrand and J. Koenemann (Eds.): IPCO 2017, LNCS 10328, pp. 1–12, 2017.
DOI: 10.1007/978-3-319-59250-3_1

minimal sets in $\{B \subseteq E : B \cap C \neq \emptyset \; \forall C \in \mathbb{F}\}$. It is well-known that $b(b(\mathbb{F})) = \mathbb{F}$ [5]. We may therefore call $\mathbb{F}, b(\mathbb{F})$ a *blocking* pair. A clutter $\mathbb{F}$ is binary if, and only if, $|B \cap C|$ is odd for all $B \in b(\mathbb{F})$ and $C \in \mathbb{F}$ [9]. Hence, if $\mathbb{F}$ is binary, then so is $b(\mathbb{F})$. Lehman's Width-Length Inequality shows that if $\mathbb{F}$ is ideal, then so is $b(\mathbb{F})$ [10]. In particular, since $\mathbb{L}_7$ and $\mathbb{O}_5$ are non-ideal, then so are $b(\mathbb{L}_7) = \mathbb{L}_7$ and $b(\mathbb{O}_5)$. Let I, J be disjoint subsets of E. Denote by $\mathbb{F}\backslash I/J$ the clutter over $E - (I \cup J)$ of minimal sets of $\{C - J : C \in \mathbb{F}, C \cap I = \emptyset\}$.[1] We say that $\mathbb{F}\backslash I/J$, and any clutter isomorphic to it, is a *minor of* $\mathbb{F}$ obtained after *deleting I* and *contracting J*. If $I \cup J \neq \emptyset$, then $\mathbb{F}\backslash I/J$ is a *proper* minor of $\mathbb{F}$. It is well-known that $b(\mathbb{F}\backslash I/J) = b(\mathbb{F})/I\backslash J$ [16]. If a clutter is binary, then so is every minor of it, and if a clutter is ideal, then so is every minor of it [17].

Let $\mathbb{F}$ be a binary clutter. Regrouping what we discussed, if $\mathbb{F}$ or $b(\mathbb{F})$ has one of $\mathbb{L}_7, \mathbb{O}_5$ as a minor, then it is non-ideal. Seymour [17] (p. 200) conjectures the converse is also true:

The flowing conjecture. Let $\mathbb{F}$ be a non-ideal binary clutter. Then $\mathbb{F}$ or $b(\mathbb{F})$ has one of $\mathbb{L}_7, \mathbb{O}_5$ as a minor.

The *two-point Fano clutter*, denoted by $\mathbb{LC}_7$, is the clutter over ground set $\{1, \ldots, 7\}$ whose sets are the lines, and their complements, of the Fano plane that intersect $\{1, 4\}$ exactly once, i.e. $\mathbb{LC}_7$ consists of $\{1, 2, 6\}, \{1, 3, 5\}, \{2, 3, 4\}, \{2, 5, 7\}$ and $\{3, 4, 5, 7\}, \{2, 4, 6, 7\}, \{1, 5, 6, 7\}, \{1, 3, 4, 6\}$. Observe that changing the two points $1, 4$ yields an isomorphic clutter. It can be readily checked that $\mathbb{LC}_7$ is binary *and* ideal. In this paper, we prove the following weakening of the flowing conjecture:

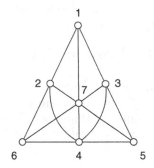

Fig. 1. The Fano plane

Theorem 1. *Let $\mathbb{F}$ be a non-ideal binary clutter. Then $\mathbb{F}$ or $b(\mathbb{F})$ has one of $\mathbb{L}_7, \mathbb{O}_5, \mathbb{LC}_7$ as a minor.*

What makes this result attractive is its relatively simple proof. The techniques used in the proof give hope of resolving the flowing conjecture. An interesting feature of the proof is the interplay between the clutter $\mathbb{F}$ and its blocker $b(\mathbb{F})$; if we fail to find one of the desired minors in the clutter, we switch to the blocker and find a desired minor there. Theorem 1 is a consequence of a stronger statement stated in the next section.

2   Preliminaries and the Main Theorem

2.1   Minimally Non-ideal Binary Clutters

A clutter is *minimally non-ideal (mni)* if it is non-ideal and every proper minor of it is ideal. Notice that every non-ideal clutter has an mni minor, and if a clutter

---

[1] Given sets A, B we denote by $A - B$ the set $\{a \in A : a \notin B\}$ and, for element a, we write $A - a$ instead of $A - \{a\}$.

is mni, then so is its blocker. Justified by this observation, instead of working with non-ideal binary clutters, we will work with mni binary clutters. The three clutters $\mathbb{L}_7, \mathbb{O}_5, b(\mathbb{O}_5)$ are mni, and the flowing conjecture predicts that these are the *only* mni binary clutters. We will need the following result of the authors:

Theorem 2 ([1]). $\mathbb{L}_7, \mathbb{O}_5$ *are the only mni binary clutters with a set of size 3.*

We will also need the following intermediate result of Alfred Lehman on mni clutters, stated only for binary clutters. Let $\mathbb{F}$ be a clutter over ground set E. Denote by $\bar{\mathbb{F}}$ the clutter of minimum size sets of $\mathbb{F}$. Denote by $M(\mathbb{F})$ the $0 - 1$ matrix whose columns are labeled by E and whose rows are the incidence vectors of the sets of $\mathbb{F}$. For an integer $r \geq 1$, a square $0 - 1$ matrix is *r-regular* if every row and every column has precisely r ones.

Theorem 3 ([2,11,15]). *Let $\mathbb{F}$ be an mni binary clutter where $n := |E(\mathbb{F})|$, and let $\mathbb{K} := b(\mathbb{F})$. Then*

(1) $M(\bar{\mathbb{F}})$ and $M(\bar{\mathbb{K}})$ are square and non-singular matrices,
(2) $M(\bar{\mathbb{F}})$ is r-regular and $M(\bar{\mathbb{K}})$ is s-regular, for some integers $r \geq 3$ and $s \geq 3$ such that $rs - n$ is even and $rs - n \geq 2$,
(3) after possibly permuting the rows of $M(\bar{\mathbb{K}})$, we have that

$$M(\bar{\mathbb{F}})M(\bar{\mathbb{K}})^\top = J + (rs - n)I = M(\bar{\mathbb{K}})^\top M(\bar{\mathbb{F}}).$$

Here, J denotes the all-ones matrix, and I the identity matrix. Given a ground set E and a set $C \subseteq E$, denote by $\chi_C \subseteq \{0,1\}^E$ the incidence vector of C. We will make use of the following corollary:

Corollary 4. *Let $\mathbb{F}$ be an mni binary clutter. Then the following statements hold:*

(1) For $C_1, C_2 \in \bar{\mathbb{F}}$, the only sets of $\mathbb{F}$ contained in $C_1 \cup C_2$ are C_1, C_2 [7,8].
(2) Choose $C_1, C_2, C_3 \in \bar{\mathbb{F}}$ and $e \in E(\mathbb{F})$ such that $C_1 \cap C_2 = C_2 \cap C_3 = C_3 \cap C_1 = \{e\}$. If C, C' are sets of $\mathbb{F}$ such that $C \cup C' \subseteq C_1 \cup C_2 \cup C_3$ and $C \cap C' \subseteq \{e\}$, then $\{C, C'\} = \{C_i, C_j\}$ for some distinct $i, j \in \{1, 2, 3\}$.

Proof. **(2)** Denote by r the minimum size of a set in $\mathbb{F}$. As $\mathbb{F}$ is binary, $C_1 \triangle C_2 \triangle C_3 \triangle C \triangle C'$ contains another set C'' of $\mathbb{F}$. Notice that $C'' \cap C \subseteq \{e\}$ and $C'' \cap C' \subseteq \{e\}$. If k many of C, C', C'' contain e, then

$$3r - 3 = |(C_1 \cup C_2 \cup C_3) - e| \geq |(C \cup C' \cup C'') - e| = |C| + |C'| + |C''| - k \geq 3r - k,$$

implying in turn that $k = 3$ and equality must hold throughout. In particular, $C, C', C'' \in \bar{\mathbb{F}}$ and $\chi_{C_1} + \chi_{C_2} + \chi_{C_3} = \chi_C + \chi_{C'} + \chi_{C''}$, so as $M(\bar{\mathbb{F}})$ is non-singular by Theorem 3 (1), we get that $\{C_1, C_2, C_3\} = \{C, C', C''\}$. □

2.2   Signed Matroids

All matroids considered in this paper are binary; we follow the notation used in Oxley [14]. Let M be a matroid over ground set E. Recall that a circuit is a minimal dependent set of M and a cocircuit is a minimal dependent set of the dual M^*. A *cycle* is the symmetric difference of circuits, and a *cocycle* is the symmetric difference of cocircuits. It is well-known that a nonempty cycle is a disjoint union of circuits ([14], Theorem 9.1.2). Let $\Sigma \subseteq E$. The pair (M, Σ) is called a *signed matroid* over ground set E. An *odd circuit of* (M, Σ) is a circuit C of M such that $|C \cap \Sigma|$ is odd.

Proposition 5 ([9,13], **also see** [4]). *The clutter of odd circuits of a signed matroid is binary. Conversely, a binary clutter is the clutter of odd circuits of a signed matroid.*

A *representation* of a binary clutter $\mathbb{F}$ is a signed matroid whose clutter of odd circuits is $\mathbb{F}$. By the preceding proposition, every binary clutter has a representation. For instance, $\mathbb{L}_7$ is represented as $(F_7, E(F_7))$, where F_7 is the Fano matroid. A *signature of* (M, Σ) is any subset of the form $\Sigma \triangle D$, where D is a cocycle of M; to *resign* is to replace (M, Σ) by $(M, \Sigma \triangle D)$. Notice that resigning does not change the family of odd cycles. We say that two signed matroids are *isomorphic* if one can be obtained from the other after a relabeling of the ground set and a resigning.

Remark 6. *Take an arbitrary element ω of F_7. Then $(F_7, E(F_7) - \omega)$ represents $\mathbb{LC}_7$.*

Proof. Suppose $E(F_7) = \{1, \ldots, 7\}$, and since F_7 is transitive, we may assume that $\omega = 7$. Consider the following representation of F_7,

$$\begin{pmatrix} 1\,0\,0\,0\,1\,1\,1 \\ 0\,1\,0\,1\,0\,1\,1 \\ 0\,0\,1\,1\,1\,0\,1 \end{pmatrix}$$

where the columns are labeled $1, \ldots, 7$ from left to right. Since $\{2, 3, 5, 6\}$ is a cocycle of F_7, $(F_7, \{1, \ldots, 6\})$ is isomorphic to $(F_7, \{1, \ldots, 6\} \triangle \{2, 3, 5, 6\}) = (F_7, \{1, 4\})$. It can be readily checked that the odd circuits of $(F_7, \{1, 4\})$ are precisely the sets of $\mathbb{LC}_7$, thereby proving the remark. □

Proposition 7 ([9,13], **also see** [8]). *In a signed matroid, the clutter of minimal signatures is the blocker of the clutter of odd circuits.*

Let I, J be disjoint subsets of E. The *minor* $(M, \Sigma) \backslash I / J$ obtained after *deleting* I and *contracting* J is the signed matroid defined as follows: if J contains an odd circuit, then $(M, \Sigma) \backslash I / J := (M \backslash I / J, \emptyset)$, and if J does not contain an odd circuit, then there is a signature Σ' of (M, Σ) disjoint from J by the preceding proposition, and we let $(M, \Sigma) \backslash I / J := (M \backslash I / J, \Sigma' - I)$. Observe that minors are defined up to resigning.

Proposition 8 ([13], **also see** [4]). *Let $\mathbb{F}$ be a binary clutter represented as (M, Σ), and take disjoint $I, J \subseteq E(\mathbb{F})$. Then $\mathbb{F} \backslash I / J$ is represented as $(M, \Sigma) \backslash I / J$.*

2.3 Hubs and the Main Theorem

Let (M, Σ) be a signed matroid, and take $e \in E(M)$. An *e-hub of* (M, Σ) is a triple (C_1, C_2, C_3) satisfying the following conditions:

(h1) C_1, C_2, C_3 are odd circuits such that, for distinct $i, j \in \{1, 2, 3\}$, $C_i \cap C_j = \{e\}$,

(h2) for distinct $i, j \in \{1, 2, 3\}$, the only nonempty cycles contained in $C_i \cup C_j$ are $C_i, C_j, C_i \triangle C_j$,

(h3) a cycle contained in $C_1 \cup C_2 \cup C_3$ is odd if and only if it contains e.

A *strict e-hub* is an e-hub (C_1, C_2, C_3) such that the following holds:

(h4) if C, C' are odd cycles contained in $C_1 \cup C_2 \cup C_3$ such that $C \cap C' = \{e\}$, then for some distinct $i, j \in \{1, 2, 3\}$, $\{C, C'\} = \{C_i, C_j\}$.

Given $I \subseteq E$, denote by $M|I$ the minor $M \setminus (E - I)$, and by $(M, \Sigma)|I$ the minor $(M, \Sigma) \setminus (E - I)$. The following is the main result of the paper:

Theorem 9. *Let* $\mathbb{F}, \mathbb{K}$ *be a blocking pair of mni binary clutters over ground set* E, *neither of which has a set of size 3. Let* (M, Σ) *represent* $\mathbb{F}$ *and let* (N, Γ) *represent* $\mathbb{K}$. *Then, for a given* $e \in E$, *the following statements hold:*

(1) (M, Σ) *has a strict e-hub* (C_1, C_2, C_3) *and* (N, Γ) *has a strict e-hub* (B_1, B_2, B_3) *where for* $i, j \in \{1, 2, 3\}$,

$$|C_i \cap B_j| \begin{cases} \geq 3 & \text{if } i = j \\ = 1 & \text{if } i \neq j, \end{cases}$$

(2) either $M|(C_1 \cup C_2 \cup C_3)$ *or* $N|(B_1 \cup B_2 \cup B_3)$ *is non-graphic,*

(3) if $M|(C_1 \cup C_2 \cup C_3)$ *is non-graphic, then* $(M, \Sigma) \setminus I/J \cong (F_7, E(F_7) - \omega)$ *for some disjoint* $I, J \subseteq E - e$, *and similarly, if* $N|(B_1 \cup B_2 \cup B_3)$ *is non-graphic, then* $(N, \Gamma) \setminus I/J \cong (F_7, E(F_7) - \omega)$ *for some disjoint* $I, J \subseteq E - e$.

Given this result, let us prove Theorem 1:

Proof (of Theorem 1). Let $\mathbb{F}$ be a non-ideal binary clutter, let $\mathbb{F}'$ be an mni minor of $\mathbb{F}$, and let $\mathbb{K}' := b(\mathbb{F}')$. If $\mathbb{F}'$ has a set of size 3, then by Theorem 2, $\mathbb{F}' \cong \mathbb{L}_7$ or $\mathbb{O}_5$. If $\mathbb{K}'$ has a set of size 3, then by Theorem 2, $\mathbb{K}' \cong \mathbb{L}_7$ or $\mathbb{O}_5$. Thus, if one of $\mathbb{F}', \mathbb{K}'$ has a set of size 3, then either $\mathbb{F}$ or $b(\mathbb{F})$ has one of $\mathbb{L}_7, \mathbb{O}_5$ as a minor. We may therefore assume that neither $\mathbb{F}'$ nor $\mathbb{K}'$ has a set of size 3. Let (M, Σ) represent $\mathbb{F}'$ and let (N, Γ) represent $\mathbb{K}'$, whose existence are guaranteed by Proposition 5. It then follows from Theorem 9 (2)–(3) that either (M, Σ) or (N, Γ) has an $(F_7, E(F_7) - \omega)$ minor. By Remark 6 and Proposition 8, we see that either $\mathbb{F}'$ or $\mathbb{K}'$ has an $\mathbb{LC}_7$ minor, implying in turn that either $\mathbb{F}$ or $b(\mathbb{F})$ has an $\mathbb{LC}_7$ minor, as required. □

In the remainder of this paper, we prove Theorem 9.

3   Proof of Theorem 9 Part (1)

Let $\mathbb{F}, \mathbb{K}$ be blocking mni binary clutters over ground set E, neither of which has a set of size 3. By Theorem 3, there are integers $r \geq 4$ and $s \geq 4$ such that $M(\mathbb{F})$ is r-regular, $M(\bar{\mathbb{K}})$ is s-regular, and after possibly permuting the rows of $M(\bar{\mathbb{K}})$, $M(\mathbb{F})M(\bar{\mathbb{K}})^\top = J + (rs - n)I = M(\bar{\mathbb{K}})^\top M(\mathbb{F})$. Thus, there is a labeling $\bar{\mathbb{F}} = \{C_1, \ldots, C_n\}$ and $\bar{\mathbb{K}} = \{B_1, \ldots, B_n\}$ so that, for all $i, j \in \{1, \ldots, n\}$,

$$(\star) \qquad |C_i \cap B_j| = \begin{cases} rs - n + 1 & \text{if } i = j \\ 1 & \text{if } i \neq j \end{cases}$$

and for all $g, h \in E$,

$$(\diamond) \qquad |\{i \in \{1, \ldots, n\} : g \in C_i, h \in B_i\}| = \begin{cases} rs - n + 1 & \text{if } g = h \\ 1 & \text{if } g \neq h. \end{cases}$$

Take an element $e \in E$. Since $rs - n \geq 2$, we may assume by $(\diamond)$ that $e \in C_i \cap B_i$ for $i \in \{1, 2, 3\}$. Recall that (M, Σ) represents $\mathbb{F}$ and that (N, Γ) represents $\mathbb{K}$. We will show that (C_1, C_2, C_3) is a strict e-hub of (M, Σ).

Claim 1. C_1, C_2, C_3 are odd circuits of (M, Σ) such that, for distinct $i, j \in \{1, 2, 3\}$, $C_i \cap C_j = \{e\}$, i.e. (h1) holds.

Proof of Claim. By definition, C_1, C_2, C_3 are odd circuits of (M, Σ). To see $C_1 \cap C_2 = \{e\}$, notice that if $f \in (C_1 \cap C_2) - e$, then $\{1, 2\} \subseteq \{i \in \{1, \ldots, n\} : f \in C_i, e \in B_i\}$, which cannot be the case as the latter set has size 1 by $(\diamond)$. Similarly, $C_2 \cap C_3 = C_3 \cap C_1 = \{e\}$. $\diamond$

Claim 2. For distinct $i, j \in \{1, 2, 3\}$, the only nonempty cycles of M contained in $C_i \cup C_j$ are $C_i, C_j, C_i \triangle C_j$, so (h2) holds.

Proof of Claim. By symmetry, we may only analyze the cycles of M contained in $C_1 \cup C_2$. By Corollary 4 (1), the only odd circuits of (M, Σ) contained in $C_1 \cup C_2$ are C_1, C_2. We first show that C_1, C_2 are the only odd cycles of (M, Σ) in $C_1 \cup C_2$. Suppose otherwise. Let A be an odd cycle different from C_1, C_2. Write C as the disjoint union of circuits $A_1, \ldots, A_k$ for some $k \geq 2$. Since $|\Sigma \cap A| = \sum_{i=1}^{k} |\Sigma \cap A_i|$ and $|\Sigma \cap A|$ is odd, we may assume that $|\Sigma \cap A_1|$ is odd, so $A_1 \in \{C_1, C_2\}$, and we may assume that $A_1 = C_1$. But then $A_2 \subseteq C_2 - e$, a contradiction as both A_2, C_2 are circuits of M. Let C be a nonempty cycle of M contained in $C_1 \cup C_2$. If C is an odd cycle of (M, Σ), then as we just showed, $C \in \{C_1, C_2\}$. Otherwise, C is an even cycle, so $C \triangle C_1$ is an odd cycle, so $C \triangle C_1 \in \{C_1, C_2\}$, implying in turn that $C = C_1 \triangle C_2$, as required. $\diamond$

Claim 3. Every odd cycle of (M, Σ) contained in $C_1 \cup C_2 \cup C_3$ uses e, so (h3) holds.

Proof of Claim. Since $s \geq 4$ and $M(\bar{\mathbb{K}})$ is s-regular, there is a $B \in \bar{\mathbb{K}} - \{B_1, B_2, B_3\}$ such that $e \in B$. Then, for each $i \in \{1, 2, 3\}$, $|B \cap C_i| = 1$ by $(\star)$, so $B \cap (C_1 \cup C_2 \cup C_3) = \{e\}$. It follows from Proposition 7 that B is a signature of (M, Σ). Thus, if C is an odd cycle of (M, Σ) contained in $C_1 \cup C_2 \cup C_3$, then $|C \cap B|$ is odd and therefore nonzero, so $e \in C$. ◇

Claim 4. *If C, C' are odd cycles of (M, Σ) contained in $C_1 \cup C_2 \cup C_3$ such that $C \cap C' = \{e\}$, then for some distinct $i, j \in \{1, 2, 3\}$, $\{C, C'\} = \{C_i, C_j\}$, so (h4) holds.*

Proof of Claim. Let D, D' be odd circuits contained in C, C', respectively. It follows from Corollary 4 (2) that, for some distinct $i, j \in \{1, 2, 3\}$, $\{D, D'\} = \{C_i, C_j\}$. Since there is no even cycle contained in $(C_1 \cup C_2 \cup C_3) - (C_i \triangle C_j)$, it follows that $D = C$ and $D' = C'$, and the claim follows. ◇

Hence, (C_1, C_2, C_3) is a strict e-hub of (M, Σ). Similarly, (B_1, B_2, B_3) is a strict e-hub of (N, Γ). This finishes the proof of Theorem 9 part (1). □

4 Hypergraphs, the Trifold, and Graphic Hubs

Let M be a binary matroid over ground set E. By definition, the cycles of M form a linear space modulo 2, so there is a $0-1$ matrix A such that the incidence vectors of the cycles in M are $\{x \in \{0, 1\}^E : Ax \equiv \mathbf{0} \pmod{2}\}$. The matrix A is referred to as a *representation of M*. Notice that elementary row operations modulo 2 applied to A yield another representation, and if $a \in \{0, 1\}^E$ belongs to the row space of A modulo 2, then $\begin{pmatrix} A \\ a^\top \end{pmatrix}$ is also a representation.

A *hypergraphic representation of M* is a representation where every column has an even number of ones. If $a^\top$ is the sum of the rows of A modulo 2, then $\begin{pmatrix} A \\ a^\top \end{pmatrix}$ is a hypergraphic representation. In particular, a binary matroid always has a hypergraphic representation. A *hypergraph* is a pair $G = (V, E)$, where V is a finite set of *vertices* and E is a family of even subsets of V, called *edges*. Note that if A is a hypergraphic representation of M, then A may be thought of as a hypergraph whose vertices are labeled by the rows and whose edges are labeled by the columns. For instance, the Fano matroid F_7 may be represented as a hypergraph on vertices $\{1, \ldots, 4\}$ and edges $\{T \subseteq \{1, \ldots, 4\} : |T| \in \{2, 4\}\}$. Denote by S_8 the binary matroid represented as the hypergraph displayed in Fig. 2, which has vertices $\{1, \ldots, 5\}$ and edges $\{1, 2\}, \{1, 3\}, \{1, 4\}, \{1, 5\}, \{2, 3\}, \{2, 4\}, \{2, 5\}, \{2, 3, 4, 5\}$. Label $\gamma := \{2, 3, 4, 5\} \in E(S_8)$. A *trifold* is any signed matroid isomorphic to $(S_8, E(S_8) - \gamma)$.

Remark 10. *A trifold has an $(F_7, E(F_7))$ minor.*

Proof. Observe that $S_8 / \gamma \cong F_7$, implying in turn that $(S_8, E(S_8) - \gamma) / \gamma \cong (F_7, E(F_7))$. □

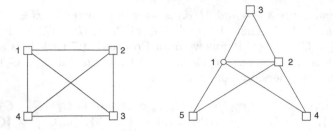

Fig. 2. The hypergraph on the left represents F_7, and the one on the right represents S_8. Line segments represent edges of size 2, and square vertices form the edges of size 4.

Given a hypergraph $G = (V, E)$ and $F \subseteq E$, let $\mathrm{odd}_G(F) := \triangle(e : e \in F) \subseteq V$. Observe that $\mathrm{odd}_G(F)$ is an even subset of V. We will make use of the following remark throughout the paper:

Remark 11. *Let M be a binary matroid over ground set $E \cup \{e\}$, where $M \setminus e$ is represented by the hypergraph $G = (V, E)$. If for some $F \subseteq E$, $F \cup \{e\}$ is a cycle of M, then the hypergraph on vertices V and edges $E \cup \{\mathrm{odd}_G(F)\}$ represents M.*

Recall that a binary matroid is graphic if it can be represented by a graph. We will also need the following result, whose proof is straight-forward:

Proposition 12. *Take a signed matroid (M, Σ), $e \in E(M)$ and an e-hub (C_1, C_2, C_3). Then there is a signature Σ' such that $\Sigma' \cap (C_1 \cup C_2 \cup C_3) = \{e\}$. Moreover, the following statements are equivalent:*

(i) $M|(C_1 \cup C_2 \cup C_3)$ is graphic,
(ii) $C_1, C_2, C_3, C_1 \triangle C_2 \triangle C_3$ are the only odd cycles contained in $C_1 \cup C_2 \cup C_3$.

5   Proof of Theorem 9 Part (2)

Let $\mathbb{F}, \mathbb{K}$ be blocking mni binary clutters over ground set E, neither of which has a set of size 3. Recall that (M, Σ) represents $\mathbb{F}$ and that (N, Γ) represents $\mathbb{K}$. Take an element $e \in E$. By Theorem 9 part (1), (M, Σ) has a (strict) e-hub (C_1, C_2, C_3) and (N, Γ) has a (strict) e-hub (B_1, B_2, B_3), where for $i \in \{1, 2, 3\}$, $|C_i \cap B_i| \geq 3$ and, for distinct $i, j \in \{1, 2, 3\}$, $C_i \cap B_j = \{e\}$. By Proposition 12, after a possible resigning of (M, Σ), we may assume that $\Sigma \cap (C_1 \cup C_2 \cup C_3) = \{e\}$. Notice further that by Proposition 7, the odd circuits of (N, Γ) are (minimal) signatures of (M, Σ). We need to show that either $M|(C_1 \cup C_2 \cup C_3)$ or $N|(B_1 \cup B_2 \cup B_3)$ is non-graphic. Suppose otherwise. Since $N|(B_1 \cup B_2 \cup B_3)$ is graphic, it follows from Proposition 12 that B_1, B_2, B_3 are the only odd circuits of (N, Γ) contained in $B_1 \cup B_2 \cup B_3$. In other words, the only sets of $\mathbb{K}$ contained in $B_1 \cup B_2 \cup B_3$ are B_1, B_2, B_3.

Claim 1. *There is an odd circuit C of (M, Σ) such that $e \notin C$ and, for each $i \in \{1, 2, 3\}$, $C \cap B_i \subseteq C_i$.*

Proof of Claim. Let B be the union of $(B_1 \cup B_2 \cup B_3) - (C_1 \cup C_2 \cup C_3)$ and $\{e\}$. Since $B_1 \cap C_1 \neq \{e\}$, it follows that $B_1 \not\subseteq B$. Similarly, $B_2 \not\subseteq B$ and $B_3 \not\subseteq B$. Thus, since the only sets of $\mathbb{K}$ contained in $B_1 \cup B_2 \cup B_3$ are B_1, B_2, B_3, we get that B does not contain a set of $\mathbb{K} = b(\mathbb{F})$. In other words, there is a set $C \in \mathbb{F}$ such that $C \cap B = \emptyset$. By definition, C is an odd circuit of (M, Σ). Clearly, $e \notin C$. Consider the intersection $C \cap B_1$. Since $C \cap B = \emptyset$, it follows that $C \cap B_1 \subseteq C_1 \cup C_2 \cup C_3$. Moreover, as $B_1 \cap C_2 = B_1 \cap C_3 = \{e\}$, we see that $C \cap B_1 \subseteq C_1$. Similarly, $C \cap B_2 \subseteq C_2$ and $C \cap B_3 \subseteq C_3$. $\Diamond$

Since $e \notin C$, we get that $C \cap \Sigma \subseteq C - (C_1 \cup C_2 \cup C_3)$, and as C is odd, it follows that $C \not\subseteq C_1 \cup C_2 \cup C_3$.

Claim 2. $(M, \Sigma)|(C_1 \cup C_2 \cup C_3 \cup C)$ *has a trifold minor.*

Proof Sketch. Let S be a minimal subset of $C - (C_1 \cup C_2 \cup C_3)$ such that (m1) $M|(C_1 \cup C_2 \cup C_3 \cup S)$ has a cycle containing S, and (m2) $|S \cap \Sigma|$ is odd. Note that S is well-defined, since $C - (C_1 \cup C_2 \cup C_3)$ satisfies both (m1)–(m2). Let

$$(M', \Sigma') := (M, \Sigma)|(C_1 \cup C_2 \cup C_3 \cup S).$$

The minimality of S implies that the elements of S are in series in M'. In particular, after a possible resigning, we may assume that $\Sigma' \cap (C_1 \cup C_2 \cup C_3 \cup S) = \{e, f\}$ for some element $f \in S$. Let

$$(M'', \{e, f\}) := (M', \Sigma')/(S - f).$$

Since B_1 is a signature for (M, Σ), and $B_1 \cap (C_1 \cup C_2 \cup C_3 \cup C) = B_1 \cap C_1$ by our choice of C, it follows that $B_1 \cap C_1$ is a signature for $(M'', \{e, f\})$. We have $M'' \backslash f = M'/(S - f) \backslash f = M' \backslash S = M|(C_1 \cup C_2 \cup C_3)$, where the second equality follows from the fact that the elements of M' in S are in series. Since $M|(C_1 \cup C_2 \cup C_3)$ is graphic, $M'' \backslash f$ may be represented as a graph $G = (V, C_1 \cup C_2 \cup C_3)$. It follows from (h2) that the circuits C_1, C_2, C_3 are pairwise vertex-disjoint except at the ends of $e = \{x, y\} \subseteq V$. By (m1), $M|(C_1 \cup C_2 \cup C_3 \cup S)$ has a cycle containing S, so M'' has a cycle $P \cup \{f\}$, for some $P \subseteq C_1 \cup C_2 \cup C_3$. By replacing P by $P \triangle C_1$, if necessary, we may assume that $e \notin P$. For each $i \in \{1, 2, 3\}$, let $P_i := P \cap C_i$ and $Q_i := C_i - (P_i \cup \{e\})$. After possibly rearranging the edges of G within each series class $C_i - e$, we may assume that each P_i is a path that starts from x. It follows from Remark 11 that M'' is represented as the hypergraph on vertices V and edges $C_1 \cup C_2 \cup C_3 \cup \{odd_G(P)\}$. We may therefore label $f = odd_G(P)$, and represent M'' with the following hypergraph

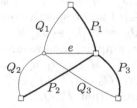

where f consists of the square vertices. Since $P \cup \{f\}$ is an odd cycle of $(M'', \{e, f\})$, it must contain an odd number of edges of the signature $B_1 \cap C_1$, implying in turn that each of P_1, Q_1 contains an odd number of edges of B_1, so $P_1 \neq \emptyset$ and $Q_1 \neq \emptyset$. Similarly, for each $i \in \{1, 2, 3\}$, $P_i \neq \emptyset$ and $Q_i \neq \emptyset$, so there are $p_i \in P_i$ and $q_i \in Q_i$. Since $\{e, p_1, p_2, p_3, q_1, q_2, q_3\}$ is a signature for $(M'', \{e, f\})$, we see that

$$(M'', \{e, f\}) \cong (M'', \{e, p_1, p_2, p_3, q_1, q_2, q_3\}).$$

Observe however that the right signed matroid has a trifold minor, obtained after contracting each $C_i - \{e, p_i, q_i\}$. As a result, $(M, \Sigma)|(C_1 \cup C_2 \cup C_3 \cup C)$ has a trifold minor. ◇

However, by Remark 10, a trifold has an $(F_7, E(F_7))$ minor, so (M, Σ) has an $(F_7, E(F_7))$ minor. As a consequence, Proposition 8 implies that $\mathbb{F}$ has an $\mathbb{L}_7$ minor. Since $\mathbb{F}$ is mni, we must have that $\mathbb{F} \cong \mathbb{L}_7$, but $\mathbb{F}$ has no set of size 3, a contradiction. This finishes the proof of Theorem 9 part (2). □

6 Non-graphic Strict Hubs

In this section, we prove the following result needed for Theorem 9 part (3):

Proposition 13. *Take a signed matroid (M, Σ), $e \in E(M)$ and a strict e-hub (C_1, C_2, C_3) such that $M|(C_1 \cup C_2 \cup C_3)$ is non-graphic. Then there exist $I \subseteq C_3 - e$ and distinct $g_1, g_2 \in (C_3 - I) - e$ where*

(1) $(C_1, C_2, C_3 - I)$ is an e-hub of $(M, \Sigma)/I$,
(2) $(M/I)|(C_1 \cup C_2 \cup \{g_i\})$ has a circuit containing g_i, for each $i \in \{1, 2\}$,
(3) $(M/I)|(C_1 \cup C_2 \cup \{g_1, g_2\})$ is non-graphic.

Proof Sketch. By Proposition 12, after a possible resigning, we may assume that $\Sigma \cap (C_1 \cup C_2 \cup C_3) = \{e\}$. Let I be a maximal subset of $C_3 - e$ such that every cycle of $M|(C_1 \cup C_2 \cup I)$ is disjoint from I. Let $(M', \{e\}) := (M, \Sigma)|(C_1 \cup C_2 \cup C_3)/I$ and $C_3' := C_3 - I$. Then (C_1, C_2, C_3') is an e-hub of $(M', \{e\})$, and as $M|(C_1 \cup C_2 \cup C_3)$ is non-graphic, it follows from Proposition 12 that M' is non-graphic. Moreover, the maximality of I implies that, for each $g \in C_3' - e$, there is a cycle D_g of $M'|(C_1 \cup C_2 \cup \{g\})$ using g, where after possibly replacing D_g by $D_g \triangle C_1$, we may assume that $e \notin D_g$. Note that $D_g \triangle C_1 \triangle C_2$ is another cycle of $M'|(C_1 \cup C_2 \cup \{g\})$ that uses g and excludes e. For each such g, refer to $D_g - g$ and $(D_g \triangle C_1 \triangle C_2) - g$ as the *outer joins of g*. Notice that an outer join intersects both C_1, C_2. As $\triangle(D_g - g : g \in C_3' - e)$ is either $C_1 - e$ or $C_2 - e$ by (h2), there exist $h_1, h_2 \in C_3' - \{e\}$ and respective outer joins J_{h_1}, J_{h_2} that cross, that is, $J_{h_1} \cap J_{h_2} \neq \emptyset$, $J_{h_1} - J_{h_2} \neq \emptyset$, $J_{h_2} - J_{h_1} \neq \emptyset$ and $J_{h_1} \cup J_{h_2} \neq C_1 \triangle C_2$. If $M'|(C_1 \cup C_2 \cup \{h_1, h_2\})$ is non-graphic, then we are done. Otherwise, it may be represented as a graph $H = (V, C_1 \cup C_2 \cup \{h_1, h_2\})$, displayed below (left figure), where $C_1 = \{e\} \cup P_1 \cup Q_1 \cup R_1$, $C_2 = \{e\} \cup P_2 \cup Q_2 \cup R_2$, $J_{h_1} = P_1 \cup P_2 \cup Q_2$, and $J_{h_2} = P_1 \cup P_2 \cup Q_1$. Notice that $P_i, Q_i, R_i \neq \emptyset$ for each $i \in \{1, 2\}$. Let

$D_1 := \{e, h_1\} \cup P_1 \cup R_2$ and $D_2 := \{e, h_2\} \cup \{P_2, R_1\}$. For $i \in \{1, 2\}$, let D_i' be a cycle of M such that $D_i \subseteq D_i' \subseteq D_i \cup I$; as $D_i' \cap \Sigma = \{e\}$, D_i' is an odd cycle of (M, Σ). Note further, for $i \in \{1, 2\}$, that D_i' is different from C_1, C_2, C_3. Thus, since (C_1, C_2, C_3) is a strict e-hub of (M, Σ) and therefore satisfies (h4), we must have that $\{e\} \subsetneq D_1' \cap D_2'$. Because $D_1 \cap D_2 = \{e\}$, there is an element $f \in I$ such that $\{e, f\} \subseteq D_1' \cap D_2'$. Consider now the minor $(M, \Sigma)/(I - f)$; note that $D_1 \cup \{f\}$ and $D_2 \cup \{f\}$ are odd cycles of this signed matroid. We may represent $M/(I - f)$ as a hypergraph $G = (V \cup \{w\}, C_1 \cup C_2 \cup \{h_1, h_2, f\})$ obtained from H by adding a vertex w, displayed below (right figure), where the square vertices form the edge h_1. Now let $J := I \triangle \{f, h_2\}$. Observe that $(M/J)|(C_1 \cup C_2 \cup \{f, h_1\})$ is non-graphic, as it has an F_7 minor obtained after contracting $P_1 \cup R_2$ and contracting each of Q_1, R_1, P_2, Q_2 to a single edge. Thus, $J \subseteq C_3' - \{e\}$ and f, h_1 satisfy (3), and it can be readily checked that they also satisfy (1)–(2).

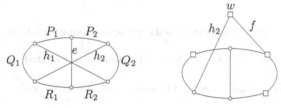

7   A Sketch of the Proof of Theorem 9 Part (3)

Let $\mathbb{F}, \mathbb{K}$ be blocking mni clutters over ground set E, neither of which has a set of size 3, where (M, Σ) represents $\mathbb{F}$ and (N, Γ) represents $\mathbb{K}$. By Theorem 9 part (1), (M, Σ) has a strict e-hub (C_1, C_2, C_3) and (N, Γ) has a strict e-hub (B_1, B_2, B_3) such that for $i \in \{1, 2, 3\}$, $|C_i \cap B_i| \geq 3$ and, for distinct $i, j \in \{1, 2, 3\}$, $C_i \cap B_j = \{e\}$. Assume further that $M|(C_1 \cup C_2 \cup C_3)$ is non-graphic. We need to show that (M, Σ) has an $(F_7, E(F_7) - \omega)$ minor going through e. By Proposition 12, after a possible resigning, we may assume that $\Sigma \cap (C_1 \cup C_2 \cup C_3) = \{e\}$. By Proposition 13, there exist $I \subseteq C_3 - e$ and distinct $g_1, g_2 \in (C_3 - I) - e$ such that (1)–(3) hold. For each $i \in \{1, 2\}$, after possibly replacing D_i by $D_i \triangle C_1$, we may assume that $e \notin D_i$; as $(C_1, C_2, C_3 - I)$ is an e-hub of $(M, \Sigma)/I$, it follows from (h2) that $D_i \cap C_1 \neq \emptyset$ and $D_i \cap C_2 \neq \emptyset$. Notice that, for each $i \in \{1, 2\}$, $B_i \cap I = \emptyset$, so B_i is a signature of $(M, \Sigma)/I$.

Claim 1. *There exists an odd circuit C of $(M, \Sigma)/I$ such that $e \notin C$ and, for each $i \in \{1, 2\}$, $C \cap B_i \subseteq C_i$.*

Let $(M', \Sigma) := (M, \Sigma)/I$. Let S be a minimal subset of $C - (C_1 \cup C_2)$ such that (m1) $M'|(C_1 \cup C_2 \cup S)$ has a cycle containing S, and (m2) $|S \cap \Sigma|$ is odd. Note that S is well-defined as $C - (C_1 \cup C_2)$ satisfies (m1)–(m2). The minimality of S implies that $S \cap \{g_1, g_2\} = \emptyset$, and the elements of S are in series in $M'|(C_1 \cup C_2 \cup \{g_1, g_2\} \cup S)$. Thus, there exists a signature Σ' of (M', Σ) such that $\Sigma' \cap (C_1 \cup C_2 \cup \{g_1, g_2\} \cup S) = \{e, f\}$, for some $f \in S$. Consider the minor

$$(M'', \{e, f\}) := (M', \Sigma')|(C_1 \cup C_2 \cup \{g_1, g_2\} \cup S)/(S - f).$$

For each $i \in \{1, 2\}$, our choice of C implies that $B_i \cap S = \emptyset$, so $B_i \cap (C_1 \cup C_2 \cup \{g_1, g_2\}) = B_i \cap C_i$ is a signature of $(M'', \{e, f\})$.

Claim 2. *If $M'' \backslash g_i$ is graphic for each $i \in \{1, 2\}$, then $(M'', \{e, f\})$ has an $(F_7, E(F_7))$ minor.*

Assume that $M'' \backslash g_i$ is graphic for each $i \in \{1, 2\}$. Then by the preceding claim, $(M'', \{e, f\})$ has an $(F_7, E(F_7))$ minor, implying in turn that (M, Σ) has an $(F_7, E(F_7))$ minor. So by Proposition 8, $\mathbb{F}$ has an $\mathbb{L}_7$ minor, and since $\mathbb{F}$ is mni, this means $\mathbb{F} \cong \mathbb{L}_7$, which cannot be as $\mathbb{F}$ has no set of size 3. Hence, by symmetry, we may assume that $M'' \backslash g_2$ is non-graphic. Thus, there exists $I \subseteq C_1 \triangle C_2$ such that $M'' \backslash g_2 / I \cong F_7$. Then $(M'', \{e, f\}) \backslash g_2 / I \cong (F_7, E(F_7) - \omega)$, and so (M, Σ) has an $(F_7, E(F_7) - \omega)$ minor going through e, as required. This finishes the proof of Theorem 9 part (3). □

References

1. Abdi, A., Guenin, B.: The minimally non-ideal binary clutters with a triangle (Submitted)
2. Bridges, W.G., Ryser, H.J.: Combinatorial designs and related systems. J. Algebra **13**, 432–446 (1969)
3. Cornuéjols, G.: Combinatorial Optimization, Packing and Covering. SIAM, Philadelphia (2001)
4. Cornuéjols, G., Guenin, B.: Ideal binary clutters, connectivity, and a conjecture of Seymour. SIAM J. Discrete Math. **15**(3), 329–352 (2002)
5. Edmonds, J., Fulkerson, D.R.: Bottleneck extrema. J. Combin. Theory Ser. B **8**, 299–306 (1970)
6. Ford, L.R., Fulkerson, D.R.: Maximal flow through a network. Can. J. Math. **8**, 399–404 (1956)
7. Guenin, B.: A characterization of weakly bipartite graphs. J. Combin. Theory Ser. B **83**, 112–168 (2001)
8. Guenin, B.: Integral polyhedra related to even-cycle and even-cut matroids. Math. Oper. Res. **27**(4), 693–710 (2002)
9. Lehman, A.: A solution of the Shannon switching game. Soc. Ind. Appl. Math. **12**(4), 687–725 (1964)
10. Lehman, A.: On the width-length inequality. Math. Program. **17**(1), 403–417 (1979)
11. Lehman, A.: The width-length inequality and degenerate projective planes. In: DIMACS, vol. 1, pp. 101–105 (1990)
12. Menger, K.: Zur allgemeinen Kurventheorie. Fundamenta Mathematicae **10**, 96–115 (1927)
13. Novick, B., Sebő, A.: On combinatorial properties of binary spaces. In: Balas, E., Clausen, J. (eds.) IPCO 1995. LNCS, vol. 920, pp. 212–227. Springer, Heidelberg (1995). doi:10.1007/3-540-59408-6_53
14. Oxley, J.: Matroid Theory, 2nd edn. Oxford University Press, New York (2011)
15. Seymour, P.D.: On Lehman's width-length characterization. In: DIMACS, vol. 1, pp. 107–117 (1990)
16. Seymour, P.D.: The forbidden minors of binary matrices. J. Lond. Math. Soc. **2**(12), 356–360 (1976)
17. Seymour, P.D.: The matroids with the max-flow min-cut property. J. Combin. Theory Ser. B **23**, 189–222 (1977)

On Scheduling Coflows

Saba Ahmadi[1], Samir Khuller[1], Manish Purohit[2], and Sheng Yang[1(✉)]

[1] University of Maryland, College Park, USA
{saba,samir,styang}@cs.umd.edu
[2] Google, Mountain View, USA
mpurohit@google.com

Abstract. Applications designed for data-parallel computation frameworks such as MapReduce usually alternate between computation and communication stages. Coflow scheduling is a recent popular networking abstraction introduced to capture such application-level communication patterns in datacenters. In this framework, a datacenter is modeled as a single non-blocking switch with m input ports and m output ports. A coflow j is a collection of flow demands $\{d_{io}^j\}_{i \in m, o \in m}$ that is said to be complete once *all* of its requisite flows have been scheduled.

We consider the offline coflow scheduling problem with and without release times to minimize the total weighted completion time. Coflow scheduling generalizes the well studied concurrent open shop scheduling problem and is thus NP-hard. Qiu, Stein and Zhong [15] obtain the first constant approximation algorithms for this problem via LP rounding and give a deterministic $\frac{67}{3}$-approximation and a randomized $(9 + \frac{16\sqrt{2}}{3}) \approx 16.54$-approximation algorithm. In this paper, we give a combinatorial algorithm that yields a deterministic 5-approximation algorithm with release times, and a deterministic 4-approximation for the case without release time.

Keywords: Coflow scheduling · Concurrent open shop

1 Introduction

Large scale data centers have emerged as the dominant form of computing infrastructure over the last decade. The success of data-parallel computing frameworks such as MapReduce [9], Hadoop [1], and Spark [19] has led to a proliferation of applications that are designed to alternate between computation and communication stages. Typically, the intermediate data generated by a computation stage needs to be transferred across different machines during a communication stage for further processing. For example, there is a "Shuffle" phase between every consecutive "Map" and "Reduce" phase in MapReduce. With an increasing reliance on parallelization, these communication stages are responsible for a large amount of data transfer in a datacenter. Chowdhury and Stoica

This work is supported by NSF grant CNS 156019.

F. Eisenbrand and J. Koenemann (Eds.): IPCO 2017, LNCS 10328, pp. 13–24, 2017.
DOI: 10.1007/978-3-319-59250-3_2

[5] introduced coflows as an effective networking abstraction to represent the collective communication requirements of a job. In this paper, we consider the problem of scheduling coflows to minimize weighted completion time and give improved approximation algorithms for this basic problem.

The communication phase for a typical application in a modern data center may contain hundreds of individual flow requests, and the phase ends only when all of these flow requests are satisfied. A coflow is defined as the collection of these individual flow requests that all share a common performance goal. The underlying data center is modeled as a single $m \times m$ *non-blocking switch* that consists of m input ports and m output ports. We assume that each port has unit capacity, i.e. it can handle at most one unit of data per unit time. Modeling the data center itself as a simple switch allows us to focus solely on the scheduling task instead of the problem of *routing* flows through the network. Each coflow j is represented as a $m \times m$ integer matrix $D^j = [d_{io}^j]$ where the entry d_{io}^j indicates the number of data units that must be transferred from input port i to output port o for coflow j. Figure 1 shows a single coflow over a 2×2 switch. For instance, the coflow depicted needs to transfer 2 units of data from input a to output b and 3 units of data from input a to output d. Each coflow j also has a weight w_j that indicates its relative importance and a release time r_j.

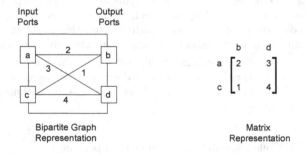

Fig. 1. An example coflow over a 2×2 switch. The figure illustrates two equivalent representations of a coflow - (i) as a weighted, bipartite graph over the set of ports, and (ii) as a $m \times m$ integer matrix.

A coflow j is available to be scheduled at its release time r_j and is said to be completed when all the flows in the matrix D^j have been scheduled. More formally, the completion time C_j of coflow j is defined as the earliest time such that for every input i and output o, d_{io}^j units of its data have been transferred from port i to port o. We assume that time is slotted and data transfer within the switch is instantaneous. Since each input port i can transmit at most one unit of data and each output port o can receive at most one unit of data in each time slot, a feasible schedule for a single time slot can be described as a matching. Our goal is to find a feasible scheduling that minimizes the total weighted completion time of the coflows, i.e. minimize $\sum_j w_j C_j$.

1.1   Related Work

Chowdhury and Stoica [5] introduced the coflow abstraction to describe the prevalent communication patterns in data centers. Since then coflow scheduling has been a topic of active research [6,7,15,20] in both the systems and theory communities. Although coflow aware network schedulers have been found to perform very well in practice in both the offline [7] and online [6] settings, no $O(1)$ approximation algorithms were known even in the offline setting until recently. Since the coflow scheduling problem generalizes the well-studied concurrent open shop scheduling problem, it is NP-hard to approximate within a factor better than $(2 - \epsilon)$ [3,17].

For the special case when all coflows have zero release time, Qiu, Stein and Zhong [15] obtain a deterministic $\frac{64}{3}$ approximation and a randomized $(8 + \frac{16\sqrt{2}}{3})$ approximation algorithm for the problem of minimizing the weighted completion time. For coflow scheduling with arbitrary release times, Qiu et al. [15] claim a deterministic $\frac{67}{3}$ approximation and a randomized $(9 + \frac{16\sqrt{2}}{3})$ approximation algorithm. However in the full version [2], we demonstrate a subtle error in their proof that deals with non-zero release times. We show that their techniques in fact only yield a deterministic $\frac{76}{3}$-approximation algorithm for coflow scheduling with release times. However their result holds for the case with equal release times.

By exploiting a connection with the well-studied concurrent open shop scheduling problem, Luo et al. [13] claim a 2-approximation algorithm for coflow scheduling when all the release times are zero. Unfortunately, as we show in the full version [2], their proof too is flawed and the result does not hold.

In a recent work, Khuller et al. [11] study coflow scheduling in the online setting where the coflows arrive online over time. Using the results of this paper (Theorem 2), they obtain an exponential time 7-competitive algorithm and a polynomial time 14-competitive algorithm.

1.2   Our Contributions

The main algorithmic contribution of this paper is a deterministic, primal-dual algorithm for the offline coflow scheduling problem with improved approximation guarantees.

Theorem 1. *There exists a deterministic, combinatorial, polynomial time 5-approximation algorithm for coflow scheduling with release times.*

Theorem 2. *There exists a deterministic, combinatorial, polynomial time 4-approximation algorithm for coflow scheduling without release times.*

Our results significantly improve upon the approximation algorithms developed by Qiu et al. [15] whose techniques yield an approximation factors of $\frac{76}{3} = 25.33$ and $(8 + \frac{16\sqrt{2}}{3}) \approx 15.54$ (see the full version [2]) respectively for the two cases. In addition, our algorithm is completely combinatorial and does not require solving a linear program. A LP-based version is also provided together with its proof, to help show the intuition behind the primal-dual one.

We also extend the primal dual algorithm by Mastrolilli et al. [14] to give a 3-approximation algorithm for the concurrent open shop problem when the jobs have arbitrary release times.

Theorem 3. *There exists a deterministic, combinatorial, polynomial time 3-approximation algorithm for concurrent open shop scheduling with release times.*

Due to space constraints, we defer all proofs to the full version [2].

1.3   Connection to Concurrent Open Shop

The coflow scheduling problem generalizes the well-studied concurrent open shop problem [4,10,12,14,18]. In the concurrent open shop problem, we have a set of m machines and each job j with weight w_j is composed of m tasks $\{t_i^j\}_{i=1}^m$, one on each machine. Let p_i^j denote the processing requirement of task t_i^j. A job j is said to be completed once all its tasks have completed. A machine can perform at most one unit of processing at a time. The goal is to find a feasible schedule that minimizes the total weighted completion time of jobs. An LP-relaxation yields a 2-approximation algorithm for concurrent open shop scheduling when all release times are zero [4,10,12] and a 3-approximation algorithm for arbitrary release times [10,12]. Mastrolilli et al. [14] show that a simple greedy algorithm also yields a 2-approximation for concurrent open shop without release times. We develop a primal-dual algorithm that yields a 3-approximation for concurrent open shop with release times.

The concurrent open shop problem can be viewed as a special case of coflow scheduling when the demand matrices D^j for all coflows j are diagonal [7,15]. At first glance, it appears that coflow scheduling is much harder than concurrent open shop. For instance, while concurrent open shop always admits an optimal permutation schedule, such a property is not be true for coflows [7]. In fact, even without release times, the best known approximation algorithm for scheduling coflows has an approximation factor of ≈ 15.54 [15], in contrast to the many 2-approximations known for the concurrent open shop problem. Surprisingly, we show that using a similar LP relaxation as for the concurrent open shop problem, we can design a primal dual algorithm to obtain a permutation of coflows such that sequentially scheduling the coflows after some post-processing in this permutation leads to provably good coflow schedules.

2   Preliminaries

We first introduce some notation to facilitate the following discussion. For every coflow j and input port i, we define the load $L_{i,j} = \sum_{o=1}^m d_{io}^j$ to be the total amount of data that coflow j needs to transmit through port i. Similarly, we define $L_{o,j} = \sum_{i=1}^m d_{io}^j$ for every coflow j and output port o. Equivalently, a coflow j can be represented by a weighted, bipartite graph $G_j = (I, O, E_j)$ where the set of input ports (I) and the set of output ports (O) form the two sides of the bipartition and an edge $e = (i,o)$ with weight $w_{G_j}(e) = d^j_{io}$ represents that the coflow j requires d^j_{io} units of data to be transferred from input port i to output port o. We will abuse notation slightly and refer to a coflow j by the corresponding bipartite graph G_j when there is no confusion.

Representing a coflow as a bipartite graph simplifies some of the notation that we have seen previously. For instance, for any coflow j, the load of j on port i is simply the weighted degree of vertex i in graph G_j, i.e., if $\mathbb{N}_{G_j}(i)$ denotes the set of neighbors of node i in the graph G_j.

$$L_{i,j} = \deg_{G_j}(i) = \sum_{o \in \mathbb{N}_{G_j}(i)} w_{G_j}(i,o) \tag{1}$$

For any graph G_j, let $\Delta(G_j) = \max_{s \in I \cup O} \deg_{G_j}(s) = \max\{\max_i L_{i,j}, \max_o L_{o,j}\}$ denote the maximum degree of any node in the graph, i.e., the load on the most heavily loaded port of coflow j.

In our algorithm, we consider coflows obtained as the union of two or more coflows. Given two weighted bipartite graphs $G_j = (I, O, E_j)$ and $G_k = (I, O, E_k)$, we define the cumulative graph $G_j \cup G_k = (I, O, E_j \cup E_k)$ to be a weighted bipartite graph such that $w_{G_j \cup G_k}(e) = w_{G_j}(e) + w_{G_k}(e)$. We extend this notation to the union of multiple graphs in the obvious manner.

2.1   Scheduling a Single Coflow

Before we present our algorithm for the general coflow scheduling problem, it is instructive to consider the problem of feasibly scheduling a *single coflow* subject to the matching constraints. Given a coflow G_j, the maximum degree of any vertex in the graph $\Delta(G_j) = \max_v \deg_G(v)$ is an obvious lower bound on the amount of time required to feasibly schedule coflow G_j. In fact, the following lemma by Qiu et al. [15] shows that this bound is always achievable for any coflow. The proof follows by repeated applications of Hall's theorem on the existence of perfect matchings in bipartite graphs.

Lemma 1 ([15]). *There exists a polynomial time algorithm that schedules a single coflow G_j in $\Delta(G_j)$ time steps.*

Lemma 1 also implicitly provides a way to decompose a bipartite graph G into two graphs G_1 and G_2 such that $\Delta(G) = \Delta(G_1) + \Delta(G_2)$. Given a time interval $(t_s, t_e]$, the following corollary uses such a decomposition to obtain a feasible coflow schedule for the given time interval by partially scheduling a coflow if necessary. We defer the proof to the full version [2].

Corollary 1. *Given a sequence of coflows $G_1, G_2, \ldots, G_n$, a start time t_s, and an end time t_e such that $t_e \geq t_s + \sum_{k=1}^{j-1} \Delta(G_k)$ and $t_e < t_s + \sum_{k=1}^{j} \Delta(G_k)$, there exists a polynomial time algorithm that finds a feasible coflow schedule for the time interval $(t_s, t_e]$ such that:*

- *coflows $G_1, G_2, \ldots, G_{j-1}$ are completely scheduled.*
- *coflow G_j is partially scheduled so that $\Delta(\tilde{G}_j) = t_s + \sum_{k=1}^{j} \Delta(G_k) - t_e$ where $\tilde{G}_j$ denotes the subset of coflow j that has not yet been scheduled.*
- *coflows $G_{j+1}, \ldots, G_n$ are not scheduled.*

2.2   Linear Programming Relaxation

By exploiting the connection with concurrent open-shop scheduling, we adapt the LP relaxation used for the concurrent open-shop problem [10,12] to formulate the following linear program as a relaxation of the coflow scheduling problem. We introduce a variable C_j for every coflow j to denote its completion time. Let $J = \{1, 2, \ldots, n\}$ denote the set of all coflows and $M = I \cup O$ denote the set of all the ports. Figure 2 shows our LP relaxation.

$$\min \sum_{j \in J} w_j C_j$$

$$\text{subject to,} \quad C_j \geq r_j + L_{i,j} \qquad\qquad\qquad \forall j \in J, \forall i \in M \qquad (2)$$

$$\sum_{j \in S} L_{i,j} C_j \geq \frac{1}{2} \left(\sum_{j \in S} L_{i,j}^2 + \left(\sum_{j \in S} L_{i,j} \right)^2 \right) \qquad \forall i \in M, \forall S \subseteq J \qquad (3)$$

Fig. 2. LP$_1$ for coflow scheduling

The first set of constraints (2) ensure that the completion time of any job j is at least its release time r_j plus the load of coflow j on any port i. The second set of constraints (3) are standard in parallel scheduling literature (e.g. [16]) and are used to effectively lower bound completion time variables. For simplicity, we define $f_i(S)$ for any subset $S \subseteq J$ and each port i as follow

$$f_i(S) = \frac{\sum_{j \in S} L_{i,j}^2 + \left(\sum_{j \in S} L_{i,j} \right)^2}{2} \qquad (4)$$

3   High Level Ideas

We use the LP above in Fig. 2 and its dual to develop a combinatorial algorithm (Algorithm 1) in Sect. 4.1 to obtain a good permutation of the coflows. This primal dual algorithm is inspired by Davis et al. [8] and Mastrolilli et al. [14]. As we show in Lemma 5, once the coflows are permuted as per this algorithm, we can bound the completion time of a coflow j in an optimal schedule in terms of $\Delta(\bigcup_{k \leq j} G_k)$, the maximum degree of the union of the first j coflows in the permutation.

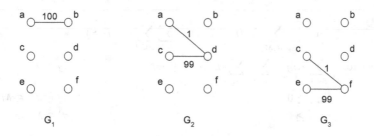

Fig. 3. Example that illustrates sequentially scheduling coflows independently can lead to bad schedules.

A naïve approach now would be to schedule each coflow independently and sequentially using Lemma 1 in this permutation. Since all coflows $k \leq j$ would need to be scheduled before starting to schedule j, the completion time of coflow j under such a scheme would be $\sum_{k \leq j} \Delta(G_k)$. Unfortunately, for arbitrary coflows we can have $\sum_{k \leq j} \Delta(G_k) \gg \Delta(\bigcup_{k \leq j} G_k)$. For instance, Fig. 3 shows three coflows such that $\Delta(G_1) + \Delta(G_2) + \Delta(G_3) = 300 > \Delta(G_1 \cup G_2 \cup G_3) = 101$.

One key insight is that sequentially scheduling coflows one after another may waste resources. Since the amount of time required to completely schedule a single coflow k only depends on the maximum degree of the graph G_k, if we augment graph G_k by adding edges such that its maximum degree does not increase, the augmented coflow can still be scheduled in the same time interval. This observation leads to the natural idea of "shifting" edges from a coflow j later in the permutation to a coflow k ($k < j$), so long as the release time of j is still respected, as such a shift does not delay coflow k further but may significantly reduce the requirements of coflow j. Consider for instance the coflows in Fig. 3 when all release times are zero; shifting the edge (c, d) from graph G_2 to G_1 and the edge (e, f) from G_3 to G_1 leaves $\Delta(G_1)$ unchanged but drastically reduces $\Delta(G_2)$ and $\Delta(G_3)$. In Algorithm 3 in Sect. 4.2, we formalize this notion of shifting edges and prove that after all such edges have been shifted, sequentially scheduling the augmented coflows leads to provably good coflow schedules.

In Sect. 6 we present an alternative approach using LP Rounding for finding a good permutation of coflows. Then we schedule the coflows using Algorithm 3 and give alternative proofs for Theorems 1 and 2.

4   Approximation Algorithm for Coflow Scheduling with Release Times

In this section we present a combinatorial 5-approximation algorithm for minimizing the weighted sum of completion times of a set of coflows with release times. Our algorithm consists of two stages. In the first stage, we design a primal-dual algorithm to find a good permutation of the coflows. In the second stage, we show that scheduling the coflows sequentially in this ordering after some postprocessing steps yields a provably good coflow schedule.

$$\max \sum_{j \in J} \sum_{i \in M} \alpha_{i,j}(r_j + L_{i,j}) + \sum_{i \in M} \sum_{S \subseteq J} \beta_{i,S} f_i(S)$$

$$\text{subject to,} \quad \sum_{i \in M} \alpha_{i,j} + \sum_{i \in M} \sum_{S/j \in S} L_{i,j} \beta_{i,S} \le w_j \qquad \forall j \in J$$

$$\alpha_{i,j} \ge 0 \qquad \forall j \in J, i \in M$$

$$\beta_{i,S} \ge 0 \qquad \forall i \in M, \forall S \subseteq J$$

Fig. 4. Dual of LP_1

4.1 Finding a Permutation of Coflows Using a Primal Dual Algorithm

Although our algorithm does not require solving a linear program, we use the linear program in Fig. 2 and its dual (Fig. 4) in the design and analysis of the algorithm.

Our algorithm works as follows. We build up a permutation of the coflows in the reverse order iteratively. Let κ be a constant that we optimize later. In any iteration, let j be the unscheduled job with the latest release time, let μ be the machine with the highest load and let L_μ be the load on machine μ. Now if $r_j > \kappa L_\mu$, we raise the dual variable $\alpha_{\mu,j}$ until the corresponding dual constraint is tight and place coflow j to be last in the permutation. But if $r_j \le \kappa L_\mu$, then we raise the dual variable $\beta_{\mu,J}$ until the dual constraint for some job j' becomes tight and place coflow j' to be last in the permutation. Algorithm 1 gives the formal description of the complete algorithm.

4.2 Scheduling Coflows According to a Permutation

We assume without loss of generality that the coflows are ordered based on the permutation given by Algorithm 1, i.e. $\sigma(j) = j$.

As we discussed in Sect. 3, naïvely scheduling the coflows sequentially in this order may not be a good idea. However, by appropriately moving edges from a coflow j to an earlier coflow k $(k < j)$, we can get a provably good schedule. The crux of our algorithm lies in the subroutine MoveEdgesBack defined in Algorithm 2.

Given two bipartite graphs G_k and G_j $(k < j)$, MoveEdgesBack greedily moves weighted edges from graph G_j to G_k so long as the maximum degree of graph G_k does not increase. The key idea behind this subroutine is that since the coflow k requires $\Delta(G_k)$ time units to be scheduled feasibly, the edges moved back can now also be scheduled in those $\Delta(G_k)$ time units for "free".

If all coflows have zero release times, then we can safely move edges of a coflow G_j to any G_k such that $k < j$. However, with the presence of arbitrary release times, we need to ensure that edges of coflow G_j do not violate their

---

Algorithm 1. Permuting Coflows

---

1 J is the set of unscheduled jobs and initially $J = \{1, 2, \cdots, n\}$;
2 Initialize $\alpha_{i,j} = 0$ for all $i \in M, j \in J$ and $\beta_{i,S} = 0$ for all $i \in M, S \subseteq J$;
3 $L_i = \sum_{j \in J} L_{ij} \ \forall i \in M$;   // load of machine i
4 **for** $k = n, n-1, \cdots, 1$ **do**
5 | $\mu(k) = \arg \max_{i \in M} L_i$;   // determine the machine with highest load
6 | $j = \arg \max_{\ell \in J} r_\ell$;   // determine job that released last
7 | **if** $r_j > \kappa \cdot L_{\mu(k)}$ **then**
8 | | $\alpha_{\mu(k),j} = (w_j - \sum_{i \in M} \sum_{S \ni j} L_{i,j}\beta_{i,S})$;
9 | | $\sigma(k) \leftarrow j$;
10 | **end**
11 | **else if** $r_{\sigma(k)} \le \kappa \cdot L_{\mu(k)}$ **then**
12 | | $j' = \arg \min_{j \in J} \left(\frac{w_j - \sum_{i \in M} \sum_{S \ni j} L_{i,j}\beta_{i,S}}{L_{\mu(k),j}} \right)$;
13 | | $\beta_{\mu(k),J} = \left(\frac{w_{j'} - \sum_{i \in M} \sum_{S \ni j'} L_{i,j'}\beta_{i,S}}{L_{\mu(k),j'}} \right)$;
14 | | $\sigma(k) \leftarrow j'$;
15 | **end**
16 | $J \leftarrow J \setminus \sigma(k)$;
17 | $L_i \leftarrow L_i - L_{i,\sigma(k)}, \forall i \in M$;
18 **end**
19 Output permutation $\sigma(1), \sigma(2), \cdots, \sigma(n)$;

Algorithm 2. The MoveEdgesBack subroutine.

---

1 **Function** *MoveEdgesBack(G_k, G_j)*
2 | **for** $e = (u, v) \in G_j$ **do**
3 | | $\delta = \min(\Delta(G_k) - \deg_{G_k}(u), \Delta(G_k) - \deg_{G_k}(v), w_{G_j}(e))$;
4 | | $w_{G_j}(e) = w_{G_j}(e) - \delta$;
5 | | $w_{G_k}(e) = w_{G_k}(e) + \delta$;
6 | **end**
7 | **return** G_k, G_j;

---

release time, i.e. they are scheduled only after they are released. Algorithm 3 describes the pseudo-code for coflow scheduling with arbitrary release times. Here q denote the number of distinct values taken by the release times of the n coflows. Further, let $t_1 < t_2 < \ldots < t_q$ be the ordered set of the release times. For simplicity, we define $t_{q+1} = T$ as a sufficiently large time horizon.

At any time step t_i, let $G'_j \subseteq G_j$ denote the subgraph of coflow j that has not been scheduled yet. We consider every ordered pair of coflows $k < j$ such that both the coflows have been released and MoveEdgesBack from graph G'_j to graph G'_k. Finally, we begin to schedule the coflows sequentially in order using Corollary 1 until all coflows are scheduled completely or we reach time t_{i+1} when a new set of coflows gets released and the process repeats.

Algorithm 3. Coflow Scheduling

---

1 $q \leftarrow$ number of distinct release times; $t_{q+1} \leftarrow T$;
2 $t_1, t_2, \ldots, t_q \leftarrow$ distinct release time in increasing order ;
3 **for** $i = 1, 2, \ldots, q$ **do**
4 $\quad$ // Each loop finds a schedule for time interval $(t_i, t_{i+1}]$
5 $\quad$ **for** $j = 1, 2, \ldots, n$ **do**
6 $\quad\quad\mid$ $G'_j \leftarrow$ unscheduled part of G_j;
7 $\quad$ **end**
8 $\quad$ **for** $k = 1, 2, \ldots, n - 1$ **do**
9 $\quad\quad$ **if** $r_k \leq t_i$ **then**
10 $\quad\quad\quad$ **for** $j = k + 1, \ldots, n$ **do**
11 $\quad\quad\quad\quad\mid$ **if** $r_j \leq t_i$ **then** $G'_k, G'_j \leftarrow$ MoveEdgesBack(G'_k, G'_j) ;
12 $\quad\quad\quad$ **end**
13 $\quad\quad$ **end**
14 $\quad$ **end**
15 $\quad$ Schedule $(G'_1, G'_2, \ldots, G'_n)$ in $(t_i, t_{i+1}]$ using Corollary 1;
16 **end**

---

5   Analysis

We first analyze Algorithm 3 and upper bound the completion time of a coflow j in terms of the maximum degree of the cumulative graph obtained by combining the first j coflows in the given permutation. For simplicity, we first state the proof when all release times are zero, then proceed to the case with non-zero release time.

5.1   Coflows with Zero Release Times

For ease of presentation we first analyze the special case when all coflows are released at time zero. In this case, we have $q = 1$ in Algorithm 3 and thus the outer *for* loop is only executed once. The following lemma shows that after the MoveEdgesBack subroutine has been executed on every ordered pair of coflows, for any coflow j, the sum of maximum degrees of graphs G'_k ($k \leq j$) is at most twice the maximum degree of the cumulative graph obtained by combining the first j coflows.

Lemma 2. *For all* $j \in \{1, 2, \ldots n\}$, $\sum_{k \leq j} \Delta(G'_k) \leq 2\Delta(\bigcup_{k \leq j} G_k)$.

Lemma 3. *Consider any coflow j and let $C_j(alg)$ denote the completion time of coflow j when scheduled as per Algorithm 3. Then $C_j(alg) \leq 2\Delta(\bigcup_{k \leq j} G_k)$.*

5.2   Coflows with Arbitrary Release Times

When the coflows have arbitrary release times, we can bound the completion time of each coflow j in terms of the maximum degree of the cumulative graph obtained by combining the first j coflows and the largest release time of all the jobs before j in the permutation.

Lemma 4. *For any coflow j, let $C_j(alg)$ denote the completion time of coflow j when scheduled as per Algorithm 3. Then $C_j(alg) \leq \max_{k \leq j} r_k + 2\Delta(\bigcup_{k \leq j} G_k)$.*

5.3   Analyzing the Primal-Dual Algorithm

We are now in a position to analyze Algorithm 1. Recall that we assume that the jobs are sorted as per the permutation obtained by Algorithm 1, i.e., $\sigma(k) = k, \forall k \in [n]$. We first give a lemma,

Lemma 5. *If there is an algorithm that generates a feasible coflow schedule such that for any coflow j, $C_j(alg) \leq a \max_{k \leq j} r_k + b\Delta(\bigcup_{k \leq j} G_k)$ for some constants a and b, then the total cost of the schedule is bounded as follows.*

$$\sum_j w_j C_j(alg) \leq (a + \frac{b}{\kappa}) \sum_{j=1}^{n} \sum_{i \in M} \alpha_{i,j} r_j + 2(a\kappa + b) \sum_{i \in M} \sum_{S \subseteq J} \beta_{i,S} f_i(S)$$

Proof Sketch. Algorithm 1 judiciously sets the dual variables such that the dual constraint for any coflow j is tight. Analyzing the cost of schedule obtained in terms of the dual variables yields the lemma. The formal proof is available in the full version [2].

Lemmas 3 and 4 along with Lemma 5 and an appropriate choice of κ now give the desired theorems. Proof in journal version.

Theorem 1. *There exists a deterministic, combinatorial, polynomial time 5-approximation algorithm for coflow scheduling with release times.*

Theorem 2. *There exists a deterministic, combinatorial, polynomial time 4-approximation algorithm for coow scheduling without release times.*

6   An Alternative Approach Using LP Rounding

This alternative approach also consists of two stages. First, we find a good permutation of coflows and after that we schedule the coflows sequentially in this ordering using Algorithm 3.

Let $\overline{C_j}$ denote the completion time of job j in an optimal **LP$_1$** solution. We assume without loss of generality that the coflows are ordered so that the following holds.

$$\overline{C_1} \leq \overline{C_2} \leq \ldots \leq \overline{C_n} \tag{5}$$

We can use the LP-constraints to provide a lower bound on $\overline{C_j}$ in terms of the maximum degree of the cumulative graph obtained by combining the first j coflows. In particular, the following lemma follows from the constraints of **LP$_1$**.

Lemma 6. *For each coflow $j = 1, 2, \ldots, n$, the following inequality holds.*

$$\overline{C_j} \geq \frac{1}{2} \max_i \left\{ \sum_{k=1}^{j} L_{i,k} \right\} = \frac{1}{2} \Delta(\bigcup_{k \leq j} G_k)$$

Lemmas 3 and 4 along with Lemma 6 give alternative proofs for Theorems 1 and 2.

References

1. https://hadoop.apache.org
2. http://cs.umd.edu/~samir/ipco17-fullversion.pdf
3. Bansal, N., Khot, S.: Inapproximability of hypergraph vertex cover and applications to scheduling problems. In: Abramsky, S., Gavoille, C., Kirchner, C., Meyer auf der Heide, F., Spirakis, P.G. (eds.) ICALP 2010. LNCS, vol. 6198, pp. 250–261. Springer, Heidelberg (2010). doi:10.1007/978-3-642-14165-2_22
4. Chen, Z.-L., Hall, N.G.: Supply chain scheduling: conflict and cooperation in assembly systems. Oper. Res. **55**(6), 1072–1089 (2007)
5. Chowdhury, M., Stoica, I. Coflow: a networking abstraction for cluster applications. In: ACM Workshop on Hot Topics in Networks, pp. 31–36. ACM (2012)
6. Chowdhury, M., Stoica, I.: Efficient coflow scheduling without prior knowledge. In: SIGCOMM, pp. 393–406. ACM (2015)
7. Chowdhury, M., Zhong, Y., Stoica, I.: Efficient coflow scheduling with varys. In: SIGCOMM, SIGCOMM 2014, pp. 443–454. ACM, New York (2014)
8. Davis, J.M., Gandhi, R., Kothari, V.H.: Combinatorial algorithms for minimizing the weighted sum of completion times on a single machine. Oper. Res. Lett. **41**(2), 121–125 (2013)
9. Dean, J., Ghemawat, S.: Mapreduce: simplified data processing on large clusters. Commun. ACM **51**(1), 107–113 (2008)
10. Garg, N., Kumar, A., Pandit, V.: Order scheduling models: hardness and algorithms. In: Arvind, V., Prasad, S. (eds.) FSTTCS 2007. LNCS, vol. 4855, pp. 96–107. Springer, Heidelberg (2007). doi:10.1007/978-3-540-77050-3_8
11. Khuller, S., Li, J., Sturmfels, P., Sun, K., Venkat, P.: Select, permute: an improved online framework for scheduling to minimize weighted completion time (2016) (Submitted)
12. Leung, J.Y.-T., Li, H., Pinedo, M.: Scheduling orders for multiple product types to minimize total weighted completion time. Discrete Appl. Math. **155**(8), 945–970 (2007)
13. Luo, S., Yu, H., Zhao, Y., Wang, S., Yu, S., Li, L.: Towards practical, near-optimal coflow scheduling for data center networks. IEEE Trans. Parallel Distrib. Syst. PP(99), 1 (2016)
14. Mastrolilli, M., Queyranne, M., Schulz, A.S., Svensson, O., Uhan, N.A.: Minimizing the sum of weighted completion times in a concurrent open shop. Oper. Res. Lett. **38**(5), 390–395 (2010)
15. Qiu, Z., Stein, C., Zhong, Y.: Minimizing the total weighted completion time of coflows in datacenter networks. In: SPAA 2015, pp. 294–303. ACM, New York (2015)
16. Queyranne, M.: Structure of a simple scheduling polyhedron. Math. Program. **58**(1–3), 263–285 (1993)
17. Sachdeva, S., Saket, R.: Optimal inapproximability for scheduling problems via structural hardness for hypergraph vertex cover. In: IEEE Conference on Computational Complexity, pp. 219–229. IEEE (2013)
18. Wang, G., Cheng, T.E.: Customer order scheduling to minimize total weighted completion time. Omega **35**(5), 623–626 (2007)
19. Zaharia, M., Chowdhury, M., Franklin, M.J., Shenker, S., Stoica, I.: Spark: cluster computing with working sets. HotCloud **10**, 10 (2010)
20. Zhao, Y., Chen, K., Bai, W., Yu, M., Tian, C., Geng, Y., Zhang, Y., Li, D., Wang, S. Rapier: integrating routing and scheduling for coflow-aware data center networks. In: INFOCOM, pp. 424–432. IEEE (2015)

Integrality Gaps of Integer Knapsack Problems

Iskander Aliev[1], Martin Henk[2], and Timm Oertel[1(✉)]

[1] Cardiff University, Cardiff, UK
{alievi,oertelt}@cardiff.ac.uk
[2] TU Berlin, Berlin, Germany
henk@math.tu-berlin.de

Abstract. We obtain optimal lower and upper bounds for the (additive) integrality gaps of integer knapsack problems. In a randomised setting, we show that the integrality gap of a "typical" knapsack problem is drastically smaller than the integrality gap that occurs in a worst case scenario.

1 Introduction

Given an integer $m \times n$ matrix A, integer vector $\boldsymbol{b} \in \mathbb{Z}^m$ and a cost vector $\boldsymbol{c} \in \mathbb{Q}^n$, consider the linear integer programming problem

$$\min\{\boldsymbol{c} \cdot \boldsymbol{x} : A\boldsymbol{x} = \boldsymbol{b}, \boldsymbol{x} \in \mathbb{Z}_{\geq 0}^n\}. \tag{1}$$

The linear programming relaxation to (1) is obtained by dropping the integrality constraint

$$\min\{\boldsymbol{c} \cdot \boldsymbol{x} : A\boldsymbol{x} = \boldsymbol{b}, \boldsymbol{x} \in \mathbb{R}_{\geq 0}^n\}. \tag{2}$$

We will denote by $IP_c(A, \boldsymbol{b})$ and $LP_c(A, \boldsymbol{b})$ the optimal values of (1) and (2), respectively.

While the problem (2) is polynomial time solvable [20], it is well known that (1) is NP-hard [14]. There are many examples, where relaxation on the integrality constraints are used to approximate, or even to solve, integer programming problems. Prominent examples can be found in the areas of cutting plane algorithms, such us Gomory cuts [15], and approximation algorithms for combinatorial problems. For further details see [3,8,28]. Therefore, a natural question is to compare the optimal values IP_c and LP_c with each other.

Suppose that (1) is feasible and bounded. The *(additive) integrality gap* $IG_c(A, \boldsymbol{b})$ is a fundamental characteristic of the problem (1), defined as

$$IG_c(A, \boldsymbol{b}) = IP_c(A, \boldsymbol{b}) - LP_c(A, \boldsymbol{b}).$$

The problem of computing bounds for the additive integrality gaps has been studied by Hoşten and Sturmfels [18], Sullivant [27], Eisenbrand and Shmonin [12] and, more recently, by Eisenbrand et al. [11]. Specifically, given a tuple $(A, \boldsymbol{c})$ one asks for the upper bounds on $IG_c(A, \boldsymbol{b})$ as $\boldsymbol{b}$ varies. In this setting, the optimal

© Springer International Publishing AG 2017
F. Eisenbrand and J. Koenemann (Eds.): IPCO 2017, LNCS 10328, pp. 25–38, 2017.
DOI: 10.1007/978-3-319-59250-3_3

bound is given by the *integer programming gap* $\mathrm{Gap}_c(A)$, defined by Hoşten and Sturmfels [18] as

$$\mathrm{Gap}_c(A) = \max_b IG_c(A, b),$$

where b ranges over integer vectors such that (1) is feasible and bounded. Note that, $\mathrm{Gap}_c(A) = 0$ for all $c \in \mathbb{Z}^n$, if and only if A is totally unimodular [25, Theorem 19.2]. Hoşten and Sturmfels [18] showed that for fixed n the value of $\mathrm{Gap}_c(A)$ can be computed in polynomial time. Eisenbrand and Shmonin [12] extended this result to integer programs in the canonical form.

Eisenbrand et al. [11] studied a closely related problem of testing upper bounds for $IG_c(A, b)$ in context of a generalised *integer rounding property*. Following [11], the tuple (A, c) with $c \in \mathbb{Z}^n$ has the *additive integrality gap of at most γ* if

$$IP_c(A, b) \leq \lceil LP_c(A, b) \rceil + \gamma$$

for each b for which the linear programming relaxation (2) is feasible.

The classical case $\gamma = 0$ corresponds to the integer rounding property and can be tested in polynomial time [25, Sect. 22.10]. The integer rounding property, in its turn, implies solvability of (1) in polynomial time [7]. The computational complexity of the problem drastically changes already for $\gamma = 1$. Eisenbrand et al. [11] showed that it is NP-hard to test whether (A, c) has additive gap of at most γ even if $m = \gamma = 1$.

A bound for the additive integrality gap in terms of A and c can be derived from the results of Cook et al. [9] on distances between optimal solutions to integer programs in canonical form and their linear programming relaxations. Let $\hat{A}$ be an integer $d \times n$ matrix and let $\hat{b}$ and c be rational vectors such that $\hat{A}x \leq \hat{b}$ has an integer solution and $\min\{c \cdot x : \hat{A}x \leq \hat{b}, x \in \mathbb{R}^n\}$ exists. Note that, in this setting $\hat{b}$ is not required to be integer. Then Corollary 2 in [9], applied in the minimisation setting, gives the bound

$$\min\{c \cdot x : \hat{A}x \leq \hat{b}, x \in \mathbb{Z}^n\} - \min\{c \cdot x : \hat{A}x \leq \hat{b}, x \in \mathbb{R}^n\} \\ \leq n\Delta(A)\|c\|_1, \tag{3}$$

where $\Delta(A)$ stands for the maximum sub-determinant of A and $\|c\|_1 = \sum_{i=1}^n |c_i|$ denotes the l_1-*norm* of c. The estimate (3) strengthened previous results of Blair and Jeroslow [4,5]. Given that $\hat{b}$ does not have to be integer, one can show that the bound (3) is essentially tight (see Remark 1). However, considering that we study linear integer programming, it is natural to assume that also $\hat{b}$ is integer, but then it is not clear whether (3) remains optimal. By studying linear integer programming problems in standard form we naturally require b and respectively $\hat{b}$ to be integer.

This paper will focus on the problem (1) with $m = 1$, to which we refer to as the *integer knapsack problem*. Note that usually the integer knapsack problem is defined in the literature as $\min\{\bar{c} \cdot x : \bar{A}x \leq b, x \in \mathbb{Z}^n_{\geq 0}\}$. However, this problem can be brought into standard form (1), by lifting the polytope by one dimension and defining $A = (\bar{A}\ 1)$ and $c = \binom{\bar{c}}{0}$. We will assume that the entries of A are positive. For the integer knapsack problem the positivity assumption guarantees that the feasible region of its linear programming relaxation (2) is bounded (or empty) for all b. Conversely, for $m = 1$ any linear problem (2) with bounded feasible region can be written with A satisfying the positivity assumption. Without loss of generality, we also assume that $n \geq 2$ and the entries of A are coprime. That is the following conditions are assumed to hold:

$$\begin{aligned} &(i)\ \ A = (a_1, \ldots, a_n), n \geq 2, a_i \in \mathbb{Z}_{>0}, i = 1, \ldots, n, \\ &(ii)\ \gcd(a_1, \ldots, a_n) = 1. \end{aligned} \tag{4}$$

For $A \in \mathbb{Z}^{1 \times n}$ we denote by $\|A\|_\infty$ its *maximum norm*, i.e., $\|A\|_\infty = \max_{i=1,\ldots,n} |a_i|$. Applying (3) with

$$\hat{A} = \begin{pmatrix} A \\ -A \\ -I_n \end{pmatrix}, \ \hat{b} = \begin{pmatrix} b \\ -b \\ 0 \end{pmatrix},$$

where I_n is the $n \times n$ identity matrix and 0 is the n dimensional zero vector, we obtain the bound

$$\mathrm{Gap}_c(A) \leq n\|A\|_\infty \|c\|_1. \tag{5}$$

How far is the bound (5) from being optimal? Does $\mathrm{Gap}_c(A)$ admit a natural lower bound? To answer these questions we will establish a link between the integer programming gaps, covering radii of simplices and Frobenius numbers. Our first result gives an upper bound on the integer programming gap that improves (5) with factor $1/n$. We also show that the obtained bound is optimal.

Theorem 1. *(i) Let A satisfy (4) and let $c \in \mathbb{Q}^n$. Then*

$$\mathrm{Gap}_c(A) \leq (\|A\|_\infty - 1)\|c\|_1. \tag{6}$$

(ii) For any positive integer k there exist A with $\|A\|_\infty = k$ satisfying (4) and $c \in \mathbb{Q}^n$ such that

$$\mathrm{Gap}_c(A) = (\|A\|_\infty - 1)\|c\|_1. \tag{7}$$

We will say that the tuple (A, c) is *generic* if for any positive $b \in \mathbb{Z}$ the linear programming relaxation (2) has a unique optimal solution. An optimal lower bound for $\mathrm{Gap}_c(A)$ with generic (A, c) can be obtained using recent results [1] on the *lattice programming gaps* associated with the group relaxations to (1).

A subset τ of $\{1, \ldots, n\}$ partitions $x \in \mathbb{R}^n$ as x_τ and $x_{\bar{\tau}}$, where x_τ consists of the entries indexed by τ and $x_{\bar{\tau}}$ the entries indexed by the complimentary set $\bar{\tau} = \{1, \ldots, n\} \setminus \tau$. Similarly, the matrix A is partitioned as A_τ and $A_{\bar{\tau}}$. Assume that (A, c) is generic and (4) holds. Then, let $\tau = \tau(A, c)$ denote the unique index of the basic variable for the optimal solution to the linear relaxation (2) with a positive $b \in \mathbb{Z}$. The index τ is well-defined. We also define $l(A, c) = c_{\bar{\tau}} - c_\tau A_\tau^{-1} A_{\bar{\tau}}$. Note that the vector $l = l(A, c)$ is positive for generic tuples (A, c).

Let ρ_d denote the *covering constant* of the standard d-dimensional simplex, defined in Sect. 2.

Theorem 2. *(i) Let A satisfy (4) and let $c \in \mathbb{Q}^n$. Suppose that (A, c) is generic. Then for $\tau = \tau(A, c)$ and $l = l(A, c)$ we have*

$$\mathrm{Gap}_c(A) \geq \rho_{n-1}(|A_\tau| l_1 \cdots l_{n-1})^{1/(n-1)} - \|l\|_1 . \tag{8}$$

(ii) For any $\epsilon > 0$, there exists a matrix A, satisfying (4) and $c \in \mathbb{Q}^n$ such that (A, c) is generic and, in the notation of part (i), we have

$$\mathrm{Gap}_c(A) < (\rho_{n-1} + \epsilon)(|A_\tau| l_1 \cdots l_{n-1})^{1/(n-1)} - \|l\|_1 . \tag{9}$$

The only known values of ρ_d are $\rho_1 = 1$ and $\rho_2 = \sqrt{3}$ (see [13]). It was proved in [2], that $\rho_d > (d!)^{1/d} > d/e$. For sufficiently large d this bound is not far from being optimal. Indeed, $\rho_d \leq (d!)^{1/d}(1 + O(d^{-1} \log d))$ (see [10,21]).

How large is the integer programming gap of a "typical" knapsack problem? To tackle this question we will utilize the recent strong results of Strömbergsson [26] (see also Schmidt [24] and references therein) on the asymptotic distribution of Frobenius numbers. The main result of this paper will show that for any $\epsilon > 2/n$ the ratio

$$\frac{\mathrm{Gap}_c(A)}{\|A\|_\infty^\epsilon \|c\|_1}$$

is bounded, on average, by a constant that depends only on dimension n. Hence, for fixed $n > 2$ and a "typical" integer knapsack problem with large $\|A\|_\infty$, its linear programming relaxation provides a drastically better approximation to the solution than in the worst case scenario, determined by the optimal upper bound (6).

For $T \geq 1$, let $Q(T)$ be the set of $A \in \mathbb{Z}^{1 \times n}$ that satisfy (4) and

$$\|A\|_\infty \leq T .$$

Let $N(T)$ be the cardinality of $Q(T)$. For $\epsilon \in (0, 1)$ let

$$N_\epsilon(t, T) = \# \left\{ A \in Q(T) : \max_{c \in \mathbb{Q}^n} \frac{\mathrm{Gap}_c(A)}{\|A\|_\infty^\epsilon \|c\|_1} > t \right\} . \tag{10}$$

In what follows, $\ll_n$ will denote the Vinogradov symbol with the constant depending on n. That is $f \ll_n g$ if and only if $|f| \leq c|g|$, for some positive constant $c = c(n)$. The notation $f \asymp_n g$ means that both $f \ll_n g$ and $g \ll_n f$ hold.

Theorem 3. *For $n \geq 3$*

$$\frac{N_\epsilon(t,T)}{N(T)} \ll_n t^{-\alpha(\epsilon,n)} \tag{11}$$

uniformly over all $t > 0$ and $T \geq 1$. Here

$$\alpha(\epsilon,n) = \frac{n-2}{(1-\epsilon)n}.$$

From (11) one can derive an upper bound on the average value of the (normalised) integer programming gap.

Corollary 4. *Let $n \geq 3$. For $\epsilon > 2/n$*

$$\frac{1}{N(T)} \sum_{A \in Q(T)} \max_{c \in \mathbb{Q}^n} \frac{\mathrm{Gap}_c(A)}{\|A\|_\infty^\epsilon \|c\|_1} \ll_n 1. \tag{12}$$

The last theorem of this paper shows that the bound in Corollary 4 is not far from being optimal. We include its proof in the Appendix.

Theorem 5. *For T large*

$$\frac{1}{N(T)} \sum_{A \in Q(T)} \max_{c \in \mathbb{Q}^n} \frac{\mathrm{Gap}_c(A)}{\|A\|_\infty^{1/(n-1)} \|c\|_1} \gg_n 1. \tag{13}$$

Hence, the optimal value of ϵ in (12) cannot be smaller than $1/(n-1)$.

Remark 1.

(i) An example due to L. Lovász [25, Sect. 17.2], with $\Delta(A) = 1$, shows that the bound (3) is best possible in this particular case. We would like to point out that by a small adaptation of Lovász's example one can show that this bound is, in all its generality, best possible up to a constant factor, i.e., the upper bound for the additive integrality gap is in $\Theta(\Delta(A)n)$. Let $\delta \in \mathbb{Z}_{>0}$ and $0 < \beta < 1$. We define

$$A = \begin{pmatrix} 1 & & & \\ -1 & 1 & & \\ & \ddots & & \\ & & -1 & 1 \\ & & & -\delta & 1 \end{pmatrix}, \quad b = \begin{pmatrix} \beta \\ \vdots \\ \beta \end{pmatrix} \quad \text{and} \quad c = \begin{pmatrix} -1 \\ \vdots \\ -1 \end{pmatrix}.$$

By construction $\Delta(A) = \delta$. The unique solution of the linear relaxation is $x^T = (\beta, 2\beta, \ldots, (n-1)\beta, (\delta(n-1)+1)\beta)$ and the unique optimal integer solution is $z^T = (0, \ldots, 0)$. Thus $\|x - z\|_\infty = (\delta(n-1)+1)\beta \approx n\Delta(A)$.

(ii) In the proof of Theorem 1 (and, subsequently, Theorem 3) we estimate the integrality gap using a covering argument that guarantees existence of a solution to (1) in an $(n-1)$-dimensional simplex of sufficiently small diameter, translated by a solution to (2). Here the diameter of the simplex is independent of c. The argument allows us, in particular, to restate Theorem 1 (i) in terms of the infinity norm:

$$\mathrm{Gap}_c(A) \leq 2\left(\|A\|_\infty - 1\right)\|c\|_\infty .$$

Depending on c this gives a stronger bound.

2 Coverings and Frobenius Numbers

In what follows, $\mathcal{K}^d$ will denote the space of all d-dimensional *convex bodies*, i.e., closed bounded convex sets with non-empty interior in the d-dimensional Euclidean space $\mathbb{R}^d$.

By $\mathcal{L}^d$ we denote the set of all d-dimensional lattices in $\mathbb{R}^d$. Given a matrix $B \in \mathbb{R}^{d \times d}$ with $\det B \neq 0$ and a set $S \subset \mathbb{R}^d$ let $BS = \{Bx : x \in S\}$ be the image of S under linear map defined by B. Then we can write $\mathcal{L}^d = \{B\mathbb{Z}^d : B \in \mathbb{R}^{d \times d}, \det B \neq 0\}$. For $\Lambda = B\mathbb{Z}^d \in \mathcal{L}^d$, $\det(\Lambda) = |\det B|$ is called the *determinant* of the lattice Λ.

Recall that the *Minkowski sum* $X + Y$ of the sets $X, Y \subset \mathbb{R}^d$ consists of all points $x + y$ with $x \in X$ and $y \in Y$. For $K \in \mathcal{K}^d$ and $\Lambda \in \mathcal{L}^d$ the *covering radius* of K with respect to Λ is the smallest positive number μ such that any point $x \in \mathbb{R}^d$ is covered by $\mu K + \Lambda$, that is

$$\mu(K, \Lambda) = \min\{\mu > 0 : \mathbb{R}^d = \mu K + \Lambda\} .$$

For further information on covering radii in the context of the geometry of numbers see e.g. Gruber [16] and Gruber and Lekkerkerker [17].

Let $\Delta = \{x \in \mathbb{R}^d_{\geq 0} : x_1 + \cdots + x_d \leq 1\}$ be the standard d-dimensional simplex. The optimal lower bound in Theorem 2 is expressed using the covering constant $\rho_d = \rho_d(\Delta)$ defined as

$$\rho_d = \inf\{\mu(\Delta, \Lambda) : \det(\Lambda) = 1\} .$$

We will be also interested in coverings of $\mathbb{Z}^d$ by lattice translates of convex bodies. For this purpose we define

$$\mu(K, \Lambda; \mathbb{Z}^d) = \min\{\mu > 0 : \mathbb{Z}^d \subset \mu K + \Lambda\} .$$

Given $A = (a_1, \ldots, a_n)$ satisfying (4) the *Frobenius number* $g(A)$ is least so that every integer $b > g(A)$ can be represented as $b = a_1 x_1 + \cdots + a_n x_n$ with nonnegative integers $x_1, \ldots, x_n$.

Kannan [19] found a nice and very useful connection between $g(A)$ and geometry of numbers. Let us consider the $(n-1)$-dimensional simplex

$$S_A = \left\{x \in \mathbb{R}^{n-1}_{\geq 0} : a_1 x_1 + \cdots + a_{n-1} x_{n-1} \leq 1\right\}$$

and the $(n-1)$-dimensional lattice

$$\Lambda_A = \left\{ \boldsymbol{x} \in \mathbb{Z}^{n-1} : a_1 x_1 + \cdots + a_{n-1} x_{n-1} \equiv 0 \bmod a_n \right\}.$$

Kannan [19] established the identities

$$\mu(S_A, \Lambda_A) = g(A) + a_1 + \cdots + a_n$$

and

$$\mu(S_A, \Lambda_A; \mathbb{Z}^{n-1}) = g(A) + a_n. \tag{14}$$

3   Proof of Theorem 1

The proof of the upper bound in part (i) will be based on two auxiliary lemmas. First we will need the following property of $\mu(K, \Lambda; \mathbb{Z}^{n-1})$.

Lemma 1. *For any $\boldsymbol{y} \in \mathbb{Z}^{n-1}$ the set $\mu(K, \Lambda; \mathbb{Z}^{n-1})K$ contains a point of the translated lattice $\boldsymbol{y} + \Lambda$.*

Proof. By the definition of $\mu(K, \Lambda; \mathbb{Z}^{n-1})$ we have $\mathbb{Z}^{n-1} \subset \mu(K, \Lambda; \mathbb{Z}^{n-1})K + \Lambda$. Therefore for any integer vector $\boldsymbol{y}$ we have $(\boldsymbol{y} + \Lambda) \cap \mu(K, \Lambda; \mathbb{Z}^{n-1})K \neq \emptyset$.   □

The next lemma gives an upper bound for the integer programming gap in terms of the Frobenius number associated with vector A.

Lemma 2. *For A satisfying (4) and $\boldsymbol{c} \in \mathbb{Q}^n$*

$$\mathrm{Gap}_{\boldsymbol{c}}(A) \leq \frac{(g(A) + \|A\|_\infty)\|\boldsymbol{c}\|_1}{\min_i a_i}. \tag{15}$$

Proof. Let b be a nonnegative integer. Consider the *knapsack polytope*

$$P(A, b) = \{\boldsymbol{x} \in \mathbb{R}^n_{\geq 0} : A\boldsymbol{x} = b\}.$$

Clearly, $P(A, b)$ is a simplex with vertices

$$(b/a_1, 0, \ldots, 0), (0, b/a_2, \ldots, 0), \ldots, (0, \ldots, 0, b/a_n)$$

and

$$P(A, b) \subset \left[0, \frac{b}{\min_i a_i} \right]^n. \tag{16}$$

Notice also that

$$bS_A = \pi_n(P(A, b)), \tag{17}$$

where $\pi_n(\cdot) : \mathbb{R}^n \to \mathbb{R}^{n-1}$ is the projection that forgets the last coordinate.

Rearranging the entries of A, if necessary, we may assume that the optimal value $LP_{\boldsymbol{c}}(A, b)$ is attained at the vertex $\boldsymbol{v} = (0, \ldots, 0, b/a_n)$ of $P(A, b)$.

If $b \leq \mu(S_A, \Lambda_A; \mathbb{Z}^{n-1})$ then (14) and (16) imply that the integrality gap is bounded by the right hand side of (15).

Suppose now that $b > \mu(S_A, \Lambda_A; \mathbb{Z}^{n-1})$. Then, in view of (17),

$$\mu(S_A, \Lambda_A; \mathbb{Z}^{n-1})S_A \subset \pi_n(P(A, b)). \tag{18}$$

Let $\Lambda(A, b) = \{x \in \mathbb{Z}^n : Ax = b\}$ be the set of integer points in the affine hyperplane $Ax = b$. There exists $y \in \mathbb{Z}^{n-1}$ such that

$$\pi_n(\Lambda(A, b)) = y + \Lambda_A. \tag{19}$$

By Lemma 1, there is a point $(z_1, \ldots, z_{n-1}) \in \pi_n(\Lambda(A, b)) \cap \mu(S_A, \Lambda_A; \mathbb{Z}^{n-1})S_A$. Hence

$$z = \left(z_1, \ldots, z_{n-1}, \frac{b}{a_n} - \frac{a_1 z_1 + \cdots + a_{n-1} z_{n-1}}{a_n}\right) \in \Lambda(A, b) \cap P(A, b) \tag{20}$$

is a feasible integer point for the knapsack problem (1).

Since $(z_1, \ldots, z_{n-1}) \in \mu(S_A, \Lambda_A; \mathbb{Z}^{n-1})S_A$, we have

$$\|v - z\|_\infty \leq \frac{\mu(S_A, \Lambda_A; \mathbb{Z}^{n-1})}{\min_i a_i} \leq \frac{g(A) + \|A\|_\infty}{\min_i a_i}, \tag{21}$$

where the last inequality follows from (14). Therefore, the integrality gap is bounded by the right hand side of (15). □

To complete the proof of part (i) we need the classical upper bound for the Frobenius number due to Schur (see Brauer [6]):

$$g(A) \leq (\min_i a_i)\|A\|_\infty - (\min_i a_i) - \|A\|_\infty. \tag{22}$$

Combining (15) and (22) we obtain (6).

To prove part (ii), we set $A = (k, \ldots, k, 1)$, $b = k - 1$ and $c = e_n$, where e_i denotes the i-th unit-vector. Note that A fulfils the conditions (4). The integer programming problem (1) has precisely one feasible, and therefore optimal, integer point, namely $(k - 1) \cdot e_n$. Thus $IP_c(A, b) = k - 1$. The corresponding linear relaxation (2) has the, in general not unique, optimal solution $\frac{k-1}{k} \cdot e_1$ with $LP_c(A, b) = 0$. Hence, $\text{Gap}_c(A) \geq IG_c(A, b) = k - 1 = (\|A\|_\infty - 1)\|c\|_1$.

4   Proof of Theorem 2

We will first establish a connection between $\text{Gap}_c(A)$ and the lattice programming gap associated with a certain lattice program.

For a vector $w \in \mathbb{Q}_{>0}^{n-1}$, a $(n-1)$-dimensional lattice $\Lambda \subset \mathbb{Z}^{n-1}$ and $r \in \mathbb{Z}^{n-1}$ consider the lattice program (also referred to as the *group problem*)

$$\min\{w \cdot x : x \equiv r(\bmod \Lambda), x \in \mathbb{R}_{\geq 0}^{n-1}\}. \tag{23}$$

Here $x \equiv r \pmod{\Lambda}$ if and only if $x - r$ is a point of Λ.

Let $m(\Lambda, \boldsymbol{w}, \boldsymbol{r})$ denote the value of the minimum in (23). The *lattice programming gap* $\mathrm{Gap}(\Lambda, \boldsymbol{w})$ of (23) is defined as

$$\mathrm{Gap}(\Lambda, \boldsymbol{w}) = \max_{\boldsymbol{r} \in \mathbb{Z}^{n-1}} m(\Lambda, \boldsymbol{w}, \boldsymbol{r}). \tag{24}$$

The lattice programming gaps were introduced and studied for sublattices of all dimensions in $\mathbb{Z}^{n-1}$ by Hoşten and Sturmfels [18].

To proceed with the proof of the part (i), we assume without loss of generality that $\tau(A, \boldsymbol{c}) = \{n\}$. Then for $\boldsymbol{l} = \boldsymbol{l}(A, \boldsymbol{c})$ the lattice programs

$$\min\{\boldsymbol{l} \cdot \boldsymbol{x} : \boldsymbol{x} \equiv \boldsymbol{r} \pmod{\Lambda_A}, \boldsymbol{x} \in \mathbb{R}_{\geq 0}^{n-1}\}, \quad \boldsymbol{r} \in \mathbb{Z}^{n-1} \tag{25}$$

are the *group relaxations* to (1).

Indeed, for any positive $b \in \mathbb{Z}$ and any integer solution $\boldsymbol{z}$ of the equation $A\boldsymbol{x} = b$ the lattice program (25) with $\boldsymbol{r} = \pi_n(\boldsymbol{z})$, is a group relaxation to (1). On the other hand, for any integer vector $\boldsymbol{r}$ the lattice program (25) is a group relaxation to (1) with $b = \pi_n(A)\boldsymbol{u}$ for a nonnegative integer vector $\boldsymbol{u}$ from $\boldsymbol{r} + \Lambda_A$.

In both cases

$$IG_c(A, b) \geq m(\Lambda_A, \boldsymbol{l}, \boldsymbol{r})$$

and, consequently,

$$\mathrm{Gap}_c(A) \geq \mathrm{Gap}(\Lambda_A, \boldsymbol{l}). \tag{26}$$

Note that for $n = 2$ we have $\mathrm{Gap}(\Lambda_A, \boldsymbol{l}) = l_1(|A_\tau| - 1)$ and thus (26) implies (8). For $n > 2$, the bound (8) immediately follows from (26) and Theorem 1.2 (i) in [1].

The proof of the part (ii) will be based on the following lemma.

Lemma 3. *Let A satisfy (4), $\boldsymbol{c} = (a_1, \ldots, a_{n-1}, 0)^t \in \mathbb{Q}^n$ and $\boldsymbol{l} = (a_1, \ldots, a_{n-1})^t \in \mathbb{Q}_{>0}^{n-1}$. Then*

$$\mathrm{Gap}_c(A) = \mathrm{Gap}(\Lambda_A, \boldsymbol{l}). \tag{27}$$

Proof. Observe that assumption (i) in (4) implies that the linear programming relaxation (2) is feasible if and only if b is nonnegative. Recall that $\Lambda(A, b) = \{\boldsymbol{x} \in \mathbb{Z}^n : A\boldsymbol{x} = b\}$ denotes the set of integer points in the affine hyperplane $A\boldsymbol{x} = b$ and $P(A, b) = \{\boldsymbol{x} \in \mathbb{R}_{\geq 0} : A\boldsymbol{x} = b\}$ denotes the knapsack polytope. Suppose that for a nonnegative b the knapsack problem (1) is feasible, with solution $\boldsymbol{y} \in \mathbb{Z}_{\geq 0}^n$. Then for $\boldsymbol{r} = \pi_n(\boldsymbol{y}) \in \mathbb{Z}_{\geq 0}^{n-1}$

$$\pi_n(\Lambda(A, b)) = \boldsymbol{r} + \Lambda_A.$$

As $c_n = 0$, the optimal value of the linear programming relaxation $LP_c(A, b) = 0$. Therefore, noting that $\boldsymbol{c} = (a_1, \ldots, a_{n-1}, 0)^t$ and $\boldsymbol{l} = \pi_n(\boldsymbol{c})$,

$$IG_c(A, b) = \min\{\boldsymbol{l} \cdot \boldsymbol{x} : \boldsymbol{x} \in \boldsymbol{r} + \Lambda_A, \boldsymbol{x} \in \pi_n(P(A, b))\}. \tag{28}$$

Since

$$\pi_n(P(A,b)) = bS_A = \{x \in \mathbb{R}_{\geq 0}^{n-1} : l \cdot x \leq b\}$$

and $l \cdot r \leq Ay = b$, the constraint $x \in \pi_n(P(A,b))$ in (28) can be removed. Consequently, we have

$$IG_c(A,b) = m(\Lambda_A, l, r).$$

Hence, by (24), we obtain

$$\text{Gap}_c(A) \leq \text{Gap}(\Lambda_A, l). \tag{29}$$

Suppose now that $\text{Gap}(\Lambda_A, l) = m(\Lambda_A, l, r_0)$. Then

$$IG_c(A, Ar_0) = m(\Lambda, l, r_0).$$

Together with (29), this implies (27).   □

As was shown in the proof of Theorem 1.1 in [1], for $l = (a_1, \ldots, a_{n-1})^t$

$$\text{Gap}(\Lambda_A, l) = g(A) + a_n.$$

Thus we obtain the following corollary.

Corollary 6. *Let* $A = (a_1, \ldots, a_n)$ *satisfy (4) and* $c = (a_1, \ldots, a_{n-1}, 0)^t$. *Then*

$$\text{Gap}_c(A) = g(A) + a_n. \tag{30}$$

For $n = 2$, we have

$$g(A) = a_1 a_2 - a_1 - a_2 \tag{31}$$

by a classical result of Sylvester (see e.g. [22]). Hence the part (ii) immediately follows from Corollary 6. For $n > 2$, noting that $|A_\tau| = a_n$, the part (ii) follows from Corollary 6 and Theorem 1.1 (ii) in [2].

5   Proof of Theorem 3

For convenience, we will work with the quantity

$$f(A) = g(A) + a_1 + \cdots + a_n$$

and the set

$$R = \{A \in \mathbb{Z}^{1 \times n} : 0 < a_1 \leq \cdots \leq a_n\}.$$

By Lemma 2, we have

$$N_\epsilon(t, T) \leq n! \# \left\{ A \in Q(T) \cap R : \frac{f(A)}{a_1 a_n^\epsilon} > t \right\}. \tag{32}$$

We may assume $t \geq 10$ since otherwise (11) follows from $N_\epsilon(t,T)/N(T) \leq 1$. We keep $t' \in [1,t]$, to be fixed later. Then, setting $s(A) = a_{n-1}a_n^{1/(n-1)}$ and noting (32), we get

$$N_\epsilon(t,T) \leq n!\, \# \left\{ A \in Q(T) \cap R : \frac{f(A)}{s(A)} > t' \text{ or } \frac{s(A)}{a_1 a_n^\epsilon} > \frac{t}{t'} \right\}$$

$$\leq n!\, \# \left\{ A \in Q(T) \cap R : \frac{f(A)}{s(A)} > t' \right\} \tag{33}$$

$$+ n!\, \# \left\{ A \in Q(T) \cap R : \frac{a_{n-1}}{a_1 a_n^{\epsilon-1/(n-1)}} > \frac{t}{t'} \right\}.$$

The first of the last two terms in (33) can be estimated using a special case of Theorem 3 in Strömbergsson [26].

Lemma 4

$$\# \left\{ A \in Q(T) \cap R : \frac{f(A)}{s(A)} > r \right\} \ll_n \frac{1}{r^{n-1}} N(T). \tag{34}$$

Proof. The inequality (34) immediately follows from Theorem 3 in [26] applied with $\mathcal{D} = [0,1]^{n-1}$. □

To estimate the last term, we will need the following lemma.

Lemma 5.

$$\# \left\{ A \in Q(T) \cap R : \frac{a_{n-1}}{a_1 a_n^{\epsilon-1/(n-1)}} > r \right\} \ll_n \frac{1}{r T^{\epsilon-1/(n-1)}} N(T). \tag{35}$$

Proof. Since $A \in R$, we have $a_{n-1} \leq a_n$. Hence

$$\# \left\{ A \in Q(T) \cap R : \frac{a_{n-1}}{a_1 a_n^{\epsilon-1/(n-1)}} > r \right\} \leq \# \left\{ A \in Q(T) \cap R : a_n^{1+1/(n-1)-\epsilon} > r a_1 \right\}.$$

Furthermore, all $A \in Q(T) \cap R$ with $a_n^{1+1/(n-1)-\epsilon} > r a_1$ are in the set

$$U = \{ A \in \mathbb{Z}^{1 \times n} : 0 < a_1 < T^{1+1/(n-1)-\epsilon}/r, 0 < a_i \leq T, i = 2, \ldots, n \}.$$

Since $\#(U \cap \mathbb{Z}^n) < T^{n+1/(n-1)-\epsilon}/r$ and $N(T) \asymp_n T^n$ (see e.g. Theorem 1 in [23]), the result follows. □

Then by (33), (34) and (35)

$$\frac{N_\epsilon(t,T)}{N(T)} \ll_n \frac{1}{(t')^{n-1}} + \frac{t'}{t T^{\epsilon-1/(n-1)}}. \tag{36}$$

Next, we will bound T from below in terms of t, similar to Theorem 3 in [26]. The upper bound of Schur (22) implies $f(A) < na_1a_n$. Thus, using (32),

$$
N_\epsilon(t,T) \leq \# \left\{ A \in Q(T) \cap R : \frac{f(A)}{a_1 a_n^\epsilon} > t \right\}
$$
$$
\leq \# \left\{ A \in Q(T) \cap R : a_n^{1-\epsilon} > \frac{t}{n} \right\}.
$$

The latter set is empty if $T \leq (t/n)^{\frac{1}{1-\epsilon}}$. Hence we may assume

$$
T > \left(\frac{t}{n} \right)^{\frac{1}{1-\epsilon}}. \tag{37}
$$

Using (36) and (37), we have

$$
\frac{N_\epsilon(t,T)}{N(T)} \ll_n \frac{1}{(t')^{n-1}} + \frac{t'}{t^{1+\frac{1}{1-\epsilon}\left(\epsilon - \frac{1}{n-1}\right)}}. \tag{38}
$$

To minimise the exponent of the right hand side of (38), set $t' = t^\beta$ and choose β with

$$
\beta(n-1) = 1 + \frac{1}{1-\epsilon} \left(\epsilon - \frac{1}{n-1} \right) - \beta. \tag{39}
$$

We get

$$
\beta = \frac{n-2}{n(n-1)(1-\epsilon)}
$$

and, by (38) and (39),

$$
\frac{N_\epsilon(t,T)}{N(T)} \ll_n t^{-\alpha(\epsilon,n)}
$$

with $\alpha(\epsilon,n) = \beta(n-1)$. The theorem is proved.

6   Proof of Corollary 4

For the upper bound we observe, that the conditions $n \geq 3$ and $\epsilon > 2/n$ imply that in (11) $\alpha(\epsilon,n) > 1$. Consider vectors $A \in Q(T)$ with

$$
e^{s-1} \leq \max_{c \in \mathbb{Q}^n} \frac{\mathrm{Gap}_c(A)}{\|A\|_\infty^\epsilon \|c\|_1} < e^s. \tag{40}
$$

The contribution of vectors satisfying (40) to the sum

$$
\sum_{A \in Q(T)} \max_{c \in \mathbb{Q}^n} \frac{\mathrm{Gap}_c(A)}{\|A\|_\infty^\epsilon \|c\|_1}
$$

on the left hand side of (12) is

$$\leq N_\epsilon(e^{s-1}, T)e^s \ll_n e^{-\alpha(\epsilon,n)s}e^s N(T),$$

where the last inequality holds by (11). Therefore

$$\frac{1}{N(T)} \sum_{A \in Q(T)} \max_{c \in \mathbb{Q}^n} \frac{\mathrm{Gap}_c(A)}{\|A\|_\infty^\epsilon \|c\|_1} \ll_n \sum_{s=1}^\infty e^{s(1-\alpha(\epsilon,n))}.$$

Finally, observe that the series

$$\sum_{s=1}^\infty e^{s(1-\alpha(\epsilon,n))}$$

is convergent for $\alpha(\epsilon, n) > 1$.

References

1. Aliev, I.: On the lattice programming gap of the group problems. Oper. Res. Lett. **43**, 199–202 (2015)
2. Aliev, I., Gruber, P.M.: An optimal lower bound for the Frobenius problem. J. Number Theor. **123**, 71–79 (2007)
3. Bertsimas, D., Weismantel, R.: Optimization Over Integers. Dynamic Ideas, Massachusetts (2005)
4. Blair, C.E., Jeroslow, R.G.: The value function of a mixed integer program: I. Discrete Math. **19**, 121–138 (1977)
5. Blair, C.E., Jeroslow, R.G.: The value function of an integer program. Math. Program. **23**, 237–273 (1982)
6. Brauer, A.: On a problem of partitions. Am. J. Math. **64**, 299–312 (1942)
7. Chandrasekaran, R.: Polynomial algorithms for totally dual integral systems and extensions. Studies on graphs and discrete programming (Brussels, 1979). Ann. Discrete Math. 11, 39–51. North-Holland, Amsterdam (1981)
8. Conforti, M., Cornuéjols, G., Zambelli, G.: Integer Programming. Graduate Texts in Mathematics, vol. 271. Springer, Heidelberg (2014)
9. Cook, W., Gerards, A.M.H., Schrijver, A., Tardos, É.: Sensitivity theorems in integer linear programming. Math. Program. **34**, 251–264 (1986)
10. Dougherty, R., Faber, V.: The degree-diameter problem for several varieties of Cayley graphs I: the abelian case. SIAM J. Discrete Math. **17**, 478–519 (2004)
11. Eisenbrand, F., Hähnle, N., Pálvölgyi, D., Shmonin, G.: Testing additive integrality gaps. Math. Program. A **141**, 257–271 (2013)
12. Eisenbrand, F., Shmonin, G.: Parametric integer programming in fixed dimension. Math. Oper. Res. **33**, 839–850 (2008)
13. Fáry, I.: Sur la densité des réseaux de domaines convexes. Bull. Soc. Math. France **78**, 152–161 (1950)
14. Garey, M.R., Johnson, D.S.: Computers and Intractability, A Guide to the Theory of NP-Completeness. A Series of Books in the Mathematical Sciences. W.H. Freeman and Co., San Francisco (1979)
15. Gomory, R.E.: Outline of an algorithm for integer solutions to linear programs. Bull. Am. Math. Soc. **64**, 275–278 (1958)

16. Gruber, P.M.: Convex and Discrete Geometry. Springer, Berlin (2007)
17. Gruber, P.M., Lekkerkerker, C.G.: Geometry of Numbers. North-Holland, Amsterdam (1987)
18. Hoşten, S., Sturmfels, B.: Computing the integer programming gap. Combinatorica **27**(3), 367–382 (2007)
19. Kannan, R.: Lattice translates of a polytope and the Frobenius problem. Combinatorica **12**, 161–177 (1992)
20. Khachiyan, L.G.: Polynomial algorithms in linear programming. USSR Comput. Math. Math. Phys. **1**(20), 53–72 (1980)
21. Marklof, J., Strömbergsson, A.: Diameters of random circulant graphs. Combinatorica **33**, 429–466 (2013)
22. Ramírez Alfonsín, J.L.: The Diophantine Frobenius Problem. Oxford Lecture Series in Mathematics and its Applications, vol. 30 (2005)
23. Schmidt, W.M.: Asymptotic formulae for point lattices of bounded determinant and subspaces of bounded height. Duke Math. J. **35**, 327–339 (1968)
24. Schmidt, W.M.: Integer matrices, sublattices of $\mathbb{Z}^m$, and Frobenius numbers. Monatsh. Math. **178**, 405–451 (2015)
25. Schrijver, A.: Theory of Linear and Integer Programming. Wiley, New Jersey (1986)
26. Strömbergsson, A.: On the limit distribution of Frobenius numbers. Acta Arith. **152**, 81–107 (2012)
27. Sullivant, S.: Small contingency tables with large gaps. SIAM J. Discrete Math. **18**, 787–793 (2005)
28. Vazirani, V.V.: Approximation Algorithms. Springer, Heidelberg (2001)

An Improved Integrality Gap
for the Călinescu-Karloff-Rabani Relaxation
for Multiway Cut

Haris Angelidakis[1], Yury Makarychev[1], and Pasin Manurangsi[2]([✉])

[1] Toyota Technological Institute at Chicago, Chicago, IL, USA
{hangel,yury}@ttic.edu
[2] University of California, Berkeley, Berkeley, CA, USA
pasin@berkeley.edu

Abstract. We construct an improved integrality gap instance for the Călinescu-Karloff-Rabani LP relaxation of the Multiway Cut problem. For $k \geqslant 3$ terminals, our instance has an integrality ratio of $6/(5 + \frac{1}{k-1}) - \varepsilon$, for every constant $\varepsilon > 0$. For every $k \geqslant 4$, this improves upon a long-standing lower bound of $8/(7 + \frac{1}{k-1})$ by Freund and Karloff [7]. Due to the result by Manokaran et al. [9], our integrality gap also implies Unique Games hardness of approximating Multiway Cut of the same ratio.

1 Introduction

In the Multiway Cut problem, we are given a weighted undirected graph and k terminal vertices. The goal is to find a set of edges of minimum total weight whose removal disconnects all the terminals. Equivalently, an optimal solution is a partition of the graph into k clusters, each containing one terminal, such that the total weight of edges across clusters is minimized. Such a partition is called a k-*way cut* and its *cost* is the total weight of edges across clusters.

Since its introduction [6], Multiway Cut has been extensively studied in the approximation algorithms community [2–8,10]. Despite this, its approximability when $k \geqslant 4$ still remains open.

When $k = 2$, Multiway Cut is simply Minimum s-t Cut, which is solvable in polynomial time. For $k \geqslant 3$, Dahlhaus et al. [6] showed that the problem becomes APX-hard and gave the first approximation algorithm for the problem, which achieves a $(2 - 2/k)$-approximation. Due to the combinatorial nature of the algorithm, several LPs were subsequently proposed but it was not until Călinescu, Karloff and Rabani's work [4] that a significant improvement in the approximation ratio was made. Their relaxation, also known as the CKR relaxation, on a graph $G = (V, E, w)$ and terminals $\{t_1, \ldots, t_k\} \subseteq V$, can be formulated as

Y. Makarychev—Supported by NSF awards CAREER CCF-1150062 and IIS-1302662.
P. Manurangsi—Supported by NSF Grants No. CCF-1540685 and CCF-1655215.
Part of this work was done while the author was visiting TTIC.

F. Eisenbrand and J. Koenemann (Eds.): IPCO 2017, LNCS 10328, pp. 39–50, 2017.
DOI: 10.1007/978-3-319-59250-3_4

$$\text{minimize} \quad \sum_{e=(u,v)\in E} w(e) \cdot \frac{1}{2}\|x^u - x^v\|_1$$

$$\text{subject to} \quad \forall u \in V, \ x^u \in \Delta_k,$$

$$\forall i \in [k], \ x^{t_i} = e^i,$$

where $[k] = \{1, \ldots, k\}$, $\Delta_k = \{(x_1, \ldots, x_k) \in [0,1]^k \mid x_1 + \cdots + x_k = 1\}$ denotes the k-simplex and e^i is the i-th vertex of the simplex, i.e. $e_i^i = 1$. Călinescu et al. [4] gave a rounding scheme for this LP that yields a $(3/2 - 1/k)$-approximation for Multiway Cut. Exploiting the geometric nature of the relaxation even further, Karger et al. [8] gave an 1.3438-approximation algorithm for the problem. They also conducted experiments that led to improvements over small k's; specifically, for $k = 3$, they gave a 12/11-approximation algorithm and proved that it is tight by constructing an integrality gap example of ratio $12/11 - \varepsilon$, for every $\varepsilon > 0$. The same result was also independently discovered by Cunningham and Tang [5]. In 2013, Buchbinder et al. [2] gave a neat 4/3-approximation algorithm to the problem for general k and additionally showed how to push the ratio down to 1.3239. This result was later improved by Sharma and Vondrák [10], who obtained an approximation ratio of 1.2965. The algorithm by Sharma and Vondrák is quite difficult, and its analysis is computer-assisted. Very recently, Buchbinder et al. [3] came up with a much simpler algorithm and analytically showed that it yields roughly the same approximation ratio.

The CKR relaxation can be used not only for designing approximation algorithms but also for proving hardness results for Multiway Cut. Manokaran et al. [9] showed, assuming the Unique Games Conjecture (UGC), that, if the integrality gap of the CKR relaxation is (at least) τ, then it is NP-hard to approximate Multiway Cut to within a factor of $\tau - \varepsilon$, for every $\varepsilon > 0$. Roughly speaking, this means that, assuming UGC, the CKR relaxation gives essentially the best approximation one can get in polynomial time. Despite this connection, few lower bounds for the integrality gap of the relaxation are known. Apart from the aforementioned $12/11 - \varepsilon$ integrality gap for $k = 3$ [5,8], the only other known lower bound is an $8/(7 + \frac{1}{k-1})$-integrality gap constructed by Freund and Karloff [7] not long after the introduction of the relaxation.

1.1   Our Contributions

We provide a new construction of integrality gap instances for Multiway Cut, which achieves an integrality gap of $6/(5 + \frac{1}{k-1}) - \varepsilon$, for every $k \geqslant 3$, as stated formally below. For every $k \geqslant 4$, our integrality gap improves on the best known integrality gap of $8/(7 + \frac{1}{k-1})$ of Freund and Karloff [7].

Theorem 1. *For every $k \geqslant 3$ and every $\varepsilon > 0$, there exists an instance $\mathcal{I}_{k,\varepsilon}$ of Multiway Cut with k terminals such that the integrality gap of the CKR relaxation for $\mathcal{I}_{k,\varepsilon}$ is at least $6/(5 + \frac{1}{k-1}) - \varepsilon$.*

Thanks to the aforementioned result of Manokaran et al. [9], the following corollary is an immediate consequence of Theorem 1.

Corollary 1. *Assuming UGC, for every $k \geqslant 3$ and every $\varepsilon > 0$, it is NP-hard to approximate Multiway Cut with k terminals to within $6/(5 + \frac{1}{k-1}) - \varepsilon$ of the optimum.*

Techniques. To see the motivation behind our construction, it is best to first gain additional intuition for the geometry of the CKR relaxation.

Geometric Interpretation of the CKR relaxation. A solution of the CKR relaxation embeds the graph into a simplex; each vertex u becomes a point $x^u \in \Delta_k$, while each edge (u, v) becomes a segment (x^u, x^v). To construct a randomized rounding scheme for the relaxation, it suffices to define a randomized k-partitioning scheme of the simplex: a distribution $\mathcal{P}$ on k-way cuts of Δ_k that separates the vertices e^i of the simplex Δ_k. We will identify each partition P in $\mathcal{P}$ with a map $P : \Delta_k \to [k]$; $P(x) = i$ if x lies in part i of the partition. Then, for every P in $\mathcal{P}$ and every $i \in [k]$, $P(e^i) = i$. Define the *maximum density* of $\mathcal{P}$ as follows,

$$\tau_k(\mathcal{P}) := \sup_{x \neq y \in \Delta_k} \frac{\Pr_{P \sim \mathcal{P}}[P(x) \neq P(y)]}{\frac{1}{2}\|x - y\|_1}.$$

In words, for any $x, y \in \Delta_k$, a random k-way cut P sampled from $\mathcal{P}$ assigns x, y to different clusters with probability at most $\tau_k(\mathcal{P}) \cdot \frac{1}{2}\|x - y\|_1$. This immediately yields the following randomized rounding scheme: pick a random $P \sim \mathcal{P}$ and place each $u \in V$ in cluster $P(x^u)$. The expected cost of the obtained solution is at most $\tau_k(\mathcal{P}) \cdot \sum_{e=(u,v) \in E} w(e) \cdot \frac{1}{2}\|x^u - x^v\|_1$. Therefore, the rounding scheme gives a $\tau_k(\mathcal{P})$-approximation for Multiway Cut with k terminals.

The above observation was implicit in [4] and was first made explicit by Karger et al. [8], who also proved the opposite inequality: for every $\varepsilon > 0$, there is an instance of Multiway Cut with k terminals whose integrality gap is at least $\tau_k^* - \varepsilon$, where τ_k^* is the minimum[1] of $\tau_k(\mathcal{P})$ among all $\mathcal{P}$'s. In other words, integrality gaps for Multiway Cut and lower bounds for τ_k^* are equivalent.

Non-Opposite Cuts. Let us consider a special class of partitions of the simplex, which we call *non-opposite cuts*. A non-opposite cut of Δ_k is a function $P : \Delta_k \to [k+1]$ such that $P(e^i) = i$ for every $i \in [k]$ and, for every $x \in \Delta_k$, $P(x) \in \mathrm{supp}(x) \cup \{k+1\}$, where $\mathrm{supp}(x) := \{i \in [k] \mid x_i \neq 0\}$ is the set of all non-zero coordinates of x. The representative case to keep in mind is when $k = 3$; in this case, a non-opposite cut partitions Δ_3 into four parts and any point on the border of Δ_3 is not assigned to the cluster corresponding to the opposite vertex of Δ_3. Figure 1 demonstrates non-opposite cuts and 3-way cuts of Δ_3.

We define τ_k for a distribution on non-opposite cuts similarly. Additionally, let $\tilde{\tau}_k^*$ be the infimum of $\tau_k(\mathcal{P})$ among all distributions $\mathcal{P}$ on non-opposite cuts of Δ_k. Observe that $\tau_k^* \leqslant \tilde{\tau}_k^*$ because each non-opposite cut P of Δ_k can be turned into a k-way cut without separating additional pairs of points by merging the

---

[1] In [8], τ_k^* is defined as the infimum of $\tau_k(\mathcal{P})$ among all $\mathcal{P}$'s but it was proved in the same work that there exists $\mathcal{P}$ that achieves the infimum.

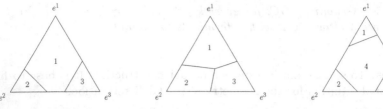

(a) A 3-way cut that is not non-opposite (b) A non-opposite 3-way cut (c) A non-opposite cut that is not 3-way

Fig. 1. Illustrations of cuts of Δ_3.

$(k+1)$-th cluster into the first cluster. However, it is not right away clear how to lower bound τ_k^* in terms of $\tilde{\tau}_k^*$. Our plan is as follows: define a "discretized" version $\tilde{\tau}_{k,n}^*$ of $\tilde{\tau}_k^*$, give a lower bound for τ_K^* in terms of $\tilde{\tau}_{k,n}^*$ for $K > k$, and then prove a lower bound for $\tilde{\tau}_{k,n}^*$.

Consider a discretization $\Delta_{k,n} := \{x \in \Delta_k \mid \forall i \in [k], x_i \text{ is a multiple of } 1/n\}$ of the simplex Δ_k. For every distribution $\mathcal{P}$ on non-opposite cuts of Δ_k, define

$$\tau_{k,n}(\mathcal{P}) := \max_{x \neq y \in \Delta_{k,n}} \frac{\Pr_{P \sim \mathcal{P}}[P(x) \neq P(y)]}{\frac{1}{2}\|x - y\|_1}$$

and $\tilde{\tau}_{k,n}^*$ as the minimum[2] of $\tau_{k,n}(\mathcal{P})$ over all distributions $\mathcal{P}$. The simple observation that allows us to construct our integrality gap is the following relation between τ_K^* and $\tilde{\tau}_{k,n}^*$: for every $K > k$,

$$\tau_K^* \geqslant \tilde{\tau}_{k,n}^* - O(kn/(K - k)). \tag{1}$$

In fact, we only use this inequality for $k = 3$ and, hence, we only sketch the proof of this case here. See the full version of this work [1] for the proof of the general case.

Let us note that the result by Freund and Karloff [7] can be seen as a proof of $\tilde{\tau}_{3,2}^* \geqslant 8/7$ (see Appendix A of the full version [1]). The bound $\tilde{\tau}_{3,2}^* \geqslant 8/7$, together with (1), immediately implies that $\tau_K^* \geqslant 8/7 - O(1/K)$. Barring the dependency on K, this is the gap proven in [7]. With this in mind, the rest of our work can be mostly seen as proving that $\tilde{\tau}_{3,n}^* \geqslant 6/5 - O(1/n)$. By selecting $n = \Theta(\sqrt{K})$, this implies that $\tau_K^* \geqslant 6/5 - O(1/\sqrt{K})$; more care can then be taken to get the right dependency on K.

We now sketch the proof of (1) for $k = 3$. Suppose that $\mathcal{P}$ is a distribution on K-way cuts such that $\tau_K(\mathcal{P}) = \tau_K^*$. We sample a non-opposite cut $\tilde{P}$ of $\Delta_{3,n}$ as follows. First, sample $P \sim \mathcal{P}$. Then, randomly select three different indices i_1, i_2, i_3 from $[K]$. Let $P_{\{i_1,i_2,i_3\}} : \Delta_3 \to [4]$ be the cut induced by P on the face with vertices i_1, i_2, i_3. More formally, let $f(i_j) = j$ for every $j \in [3]$ and $g(x) = x_1 e^{i_1} + x_2 e^{i_2} + x_3 e^{i_3}$ for every $x \in \Delta_3$. Define $P_{\{i_1,i_2,i_3\}}$ by

---

[2] The minimum exists since there is only a finite number of k-way cuts of the discretized simplex $\Delta_{k,n}$.

$$P_{\{i_1,i_2,i_3\}}(x) = \begin{cases} f(P(g(x))), & \text{if } P(g(x)) \in \{i_1,i_2,i_3\}, \\ 4, & \text{otherwise.} \end{cases}$$

If $P_{\{i_1,i_2,i_3\}}$ is a non-opposite cut, we simply let $\tilde{P} = P_{\{i_1,i_2,i_3\}}$. Otherwise, we "fix" $P_{\{i_1,i_2,i_3\}}$ by changing its value at every point x that violates the non-opposite cut condition; we call such points "bad". Specifically, we let

$$\tilde{P}(x) = \begin{cases} P_{\{i_1,i_2,i_3\}}(x), & \text{if } P_{\{i_1,i_2,i_3\}}(x) \in \text{supp}(x) \cup \{4\}, \\ 4, & \text{otherwise.} \end{cases}$$

It is obvious that $\tilde{P}$ is a non-opposite cut. Moreover, if any two points $x, y \in \Delta_{3,n}$ are separated in $\tilde{P}$, then either $P_{\{i_1,i_2,i_3\}}(x) \neq P_{\{i_1,i_2,i_3\}}(y)$ or exactly one of x, y is bad. From the definition of $P_{\{i_1,i_2,i_3\}}$, the former implies $P(g(x)) \neq P(g(y))$, which happens with probability at most $\tau_K^* \cdot \frac{1}{2}\|x - y\|_1$. Hence, we are left to show that the probability that any $x \in \Delta_{3,n}$ is bad is at most $O(1/K)$; this immediately yields the intended bound since $\|x - y\|_1 \geqslant 2/n$ for every $x \neq y \in \Delta_{3,n}$.

If x is bad, then $P_{\{i_1,i_2,i_3\}}(x) \in [3] \setminus \text{supp}(x)$, which implies that $|\text{supp}(x)| = 2$. Assume w.l.o.g. that $\text{supp}(x) = \{1, 2\}$. We want to bound $\Pr[P_{\{i_1,i_2,i_3\}}(x) = 3]$. Fix P, i_1 and i_2 (so that only i_3 is random). Note that now the value of $P(g(x))$ is also fixed, since $x_3 = 0$. Finally, observe that $P_{\{i_1,i_2,i_3\}}(x) = 3$ iff $i_3 = P(g(x))$. Since i_3 is a random element of $[K] \setminus \{i_1, i_2\}$, this happens with probability at most $1/(K - 2)$, completing our proof sketch.

Lower Bound for $\tilde{\tau}_{3,n}^*$. To lower bound $\tilde{\tau}_{3,n}^*$, it suffices to construct a weighted undirected graph with vertices $\Delta_{3,n}$ such that the CKR LP has a small value but every non-opposite cut has a large cost. Similar to Karger et al.'s integrality gap instance for $k = 3$ [8], a crucial component in our proof is a characterization of candidate optimal non-opposite cuts. This allows us to restrict our attention to certain types of cuts, for which it is easier to prove lower bounds.

2   Preliminaries and Notation

Henceforth, we consider graphs $(\Delta_{k,n}, E_{k,n})$ on $\Delta_{k,n}$ with the edge set $E_{k,n} := \{(x, y) \mid x, y \in \Delta_{k,n}, \|x - y\|_1 = 2/n\}$ for some k and n. The terminals are the k vertices of the simplex Δ_k. To avoid confusion, we refer to the simplex vertices simply as *vertices* and to the vertices of the graph as *points*.

We denote the value of the LP solution in which each point is assigned to itself with respect to $w : E_{k,n} \to \mathbb{R}_{\geq 0}$ by

$$\mathcal{L}(w) := \sum_{(x,y) \in E_{k,n}} w(x,y) \cdot \frac{1}{2}\|x - y\|_1 = \frac{1}{n} \sum_{(x,y) \in E_{k,n}} w(x,y).$$

For every cut $P : \Delta_{k,n} \to \mathbb{N}$, we denote its cost with respect to w by

$$\mathcal{C}(P, w) := \sum_{(x,y) \in E_{k,n}} w(x,y) \cdot \mathbb{1}[P(x) \neq P(y)],$$

where $\mathbb{1}[P(x) \neq P(y)]$ is the indicator variable of the event $P(x) \neq P(y)$.

3   A Lower Bound on $\tilde{\tau}_{3,n}^*$

We will first construct an integrality gap on the graph $(\Delta_{3,n}, E_{3,n})$ such that its CKR LP value is small but every non-opposite cut has a large cost. The main result of this section is that $\tilde{\tau}_{3,n}^* \geqslant 6/5 - O(1/n)$, equivalently stated as follows.

Lemma 1. *For every n divisible by 3, there is $w : E_{3,n} \to \mathbb{R}_{\geq 0}$ such that $\mathcal{L}(w) = 5/6 + O(1/n)$ and, for every non-opposite cut P of $(\Delta_{3,n}, E_{3,n})$, $\mathcal{C}(P, w) \geqslant 1$.*

To prove Lemma 1, we start by characterizing optimal non-opposite cuts.

3.1   A Characterization of Non-opposite Cuts of $\Delta_{3,n}$

We now characterize non-opposite cuts in $(\Delta_{3,n}, F_{3,n})$; we do so by extending a characterization of 3-way cuts by Karger et al. [8].

To characterize 3-way cuts, Karger et al. [8] consider the dual graph of an augmented version of $(\Delta_{3,n}, E_{3,n})$, in which each simplex vertex has an edge heading out infinitely. This augmentation creates three outer faces O_1, O_2 and O_3 (opposite to e^1, e^2 and e^3 respectively). For convenience, we disregard the edges among O_1, O_2, O_3 in the dual. Figure 2a and b illustrate an augmented graph and its planar dual. They are reproduced from Figs. 1 and 2 in [8].

A cut in the original graph can be viewed as a collection of edges in the dual graph; an edge in the dual graph between two faces corresponding to the shared edge of the faces being cut. With this interpretation, Karger et al. give the following characterization for candidate optimal cuts of any weight function.

Observation 2 ([8]). *For every $w : E_{3,n} \to \mathbb{R}_{\geq 0}$, there exists a least-cost 3-way cut P that is of one of the following forms:*

- *P contains three non-intersecting paths from a triangle to O_1, O_2 and O_3. Such P is called a ball cut. (See Fig. 2c, reproduced from Fig. 3 of [8].)*
- *P contains two non-intersecting paths among O_1, O_2 and O_3. Such P is called a 2-corner cut. (See Fig. 2d, reproduced from Fig. 4 of [8].)*

In other words, to prove that every 3-way cut incurs large cost against w, it suffices to consider only ball cuts and 2-corner cuts, which are more convenient to work with. Karger et al. take advantage of this when constructing their gap for Multiway Cut with three terminals. We make the following analogous observation for non-opposite cuts. Since the proof is straightforward, we omit it here.

Observation 3. *For every $w : E_{3,n} \to \mathbb{R}_{\geq 0}$, there exists a least-cost non-opposite cut P that is of one of the following forms:*

- *P is a ball cut.*
- *P contains three non-intersecting paths among O_1, O_2 and O_3. Such P is called a 3-corner cut or simply a corner cut. (See Fig. 2e.)*

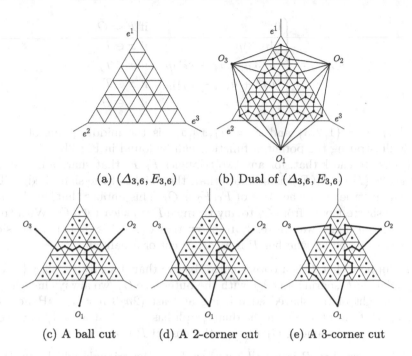

(a) $(\Delta_{3,6}, E_{3,6})$ (b) Dual of $(\Delta_{3,6}, E_{3,6})$

(c) A ball cut   (d) A 2-corner cut   (e) A 3-corner cut

Fig. 2. The augmented $(\Delta_{3,6}, E_{3,6})$, its dual and various types of cuts.

3.2   The Integrality Gap

With the characterization in place, we are now ready to prove Lemma 1.

Proof of Lemma 1. To construct the instance, we first divide the vertex set into *corner triangles* T_1, T_2, T_3 and a *middle hexagon* H as follows[3]. H contains all points $x \in \Delta_{3,n}$ such that $x_1, x_2, x_3 \leqslant 2/3$, whereas each T_i contains all the points x's such that $x_i \geqslant 2/3$. Note that this is not a partition since H and T_i share the line $x_i = 2/3$, but this notation will be more convenient for us.

Every edge in H, including its border, has weight $\rho := 1/(2n)$. For each i, every non-border edge in T_i not parallel to the opposite side of e^i also has weight ρ, whereas the non-border edges parallel to the opposite side of e^i have weight zero. Finally, for each of the two borders of T_i containing e^i, the edge closest to e^i is assigned weight $(n/3)\rho$, the second closest is assigned $(n/3 - 1)\rho$, and so on. An illustration of the construction is shown in Fig. 3a.

It is easy to check that $\mathcal{L}(w) = 5/6 + O(1/n)$. We will next prove that, for any non-opposite cut P, $\mathcal{C}(P, w) \geqslant 1$. From Observation 3, we can assume that P is either a ball cut or a corner cut. Recall that each node in the dual graph is either O_1, O_2, O_3 or a triangle. We represent each triangle by its median. For each i, we define the potential function Φ_i on O_i and all the triangles as follows.

---

[3] This terminology is from [5] but our instance differs significantly from theirs.

$$\Phi_i(F) = \begin{cases} 0 & \text{if } F = O_i, \\ (4n/3)\rho & \text{if } x \in T_i, \\ (n/3 + n(x_i - x_\ell))\rho & \text{if } x \in T_j, \\ (n/3 + n(x_i - x_j))\rho & \text{if } x \in T_\ell, \\ \lceil 2nx_i \rceil \rho & \text{if } x \in H, \end{cases}$$

where $\{i, j, \ell\} = \{1, 2, 3\}$ and $x = (x_1, x_2, x_3)$ is the middle point of F. An example illustrating the potential function can be found in Fig. 3b.

It is easy to check that, for any two triangles F_1, F_2 that share an edge, the difference $|\Phi_i(F_1) - \Phi_i(F_2)|$ is no more than the weight of the shared edge. The same remains true even when one of F_1, F_2 is O_i. This implies that, in the dual graph, the shortest path from O_i to any triangle F is at least $\Phi_i(F)$. With these observations in mind, we are ready to show that $\mathcal{C}(P, w) \geqslant 1$. Let us consider the two cases based on whether P is a corner cut or a ball cut.

Case 1. Suppose that P is a corner cut. Observe that, for any $j \in [3] \setminus \{i\}$ and for any triangle F sharing an edge with the outer face O_j, we always have $\Phi_i(F)$ plus the weight of the shared edge being at least $(2n/3)\rho = 1/3$. Hence, the shortest path from O_i to O_j in the dual graph has weight at least $1/3$. Since P contains three paths among O_1, O_2, O_3, the cost of P is at least 1.

Case 2. Suppose that P is a ball cut. Let F be the triangle which the three paths to O_1, O_2, O_3 in P originate from and x be the median of F. To show that $\mathcal{C}(P, w) \geqslant 1$, it is enough to show that the total length of the shortest paths from F to O_1, O_2 and O_3 is at least one. Since these shortest paths are lower bounded by the potential functions, we only need to prove that $\Phi_1(F) + \Phi_2(F) + \Phi_3(F) \geqslant 1$.

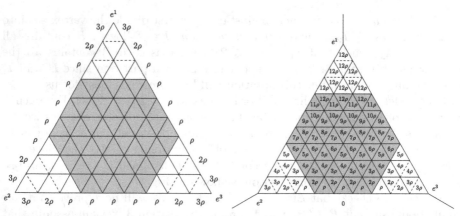

(a) Weights of border edges are shown. For non-border edges, the dashed edges have weight 0 whereas others have weight ρ.

(b) The potential function Φ_1.

Fig. 3. The integrality gap instance and a potential function when $n = 9$ are shown in (a) and (b) respectively. The middle hexagon H is shaded.

If $x \in H$, then $\sum_{i \in [3]} \Phi_i(F) = \sum_{i \in [3]} \lceil 2nx_i \rceil \rho \geqslant \sum_{i \in [3]} (2nx_i)\rho = 1$. Otherwise, if $x \notin H$, assume without loss of generality that $x \in T_1$. We also have $\sum_{i \in [3]} \Phi_i(F) = (4n/3)\rho + (n/3 + n(x_2 - x_3))\rho + (n/3 + n(x_3 - x_2))\rho = 1$.

In both cases, we have $\mathcal{C}(P, w) \geqslant 1$, thus completing the proof of Lemma 1. $\qquad\square$

4   Proof of the Main Theorem

We will now construct an integrality gap for Multiway Cut with $k \geqslant 3$ terminals on the graph $(\Delta_{k,n}, E_{k,n})$ and prove our main theorem, which is stated below. Clearly, by taking n sufficiently large, Theorem 4 implies Theorem 1.

Theorem 4. *For every n divisible by 3 and $k \geqslant 3$, there exists $\tilde{w} : E_{k,n} \to \mathbb{R}_{\geqslant 0}$ such that $\mathcal{L}(\tilde{w}) \leqslant (5 + \frac{1}{k-1})/6 + O(1/n)$ and, for every k-way cut P, $\mathcal{C}(P, \tilde{w}) \geqslant 1$.*

4.1   An Integrality Gap for $\Delta_{k,n}$ from $\Delta_{3,n}$

To prove Theorem 4, we first construct a simple integrality gap for $\Delta_{k,n}$ from our gap for $\Delta_{3,n}$ in Sect. 3; this is similar to the proof of $\tilde{\tau}^*_{3,n} - O(n/k) \leqslant \tau^*_k$ sketched in the introduction, although we will need to be more precise in order to get the right dependency on k. We will later tweak this instance slightly to arrive at our final integrality gap for Multiway Cut. We start by proving the following proposition, which is crucial in the analysis of the gap.

Proposition 1. *Given arbitrary $k \geqslant 3$ and n, let P be a k-way cut of $\Delta_{k,n}$. For every $\{i_1, i_2, i_3\} \subseteq [k]$, let $P_{\{i_1,i_2,i_3\}}$ be defined as in the introduction. Moreover, for each $\{i, j\} \subseteq [k]$, let us define $D_{i,j}(P)$ to be the set of all clusters that the points on the line between e^i and e^j (inclusive) are assigned to, i.e., $D_{i,j}(P) = \{P(x) \mid x \in \Delta_{k,n}, supp(x) \subseteq \{i, j\}\}$. Finally, let $D(P) = \mathbb{E}_{\{i,j\} \subseteq [k]}[|D_{i,j}(P)|]$. If i_1, i_2, i_3 are three randomly selected distinct elements of $[k]$, then*

$$\Pr_{i_1,i_2,i_3}[P_{\{i_1,i_2,i_3\}} \text{ is non-opposite}] \geqslant 1 - 3(D(P) - 2)/(k - 2).$$

Proof. If $P_{\{i_1,i_2,i_3\}}$ is not non-opposite, there is x such that $supp(x) \subseteq \{i_1, i_2, i_3\}$ and $P(x) \in \{i_1, i_2, i_3\} \setminus supp(x)$. Such x must have $|supp(x)| = 2$. Thus,

$$\Pr_{i_1,i_2,i_3}[P_{\{i_1,i_2,i_3\}} \text{ is not non-opposite}]$$

$$\leqslant \sum_{\{j_1,j_2\} \subseteq \{i_1,i_2,i_3\}} \Pr_{i_1,i_2,i_3}[\exists x, supp(x) = \{j_1, j_2\} \text{ and } P(x) \in \{i_1, i_2, i_3\} \setminus \{j_1, j_2\}]$$

$$= 3 \Pr_{i_1,i_2,i_3}[\exists x, supp(x) = \{i_1, i_2\} \text{ and } P(x) = i_3],$$

where the last equality comes from symmetry. Now, let us fix i_1, i_2. There is an x with $supp(x) = \{i_1, i_2\}$ such that $P(x) = i_3$ if and only if $i_3 \in D_{i_1,i_2}(P)$.

Observe that $i_1, i_2 \in D_{i_1,i_2}(P)$. Hence, the probability that i_3, a uniformly random element from $[k]\setminus\{i_1, i_2\}$, lies in $D_{i_1,i_2}(P)$ is exactly $(|D_{i_1,i_2}(P)|-2)/(k-2)$. Thus,

$$\Pr_{i_1,i_2,i_3}[\exists x, \operatorname{supp}(x) = \{i_1, i_2\} \text{ and } P(x) = i_3] = \underset{i_1,i_2}{\mathbb{E}}[(|D_{i_1,i_2}(P)| - 2)/(k-2)]$$
$$= (D(P) - 2)/(k-2).$$

Hence, $\Pr_{i_1,i_2,i_3}[P_{\{i_1,i_2,i_3\}} \text{ is non-opposite}] \geq 1 - 3(D(P) - 2)/(k-2)$. $\qquad\square$

Another component in our analysis is the observation that, even when a 3-way cut is non-opposite, it still incurs some cost against w (defined in Lemma 1).

Observation 5. *For every 3-way cut P of $(\Delta_{3,n}, E_{3,n})$, $\mathcal{C}(P, w) \geq 2/3$.*

To see that the observation is true, recall Observation 2; we only need to consider ball cuts and 2-corner cuts. Moreover, we showed in Subsect. 3.2 that the cost of any ball cut and 3-corner cut against w is at least one. This implies that the cost of any 2-corner cut is at least $2/3$, meaning that Observation 5 is true.

We are now ready to describe and analyze the integrality gap instance for $(\Delta_{k,n}, E_{k,n})$.

Lemma 2. *For every n divisible by 3 and $k \geq 3$, there is $\overline{w} : E_{k,n} \to \mathbb{R}_{\geq 0}$ such that $\mathcal{L}(\overline{w}) = 5/6 + O(1/n)$ and, for every k-way cut P of $(\Delta_{k,n}, E_{k,n})$, $\mathcal{C}(P, \overline{w}) \geq 1 - (D(P) - 2)/(k-2)$.*

Proof. Our instance is created by embedding the weight w from Lemma 1 to every triangular face of $\Delta_{k,n}$. Specifically, for each $\{i_1, i_2, i_3\} \in \binom{[k]}{3}$, the weight $\overline{w}_{\{i_1,i_2,i_3\}} : E_{k,n} \to \mathbb{R}_{\geq 0}$ of the gap embedded to the face of $e^{i_1}, e^{i_2}, e^{i_3}$ is

$$\overline{w}_{\{i_1,i_2,i_3\}}(x,y) = \begin{cases} w((x_{i_1}, x_{i_2}, x_{i_3}), (y_{i_1}, y_{i_2}, y_{i_3})) & \text{if } \operatorname{supp}(x), \operatorname{supp}(y) \subseteq \{i_1, i_2, i_3\}, \\ 0 & \text{otherwise.} \end{cases}$$

Finally, let $\overline{w} = \mathbb{E}_{i_1,i_2,i_3}[\overline{w}_{\{i_1,i_2,i_3\}}]$ where the expectation is over all random distinct $i_1, i_2, i_3 \in [k]$. Note that $\mathcal{L}(\overline{w}) = \mathcal{L}(w) = 5/6 + O(1/n)$.

Next, consider any k-way cut P. Observe that $\mathcal{C}(P, \overline{w}) = \mathbb{E}_{i_1,i_2,i_3}[\mathcal{C}(P, \overline{w}_{\{i_1,i_2,i_3\}})] \geq \mathbb{E}_{i_1,i_2,i_3}[\mathcal{C}(P_{\{i_1,i_2,i_3\}}, w)]$. Let $E_{\{i_1,i_2,i_3\}}$ be the event that $P_{\{i_1,i_2,i_3\}}$ is non-opposite. We can lower bound $\mathcal{C}(P, \overline{w})$ further as follows.

$$\underset{i_1,i_2,i_3}{\mathbb{E}}[\mathcal{C}(P_{\{i_1,i_2,i_3\}}, w)] = \underset{i_1,i_2,i_3}{\Pr}[E_{\{i_1,i_2,i_3\}}]\underset{i_1,i_2,i_3}{\mathbb{E}}[\mathcal{C}(P_{\{i_1,i_2,i_3\}}, w) \mid E_{\{i_1,i_2,i_3\}}]+$$
$$\underset{i_1,i_2,i_3}{\Pr}[\neg E_{\{i_1,i_2,i_3\}}]\underset{i_1,i_2,i_3}{\mathbb{E}}[\mathcal{C}(P_{\{i_1,i_2,i_3\}}, w) \mid \neg E_{\{i_1,i_2,i_3\}}]$$

(Lemma 1, Observation 5) $\geq \Pr[E_{\{i_1,i_2,i_3\}}] + (2/3)\Pr[\neg E_{\{i_1,i_2,i_3\}}]$

(Proposition 1) $\geq 1 - (D(P) - 2)/(k-2)$. $\qquad\square$

4.2   The Final Integrality Gap

Since $\overline{w}$ does not work well against P with large $D(P)$, we need another integrality gap instance to deal with this case. This instance is constructed simply by distributing the weight equally on simplex edges. Its properties are stated below; we defer the straightforward analysis of this instance to the full version [1].

Lemma 3. *For every $n \geqslant 2$ and every $k \geqslant 2$, there exists $w' : E_{k,n} \to \mathbb{R}_{\geq 0}$ such that $\mathcal{L}(w') = 1$ and, for every k-way cut P, $\mathcal{C}(P, w') \geqslant D(P) - 1$.*

Our final weight $\tilde{w}$ is simply an appropriate linear combination of $\overline{w}$ and w'.

Proof of Theorem 4. Let $\tilde{w} = \left(\frac{k-2}{k-1}\right) \overline{w} + \left(\frac{1}{k-1}\right) w'$. Note that $\mathcal{L}(\tilde{w}) = \left(\frac{k-2}{k-1}\right) \mathcal{L}(\overline{w}) + \left(\frac{1}{k-1}\right) \mathcal{L}(w') = \frac{5 + \frac{1}{k-1}}{6} + O(1/n)$. Moreover, for any k-way cut P,

$$\mathcal{C}(P, \tilde{w}) \geqslant \left(\frac{k-2}{k-1}\right) \left(1 - \frac{D(P) - 2}{k - 2}\right) + \left(\frac{1}{k-1}\right) (D(P) - 1) = 1$$

where the inequality comes from Lemmas 2 and 3.   □

5   Conclusion

We construct integrality gap instances of ratio $6/(5 + \frac{1}{k-1}) - \varepsilon$ for the CKR relaxation of Multiway Cut with $k \geqslant 3$ terminals. Thanks to Manokaran et al.'s result [9], this implies UGC-hardness of approximating Multiway Cut with similar ratios. Our construction is based on the observation that $\tilde{\tau}^*_{3,n} - O(n/K) \leqslant \tau^*_K$; from there, we extend Karger et al.'s characterization of 3-way cuts [8] to non-opposite cuts, leading to a lower bound of $6/5 - O(1/n)$ for $\tilde{\tau}^*_{3,n}$. We also observe that Freund and Karloff's integrality gap [7] is subsumed by our approach.

Even with our result, the approximability of Multiway Cut is far from resolved; the best known approximation ratio [10] is 1.2965 as $k \to \infty$ whereas our gap is only 1.2. Note also that a modification of Karger et al.'s algorithm shows that $\tilde{\tau}^*_3 \leqslant 1.2$ (see Appendix C of [1]), meaning that our lower bound for $\tilde{\tau}^*_3$ is the best possible. Thus, to construct better gaps, one likely needs to exploit properties of higher-dimensional simplexes, which seems challenging as all known gaps, including ours, only deal with Δ_3, whose cuts can be conveniently characterized.

References

1. Angelidakis, H., Makarychev, Y., Manurangsi, P.: An improved integrality gap for the Calinescu-Karloff-Rabani relaxation for multiway cut. CoRR abs/1611.05530 (2016). http://arxiv.org/abs/1611.05530
2. Buchbinder, N., Naor, J., Schwartz, R.: Simplex partitioning via exponential clocks and the multiway cut problem. In: Proceedings of the 45th ACM Symposium on Theory of Computing, STOC, pp. 535–544 (2013)

3. Buchbinder, N., Schwartz, R., Weizman, B.: Simplex transformations and the multiway cut problem. In: Proceedings of the 28th ACM-SIAM Symposium on Discrete Algorithms, SODA, pp. 2400–2410 (2017)
4. Călinescu, G., Karloff, H.J., Rabani, Y.: An improved approximation algorithm for multiway cut. J. Comput. Syst. Sci. **60**(3), 564–574 (2000)
5. Cunningham, W.H., Tang, L.: Optimal 3-terminal cuts and linear programming. In: Cornuéjols, G., Burkard, R.E., Woeginger, G.J. (eds.) IPCO 1999. LNCS, vol. 1610, pp. 114–125. Springer, Heidelberg (1999). doi:10.1007/3-540-48777-8_9
6. Dahlhaus, E., Johnson, D.S., Papadimitriou, C.H., Seymour, P.D., Yannakakis, M.: The complexity of multiterminal cuts. SIAM J. Comput. **23**(4), 864–894 (1994)
7. Freund, A., Karloff, H.: A lower bound of $8/(7+\frac{1}{k-1})$ on the integrality ratio of the Călinescu-Karloff-Rabani relaxation for multiway cut. Inf. Process. Lett. **75**(1–2), 43–50 (2000)
8. Karger, D.R., Klein, P.N., Stein, C., Thorup, M., Young, N.E.: Rounding algorithms for a geometric embedding of minimum multiway cut. Math. Oper. Res. **29**(3), 436–461 (2004)
9. Manokaran, R., Naor, J., Raghavendra, P., Schwartz, R.: SDP gaps and UGC hardness for multiway cut, 0-extension, and metric labeling. In: Proceedings of the 40th ACM Symposium on Theory of Computing, STOC, pp. 11–20 (2008)
10. Sharma, A., Vondrák, J.: Multiway cut, pairwise realizable distributions, and descending thresholds. In: Proceedings of the 46th ACM Symposium on Theory of Computing, STOC, pp. 724–733 (2014)

Approximation of Corner Polyhedra with Families of Intersection Cuts

Gennadiy Averkov[1], Amitabh Basu[2], and Joseph Paat[2(✉)]

[1] Faculty of Mathematics, Institute of Mathematical Optimization,
University of Magdeburg, Magdeburg, Germany
[2] Department of Applied Mathematics and Statistics,
Johns Hopkins University, Baltimore, USA
jspaat@gmail.com

Abstract. We study the problem of approximating the corner polyhedron using intersection cuts derived from families of lattice-free sets. In particular, we look at the problem of characterizing families that approximate the corner polyhedron up to a constant factor in fixed dimension n (the constant depends on n). The literature already contains several results in this direction. In this paper, we use the maximum number of facets of a lattice-free set in a family as a measure of its complexity and precisely characterize the level of complexity of a family required for constant factor approximations. As one of the main results, we show that for each natural number n, a corner polyhedron for n integer variables is approximated by intersection cuts from lattice-free sets with at most i facets up to a constant factor (depending only on n) if $i > 2^{n-1}$ and that no such approximation is possible if $i \leq 2^{n-1}$. When the approximation factor is allowed to depend on the denominator of the underlying fractional point of the corner polyhedron, we show that the threshold is $i > n$ versus $i \leq n$. The tools introduced for proving such results are of independent interest for studying intersection cuts.

1 Introduction

Given $n, k \in \mathbb{N}$, a matrix $R := (r_1, \ldots, r_k) \in \mathbb{R}^{n \times k}$ with columns $r_1, \ldots, r_k \in \mathbb{R}^n$, and a vector $f \in \mathbb{R}^n \backslash \mathbb{Z}^n$, the set

$$\mathrm{Cor}(R, f) := \mathrm{conv}\left\{ s \in \mathbb{R}^k_{\geq 0} \ : \ f + \sum_{i=1}^k s_i r_i \in \mathbb{Z}^n \right\}$$

has been studied in the integer programming literature, as a framework for deriving *cutting planes (cuts)* for general mixed-integer programs; see Chap. 6 of [10] for a detailed discussion. When both R and f are rational, the well-known Meyer's theorem (see [14]) implies that $\mathrm{Cor}(R, f)$ is a rational polyhedron. In the case of rational (R, f), we will refer to $\mathrm{Cor}(R, f)$ as the *corner polyhedron for (R, f)*. The original definition of the corner polyhedron going back to [12]

A. Basu and J. Paat—Supported by the NSF grant CMMI1452820.

F. Eisenbrand and J. Koenemann (Eds.): IPCO 2017, LNCS 10328, pp. 51–62, 2017.
DOI: 10.1007/978-3-319-59250-3_5

involved the condition $s \in \mathbb{Z}_{\geq 0}^k$, but the term has since been used with s allowed to take mixed-integer values (see Chap. 6 of [10] and the references therein).

An inequality description of $\mathrm{Cor}(R, f)$ can be obtained via gauge functions of lattice-free sets. A set $B \subseteq \mathbb{R}^n$ is *lattice-free* if B is n-dimensional, closed, convex, and the interior of B does not contain points of $\mathbb{Z}^n$. A lattice-free set is called *maximal* if it is not a proper subset of another lattice-free set[1]. Let $B \subseteq \mathbb{R}^n$ be a closed and convex set with $0 \in \mathrm{int}(B)$. The *gauge function* $\psi_B : \mathbb{R}^n \to \mathbb{R}$ of B is $\psi_B(r) := \inf \{\lambda > 0 : r \in \lambda B\}$.

Given a lattice-free set B with $f \in \mathrm{int}(B)$, the *intersection cut for* (R, f) *generated by* B (or the *B-cut* of (R, f)) is

$$C_B(R, f) := \left\{s \in \mathbb{R}_{\geq 0}^k : \sum_{i=1}^k s_i \psi_{B-f}(r_i) \geq 1\right\}.$$

In the degenerate case where $f \in \mathbb{R}^n \backslash \mathrm{int}(B)$, we define $C_B(R, f) := \mathbb{R}_{\geq 0}^k$. Given a family $\mathcal{B}$ of lattice-free subsets of $\mathbb{R}^n$ we call the set

$$C_{\mathcal{B}}(R, f) := \bigcap_{B \in \mathcal{B}} C_B(R, f)$$

the *$\mathcal{B}$-closure* of (R, f). If the family $\mathcal{B}$ is empty, we define $C_{\mathcal{B}}(R, f) = \mathbb{R}_{\geq 0}^k$.

Intersection cuts can be partially ordered by set inclusion. If B_1, B_2 are lattice-free sets then the inclusion $B_1 \subseteq B_2$ implies $C_{B_2}(R, f) \subseteq C_{B_1}(R, f)$ for all (R, f). Hence maximal lattice-free sets produce the strongest cuts [8,9]. Furthermore, all lattice-free sets are contained in maximal lattice-free sets, and all maximal lattice-free sets are polyhedra [13]. Therefore our focus can be directed towards intersection cuts from polyhedra.

Definition 1 ($\mathcal{L}_i^n$, i-hedral closures, and $\mathcal{L}_*^n$). *For $i \in \mathbb{N}$, let $\mathcal{L}_i^n$ denote the family of all lattice-free (not necessarily maximal) polyhedra in $\mathbb{R}^n$ with at most i facets; we call $C_{\mathcal{L}_i^n}(R, f)$ the i-hedral closure of (R, f). Let $\mathcal{L}_*^n$ denote the family of all lattice-free (not necessarily maximal) polyhedra in $\mathbb{R}^n$.*

For every $\mathcal{B}$, the $\mathcal{B}$-closure $C_{\mathcal{B}}(R, f)$ is a relaxation of $\mathrm{Cor}(R, f)$, which means the inclusion $\mathrm{Cor}(R, f) \subseteq C_{\mathcal{B}}(R, f)$ holds for every choice of (R, f). Furthermore, the equality $\mathrm{Cor}(R, f) = C_{\mathcal{B}}(R, f)$ is attained when $\mathcal{B}$ contains all maximal lattice-free polyhedra and (R, f) is rational [16]. This implies that one approach to computing $\mathrm{Cor}(R, f)$ for rational (R, f) is to classify maximal lattice-free sets and compute cuts using the corresponding gauge functions. Recent work has focused on this classification [1,3,4,11]. The classification was given for $n = 2$ in [11], but a classification is not known even for $n = 3$. Furthermore, even if such a classification was available for an arbitrary dimension n, the respective gauge functions would be difficult to compute, in general. In fact, the number i of facets of an arbitrary maximal lattice-free polyhedron $B \subseteq \mathbb{R}^n$ can be as large as 2^n, while the computation of the respective gauge function would

---

[1] Some sources do not impose the condition $\dim(B) = n$ in the definition of maximal lattice-free sets, but the case $\dim(B) < n$ is not needed for this paper.

require evaluation of i scalar products if B is not required to have any particular structure.

In light of these difficulties, instead of fully describing $\text{Cor}(R, f)$ by classifying lattice-free sets, one can aim to find a small and simple family of lattice-free sets whose closure approximates $\text{Cor}(R, f)$ within a desired tolerance [2,5,7]. In other words, for a fixed $n \in \mathbb{N}$, one can search for a simple family $\mathcal{B}$ and a constant $\alpha \geq 1$ (potentially depending on n) such that the inclusions

$$\text{Cor}(R, f) \subseteq C_{\mathcal{B}}(R, f) \subseteq \tfrac{1}{\alpha} \text{Cor}(R, f)$$

hold for all (R, f) (it is well known that $\text{Cor}(R, f) \subseteq \tfrac{1}{\alpha} \text{Cor}(R, f)$ for $\alpha \geq 1$). The inclusion $\text{Cor}(R, f) \subseteq C_{\mathcal{B}}(R, f)$ always holds, so we are led to consider the following main question:

Question 1. *Let $\mathcal{B}$ and $\mathcal{L}$ be families of lattice-free sets in $\mathbb{R}^n$. Under what conditions does there exist some $\alpha > 1$ such that the inclusion $C_{\mathcal{B}}(R, f) \subseteq \tfrac{1}{\alpha} C_{\mathcal{L}}(R, f)$ holds for all pairs (R, f)? Moreover, for a fixed $f \in \mathbb{Q}^n \backslash \mathbb{Z}^n$, when does there exist α such that $C_{\mathcal{B}}(R, f) \subseteq \tfrac{1}{\alpha} C_{\mathcal{L}}(R, f)$ holds for all rational R?*

If such an α exists, then the $\mathcal{B}$-closure approximates the $\mathcal{L}$-closure within a factor of α, that is the $\mathcal{B}$-closure provides a finite approximation of the $\mathcal{L}$-closure for all choices of (R, f) (or for a fixed f and all R). In this paper, we focus on answering Question 1. Since for rational (R, f), the corner polyhedron of (R, f) coincides with $C_{\mathcal{L}^n_*}(R, f)$, we are particularly interested in studying the case of $\mathcal{L} = \mathcal{L}^n_*$. On the other hand, as the number of facets is a natural measure for describing the complexity of maximal lattice-free sets, we are interested in the case $\mathcal{B} = \mathcal{L}^n_i$ with $i \in \mathbb{N}$.

Notation and Preliminaries. For background information on convex sets, polyhedra, and integer programming, see for example [10,15].

We use $\mathbb{N}$ to denote the set of all positive integers and $[m] := \{1, \ldots, m\}$ for $m \in \mathbb{N}$. For $X \subseteq \mathbb{R}^n$, we use $\text{cone}(X)$, $\text{conv}(X)$, $\text{int}(X)$, $\text{relint}(X)$ to denote the conic hull, the convex hull, the interior, and the relative interior of X, respectively. For $i \in [n]$, the vector $e_i \in \mathbb{R}^n$ denotes the i-th standard basis vector. The value $n \in \mathbb{N}$ will always denote the dimension of the ambient space $\mathbb{R}^n$, $k \in \mathbb{N}$ will always represent the number of columns of R, and $r_1, \ldots, r_k$ will always denote the columns of R. Stating that a condition holds for every R means that the condition holds for every $R \in \bigcup_{k=1}^{\infty} \mathbb{R}^{n \times k}$. Stating that a condition holds for every (R, f) means the condition holds for every $R \in \bigcup_{k=1}^{\infty} \mathbb{R}^{n \times k}$ and $f \in \mathbb{R}^n \backslash \mathbb{Z}^n$. Due to space constraints, most proofs appear exclusively in the journal version of this paper.

2   Summary of Results

For $\alpha \geq 1$, we call $\tfrac{1}{\alpha} C_{\mathcal{B}}(R, f)$ the α-*relaxation* of the cut $C_{\mathcal{B}}(R, f)$. Analogously, for a family of lattice-free sets $\mathcal{B}$, we call $\tfrac{1}{\alpha} C_{\mathcal{B}}(R, f)$ the α-*relaxation* of the $\mathcal{B}$-closure $C_{\mathcal{B}}(R, f)$. Using α-relaxations, the relative strength of cuts and closures can be quantified naturally as follows. For $f \in \mathbb{R}^n \backslash \mathbb{Z}^n$ and lattice-free subsets B and L of $\mathbb{R}^n$, we define

$$\rho_f(B, L) := \inf \left\{ \alpha > 0 \, : \, C_B(R, f) \subseteq \tfrac{1}{\alpha} C_L(R, f) \; \forall R \right\}. \tag{1}$$

The value $\rho_f(B, L)$ quantifies up to what extent $C_B(R, f)$ can 'replace' $C_L(R, f)$. For $\alpha \geq 1$, the inclusion $C_B(R, f) \subseteq \tfrac{1}{\alpha} C_L(R, f)$ says that the cut $C_B(R, f)$ is at least as strong as the α-relaxation of the cut $C_L(R, f)$. For $\alpha < 1$, the previous inclusion says that not just $C_B(R, f)$ but also the $\tfrac{1}{\alpha}$-relaxation of the cut $C_B(R, f)$ is at least as strong as the cut $C_L(R, f)$. Thus, if $\rho_f(B, L) \leq 1$, the B-cut of (R, f) is stronger than the L-cut of (R, f) for every R, and the value $\rho_f(B, L)$ quantifies how much stronger they are. If $1 < \rho_f(B, L) < \infty$, then the B-cut of (R, f) is not stronger than the L-cut of (R, f) but stronger than the α-relaxation of the L-cut for some $\alpha > 0$ independent of R, where the value $\rho_f(B, L)$ quantifies up to what extent the L-cut should be relaxed. If $\rho_f(B, L) = \infty$, then $C_B(R, f)$ cannot 'replace' $C_L(R, f)$ as there is no $\alpha \geq 1$ independent of R such that $C_B(R, f)$ is stronger than the α-relaxation of $C_L(R, f)$.

In addition to comparing the cuts coming from two lattice-free sets, we want to compare the relative strength of a family $\mathcal{B}$ to a single set L, and the relative strength of two families $\mathcal{B}$ and $\mathcal{L}$. We consider these comparisons when f is fixed or arbitrary. For the case of a fixed $f \in \mathbb{R}^n \backslash \mathbb{Z}^n$ we introduce the functional

$$\rho_f(\mathcal{B}, L) := \inf \left\{ \alpha > 0 \, : \, C_{\mathcal{B}}(R, f) \subseteq \tfrac{1}{\alpha} C_L(R, f) \; \forall R \right\},$$

which compares $\mathcal{B}$-closures to L-cuts for a fixed f. We also introduce

$$\rho_f(\mathcal{B}, \mathcal{L}) := \inf \left\{ \alpha > 0 \, : \, C_{\mathcal{B}}(R, f) \subseteq \tfrac{1}{\alpha} C_{\mathcal{L}}(R, f) \; \forall R \right\}$$

for comparing $\mathcal{B}$-closures to $\mathcal{L}$-closures for a fixed f. The analysis of $\rho_f(\mathcal{B}, \mathcal{L})$ can be reduced to the analysis of $\rho_f(\mathcal{B}, L)$ for $L \in \mathcal{L}$, since one obviously has

$$\rho_f(\mathcal{B}, \mathcal{L}) = \sup \left\{ \rho_f(\mathcal{B}, L) \, : \, L \in \mathcal{L} \right\}. \tag{2}$$

For the analysis in the case of varying f, we introduce the functionals:

$$\rho(\mathcal{B}, L) := \sup \left\{ \rho_f(\mathcal{B}, L) \, : \, f \in \mathbb{R}^n \backslash \mathbb{Z}^n \right\},$$
$$\rho(\mathcal{B}, \mathcal{L}) := \sup \left\{ \rho_f(\mathcal{B}, \mathcal{L}) \, : \, f \in \mathbb{R}^n \backslash \mathbb{Z}^n \right\}.$$

Observe that

$$\rho(\mathcal{B}, L) = \sup \left\{ \rho_f(\mathcal{B}, L) \, : \, f \in \mathrm{int}(L) \right\}, \tag{3}$$
$$\rho(\mathcal{B}, \mathcal{L}) = \sup \left\{ \rho_f(\mathcal{B}, L) \, : \, f \in \mathrm{int}(L), \, L \in \mathcal{L} \right\}. \tag{4}$$

The functional $\rho(\mathcal{B}, \mathcal{L})$ was introduced in [5, Sect. 1.2], where the authors initiated a systematic study for the case of $n = 2$. In the case that (R, f) is rational, since $C_{\mathcal{L}_*^n}(R, f) = \mathrm{Cor}(R, f)$, the value $\rho(\mathcal{B}, \mathcal{L}_*^n) \geq 1$ describes how well $C_{\mathcal{B}}(R, f)$ approximates $\mathrm{Cor}(R, f)$.

Our first main result examines i-hedral closures using the functional $\rho(\mathcal{B}, \mathcal{L})$.

Theorem 1. *Let* $i \in \{2, \ldots, 2^n\}$. *Then* $\rho(\mathcal{L}_i^n, \mathcal{L}_{i+1}^n) = \infty$ *if* $i \leq 2^{n-1}$ *and* $\rho(\mathcal{L}_i^n, \mathcal{L}_*^n) \leq 4 \, \mathrm{Flt}(n)$ *if* $i > 2^{n-1}$, *where* $\mathrm{Flt}(n)$ *is the flatness constant[2].*

Another way to examine the relative strength of i-hedral closures is with the functional $\rho_f(\mathcal{B}, \mathcal{L})$ for a fixed $f \in \mathbb{R}^n \backslash \mathbb{Z}^n$. As $\rho_f(\mathcal{L}_i^n, \mathcal{L}_*^n) < \rho(\mathcal{L}_i^n, \mathcal{L}_*^n)$ for every $f \in \mathbb{R}^n \backslash \mathbb{Z}^n$ and $i \in \mathbb{N}$, Theorem 1 immediately implies that $\rho_f(\mathcal{L}_i^n, \mathcal{L}_*^n) < \infty$ for $i > 2^{n-1}$. It turns out that in the case of a fixed rational f, the finiteness $\rho_f(\mathcal{L}_i^n, \mathcal{L}_*^n) < \infty$ holds for every $i > n$ (that is, already starting from $i = n + 1$). The *denominator* of f is the minimal $s \in \mathbb{N}$ such that $sf \in \mathbb{Z}^n$.

Theorem 2. *Let* $f \in \mathbb{Q}^n \backslash \mathbb{Z}^n$ *and* $i \in \{2, \ldots, 2^n\}$. *Then* $\rho_f(\mathcal{L}_i^n, \mathcal{L}_{i+1}^n) = \infty$ *for every* $i \leq n$ *and* $\rho_f(\mathcal{L}_i^n, \mathcal{L}_*^n) < \mathrm{Flt}(n) 4^{n-1} s$ *for every* $i > n$, *where* s *is the denominator of* f.

In light of Theorems 1 and 2, upper bounds on $\rho_f(\mathcal{L}_i^n, \mathcal{L}_*^n)$ necessarily depend on f for $n < i \leq 2^{n-1}$. An important point to note is that Theorem 2 assumes rationality of f. Rationality on f or R is not required for the other results in this paper. The finite approximation directions of Theorems 1 and 2 are provided in Sect. 5.1, and the inapproximability results are shown in Sect. 5.2.

Our main tool used in proving Theorems 1 and 2 is Theorem 3, which is geometric in nature. We set up some notation necessary to state the result.

Definition 2 ($\mathcal{C}_f$ and $\mathcal{B}_f$). *For* $f \in \mathbb{R}^n \backslash \mathbb{Z}^n$, *let* $\mathcal{C}_f$ *be the collection of all closed, full-dimensional, convex sets with* f *in the interior. For a family of lattice-free sets* $\mathcal{B}$, *define* $\mathcal{B}_f := \mathcal{C}_f \cap \mathcal{B}$.

Theorem 3 requires a metric on the space $\mathcal{C}_f$. For $f \in \mathbb{R}^n \backslash \mathbb{Z}^n$, we require a topology that allows us to take limits of the inequalities defining intersection cuts, that is, we need a topology that considers the convergence of gauge functions. To this end, we say that a sequence of sets B_t converges to B in *the* f-*metric* if $f \in \mathrm{int}(B)$, $f \in \mathrm{int}(B_t)$ for $t \in \mathbb{N}$, and $\psi_{B_t - f}$ converges to $\psi_{B - f}$ pointwise as $t \to \infty$. The formal definition of the f-metric appears in Sect. 3. We believe that this topology on the collection of lattice-free sets could be useful in future research. We use cl_f to denote the closure operator with respect to the f-metric.

Theorem 3 (Geometric One-for-all Theorem for two families). *Let* $\mathcal{B}$ *be a family of lattice-free subsets of* $\mathbb{R}^n$. *Let* $\mathcal{L} \subseteq \mathcal{L}_*^n$ *be such that there is a constant* $N \in \mathbb{N}$ *satisfying the following condition: every* $L \in \mathcal{L}$ *has a representation* $L = \mathrm{conv}(V) + \mathrm{cone}(W)$ *using a nonempty finite subset* V *of* $\mathbb{R}^n$ *and a finite (possible empty) subset* W *of* $\mathbb{R}^n \backslash \{0\}$ *such that* $|V| + |W| + 1 \leq N$ *holds. Then the following hold:*

(a) *Suppose* $\mathrm{cl}_f(\mathcal{B}_f) = \mathcal{B}_f$ *for a fixed* $f \in \mathbb{R}^n \backslash \mathbb{Z}^n$. *Then* $\rho_f(\mathcal{B}, \mathcal{L}) < \infty$ *if and only if there exists* $\mu \in (0, 1)$ *such that for every* $L \in \mathcal{L}_f$, *there exists some* $B \in \mathcal{B}$ *satisfying* $B \supseteq \mu L + (1 - \mu) f$.

(b) *Suppose* $\mathrm{cl}_f(\mathcal{B}_f) = \mathcal{B}_f$ *for all* $f \in \mathbb{R}^n \backslash \mathbb{Z}^n$. *Then* $\rho(\mathcal{B}, \mathcal{L}) < \infty$ *if and only if there exists* $\mu \in (0, 1)$ *such that for every* $f \in \mathbb{R}^n \backslash \mathbb{Z}^n$ *and each* $L \in \mathcal{L}_f$, *there exists some* $B \in \mathcal{B}$ *satisfying* $B \supseteq \mu L + (1 - \mu) f$.

---

[2] The flatness constant $\mathrm{Flt}(n)$ is known to be upper bounded by $n^{5/2}$ [6, p. 317].

Let $\mathcal{B}$ and $\mathcal{L}$ be families of lattice-free sets. A $\mathcal{B}$-closure approximates an $\mathcal{L}$-closure if and only if the $\mathcal{B}$-closure approximates the L-cut for each $L \in \mathcal{L}$. The somewhat surprising message of Theorem 3 is that in order for the $\mathcal{B}$-closure to approximate an L-cut for some $L \in \mathcal{L}$, it is necessary that there exists a *single* $B \in \mathcal{B}$ such that the corresponding B-cut approximates the L-cut. So with a view towards constant factor approximations of L-cuts, there is no synergy of all B-cuts for $B \in \mathcal{B}$ that contributes to the approximation. In Theorem 3, which is proved in Sect. 4, the above informal message is expressed rigorously in convenient geometric terms.

3   The f-metric

Theorem 3 requires one to consider the closure of a family $\mathcal{B}_f$ under the f-metric. The following example shows it is not always sufficient to consider only $\mathcal{B}_f$ when examining the approximation functional $\rho_f(\mathcal{B}, L)$.

Example. *Let $L \subseteq \mathbb{R}^2$ be a lattice-free split (see [10, p. 196]). Choose $f \in \mathrm{int}(L)$ and let $\mathcal{B} := \mathcal{L}_3^2 \cap \mathcal{C}_f$ be the set of all maximal lattice-free triangles containing f in the interior. Since L is a split, there is a nonzero vector r in the lineality space of L. The intersection cut $C_L((r), f)$ is an empty set while $C_B((r), f)$ is nonempty for each $B \in \mathcal{B}$. Hence $\rho_f(B, L) = \infty$ for each $B \in \mathcal{B}$. However, it is not hard to see that $\rho_f(\mathcal{B}, L) \leq 1$, see also [7, Theorem 1.4].*   ◇

The issue in this example is that L is a 'limit point' of $\mathcal{B}$, but $L \notin \mathcal{B}$. Examples such as this motivate the use of a metric such that these 'limit points' can be considered.

For $f \in \mathbb{R}^n \backslash \mathbb{Z}^n$, recall that $\mathcal{C}_f$ is the collection all closed, full-dimensional, convex sets in $\mathbb{R}^n$ that contain f in their interior. We define the *f-metric* d_f : $\mathcal{C}_f \times \mathcal{C}_f \to \mathbb{R}_{\geq 0}$ on $\mathcal{C}_f$ to be $d_f(B_1, B_2) := d_H((B_1 - f)^\circ, (B_2 - f)^\circ)$, where $B_1, B_2 \in \mathcal{C}_f$, $(B_i - f)^\circ$ denotes the polar of $B_i - f$ for $i = 1, 2$, and d_H denotes the Hausdorff metric[3]. Since f is in the interior of B_1 and B_2, the sets $(B_1 - f)^\circ$ and $(B_2 - f)^\circ$ are compact, which shows that $d_f(B_1, B_2)$ is well-defined.

4   One-for-all Theorems and Proof of Theorem 3

For proving Theorem 3, we first derive an analogous result about approximation of a single set L by a family $\mathcal{B}$ in the case of a fixed f.

Theorem 4 (One-for-all Theorem for a family $\mathcal{B}$ and a set L). *Let $f \in \mathbb{R}^n \backslash \mathbb{Z}^n$, let $\mathcal{B}$ be a family of lattice-free subsets of $\mathbb{R}^n$ satisfying $\mathrm{cl}_f(\mathcal{B}_f) = \mathcal{B}_f$, and let L be a lattice-free polyhedron given by $L = \mathrm{conv}(V) + \mathrm{cone}(W)$, where V*

---

[3] For a set $B \subseteq \mathbb{R}^n$, the *polar* of B is $B^\circ := \{r \in \mathbb{R}^n : r \cdot x \leq 1 \; \forall x \in B\}$. The *Hausdorff metric* is defined on the family of compact sets of $\mathbb{R}^n$ as follows: $d_H(A, B)$ for compact sets A, B is the minimum $\gamma > 0$ such that $A \subseteq B + D(0, \gamma)$ and $B \subseteq A + D(0, \gamma)$, where $D(0, \gamma)$ is the closed ball of radius γ around the origin.

is a nonempty finite subset of $\mathbb{R}^n$ and W is a finite (possibly empty) subset of $\mathbb{R}^n \backslash \{0\}$. Then

$$\frac{1}{|V|+|W|+1} \inf_{B \in \mathcal{B}} \rho_f(B, L) \leq \rho_f(\mathcal{B}, L) \leq \inf_{B \in \mathcal{B}} \rho_f(B, L). \tag{5}$$

Proof (Sketch). Note that for every $B \in \mathcal{B}$, $\rho_f(\mathcal{B}, L) \leq \rho_f(B, L)$ and so the right inequality of (5) holds. For the left inequality of (5), first suppose that L is a polytope, i.e. $W = \emptyset$. We may assume $f \in \text{int}(L)$, otherwise all the functionals evaluate to 0. We may also assume $\rho_f(\mathcal{B}, L) < \infty$, which implies $\mathcal{B}_f \neq \emptyset$. Using the fact that $\rho_f(\mathcal{B}, L) < \infty$, one can show that for $\epsilon^* := 1/((|V|+1)\rho_f(\mathcal{B}, L))$, there is some $B \in \mathcal{B}_f \subseteq \mathcal{B}$ satisfying $\epsilon^* L + (1 - \epsilon^*) f \subseteq B$. Therefore $C_B(R, f) \subseteq \epsilon^* C_L(R, f)$ for all R. From (1), $\rho_f(B, L) \leq (|V|+1)\rho_f(\mathcal{B}, L)$. This yields

$$\frac{1}{|V|+1} \inf_{B \in \mathcal{B}} \rho_f(B, L) \leq \frac{1}{|V|+1}\rho_f(B, L) \leq \rho_f(\mathcal{B}, L).$$

If L is not a polytope, we can restrict ourselves to the polytope case by representing L using a sequence of polytopes and employing a limiting argument. A complete proof appears in the journal version of this paper. □

An immediate corollary of Theorem 4 handles the case of arbitrary f.

Corollary 1. *Let $\mathcal{B}$ and L be as in Theorem 4. Further assume that $\text{cl}_f(\mathcal{B}_f) = \mathcal{B}_f$ for all $f \in \mathbb{R}^n \backslash \mathbb{Z}^n$. Then*

$$\frac{1}{|V|+|W|+1} \sup_{f \in \mathbb{R}^n \backslash \mathbb{Z}^n} \inf_{B \in \mathcal{B}} \rho_f(B, L) \leq \rho(\mathcal{B}, L) \leq \sup_{f \in \mathbb{R}^n \backslash \mathbb{Z}^n} \inf_{B \in \mathcal{B}} \rho_f(B, L). \tag{6}$$

Corollary 1 and Theorem 4 together give a more general One-for-all type result where $\mathcal{L}$ is a family of sets.

Theorem 5 (One-for-all Theorem for two families). *Let $\mathcal{B}$, $\mathcal{L}$, and N be as in Theorem 3. Then the following hold:*

(a) Let $f \in \mathbb{R}^n \backslash \mathbb{Z}^n$. If $\text{cl}_f(\mathcal{B}_f) = \mathcal{B}_f$ then

$$\frac{1}{N} \sup_{L \in \mathcal{L}} \inf_{B \in \mathcal{B}} \rho_f(B, L) \leq \rho_f(\mathcal{B}, \mathcal{L}) \leq \sup_{L \in \mathcal{L}} \inf_{B \in \mathcal{B}} \rho_f(B, L). \tag{7}$$

(b) If $\text{cl}_f(\mathcal{B}_f) = \mathcal{B}_f$ for all $f \in \mathbb{R}^n \backslash \mathbb{Z}^n$, then

$$\frac{1}{N} \sup_{L \in \mathcal{L}, f \in \text{int}(L)} \inf_{B \in \mathcal{B}} \rho_f(B, L) \leq \rho(\mathcal{B}, \mathcal{L}) \leq \sup_{L \in \mathcal{L}, f \in \text{int}(L)} \inf_{B \in \mathcal{B}} \rho_f(B, L). \tag{8}$$

Note that (7) follows from (5), and (8) follows from (6). Theorem 5 shows that finiteness of $\rho_f(\mathcal{B}, \mathcal{L})$ and $\rho(\mathcal{B}, \mathcal{L})$ depends on the individual values $\rho_f(B, L)$ for sets $B \in \mathcal{B}$ and $L \in \mathcal{L}$. The following proposition gives a geometric characterization of $\rho_f(B, L)$; this combined with Theorem 5 implies Theorem 3.

Proposition 1. *Let $f \in \mathbb{R}^n \backslash \mathbb{Z}^n$, and let B and L be lattice-free subsets of $\mathbb{R}^n$. Then*

$$\rho_f(B, L) = \inf \left\{ \alpha > 0 : B \supseteq \tfrac{1}{\alpha}(L - f) + f \right\}$$

if $f \in \text{int}(L)$ and $\rho_f(B, L) = 0$, otherwise.

5 The Relative Strength of i-hedral Closures

From Theorem 3, identifying the values of i yield $\rho(\mathcal{L}_i^n, \mathcal{L}_*^n) < \infty$ or $\rho_f(\mathcal{L}_i^n, \mathcal{L}_*^n) < \infty$ can be done by analyzing the structure of polyhedra with at most i facets. In order to help with this geometric analysis, we make use of the flatness constant. For every nonempty subset X of $\mathbb{R}^n$, the *width function* $w(X, \cdot) : \mathbb{R}^n \to [0, \infty]$ of X is defined to be $w(X, u) := \sup_{x \in X} x \cdot u - \inf_{x \in X} x \cdot u$. The value $w(X) := \inf_{u \in \mathbb{Z}^n \setminus \{0\}} w(X, u)$ is called the *lattice width* of X.

Theorem 6 (Flatness Theorem). *The* flatness constant in dimension n *is* finite, *i.e.* $\mathrm{Flt}(n) := \sup \{w(B) : B \text{ lattice-free set in } \mathbb{R}^n\} < \infty$.

For upper bounds on $\mathrm{Flt}(n)$ see, for example, [6, p. 317]. Theorem 3 also requires $\mathrm{cl}_f(\mathcal{L}_i^n \cap \mathcal{C}_f) = \mathcal{L}_i^n \cap \mathcal{C}_f$ for every $f \in \mathbb{R}^n \setminus \mathbb{Z}^n$.

Proposition 2. *Let* $i \in \mathbb{N}$ *and* $f \in \mathbb{R}^n \setminus \mathbb{Z}^n$. *Then* $\mathrm{cl}_f(\mathcal{L}_i^n \cap \mathcal{C}_f) = \mathcal{L}_i^n \cap \mathcal{C}_f$.

5.1 On the Approximability of i-hedral Closures

For Propositions 3 and 4 below, most of the proof is contained in Lemma 1.

Lemma 1. *Let* $L \in \mathcal{L}_*^n$ *and* $f \in \mathrm{int}(L)$. *Assume there exist values* $m \in \mathbb{N}$ *and* $t \in \mathbb{Z}$, *and a maximal lattice-free set* $D \in \mathcal{L}_m^{n-1}$ *such that* $L \cap (\mathbb{R}^{n-1} \times \{t\}) \subseteq D \times \{t\}$. *Assume there exists* $\gamma \in (0, 1]$ *such that* $w(L', e_n) \leq 1$ *and* $L' \cap (\mathbb{R}^{n-1} \times \{t\})$ *is nonempty, where* $L' := \gamma(L - f) + f$. *Then there exists a* $B \in \mathcal{L}_{m+1}^n$ *such that* $\frac{1}{4}\gamma(L - f) + f \subseteq B$.

Proposition 3. $\rho(\mathcal{L}_i^n, \mathcal{L}_*^n) \leq 4 \mathrm{Flt}(n)$ *for* $i > 2^{n-1}$.

Proof. It suffices to consider the case $i = 2^{n-1} + 1$, as every $\mathcal{L}_i^n$ with $i > 2^{n-1}$ contains $\mathcal{L}_{2^{n-1}+1}^n$ as a subset. The assertion is trivial for $n = 1$, and so we assume that $n \geq 2$. Using the definitions of the functionals $\rho_f(B, L)$ and $\rho_f(\mathcal{B}, L)$, and Eq. (4), it follows that

$$\rho(\mathcal{L}_i^n, \mathcal{L}_*^n) \leq \sup_{L \in \mathcal{L}_*^n, f \in \mathrm{int}(L)} \inf_{B \in \mathcal{L}_i^n} \rho_f(B, L).$$

Let $L \in \mathcal{L}_*^n$ and $f \in \mathrm{int}(L)$. From the previous inequality, it is enough to show that there exists a $B \in \mathcal{L}_i^n$ such that $\rho_f(B, L) \leq 4 \mathrm{Flt}(n)$. From Proposition 1, this condition is equivalent to the geometric condition $\frac{1}{4}(L' - f) + f \subseteq B$, where

$$L' := \tfrac{1}{\mathrm{Flt}(n)}(L - f) + f.$$

Thus we aim to find a $B \in \mathcal{L}_i^n$ such that $\frac{1}{4}(L' - f) + f \subseteq B$.

By Theorem 6, there exists a $u \in \mathbb{Z}^n \setminus \{0\}$ such that $w(L, u) \leq \mathrm{Flt}(n)$. After a unimodular transformation, we may assume $u = e_n$. For $t \in \mathbb{Z}$, let $U_t := \mathbb{R}^{n-1} \times \{t\}$. If $\frac{1}{4}(L' - f) + f \subseteq \mathrm{conv}(U_t \cup U_{t+1})$ for some $t \in \mathbb{Z}$, then setting $B := \mathrm{conv}(U_t \cup U_{t+1})$ yields the desired result. Otherwise fix $t \in \mathbb{Z}$ such that

$$\emptyset \neq (\tfrac{1}{4}(\mathrm{int}(L') - f) + f) \cap U_t \subseteq \mathrm{int}(L') \cap U_t \subseteq \mathrm{int}(L) \cap U_t.$$

Since L is lattice-free and $\mathrm{int}(L) \cap U_t \neq \emptyset$, the set $\{x \in \mathbb{R}^{n-1} : (x,t) \in L\}$ is lattice-free. The latter set is a subset of a maximal lattice-free polyhedron $D \subseteq \mathbb{R}^{n-1}$. We have thus shown that the assumptions of Lemma 1 are fulfilled with $m = 2^{n-1}$. Applying Lemma 1, we get the desired conclusion. $\qquad\square$

Proposition 4. *Let $f \in \mathbb{Q}^n \backslash \mathbb{Z}^n$ and let $s \in \mathbb{N}$ be the denominator of f. Let $i \in \mathbb{N}$ with $i \geq n+1$. Then $\rho_f(\mathcal{L}_i^n, \mathcal{L}_*^n) \leq \mathrm{Flt}(n)4^{n-1}s$.*

Proof. For each $L \in \mathcal{L}_*^n \cap \mathcal{C}_f$, we introduce two homothetical copies of L:

$$L' := \tfrac{1}{\mathrm{Flt}(n)4^{n-2}s}(L - f) + f,$$
$$L'' := \tfrac{1}{4}(L' - f) + f = \tfrac{1}{\mathrm{Flt}(n)4^{n-1}s}(L - f) + f.$$

Using the definitions of $\rho_f(B, L)$ and $\rho_f(\mathcal{B}, L)$, and Eq. (2), it follows that

$$\rho_f(\mathcal{L}_i^n, \mathcal{L}_*^n) \leq \sup_{L \in \mathcal{L}_*^n} \inf_{B \in \mathcal{L}_i^n} \rho_f(B, L).$$

From the previous inequality and Proposition 1, it is enough to show that for an arbitrary $L \in \mathcal{L}_*^n$ with $f \in \mathrm{int}(L)$, there exists a $B \in \mathcal{L}_i^n$ such that $L'' \subseteq B$. We verify this by induction on n. The assertion is clear for $n = 1$ by setting $B = L$. Consider $n \geq 2$ such that for every $f' \in \mathbb{Q}^{n-1}/\mathbb{Z}^{n-1}$ with denominator s and for every $\bar{L} \in \mathcal{L}_*^{n-1}$ there exists $\bar{B} \in \mathcal{L}_n^{n-1}$ satisfying $\bar{L}'' \subseteq \bar{B}$.

Let $L \in \mathcal{L}_*^n$ with $f \in \mathrm{int}(L)$. Let $u \in \mathbb{Z}^n \backslash \{0\}$ be the primitive vector for which the lattice width of L is attained. One has $u \cdot f \in \frac{1}{s}\mathbb{Z}$. After a unimodular transformation, we may assume $u = e_n$ (recall that unimodular transformations do not change the denominator of rational vectors). For $t \in \mathbb{Z}$ let $U_t := \mathbb{R}^{n-1} \times \{t\}$. Since $w(L', e_n) \leq 1$, there is some $m \in \mathbb{Z}$ such that $L' \subseteq \mathrm{conv}(U_{m-1}, U_{m+1})$. We may assume that $m = 0$ and so $L' \subseteq \mathrm{conv}(U_{-1}, U_1)$.

Consider cases on the integrality of f_n. First suppose $f_n \notin \mathbb{Z}$. Without loss of generality, we may assume that $f_n \in (0,1)$. Thus $f \in \mathbb{R}^{n-1} \times [\frac{1}{s}, 1 - \frac{1}{s}]$. Furthermore, $w(L'', e_n) \leq \frac{1}{s}$ by the choice of L''. Consequently L'' is a subset of the lattice-free split $B := \mathbb{R}^{n-1} \times [0,1]$.

For the case when $f_n \in \mathbb{Z}$, we use the induction assumption. Observe $f = (f', 0)$, where $f' \in \mathbb{Q}^{n-1} \backslash \mathbb{Z}^{n-1}$ has the same denominator as f. Also, the set $\{x \in \mathbb{R}^{n-1} : (x,0) \in L\}$ is a lattice-free polyhedron in U_0. Thus there is a $(n-1)$-dimensional maximal lattice-free set M_0 such that $\{x \in \mathbb{R}^{n-1} : (x,0) \in L\} \subseteq M_0 \in \mathcal{L}_{2^{n-1}}^{n-1}$.

Applying the induction assumption to M_0, we obtain a lattice-free set $D \in \mathcal{L}_n^{n-1}$ with respect to the lattice $\mathbb{Z}^{n-1} \times \{0\}$ satisfying $\frac{1}{\mathrm{Flt}(n-1)4^{n-2}s}(M_0 - f') + f' \subseteq D$. From the fact that $\mathrm{Flt}(n) \geq \mathrm{Flt}(n-1)$, one also has

$$L' \cap U_0 \subseteq \left(\tfrac{1}{\mathrm{Flt}(n)4^{n-2}s}(M_0 - f') + f'\right) \times \{0\} \subseteq D \times \{0\}.$$

Since $f_n = 0$, the set $L' \cap U_0$ is nonempty. Observe that $L', \gamma = 1, t = 0$, and D satisfy the assumptions of Lemma 1. Thus there is some $B \in \mathcal{L}_{n+1}^n$ such that $L'' = \frac{1}{4}L' + \frac{3}{4}f \subseteq B$, as desired. $\qquad\square$

5.2   On the Inapproximability of i-hedral Closures

The following lemmas help prove $\rho(\mathcal{L}_i^n, \mathcal{L}_{i+1}^n) = \infty$ for $2 \le i \le 2^{n-1}$.

Lemma 2. *Let $M \subseteq \mathbb{R}^n$ be a maximal lattice-free polyhedron with m facets and let $c \in \mathrm{int}(M)$. Then there exists an $\epsilon \in (0,1)$ such that every lattice-free polyhedron containing $(1 - \epsilon)M + \epsilon c$ has at least m facets.*

Lemma 3. *Let $i \in \mathbb{N}$ such that $2 \le i \le 2^n$. Then there exists an n-dimensional maximal lattice-free set with exactly i facets.*

Proposition 5. $\rho(\mathcal{L}_i^n, \mathcal{L}_{i+1}^n) = \infty$ *for $i \le 2^{n-1}$.*

Proof. Assuming $\rho(\mathcal{L}_i^n, \mathcal{L}_{i+1}^n) < \infty$, we derive a contradiction. Let $M \in \mathcal{L}_i^{n-1} \backslash \mathcal{L}_{i-1}^{n-1}$ be maximal lattice-free in $\mathbb{R}^{n-1}$ as guaranteed by Lemma 3. Since M is maximal, after applying an appropriate unimodular transformation we may write $M = D \times \mathbb{R}^{n-m-1}$ with $D \subseteq \mathbb{R}^m$ a maximal lattice-free polytope in $\mathbb{R}^m$ for $1 \le m \le n-1$ and D having i facets (see [13]). We consider D as embedded in $\mathbb{R}^m \times \{0\}^{n-m}$ and M in $\{x \in \mathbb{R}^n : x_{m+1} = 0\}$. For $z \in \mathrm{relint}(M)$, let $\varepsilon(z)$ denote the value obtained from Lemma 2 by setting $c = z$.

Fix $z = (z_D, 0) \in \mathrm{relint}(M)$ with $z_D \in D$. For $\epsilon > 0$, consider the polyhedron $L^\epsilon := \mathrm{conv}\left(\{z + \epsilon e_{m+1}\} \cup \left((1 + \frac{1}{\epsilon})(D - z) + z - e_{m+1}\right)\right) \subseteq \mathbb{R}^{m+1}$. Note $L^\epsilon \times \mathbb{R}^{n-m-1} \in \mathcal{L}_{i+1}^n$ is maximal lattice-free with exactly $i + 1$ facets. Theorem 3(b) and Proposition 2 imply the existence of $\mu \in (0,1)$ independent of $\varepsilon > 0$ such that for every $f \in L^\epsilon \times \mathbb{R}^{n-m-1}$ some $B_\epsilon \in \mathcal{L}_i^n$ satisfies $\mu(L^\epsilon \times \mathbb{R}^{n-m-1}) + (1 - \mu)f \subseteq B_\epsilon$. Let $\epsilon > 0$ and $\gamma \in (0, \epsilon)$ be chosen such that

(a) $\varepsilon(1 - \mu) - \mu < 0$, and
(b) $\mu(1 + \frac{\gamma}{\epsilon}(\frac{1-\mu}{\mu})) > 1 - \varepsilon(z)$.

For example, one can choose $\epsilon = \frac{1}{2}(\frac{\mu}{1-\mu})$ and $\gamma = \max\{\frac{\epsilon}{2}, \frac{\epsilon}{2}(1 + \frac{1-\varepsilon(z)-\mu}{1-\mu})\}$. Choose $f = z + \gamma e_{m+1}$.

Note that $D \cup \{z_D, f\} \subseteq \mathbb{R}^{m+1} \times \{0\}^{n-m-1}$. Conserving notation, we use $D, z_D,$ and f to denote the respective projections of $D, z_D,$ and f onto $\mathbb{R}^{m+1}$. With our choice of f, $L' := \mu L^\epsilon + (1 - \mu)f = \mathrm{conv}(\{a\} \cup \Delta)$ is a pyramid in $\mathbb{R}^{m+1}$ with apex $a := z_D + (\mu\epsilon + (1 - \mu)\gamma)e_{m+1}$ and base $\Delta := \mu\left(1 + \frac{1}{\epsilon}\right)(D - z_D) + z_D + (\gamma(1 - \mu) - \mu)e_{m+1}$. From (a) and the fact that $\gamma < \epsilon$, we obtain that $\gamma(1 - \mu) < \mu$, i.e., $\gamma(1 - \mu) - \mu < 0$. Thus the base of L' is below the hyperplane $\mathbb{R}^m \times \{0\}$.

For $\lambda \in \mathbb{R}$, define $L_\lambda^\epsilon := L^\epsilon \cap (\mathbb{R}^m \times \{\lambda\})$ and $L_\lambda' := L' \cap (\mathbb{R}^m \times \{\lambda\})$.

Claim 1. *Let $\lambda \in [-1, 0]$. Then $L_\lambda^\epsilon = \left(1 - \frac{\lambda}{\epsilon}\right)(D - z_D) + z_D + \lambda e_{m+1}$.*

Proof of Claim. Using the definitions of L_λ^ϵ and L^ϵ, we see $x \in L_\lambda^\epsilon$ if and only if

$$\text{iff } x_{m+1} = \lambda \text{ and } x \in L_{m+1}^\epsilon$$
$$\text{iff } x = \left(1 - \frac{\lambda}{\epsilon}\right)(y - z_D) + z_D + \lambda e_{m+1}, \text{ for } y \in D$$
$$\text{iff } x \in \left(1 - \frac{\lambda}{\epsilon}\right)(D - z_D) + z_D + \lambda e_{m+1}.$$

◇

Define $\beta := \frac{-\gamma(1-\mu)}{\mu}$. From (a) and the fact that $\gamma < \epsilon$, we see that $\beta \in (-1, 0)$.

Claim 2. $L_0' = \mu L_\beta^\epsilon + (1 - \mu)f$.

Proof of Claim. L_0' is the set of points $y \in L'$ with $y_{m+1} = 0$. L' is the set of points that can be written as $\mu x + (1 - \mu)f$ for $x \in L^\epsilon$. All points of this form that have 0 in the last coordinate must satisfy $x_{m+1} = -\frac{\gamma(1-\mu)}{\mu} = \beta$. Thus L' is the set of points that can be written as $\mu x + (1 - \mu)f$ for some $x \in L_\beta^\epsilon$.   ◇

We now follow this sequence of equalities:

$$L_0' = \mu L_\beta^\epsilon + (1 - \mu)f \qquad \text{from Claim 2}$$
$$= \mu((1 - \tfrac{\beta}{\epsilon})(D - z_D) + z_D + \beta e_{m+1}) + (1 - \mu)(z_D + \gamma e_{m+1}) \quad \text{from Claim 1}$$
$$= \mu(1 - \tfrac{\beta}{\epsilon})D + (1 - \mu(1 - \tfrac{\beta}{\epsilon}))z_D.$$

From (a) and (b), $\mu(1 - \frac{\beta}{\epsilon}) > 1 - \varepsilon(z) = 1 - \varepsilon(z_D)$. Thus the definition of $\varepsilon(z_D)$ implies that any lattice-free polyhedron containing L_0' requires i facets. In particular, B_ϵ must have at least i facets since the cross-section of B_ϵ by the hyperplane $\mathbb{R}^{m+1} \times \{0\}$ contains L_0'. Since $B_\epsilon \in \mathcal{L}_i^n$, B_ϵ must have exactly i facets. However, for small enough ϵ, $w(\Delta) > \mathrm{Flt}(m)$. This would imply that B_ϵ is not a cylinder since it must contain Δ. Therefore B_ϵ must have a full-dimensional recession cone. However, this contradicts that B_ϵ is lattice-free. □

Proving $\rho_f(\mathcal{L}_i^n, \mathcal{L}_*^n) = \infty$ requires us to identify, for every $\mu \in (0, 1)$, some $L \in \mathcal{L}_k^n$ with $k > i$ satisfying $B \not\supseteq \mu L + (1 - \mu)f$ for every $B \in \mathcal{L}_i^n$. However, unlike in the proof of Proposition 5 where we first fix L and then choose f in the interior of L, proving $\rho_f(\mathcal{L}_i^n, \mathcal{L}_*^n) = \infty$ requires us to construct L for any arbitrary fixed f. The next result helps us overcome this complication.

Lemma 4. *Let $f \in \mathbb{Q}^n \setminus \mathbb{Z}^n$ and $\mu \in (0, 1)$. Then*

(a) *There exists a maximal lattice-free simplex $L \in \mathcal{L}_{n+1}^n \cap \mathcal{C}_f$ such that, for some choice of $n+1$ integer points $z_1, \ldots, z_{n+1}$ in the relative interior of the $n+1$ distinct facets of L, the following is fulfilled: every closed half-space disjoint from $\mathrm{int}(\mu L + (1 - \mu)f)$ contains at most one point of the set $\{z_1, \ldots, z_{n+1}\}$.*

(b) *For every simplex L in (a), $B \not\supseteq \mu L + (1 - \mu)f$ for every $B \in \mathcal{L}_n^{n+1} \cap \mathcal{C}_f$.*

Proposition 6. *$\rho_f(\mathcal{L}_i^n, \mathcal{L}_{i+1}^n) = \infty$ for each $f \in \mathbb{Q}^n \setminus \mathbb{Z}^n$ and every $i \leq n$.*

Proof. From Theorem 3, it suffices to show that for all $i \in \{1, \ldots, n\}$, $f \in \mathbb{Q}^n \setminus \mathbb{Z}^n$ and $\mu \in (0, 1)$, there exists $L \in \mathcal{L}_{i+1}^n \cap \mathcal{C}_f$ satisfying $B \not\supseteq \mu L + (1 - \mu)f$ for all $B \in \mathcal{L}_i^n \cap \mathcal{C}_f$. For $i = n$, the assertion follows by choosing L as in Lemma 4.

Consider the case $i < n$. After applying an appropriate unimodular transformation we may assume that $f = (f', 0, \ldots, 0) \in \mathbb{R}^n$ for some $f' \in \mathbb{Q}^i \setminus \mathbb{Z}^i$. Application of Lemma 4 in dimension i yields the existence of a maximal lattice-free simplex $L' \in \mathcal{L}_{i+1}^i$ such that $B' \not\supseteq \mu L' + (1-\mu)f'$ holds for every $B' \in \mathcal{L}_i^i$. We choose $L = L' \times \mathbb{R}^{n-i}$ and show that $B \not\supseteq \mu L + (1 - \mu)f$ for every $B \in \mathcal{L}_i^n$. Note that $\mu L + (1 - \mu)f$ contains the affine space $A := \{f'\} \times \mathbb{R}^{n-i}$. If $B \not\supseteq A$, we get $B \not\supseteq \mu L + (1 - \mu)f$. Otherwise, $B \supseteq A$ and thus B can be represented as $B = B' \times \mathbb{R}^{n-i}$ with $B' \in \mathcal{L}_i^i$. In this case, $B \not\supseteq \mu L + (1 - \mu)f$ since $(B \cap \mathbb{R}^i) \not\supseteq \mu L' + (1 - \mu)f'$.   □

References

1. Andersen, K., Louveaux, Q., Weismantel, R., Wolsey, L.A.: Inequalities from two rows of a simplex tableau. In: Fischetti, M., Williamson, D.P. (eds.) IPCO 2007. LNCS, vol. 4513, pp. 1–15. Springer, Heidelberg (2007). doi:10.1007/978-3-540-72792-7_1
2. Andersen, K., Wagner, C., Weismantel, R.: On an analysis of the strength of mixed-integer cutting planes from multiple simplex tableau rows. SIAM J. Optim. **20**(2), 967–982 (2009)
3. Averkov, G., Krümpelmann, J., Weltge, S.: Notions of maximality for integral lattice-free polyhedra: the case of dimension three (2015). http://arxiv.org/abs/1509.05200
4. Averkov, G., Wagner, C., Weismantel, R.: Maximal lattice-free polyhedra: finiteness and an explicit description in dimension three. Math. Oper. Res. **36**(4), 721–742 (2011)
5. Awate, Y., Cornuéjols, G., Guenin, B., Tuncel, L.: On the relative strength of families of intersection cuts arising from pairs of tableau constraints in mixed integer programs. Math. Program. **150**, 459–489 (2014)
6. Barvinok, A.: A Course in Convexity. American Mathematical Society (2002)
7. Basu, A., Bonami, P., Cornuéjols, G., Margot, F.: On the relative strength of split, triangle and quadrilateral cuts. Math. Program. Ser. A **126**, 281–314 (2009)
8. Basu, A., Conforti, M., Cornuéjols, G., Zambelli, G.: Maximal lattice-free convex sets in linear subspaces. Math. Oper. Res. **35**, 704–720 (2010)
9. Borozan, V., Cornuéjols, G.: Minimal valid inequalities for integer constraints. Math. Oper. Res. **34**, 538–546 (2009)
10. Conforti, M., Cornuéjols, G., Zambelli, G.: Integer programming, vol. 271. Springer, Switzerland (2014)
11. Dey, S.S., Wolsey, L.A.: Lifting integer variables in minimal inequalities corresponding to lattice-free triangles. In: Lodi, A., Panconesi, A., Rinaldi, G. (eds.) IPCO 2008. LNCS, vol. 5035, pp. 463–475. Springer, Heidelberg (2008). doi:10.1007/978-3-540-68891-4_32
12. Gomory, R.E.: Some polyhedra related to combinatorial problems. Linear Algebra Appl. **2**(4), 451–558 (1969)
13. Lovász, L.: Geometry of numbers and integer programming. In: Iri, M., Tanabe, K. (eds.) Mathematical Programming: State of the Art, pp. 177–201. Mathematical Programming Society (1989)
14. Meyer, R.: On the existence of optimal solutions to integer and mixed-integer progamming problems. Math. Program. **7**, 223–235 (1974)
15. Schneider, R.: Convex Bodies: The Brunn-Minkowski Theory, vol. 44. Cambridge University Press, Cambridge (2014)
16. Zambelli, G.: On degenerate multi-row gomory cuts. Oper. Res. Lett. **37**(1), 21–22 (2009)

The Structure of the Infinite Models
in Integer Programming

Amitabh Basu[1], Michele Conforti[2], Marco Di Summa[2($\boxtimes$)], and Joseph Paat[1]

[1] Department of Applied Mathematics and Statistics,
The Johns Hopkins University, Baltimore, USA
[2] Dipartimento di Matematica,
Università degli Studi di Padova, Padua, Italy
disumma@math.unipd.it

Abstract. The infinite models in integer programming can be described as the convex hull of some points or as the intersection of halfspaces derived from valid functions. In this paper we study the relationships between these two descriptions. Our results have implications for finite dimensional corner polyhedra. One consequence is that nonnegative continuous functions suffice to describe finite dimensional corner polyhedra with rational data. We also discover new facts about corner polyhedra with non-rational data.

1 Introduction

Let $b \in \mathbb{R}^n \backslash \mathbb{Z}^n$. The *mixed-integer infinite group relaxation* M_b is the set of all pairs of functions (s, y) with $s : \mathbb{R}^n \to \mathbb{R}_+$ and $y : \mathbb{R}^n \to \mathbb{Z}_+$ having finite support (that is, $\{r : s(r) > 0\}$ and $\{p : y(p) > 0\}$ are finite sets) satisfying

$$\sum_{r \in \mathbb{R}^n} r s(r) + \sum_{p \in \mathbb{R}^n} p y(p) \in b + \mathbb{Z}^n. \tag{1.1}$$

M_b is a subset of the infinite dimensional vector space $\mathbb{R}^{(\mathbb{R}^n)} \times \mathbb{R}^{(\mathbb{R}^n)}$, where $\mathbb{R}^{(\mathbb{R}^n)}$ denotes the set of finite support functions from $\mathbb{R}^n$ to $\mathbb{R}$ (similarly, $\mathbb{R}_+^{(\mathbb{R}^n)}$ will denote the set of finite support functions from $\mathbb{R}^n$ to $\mathbb{R}$ that are nonnegative). We will work with this vector space throughout the paper. A tuple (ψ, π, α), where $\psi, \pi : \mathbb{R}^n \to \mathbb{R}$ and $\alpha \in \mathbb{R}$, is a *valid tuple for M_b* if

$$\sum_{r \in \mathbb{R}^n} \psi(r) s(r) + \sum_{p \in \mathbb{R}^n} \pi(p) y(p) \geq \alpha \tag{1.2}$$

for every $(s, y) \in M_b$. Since for $\lambda > 0$ the inequalities (1.2) associated with (ψ, π, α) and $(\lambda\psi, \lambda\pi, \lambda\alpha)$ are equivalent, from now on we assume $\alpha \in \{-1, 0, 1\}$.

The set of functions $y : \mathbb{R}^n \to \mathbb{Z}_+$ such that $(0, y) \in M_b$ will be called the *pure integer infinite group relaxation I_b*. In other words, $I_b = \{y : (0, y) \in M_b\}$.

A. Basu and J. Paat—Supported by the NSF grant CMMI1452820.
M. Conforti and M. Di Summa—Supported by the grant "Progetto di Ateneo 2013".

© Springer International Publishing AG 2017
F. Eisenbrand and J. Koenemann (Eds.): IPCO 2017, LNCS 10328, pp. 63–74, 2017.
DOI: 10.1007/978-3-319-59250-3_6

By definition, $I_b \subseteq \mathbb{R}^{(\mathbb{R}^n)}$. However, when convenient we will see I_b as a subset of M_b. A tuple (π, α), where $\pi : \mathbb{R}^n \to \mathbb{R}$ and $\alpha \in \mathbb{R}$, is called a *valid tuple for* I_b if

$$\sum_{p \in \mathbb{R}^n} \pi(p)y(p) \geq \alpha \tag{1.3}$$

for every $y \in I_b$. Again, we will assume $\alpha \in \{-1, 0, 1\}$.

Models M_b and I_b were defined by Gomory and Johnson in a series of papers [10–12,14] as a template to generate valid inequalities, derived from (1.2) and (1.3), for general integer programs. They have been the focus of extensive research, as summarized, e.g., in [3,4] and [6, Chap. 6].

Our Results. One would expect that the intersection of (1.2) for all valid tuples for M_b would be equal to $\mathrm{conv}(M_b)$, where $\mathrm{conv}(\cdot)$ denotes the convex hull operator. However, this is not true: this intersection is a strict superset of $\mathrm{conv}(M_b)$. One of our main results (Theorem 2.13) shows that the intersection of all valid tuples for M_b is, in fact, the closure of $\mathrm{conv}(M_b)$ under a norm topology on $\mathbb{R}^{(\mathbb{R}^n)} \times \mathbb{R}^{(\mathbb{R}^n)}$ that was first defined by Basu et al. [2]. We then give an explicit characterization that shows that this closure coincides with $\mathrm{conv}(M_b) + (\mathbb{R}_+^{(\mathbb{R}^n)} \times \mathbb{R}_+^{(\mathbb{R}^n)})$. A similar phenomenon happens for I_b (Theorem 2.14).

A valid tuple (ψ, π, α) for M_b is minimal if there does not exist a pair of functions (ψ', π') different from (ψ, π), with $(\psi', \pi') \leq (\psi, \pi)$, such that (ψ', π', α) is a valid tuple for M_b. Our main tool is a characterization of the minimal tuples (Theorem 2.4) that extends a result of Johnson (see, e.g., Theorem 6.34 in [6]), that was obtained under the assumption that $\pi \geq 0$. The main novelty of our result over Johnson's is that minimality of a valid tuple (ψ, π, α) *implies* nonnegativity of π (no need to assume it). Moreover, π has to be continuous (in fact, it is Lipschitz continuous.)

Most of the prior literature on valid tuples (π, α) for I_b proceeds under the restrictive assumption that π is nonnegative (in fact, Gomory and Johnson included the assumption $\pi \geq 0$ in their original definition of valid tuple for I_b). This assumption has been criticized in more recent work on I_b, as there are valid functions not satisfying $\pi \geq 0$. In this paper, we prove that every valid tuple for I_b has an equivalent representation (π, α) where $\pi \geq 0$. More specifically, we show that for every valid tuple (π, α), there exist $\theta : \mathbb{R}^n \to \mathbb{R}$ and $\beta \in \mathbb{R}$ such that both (θ, β), $(-\theta, -\beta)$ are valid tuples and the valid tuple $(\pi', \alpha') = (\pi + \theta, \alpha + \beta)$ satisfies $\pi' \geq 0$ (Theorem 3.5). This settles an open question in [3, Open Question 2.5]. Being able to restrict to nonnegative valid tuples without loss of generality has the added advantage that nonnegative minimal valid tuples form a *compact, convex set* under the natural product topology on functions. Thus, one approach to understanding valid tuples is to understand the extreme points of this compact convex set, which are termed *extreme functions/tuples* in the literature. While this approach was standard for the area, our result about nonnegative valid tuples now gives a rigorous justification for this.

A valid tuple (π, α) for I_b is liftable if there there exists $\psi : \mathbb{R}^n \to \mathbb{R}$ such that (ψ, π, α) is a valid tuple for M_b. Minimal valid tuples (π, α) that are liftable are a strict subset of minimal valid tuples, as we show that such π have to be nonnegative and Lipschitz continuous (Proposition 2.6 and Remark 2.7). This has some consequences for finite dimensional corner polyhedra that have rational data, which are sets of the form $\mathrm{conv}(I_b) \cap \{y : y_r = 0, r \in \mathbb{R}^n \backslash P\}$, where P is a finite subset of $\mathbb{Q}^n$. Corollary 4.4 shows that inequalities (1.3) associated with liftable tuples, when restricted to the space $\{y : y_r = 0, r \in \mathbb{R}^n \backslash P\}$, suffice to provide a complete inequality description for such corner polyhedra. Literature on valid tuples contains constructions of families of extreme valid tuples (π, α) such that π is discontinuous [7,8,13,15,17,18] (or continuous but not Lipschitz continuous [15]). Our result above shows that such functions may be disregarded, if one is interested in valid inequalities or facets of rational corner polyhedra. Similarly, valid tuples (π, α) where $\pi \not\geq 0$ are also superfluous for such polyhedra. This is interesting, in our opinion, as it shows that such extreme tuples are redundant within the set of valid tuples, as far as rational corner polyhedra are concerned. Some further characterizations of rational corner polyhedra are derived in Theorem 4.3.

Crucial to the proof of the above result on rational corner polyhedra is our characterization of the equations defining the affine hull of $\mathrm{conv}(I_b)$, which extends a result in [3]. This characterization is also essential in understanding the recession cone of $\mathrm{conv}(I_b) \cap \{y : y_r = 0, r \in \mathbb{R}^n \backslash P\}$, where P is a finite subset of $\mathbb{R}^n$. We use this to prove that $\mathrm{conv}(I_b) \cap \{y : y_r = 0, r \in \mathbb{R}^n \backslash P\}$ is a polyhedron, even if $P \cup \{b\}$ contains non-rational vectors (Theorem 4.2).

Due to space constraints, all missing proofs will appear in the journal version of this extended abstract.

2   The Structure of $\mathrm{conv}(M_b)$ and $\mathrm{conv}(I_b)$

A valid tuple (ψ, π, α) for M_b is said to be *minimal* if there does not exist a pair of functions (ψ', π') different from (ψ, π), with $(\psi', \pi') \leq (\psi, \pi)$, such that (ψ', π', α) is a valid tuple for M_b. Similarly, we say that a valid tuple (π, α) for I_b is *minimal* if there does not exist a function π' different from π, with $\pi' \leq \pi$, such that (π', α) is a valid tuple for I_b.

Remark 2.1. An application of Zorn's lemma (see, e.g., [5, Proposition A.1]) shows that, given a valid tuple (ψ, π, α) for M_b, there exists a minimal valid tuple (ψ', π', α) for M_b with $\psi' \leq \psi$ and $\pi' \leq \pi$. Similarly, given a valid tuple (π, α) for I_b, there exists a minimal valid tuple (π', α) for I_b with $\pi' \leq \pi$. We will use this throughout the paper.

Given a tuple (ψ, π, α), we define

$$H_{\psi,\pi,\alpha} = \left\{ (s,y) \in \mathbb{R}^{(\mathbb{R}^n)} \times \mathbb{R}^{(\mathbb{R}^n)} : \sum_{r \in \mathbb{R}^n} \psi(r)s(r) + \sum_{p \in \mathbb{R}^n} \pi(p)y(p) \geq \alpha \right\}.$$

A valid tuple (ψ, π, α) for M_b is *trivial* if $\mathbb{R}_+^{(\mathbb{R}^n)} \times \mathbb{R}_+^{(\mathbb{R}^n)} \subseteq H_{\psi,\pi,\alpha}$. This happens if and only if $\psi \geq 0$, $\pi \geq 0$ and $\alpha \in \{0, -1\}$. Similarly, a valid tuple (π, α) for I_b is *trivial* if $\pi \geq 0$ and $\alpha \in \{0, -1\}$.

A function $\phi : \mathbb{R}^n \to \mathbb{R}$ is *subadditive* if $\phi(r_1) + \phi(r_2) \geq \phi(r_1 + r_2)$ for every $r_1, r_2 \in \mathbb{R}^n$, and is *positively homogenous* if $\phi(\lambda r) = \lambda \phi(r)$ for every $r \in \mathbb{R}^n$ and $\lambda \geq 0$. If ϕ is subadditive and positive homogenous, then ϕ is called *sublinear*.

Proposition 2.2. *Let* (ψ, π, α) *be a minimal valid tuple for* M_b*. Then* ψ *is sublinear and* $\pi \leq \psi$.

Lemma 2.3. *Suppose* $\pi : \mathbb{R}^n \to \mathbb{R}$ *is subadditive and* $\sup_{\varepsilon > 0} \frac{\pi(\varepsilon r)}{\varepsilon} < \infty$ *for all* $r \in \mathbb{R}^n$. *Define* $\psi(r) = \sup_{\varepsilon > 0} \frac{\pi(\varepsilon r)}{\varepsilon}$. *Then* ψ *is sublinear and* $\pi \leq \psi$.

Proof. Since π is subadditive, ψ is readily checked to be subadditive as well. The fact that $\pi \leq \psi$ follows by taking $\varepsilon = 1$. Finally, positive homogeneity of ψ follows from the definition of ψ. $\square$

Theorem 2.4. *Let* $\psi : \mathbb{R}^n \to \mathbb{R}$, $\pi : \mathbb{R}^n \to \mathbb{R}$ *be any functions, and* $\alpha \in \{-1, 0, 1\}$. *Then* (ψ, π, α) *is a nontrivial minimal valid tuple for* M_b *if and only if the following hold:*

(a) π *is subadditive;*
(b) $\psi(r) = \sup_{\varepsilon > 0} \frac{\pi(\varepsilon r)}{\varepsilon} = \lim_{\varepsilon \to 0+} \frac{\pi(\varepsilon r)}{\varepsilon} = \limsup_{\varepsilon \to 0+} \frac{\pi(\varepsilon r)}{\varepsilon}$ *for every* $r \in \mathbb{R}^n$;
(c) π *is Lipschitz continuous with Lipschitz constant* $L := \max_{\|r\|=1} \psi(r)$;
(d) $\pi \geq 0$, $\pi(z) = 0$ *for every* $z \in \mathbb{Z}^n$, *and* $\alpha = 1$;
(e) *(symmetry condition)* π *satisfies* $\pi(r) + \pi(b - r) = 1$ *for all* $r \in \mathbb{R}^n$.

The above theorem can be deduced from a result of Yıldız and Cornuéjols [19, Theorem 37] by using the characterization of the nontrivial minimal valid tuples for I_b due to Gomory and Johnson (see, e.g., [6, Theorem 6.22]).

Corollary 2.5. *Let* (π, α) *be a nontrivial minimal valid tuple for* I_b *such that* $\sup_{\varepsilon > 0} \frac{\pi(\varepsilon r)}{\varepsilon} < \infty$ *for every* $r \in \mathbb{R}^n$. *Define* $\psi(r) = \sup_{\varepsilon > 0} \frac{\pi(\varepsilon r)}{\varepsilon}$. *Then* (ψ, π, α) *satisfies conditions (a)–(e) of Theorem 2.4 and therefore is a nontrivial minimal valid tuple for* M_b.

  Conversely, if (ψ, π, α) *is a nontrivial minimal valid tuple for* M_b*, then* (π, α) *is a nontrivial minimal valid tuple for* I_b.

Proof. Since (π, α) is minimal, the same argument as in the proof of Proposition 2.2 shows that π is subadditive. Let ψ be defined as above. Following the proof of Theorem 2.4 it can be checked that minimality and nontriviality of (π, α) suffice to show that (ψ, π, α) satisfies (a)–(e), and therefore (ψ, π, α) is a nontrivial minimal valid tuple for M_b.

For the converse, we use a theorem of Gomory and Johnson (see, e.g., [6, Theorem 6.22]) stating that if $(\pi, 1)$ is a nontrivial valid tuple with $\pi \geq 0$, then $(\pi, 1)$ is minimal if and only if π is subadditive, $\pi(z) = 0$ for every $z \in \mathbb{Z}^n$, and π satisfies the symmetry condition. Let (ψ, π, α) be a nontrivial minimal valid tuple for M_b. By Theorem 2.4, $\pi \geq 0$, $\alpha = 1$, π is subadditive, $\pi(z) = 0$ for every $z \in \mathbb{Z}^n$, and π satisfies the symmetry condition. Therefore, by the above theorem, (π, α) is a nontrivial minimal valid tuple for I_b. $\square$

A valid tuple (π, α) for I_b is called *liftable* if there exists a function $\psi : \mathbb{R}^n \to \mathbb{R}$ such that (ψ, π, α) is a valid tuple for M_b.

Proposition 2.6. *Let (π, α) be a nontrivial valid tuple for I_b. Then (π, α) is liftable if and only if there exists a minimal valid tuple (π', α) such that $\pi' \leq \pi$ and $\sup_{\varepsilon > 0} \frac{\pi'(\varepsilon r)}{\varepsilon} < \infty$ for every $r \in \mathbb{R}^n$. In this case, defining $\psi(r) = \sup_{\varepsilon > 0} \frac{\pi'(\varepsilon r)}{\varepsilon}$ gives a valid tuple (ψ, π', α) for M_b satisfying conditions (a)–(e) of Theorem 2.4.*

Proof. If (π, α) is nontrivial and liftable, then there exists ψ such that (ψ, π, α) is a valid tuple for M_b. Let (ψ', π', α) be a minimal valid tuple with $\psi' \leq \psi$ and $\pi' \leq \pi$. Since (π, α) is nontrivial, so is (ψ', π', α). By Theorem 2.4, $\alpha = 1$, $\pi' \geq 0$, and $\sup_{\varepsilon > 0} \frac{\pi'(\varepsilon r)}{\varepsilon} < \infty$ for every $r \in \mathbb{R}^n$. By Corollary 2.5, (π', α) is minimal.

Conversely, let (π, α) be a nontrivial valid tuple for I_b, and let $\pi' \leq \pi$ be such that (π', α) is minimal (and nontrivial) and $\psi(r) := \sup_{\varepsilon > 0} \frac{\pi'(\varepsilon r)}{\varepsilon}$ is finite for every $r \in \mathbb{R}^n$. By Corollary 2.5, (ψ, π', α) is a nontrivial minimal valid tuple for M_b, and therefore (π', α) is liftable. Since $\pi \geq \pi'$, (π, α) is liftable as well. □

Remark 2.7. Let (π, α) be a nontrivial minimal valid tuple for I_b that is liftable. It follows from Proposition 2.6 (with $\pi' = \pi$) that $\psi(r) := \sup_{\varepsilon > 0} \frac{\pi(\varepsilon r)}{\varepsilon}$ is finite for all $r \in \mathbb{R}^n$, and (ψ, π, α) is a minimal valid tuple for M_b that satisfies conditions (a)–(e) of Theorem 2.4. Therefore π is Lipschitz continuous and $\pi \geq 0$. There are nontrivial minimal valid tuples (π, α) for I_b for which π is not continuous, or π is continuous but not Lipschitz continuous, see the construction in [15, Sect. 5]. There are also nontrivial minimal valid tuples (π, α) for I_b with $\pi \not\geq 0$. None of these minimal tuples is liftable.

2.1 The Closure of conv(M_b)

Lemma 2.8. *The following sets coincide:*

(a) $\left(\mathbb{R}_+^{(\mathbb{R}^n)} \times \mathbb{R}_+^{(\mathbb{R}^n)}\right) \cap \bigcap \{H_{\psi, \pi, \alpha} : (\psi, \pi, \alpha) \text{ valid tuple}\}$

(b) $\left(\mathbb{R}_+^{(\mathbb{R}^n)} \times \mathbb{R}_+^{(\mathbb{R}^n)}\right) \cap \bigcap \{H_{\psi, \pi, \alpha} : (\psi, \pi, \alpha) \text{ nontrivial valid tuple}\}$

(c) $\left(\mathbb{R}_+^{(\mathbb{R}^n)} \times \mathbb{R}_+^{(\mathbb{R}^n)}\right) \cap \bigcap \{H_{\psi, \pi, \alpha} : (\psi, \pi, \alpha) \text{ minimal nontrivial valid tuple}\}$

(d) $\left(\mathbb{R}_+^{(\mathbb{R}^n)} \times \mathbb{R}_+^{(\mathbb{R}^n)}\right) \cap \bigcap \{H_{\psi, \pi, \alpha} : (\psi, \pi, \alpha) \text{ minimal nontrivial valid tuple}, \psi, \pi \geq 0, \alpha = 1\}.$

Proof. The equivalence of (a) and (b) follows from the definition of nontrivial valid tuple. The sets (b) and (c) coincide by Remark 2.1. Finally, Theorem 2.4 shows that (c) is equal to (d). □

From now on, we denote by Q_b the set(s) in Lemma 2.8.

While $\text{conv}(M_b) \subseteq Q_b$, this containment is strict, as shown in Remark 4.6. However, Theorem 2.13 below proves that, under an appropriate topology, the closure of $\text{conv}(M_b)$ is exactly Q_b. In order to show this result, we need the following lemma, that may be of independent interest.

Lemma 2.9. *If $C \subseteq \mathbb{R}^n_+$ is closed, then $\mathrm{conv}(C) + \mathbb{R}^n_+$ is also closed.*

Define the following norm on $\mathbb{R}^{(\mathbb{R}^n)} \times \mathbb{R}^{(\mathbb{R}^n)}$, which was first introduced in [2]:

$$|(s,y)|_* = |s(0)| + \sum_{r \in \mathbb{R}^n} \|r\| |s(r)| + |y(0)| + \sum_{p \in \mathbb{R}^n} \|p\| |y(p)|.$$

Define $\mathrm{cl}(\cdot)$ as the closure operator with respect to the topology induced by $|(\cdot, \cdot)|_*$. For any two functions $\psi : \mathbb{R}^n \to \mathbb{R}$, $\pi : \mathbb{R}^n \to \mathbb{R}$, we define a linear functional $F_{\psi, \pi}$ on the space $\mathbb{R}^{(\mathbb{R}^n)} \times \mathbb{R}^{(\mathbb{R}^n)}$ as follows:

$$F_{\psi, \pi}(s, y) = \sum_{r \in \mathbb{R}^n} \psi(r) s(r) + \sum_{p \in \mathbb{R}^n} \pi(p) y(p). \tag{2.1}$$

Lemma 2.10. *Under the $|(\cdot, \cdot)|_*$ norm, the linear functional $F_{\psi, \pi}$ is continuous if $(\psi, \pi, 1)$ is a nontrivial minimal valid tuple for M_b.*

Lemma 2.11. *Under the topology induced by $|(\cdot, \cdot)|_*$, the set Q_b is closed.*

For any subsets $R, P \subseteq \mathbb{R}^n$, define

$$V_{R,P} = \left\{ (s, y) \in \mathbb{R}^{(\mathbb{R}^n)} \times \mathbb{R}^{(\mathbb{R}^n)} : s(r) = 0 \ \forall r \notin R, \ y(p) = 0 \ \forall p \notin P \right\}.$$

When convenient, we will see $V_{R,P}$ as a subset of $\mathbb{R}^R \times \mathbb{R}^P$ by dropping the variables set to 0. Similarly, V_P will denote $\{y \in \mathbb{R}^{(\mathbb{R}^n)} : y(p) = 0 \ \forall p \notin P\}$.

Lemma 2.12. *For any $R, P \subseteq \mathbb{R}^n$, $V_{R,P}$ is a closed subspace of $\mathbb{R}^{(\mathbb{R}^n)} \times \mathbb{R}^{(\mathbb{R}^n)}$.*

Theorem 2.13. $Q_b = \mathrm{cl}(\mathrm{conv}(M_b)) = \mathrm{conv}(M_b) + (\mathbb{R}^{(\mathbb{R}^n)}_+ \times \mathbb{R}^{(\mathbb{R}^n)}_+).$

Proof. We first show that $Q_b \supseteq \mathrm{cl}(\mathrm{conv}(M_b))$. By definition, Q_b is convex, and by Lemma 2.11, Q_b is closed. Thus, it suffices to show that $Q_b \supseteq M_b$. This follows from the fact that $M_b \subseteq \mathbb{R}^{(\mathbb{R}^n)}_+ \times \mathbb{R}^{(\mathbb{R}^n)}_+$ and every inequality that defines Q_b is valid for M_b.

We next show that $Q_b \subseteq \mathrm{cl}(\mathrm{conv}(M_b))$. Consider a point $(s, y) \notin \mathrm{cl}(\mathrm{conv}(M_b))$. By the Hahn-Banach theorem, there exists a continuous linear functional that separates (s, y) from $\mathrm{cl}(\mathrm{conv}(M_b))$. In other words, there exist two functions $\psi, \pi : \mathbb{R}^n \to \mathbb{R}$ and a real number α such that $F_{\psi, \pi}(s, y) < \alpha$ and $\mathrm{cl}(\mathrm{conv}(M_b)) \subseteq H_{\psi, \pi, \alpha}$, implying that (ψ, π, α) is a valid tuple for M_b. Thus $(s, y) \notin Q_b$.

We now show that $\mathrm{conv}(M_b) + (\mathbb{R}^{(\mathbb{R}^n)}_+ \times \mathbb{R}^{(\mathbb{R}^n)}_+) \subseteq Q_b$. Consider any point $(s_1, y_1) + (s_2, y_2)$, where $(s_1, y_1) \in \mathrm{conv}(M_b)$ and $s_2 \geq 0, y_2 \geq 0$. Since Q_b can be written as the set (d) in Lemma 2.8 and $\mathrm{conv}(M_b) \subseteq \mathbb{R}^{(\mathbb{R}^n)}_+ \times \mathbb{R}^{(\mathbb{R}^n)}_+$, we just need to verify that $(s_1, y_1) + (s_2, y_2) \in H_{\psi, \pi, 1}$ for all valid $\psi, \pi \geq 0$. This follows because $(s_1, y_1) \in H_{\psi, \pi, 1}$ and (s_2, y_2) and ψ, π are all nonnegative.

We finally show that $\mathrm{conv}(M_b) + (\mathbb{R}^{(\mathbb{R}^n)}_+ \times \mathbb{R}^{(\mathbb{R}^n)}_+) \supseteq Q_b$. Consider $(s^*, y^*) \notin \mathrm{conv}(M_b) + (\mathbb{R}^{(\mathbb{R}^n)}_+ \times \mathbb{R}^{(\mathbb{R}^n)}_+)$. We prove that $(s^*, y^*) \notin Q_b$. This is obvious when

$(s^*, y^*) \notin \mathbb{R}_+^{(\mathbb{R}^n)} \times \mathbb{R}_+^{(\mathbb{R}^n)}$. Therefore we assume $s^* \geq 0$, $y^* \geq 0$. Let $R \subseteq \mathbb{R}^n$ be a finite set containing the support of s^* and satisfying $\text{cone}(R) = \mathbb{R}^n$ (where $\text{cone}(R)$ denotes the conical hull of R), and let $P \subseteq \mathbb{R}^n$ be a finite set containing the support of y^*. Then $(s^*, y^*) \notin \text{conv}(M_b \cap V_{R,P}) + (\mathbb{R}_+^R \times \mathbb{R}_+^P)$. (We use the same notation (s^*, y^*) to indicate the restriction of (s^*, y^*) to $\mathbb{R}^R \times \mathbb{R}^P$.) Since $M_b \cap V_{R,P}$ is the inverse image of the closed set $b + \mathbb{Z}^n$ under the linear transformation given by the matrix (R, P), $M_b \cap V_{R,P}$ is closed in the usual finite dimensional topology of $V_{R,P}$. Therefore, by Lemma 2.9, $\text{conv}(M_b \cap V_{R,P}) + (\mathbb{R}_+^R \times \mathbb{R}_+^P)$ is closed as well. This implies that there exists a valid inequality in $\mathbb{R}^R \times \mathbb{R}^P$ separating (s^*, y^*) from $\text{conv}(M_b \cap V_{R,P}) + (\mathbb{R}_+^R \times \mathbb{R}_+^P)$. Since the recession cone of $\text{conv}(M_b \cap V_{R,P}) + (\mathbb{R}_+^R \times \mathbb{R}_+^P)$ contains $(\mathbb{R}_+^R \times \mathbb{R}_+^P)$ and because $s^*, y^* \geq 0$, this valid inequality is of the form $\sum_{r \in R} h(r)s(r) + \sum_{p \in P} d(p)y(p) \geq 1$ where $h(r) \geq 0$ for $r \in R$ and $d(p) \geq 0$ for $p \in P$.

Now define the functions

$$\psi(r) = \inf \left\{ \sum_{r' \in R} h(r')s(r') : r = \sum_{r' \in R} r's(r'), \, s : R \to \mathbb{R}_+ \right\},$$

$$\pi(p) = \inf \left\{ \sum_{r' \in R} h(r')s(r') + \sum_{p' \in P} d(p')y(p') : \right.$$
$$\left. p = \sum_{r' \in R} r's(r') + \sum_{p' \in P} p'y(p'), \, s : R \to \mathbb{R}_+, \, y : P \to \mathbb{Z}_+ \right\}.$$

Since $\text{cone}(R) = \mathbb{R}^n$, ψ and π are well-defined functions. As the sum only involves nonnegative terms, $\psi, \pi \geq 0$. It can be checked that $(\psi, \pi, 1)$ is a valid tuple for M_b, and since $(s^*, y^*) \notin H_{\psi, \pi, 1}$, we have $(s^*, y^*) \notin Q_b$. $\square$

2.2   The Closure of $\text{conv}(I_b)$

In the following, we see $\mathbb{R}^{(\mathbb{R}^n)}$ as a topological vector subspace of the space $\mathbb{R}^{(\mathbb{R}^n)} \times \mathbb{R}^{(\mathbb{R}^n)}$ endowed with the topology induced by the norm $|(\cdot, \cdot)|_*$. With a slight abuse of notation, for any $y \in \mathbb{R}^{(\mathbb{R}^n)}$, $|y|_* = |y(0)| + \sum_{p \in \mathbb{R}^n} \|p\| |y(p)|$. Also, given $\pi : \mathbb{R}^n \to \mathbb{R}$ and $\alpha \in \mathbb{R}$, we let $H_{\pi, \alpha} = \{ y \in \mathbb{R}^{(\mathbb{R}^n)} : \sum_{p \in \mathbb{R}^n} \pi(p)y(p) \geq \alpha \}$.

We define $G_b = \{ y \in \mathbb{R}^{(\mathbb{R}^n)} : (0, y) \in Q_b \}$. Since Q_b can be written as the set (d) in Lemma 2.8, by Corollary 2.5 we have that

$$G_b = \mathbb{R}_+^{(\mathbb{R}^n)} \cap \bigcap \{ H_{\pi, \alpha} : (\pi, \alpha) \text{ minimal nontrivial liftable tuple} \}. \quad (2.2)$$

Similar to the mixed-integer case, $\text{conv}(I_b) \subsetneq G_b$ (this will be shown in Remark 3.3).

Theorem 2.14. $G_b = \text{cl}(\text{conv}(I_b)) = \text{conv}(I_b) + \mathbb{R}_+^{(\mathbb{R}^n)}$.

Proof. By Theorem 2.13, $Q_b = \text{conv}(M_b) + (\mathbb{R}_+^{(\mathbb{R}^n)} \times \mathbb{R}_+^{(\mathbb{R}^n)})$. Since the inequality $s \geq 0$ is valid for M_b, by taking the intersection with the subspace $\{(s, y) : s = 0\}$ we obtain the equality $G_b = \text{conv}(I_b) + \mathbb{R}_+^{(\mathbb{R}^n)}$. Furthermore, since G_b coincides with the intersection of the closed set Q_b with the closed subspace defined by $s = 0$ (this subspace is closed by Lemma 2.12), G_b is a closed set. Therefore, $\text{conv}(I_b) + \mathbb{R}_+^{(\mathbb{R}^n)}$ is a closed set, and we have $\text{cl}(\text{conv}(I_b)) \subseteq \text{conv}(I_b) + \mathbb{R}_+^{(\mathbb{R}^n)}$.

It remains to show that $\mathrm{conv}(I_b) + \mathbb{R}_+^{(\mathbb{R}^n)} \subseteq \mathrm{cl}(\mathrm{conv}(I_b))$. To prove this, it suffices to show that for every $\bar{y} \in I_b$ and $r \in \mathbb{R}^n$, the point $\bar{y} + \hat{y}_r$, where $\hat{y}_r(r) = 1$ and $\hat{y}_r(p) = 0$ for $p \neq r$, is the limit of a sequence of points in $\mathrm{conv}(I_b)$ with respect to our topology. So fix $\bar{y} \in I_b$ and $r \in \mathbb{R}^n$. For every integer $k \geq 1$, there exist $q_k \in \mathbb{Z}^n$ and a real number $\lambda_k \geq 1$ such that $\|q_k - \lambda_k r\| < \frac{1}{k}$. Define y_k by setting $y_k(r) = y_k\left(\frac{q_k - \lambda_k r}{\lambda_k}\right) = 1$, and $y_k(p) = 0$ for $p \neq r$. Since $\sum_{p \in \mathbb{R}^n} p \cdot (\lambda_k y_k(p)) = q_k \in \mathbb{Z}^n$, every point of the form $\bar{y} + \lambda_k y_k$ is in I_b. Since $\lambda_k \geq 1$ for every $k \geq 1$, we have $\bar{y} + y_k = \frac{\lambda_k - 1}{\lambda_k}\bar{y} + \frac{1}{\lambda_k}(\bar{y} + \lambda_k y_k) \in \mathrm{conv}(I_b)$. Furthermore, $\|y_k - \hat{y}_r\|_* = \left\|\frac{q_k - \lambda_k r}{\lambda_k}\right\| < \frac{1}{k}$. Therefore, the sequence of points $\bar{y} + y_k$ converges to $\bar{y} + \hat{y}_r$ as $k \to \infty$. □

3   Affine Hulls and Nonnegative Representation of Valid Tuples

In any vector space (possibly infinite dimensional) the affine hull of any subset C can be equivalently described as the set of affine combinations of points in C or the intersection of all hyperplanes containing C. The next proposition shows that there is no hyperplane containing M_b.

Proposition 3.1. $\mathrm{aff}(M_b) = \mathbb{R}^{(\mathbb{R}^n)} \times \mathbb{R}^{(\mathbb{R}^n)}$.

The characterization of $\mathrm{aff}(I_b)$ is more involved and requires some preliminary notions. A function $\theta : \mathbb{R}^n \to \mathbb{R}$ is said to be *additive* if $\theta(u + v) = \theta(u) + \theta(v)$ for all $u, v \in \mathbb{R}^n$; see [1] for a survey on this family of functions. The following result is an immediate extension of a result of Basu, Hildebrand and Köppe (see [3, Propositions 2.2–2.3]).

Proposition 3.2. *The affine hull of I_b is described by the equations*

$$\sum_{p \in \mathbb{R}^n} \theta(p)y(p) = \theta(b) \tag{3.1}$$

for all additive functions $\theta : \mathbb{R}^n \to \mathbb{R}$ such that $\theta(p) = 0$ for every $p \in \mathbb{Q}^n$.

Remark 3.3. Proposition 3.2 shows that $\mathrm{conv}(I_b)$ is contained in some hyperplane; thus, $\mathrm{conv}(I_b) \subsetneq \mathrm{conv}(I_b) + \mathbb{R}_+^{(\mathbb{R}^n)} = G_b$, by Theorem 2.14.

Proposition 3.4. *Let P be a finite subset of $\mathbb{R}^n$. Then $\mathrm{aff}(I_b) \cap V_P$ is a rational affine subspace of $\mathbb{R}^P$, i.e., there exist a natural number $m \leq |P|$, a rational matrix $\Theta \in \mathbb{Q}^{m \times |P|}$ and a vector $d \in \mathbb{R}^m$ such that $\mathrm{aff}(I_b) \cap V_P = \{s \in \mathbb{R}^P : \Theta s = d\}$. Moreover, $\mathrm{aff}(I_b) \cap V_P = V_P$ if and only if $P \subseteq \mathbb{Q}^n$.*

3.1   Sufficiency of Nonnegative Functions to Describe conv(I_b)

As mentioned in the introduction, to the best of our knowledge the study of valid tuples for I_b in prior literature is restricted to nonnegative valid tuples, with the exception of [4]. The standard justification behind this assumption is the fact that valid tuples are nonnegative on the rational vectors. Since in practice we are interested in finite dimensional faces of $\text{conv}(I_b)$ that correspond to rational vectors, such an assumption seems reasonable. However, no mathematical evidence exists in the literature that a complete inequality description of these faces can be obtained from the nonnegative valid tuples only.[1] We prove below that any valid tuple is equivalent to a nonnegative valid tuple, modulo the affine hull. This gives the first proof of the above assertion and puts the nonnegativity assumption on a sound mathematical foundation. Later we will show that even a smaller class of nonnegative valid tuples suffices to describe the finite dimensional faces of $\text{conv}(I_b)$ that correspond to rational vectors, in particular the nontrivial minimal liftable tuples suffice (Corollary 4.4).

Theorem 3.5. *For every valid tuple (π, α) for I_b, there exists a unique additive function $\theta : \mathbb{R}^n \to \mathbb{R}$ such that $\theta(p) = 0$ for every $p \in \mathbb{Q}^n$ and the valid tuple $(\pi', \alpha') = (\pi + \theta, \alpha + \theta(b))$ satisfies $\pi' \geq 0$.*

This answers Open Question 2.5 in [3].

4   Recession Cones and Canonical Faces

A *canonical face* of $\text{conv}(M_b)$ is a face of the form $F = \text{conv}(M_b) \cap V_{R,P}$ for some $R, P \subseteq \mathbb{R}^n$. If R and P are finite, F is a *finite canonical face* of $\text{conv}(M_b)$. The same definitions can be given for $\text{conv}(I_b)$. The corner polyhedra defined by Gomory and Johnson [10–12] are precisely the finite canonical faces of $\text{conv}(I_b)$.

The notion of recession cone of a closed convex set is standard (see, e.g., [16]). We extend it to general convex sets in general vector spaces (possibly infinite dimensional) in the following way. Let V be a vector space and let $C \subseteq V$ be a convex set. For any $x \in C$, define

$$C_\infty(x) = \{r \in V : x + \lambda r \in C \text{ for all } \lambda \geq 0\}.$$

We define the *recession cone* of a nonempty convex set C as $\text{rec}(C) = \bigcap_{x \in C} C_\infty(x)$. Theorem 2.13 yields the following result.

Corollary 4.1. *Let $F = \text{conv}(M_b) \cap V_{R,P}$ be a canonical face of $\text{conv}(M_b)$. Then F is a face of $\text{cl}(\text{conv}(M_b))$ if and only if $F + (\mathbb{R}_+^R \times \mathbb{R}_+^P) = F$, i.e., $\text{rec}(F)$ is the nonnegative orthant.*

Proof. By Theorem 2.13,

$$\text{cl}(\text{conv}(M_b)) \cap V_{R,P} = \left(\text{conv}(M_b) + (\mathbb{R}_+^{(\mathbb{R}^n)} \times \mathbb{R}_+^{(\mathbb{R}^n)})\right) \cap V_{R,P}$$
$$= (\text{conv}(M_b) \cap V_{R,P}) + (\mathbb{R}_+^R \times \mathbb{R}_+^P)$$
$$= F + (\mathbb{R}_+^R \times \mathbb{R}_+^P)$$

The results follows from the observation that F is a face of $\text{cl}(\text{conv}(M_b))$ if and only if $F = \text{cl}(\text{conv}(M_b)) \cap V_{R,P}$.   □

---

[1] Such results are obtainable in the case $n = 1$ by more elementary means such as interpolation. We are unaware of a way to establish these results for general $n \geq 2$ without using the technology developed in this paper.

Define L to be the linear space parallel to the affine hull of $\mathrm{conv}(I_b)$; Proposition 3.2 shows that L is the set of all $y \in \mathbb{R}^{(\mathbb{R}^n)}$ that satisfy $\sum_{p \in \mathbb{R}^n} \theta(p)y(p) = 0$ for all additive functions $\theta : \mathbb{R}^n \to \mathbb{R}$ such that $\theta(p) = 0$ for all $p \in \mathbb{Q}^n$. For any $P \subseteq \mathbb{R}^n$, define the face $C^P = \mathrm{conv}(I_b) \cap V_P$ of $\mathrm{conv}(I_b)$.

Theorem 4.2. *For every finite subset $P \subseteq \mathbb{R}^n$ such that $C^P \neq \emptyset$, the following are all true:*

(a) the face $C^P = \mathrm{conv}(I_b) \cap V_P$ is a rational polyhedron in $\mathbb{R}^P$;
(b) every extreme ray of C^P is spanned by some $r \in \mathbb{Z}_+^P$ such that $\sum_{p \in P} pr(p) \in \mathbb{Z}^n$;
(c) $\mathrm{rec}(C^P) = L \cap \mathbb{R}_+^{(\mathbb{R}^n)} \cap V_P = (L \cap V_P) \cap \mathbb{R}_+^P$.

Proof. By dropping variables set to zero, $I_b \cap V_P$ is the set of vectors $y \in \mathbb{Z}_+^P$ such that $\sum_{p \in P} py(p) \in b + \mathbb{Z}^n$. We say that a feasible point $y \in I_b \cap V_P$ is minimal if there is no feasible point $y' \neq y$ such that $y' \leq y$. Every vector $d \in \mathbb{Z}_+^P$ such that $\sum_{p \in P} pd(p) \in \mathbb{Z}^n$ is called a *ray*. A ray d is minimal if there is no ray $d' \neq d$ such that $d' \leq d$.

We claim that every feasible point y is the sum of a minimal feasible point and a nonnegative integer combination of minimal rays. To see this, as long as there is a ray d such that $d \leq y$, replace y with $y - d$. Note that this operation can be repeated only a finite number of times. Denote by $\bar{y}$ the feasible point obtained at the end of this procedure. Then y is the sum of $\bar{y}$ and a nonnegative integer combination of rays. We observe that $\bar{y}$ is minimal: if not, there would exist a feasible point $y' \neq \bar{y}$ such that $y' \leq \bar{y}$; but then the vector $d := \bar{y} - y'$ would be a ray satisfying $d \leq \bar{y}$, contradicting the fact that the procedure has terminated. Therefore y is the sum of a minimal feasible point $\bar{y}$ and a nonnegative integer combination of rays. Since every ray is a nonnegative integer combination of minimal rays (argue as above), we conclude that y is the sum of a minimal feasible point and a nonnegative integer combination of minimal rays.

By the Gordan–Dickson lemma (see, e.g., [9]), the set of minimal feasible points and the set of minimal rays are both finite. Let Y be the set of points that are the sum of a minimal feasible point and a nonnegative integer combination of minimal rays. Thus, there exist finite sets $E \subseteq \mathbb{Z}_+^P$ and $R \subseteq \mathbb{Z}_+^P$ such that $Y = E + \mathrm{integ.cone}(R)$, where $\mathrm{integ.cone}(R)$ denotes the set of all nonnegative integer combinations of vectors in R. So $\mathrm{conv}(Y) = \mathrm{conv}(E + \mathrm{integ.cone}(R)) = \mathrm{conv}(E) + \mathrm{conv}(\mathrm{integ.cone}(R)) = \mathrm{conv}(E) + \mathrm{cone}(R)$. Hence, $\mathrm{conv}(Y)$ is a rational polyhedron, by the Minkowski-Weyl Theorem [6, Theorem 3.13]. The above observation proves that $I_b \cap V_P \subseteq Y$. On the other hand, by using the fact that if y is a feasible point and d is a ray then $y + d$ is a feasible point, one readily verifies that $Y \subseteq I_b \cap V_P$. Then $I_b \cap V_P = Y$ and therefore $\mathrm{conv}(I_b) \cap V_P = \mathrm{conv}(I_b \cap V_P) = \mathrm{conv}(Y)$. Hence, $\mathrm{conv}(I_b) \cap V_P$ is a rational polyhedron.

The above analysis proves (a) and (b) simultaneously. We now prove (c).

We first show that $\mathrm{rec}(C^P) \subseteq L \cap \mathbb{R}_+^{(\mathbb{R}^n)} \cap V_P$. Consider any $\bar{d} \in \mathrm{rec}(C^P)$. By part (b), $\bar{d}$ is a nonnegative combination of vectors $d \in \mathbb{Z}_+^P$ such that

$\sum_{p \in P} pr(p) \in \mathbb{Z}^n$. Observe that each such $d \in L$. Thus, $\bar{d} \in L$ since L is a linear space. Therefore, $\mathrm{rec}(C^P) \subseteq L \cap \mathbb{R}_+^{(\mathbb{R}^n)} \cap V_P$.

We now want to establish that $L \cap \mathbb{R}_+^{(\mathbb{R}^n)} \cap V_P \subseteq \mathrm{rec}(C^P)$. First, consider any $d \in L \cap \mathbb{R}_+^{(\mathbb{R}^n)} \cap V_P$ such that $d \in \mathbb{Q}^P$, i.e., d has only rational coordinates. Let $\lambda > 0$ be such that $\bar{d} = \lambda d \in \mathbb{Z}_+^P$. We claim that $\sum_{p \in P} p\bar{d}(p) \in \mathbb{Q}^n$. Otherwise, there exists[2] an additive function $\theta : \mathbb{R}^n \to \mathbb{R}$ such that $0 \neq \theta(\sum_{p \in P} p\bar{d}(p)) = \sum_{p \in P} \theta(p)\bar{d}(p) = \lambda \sum_{p \in P} \theta(p)d(p)$, which violates the fact that $d \in L$. Since $\sum_{p \in P} p\bar{d}(p) \in \mathbb{Q}^n$, there exists a positive scaling $\tilde{d}$ of d such that $\sum_{p \in P} p\tilde{d}(p) \in \mathbb{Z}^n$. It is easy to verify that $\tilde{d} \in \mathrm{rec}(C^P)$ and therefore $d \in \mathrm{rec}(C^P)$. This shows that all rational vectors in $L \cap \mathbb{R}_+^{(\mathbb{R}^n)} \cap V_P$ are in $\mathrm{rec}(C^P)$. Since, by Proposition 3.4, $L \cap V_P$ is a rational subspace, $L \cap \mathbb{R}_+^{(\mathbb{R}^n)} \cap V_P \subseteq \mathrm{rec}(C^P)$.   □

Theorem 4.3. *Let $P \subseteq \mathbb{R}^n$ be finite such that $C^P \neq \emptyset$. Then the following are equivalent:*

(a) $P \subseteq \mathbb{Q}^n$;
(b) $\mathrm{rec}(C^P) = \mathbb{R}_+^P$;
(c) the dimension of C^P is $|P|$;
(d) $C^P = G_b \cap V_P$.

Proof. (a) is equivalent to (b) by Proposition 3.4 and Theorem 4.2. (b) is equivalent to (c) by Proposition 3.4. The equivalence of (a) and (d) follows from the equivalence of (a) and (b), Corollary 4.1 and Theorem 2.14.

□

By applying (2.2) to condition (d) in Theorem 4.3, we get

Corollary 4.4. *A finite dimensional corner polyhedron C^P can be expressed as $C^P = \{y \in \mathbb{R}_+^P : \sum_{p \in P} \pi(p)y(p) \geq 1, \ (\pi, 1) \text{ is minimal and liftable}\}$ if and only if $P \subseteq \mathbb{Q}^n$.*

Example 4.5. There are finite dimensional faces of $\mathrm{conv}(M_b)$ that are not closed. Let $n = 1$, $b \in \mathbb{Q}$, $\omega \in \mathbb{R} \setminus \mathbb{Q}$, $R = \{-1\}$, $P = \{b, \omega\}$. Consider the point $(\bar{s}, \bar{y})$ defined by $\bar{s}(-1) = 0$ and $\bar{y}(b) = \bar{y}(\omega) = 1$. Note that $(\bar{s}, \bar{y}) \notin \mathrm{conv}(M_b) \cap V_{R,P}$, as the only point in M_b satisfying $s(-1) = 0$ and $y(b) \leq 1$ has $y(b) = 1$, $y(\omega) = 0$.

We now show that $(\bar{s}, \bar{y}) \in \mathrm{cl}(\mathrm{conv}(M_b) \cap V_{R,P})$ by constructing for every $\varepsilon > 0$ a point in $\mathrm{conv}(M_b) \cap V_{R,P}$ whose Euclidean distance from $(\bar{s}, \bar{y})$ is at most ε. So fix $\varepsilon > 0$. Let $\hat{y}(\omega)$ be a positive integer such that the fractional part of $\omega\hat{y}(\omega)$ is at most ε. Let $\hat{s}(-1)$ be equal to this fractional part, and $\hat{y}(b) = 1$. Then $(\hat{s}, \hat{y}) \in M_b \cap V_{R,P}$. By taking a suitable convex combination of $(\hat{s}, \hat{y})$ and the point of $M_b \cap V_{R,P}$ defined by $y(b) = 1$, $s(-1) = y(\omega) = 0$, we find a point in $\mathrm{conv}(M_b) \cap V_{R,P}$ whose distance from $(\bar{s}, \bar{y})$ is at most ε.

Remark 4.6. Since $Q_b = \mathrm{cl}(\mathrm{conv}(M_b))$ by Theorem 2.13, for every $R, P \subseteq \mathbb{R}^n$ the set $Q_b \cap V_{R,P}$ is closed by Lemma 2.12. The previous example gives sets R, P such that $\mathrm{conv}(M_b) \cap V_{R,P}$ is not closed. Thus $\mathrm{conv}(M_b)$ is a strict subset of Q_b.

---

[2] For an explicit construction of such a function, see the journal version of the paper.

References

1. Aczél, J., Dhombres, J.G.: Functional Equations in Several Variables. Encyclopedia of Mathematics and Its Applications, vol. 31. Cambridge University Press (1989)
2. Basu, A., Conforti, M., Cornuéjols, G., Zambelli, G.: Maximal lattice-free convex sets in linear subspaces. Math. Oper. Res. **35**, 704–720 (2010)
3. Basu, A., Hildebrand, R., Köppe, M.: Light on the infinite group relaxation I: foundations and taxonomy. 4OR **14**(1), 1–40 (2016)
4. Basu, A., Hildebrand, R., Köppe, M.: Light on the infinite group relaxation II: sufficient conditions for extremality, sequences, and algorithms. 4OR **14**(2), 1–25 (2016)
5. Basu, A., Paat, J.: Operations that preserve the covering property of the lifting region. SIAM J. Optim. **25**(4), 2313–2333 (2015)
6. Conforti, M., Cornuéjols, G., Zambelli, G.: Integer Programming, vol. 271. Springer, Switzerland (2014)
7. Dash, S., Günlük, O.: Valid inequalities based on simple mixed-integer sets. Math. Program. **105**, 29–53 (2006)
8. Dey, S.S., Richard, J.P.P., Li, Y., Miller, L.A.: On the extreme inequalities of infinite group problems. Math. Program. **121**(1), 145–170 (2009)
9. Dickson, L.E.: Finiteness of the odd perfect and primitive abundant numbers with n distinct prime factors. Am. J. Math. **35**(4), 413–422 (1913)
10. Gomory, R.E.: Some polyhedra related to combinatorial problems. Linear Algebra Appl. **2**(4), 451–558 (1969)
11. Gomory, R.E., Johnson, E.L.: Some continuous functions related to corner polyhedra, I. Math. Program. **3**, 23–85 (1972)
12. Gomory, R.E., Johnson, E.L.: Some continuous functions related to corner polyhedra, II. Math. Program. **3**, 359–389 (1972)
13. Hildebrand, R.: Algorithms and cutting planes for mixed integer programs. Ph.D. thesis, University of California, Davis, June 2013
14. Johnson, E.L.: On the group problem for mixed integer programming. Math. Program. Study **2**, 137–179 (1974)
15. Köppe, M., Zhou, Y.: An electronic compendium of extreme functions for the gomory-johnson infinite group problem. Oper. Res. Lett. **43**(4), 438–444 (2015)
16. Lemaréchal, C.: Convex Analysis and Minimization Algorithms I. Grundlehren der mathematischen Wissenschaften, vol. 305 (1996)
17. Letchford, A.N., Lodi, A.: Strengthening Chvátal-Gomory cuts and gomory fractional cuts. Oper. Res. Lett. **30**(2), 74–82 (2002)
18. Miller, L.A., Li, Y., Richard, J.P.P.: New inequalities for finite and infinite group problems from approximate lifting. Naval Res. Logistics (NRL) **55**(2), 172–191 (2008)
19. Yıldız, S., Cornuéjols, G.: Cut-generating functions for integer variables. Math. Oper. Res. **41**, 1381–1403 (2016)

Mixed-Integer Linear Representability, Disjunctions, and Variable Elimination

Amitabh Basu[1]([⊠]), Kipp Martin[2], Christopher Thomas Ryan[2], and Guanyi Wang[3]

[1] Department of Applied Mathematics and Statistics,
Johns Hopkins University, Baltimore, USA
basu.amitabh@jhu.edu
[2] Booth School of Business, University of Chicago, Chicago, USA
[3] Industrial and Systems Engineering, Georgia Institute of Technology, Atlanta, USA

Abstract. Jeroslow and Lowe gave an exact geometric characterization of subsets of $\mathbb{R}^n$ that are projections of mixed-integer linear sets, a.k.a MILP-representable sets. We give an alternate algebraic characterization by showing that a set is MILP-representable *if and only if* the set can be described as the intersection of finitely many *affine Chvátal inequalities*. These inequalities are a modification of a concept introduced by Blair and Jeroslow. This gives a sequential variable elimination scheme that, when applied to the MILP representation of a set, explicitly gives the affine Chvátal inequalities characterizing the set. This is related to the elimination scheme of Wiliams and Williams-Hooker, who describe projections of integer sets using *disjunctions* of affine Chvátal systems. Our scheme extends their work in two ways. First, we show that disjunctions are unnecessary, by showing how to find the affine Chvátal inequalities that cannot be discovered by the Williams-Hooker scheme. Second, disjunctions of Chvátal systems can give sets that are *not* projections of mixed-integer linear sets; so the Williams-Hooker approach does not give an exact characterization of MILP representability.

1 Introduction

Researchers are interested in characterizing sets that are projections of mixed-integer sets described by linear constraints. Such sets have been termed *MILP-representable sets*; see [8] for a thorough survey. Knowing which sets are MILP-representable is important because of the prevalence of good algorithms and software for solving MILP formulations. Therefore, if one encounters an application that can be modeled using MILP-representable sets, then this sophisticated technology can be used to solve the application.

A seminal result of Jeroslow and Lowe [4] provides a geometric characterization of MILP-representable sets as the sum of a finitely generated monoid, and a disjunction of finitely many polytopes (see Theorem 1 below for a precise statement). An algebraic approach based on an explicit elimination scheme for

---

A. Basu—Supported by the NSF grant CMMI1452820.

F. Eisenbrand and J. Koenemann (Eds.): IPCO 2017, LNCS 10328, pp. 75–85, 2017.
DOI: 10.1007/978-3-319-59250-3_7

integer variables was first developed by Williams in [9–11]. Williams (and later Williams and Hooker) adapted the Fourier-Motzkin elimination approach for linear inequalities to handle integer variables. Balas, in [1], also explores how to adapt Fourier-Motzkin elimination in the case of binary variables. In both instances, there is a need to introduce *disjunctions* of inequalities that involve either rounding operations or congruence relations. We emphasize that the geometric approach of Jeroslow-Lowe and the algebraic approach of Williams-Hooker-Balas both require the use of disjunctions.

Our point of departure is that we provide a constructive algebraic characterization of MILP-representability that does not need disjunctions, but instead makes use of *affine Chvátal inequalities*, i.e. affine linear inequalities with rounding operations (for a precise definition see Definition 1 below). We show that MILP-representable sets are exactly those sets that satisfy a finite system of affine Chvátal inequalities. In contrast, Williams and Hooker [9–11] require *disjunctions* of systems of affine Chvátal inequalities. Another disadvantage in their work is the following: there exist sets given by disjunctions of affine Chvátal systems that are *not* MILP-representable. Finally, our proof of the non-disjunctive characterization is constructive and implies a sequential variable elimination scheme for mixed-integer linear sets (see Sect. 5).

We thus simultaneously show three things: (1) disjunctions are not necessary for MILP-representability (if one allows affine Chvátal inequalities), an operation that shows up in both the Jeroslow-Lowe and the Williams-Hooker approaches, (2) our algebraic characterization comes with a variable elimination scheme, which is an advantage, in our opinion, to the geometric approach of Jeroslow-Lowe, and (3) our algebraic characterization is exact, as opposed to the algebraic approach of Williams-Hooker, whose algebraic descriptions give a strictly larger family of sets than MILP-representable sets.

Our algebraic characterization could be useful to obtain other insights into the structure of MIP representable sets that is not apparent from the geometric perspective. As an illustration, we resolve an open question posed in Ryan [6] on the representability of integer monoids using our characterization. Theorem 1 in [6] shows that every finitely-generated integer monoid can be described as a finite system of Chvátal inequalities but leaves open the question of how to construct the associated Chvátal functions via elimination. Ryan states that the elimination methods of Williams in [9,10] do not address her question because of the introduction of disjunctions. Our work provides a constructive approach for finding a Chvátal inequality representation of finitely-generated integer monoids using elimination.

Our new algebraic characterization could also lead to novel algorithmic ideas where researchers optimize by directly working with affine Chvatal functions, rather than using traditional branch-and-cut/cutting plane type of methods for mixed-integer optimization.

2   Preliminaries

$\mathbb{Z}, \mathbb{Q}, \mathbb{R}$ denote the set of integers, rational numbers and reals, respectively. Any of these sets subscripted by a plus means the nonnegative elements of that set. For instance, $\mathbb{Q}_+$ is the set of nonnegative rational numbers. The projection operator proj_Z where $Z \subseteq \{x_1, \ldots, x_n\}$ projects a vector $x \in \mathbb{R}^n$ onto the coordinates in Z. Following [4] we say a set $S \subseteq \mathbb{R}^n$ is *mixed integer linear representable* (or MILP-representable) if there exists rational matrices A, B, C and a rational vector d such that

$$S = \text{proj}_x \{(x, y, z) \in \mathbb{R}^n \times \mathbb{R}^p \times \mathbb{Z}^q : Ax + By + Cz \geq d\}. \tag{1}$$

The following is the main result from [4] stated as Theorem 4.47 in [3]:

Theorem 1. *A set $S \subset \mathbb{R}^n$ is MILP-representable if and only if there exists rational polytopes $P_1, \ldots, P_k \subseteq \mathbb{R}^n$ and vectors $r^1, \ldots, r^t \in \mathbb{Z}^n$ such that*

$$S = \bigcup_{i=1}^{k} P_i + \text{intcone} \{r^1, \ldots, r^t\}, \tag{2}$$

where $\text{intcone} \{r^1, \ldots, r^t\}$ *denotes the set of nonnegative integer linear combinations of $r^1, \ldots, r^t$.*

The ceiling operator $\lceil a \rceil$ gives the smallest integer no less than $a \in \mathbb{R}$. *Chvátal functions*, first introduced by [2], are obtained by taking linear combinations of linear functions and using the ceiling operator. We extend this original definition to allow for affine linear functions, as opposed to homogenous linear functions. Consequently, we term our functions *affine Chvátal functions*. We use the concept of finite binary trees from [5] to formally define these functions.

Definition 1. *An affine Chvátal function $f : \mathbb{R}^n \to \mathbb{R}$ is constructed as follows. We are given a finite binary tree where each node of the tree is either: (i) a leaf, which corresponds to an affine linear function on $\mathbb{R}^n$ with rational coefficients; (ii) has one child with corresponding edge labeled by either a $\lceil \cdot \rceil$ or a number in $\mathbb{Q}_+$, or (iii) has two children, each with edges labelled by a number in $\mathbb{Q}_+$.*

The function f is built as follows. Start at the root node and (recursively) form functions corresponding to subtrees rooted at its children. If the root has a single child whose subtree is g, then either (a) $f = \lceil g \rceil$ if the corresponding edge is labeled $\lceil \cdot \rceil$ or (b) $f = \alpha g$ if the corresponding edge is labeled by $a \in \mathbb{Q}_+$. If the root has two children with corresponding edges labeled by $a \in \mathbb{Q}_+$ and $b \in \mathbb{Q}_+$ then $f = ag + bh$ where g and h are functions corresponding to the respective children of the root.[1]

*The **depth** of a binary tree representation T of an affine Chvátal function is the length of the longest path from the root to a node in T, and $cc(T)$ denotes the **ceiling count** of T, i.e., the total number of edges of T labelled $\lceil \cdot \rceil$.*

---

[1] The original definition of Chvátal function in [2] does not employ binary trees. Ryan shows the two definitions are equivalent in [5].

The original definition of *Chvátal function* in the literature requires the leaves of the binary tree to be linear functions, and the domain of the function to be $\mathbb{Q}^n$ (see [2,5,6]). Our definition above allows for *affine* linear functions at the leaves, and the domain of the functions to be $\mathbb{R}^n$. We use the term *Chvátal function*, as opposed to *affine* Chvátal function, to refer to the setting where the leaves are linear functions. In this paper, the domain of all functions is $\mathbb{R}^n$.

An inequality $f(x) \leq b$, where f is an affine Chvátal function and $b \in \mathbb{R}$, is called an *affine Chvátal inequality*. A *mixed-integer Chvátal (MIC) set* is a mixed-integer set described by finitely many affine Chvátal inequalities. That is, a set S is a mixed integer Chvátal set if there exist affine Chvátal functions f_i and $b_i \in \mathbb{R}$ for $i = 1, \ldots, m$ such that $S = \{(x, z) \in \mathbb{R}^n \times \mathbb{Z}^q : f_i(x, z) \leq b_i \text{ for } i = 1, \ldots, m\}$. A set S is a *disjunctive mixed-integer Chvátal (DMIC) set* if there exist affine Chvátal functions f_{ij} and $b_{ij} \in \mathbb{R}$ for $i = 1, \ldots, m$ and $j = 1, \ldots, t$ such that $S = \bigcup_{j=1}^{t} \{(x, z) \in \mathbb{R}^n \times \mathbb{Z}^q : f_{ij}(x, z) \leq b_{ij} \text{ for } i = 1, \ldots, m\}$.

3   MILP-representable Sets as DMIC Sets

From the perspective of MILP-representability, the following result summarizes the work in [9–11] that relates affine Chvátal functions and projections of integer variables.

Theorem 2. *Every MILP-representable set is a DMIC set.*

Theorem 2 is not explicitly stated in [9–11], even though it summarizes the main results of these papers, for two reasons: (i) the development in [11] works with linear congruences and inequalities as constraints and not affine Chvátal inequalities and (ii) they only treat the pure integer case. These differences are only superficial. For (i), an observation due to Ryan in [6] shows that congruences can always be expressed equivalently as affine Chvátal inequalities. For (ii), continuous variables (the y variables in (1)) can first be eliminated using Fourier-Motzkin elimination, which introduces no complications.

The converse of Theorem 2 is not true. As the following example illustrates, not every DMIC set is MILP-representable.

Example 1. Consider the set $E := \{(\lambda, 2\lambda) : \lambda \in \mathbb{Z}_+\} \cup \{(2\lambda, \lambda) : \lambda \in \mathbb{Z}_+\}$ as illustrated in Fig. 1. This set is a DMIC set because it can be expressed as $E = \{x \in \mathbb{Z}_+^2 : 2x_1 - x_2 = 0\} \cup \{x \in \mathbb{Z}_+^2 : x_1 - 2x_2 = 0\}$.

E is not the projection of any mixed integer linear program. Indeed, by Theorem 1 every MILP-representable set has the form (2). Suppose E has such a form. Consider the integer points in E of the form $(\lambda, 2\lambda)$ for $\lambda \in \mathbb{Z}_+$. There are infinitely many such points and so cannot be captured inside of the finitely-many polytopes P_k in (2). Thus, the ray $\lambda(1, 2)$ for $\lambda \in \mathbb{Z}_+$ must lie inside intcone$\{r^1, \ldots, r^t\}$. Identical reasoning implies the ray $\lambda(2, 1)$ for $\lambda \in \mathbb{Z}_+$ must also lie inside intcone$\{r^1, \ldots, r^t\}$. But then, every conic integer combination of these two rays must lie in E. Observe that $(3, 3) = (2, 1) + (1, 2)$ is one such integer combination but $(3, 3) \notin E$. We conclude that E cannot be represented in the form (2) and hence E is not MILP-representable.

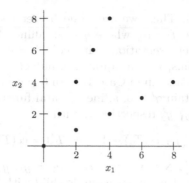

Fig. 1. A DMIC set that is not MILP-representable.

4 Characterization of MILP-representable Sets as MIC Sets

In this section we characterize MILP-representable sets as MIC sets. This is achieved in two steps across two subsections. The main results are Theorems 4 and 6, which are converses of each other.

4.1 MIC Sets are MILP-representable

We show how to "lift" a MIC set to a mixed-integer linear set. The idea is simple – replace ceiling operators with additional integer variables. However, we need to work with an appropriate representation of an affine Chvátal function in order to implement this idea. The next result provides the correct representation.

Theorem 3. *For every affine Chvátal function f represented by a binary tree T, one of the following cases hold:*

Case 1: $cc(T) = 0$, *which implies that f is an affine linear function.*
Case 2: $f = \gamma \lceil g_1 \rceil + g_2$, *where $\gamma > 0$ and g_1, g_2 are affine Chvátal functions such that there exist binary tree representations T_1, T_2 for g_1, g_2 respectively, with $cc(T_1) + cc(T_2) + 1 \leq cc(T)$.*

Proof. We use induction on the depth of the binary tree T. For the base case, if T has depth 0, then $cc(T) = 0$ and we are in Case 1. The inductive hypothesis assumes that for some $k \geq 0$, every affine Chvátal function f with a binary tree representation T of depth less or equal to k, can be expressed in Case 1 or 2.

For the inductive step, consider an affine Chvátal function f with a binary tree representation T of depth $k + 1$. If the root node of T has a single child, let T' be the subtree of T with root node equal to the child of the root node of T. We now consider two cases: the edge at the root node is labeled with a $\lceil \cdot \rceil$, or the edge is labeled with a scalar $\alpha > 0$. In the first case, $f = \lceil g \rceil$ where g is an affine Chvátal function which has T' as a binary tree representation.

Also, $cc(T') + 1 = cc(T)$. Thus, we are done by setting $g_1 = g$, $g_2 = 0$ and $\gamma = 1$. In the second case, $f = \alpha g$ where g is an affine Chvátal function which has T' as a binary tree representation, with $cc(T') = cc(T)$. Note that T' has smaller depth than T. Thus, we can apply the induction hypothesis on g with representation T'. If this ends up in Case 1, then $0 = cc(T') = cc(T)$ and f is in Case 1. Otherwise, we obtain $\gamma' > 0$, affine Chvátal functions g'_1, g'_2, and binary trees T'_1, T'_2 representing g'_1, g'_2 respectively, with

$$cc(T'_1) + cc(T'_2) + 1 \leq cc(T') = cc(T) \tag{3}$$

such that $g = \gamma' \lceil g'_1 \rceil + g'_2$. Now set $\gamma = \alpha \gamma'$, $g_1 = g'_1$, $g_2 = \alpha g'_2$, $T_1 = T'_1$ and T_2 to be the tree whose root node has a single child with T'_2 as the subtree, and the edge at the root labeled with α. Note that $cc(T_2) = cc(T'_2)$. Also, observe that T_1, T_2 represents g_1, g_2 respectively. Combined with (3), we obtain that $cc(T_1) + cc(T_2) + 1 \leq cc(T)$.

If the root node of T has two children, let S_1, S_2 be the subtrees of T with root nodes equal to the left and right child, respectively, of the root node of T. Then, $f = \alpha h_1 + \beta h_2$, where $\alpha, \beta > 0$ and h_1, h_2 are affine Chvátal functions with binary tree representations S_1, S_2 respectively. Also note that the depths of S_1, S_2 are both strictly less than the depth of T, and

$$cc(S_1) + cc(S_2) = cc(T) \tag{4}$$

By the induction hypothesis applied to h_1 and h_2 with representations S_1, S_2, we can assume both of them end up in Case 1 or 2 of the statement of the theorem. If both of them are in Case 1, then $cc(S_1) = cc(S_2) = 0$, and by (4), $cc(T) = 0$. So f is in Case 1.

Thus, we may assume that h_1 or h_2 (or both) end up in Case 2. There are three subcases, (i) h_1, h_2 are both in Case 2, (ii) h_1 is Case 2 and h_2 in Case 1, or (iii) h_2 in Case 2 and h_1 in Case 1. We analyze subcase (i), the other two subcases are analogous. This implies that there exists $\gamma' > 0$, and affine Chvátal functions g'_1 and g'_2 such that $h_1 = \gamma' \lceil g'_1 \rceil + g'_2$, and there exist binary tree representations T'_1, T'_2 for g'_1, g'_2 respectively, such that

$$cc(T'_1) + cc(T'_2) + 1 \leq cc(S_1). \tag{5}$$

Now set $\gamma = \alpha \gamma'$, $g_1(x) = g'_1(x)$ and $g_2(x) = \alpha g'_2(x) + \beta h_2(x)$. Then $f = \gamma \lceil g_1 \rceil + g_2$. Observe that g_2 has a binary tree representation T_2 such that the root node of T_2 has two children: the subtrees corresponding to these children are T'_2 and S_2, and the edges at the root node of T_2 are labeled by α and β respectively. Therefore,

$$cc(T_2) \leq cc(T'_2) + cc(S_2). \tag{6}$$

Moreover, we can take $T_1 = T'_1$ as the binary tree representation of g_1. We observe that

$$cc(T_1) + cc(T_2) + 1 \leq cc(T'_1) + cc(T'_2) + cc(S_2) + 1$$
$$\leq cc(S_1) + cc(S_2) = cc(T)$$

where the first inequality is from the fact that $T_1 = T_1'$ and (6), the second inequality is from (5) and the final equation is (4).   □

For a system of affine Chvátal inequalities where each affine Chvátal function is represented by a binary tree, the *total ceiling count of this representation* is the sum of the ceiling counts of all these binary trees. The next lemma shows how to reduce the total ceiling count of a MIC set by one, in exchange for an additional integer variable.

Lemma 1. *Given a system $C = \{x \in \mathbb{R}^n \times \mathbb{Z}^q : f_i(x) \le b_i\}$ of affine Chvátal inequalities with a total ceiling count $c \ge 1$, there exists a system $P = \{(x, z) \in \mathbb{R}^n \times \mathbb{Z}^q \times \mathbb{Z} : f_i'(x) \le b_i'\}$ of affine Chvátal inequalities with a total ceiling count of at most $c - 1$, and $C = \mathrm{proj}_x(P)$.*

Proof. Since $c \ge 1$, at least one of the f_i is given with a binary tree representation T with strictly positive ceiling count. Without loss of generality we assume it is f_1. This means f_1, along with its binary tree representation T, falls in Case 2 of Theorem 3. Therefore, one can write f as $f_1 = \gamma\lceil g_1 \rceil + g_2$, with $\gamma > 0$, and g_1, g_2 are affine Chvátal functions such that there exist binary tree representations T_1, T_2 for g_1, g_2 respectively, with $\mathrm{cc}(T_1) + \mathrm{cc}(T_2) + 1 \le \mathrm{cc}(T)$. Dividing by γ on both sides, the inequality $f_1(x) \le b_1$ is equivalent to $\lceil g_1(x) \rceil + (1/\gamma)g_2(x) \le b_1/\gamma$. Moving $(1/\gamma)g_2(x)$ to the right hand side, we get $\lceil g_1(x) \rceil \le -(1/\gamma)g_2(x) + b_1/\gamma$. This inequality is easily seen to be equivalent to two inequalities, involving an extra integer variable $z \in \mathbb{Z}$: $\lceil g_1(x) \rceil \le z \le -(1/\gamma)g_2(x) + b_1/\gamma$, which, in turn is equivalent to $g_1(x) \le z \le -(1/\gamma)g_2(x) + b_1/\gamma$, since $z \in \mathbb{Z}$. Therefore, we can replace the constraint $f_1(x) \le b_1$ with the two constraints

$$g_1(x) - z \le 0, \tag{7}$$
$$(1/\gamma)g_2(x) + z \le b_1/\gamma \Leftrightarrow g_2(x) + \gamma z \le b_1 \tag{8}$$

as long as we restrict $z \in \mathbb{Z}$. Note that the affine Chvátal functions on the left hand sides of (7) and (8) have binary tree representations with ceiling count equal to $\mathrm{cc}(T_1)$ and $\mathrm{cc}(T_2)$ respectively. Since $\mathrm{cc}(T_1) + \mathrm{cc}(T_2) + 1 \le \mathrm{cc}(T)$, the total ceiling count of the new system is at least one less than the total ceiling count of the previous system.   □

The key result of this subsection is an immediate consequence.

Theorem 4. *Every MIC set is MILP-representable.*

Proof. Consider any system of affine Chvátal inequalities describing the MIC set, with total ceiling count $c \in \mathbb{N}$. Apply Lemma 1 at most c times to get the desired result.   □

4.2   MIP-representable Sets Are MIC Sets

We now turn to showing the converse of Theorem 4, that every MILP-representable set is a MIC set (Theorem 6 below). This direction leverages some established theory in integer programming, in particular,

Theorem 5. *For any rational $m \times n$ matrix A, there exists a finite set of Chvátal functions $f_i : \mathbb{R}^m \to \mathbb{R}$, $i \in I$ with the following property: for every $b \in \mathbb{R}^m$, $\{x \in \mathbb{Z}^n : Ax \geq b\}$ is nonempty if and only if $f_i(b) \leq 0$ for all $i \in I$. Moreover, these functions can be explicitly constructed from the matrix A.*

The above result is quite similar to Corollary 23.4 in [7]. This result from [7] was originally obtained by Blair and Jeroslow in [2, Theorem 5.1]). This work in turn builds on seminal work on integer programming duality by Wolsey in [12]. Wolsey showed that the family of subadditive functions suffices to give a result like Theorem 5; Blair and Jeroslow improved this to show that the smaller family of Chvatal functions suffice. The main difference between Corollary 23.4 in [7] and our result here is that we allow the right hand side b to be non rational. This difference is indispensable in our analysis (see the proof of Theorem 6). Although our proof of Theorem 5 is conceptually similar to the approach in [7], we need to handle some additional technicalities related to irrationality. We omit this analysis from this extended abstract; it will be included in the full version. The following lemma is easy to verify.

Lemma 2. *Let $T : \mathbb{R}^{n_1} \to \mathbb{R}^{n_2}$ be an affine transformation involving rational coefficients, and let $f : \mathbb{R}^{n_2} \to \mathbb{R}$ be an affine Chvátal function. Then $f \circ T : \mathbb{R}^{n_1} \to \mathbb{R}$ can be expressed as $f \circ T(x) = g(x)$ for some affine Chvátal function $g : \mathbb{R}^{n_1} \to \mathbb{R}$.*

Theorem 6. *Every MILP-representable set is a MIC set. Moreover, given an explicit inequality description of the MILP-representable set, the MIC set can be obtained algorithmically.*

Proof. Let $m, n, p, q \in \mathbb{N}$. Let $A \in \mathbb{Q}^{m \times n}, B \in \mathbb{Q}^{m \times p}, C \in \mathbb{Q}^{m \times q}$ be any rational matrices, and let $d \in \mathbb{Q}^m$. Define $\mathcal{F} = \{(x, y, z) \in \mathbb{R}^n \times \mathbb{R}^p \times \mathbb{Z}^q : Ax + By + Cz \geq d\}$. It suffices to show that the projection of $\mathcal{F}$ onto the x space is a MIC set.

By applying Fourier-Motzkin elimination on the y variables, we obtain rational matrices A', C' with m' rows for some natural number m', and a vector $d' \in \mathbb{Q}^{m'}$ such that the projection of $\mathcal{F}$ onto the (x, z) space is given by $\overline{\mathcal{F}} := \{(x, z) \in \mathbb{R}^n \times \mathbb{Z}^q : A'x + C'z \geq d'\}$.

Let $f_i : \mathbb{R}^{m'} \to \mathbb{R}$, $i \in I$ be the set of Chvátal functions obtained by applying Theorem 5 to the matrix C'. It suffices to show that the projection of $\overline{\mathcal{F}}$ onto the x space is $\hat{\mathcal{F}} := \{x \in \mathbb{R}^n : f_i(d' - A'x) \leq 0, \ i \in I\}$ since for every $i \in I$, $f_i(d' - A'x) \leq 0$ can be written as $g_i(x) \leq 0$ for some affine Chvátal function g_i, by Lemma 2.[2] This follows from the following sequence of equivalences.

$$x \in \text{proj}_x(\mathcal{F}) \Leftrightarrow x \in \text{proj}_x(\overline{\mathcal{F}})$$
$$\Leftrightarrow \exists z \in \mathbb{Z}^q \text{ such that } (x, z) \in \overline{\mathcal{F}}$$
$$\Leftrightarrow \exists z \in \mathbb{Z}^q \text{ such that } C'z \geq d' - A'x$$
$$\Leftrightarrow f_i(d' - A'x) \leq 0 \text{ for all } i \in I \quad \text{(By Theorem 5)}$$
$$\Leftrightarrow x \in \hat{\mathcal{F}}. \quad \text{(By definition of } \hat{\mathcal{F}}) \qquad \square$$

---

[2] This is precisely where we need to allow the arguments of the f_i's to be non rational because the vector $d' - A'x$ that arise from all possible x is sometimes non rational.

Remark 1. We note in the proof of Theorem 6 that if the right hand side d of the mixed-integer set is 0, then the affine Chvátal functions g_i are actually Chvátal functions. This follows from the fact that the function g in Lemma 2 is a Chvátal function if f is a Chvátal function and T is a linear transformation.

5 An Sequential Variable Elimination Scheme for Mixed-Integer Chvátal Sytems

Ryan shows (see Theorem 1 in [6]) that Y is a finitely generated integral monoid if and only if there exist Chvátal functions $f_1, \ldots, f_p$ such that $Y = \{b : f_i(b) \leq 0, i = 1, \ldots, p\}$. By definition, a finitely generated integral monoid Y is MILP representable since $Y = \{b : b = Ax, x \in \mathbb{Z}_+^n\}$ where A is an $m \times n$ integral matrix. Thus, an alternate proof of Ryan's characterization follows from Theorems 4 and 6 and Remark 1.

Ryan [6] further states that "It is an interesting open problem to find an elimination scheme to construct the Chvátal constraints for an arbitrary finitely generated integral monoid." The results of Sect. 4 provide such an elimination scheme, as we show below.

A number of authors have studied sequential projection algorithms for linear integer programs [1,9–11]. However, their sequential projection algorithms *do not* resolve Ryan's open question because they do not generate the Chvátal functions $f_i(b)$ required to describe Y. Below, we show that, in fact, all these schemes have to *necessarily* resort to the use of disjunctions because they try to adapt the classical Fourier-Motzkin procedure and apply it to the system $b = Ax, x \in \mathbb{Z}_+^n$.

Our resolution to Ryan's open question hinges on the observation that the Chvátal functions that define Y can be generated if certain redundant linear inequalities are added to those generated by the Fourier-Motzkin procedure, and then the ceiling operator is applied to these redundant inequalities. We illustrate the idea with Example 2 below and then outline the general procedure. Rather than work with $Ax = b$, $x \in \mathbb{Z}_+^n$ we work with the system $Ax \geq b$ and $x \in \mathbb{Z}^n$.

Example 2. Let $\mathcal{B}$ denote the set of all $b = (b_1, \ldots, b_5) \in \mathbb{R}^5$ such that there exist $x_1, x_2, x_3 \in \mathbb{Z}$ satisfying the following inequalities.

$$
\begin{aligned}
-x_1 + \tfrac{1}{2}x_2 - \tfrac{1}{10}x_3 &\geq b_1 \\
x_1 - \tfrac{1}{4}x_2 \quad\quad &\geq b_2 \\
-x_2 + x_3 &\geq b_3 \\
x_3 &\geq b_4 \\
-x_3 &\geq b_5
\end{aligned}
\tag{9}
$$

Performing Fourier-Motzkin elimination on the linear relaxation of (9) gives

$$
0 \geq 2b_1 + 2b_2 + \tfrac{1}{2}b_3 + \tfrac{3}{10}b_5 \tag{10}
$$

$$
0 \geq \tfrac{1}{10}b_4 + \tfrac{1}{10}b_5. \tag{11}
$$

Unfortunately, there is no possible application of the ceiling operator to any combination of terms in these two inequalities that results in affine Chvátal functions that characterize $\mathcal{B}$. In particular, $b^1 = (0,0,0,1,-1) \notin \mathcal{B}$ while $b^2 = (-1,0,0,1,-1) \in \mathcal{B}$. Consider b^1. This forces $x_3 = 1$ and the only feasible values for x_1 are $1/10 \le x_1 \le 4/10$. Therefore, for this set of b values applying the ceiling operator to some combination of terms in (10)–(11) must result in either (10) positive or (11) positive. Since $b_1 = b_2 = b_3 = 0$ and $b_5 = -1$ there is no ceiling operator that can be applied to any term in (10) to make the right hand side positive. Hence a ceiling operator must be applied to (11) in order to make the right hand side positive for $b_4 = 1$ and $b_5 = -1$. However, consider b^2. For this right-hand-side, $x_1 = x_2 = x_3 = 1$ is feasible. Since we still have $b_4 = 1$ and $b_5 = -1$, the ceiling operator applied to (11) will incorrectly conclude that there is no integer solution with b^2.

However, Fourier-Motzkin elimination *will* work in conjunction with ceiling operations *appropriate redundant inequalities are added*. Consider the inequality $x_1 \ge b_1 + 2b_2 + \frac{1}{10}b_4$ which is redundant to (9). Integrality of x_1 implies $x_1 \ge \lceil b_1 + 2b_2 + \frac{1}{10}b_4 \rceil$. Applying Fourier-Motzkin elimination to (9) along with $x_1 \ge \lceil b_1 + 2b_2 + \frac{1}{10}b_4 \rceil$ generates the additional inequality $0 \ge b_1 + \frac{1}{2}b_3 + \lceil b_1 + 2b_2 + \frac{1}{10}b_4 \rceil + \frac{4}{10}b_5$, which separates b^1 and b^2.

Our proofs in Sect. 4 give a general method to systematically add the necessary redundant constraints, such as $x_1 \ge b_1 + 2b_2 + \frac{1}{10}b_4$ in Example 2. This results in the following variable elimination scheme: at iterative step k maintain a MIC set with variables $(x_k, \ldots, x_n)$ that is indeed the true projection of the original set onto these variables. By Theorem 4, this MIC set is MILP representable with a set of variables $(x_k, \ldots, x_n, z)$. Then by Theorem 6 we can project out variable x_k and additional auxiliary z variables that were used to generate the MILP representation and obtain a new MIC in only variables $(x_{k+1}, \ldots, x_n)$. *The key point is that adding these auxiliary variables and then using Theorem 6 introduces the necessary redundant inequalities*. Repeat until all variables are eliminated and a MIC set remains in the b variables. This positively answers the question of Ryan [6] and provides a projection algorithm in a similar vein to Williams [9–11] and Balas [1] but *without* use of disjunctions.

References

1. Balas, E.: Projecting systems of linear inequalities with binary variables. Ann. Oper. Res. **188**(1), 19–31 (2011)
2. Blair, C., Jeroslow, R.: The value function of an integer program. Math. Program. **23**, 237–273 (1982)
3. Conforti, M., Cornuéjols, G., Zambelli, G.: Integer Programming. Springer, Heidelberg (2014)
4. Jeroslow, R., Lowe, J.: Modelling with integer variables. In: Korte, B., Ritter, K. (eds.) Mathematical Programming at Oberwolfach II, vol. 22, pp. 167–184. Springer, Heidelberg (1984)
5. Ryan, J.: Integral monoid duality models. Technical report, Cornell University Operations Research and Industrial Engineering (1986)

 6. Ryan, J.: Decomposing finitely generated integral monoids by elimination. Linear Algebra Appl. **153**, 209–217 (1991)
 7. Schrijver, A.: Theory of Linear and Integer Programming. Wiley, New York (1986)
 8. Vielma, J.P.: Mixed integer linear programming formulation techniques. SIAM Rev. **57**(1), 3–57 (2015)
 9. Williams, H.P.: Fourier-Motzkin elimination extension to integer programming problems. J. Comb. Theor. A **21**, 118–123 (1976)
10. Williams, H.P.: The elimination of integer variables. J. Oper. Res. Soc. **43**, 387–393 (1992)
11. Williams, H., Hooker, J.: Integer programming as projection. Discrete Optim. **22**, 291–311 (2016)
12. Wolsey, L.: The b-hull of an integer program. Discrete Appl. Math. **3**(3), 193–201 (1981)

Deterministic Fully Dynamic Approximate Vertex Cover and Fractional Matching in $O(1)$ Amortized Update Time

Sayan Bhattacharya[1]($\boxtimes$), Deeparnab Chakrabarty[2], and Monika Henzinger[3]

[1] University of Warwick, Coventry, UK
s.bhattacharya@warwick.ac.uk
[2] Department of Computer Science, Dartmouth College,
6211 Sudikoff Lab, Hanover, NH 03755, USA
deeparnab.chakrabarty@dartmouth.edu
[3] University of Vienna, Vienna, Austria
monika.henzinger@univie.ac.at

Abstract. We consider the problems of maintaining approximate maximum matching and minimum vertex cover in a dynamic graph. Starting with the seminal work of Onak and Rubinfeld [STOC 2010], this problem has received significant attention in recent years. Very recently, extending the framework of Baswana, Gupta and Sen [FOCS 2011], Solomon [FOCS 2016] gave a randomized 2-approximation dynamic algorithm for this problem that has amortized update time of $O(1)$ with high probability. We consider the natural open question of derandomizing this result. We present a new *deterministic* fully dynamic algorithm that maintains a $O(1)$-approximate minimum vertex cover and maximum fractional matching, with an amortized update time of $O(1)$. Previously, the best deterministic algorithm for this problem was due to Bhattacharya, Henzinger and Italiano [SODA 2015]; it had an approximation ratio of $(2 + \epsilon)$ and an amortized update time of $O(\log n/\epsilon^2)$. Our result can be generalized to give a fully dynamic $O(f^3)$-approximation algorithm with $O(f^2)$ amortized update time for the hypergraph vertex cover and fractional matching problems, where every hyperedge has at most f vertices.

1 Introduction

Computing a maximum cardinality matching is a fundamental problem in computer science with applications, for example, in operations research, computer science, and computational chemistry. In many of these applications the underlying graph can change. Thus, it is natural to ask how quickly a maximum matching can be maintained after a change in the graph. As nodes usually change less

---

D. Chakrabarty—Work done while the author was at Microsoft Research, India.

M. Henzinger—The research leading to these results has received funding from the European Research Council under the European Union's Seventh Framework Programme (FP7/2007-2013)/ERC Grant Agreement number 340506.

F. Eisenbrand and J. Koenemann (Eds.): IPCO 2017, LNCS 10328, pp. 86–98, 2017.
DOI: 10.1007/978-3-319-59250-3_8

frequently than edges, dynamic matching algorithms usually study the problem where edges are inserted and deleted, which is called the *(fully) dynamic matching problem*. The goal of a dynamic matching algorithm is to maintain either an actual matching (called the *matching version*) or the *value* of the matching (called the *value version*) as efficiently as possible.

Unfortunately, the problem of maintaining even just the value of the maximum cardinality matching is hard: There is a conditional lower bound that shows that no (deterministic or randomized) algorithm can achieve at the same time an amortized update time of $O(m^{1/2-\epsilon})$ and a query (for the size of the matching) time of $O(m^{1-\epsilon})$ for any small $\epsilon > 0$ [9] (see [1] for conditional lower bounds using different assumptions). The best upper bound is Sankowski's randomized algorithm [14] that solves the value problem in time $O(n^{1.495})$ per update and $O(1)$ per query. Thus, it is natural to study the dynamic *approximate* maximum matching problem, and there has been a large body [2,5,6,8,11,12,15] of work on it and its dual, the approximate vertex cover problem, in the last few years.

Dynamic algorithms can be further classified into two types: Algorithms that require an *oblivious* (aka *non-adaptive*) adversary, i.e., an adversary that does *not* base future updates and queries on the answers to past queries, and algorithms that work even for an adaptive adversary. The earlier kind of algorithms are less general than the later. Unfortunately, all randomized dynamic approximate matching and vertex cover algorithms so far do not work for an adaptive adversary [2,12,15]. Solomon [15] gives the best such randomized algorithm: It achieves $O(1)$ amortized update time (with high probability) and maintains a 2-approximate maximum matching and a 2-approximate minimum vertex cover.

We present the first *deterministic* algorithm that maintains an $O(1)$ approximation to the size of the maximum matching in $O(1)$ amortized update time. We also maintain an $O(1)$-approximate vertex cover in the same update time. Note that this is the first *deterministic* dynamic algorithm with constant update time for any non-trivial dynamic graph problem. This is significant as for other dynamic problems such as the dynamic connectivity problem or the dynamic planarity testing problem there are non-constant lower bounds in the cell probe model on the time per operation [10,13]. Thus, we show that no such lower bound can exist for the dynamic approximate matching problem.

There has been prior work on deterministic algorithms for dynamic approximate matching, but they all have $\Omega(\text{poly}(\log n))$ update time. One line of work concentrated on reducing the approximation ratio as much as possible, or at least below 2: Neiman and Solomon [11] achieved an update time $O(\sqrt{m})$ for maintaining a 3/2-approximate maximum matching and 2-approximate minimum vertex cover. This result was improved by Gupta and Peng [8] who gave an algorithm with update time $O(\sqrt{m}/\epsilon^2)$ for maintaining a $(1+\epsilon)$-approximate maximum matching. Recently, Bernstein and Stein [3] gave an algorithm with $O(m^{1/4}/\epsilon^2)$ amortized update time for maintaining a $(3/2 + \epsilon)$-approximate maximum matching. Another line of work, and this paper fits in this line, concentrated on getting a constant approximation while reducing the update time to polylogarithmic: Bhattacharya, Henzinger and Italiano [5] achieved an

$O(\log n/\epsilon^2)$ update time for maintaining a $(2 + \epsilon)$-approximate maximum *fractional* matching and a $(2+\epsilon)$-approximate minimum vertex cover. Note that any fractional matching algorithm solves the value version of the dynamic matching problem while degrading the approximation ratio by a factor of $3/2$. Thus, the algorithm in [5] maintains a $(3 + \epsilon)$-approximation of the value of the maximum matching. The fractional matching in this algorithm was later "deterministically rounded" by Bhattacharya, Henzinger and Nanongkai [6] to achieve a $O(\text{poly}(\log n, 1/\epsilon))$ update time for maintaining a $(2+\epsilon)$-approximate maximum matching.

Our method also generalizes to the hypergraph vertex (set) cover and hypergraph fractional matching problem which was considered by [4]. In this problem the hyperedges of a hypergraph are inserted and deleted over time. f indicates the maximum cardinality of any hyperedge. The objective is to maintain a hypergraph vertex cover, that is, a set of vertices that hit every hyperedge. Similarly a fractional matching in the hypergraph is a fractional assignment (weights) to the hyperedges so that the total weight faced by any vertex is at most 1. We give an $O(f^3)$-approximate algorithm with amortized $O(f^2)$ update time.

Our Techniques. Our algorithm builds and simplifies the framework of hierarchical partitioning of vertices proposed by Onak and Rubinfeld [12], which was later enhanced by Bhattacharya, Henzinger and Italiano [5] to give a deterministic fully-dynamic $(2 + \epsilon)$-approximate vertex cover and maximum matching in $O(\log n/\epsilon^2)$-amortized update time. The hierarchical partition divides the vertices into $O(\log n)$-many levels and maintains a *fractional matching* and vertex cover. To prove that the approximation factor is good, Bhattacharya et al. [5] also maintain approximate complementary slackness conditions. An edge insertion or deletion can disrupt these conditions (and indeed at times the feasibility of the fractional matching), and a *fixing procedure* maintains various invariants. To argue that the update time is bounded, [5] give a rather involved potential function argument which proves that the update time bounded by $O(L)$, the number of levels, and is thus $O(\log n)$. It seems unclear whether the update time can be argued to be a constant or not.

Our algorithm is similar to that in Bhattacharya et al. [5], except that we are a bit stricter when we fix nodes. As in [5], whenever an edge insertion or deletion or a previous update violates an invariant condition, we move nodes across the partitioning (incurring update costs), but after a node is fixed we often ensure it satisfies a stronger condition than what the invariant requires. For example, suppose a node v violates the upper bound of a fractional matching, that is, the total fractional weight it faces becomes larger than 1, then the fixing subroutine will at the end ensure that the final weight the node faces is significantly less than 1. Intuitively, this slack allows us to make a charging argument of the following form – if this node violates the upper bound again, then a lot of "other things" must have occurred to increase its weight (for instance, maybe edge insertions have occurred). Such a charging argument, essentially, allows us to bypass the $O(\log n)$-update time to an $O(1)$-update time. The flip side of the slack is that our complementary slackness conditions become weak, and therefore instead of a $2 + \varepsilon$-approximation we can only ensure an $O(1)$-approximation. The same technique easily generalizes to the hypergraph setting. It would be interesting to see other scenarios where approximation ratios can be slightly traded in for huge improvements in the update time.

Remark. Independently of our work, Gupta et al. [7] achieved a $O(f^3)$ approximation algorithm for maximum fractional matching and minimum vertex cover in a hypergraph in $O(f^2)$ amortized update time. Here, the symbol f denotes the maximum number of nodes that can be incident on a hyperedge. Due to space limitations, some proofs are deferred to the full version of the paper.

2   Notations and Preliminaries

Since the hypergraph result implies the graph result, henceforth we consider the former problem. The input hypergraph $\mathcal{G} = (V, E)$ has $|V| = n$ nodes. Initially, the set of hyperedges is empty, i.e., $E = \emptyset$. Subsequently, an adversary inserts or deletes hyperedges in the hypergraph $\mathcal{G} = (V, E)$. The node-set V remains unchanged with time. Each hyperedge contains at most f nodes. We say that f is the maximum *frequency* of a hyperedge. If a hyperedge e has a node v as one of its endpoints, then we write $v \in e$. For every node $v \in V$, we let $E_v = \{e \in E : v \in e\}$ denote the set of hyperedges that are incident on v. In this fully dynamic setting, our goal is to maintain an approximate maximum fractional matching and an approximate minimum vertex cover in $\mathcal{G}$. The main result of this paper is summarized in Theorem 1. Throughout the rest of the paper, we fix two parameters α, β as follows.

$$\beta = 17, \text{ and } \alpha = 1 + 36f^2\beta^2. \tag{1}$$

Theorem 1. *We can maintain an $O(f^3)$ approximate maximum fractional matching and an $O(f^3)$ approximate minimum vertex cover in the input hypergraph $\mathcal{G} = (V, E)$ in $O(f^2)$ amortized update time.*

We will maintain a hierarchical partition of the node-set V into $L + 1$ levels $\{0, \ldots, L\}$, where $L = \lceil f \cdot \log_\beta n \rceil + 1$. We let $\ell(v) \in \{0, \ldots, L\}$ denote the *level* of a node $v \in V$. We define the *level* of a hyperedge $e \in E$ to be the maximum level among its endpoints, i.e., $\ell(e) = \max_{v \in e} \ell(v)$. The levels of nodes (and therefore hyperedges) induce the following weights on hyperedges: $w(e) := \beta^{-\ell(e)}$ for every hyperedge $e \in E$. For all nodes $v \in V$, let $W_v := \sum_{e \in E_v} w(e)$ be the total weight received by v from its incident hyperedges. We will satisfy the following invariant after processing a hyperedge insertion or deletion.

Invariant 2. *Every node $v \in V$ at level $\ell(v) > 0$ has weight $1/(\alpha\beta^2) < W_v < 1$. Every node $v \in V$ at level $\ell(v) = 0$ has weight $0 \leq W_v \leq 1/\beta^2$.*

Corollary 3. *The nodes in levels $\{1, \ldots, L\}$ form a vertex cover in $\mathcal{G}$.*

Proof. Suppose that there is a hyperedge $e \in E$ with $\ell(v) = 0$ for all $v \in e$. Then we also have $\ell(e) = 0$ and $w(e) = 1/\beta^{\ell(e)} = 1/\beta^0 = 1$. So for every node $v \in e$, we get: $W_v \geq w(e) = 1$. This violates Invariant 2.

Invariant 2 ensures that $w(e)$'s form a fractional matching satisfying approximate complementary slackness conditions with the vertex cover defined in Corollary 3.

Theorem 4. *In our algorithm, the hyperedge weights $\{w(e)\}$ form a $f\alpha\beta^2$-approximate maximum fractional matching, and the nodes in levels $\{1, \ldots, L\}$ form a $f\alpha\beta^2$-approximate minimum vertex cover.*

For each node v, let $W_v^+ := \sum_{e \in E_v : \ell(e) > \ell(v)} w(e)$ be the total *up-weight* received by v, that is, weight from those incident hyperedges whose levels are strictly greater than $\ell(v)$. For all levels $i \in [0, L]$, let $W_{v \to i}$ and $W_{v \to i}^+$ respectively denote the values of W_v and W_v^+ if the node v were to go to level i and the levels of all the other nodes were to remain unchanged. More precisely, for every hyperedge $e \in E$ and node $v \in e$, let $\ell_v(e) = \max_{u \in e : u \neq v} \ell(u)$ be the maximum level among the endpoints of e that are distinct from v. Then $W_{v \to i} := \sum_{e \in E_v} \beta^{-\max(\ell_v(e), i)}$ and $W_{v \to i}^+ := \sum_{e \in E_v : \ell_v(e) > i} \beta^{-\ell_v(e)}$. We maintain a notion of time such that in each time step the algorithm performs *one* elementary operation. Let $W_v(t)$ denote the weight (resp, up-weight) faced by v *right before* the operation at time t. Similarly define $W_{v \to i}(t)$, $W_v^+(t)$, and $W_{v \to i}^+(t)$.

Before the insertion/deletion of a hyperedge in $\mathcal{G}$, all nodes satisfy Invariant 2. When a hyperedge is inserted (resp. deleted), it increases (resp. decreases) the weights of its endpoints. Accordingly, one or more endpoints can violate Invariant 2 after the insertion/deletion of a hyperedge. Our algorithm *fixes* these nodes by changing their levels, which may lead to new violations, and so on and so forth. To describe the algorithm, we need to define certain *states* of the nodes.

Definition 5. *A node $v \in V$ is* Down-Dirty *iff $\ell(v) > 0$ and $W_v \leq 1/(\alpha\beta^2)$. A node $v \in V$ is* Up-Dirty *iff either $\{\ell(v) = 0, W_v > 1/\beta^2\}$ or $\{\ell(v) > 0, W_v \geq 1\}$. A node is* Dirty *if it is either* Down-Dirty *or* Up-Dirty*. Note that Invariant 2 is satisfied if and only if no node is* Dirty*.*

Definition 6. *A node $v \in V$ is* Super-Clean *iff either (1) We have $\ell(v) = 0$ and $W_v \leq 1/\beta^2$, or (2) We have $\ell(v) > 0$, $1/\beta^2 < W_v \leq 1/\beta$, and $W_v^+ \leq 1/\beta^2$.*

Note that a Super-Clean node v with $\ell(v) > 0$ has a stronger upper bound on the weight W_v it faces and also an even stronger upper bound on the up-weight W_v^+ it faces. At a high level, one of our subroutines will lead to Super-Clean nodes, and the *slack* in the parameters is what precisely allows us to perform an amortized analysis in the update time.

Data Structures. For all nodes $v \in V$ and levels $i \in [0, L]$, let $E_{v,i} := \{e \in E_v : \ell(e) = i\}$ be the set of hyperedges incident on v that are at level i. Note that $E_{v,i} = \emptyset$ for all $i < \ell(v)$. We maintain the following data structures. (1) For every level $i \in [0, L]$ and node $v \in V$, we store the set of hyperedges $E_{v,i}$ as a doubly linked list, and maintain a counter that stores the number of hyperedges in $E_{v,i}$. (2) For every node $v \in V$, we store the weights W_v and W_v^+, its level $\ell(v)$ and an indicator variable for each of the states DOWN-DIRTY, UP-DIRTY, DIRTY and SUPER-CLEAN. (3) For each hyperedge $e \in E$, we store the values of its level $\ell(e)$ and therefore its weight $w(e)$. Finally, using appropriate pointers, we ensure that a hyperedge can be inserted into or deleted from any given linked list in constant time. We now state two lemmas that will be useful in analysing the update time of our algorithm.

Lemma 7. *Suppose that a node v is currently at level $\ell(v) = i \in [0, L-1]$ and we want to move it to some level $j \in [i+1, L]$. Then it takes $O(f \cdot |\{e \in E_v : \ell_v(e) < j\}|)$ time to update the relevant data structures.*

Proof. If a hyperedge e is not incident on the node v, then the data structures associated with e are not affected as v moves up from level i to level j. Further, among the hyperedges $e \in E_v$, only the ones with $\ell_v(e) < j$ get affected (i.e., the data structures associated with them need to be changed) as v moves up from level i to level j. Finally, for every hyperedge that gets affected, we need to spend $O(f)$ time to update the data structures for its f endpoints.

Lemma 8. *Suppose that a node v is currently at level $\ell(v) = i \in [1, L]$ and we want to move it down to some level $j \in [0, i-1]$. Then it takes $O(f \cdot |\{e \in E_v : \ell_v(e) \leq i\}|)$ time to update the relevant data structures.*

Proof. Similar to the proof of Lemma 7.

3   Handling the Insertion/Deletion of a Hyperedge

Initially, the graph $\mathcal{G}$ is empty, every node is at level 0, and Invariant 2 holds. By induction, we will ensure that the following property is satisfied just before the insertion/deletion of a hyperedge.

Proposition 9. *No node $v \in V$ is DIRTY.*

Insertion of a hyperedge e. When a hyperedge e is inserted into the input graph, it is assigned a level $\ell(e) = \max_{v \in e} \ell(v)$ and a weight $w(e) = \beta^{-\ell(e)}$. The hyperedge gets inserted into the linked lists $E_{v,\ell(e)}$ for all nodes $v \in e$. Furthermore, for every node $v \in e$, the weights W_v increases by $w(e)$. For every endpoint $v \in e$, if $\ell(v) < \ell(e)$, then the weight W_v^+ increases by $w(e)$. As a result of these operations, one or more endpoints of e can now become UP-DIRTY and Property 9 might no longer be satisfied. Hence, in order to restore Property 9 we call the subroutine described in Fig. 1.

Deletion of a hyperedge e. When a hyperedge e is deleted from the input graph, we erase all the data structures associated with it. We remove the hyperedge from the linked lists $E_{v,\ell(e)}$ for all $v \in e$, and erase the values $w(e)$ and

01.	WHILE the set of DIRTY nodes is nonempty
02.	IF there exists some UP-DIRTY node v:
03.	FIX-UP-DIRTY(v).
04.	ELSE IF there exists some DOWN-DIRTY node v:
05.	FIX-DOWN-DIRTY(v).

Fig. 1. FIX-DIRTY($\cdot$)

$\ell(e)$. For every node $v \in e$, the weight W_v decreases by $w(e)$. Further, for every endpoint $v \in e$, if $\ell(v) < \ell(e)$, then we decrease the weight W_v^+ by $w(e)$. As a result of these operations, one or more endpoints of e can now become DOWN-DIRTY, and Property 9 might get violated. Hence, in order to restore Property 9 we call the subroutine described in Fig. 1.

The algorithm is simple – as long as some DIRTY node remains, it runs either FIX-UP-DIRTY or FIX-DOWN-DIRTY to take care of UP-DIRTY and DOWN-DIRTY nodes respectively. One crucial aspect is that we prioritize UP-DIRTY nodes over DOWN-DIRTY ones.

FIX-DOWN-DIRTY (v): Suppose that $\ell(v) = i$ when the subroutine is called at time t. By definition, we have $i > 0$ and $W_v(t) \leq 1/(\alpha\beta^2)$. We need to increase the value of W_v if we want to ensure that v no longer remains DIRTY. This means that we should decrease the level of v, so that some of the hyperedges incident on v can increase their weights. Accordingly, we find the *largest* possible level $j \in \{1, \ldots, (i - 1)\}$ such that $W_{v \to j}(t) > 1/\beta^2$, and move the node v down to this level j. If no such level exists, that is, if even $W_{v \to 1}(t) \leq 1/\beta^2$, then we move the node v down to level 0. Note that in this case there is no hyperedge $e \in E_v$ with $\ell_v(e) = 0$ for such a hyperedge would have $w(e) = \beta^{-1} > 1/\beta^2$ when v is moved to level 1. In particular, we get $W_{v \to 0}(t) = W_{v \to 1}(t)$.

Claim 10. *FIX-DOWN-DIRTY (v) makes the node v SUPER-CLEAN.*

FIX-UP-DIRTY (v): Suppose that $\ell(v) = i$ when the subroutine is called at time t. At this stage, we have either $\{i = 0, W_v(t) > 1/\beta^2\}$ or $\{i > 1, W_v(t) \geq 1\}$. We need to increase the level of v so as to reduce the weight faced by it. Accordingly, we find the *smallest* possible level $j \in \{i + 1, \ldots, L\}$ where $W_{v \to j}(t) \leq 1/\beta$ and move v up to level j. Such a level j always exists because $W_{v \to L}(t) \leq n^f \cdot \beta^{-L} \leq 1/\beta$.

Claim 11. *After a call to the subroutine* FIX-UP-DIRTY *(v) at time t, we have* $1/\beta^2 < W_v \leq 1/\beta$.

It is clear that if and when FIX-DIRTY() terminates, we are in a state which satisfies Invariant 2. In the next section we show that after T hyperedge insertions and deletions, the total update time is indeed $O(f^2 \cdot T)$ and so our algorithm has $O(f^2)$-amortized update time.

4 Analysis of the Algorithm

Starting from an empty graph $\mathcal{G} = (V, E)$, fix any sequence of T *updates*. The term "update" refers to the insertion or deletion of a hyperedge in $\mathcal{G}$. We show that the total time taken by our algorithm to handle this sequence of updates is $O(f^2 \cdot T)$. We also show that our algorithm has an approximation ratio of $O(f^3)$.

Relevant counters. We define three counters C^{up}, C^{down} and I^{down}. The first two counters account for the time taken to update the data structures while the third accounts for the time taken to find the index j in both FIX-DOWN-DIRTY(v) and FIX-UP-DIRTY(v). Initially, when the input graph is empty, all the three counters are set to zero. Subsequently, we increment these counters as follows.

1. Suppose node v moves from level i to level $j > i$ upon a call of FIX-UP-DIRTY(v). Then for every hyperedge $e \in E_v$ with $\ell_v(e) \leq j-1$, we increment C^{up} by one.
2. Suppose node v moves from level i to level $j < i$ upon a call of FIX-DOWN-DIRTY(v). Then for every hyperedge $e \in E_v$ with $\ell_v(e) \leq i$, we increment the value of C^{down} by one. Furthermore, we increment the value of I^{down} by β^{i-2}/α.

The next lemma upper bounds the total time taken by our algorithm in terms of the values of these counters. The proof of Lemma 12 appears in Sect. 4.5.

Lemma 12. *Our algorithm takes $\Theta(f \cdot (C^{up} + C^{down} + T) + f^2 I^{down})$ time to handle a sequence of T updates.*

We will show that $C^{up} = \Theta(f) \cdot T$ and $C^{down} + I^{down} = O(1) \cdot T$, which will imply an amortized update time of $O(f^2)$ for our algorithm. Towards this end, we now prove three lemmas that relate the values of these three counters.

Lemma 13. *We have: $C^{down} \leq I^{down}$.*

Lemma 14. *We have: $I^{down} \leq \frac{f}{\alpha-1} \cdot (T + C^{up})$.*

Lemma 15. *We have: $C^{up} \leq 9f\beta^2 \cdot (T + C^{down})$.*

The proofs of Lemmas 13, 14 and 15 appear in Sects. 4.2, 4.3 and 4.4 respectively. All these three proofs use the concepts of epochs, jumps and phases as defined in Sect. 4.1. The main result of our paper (see Theorem 1) now follows from Theorem 4, Lemmas 12 and 16.

Lemma 16 (Corollary to Lemma 13, 14, and 15). *We have: $C^{up} = \Theta(f) \cdot T$ and $C^{down} + I^{down} = \Theta(1) \cdot T$.*

Proof. Replacing C^{down} in the RHS of Lemma 15 by the upper bounds from Lemmas 13 and 14, we get:

$$
\begin{aligned}
C^{up} &\leq (9f\beta^2) \cdot T + (9f\beta^2) \cdot C^{down} \\
&\leq (9f\beta^2) \cdot T + (9f\beta^2) \cdot I^{down} \\
&\leq (9f\beta^2) \cdot T + \frac{(9f\beta^2)f}{(\alpha - 1)} \cdot (T + C^{up}) \\
&\leq (9f\beta^2) \cdot T + (1/4) \cdot T + (1/4) \cdot C^{up} \qquad \text{(see Eq. (1))}
\end{aligned}
$$

Rearranging the terms in the above inequality, we get: $(3/4) \cdot C^{up} \leq (9f\beta^2 + 1/4) \cdot T = (36f\beta^2 + 1) \cdot (T/4)$. Multiplying both sides by $(4/3)$, we get: $C^{up} \leq (12f\beta^2 + 1/3) \cdot T \leq (13f\beta^2)T$. Since $\beta = 17$, we get:

$$
C^{up} \leq \Theta(f) \cdot T \tag{2}
$$

Since $\alpha = \Theta(f^2)$, Lemmas 13 and 14 and Eq. (2) imply that:

$$
C^{down} \leq I^{down} \leq \Theta(1) \cdot T \tag{3}
$$

4.1 Epochs, Jumps and Phases

Fix any node $v \in V$. An *epoch* of v is a maximal time-interval during which the node stays at the same level. An epoch ends when either (a) the node v moves up to a higher level due to a call to FIX-UP-DIRTY, or (b) the node v moves down to a lower level due to a call to the subroutine FIX-DOWN-DIRTY. These events called *jumps*. Accordingly, there are UP-JUMPS and DOWN-JUMPS. Next, we define a *phase* of a node to be a maximal sequence of consecutive epochs where the levels of the node keep on increasing. The phase of a node v is denoted by Φ_v. Suppose that a phase Φ_v consists of k consecutive epochs of v at levels $i_1, \ldots, i_k \in \{0, 1, \ldots, L\}$. Then we have: $i_1 < i_2 < \cdots < i_k$. By definition, the epoch immediately before Φ_v must have level larger than i_1 implying FIX-DOWN-DIRTY(v) landed v at level i_1. Similarly, the epoch subsequent to i_k is smaller than i_k implying FIX-DOWN-DIRTY(v) is called again.

4.2 Proof of Lemma 13

Suppose that a node v moves down from (say) level j to level $i < j$ at time (say) t due to a call to the subroutine FIX-DOWN-DIRTY(v). Let Δ^{down} and Δ_I^{down} respective denote the increase in the counters C^{down} and I^{down} due to this event. We will show that $\Delta^{down} \leq \Delta_I^{down}$, which will conclude the proof of the lemma. By definition, we have:

$$
\Delta_I^{down} = \beta^{i-2}/\alpha \tag{4}
$$

Let $X = \{e \in E_v : \ell_v(e) \leq i\}$ be the set of hyperedges incident on v that contribute to the increase in C^{down} due to the DOWN-JUMP of v at time t. We have: $|X| = \Delta^{down}$. Each edge $e \in X$ contributes β^{-i} towards the node-weight $W_{v \to i}(t)$. Thus, we get: $|X| \cdot \beta^{-i} \leq W_{v \to i}(t) \leq 1/(\alpha\beta^2)$. The last inequality holds since v is DOWN-DIRTY in the beginning of time-step t. Rearranging the terms, we get: $\Delta^{down} = |X| \leq \beta^{i-2}/\alpha$. The lemma now follows from Eq. (4).

4.3 Proof of Lemma 14

Suppose we call FIX-DOWN-DIRTY(v) at some time t_2. Let $\ell(v) = i$ just before the call, and let $[t_1, t_2]$ be the epoch with level of v being i. Let $X := \{e \in E_v : \ell_v(e) \leq i\}$ at time t_2. By definition, I^{down} increases by β^{i-2}/α during the execution of FIX-DOWN-DIRTY(v); let us call this increase Δ_I^{down}. Thus:

$$\Delta_I^{down} = \beta^{i-2}/\alpha \tag{5}$$

Consider the time between $[t_1, t_2]$ and let us address how W_v can decrease in this time while v's level is fixed at i. Either some hyperedge incident on v is deleted, or some hyperedge $e \in E_v$ incident on it decreases its weight. In the latter case, the level $\ell(e)$ of such an hyperedge e must increase above i. Let Δ^T denote the number of hyperedge deletions incident on v during the time-interval $[t_1, t_2]$. Let Δ^{up} denote the increase in the value of C^{up} during the time-interval $[t_1, t_2]$ *due to the hyperedges incident on v*. Specifically, at time t_1, we have $\Delta^T = \Delta^{up} = 0$. Subsequently, during the time-interval $[t_1, t_2]$, we increase the value of Δ^{up} by one each time we observe that a hyperedge $e \in E_v$ increases its level $\ell(e)$ to something larger than i. Note that $\ell(v) = i$ throughout the time-interval $[t_1, t_2]$. Hence, each time we observe an unit increase in $\Delta^T + \Delta^{up}$, this decreases the value of W_v by at most β^{-i}. Just before time t_1, the node v made either an UP-JUMP, or a DOWN-JUMP. Hence, Claims 10 and 11 imply that $W_{v \to i}(t_1) > 1/\beta^2$. As $W_v(t_2) \leq 1/(\alpha\beta^2)$ at time t_2, we infer that W_v has dropped by at least $(1 - 1/\alpha) \cdot \beta^{-2}$ during the time-interval $[t_1, t_2]$. In order to account for this drop in W_v, the value of $\Delta^T + \Delta^{up}$ must have increased by at least $(1 - 1/\alpha) \cdot \beta^{-2}/\beta^{-i} = (1 - 1/\alpha) \cdot \beta^{i-2}$. Since $\Delta^T = \Delta^{up} = 0$ at time t_1, at time t_2 we get: $\Delta^T + \Delta^{up} \geq (1 - 1/\alpha) \cdot \beta^{i-2}$. Hence, (5) gives us:

$$\Delta_I^{down} \leq (\alpha - 1)^{-1} \cdot (\Delta^T + \Delta^{up}) \tag{6}$$

Each time the value of I^{down} increases due to FIX-DOWN-DIRTY on some node, inequality (6) applies. If we sum all these inequalities, then the left hand side (LHS) will be exactly equal to the final value of I^{down}, and the right hand side (RHS) will be at most $(\alpha - 1)^{-1} \cdot (f \cdot T + (f - 1) \cdot C^{up})$. The factor f appears in front of T because each hyperedge deletion can contribute f times to the sum $\sum \Delta^T$, once for each of its endpoints. Similarly, the factor $(f - 1)$ appears in front of C^{up} because whenever the level of an hyperedge e moves up due to the increase in the level $\ell(v)$ of some endpoint $v \in e$, this contributes at most $(f - 1)$ times to the sum $\sum \Delta^{up}$, once for every other endpoint $u \in e, u \neq v$. Since LHS $\leq$ RHS, we get: $I^{down} \leq (\alpha - 1)^{-1} \cdot (f \cdot T + (f - 1) \cdot C^{up}) \leq (f/(\alpha - 1)) \cdot (T + C^{up})$.

4.4 Proof Sketch of Lemma 15

In this extended abstract we give a sketch; the full proof can be found in the full version. For a vertex v, we fix a phase Φ_v and we account for the total up-movement ΔC^{up} in this phase. Note that a phase could have many up-jumps. After a node performs an up-jump, its total weight is bounded and this in turn bounds the number of hyperedges that have contributed to ΔC^{up}. For all but the last up-jump, this can be charged to incident hyperedges being inserted or some neighboring vertices making a down jump; the last one is an outlier since the node may never become UP-DIRTY at that level. To argue about the last level, we focus on two cases. One, when the last level is not much larger than the last-but-one, in which case we can charge the last level to the remaining. Otherwise, the last up-jump is a large jump. But this implies that v must fact significant W_v^+ weight just before the last up-jump. This was, in particular, not true at the beginning of the phase. So either some hyperedges have been inserted or some vertices have made down-jumps.

4.5   Proof of Lemma 12

For technical reasons, we assume that we end with the empty graph as well. This is without loss of generality due to the following reason. Suppose we made T updates and the current graph is G. At this point, the graph has $T' \leq T$ edges. Suppose the time taken by our algorithm till now is T_1. Now delete all the T' edges, and let the time taken by our algorithm to take care of these T' updates be T_2. If $T_1 + T_2 = \Theta(f^2(T + T')) = \Theta(f^2 T)$, then $T_1 = \Theta(T)$ as well. Therefore, we assume we end with an empty graph.

When a hyperedge e is inserted into or deleted from the graph, we take $O(f)$ time to update the relevant data structures for its f endpoints. The rest of the time is spent in implementing the WHILE loop in Fig. 1. We take care of the two subroutines separately.

Case 1. The subroutine FIX-DOWN-DIRTY(v) is called which moves the node v from level i to level $j < i$ (say). We need to account for the time to find the relevant index j and the time taken to update the relevant data structures. By Lemma 8, the time taken for the latter is proportional $\Theta(f \cdot \Delta C^{down})$. Further, the value of C^{up} remains unchanged. For finding the index $j < i$, it suffices to focus on the edges $E_{v,i} = \{e \in E_v : \ell_v(e) \leq i\}$ since these are the only edges that change weight as v goes down. Therefore, this takes time $\Theta(|\{e \in E_v : \ell_v(e) \leq i\}|)$. Since each of these edges had $w(e) = \beta^{-i}$ and since $W_v \leq \frac{1}{\alpha\beta^2}$ before the FIX-DOWN-DIRTY(v) call, we have $|\{e \in E_v : \ell_v(e) \leq i\}| \leq \beta^{i-2}/\alpha$ which is precisely ΔI^{down}. Therefore, the time taken to find the index j is $\Theta(\Delta I^{down})$.

Case 2. The subroutine FIX-UP-DIRTY(v) is called which moves the node v from level i to level $j > i$, say. Once again, we need to account for the time to find the relevant index j and the time taken to update the relevant data structures, and once again by Lemma 7 the time taken for the latter is $\Theta(f \cdot \Delta C^{up})$. Further, the value of C^{down} remains unchanged. We now account for the time taken to find the index j.

Claim 17. *j can be found in time $\Theta(j - i)$.*

Proof. To see this note that for $k \geq i$,

$$W_{v \to k}(t) = \sum_{\ell \geq k} \sum_{e \in E_{v,\ell}} w(e) + \sum_{\ell < k} \frac{1}{\beta^{k-\ell}} \sum_{e \in E_{v,\ell}} w(e)$$

since (a) edges not incident on v are immaterial, (b) the edges incident on v whose levels are already $\geq k$ do not change their weight, and (c) edges whose levels are $\ell < k$ have their weight go from $\beta^{-\ell}$ to β^{-k}. The above implies that for $k \geq i$,

$$W_{v \to (k+1)}(t) = W_{v \to k}(t) - \left(1 - \tfrac{1}{\beta}\right) \sum_{e \in E_{v,k}} w(e) = W_{v \to k}(t) - \left(1 - \tfrac{1}{\beta}\right)|E_{v,k}| \cdot \beta^{-k}$$

That is, $W_{v \to (k+1)}$ can be evaluated from $W_{v \to k}(t)$ in $\Theta(1)$ time since we store $|E_{v,k}|$ in our data structure. The claim follows.

Note that the LHS of Claim 17 can be as large as $\Theta(\log n)$. To account for the movement, we again fix a vertex v and a phase Φ_v where the level of v changes from i_1 to say i_k. The total time for finding indices is $\Theta(i_k - i_1)$. After this, there must be a DOWN-JUMP due to a call to FIX-DOWN-DIRTY(v) since the final graph is empty. Thus, we can charge the time taken in finding indices in this phase Φ_v to ΔI^{down} in the FIX-DOWN-DIRTY(v) call right at the end of this phase. We can do so since $\Delta I^{down} = \beta^{i_k - 2}/\alpha = \tfrac{1}{f^2}\Theta(i_k)$ since $\beta = \Theta(1)$ and $\alpha = \Theta(f^2)$ by (1). Therefore, the total time taken to find indices in the FIX-UP-DIRTY(v) calls in all is at most $f^2 I^{down}$.

In sum, the total time taken to initialize update data structures is at most $\Theta\left(f \cdot \left(C^{up} + C^{down} + T\right)\right)$ and the total time taken to find indices is at most $\Theta(f^2 \cdot I^{down})$. This proves Lemma 12.

References

1. Abboud, A., Williams, V.V.: Popular conjectures imply strong lower bounds for dynamic problems. In: FOCS (2014)
2. Baswana, S., Gupta, M., Sen, S.: Fully dynamic maximal matching in $O(\log n)$ update time. In: FOCS (2011)
3. Bernstein, A., Stein, C.: Faster fully dynamic matchings with small approximation ratios. In: SODA (2016)
4. Bhattacharya, S., Henzinger, M., Italiano, G.F.: Design of dynamic algorithms via primal-dual method. In: Halldórsson, M.M., Iwama, K., Kobayashi, N., Speckmann, B. (eds.) ICALP 2015. LNCS, vol. 9134, pp. 206–218. Springer, Heidelberg (2015). doi:10.1007/978-3-662-47672-7_17
5. Bhattacharya, S., Henzinger, M., Italiano, G.F.: Deterministic fully dynamic data structures for vertex cover and matching. In: SODA (2015)
6. Bhattacharya, S., Henzinger, M., Nanongkai, D.: New deterministic approximation algorithms for fully dynamic matching. In: STOC (2016)
7. Gupta, A., Krishnaswamy, R., Kumar, A., Panigrahi, D.: Online and dynamic algorithms for set cover. In: STOC (2017)
8. Gupta, M., Peng, R.: Fully dynamic $(1 + \epsilon)$-approximate matchings. In: FOCS (2013)
9. Henzinger, M., Krinninger, S., Nanongkai, D., Saranurak, T.: Unifying and strengthening hardness for dynamic problems via the online matrix-vector multiplication conjecture. In: STOC (2015)
10. Henzinger, M.R., Fredman, M.L.: Lower bounds for fully dynamic connectivity problems in graphs. Algorithmica **22**(3), 351–362 (1998)

11. Neiman, O., Solomon, S.: Simple deterministic algorithms for fully dynamic maximal matching. In: STOC (2013)
12. Onak, K., Rubinfeld, R.: Maintaining a large matching and a small vertex cover. In: STOC (2010)
13. Patrascu, M.: Lower bounds for dynamic connectivity. In: Encyclopedia of Algorithms, pp. 1162–1167 (2016)
14. Sankowski, P.: Faster dynamic matchings and vertex connectivity. In: SODA (2007)
15. Solomon, S.: Fully dynamic maximal matching in constant update time. In: FOCS (2016)

Cutting Planes from Wide Split Disjunctions

Pierre Bonami[1], Andrea Lodi[2(✉)], Andrea Tramontani[3], and Sven Wiese[4]

[1] CPLEX Optimization, IBM Spain, Madrid, Spain
pierre.bonami@es.ibm.com
[2] CERC - École Polytechnique de Montreal, Montreal, Canada
andrea.lodi@polymtl.ca
[3] CPLEX Optimization, IBM Italy, Bologna, Italy
andrea.tramontani@it.ibm.com
[4] DEI - University of Bologna, and OPTIT srl, Bologna, Italy
sven.wiese@unibo.it

Abstract. In this paper, we discuss an extension of split cuts that is based on widening the underlying disjunctions. That the formula for deriving intersection cuts based on splits can be adapted to this case has been known for a decade now. For the first time though, we present applications and computational results. We further provide some theory that supports our findings, discuss extensions with respect to cut strengthening procedures and present some ideas on how to use the wider disjunctions also in branching.

1 Introduction

As many authors have noted before us, cutting planes are nowadays an essential ingredient in virtually all Mixed Integer Linear Programming (MILP) codes. Several classes of cutting planes are derived from disjunctions whose validity can be easily verified by the integrality requirements of the underlying MILP. For example, split cuts [6] make use of the fact that inside the feasible set, the dot product of the MILP's integer variables with an integral vector never maps to a fractional value, i.e., into the open interval between any two consecutive integers. In this paper, we study situations in which more information is available, and we can exclude the whole interval between two not necessarily consecutive integers from the feasible set.

Our study is motivated by simple observations regarding the modeling techniques used in Constraint Programming (CP). As an illustrative example, consider a sudoku game as in Fig. 1. In CP, we are allowed to work with finite domains, and the domain of the variable $y_{5,5}$, that is to model the number in the 5th row of the 5th column, can simply be written as $D(y_{5,5}) = \{1, 2, 5, 6, 8\}$. In MILP instead, we would write $1 \leq y_{5,5} \leq 8$ and $y_{5,5} \in \mathbb{Z}$, but we have not yet accounted for the constraint $y_{5,5} \notin \{3, 4, 7\}$, which in this direct form is not foreseen in the modeling tools provided by MILP. Of course, all integer programmers would object and (correctly) assert that it is an easy exercise to introduce auxiliary variables that impose this condition. This is what has been done in integer programming for roughly 50 years now and is thus well-proven practice.

© Springer International Publishing AG 2017
F. Eisenbrand and J. Koenemann (Eds.): IPCO 2017, LNCS 10328, pp. 99–110, 2017.
DOI: 10.1007/978-3-319-59250-3_9

Yet, it highlights exactly the point we want to make
with our admittedly informal example. In order to
express the condition that a variable be different
from a value in the "interior" of its domain, we have
to use auxiliary variables. In CP, such constraints
are routinely expressed and, much more important,
also exploited algorithmically, e.g., through filtering and propagation. Our aim here is to analyze
whether and how we can exploit such explicit representations in MILP.

In the above case, one could also say that $y_{5,5}$
has holes in its domains. For example, it is allowed
to take the values 2 and 5, but nothing in between.
More formally, this can be expressed by the
disjunction

Fig. 1. A 3×3 sudoku

$$y_{5,5} \leq 2 \vee y_{5,5} \geq 5. \tag{1}$$

As stated earlier, in the theory of classical split cuts, the right-hand sides of
such two disjunctive terms are always consecutive integers. One special case of
split cuts are intersection cuts [3] from split sets, for which a closed form formula exists, and that this formula can be easily extended to disjunctions with
non-consecutive right-hand sides has already been proven in [2]. Nevertheless,
in more than ten years since, nobody has ever applied it in practice in order
to conduct computational results. While the authors in [2] use the term *general
split disjunctions*, we will call constructs like (1) *wide split disjunctions*, and the
resulting cutting planes *wide split cuts*. Our contribution is to revive the aforementioned formula of [2]. We focus on computational aspects and experiments.
We are interested, in particular, in recognizing examples where wide split disjunctions occur and in exploring to what extent the use of wide split cuts in
practical MILP codes can be advantageous. In addition, we back our findings up
by some theoretical observations.

The rest of the paper is organized as follows. We provide examples of the
validity of wide split disjunctions in Sect. 2, and show that exploiting this "hole
information" algorithmically through cutting planes can be advantageous in
Sect. 3. Finally, in Sect. 4, we examine the combination of wide split cuts with
branching on wide split disjunctions inside a branch-and-cut tree.

Throughout the text, we use the notation $[n] := \{1, \ldots, n\}$ for any positive
integer n.

2 Validity of Wide Split Disjunctions

For the moment, we assume the existence of an underlying MILP

$$\min \quad c^T x \tag{2}$$
$$\text{s.t.} \quad Ax = b \tag{3}$$
$$x \in \mathbb{R}_+^{n-p} \times \mathbb{Z}_+^p. \tag{4}$$

Wide split disjunctions are of the form

$$\pi^T x \le \pi_l \lor \pi^T x \ge \pi_u, \tag{5}$$

where the triple (π^T, π_l, π_u) belongs to the set

$$\Pi := \{(\pi^T, \pi_l, \pi_u) \in \mathbb{Z}^{n+2} \mid \pi_i = 0 \ \forall \, i = 1, \dots, n - p\}.$$

A natural question that arises is the one about the validity of a wide split disjunction for a MILP. While it is easy to see that any solution to (2)–(4) satisfies any split disjunction with consecutive right-hand sides, in the following called *ordinary* split, this is not necessarily true for wide splits. Finding wide split cuts in MILPs is very difficult by itself: While it is not our main focus in this paper, we provide here some examples where wide splits can be found.

2.1   Certifying Split Validity by Primal Information

We first give examples in which wide split disjunctions in the form of holes in the domains of variables are implied by the constraints present in an underlying MILP. This happens, e.g., when some modeling tricks with auxiliary binary variables as mentioned in the introduction are applied. The first of these tricks involves big-M constraints.

Consider an integer variable y and the set of constraints

$$l^j - y \le (1 - x_j) \cdot (l^j - l^1) \quad \forall \, j \in [m],$$

$$y - u^j \le (1 - x_j) \cdot (u^L - u^j) \quad \forall \, j \in [m], \quad \sum_{j=1}^{m} x_j = 1,$$

and assume $u^{j-1} < l^j$. In any feasible solution, y will lie in exactly one of the intervals $[l^j, u^j]$. It is easy to check that the above set of constraints implies validity of the simple wide split disjunctions $y \le u^{j-1} \lor y \ge l^j$, $j = 2, \dots, m$. Clearly, if $u^{j-1} + 1 < l^j$ for some j, there is at least one non-ordinary wide split disjunction. The above constraint structure can be found, e.g., in straightforward MILP formulations for the Traveling Salesman Problem with Multiple Time Windows (TSPMTW) [5].

A GUB-link constraint is characterized by a pair of equations,

$$y = \sum_{j=1}^{m} \lambda_j x_j, \quad \sum_{j=1}^{m} x_j = 1. \tag{6}$$

Here, the wide split disjunctions $y \le \lambda_{j-1} \lor y \ge \lambda_j$, $j = 2, \dots, m$ are valid. Clearly, if the λ_j are non-consecutive integers, there are non-ordinary wide split disjunctions. Such structures can be found in time-indexed MILP formulations for scheduling problems, see, e.g., [7].

2.2   Certifying Split Validity by Dual Information

In the previous section, we deduced the validity of wide split disjunctions by primal information: every (primal) feasible solution was assured to satisfy them. Now instead, we allow that disjunctions are imposed that not every primal feasible solution satisfies. In the following MILP, however, we will see that for every solution that is excluded in this way, there is at least one other feasible solution with identical objective value. The resulting problem and the original MILP are thus equivalent in the sense that they have the same optimal objective value. Hence, we can characterize these domains to be derived from dual information.

The MILP we are going to analyze is a formulation for the so-called Lazy Bureaucrat Problem (LBP), and strictly speaking, it is a pure Integer Linear Program (ILP). The LBP can be seen as a lazy counterpart of the classical Knapsack Problem [11]. Similar to therein, we are given a set of items $i \in [n]$ with non-negative profits p_i and non-negative weights w_i, both of which we assume to be integral, and a knapsack with capacity C. The objective is to pack a subset of items into the knapsack such that:

- the profit of all packed items is minimized,
- their weight does not exceed the capacity,
- but adding any non-packed item would exceed it.

The task is thus to find a so-called maximal packing with minimum profit. In [8], several ILP formulations for LBP are proposed. We present here the most promising one according to [8]. Assuming that the items are ordered increasingly according to their weight, i.e., $w_i \leq w_j$ for $i < j$, the critical item is defined as the first item that exceeds the capacity in a complete packing, $i_c := \min\{i \in [n] \mid \sum_{j \leq i} w_j > C\}$. The critical weight is $w_c := w_{i_c}$. A valid formulation is then given by

$$\min \quad \sum_{i=1}^{n} p_i x_i \tag{7}$$

$$\text{s.t.} \quad \sum_{i=1}^{n} w_i x_i \leq C \tag{8}$$

$$\sum_{i=1}^{n} w_i x_i + z \geq C + 1 \tag{9}$$

$$z \leq w_c - (w_c - w_i)(1 - x_i) \qquad \forall\, i \in [i_c] \tag{10}$$

$$(x, z) \in \{0, 1\}^n \times \mathbb{Z}_+. \tag{11}$$

The variable z models the weight of the smallest item left out of the packing, and it is easy to construct examples of feasible solutions with $z \notin \{w_i \mid i \leq i_c\}$. Yet, we have the following simple result.

Lemma 1. *For every feasible solution $(\bar{x}, \bar{z})$ of (7)–(11), there is another feasible solution $(\bar{x}, \tilde{z})$ with the same cost and $\tilde{z} \in \{w_i \mid i \leq i_c\}$.*

Proof. Denote the packing corresponding to $(\bar{x}, \bar{z})$ by $A := \{i \in [n] \mid \bar{x}_i = 1\}$. Further, let $\bar{i} := \min\{i \in [n] \mid i \notin A\}$. Clearly, $\bar{i} \leq i_c$, and from (10) we get $\bar{z} \leq w_{\bar{i}}$. Because increasing the value of $\bar{z}$ does not violate (9), $(\bar{x}, \tilde{z})$ with $\tilde{z} := w_{\bar{i}} \in \{w_i \mid i \leq i_c\}$ is feasible and clearly has the same cost as $(\bar{x}, \bar{z})$. $\square$

In the end, if the weights are non-consecutive integers, we can consider a version of the problem with valid wide split disjunctions.

3   Cut Derivation and Computation

In this section, before coming to our main contributions, we first recall how to algebraically derive wide split cuts. We denote the points of the feasible region of (2)–(4), that in addition satisfy the wide split disjunction (5), by $\mathcal{F}$. Also, we denote the feasible region of the LP relaxation of (2)–(4) by P, i.e., $P = \{x \in \mathbb{R}^n \mid Ax = b, x \geq 0\}$. Thus, we set

$$P_l^{(\pi, \pi_l)} := \{x \in P \mid \pi^T x \leq \pi_l\}, \quad P_u^{(\pi, \pi_u)} := \{x \in P \mid \pi^T x \geq \pi_u\},$$

and $P^{(\pi, \pi_l, \pi_u)} := \mathrm{conv}(P_l^{(\pi, \pi_l)} \cup P_u^{(\pi, \pi_u)})$. By definition, a wide split cut is any linear inequality that is valid for $P^{(\pi, \pi_l, \pi_u)}$. Since $\mathcal{F} \subseteq P^{(\pi, \pi_l, \pi_u)}$, any wide split cut is also valid for $\mathcal{F}$. We pick up on what has been shown in [2], i.e., how to derive wide split cuts as intersection cuts, but we note that it is also possible to derive wide split cuts in a lift-and-project fashion, see [13, Sect. 4.2.2].

3.1   Intersection Cuts from Wide Split Disjunctions

We assume that the LP relaxation of (2)–(4) has been solved to the point $\hat{x}$ by means of the simplex method and that we are given the optimal basis $B \subseteq [n]$ and simplex tableau $T = T(B)$ of P,

$$x_i = \hat{x}_i + \sum_{j \in N} r_j^i x_j \quad \forall i \in B, \quad x_i \geq 0 \quad \forall i \in [n], \tag{12}$$

where the index set of variables is partitioned into basic and non-basic variables, $N = [n] \backslash B$, respectively. We state the aforementioned result of [2].

Proposition 1. *Assume that $\hat{x}$ violates the wide split disjunction (π^T, π_l, π_u), i.e., $\pi_l < \pi^T \hat{x} < \pi_u$, and $\forall\, j \in N$, define $f_j := \pi_j + \sum_{i \in B} \pi_i r_j^i$. A valid inequality for $P^{(\pi, \pi_l, \pi_u)}$ is then given by*

$$\sum_{j \in N} \max\left\{\frac{-f_j}{\pi^T \hat{x} - \pi_l}, \frac{f_j}{\pi_u - \pi^T \hat{x}}\right\} x_j \geq 1. \tag{13}$$

The valid inequality of Proposition 1 is clearly a wide split cut, and it is easy to check that it is violated by $\hat{x}$. If in (12) one relaxes the non-negativity on the basic variables, then the resulting set can be shown to be a translated polyhedral

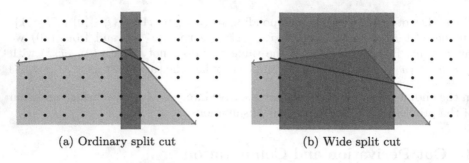

(a) Ordinary split cut   (b) Wide split cut

Fig. 2. Intersection cuts in the plane

cone, often denoted by $P(B)$. Geometrically, (13) can be obtained by computing the intersection points of the extreme rays of $P(B)$ with the boundary of the split set $S := \{x \in \mathbb{R}^n \mid \pi_l \le \pi^T x \le \pi_u\}$. Figure 2 (a) and (b) depict an example of $P(B)$, and the split sets with corresponding cuts of an ordinary split disjunction and a wide split disjunction, respectively, in a two-dimensional basic space. We now give an algebraic example of Proposition 1.

Example 1. Consider the MILP with GUB-Link constraints

$$\min \quad -x_2 \tag{14}$$
$$\text{s.t.} \quad x_1 + x_2 \le 6 \tag{15}$$
$$-x_1 + x_2 \le 0 \tag{16}$$
$$x_1 = x_3 + 2x_4 + 4x_5 + 8x_6 \tag{17}$$
$$1 = x_3 + x_4 + x_5 + x_6 \tag{18}$$
$$(x_1, x_2, x_3, x_4, x_5, x_6) \in \mathbb{Z}_+^6. \tag{19}$$

Note that variables x_3, x_4, x_5, x_6 are implicitly binary constrained due to (18). Therefore, it is easy to see that $x_1 \le 2 \lor x_1 \ge 4$ is a valid (simple) wide split disjunction. Introducing slack variables s_1 and s_2 for (15) and (16), respectively, one can check that the basic row of x_1 in the optimal simplex tableau is $x_1 = 3 - \frac{1}{2}s_1 + \frac{1}{2}s_2$, showing that $\hat{x}_1 = 3$ and that the wide split disjunction is violated. This equation can be used together with (13) to calculate the wide split cut $s_1 + s_2 \ge 2$, that, after substituting slacks, is equivalent to $x_2 \le 2$. One can check that the wide split cut is a facet of the integer hull of the MILP in the (x_1, x_2)-space. Clearly, in the optimal solution of the LP relaxation, there will be two fractional binary variables, one between x_3 and x_4, and one between x_5 and x_6. Assume that $0 < \hat{x}_4 < 1$ and $0 < \hat{x}_5 < 1$. The two intersection cuts corresponding to these two violated split disjunctions can be checked to be $x_2 - 3x_3 - 4x_6 \le 2$ and $x_2 - x_3 - 6x_6 \le 2$. Both cut off the optimal LP-solution, but are dominated by the wide split cut.   □

In the above example, the advantage of wide splits over ordinary splits lies in the fact that no ordinary split cut with the same split vector would have been separated, because $\hat{x}_1$ is not fractional. We highlight though that even in situations where the split vector gives rise to a fractional value, a theoretical dominance of wide split cuts as in Proposition 1 over corresponding ordinary split cuts can be shown, see [13, Proposition 4.7].

We showed in Example 1 that in the case of GUB-Links, a wide split cut can dominate simple ordinary split cuts on the involved binary variables. In this special case however, there is another type of ordinary but non-simple split cuts, that seems superior to the simple one, and that we analyze in the following. Formally, given the GUB-link (6), let W denote the wide split disjunction $y \leq \lambda_\ell \vee y \geq \lambda_{\ell+1}$. Also, assume the λ_j to be ordered increasingly and let Q denote the ordinary split disjunction $\sum_{j \in L} x_j \leq 0 \vee \sum_{j \in L} x_j \geq 1$ with $L = \{1, \ldots, \ell\}$. The relationship between the cuts obtained by both disjunctions is settled by the two following lemmata.

Lemma 2. *Let $C(D)$ denote the split closure, i.e., the set of all split cuts that can be obtained from a split disjunction D (see, e.g., [2]). Then, $C(Q)$ dominates $C(W)$, i.e., $C(Q) \subseteq C(W)$.*

Proof. Define $W_0 := \{(x,y) \in P \mid y \leq \lambda_\ell\}$, $W_1 := \{(x,y) \in P \mid y \geq \lambda_{\ell+1}\}$, $Q_0 := \{(x,y) \in P \mid \sum_{j \in L} x_j \leq 0\}$, and $Q_1 := \{(x,y) \in P \mid \sum_{j \in L} x_j \geq 1\}$. The closure $C(Q)$ is simply $C(Q) = \mathrm{conv}(Q_0 \cup Q_1)$, while $C(W) = \mathrm{conv}(W_0 \cup W_1)$. It holds that $\sum_{j \in L} x_j \geq 1$ implies $y \leq \lambda_\ell$, therefore $Q_1 \subseteq W_0$. Also, $\sum_{j \in L} x_j \leq 0$ implies $y \geq \lambda_{\ell+1}$, and thus $Q_0 \subseteq W_1$. The claim follows. □

Lemma 3. *The dominance in Lemma 2 only applies to closures. In particular, given an optimal tableau T, the wide split cut associated with T and W is not necessarily dominated by the split cut associated with T and Q, and it could even happen that the reverse is true (i.e., the cut from W dominates the cut from Q).*

Proof. In Example 1, assuming again that x_4 and x_5 are basic in the optimal simplex solution, the split Q is given by $x_3 + x_4 \leq 0 \vee x_3 + x_4 \geq 1$, and is violated. One can show that the intersection cut from the above binarization split is $x_2 - x_3 - 4x_6 \leq 2$. This cut is still weaker than the wide split cut: point $(3, 3, \frac{1}{5}, \frac{3}{5}, 0, \frac{1}{5})$ satisfies it. □

3.2 Computation

We now present the results of computational experiments with wide split cuts applied to the LBP introduced in Sect. 2.2. Recall that the valid wide split disjunctions in this case are simple ones, i.e., we have holes in the domains of integer variables. We are particularly interested in testing the computational advantage of wide split cuts over corresponding cuts from ordinary splits. Therefore, we include intersection cuts from simple, ordinary splits in our experiments. However, we do not only include such cuts corresponding to fractional values that fall into a hole, but also to fractional values that do not, and in particular corresponding to all integer and binary variables in the model. That is, we do not restrict to the simple split vectors that appear in valid wide split disjunctions. This will give us more flexibility in testing different cut separation strategies.

In general, we perform several rounds of separation. In a single round, we solve the LP relaxation of the underlying MILP to optimality, separate cuts according to the chosen strategy, add all separated cuts to the model and solve again. The strategies we test are

- w/o (without wide splits): For each basic fractional binary or integer variable, we compute an intersection cut from an ordinary split.
- w (with wide splits): For each basic integer variable, we compute a wide split cut if its value lies in a hole, or otherwise an intersection cut from an ordinary split, if its value is fractional. In addition, for each basic fractional binary variable, we compute an intersection cut from an ordinary split.
- o (only wide splits): For each basic integer variable, we compute a wide split cut if its value lies in a hole.

The procedure has been coded in C/C++ with CPLEX 12.6.1 as LP solver. Throughout the following, we usually report the percentage of the initial dual gap that is closed by the separated cuts. Also, the total number of cuts that have been generated is sometimes shown. In order to create test instances, we used *class 4* of the knapsack instance generator presented in [10] and available at [12]. As in [8], this led to a total of 54 instances with different combinations of the parameters.

Table 1 shows an excerpt (and mean values) of those 25 out of the 54 instances in which we can close significantly more gap with strategy w than with w/o in 10 rounds of separation. That means that the separation of wide split cuts on top of ordinary split cuts is highly advanta-

Table 1. Separation of wide split cuts on LBP instances

Instance	% gap closed			#cuts		
	w/o	w	o	w/o	w	o
10000-4-100-75	9.21	12.53	11.73	139	62	5
10000-4-10-25	32.97	34.28	28.40	46	27	2
10000-4-10-50	79.12	100.00	100.00	1	1	1
⋮	⋮			⋮		
1000-4-40-75	8.92	42.90	41.38	150	53	5
1000-4-500-75	2.97	5.10	4.89	134	73	6
1000-4-50-50	13.01	17.12	15.26	113	60	4
Mean	19.81	31.95	28.13	89.92	49.16	3.28

geous. Interestingly, on these instances the separation of only wide split cuts (strategy o) already leads to significantly more gap closure than strategy w/o, and almost reaches the one of strategy w. A (positive) side effect of wide split cuts is the reduction of the number of cuts that are separated in total. Apart from closing more gap, strategy w also decreases this number significantly with respect to strategy w/o. This is a desirable effect since the number of constraints in an LP, to which separated cutting planes have to be added in order to benefit from closing additional gap, influences the computational effort when solving it.

Remarkably, the significant gap closed by strategy o requires a small number of separated cuts. We note that the side effect of reducing the number of cuts also persists in instances that do not benefit significantly from the separation of wide split cuts in terms of the gap closed. In particular, in 14 out of the remaining 29 instances, the same gap can be closed with a significantly lower number of separated cuts. In the remaining 15 instances, strategy o is often highly competitive and requires a small number of cuts in total to close almost the same gap as strategies w/o and w.

3.3   Cut Strengthening

Ordinary split cuts can be seen to exploit the integrality requirements on the basic variables. In fact, since all non-basic variables are zero in the optimal simplex solution $\hat{x}$, the split violation is determined by the basic variables only. Intuitively, deriving a cutting plane by taking into account the integrality requirements of the non-basic variables as well should result in stronger cutting planes. This idea leads to the integer strengthening principle of intersection cuts from split disjunctions, outlined for example in [1], or recovered from the concept of monoidal strengthening introduced in [4]. An extension of this concept to wide split disjunctions is presented in [13, Sect. 4.2.3]. It turns out, however, that this strengthening is rather weak for wide split cuts, meaning that the improvement of the strengthened cutting plane is marginal. We experienced this weakness computationally: substituting every wide split cut in the experiments in Table 1 by its strengthened version leads to a negligible improvement in almost all cases. Again in [13, Sect. 4.2.3] is developed some interesting and promising theory on the strengthening of wide split cuts when holes in domains are distributed regularly. However, the detection of regularly distributed holes and the corresponding strengthening is outside the scope of this paper.

The aforementioned strengthening principle of intersection cuts from ordinary split disjunctions can be shown to give precisely one of Gomory's Mixed Integer (GMI) cuts [9], see [1]. While the strengthening of wide split cuts is relatively weak, GMI cuts are generally considered to be able to give quite strong improvements. This can be seen as some kind of dilemma. Whenever we have a violated wide split disjunction, it is not clear whether the best strategy is to separate a wide split cut that can then be strengthened only weakly, or to weaken the disjunction to an ordinary split and separate a cut that can then be strengthened strongly to a GMI cut. To analyze this dilemma computationally, we introduce the new cut generation strategy w-g (with GMI cuts instead of wide splits): equal to w, except that for every basic variable that violates a wide split disjunction (in which case with strategy w we separate a

Table 2. GMI vs. wide split cuts on LBP instances

Instance	% gap closed	
	w	w-g
10000-4-100-75	12.53	9.21
10000-4-10-25	34.28	32.96
10000-4-10-75	42.21	34.31
⋮	⋮	⋮
1000-4-40-75	42.90	8.55
1000-4-500-75	5.10	3.04
1000-4-50-50	17.12	13.01
Mean	29.11	17.07

wide split cut), we separate a GMI cut. Table 2 compares the gap closed by strategies w and w-g on the instances of Table 1. Strategy w beats w-g, meaning that GMI cuts are not able to close the gap that can be closed by exploiting the wide split disjunction on these instances.

3.4   MIPLIB2010 Instances

In order to conclude this section, we performed a preliminary experiment on five MIPLIB2010 instances in which, by inspecting the cliques explicitly present in the problem, GUB-Link constraints with an auxiliary integer variable containing holes in its domain can be defined. Table 3 reports the gap closed by adding five rounds of wide split cuts at the end of CPLEX 12.6.1's root node executed in either default

Table 3. Wide split cuts on top of CPLEX cuts

Instance	% gap closed			
	cpx_d	cpx_d+w	cpx_a	cpx_a+w
sp97ar	16.71	18.94	16.71	19.64
leo2	23.57	25.00	23.75	25.56
ns1830653	37.80	37.80	39.65	40.34
blp-ar98	77.86	78.40	80.98	81.27
n3div36	48.74	48.77	49.05	49.05
geo_mean	35.51	36.90	36.24	38.14

(cpx_d) or aggressive cuts (cpx_a) mode. This admittedly limited experiment shows that there seems to be a value in adding wide split cuts on top of CPLEX 12.6.1's existing cut separation procedures, i.e., the information associated with the holes of the integer variables does not seem to be recovered by standard cuts.

4   Branching on Wide Split Disjunctions in a Search Tree

In this final section we present some additional considerations and experimental algorithms on how wide split disjunctions can be useful not only for deriving cutting planes and thus strengthening the LP relaxation of a MILP, but also for solving the MILP to optimality. We assume to have a MILP as in (2)–(4) with the additional condition that some integer variables satisfy simple wide split disjunctions, which is imposed by means of the mixed integer constraints

$$x + Ex_a = g, \quad x_a \in \{0,1\}^k. \tag{20}$$

These constraints can be thought of as big-M or GUB-Link constraints including the auxiliary binary variables x_a. In order to solve the resulting MILP given by (2)–(4) and (20), we explore the following three strategies:

– B&C-F: Apply a MILP solver to the full model (2)–(4), (20).
– B&C-R: Apply a MILP solver to the relaxed model (2)–(4), but whenever an incumbent is found, check for satisfaction of the wide split disjunctions. If some disjunction (e_i^T, π_l, π_u) for x_i is violated, branch with π_l and π_u as new upper and lower bound, respectively: $x_i \leq \pi_l$ OR $x_i \geq \pi_u$.
– B&C-R+cuts: the same as B&C-R with the separation of r rounds of wide split cuts at the root node.

Of course, all three strategies are ways of solving model (2)–(4), (20) exactly. The motivation behind B&C-R is that the model size is reduced with respect to B&C-F, depending on the number of holes, more or less significantly. An additional advantage of B&C-R+cuts is that information used for strengthening the LP relaxation contained in (20) is potentially preserved by wide split cuts.

All three procedures have been implemented in C/C++ with CPLEX 12.6.1 as a MILP-solver. B&C-R and B&C-R+cuts can be implemented using incumbent- and branch-callbacks. We further used randomly generated test instances. In particular, we took all MIP or IP instances from MIPLIB2010 (with problem status *easy* as per January 2016), and randomly generated holes, distributed in the domains of the involved integer variables. Here, we discuss an excerpt of the outcome for 25 such random instances based on the MIPLIB2010 one neos-555424. Table 4 shows solution times in seconds and number of branch-and-bound nodes explored for all three strategies in that case. A time limit of two hours was imposed, the number of rounds r was set to 10 and (20) were encoded by big-M constraints. We immediately note that B&C-R hits the time limit in almost all cases. Essentially, the algorithm keeps finding incumbent solutions that are then rejected. Remarkably, the separation of wide split cuts in B&C-R+cuts consistently avoids this phenomenon. In the second instance for example, a total of two separated cuts is enough to avoid hitting the time limit. More importantly, B&C-R+cuts clearly beats B&C-F, showing how the exploitation of wide split disjunctions in branching and through cuts can lead to significant computational advantages.

The picture of Table 4 does not persist throughout all variations of all the original MIPLIB2010 instances we tested. There are cases in which already B&C-R wins over B&C-F, but also cases in which B&C-F is clearly the winning strategy. In general, there seem to be problems that suffer the step from B&C-F to B&C-R, like in the case of neos-555424, and others that do not. A problem from the literature that belongs to the latter class is the TSPMTW, where B&C-R

Table 4. Comparison of B&C-F, B&C-R and B&C-R+cuts on neos-555424 instances

Instance	B&C-F		B&C-R		B&C-R+cuts	
	time	nodes	time	nodes	time	nodes
1-1	540.78	66302	∞	∞	287.02	39925
1-2	930.40	108000	∞	∞	134.33	19451
1-3	384.16	39605	159.64	25244	158.93	25244
⋮	⋮		⋮		⋮	
5-3	86.06	15599	∞	∞	33.30	7803
5-4	4339.04	137497	∞	∞	17.70	3788
5-5	37.43	9077	∞	∞	6.61	1326
Mean	621.67	51667.00	-	-	107.55	16357.04

led to an average reduction in computing times of around 20% with respect to B&C-F on our testbed. However, B&C-R+cuts does not lead to an additional improvement. This is probably due to the fact that wide split cuts act on the scheduling component of the TSPMTW, leaving the LP relaxation at the root node still very weak due to the TSP component. A future direction of the considerations in this section is therefore the experimentation of the separation of wide split cuts at nodes inside the search tree. Also, there is clearly a more efficient way of replicating strategy B&C-R, that is, by incorporating the branching on wide split disjunctions directly in the branching process of the MILP-solver instead of applying it only when incumbent solutions are found. However, both aspects are outside the scope of this paper.

References

1. Andersen, K., Cornuéjols, G., Li, Y.: Reduce-and-split cuts: Improving the performance of mixed-integer Gomory cuts. Manag. Sci. **51**, 1720–1732 (2005)
2. Andersen, K., Cornuéjols, G., Li, Y.: Split closure and intersection cuts. Math. Program. **102**, 457–493 (2005)
3. Balas, E.: Intersection cuts-a new type of cutting planes for integer programming. Oper. Res. **19**, 19–39 (1971)
4. Balas, E., Jeroslow, R.G.: Strengthening cuts for mixed integer programs. Eur. J. Oper. Res. **4**, 224–234 (1980)
5. Belhaiza, S., Hansen, P., Laporte, G.: A hybrid variable neighborhood tabu search heuristic for the vehicle routing problem with multiple time windows. Comput. Oper. Res. **52**, 269–281 (2014)
6. Cook, W., Kannan, R., Schrijver, A.: Chvátal closures for mixed integer programming problems. Math. Program. **47**, 155–174 (1990)
7. Coughlan, E.T., Lübbecke, M.E., Schulz, J.: A Branch-and-price algorithm for multi-mode resource leveling. In: Festa, P. (ed.) SEA 2010. LNCS, vol. 6049, pp. 226–238. Springer, Heidelberg (2010). doi:10.1007/978-3-642-13193-6_20
8. Furini, F., Ljubić, I., Sinnl, M.: ILP and CP formulations for the lazy bureaucrat problem. In: Michel, L. (ed.) CPAIOR 2015. LNCS, vol. 9075, pp. 255–270. Springer, Cham (2015). doi:10.1007/978-3-319-18008-3_18
9. Gomory, R.E.: An algorithm for integer solutions to linear programs. In: Graves, R.L., Wolfe, P. (eds.) Recent Advances in Mathematical Programming, pp. 269–302. McGraw-Hill, New York (1963)
10. Martello, S., Pisinger, D., Toth, P.: Dynamic programming and strong bounds for the 0–1 knapsack problem. Manag. Sci. **45**, 414–424 (1999)
11. Martello, S., Toth, P.: Knapsack Problems: Algorithms and Computer Implementations. Wiley, UK (1990)
12. Pisinger, D.: http://www.diku.dk/pisinger/codes.html. Accessed 19 Feb 2016
13. Wiese, S: On the interplay of Mixed Integer Linear, Mixed Integer Nonlinear and Constraint Programming. Ph.D. thesis, University of Bologna (2016)

The Saleman's Improved Tours
for Fundamental Classes

Sylvia Boyd[1](✉) and András Sebő[2]

[1] SEECS, University of Ottawa, Ottawa K1N 6N5, Canada
sylvia@site.uottawa.ca
[2] CNRS, Laboratoire G-SCOP, Univ. Grenoble Alpes, Grenoble, France

Abstract. Finding the exact integrality gap α for the LP relaxation of the metric Travelling Salesman Problem (TSP) has been an open problem for over thirty years, with little progress made. It is known that $4/3 \leq \alpha \leq 3/2$, and a famous conjecture states $\alpha = 4/3$. For this problem, essentially two "fundamental" classes of instances have been proposed. This fundamental property means that in order to show that the integrality gap is at most ρ for all instances of metric TSP, it is sufficient to show it only for the instances in the fundamental class.

However, despite the importance and the simplicity of such classes, no apparent effort has been deployed for improving the integrality gap bounds for them. In this paper we take a natural first step in this endeavour, and consider the 1/2-integer points of one such class. We successfully improve the upper bound for the integrality gap from 3/2 to 10/7 for a superclass of these points, as well as prove a lower bound of 4/3 for the superclass.

Our methods involve innovative applications of tools from combinatorial optimization which have the potential to be more broadly applied.

Keywords: TSP · Approximation · Cubic graphs · LP · Integrality gap

1 Introduction

Given the complete graph $K_n = (V_n, E_n)$ on n nodes with non-negative edge costs $c \in \mathbb{R}^{E_n}$, the *Traveling Salesman Problem* (henceforth TSP) is to find a Hamiltonian cycle of minimum cost in K_n. When the costs satisfy the triangle inequality, i.e. $c_{ij} + c_{jk} \geq c_{ik}$ for all $i, j, k \in V_n$, the problem is called the *metric* TSP. If the metric is defined by the shortest (cardinality) paths of a graph, then it is called a *graph-metric*; the TSP specialized to graph-metrics is the *graph-TSP*.

For $G = (V, E)$, $x \in \mathbb{R}^E$ and $F \subseteq E$, $x(F) := \sum_{e \in F} x_e$; for $U \subseteq V$, $\delta(U) := \delta_G(U) := \{uv \in E : u \in U, v \in V \backslash U\}$; $E[U] := \{uv \in E : u \in U, v \in U\}$.

S. Boyd—Partially supported by the National Sciences and Engineering Research Council of Canada. This work was done during visits in Laboratoire G-SCOP, Grenoble; support from the CNRS and Grenoble-INP is gratefully acknowledged.
A. Sebő—Supported by LabEx PERSYVAL-Lab (ANR 11-LABX-0025), équipe-action GALOIS.

© Springer International Publishing AG 2017
F. Eisenbrand and J. Koenemann (Eds.): IPCO 2017, LNCS 10328, pp. 111–122, 2017.
DOI: 10.1007/978-3-319-59250-3_10

A natural linear programming relaxation for the TSP is the following *subtour LP*:

$$\text{minimize } cx \tag{1}$$
$$\text{subject to: } x(\delta(v)) = 2 \qquad \text{for all } v \in V_n, \tag{2}$$
$$x(\delta(S)) \geq 2 \qquad \text{for all } \emptyset \neq S \subsetneq V_n, \tag{3}$$
$$0 \leq x_e \leq 1 \qquad \text{for all } e \in E_n. \tag{4}$$

For a given cost function $c \in \mathbb{R}^{E_n}$, we use $LP(c)$ to denote the optimal solution value for the subtour LP and $OPT(c)$ to denote the optimal solution value for the TSP. The polytope associated with the subtour LP, called the *subtour elimination polytope* and denoted by S^n, is the set of all vectors x satisfying the constraints of the subtour LP, i.e. $S^n = \{x \in \mathbb{R}^{E_n} : x \text{ satisfies } (2), (3), (4)\}$.

The metric TSP is known to be NP-hard. One approach taken for finding reasonably good solutions is to look for a *ρ-approximation algorithm* for the problem, i.e. a polynomial-time algorithm that always computes a solution of value at most ρ times the optimum. Currently the best such algorithm known for the metric TSP is the algorithm due to Christofides [7] for which $\rho = \frac{3}{2}$. Although it is widely believed that a better approximation algorithm is possible, no one has been able to improve upon Christofides algorithm in four decades. For arbitrary nonnegative costs not constrained by the triangle inequality there does not exist a ρ-approximation algorithm for any $\rho \in \mathbb{R}$, unless $P = NP$, since such an algorithm would be able to decide if a given graph is Hamiltonian.

For an approximation guarantee of a minimization problem one needs lower bounds for the optimum, often provided by linear programming. For the TSP a commonly used lower bound is $LP(c)$. Then finding a solution of objective value at most $\rho LP(c)$ in polynomial time implies a ρ-approximation algorithm. The theoretically best possible bound for ρ is the *integrality gap* α for the subtour LP, which is the worst-case ratio between $OPT(c)$ and $LP(c)$ over all metric cost functions c.

It is known that $\alpha \leq \frac{3}{2}$ [19,20], however no example for which the ratio is greater than $\frac{4}{3}$ is known. In fact, a famous conjecture, often referred to as the $\frac{4}{3}$ *Conjecture*, states the following:

Conjecture 1. The integrality gap for the subtour LP is at most $\frac{4}{3}$.

Well-known examples show that α is at least $\frac{4}{3}$. In almost thirty years, there have been no improvements made for the upper bound of $\frac{3}{2}$ or lower bound of $\frac{4}{3}$ for the integrality gap for the subtour LP.

The definition of the integrality gap can be reformulated in terms of a containment relation between two polyhedra that do not depend on the objective function and involve only a sparse subset of (less than $2n$) edges, which is well-known, but not always exploited. We will not only use it here, but it is the very tool that we need.

Define a *tour* to be the edge-set of a spanning Eulerian (connected with all degrees even) multi-subgraph of K_n. If none of the multiplicities can be decreased, then all multiplicities are at most two; however, there are some technical advantages to allowing higher multiplicities. Given a metric cost function, a tour can always be shortcut to a Hamiltonian cycle of the same cost or less.

For any multi-set $J \subseteq E_n$, the *incidence vector of J*, denoted by χ^J, is the vector in $\mathbb{R}^{E_n}$ for which χ_e^J is equal to the number of copies of edge e in J for all $e \in E_n$.

Showing for some constant $\rho \in \mathbb{N}$ that ρx is a convex combination of incidence vectors of tours for each $x \in S^n$ gives an upper bound of ρ on the integrality gap for the subtour LP: it implies that for any cost function $c \in \mathbb{R}^{E_n}$ for which $cx = LP(c)$, at least one of the tours in the convex combination has cost at most $\rho(cx) = \rho LP(c)$. If the costs are metric, this tour can be shortcut to a TSP solution of cost at most $\rho LP(c)$, giving a ratio of $OPT(c)/LP(c) \leq \rho$. The essential part "(ii) implies (i)" of the following theorem asserts that the converse is also true: if ρ is at least the integrality gap then $\rho S^n := \{y \in \mathbb{R}^{E_n} : y = \rho x, x \in S^n\}$ is a subset of the convex hull of incidence vectors of tours:

Theorem 1 [6]. *Let $K_n = (V_n, E_n)$ be the complete graph on n nodes and let $\rho \in \mathbb{R}, \rho \geq 1$. The following statements are equivalent:*

(i) For any weight function $c : E_n \to \mathbb{R}_+ : OPT(c) \leq \rho LP(c)$.
(ii) For any $x \in S^n$, ρx is in the convex hull of incidence vectors of tours.
(iii) For any vertex x of S^n, ρx is in the convex hull of incidence vectors of tours.

So Conjecture 1 can also be reformulated as follows:

Conjecture 2. The polytope $\frac{4}{3} S^n$ is a subset of the convex hull of the incidence vectors of tours.

Given a vector $x \in S^n$, the *support graph* $G_x = (V_n, E_x)$ of x is defined with $E_x = \{e \in E_n : x_e > 0\}$. We call a point $x \in S_n$ $\frac{1}{2}$-*integer* if $x_e \in \{0, \frac{1}{2}, 1\}$ for all $e \in E_n$. For such a vector we call the edges $e \in E_n$ $\frac{1}{2}$-*edges* if $x_e = \frac{1}{2}$ and 1-*edges* if $x_e = 1$. Note that the 1-edges form a set of disjoint paths that we call 1-*paths* of x, and the $\frac{1}{2}$-edges form a set of edge-disjoint cycles we call the $\frac{1}{2}$-*cycles* of x. Cycles and paths are simple (without repetition of nodes) in this article.

For Conjecture 2, it seems that $\frac{1}{2}$-integer vertices play an important role (see [1,5,14]). In fact it has been conjectured by Schalekamp et al. [14] that a subclass of these $\frac{1}{2}$-integer vertices are the ones that give the biggest gap. Here we state their conjecture more broadly:

Conjecture 3. The integrality gap for the subtour LP is reached on $\frac{1}{2}$-integer vertices.

Very little progress has been made on the above conjectures, even though they have been around for a long time and have been well-studied. For the special case of graph-TSP an upper bound of $\frac{7}{5}$ is known for the integrality gap [17]. Conjecture 2 has been verified for the so-called *triangle vertices* $x \in S^n$ for which the values are $\frac{1}{2}$-integer, and the $\frac{1}{2}$-edges form triangles in the support graph [3]. The lower bound of $\frac{4}{3}$ for the integrality gap is provided by triangle vertices with just two triangles.

A concept first introduced by Carr and Ravi [5] (for the 2-edge-connected subgraph problem) is that of a *fundamental class*, which is a class of points F in the subtour elimination polytope with the following property: showing that ρx is in the convex hull of incidence vectors of tours for all vertices $x \in F$ implies the same holds for *all* vertices of the polytope, and thus implies that the integrality gap for the subtour LP is at most ρ.

Two main classes of such vertices have been introduced, one by Carr and Vempala [6], the other by Boyd and Carr [3]. In this paper we will focus on the latter one, i.e. we define a *Boyd-Carr point* to be a point $x \in S^n$ that satisfies the following conditions:

(i) The support graph G_x of x is cubic and 3-edge connected.
(ii) In G_x, there is exactly one 1-edge incident to each node.
(iii) The fractional edges of G_x form disjoint 4-cycles.

A *Carr-Vempala point* is one that satisfies (i), (ii) and instead of (iii) the fractional edges form a Hamiltonian cycle.

Despite their significance and simplicity, no effort has been deployed to exploring new integrality gap bounds for these classes, and no improvement on the general $\frac{3}{2}$ upper bound on the integrality gap has been made for them, not even for special cases. A natural first step in this endeavour is to try to improve the general bounds for the special case of $\frac{1}{2}$-integer Boyd-Carr or Carr-Vempala points.

In this paper we improve the upper bound for the integrality gap from $\frac{3}{2}$ to $\frac{10}{7}$ for $\frac{1}{2}$-integer Boyd-Carr points. In fact we prove this for a superclass of these points. Replacing the 1-edges by paths of arbitrary length between their two endpoints, we get all the $\frac{1}{2}$-integer vectors of S^n for which the $\frac{1}{2}$-edges form disjoint 4-cycles, or *squares* in the support graph. We call these *square points*. We also show that square points contain a subclass for which the integrality gap is at least $\frac{4}{3}$. Note that this subclass is not in the class of $\frac{1}{2}$-integer vertices conjectured by Schalekamp et al. [14] to give the biggest integrality ratio, which makes the class of square points interesting with respect to this conjecture.

In the endeavour to find improved upper bounds on the integrality gap we examine the structure of the support graphs of Boyd-Carr points, which we call *Boyd-Carr graphs*. We show that they are all Hamiltonian, an important ingredient of our bounding of their integrality gap. The proof uses a simple and nice theorem by Kotzig [12] on Eulerian trails with forbidden transitions. An *Eulerian trail* in a graph is a closed walk containing each of its edges exactly once. Note that contrary to tours, it is more than just an edge-set.

Similarly, *Carr-Vempala graphs* are the support graphs of Carr-Vempala points. These are by definition Hamiltonian.

In Sect. 2.1 we show a first, basic application of these ideas, where some parts of the difficulties do not occur. We prove that all edges can be uniformly covered 6/7 times by tours in the support graphs of both fundamental classes. This is better than the conjectured general bound 8/9 that would follow for arbitrary cubic graphs from Conjecture 2.

Another new way of using classical combinatorial optimization for the TSP occurs in Sect. 2.2, where we use an application of Edmonds' matroid intersection theorem to write the optimum x of the subtour elimination polytope as the convex hull of incidence vectors of "rainbow" spanning trees in edge-coloured graphs. The idea of using spanning trees with special structures to get improved results has recently been used successfully in [10] for graph-TSP, and in [11,18] for a related problem, namely the metric $s - t$ path TSP. However, note that we obtain and use our trees in a completely different way.

Our main results concerning the integrality ratio of $\frac{1}{2}$-integer Boyd-Carr points are proved in Sect. 3. We conclude that section by outlining a potential strategy for using the Carr-Vempala points of [6] for proving the $\frac{4}{3}$ Conjecture.

2 Polyhedral Preliminaries and Other Useful Tools

In this section we will discuss some useful and powerful tools that we will need in the proof of our main result in Sect. 3. We begin with some preliminaries.

Given a graph $G = (V, E)$ with a node in V labelled 1, a 1-*tree* is a subset F of E such that $|F \cap \delta(1)| = 2$ and $F \backslash \delta(1)$ forms a spanning tree on $V \backslash \{1\}$. The convex hull of the incidence vectors of 1-trees of G, which we will refer to as the 1-*tree polytope* of the graph G, is given by the following [13]:

$$\{x \in \mathbb{R}^E : x(\delta(1)) = 2, \ x((E[U])) \leq |U| - 1 \text{ for all } \emptyset \neq U \subseteq V \backslash \{1\},$$
$$0 \leq x_e \leq 1 \text{ for all } e \in E, x(E) = |V|\} \tag{5}$$

It is well-known that the 1-trees of a connected graph satisfy the basis axioms of a matroid (see [13]).

Given $G = (V, E)$ and $T \subseteq V$, $|T|$ even, a T-*join* of G is a set $J \subseteq E$ such that T is the set of odd degree nodes of the graph (V, J). A cut $C = \delta(S)$ for some $S \subset V$ is called a T-*cut* if $|S \cap T|$ is odd. We say that a vector *majorates* another if it is coordinatewise greater than or equal to it. The set of all vectors x that majorate some vector y in the convex hull of incidence vectors of T-joins of G is given by the following [9]:

$$\{x \in \mathbb{R}^E : x(C) \geq 1 \text{ for each } T\text{-cut } C, x_e \geq 0 \text{ for al } e \in E\}. \tag{6}$$

This is the T-*join polyhedron* of the graph G.

The following two results are well-known (see [19,20]), but we include the proofs as they illustrate the methods we will use:

Lemma 1 [19,20]. *If $x \in S^n$, then (i) it is a convex combination of incidence vectors of 1-trees of K_n, and (ii) $x/2$ majorates a convex combination of incidence vectors of T-joins of K_n for every $T \subseteq V_n$, $|T|$ even.*

Proof. By using the Eq. (2) of the subtour LP, we see that $x(E_n) = |V_n|$ and that the inequalities (3) can be replaced by $x(E_n[S]) \leq |S| - 1$, for all $\emptyset \neq S \subsetneq V_n$. Thus $x \in S^n$ satisfies all of the constraints of the 1-tree polytope for K_n and (i) of the lemma follows. To check (ii), note that $x/2$ satisfies the constraints of the T-join polyhedron of K_n for all $T \subseteq V_n$, $|T|$ even (in fact $x(C)/2 \geq 1$ on every cut C), that is, it majorates a convex combination of incidence vectors of T-joins.   □

Theorem 2 [19,20]. *If $x \in S^n$, $\frac{3}{2}x$ is in the convex hull of incidence vectors of tours.*

Proof. By (i) of Lemma 1, x is a convex combination of incidence vectors of 1-trees of K_n. Let F be any 1-tree of such a convex combination, and T_F be the set of odd degree nodes in the graph (V_n, F). Then by (ii) of Lemma 1, $x/2$ majorates a convex combination of incidence vectors of T_F-joins. So $\chi^F + x/2$ majorates a convex combination of incidence vectors of tours, and taking the average with the coefficients of the convex combination of 1-trees, we get that $x + x/2$ majorates a convex combination of incidence vectors of tours. Since adding 2 to the multiplicity of any edge in a tour results in another tour, it follows that $\frac{3}{2}x$ is a convex combination of incidence vectors of tours.   □

The tools of the following two subsections are new for the TSP and appear to be very useful.

2.1   Eulerian Trails with Forbidden Bitransitions

Let $G = (V, E)$ be a connected 4-regular multigraph. For any node $v \in V$, a *bitransition* (at v) means a partition of $\delta(v)$ into two pairs of edges. Clearly every Eulerian trail of G *uses* exactly one bitransition at every node, meaning the two disjoint pairs of consecutive edges of the trail at the node. There are 3 bitransitions at every node and the simple theorem below, which follows from a nice result due to Kotzig [12], states that we can forbid one of these and still have an allowed Eulerian trail. As we will show, this provides Hamiltonian cycles containing all the 1-edges of square points.

Theorem 3 [12]. *Let $G = (V, E)$ be a 4-regular connected multigraph with a forbidden bitransition for every $v \in V$. Then G has an Eulerian trail not using the forbidden bitransition of any node.*

Lemma 2. *Let x be any square point, and let $G_x = (V_n, E_x)$ be its support graph. Then G_x has a Hamiltonian cycle H that contains all the 1-edges of G_x.*

Proof. Shrinking all the 1/2-squares of G_x and replacing each path of 1-edges by a single edge, we obtain a 4-regular connected multigraph $G' = (V', E')$ whose edges are precisely the 1-paths of G_x and whose nodes are precisely the squares of G_x. To each contracted square we associate the forbidden bitransition consisting of the pairs of 1-edges incident with the square in G_x which are diagonally opposite to each other, as shown in Fig. 1. By Theorem 3, there is an Eulerian trail K of G' that does not use these forbidden bitransitions. Consecutive edges in K at each node in G' are thus joined by a set of parallel edges in the corresponding square in G_x, and by adding these edges to K and replacing the edges in K with their corresponding 1-paths in G_x, we obtain the desired Hamiltonian cycle for G_x.   □

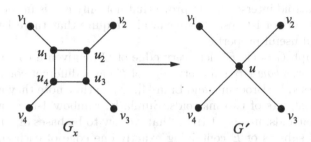

Fig. 1. Shrinking a square in G_x to node u; forbidden: $\{(uv_1, uv_3), (uv_2, uv_4)\}$.

The exhibited connection of Eulerian graphs with forbidden bitransitions sends us to a link on delta-matroids [2] with well-known optimization properties that we wish to explore in a forthcoming work. We content ourselves in this section by providing a simple first application of Lemma 2 which shows a basic idea we will use in the proof of our main result in Sect. 3, without the additional difficulty of the more refined application.

Given a graph $G = (V, E)$ and a value k, we call $y \in \mathbb{R}^{E_{|V|}}$ the *everywhere k vector for G* if $y_e = k$ for all edges $e \in E$ and $y_e = 0$ for all the other edges in the complete graph $K_{|V|}$.

Theorem 4. *If $G = (V, E)$ is cubic, 3-edge-connected and Hamiltonian, so in particular if it is a Boyd-Carr or Carr-Vempala graph, then the everywhere 6/7 vector for G is a convex combination of incidence vectors of tours.*

Proof. Let H be a Hamiltonian cycle of G, and let $M := E \backslash H$ be the perfect matching complementary to H. It can be easily seen that the point $x \in \mathbb{R}^{E_{|V|}}$ defined by $x_e = 1$ if $e \in M$, $x_e = 1/2$ if $e \in H$ and $x_e = 0$ otherwise is in the subtour elimination polytope $S^{|V|}$. By Theorem 2, $\frac{3}{2}x$ is then a convex combination of incidence vectors of tours.

Now take the convex combination $t := \frac{3}{7}\chi^H + \frac{4}{7}\frac{3}{2}x$. Then for edges $e \in M$ we have $t_e = 0 + \frac{4}{7}\frac{3}{2} = \frac{6}{7}$. For edges $e \in H$ we have $t_e = \frac{3}{7} + \frac{4}{7}\frac{3}{2}\frac{1}{2} = \frac{6}{7}$, and $x_e = 0$ for all edges e not in G, finishing the proof. The additional statement follows from the Hamiltonicity of the graphs (by Lemma 2 for Boyd-Carr, and by definition for Carr-Vempala).   □

Note that $\frac{6}{7} < \frac{8}{9}$, where $\frac{8}{9}$ is the value one gets from Conjecture 2 applied to the everywhere $\frac{2}{3}$ vector for G, which is feasible for $S^{|V|}$. However, the problem of whether the everywhere $\frac{8}{9}$ vector is a convex combination of incidence vectors of tours remains open for general cubic 3-edge-connected graphs [16], while the corresponding problem for the $s - t$ path TSP has been solved [18].

2.2   Rainbow 1-trees

We now use matroid intersection to prove that not only is x is in the convex hull of incidence vectors of 1-trees, but we can also require that these 1-trees satisfy some additional useful properties.

Given a graph $G = (V, E)$, let every edge of G be given a colour. We call a 1-tree F of G a *rainbow 1-tree* if every edge of F has a different colour. Rainbow trees are discussed by Broersma and Li in [4], where they note they are the common independent sets of two matroids. Similarly, rainbow 1-trees are common bases of two matroids, namely 1-trees, that we saw to be bases of a matroid (see after (5)), and subsets of E containing exactly one edge of each colour, which are bases of a *partition matroid* [15]. Luckily, $\frac{1}{2}$-integer points of $x \in S^n$ will be readily checked to be in the intersection of the convex hulls of each of these two sets of matroid bases. A Corollary of Edmonds' matroid intersection theorem [8] then presents x as a convex combination of rainbow 1-trees:

Theorem 5. *Let $x \in S^n$ be $\frac{1}{2}$-integer, and let $\mathcal{P}$ be any partition of the $\frac{1}{2}$-edges into pairs. Then x is in the convex hull of incidence vectors of 1-trees that each contain exactly one edge from each pair in $\mathcal{P}$.*

Proof. Let $G_x = (V_n, E_x)$ be the support graph of x. Consider the partition matroid defined on E_x by the partition $\mathcal{P} \cup \{\{e\} : e \in E_x,\ e \text{ is a 1-edge}\}$. By Lemma 1, x is in the convex hull of incidence vectors of 1-trees in E_x; since $x(Q) = 1$ for every class Q of the defined partition matroid, it is also in the convex hull of its bases. Thus by [15, Corollary 41.12d], x is in the convex hull of incidence vectors of the common bases of the two matroids.   □

3   Improved Bounds for 1/2-Integer Points

In this section we show that $\frac{10}{7}x$ is a convex combination of incidence vectors of tours for all square points $x \in S^n$, and thus for all $\frac{1}{2}$-integer Boyd-Carr points x as well. We also analyze the possibility of a similar proof for Carr-Vempala points. We begin by stating two properties which we will later prove to be sufficient to guarantee this for *any* $\frac{1}{2}$-integer vector x in S^n:

(A) The support graph G_x of x has a Hamiltonian cycle H.
(B) Vector x is a convex combination of incidence vectors of 1-trees of K_n, each containing exactly two edges in every cut of G_x consisting of four $\frac{1}{2}$-edges in H.

We will use χ^H of (A) as part of the convex combination for $\frac{10}{7}x$, which is globally good, since H has only n edges, but the $\frac{1}{2}$-edges of H have too high a value (equal to 1), contributing too much in the convex combination. To compensate for this, property (B) ensures that x is not only a convex combination of 1-trees, but these 1-trees are even for certain edge cuts $\delta(S)$, allowing us to use a value essentially less than the $\frac{x}{2} = \frac{1}{4}$ for $\frac{1}{2}$ edges in H for the corresponding T-join. The details of how to ensure we still remain feasible for the T-join polyhedron overall will be given in the proof of Theorem 6.

While condition (A) may look at first sight impossibly difficult to meet, Lemma 2 shows that one can count on the bonus of the naturally arising properties: any square point x satisfies property (A), and the additional property stated in this lemma together with the "rainbow 1-tree decomposition" of Theorem 5 will also imply (B) for square points. The reason we care about the somewhat technical property (B) instead of its more natural consequences is future research: in a new situation we may have to use the most general condition.

Lemma 3. *Let x be any square point. Then x satisfies both (A) and (B).*

Proof. Point x satisfies Property (A) by Lemma 2. Moreover, by the additional statement in this lemma, H contains all the 1-edges in G_x: it follows that H contains a perfect matching from each square of G_x.

Define $\mathcal{P}$ to be the partition of the set of $\frac{1}{2}$-edges of G_x into pairs whose classes are the perfect matchings of squares. Then by Theorem 5, x is in the convex hull of incidence vectors of 1-trees that contain exactly one edge from each pair $P \in \mathcal{P}$. Property (B) follows, since every cut that contains four $\frac{1}{2}$-edges of H is partitioned by two classes $P_1, P_2 \in \mathcal{P}$ by the preceding first paragraph of this proof, and both P_1 and P_2 are met by exactly one edge of each tree of the just constructed convex combination. $\square$

Next we prove that properties (A) and (B) are sufficient to guarantee that $\frac{10}{7}x$ is a convex combination of incidence vectors of tours for *any* $\frac{1}{2}$-integer point of S^n. Recall that properties (A) and (B) are more general than what we need for square points; the condition of the theorem we prove does not require that the Hamiltonian cycle for property (A) contains the 1-edges of G_x, as Lemma 2 asserts for square points. However, we keep the generality of (A) and (B) to remain open to eventual posterior demands of future research:

Theorem 6. *Let $x \in S^n$ be a $\frac{1}{2}$-integer point satisfying properties (A) and (B). Then $\frac{10}{7}x$ is in the convex hull of incidence vectors of tours.*

Proof. Let H be the Hamiltonian cycle of (A) and let $G_x = (V_n, E_x)$ be the support graph of x. Let the 1-trees in the convex combination for property (B) be F_i, $i = 1, 2, ..., k$, and for each tree F_i let T_{F_i} be the set of odd degree nodes in the graph (V_n, F_i). Consider the vector $y \in \mathbb{R}^{E_n}$ defined as follows:

$$y_e = \begin{cases} \frac{1}{6} & \text{if } x_e = \frac{1}{2} \text{ and } e \in H, \\ \frac{1}{3} & \text{if } x_e = \frac{1}{2} \text{ and } e \notin H, \\ \frac{1}{2} & \text{if } x_e = 1 \text{ and } e \in H, \\ \frac{2}{3} & \text{if } x_e = 1 \text{ and } e \notin H, \\ 0 & \text{if } x_e = 0. \end{cases}$$

Claim: Vector y is in the T_{F_i}-join polyhedron for K_n for $i = 1, \ldots, k$.
Let C be a T_{F_i}-cut in K_n for some $i \in \{1, \ldots, k\}$.

Case 1: Cut C contains a 1-edge of G_x. If C contains another 1-edge then $y(C) \geq 1$, as required. Otherwise it contains exactly one 1-edge. Since $x(C) \geq 2$, and the $\frac{1}{2}$-edges in G_x form edge-disjoint cycles, C contains an even (non-zero) number of $\frac{1}{2}$-edges, thus $|C \cap E_x|$ is odd. Since $|H \cap C|$ is even and non-zero, at least one edge e of $C \cap E_x$ is not in H. If e is the single 1-edge in C, then $y(C) \geq \frac{2}{3} + \frac{1}{6} + \frac{1}{6} = 1$. If e is a $\frac{1}{2}$-edge, then $y(C) \geq \frac{1}{2} + \frac{1}{3} + \frac{1}{6} = 1$.

Case 2: Cut C does not contain a 1-edge of G_x. Again using the facts that $x(C) \geq 2$ and the $\frac{1}{2}$-edges in G_x form edge-disjoint cycles, we have that C contains an even number of $\frac{1}{2}$-edges, and $|C \cap E_x| \geq 4$. If $|C \cap E_x| \geq 6$, then $y(C) \geq 6(\frac{1}{6}) = 1$. Otherwise we have $|C \cap E_x| = 4$ and $|C \cap H| = 2$ or 4. If $|C \cap H| = 2$, then $y(C) = 2(\frac{1}{6}) + 2(\frac{1}{3}) = 1$. If $|C \cap H| = 4$, then by property (B), tree F_i has exactly two edges in $C \cap E_x$ (and thus in C as well), which means that $|C \cap F_i|$ is even. Thus C is not a T_{F_i}-cut, so $y(C) \geq 1$ is not required. This completes the proof of the claim.

Using the claim, it follows that $\chi^{F_i} + y$ is in the convex hull of incidence vectors of tours for all $i = 1, \ldots, k$, and therefore $x + y$ is in the convex hull of incidence vectors of tours. Now $z := \frac{1}{7}\chi^H + \frac{6}{7}(x + y)$ is also in the convex hull of incidence vectors of tours, and $z = \frac{10}{7}x$ is easy to check: indeed, the value of $\chi_e^H + 6x_e + 6y_e$ ($e \in E_x$) is apparent from the definition of y_e (above the claim): this value is 5 if $x_e = \frac{1}{2}$, and 10 if $x_e = 1$. □

Our main result is an immediate corollary of this theorem:

Theorem 7. *Let x be a square point. Then $\frac{10}{7}x$ is in the convex hull of incidence vectors of tours. In particular, this holds if x is a $\frac{1}{2}$-integer Boyd-Carr point.*

Proof. By Theorem 6 it is enough to make sure that x satisfies properties (A) and (B), which is exactly the assertion of Lemma 3. □

We can also show that square points are worst-case with respect to Conjecture 2, in that they have an integrality gap of at least $\frac{4}{3}$. Consider the subclass of square points we call k-*donuts*, $k \in \mathbb{Z}$, $k \geq 2$, defined as follows: the support graph $G_x = (V_n, E_x)$ consists of k $\frac{1}{2}$-squares arranged in a circular donut fashion, where the squares are joined by 1-paths, each of length k. In other words, G_x consists of an outer cycle C_{out} and inner cycle C_{in}, both consisting of k paths of $k + 1$ edges, the last of which is a $\frac{1}{2}$-edge, and the others are 1-edges. There are

$2k$ $\frac{1}{2}$-edges between the two cycles so that the $\frac{1}{2}$-edges form squares. In Fig. 2 the support graph of a 4-donut is shown. In the figure, dashed edges represent $\frac{1}{2}$-edges and solid edges represent 1-edges.

We define the cost of each edge in E_x to be 1, except for the $\frac{1}{2}$-edges in each of C_{out} and C_{in} which are defined to have cost k (see the figure, where only edges of cost k are labelled). The costs of other edges of K_n are defined by the metric closure (cost of shortest paths in G_x). For these defined costs $c^{(k)}$, we have $OPT(c^{(k)}) = 4k^2 - 2k + 2$ and $LP(c^{(k)}) = 3k^2 + k$, thus $\lim_{k\to\infty} \frac{OPT(c^{(k)})}{LP(c^{(k)})} = \frac{4}{3}$. Along with Theorem 7, this gives the following:

Corollary 1. *The integrality gap for square points lies between $\frac{4}{3}$ and $\frac{10}{7}$.*

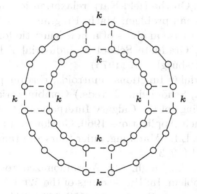

Fig. 2. Graph G_x for a k-donut x, k=4.

We finally discuss the structure of Carr-Vempala points.

Note that for the Boyd-Carr points that have been our focus, the transformation used from general vertices $x \in S^n$ to these Boyd-Carr points does not completely preserve the denominators. In particular, $\frac{1}{2}$-integer vertices of S^n get transformed into Boyd-Carr points x^* with x_e^* values in $\{1, \frac{1}{2}, \frac{3}{4}, \frac{1}{4}, 0\}$. However, for the Carr-Vempala points, general $\frac{1}{2}$-integer vertices of S^n lead to $\frac{1}{2}$-integer Carr-Vempala vertices. In fact we have the following theorem which, if Conjecture 3 is true, would provide a nice approach for proving Conjecture 2, since it is given for free that Carr-Vempala vertices satisfy property (A):

Theorem 8. *If ρx is in the convex hull of incidence vetors of tours for each $\frac{1}{2}$-integer Carr-Vempala point $x \in S^n$, then ρx is in the convex hull of incidence vectors of tours for every $\frac{1}{2}$-integer point $x \in S^n$.*

In light of these results and conjectures it seems worthwhile to study further "fundamental classes" and the role of $\frac{1}{2}$-integer points.

Acknowledgements. We are indebted to Michel Goemans for an email from his sailboat with a pointer to Theorem 5; to Alantha Newman, Frans Schalekamp, Kenjiro Takazawa and Anke van Zuylen for helpful discussions.

References

1. Benoit, G., Boyd, S.: Finding the exact integrality gap for small traveling salesman problems. Math. Oper. Res. **33**(4), 921–931 (2008)
2. Bouchet, A., Cunningham, W.: Delta-matroids, jump systems, and bisubmodular polyhedra. SIAM J. Discrete Math. **8**, 17–32 (1995)
3. Boyd, S., Carr, R.: Finding low cost TSP and 2-matching solutions using certain half-integer subtour vertices. Discrete Optim. **8**, 525–539 (2011)
4. Broersma, H., Li, X.: Spanning trees with many or few colors in edge-colored graphs. Discussiones Math. Graph Theory **17**, 259–269 (1997)
5. Carr, R., Ravi, R.: A new bound for the 2-edge connected subgraph problem. In: Bixby, R.E., Boyd, E.A., Ríos-Mercado, R.Z. (eds.) IPCO 1998. LNCS, vol. 1412, pp. 112–125. Springer, Heidelberg (1998). doi:10.1007/3-540-69346-7_9
6. Carr, R., Vempala, S.: On the Held-Karp relaxation for the asymmetric and symmetric travelling salesman problem. Math. Program. A **100**, 569–587 (2004)
7. Christofides, N.: Worst case analysis of a new heuristic for the traveling salesman problem, Report 388, Graduate School of Industrial Administration, Carnegie-Mellon University, Pittsburgh, PA (1976)
8. Edmonds, J.: Submodular functions, matroids, certain polyhedra. In: Guy, R., Hanani, H., Sauer, N., Schönheim, J. (eds.) Combinatorial Structures and Their Applications; Proceedings of the Calgary International Conference on Combinatorial Structures and Their Applications 1969, Gordon and Breach, New York (1970)
9. Edmonds, J., Johnson, E.L.: Matching, euler tours and the chinese postman. Math. Program. **5**(1), 88–124 (1973)
10. Gharan, S.O., Saberi, A., Singh, M.: A randomized rounding approach to the traveling salesman problem. In: Proceedings of the 52nd Annual IEEE Symposium on Foundations of Computer Science, pp. 550–559 (2011)
11. Gottschalk, C., Vygen, J.: Better s-t -tours by Gao trees. In: Louveaux, Skutella (eds.) Proceedings of the 18th IPCO Conference, Liège, Belgium (2016)
12. Kotzig, A.: Moves without forbidden transitions in a graph. Mat. Casopis Sloven, Akad. Vied **18**, 76–80 (1968)
13. Lawler, E.L., Lenstra, J.K., Rinnooy Kan, A.H.G., Shmoys, D.B. (eds.) The Traveling Salesman Problem - A Guided Tour of Combinatorial Optimization. Wiley, Chichester (1985)
14. Schalekamp, F., Williamson, D., van Zuylen, A.: 2-matchings, the traveling salesman problem, and the subtour LP: a proof of the Boyd-Carr conjecture. Math. Oper. Res. **39**(2), 403–417 (2014)
15. Schrijver, A.: Combinatorial Optimization. Springer, Heidelberg (2003)
16. Sebő, A., Benchetrit, Y., Stehlik, M.: Problems about uniform covers, with tours and detours. Matematisches Forschungsinstitut Oberwolfach Report No. 51/2014, pp. 2912–2915 (2015). doi:10.4171/OWR/2014/51
17. Sebő, A., Vygen, J.: Shorter tours by nice ears. Combinatorica **34**, 597–629 (2014)
18. Sebő, A., van Zuylen, A.: Paths, The salesman's improved paths: a 3/2+1/34 approximation. In: Foundations of Computer Science, (FOCS 2016), October 2016
19. Shmoys, D., Williamson, D.: Analysis of the Held-Karp TSP bound: a monotoncity property with application. Inf. Process. Lett. **35**, 281–285 (1990)
20. Wolsey, L.: Heuristic analysis, linear programming and branch and bound. Math. Program. Study **13**, 121–134 (1980)

The Heterogeneous Capacitated
k-Center Problem

Deeparnab Chakrabarty[1], Ravishankar Krishnaswamy[2], and Amit Kumar[3(✉)]

[1] Department of Computer Science, Dartmouth College, Hanover, USA
deeparnab.chakrabarty@dartmouth.edu
[2] Microsoft Research India, Bengaluru, India
ravishankar.k@gmail.com
[3] Department of Computer Science and Engineering, IIT Delhi, New Delhi, India
amitk@cse.iitd.ac.in

Abstract. In this paper we initiate the study of the *heterogeneous capacitated k-center problem*: we are given a metric space $X = (F \cup C, d)$, and a collection of capacities. The goal is to open each capacity at a unique facility location in F, and also to assign clients to facilities so that the number of clients assigned to any facility is at most the capacity installed; the objective is then to minimize the maximum distance between a client and its assigned facility. If all the capacities c_i's are identical, the problem becomes the well-studied *uniform capacitated k-center problem* for which constant-factor approximations are known [7,22]. The additional choice of determining which capacity should be installed in which location makes our problem considerably different from this problem and the non-uniform generalizations studied thus far in literature. In fact, one of our contributions is in relating the heterogeneous problem to special-cases of the classical *santa-claus problem*. Using this connection, and by designing new algorithms for these special cases, we get the following results for Heterogeneous Cap-k-Center.

- A quasi-polynomial time $O(\log n/\epsilon)$-approximation where every capacity is violated by $(1 + \epsilon)$ factor.
- A polynomial time $O(1)$-approximation where every capacity is violated by an $O(\log n)$ factor.

We get improved results for the *soft-capacities* version where we can place multiple facilities in the same location.

1 Introduction

The capacitated k-center problem is a classic optimization problem where a finite metric space (X, d) needs to be partitioned into k clusters so that every cluster has cardinality at most some specified value L, and the objective is to

---

D. Chakrabarty—This work was done while the author was at Microsoft Research India.

A. Kumar—This work was done while the author was visiting Microsoft Research India.

F. Eisenbrand and J. Koenemann (Eds.): IPCO 2017, LNCS 10328, pp. 123–135, 2017.
DOI: 10.1007/978-3-319-59250-3_11

minimize the maximum intra-cluster distance. This problem was introduced by Bar-Ilan et al. [7] and has many applications [26–28] such as resource allocation in networks (e.g., where to place servers to minimize the total load and client latency), vehicle routing, etc. The basic problem is *homogeneous* in the sizes of the clusters, that is, it has the same cardinality constraint L for each cluster. In certain applications however, a *heterogeneous* version of the problem where we have different cardinality constraints for the clusters might be more applicable. For example, consider the following problem: given a set of demands/clients in a network, and a collection of heterogeneous machines/servers of different capacities, where do we place the servers and how do we assign the clients to servers so that machine capacities are respected and the maximum client-server distance is minimized. Motivated by such applications, we study the worst-case complexity of this heterogenous version in this paper[1].

Definition 1 (The Heterogeneous Cap-k-Center Problem[2]). *We are given a metric space* $(X = F \cup C, d)$ *where* C *and* F *represent the clients and facility locations. We are also given a collection of heterogeneous capacities:* (k_1, c_1), $(k_2, c_2), \ldots, (k_P, c_P)$ *with* k_i *copies of capacity* c_i. *The objective is to install these capacities at unique locations* $F' \subseteq F$, *and find an assignment* $\phi : C \to F'$ *of clients to these locations, such that for any* $i \in F'$ *the number of clients* j *with* $\phi(j) = i$ *is at most the capacity installed at* i, *and* $\max_{j \in C} d(j, \phi(j))$ *is minimized. A weaker version, which we call* Heterogeneous Cap-k-Center *with soft capacities, allows multiple capacities to be installed at the same location.*

Note that when all $c_p = L$ and $\sum_p k_p = k$, we get back the usual capacitated k-center problem. The Heterogeneous Cap-k-Center problem is relevant in many applications where the resources available are heterogenous. The machine placement problem is one example which has applications in network scheduling [20,30] and distributed databases [27,32]. Another example is that of vehicle routing problems with fleets of different speeds [17]. A third relevant application may be clustering; often clusters of equal sizes are undesirable [18] and explicitly introducing heterogeneous constraints might lead to desirable clusters.

For the homogeneous (uniform capacities) problem, Bar-Ilan et al. [7] gave a 10-approximation which was improved to a 6-factor approximation by Khuller and Sussmann [22]. One cannot get a better than 2-approximation even for the *uncapacitated k-center problem* [19]. Recently, some works [1,13] study the *non-uniform capacitated k-center problem* and get constant-factor approximations: in this problem, every facility location $v \in F$ has a *pre-determined* capacity c_v if opened (and 0 otherwise), and the objective is again to minimize the maximum distance of a client to its assigned open facility while opening k facilities.

---

[1] Due to the page limits, we have provided only a very high-level overview of our algorithms and completely omitted the details and all the proofs. We recommend reading the full version available online [12].

[2] Technically, we should call our problem the Heterogeneous Capacitated k-Supplier Problem since we can only open centers in F. However, we avoid making this distinction throughout this paper.

We remark that the non-uniform version and our heterogeneous version seem unrelated in the sense that none is a special case of the other, and moreover, they present different sets of technical challenges to overcome.

1.1   Main Results

As mentioned above, both the uniform and the non-uniform capacitated k-center problems described above admit $O(1)$-approximation. In contrast, we show using a simple reduction that, assuming $P \neq NP$, *no non-trivial approximation exists* for even the soft-version of **Heterogeneous Cap-k-Center**, unless we *violate the capacities*. This observation (a) highlights the technical differences between our problem and these versions studied previously, and (b) motivates us to look at *bicriteria approximations*: an (a, b)-bicriteria approximation approximates the distance objective by a factor of a, while violating capacities by a factor of b.

Theorem 1. *Fix an $\varepsilon > 0$. There exists an $(O(\log n/\varepsilon), (1 + \varepsilon))$-bicriteria approximation algorithm for the **Heterogeneous Cap-k-Center** problem running in time $C_\varepsilon^{\tilde{O}(\log^3 n)}$ for a constant C_ε depending only on ϵ. For **Heterogeneous Cap-k-Center** with soft capacities, there exists an $(O(\log n/\varepsilon), (1 + \varepsilon))$-bicriteria approximation algorithm running in time $n^{O(1/\epsilon)}$.*

We prove the above theorem by reducing the **Heterogeneous Cap-k-Center** problems to a class of *max-min allocation problems* for which we design good algorithms (details appear below). Our next set of results, which also forms one of the main technical contributions of the paper, aims at reducing the logarithmic factor in the approximation to the distance.

Theorem 2. *There is a polynomial time $(O(1), O(\log n))$-bicriteria approximation algorithm for the **Heterogeneous Cap-k-Center** problem.*

Theorem 3. *For any $\delta > 0$, there is a polynomial time $(\tilde{O}(1/\delta), 2+\delta)$-bicriteria approximation for **Heterogeneous Cap-k-Center** problem with soft capacities.*

Connection to Non-uniform Max-Min Allocation Problems. One main finding of this paper is the connection of **Heterogeneous Cap-k-Center** to the *non-uniform* max-min allocation (also known as Santa Claus [6]) problem. We now define these max-min allocation problems using scheduling parlance.

Definition 2 ($Q||C_{min}$ and $Q|f_i|C_{min}$). *In the[3] $Q||C_{min}$ problem, one is given m machines with demands $D_1, \dots, D_m$ and n jobs with capacities $c_1, \dots, c_n$, and the objective is to find an assignment of the jobs to machines satisfying each demand (i.e., the total capacity of jobs assigned to machine i must be at least D_i). In the cardinality constrained non-uniform max-min allocation problem, denoted as the $Q|f_i|C_{min}$ problem, each machine further comes with a cardinality constraint f_i, and a feasible solution cannot allocate more than f_i jobs to machine i. The objective remains the same. An α-approximate feasible solution assigns each machine i total capacity at least D_i/α.*

---

[3] (Ab)using Graham's notation.

To show one side of the connection, we sketch how these problems arise as special cases of Heterogeneous Cap-k-Center, even with soft capacities.

Remark 1 (Reduction from $Q|f_i|C_{min}$). Given an instance $\mathcal{I}$ of $Q|f_i|C_{min}$, construct the instance of Heterogeneous Cap-k-Center as follows. The capacities available to us are precisely the capacities of the jobs in $\mathcal{I}$. The metric space is divided into m groups $(F_1 \cup C_1), \ldots, (F_m \cup C_m)$ such that the distance between nodes in any group is 0 and across groups is 1. Furthermore, for $1 \le i \le m$, $|F_i| = f_i$ and $|C_i| = D_i$. Observe that the Heterogeneous Cap-k-Center instance has a 0-cost, capacity-preserving solution iff $\mathcal{I}$ has a feasible assignment.

Indeed, the strong NP-hardness[4] of $Q|f_i|C_{min}$ and $Q||C_{min}$ shows that we cannot get true approximations for Heterogeneous Cap-k-Center, with or without soft-capacities. As mentioned before, our main technical contribution is in showing a connection in the reverse direction as well. Indeed, our algorithms in Theorems 1 to 3 use the following results we obtain for $Q|f_i|C_{min}$ and $Q||C_{min}$.

Theorem 4. *There is a QPTAS for the $Q|f_i|C_{min}$ problem.*

Theorem 5. *There is a poly-time logarithmic approximation algorithm for $Q|f_i|C_{min}$.*

Theorem 6. *There is a simple greedy 2-approximation algorithm for $Q||C_{min}$.*

To our knowledge, $Q|f_i|C_{min}$ has not been explicitly studied in the literature. However, in a straightforward manner one can reduce $Q|f_i|C_{min}$ to *non-uniform*, restricted-assignment max-min allocation problem (which we denote as $Q|restr|C_{min}$) where, instead of the cardinality constraint dictated by f_i, we restrict jobs to be assigned only to a subset of the machines.[5] Clearly $Q|restr|C_{min}$ is a special case of the general max-min allocation problem [10] and therefore for any $\epsilon > 0$, there are $n^{O(1/\epsilon)}$-time algorithms achieving $O(n^\epsilon)$-approximation. We do not know of any better approximations for $Q|restr|C_{min}$. The *uniform* version $P|restr|C_{min}$ where all demands are the same [6] has several $O(1)$-approximations [3,15,29]. However all these algorithms use the configuration LP, which unfortunately has an integrality gap of $\Omega(\sqrt{n})$ for the non-uniform version $Q|restr|C_{min}$ (see the full version for details).

1.2 Outline of Techniques

As mentioned before, we obtain our results by reducing Heterogeneous Cap-k-Center to the $Q|f_i|C_{min}$ problem (complementing the *from* reduction discussed in

---

[4] A simple reduction from 3-dimensional matching shows NP-hardness of $Q|f_i|C_{min}$ and $Q||C_{min}$ even when the demands and capacities are polynomially bounded.

[5] The reduction proceeds as follows: for every machine i and job j, restrict j to be assigned to i iff $c_j \ge D_i/2f_i$. It is not hard to see that a ρ-approximation for the $Q|restr|C_{min}$ implies a 2ρ-approximation for the $Q|f_i|C_{min}$ instance.

Remark 1). We provide two reductions – the first incurs logarithmic approximation to the cost but uses black-box algorithms for $Q|f_i|C_{min}$, the second incurs $O(1)$-approximation to the cost but uses "LP-based" algorithms for $Q|f_i|C_{min}$.

Warm-up: Weak Decompostion. Given a Heterogeneous Cap-k-Center instance, suppose we *guess* the optimal objective value, which we can assume to be 1 after scaling. Then, we construct a graph connecting client j with facility location i iff $d(i,j) \leq 1$. Then, starting at an arbitrary client and using a simple region-growing technique (like those used for the graph cut problems [16,23]), we can find a set of clients J_1 of along with their neighboring facility locations $T_1 = \Gamma(J_1)^6$, such that: (a) the diameter of J_1 is $O(\log n/\epsilon)$, and (b) the number of additional clients in the boundary $|\Gamma(T_1)\backslash J_1|$ is at most $\epsilon|J_1|$. Now, we simply *delete* these boundary clients and charge them to J_1, incurring a capacity violation of $(1 + \epsilon)$. Moreover, note that in an optimal solution, *all* the clients in J_1 *must be* assigned to facilities opened in T_1. Using this fact, we define our first demand in the $Q|f_i|C_{min}$ instance by $D_1 = |J_1|$ and $f_1 = |T_1|$. Repeating this process, we get a collection of pairs $\{(J_i, T_i)\}$ which naturally defines our $Q|f_i|C_{min}$ instance. It is then easy to show that an α-approximation to this instance then implies an $(O(\log n/\epsilon), \alpha(1 + \epsilon))$-bicriteria algorithm for Heterogeneous Cap-k-Center.

LP-Based Strong Decompostion. To get $O(1)$-approximations, we resort to linear programming relaxations. Indeed, one can write the natural LP relaxation (L1)–(L6) described in Sect. 2 – the relaxation has y_{ip} variables which denote opening a facility with capacity c_p at i. Armed with a feasible solution to the LP, we prove a *stronger decomposition theorem* (Theorem 7): we show that we can delete a set of clients C_{del} which can be charged to the remaining ones, and then partition the remaining clients and facilities into *two* classes. One class $\mathcal{T}$ is the so-called *complete neighborhood sets* of the form $\{(J_i, T_i)\}$ with $\Gamma(J_i) \subseteq T_i$ as described above — we define our $Q|f_i|C_{min}$ instance using these sets. The other class $\mathcal{S}$ is of, what we call, *roundable* sets (Definition 3). Roundable sets have "enough" y-mass such that installing as many capacities as prescribed by the LP (rounded down to the nearest integer) supports the total demand incident on the set (with a $(1 + \epsilon)$-factor capacity violation). Moreover, the diameter of any of these sets constructed is $\tilde{O}(1/\epsilon)$.

Technical Roadblock. It may seem that the above decomposition theorem implies a reduction to the $Q|f_i|C_{min}$ problem – the class $\mathcal{T}$ defines a $Q|f_i|C_{min}$ instance and we can use black-box algorithms, while the roundable sets in $\mathcal{S}$ are taken care of almost by definition. The nub of the problem lies in the *supply* of capacities to each of these classes. Indeed, the $Q|f_i|C_{min}$ instance formed from $\mathcal{T}$ must have a solution if the Heterogeneous Cap-k-Center problem is feasible, *but only if all the k_p copies of capacity c_p are available to it*. However, we have already used up some of these copies to take care of the $\mathcal{S}$ sets, and what we actually have available for $\mathcal{T}$ is what the *LP prescribes*. In fact, this natural LP

---

^{6} For $S \subseteq C \cup F$, $\Gamma(S)$ denotes the neighboring vertices of S.

relaxation (and the natural LP for $Q|f_i|C_{min}$) have arbitrarily bad integrality gaps, even for bicriteria algorithms.

The Supply Polyhedra. We circumvent this issue in the following manner: the above method would be fine if the supply of facility capacities prescribed by the LP to the complete-neighborhood sets in $\mathcal{T}$ can approximately satisfy the demands in the corresponding $Q|f_i|C_{min}$ instance. This motivates us to define *supply polyhedra* for $Q|f_i|C_{min}$. Informally, the supply polyhedron (Definition 6) of a $Q|f_i|C_{min}$ instance is supposed to capture all the vectors $(s_1, \ldots, s_n)$ such that s_j copies of capacity c_j can approximately satisfy the demands of all the machines. Conversely, any vector in this polyhedron should also be a feasible (or approximately feasible) supply vector for this instance.

If such an object $\mathcal{P}$ existed, then we could strengthen our natural LP relaxation as follows. For *every* collection $\mathcal{T}$ of complete-neighborhood sets, we add a constraint (described as (L7)) stating that the fractional capacity allocated to the facilities in $\mathcal{T}$ should lie in the supply polyhedron of the corresponding $Q|f_i|C_{min}$ instance. Note that this LP has exponentially many constraints, and it is not clear how to solve it. However, we can use the "round-and-cut" framework (of inferring a separating hyperplane if our rounding fails, and then using the ellipsoid algorithm overall) exploited earlier in many papers [2,8,9,14,24,25].

Using this decomposition, in Theorem 8, we effectively reduce **Heterogeneous Cap-k-Center** to the task of designing good supply polyhedra for $Q|f_i|C_{min}$.

Supply Polyhedron for $Q|f_i|C_{min}$ and $Q||C_{min}$. Do good supply polyhedra exist for $Q|f_i|C_{min}$ or even the simpler $Q||C_{min}$ problem? On the positive side, we can show that the natural assignment LP is a 2-approximate supply polyhedron for $Q||C_{min}$. For $Q|f_i|C_{min}$ we describe a supply polyhedron based on the *configuration LP* and prove that it is $O(\log D)$-approximate (Theorem 9) where D is the ratio of maximum and minimum demand. This along with our strong decomposition proves Theorem 2. We note that this also implies a *polynomial time* $O(\log D)$-approximation algorithm for the $Q|f_i|C_{min}$ problem, improving considerably over the guarantees implied by the current santa-claus algorithms. We complement this by showing (in the full version) that the integrality gap of the configuration LP is $\Omega(\log D/\log\log D)$, using which we also show a *lower-bound* on the approximation factor possible using supply polyhedra: any supply polyhedra for $Q|f_i|C_{min}$ must violate the demands by $\Omega(\log D/\log\log D)$. This shows that our approach inherently needs to violate capacities by this factor.

1.3 Related Work

Capacitated Location problems have a rich literature although most of the work has focused on versions where each facility arrives with a predetermined capacity and the decision process is to whether open a facility or not. We have already mentioned the state of the art for capacitated k-center problems. For the capacitated facility location problem a 5-approximation is known via local search [5], while more recently an $O(1)$-approximate *LP-based* algorithm was proposed [2].

All these are true approximation algorithms in that they do not violate capacities. It is an outstanding open problem to obtain true approximations for the capacitated k-median problem. The best known algorithm is the recent work of Demirci and Li [14] who for any $\epsilon > 0$ give a ploy($1/\epsilon$)-approximate algorithm violating the capacities by $(1 + \epsilon)$-factor. The technique of this algorithm and its precursors [2, 24, 25] are similar to ours in that they follow the round-and-cut strategy to exploit exponential sized linear programming relaxations.

The $Q|f_i|C_{min}$ problem is a cardinality constrained max-min allocation problem. There has been some work in the scheduling literature on cardinality-constrained min-max problem. When all the machines are identical, the problem is called the k_i-partitioning problem [4]. When the number of machines is a constant, Woeginger [34] gives a FPTAS for the problem, and the best known result is a 1.5-approximation due to Kellerer and Kotov [21]. To the best of our knowledge, the related machines case where machines have different speeds has not been looked at. When the machines are unrelated, Saha and Srinivasan [31] showed a 2-approximation; in fact this follows from the Shmoys-Tardos rounding of the assignment LP [33].

As we have discussed above, the Heterogeneous Cap-k-Center problem behaves rather differently than the usual homogeneous capacitated k-center problem. This distinction in complexity when we have heterogeneity in resource is a curious phenomenon which deserves more attention. A previous work [11] of the first two authors (with P. Goyal) looked at the (uncapacitated) k-center problem where the heterogeneity was in the radius of the balls covering the metric space. As in our work, even for that problem one needs to resort to bicriteria algorithms where the two criteria are cost and *number* of centers opened. That paper gives an $(O(1), O(1))$-bicriteria approximation algorithm. In contrast, we do not wish to violate the number of capacities available at all (in fact, the problem is considerably easier if we are allowed to do so – we do not expand on this any further).

2 Preliminaries

Given a Heterogeneous Cap-k-Center instance, we start by guessing OPT. We either prove OPT is infeasible, or find an (a, b)-bicriteria approximate allocation of clients to facilities. We define the bipartite graph $G = (F \cup C, E)$ where $(i, j) \in E$ iff $d(i, j) \leq$ OPT. If OPT is feasible, then the following assignment LP (L1)–(L6) must have a feasible solution. In this LP, we have opening variables y_{ip} for every $i \in F, p \in [P]$ indicating whether we open a facility with capacity c_p at location i. Recall that the capacities available to us are $c_1, c_2, \ldots, c_P$ – a facility with capacity c_p installed on it will be referred to as a *type p facility*. We have connection variables x_{ijp} indicating the fraction to which client $j \in C$ connects to a facility at location i where a type p facility has been opened. We force $x_{ijp} = 0$ for all pairs i, j and type p such that $d(i, j) >$ OPT.

$$\forall j \in C, \sum_{i \in F} \sum_{p \in [P]} x_{ijp} \geq 1 \tag{L1}$$

$$\forall i \in F, p \in [P], \sum_{j \in C} x_{ijp} \leq c_p y_{ip} \tag{L2}$$

$$\forall p \in [P], \sum_{i \in F} y_{ip} \leq k_p \tag{L3}$$

$$\forall i \in F, j \in C, p \in [P], \qquad\qquad x_{ijp} \leq y_{ip} \tag{L4}$$

$$\forall i \in F, \qquad\qquad \sum_{p \in [P]} y_{ip} \leq 1 \tag{L5}$$

$$\forall i \in F, j \in C, p \in [P], \qquad\qquad x_{ijp}, y_{ip} \geq 0 \tag{L6}$$

We say a solution (x, y) is (a, b)-feasible if it satisfies (L1), (L3)–(L6), and (L2) with the RHS replaced by $bc_p y_{ip}^{\text{int}}$, and $x_{ijp} > 0$ only if $d(i, j) \leq a \cdot \text{OPT}$.

Claim. Given an (a, b)-feasible solution (x, y^{int}) where $y_{ip}^{\text{int}} \in \{0, 1\}$, we can get an (a, b)-approximate solution to the Heterogeneous Cap-k-Center problem.

We remark that as it is, the LP has an unbounded integrality gap for Heterogeneous Cap-k-Center, and indeed, the gap instances also happen to be of the $Q|f_i|C_{min}$ variety. So we strengthen it by adding some additional constraints which we explain later. However, since our strong decomposition theorem merely uses these y_{ip} and x_{ijp} values, we present that first.

Definition 3 (Roundable Sets). *A set of facilities $S \subseteq F$ is said to be (a, b)-roundable w.r.t (x, y) if*

(a) $\text{diam}_G(S) \leq a$
(b) *there exists a rounding $y_{ip}^{\text{int}} \in \{0, 1\}$ for all $i \in S, p \in [P]$ such that*
   1. $\sum_{q \geq p} \sum_{i \in S} y_{iq}^{\text{int}} \leq \lfloor \sum_{q \geq p} \sum_{i \in S} y_{iq} \rfloor$ *for all p, and*
   2. $\sum_{j \in C} \sum_{i \in S, p \in [P]} x_{ijp} \leq b \cdot \sum_{i \in S} \sum_{p \in [P]} c_p y_{ip}^{\text{int}}$

So if we can partition the facilities into roundable sets with reasonable parameters, we would be done. It turns out that sets which are not roundable have a *non-expanding structure*, and indeed we define our $Q|f_i|C_{min}$ instance over such sets. The following definition comes handy in this case.

Definition 4 (Complete Neighborhood Sets). *A subset $T \subseteq F$ of facilities is called a* complete neighborhood *if there exist clients $J \subseteq C$ such that $\Gamma(J) \subseteq T$. In this case J is said to be* responsible *for T. Additionally, a complete neighborhood T is said to be an α-complete neighborhood if $\text{diam}(T) \leq \alpha$.*

If we find a complete neighborhood T of facilities with a set J of clients responsible for it, then we know that the optimal solution *must satisfy* all the demand in J by suitably opening facilities of sufficient capacity in T. Thus, if we can partition the entire instance into a collection $\mathcal{T} = (T_1, \ldots, T_m)$ of disjoint α-complete neighborhood sets with J_i responsible for T_i, we can define an instance $\mathcal{I}$ of $Q|f_i|C_{min}$ with m machines with demands $D_i = |J_i|$ and cardinality constraint $f_i = |T_i|$, and there are n_i jobs of capacities c_i for $1 \leq i \leq P$.

Our next definition is that of (τ, ρ)-*deletable clients* that can be removed from the instance since they can be "ρ-charged" to the remaining clients that are at most τ-away. In other words, such clients (with constant values of τ and ρ) can be safely removed at the expense of violating capacities and increasing objective value by small factors.

Definition 5 (Deletable Clients). *A subset $C_{\text{del}} \subseteq C$ of clients is (τ, ρ)-deletable if there exists a mapping $\phi_{j,j'} \in [0,1]$ for $j \in C_{\text{del}}$ and $j' \in C \setminus C_{\text{del}}$ satisfying (a) $\sum_{j' \in C \setminus C_{\text{del}}} \phi_{j,j'} = 1$ for all $j \in C_{\text{del}}$, and (b) $\sum_{j \in C_{\text{del}}} \phi_{j,j'} \leq \rho$ for all $j' \in C \setminus C_{\text{del}}$. Furthermore, $\phi_{j,j'} > 0$ only if $d(j,j') \leq \tau \cdot \mathsf{OPT}$.*

We now state our decomposition result.

Theorem 7 (Decomposition Theorem). *Given a feasible solution (x, y) to $LP(L1)$–$(L6)$, and $\delta > 0$, there is a polynomial time algorithm which finds a solution x satisfying (L2) and (L4), and a decomposition as follows.*

1. *The facility set F is partitioned into two families $\mathcal{S} = (S_1, S_2, \ldots, S_K)$ and $\mathcal{T} = (T_1, T_2, \ldots, T_L)$ of mutually disjoint subsets. The client set C is partitioned into three disjoint subsets $C = C_{\text{del}} \cup C_{\text{black}} \cup C_{\text{blue}}$ where C_{del} is a $(\tilde{O}(1/\delta), \delta)$-deletable subset.*
2. *Each $S_k \in \mathcal{S}$ is $(\tilde{O}(1/\delta), (1 + \delta))$-roundable with respect to (x, y), and moreover, each client in C_{blue} satisfies $\sum_{i \in \mathcal{S}, p} \mathsf{x}_{ijp} \geq 1 - \frac{\delta}{100}$.*
3. *Each T_ℓ is a $\tilde{O}(1/\delta)$-complete neighborhood with a corresponding set J_ℓ of clients responsible for it, and $C_{\text{black}} = \cup_{\ell=1}^{L} J_\ell$.*

In general, our decomposition theorem only ensures that we can partition the instance into sets which are either roundable or are complete neighborhoods (after removing the deletable clients), and the crux of the rounding algorithm lies in combining the two cases while meeting the k_i bounds for all capacities.

Our final ingredient is that of supply polyhedra. Recall that an instance of $Q|f_i|C_{min}$ has m machines M with demands $D_1, \ldots, D_m$ and cardinality constraints $f_1, \ldots, f_m$, and n jobs J with capacities $c_1, \ldots, c_n$ respectively. Now, we generalize this in the following manner: A *supply vector* $(s_1, \ldots, s_n)$ where each s_j is a non-negative integer is called *feasible* for this instance if the ensemble formed by s_j copies of jobs of capacity c_j can satisfy all the demands. The *supply polyhedra* then desires to capture these feasible supply vectors.

Definition 6 (Supply Polyhedron). *Given an instance $\mathcal{I}$ for a max-min allocation problem, a polyhedron $\mathcal{P}(\mathcal{I})$ is called an α-approximate supply polyhedron if (a) all feasible supply vectors lie in $\mathcal{P}(\mathcal{I})$, and (b) given any non-negative integer vector $(s_1, \ldots, s_n) \in \mathcal{P}(\mathcal{I})$ there exists an assignment of the s_j jobs of capacity c_j to the machines such that machine i receives capacity $\geq D_i/\alpha$.*

Ideally, we would like *exact* supply polyhedra, and one choice would be the convex hull of all the feasible supply vectors; indeed this is the tightest polytope satisfying condition (a). Unfortunately, there are instances of

$Q|f_i|C_{min}$ where the convex hull contains infeasible integer points for which $\alpha = \Theta(\log n / \log \log n)$.

3 Heterogeneous Cap-k-Center via Supply Polyhedra

In this section, we prove the following theorem.

Theorem 8. *Suppose there exists β-approximate supply polyhedra for all instances of $Q|f_i|C_{min}$ (resp., $Q||C_{min}$) which have γ-approximate separation oracles. Then for any $\delta \in (0,1)$, there is an $\left(\tilde{O}(1/\delta), \gamma\beta(1 + 5\delta)\right)$-bicriteria approximation algorithm for Heterogeneous Cap-k-Center (resp., with soft capacities).*

Our results for Heterogeneous Cap-k-Center follow from Theorem 8 and results about supply polyhedra. For example, Theorem 2 follows from Theorem 8 (using $\delta = 0.5$, say) and Theorem 9, and also noting that $D_{max}/D_{min} \leq n$ in our reduction. The proof of Theorem 8 is based on the decomposition theorem.

Proof (**Proof Sketch of Theorem 8**). Let us first describe an approach which fails. Let (x, y) be a feasible solution to LP (L1)–(L6), and apply Theorem 7. Although the sets in $\mathcal{S}$ by definition are roundable which takes care of the clients in C_{blue}, the issue arises in assigning clients of C_{black}. In particular, $y_p^T := \sum_{i \in T} y_{ip}$ for all $1 \leq p \leq P$ which indicates the "supply" of capacity c_p available for the C_{black} clients. However, this may not be enough for serving all these clients (even with violation). That is, the vector y^T may not lie in the (approximate) supply polyhedra of the $Q|f_i|C_{min}$ instance defined by T. That we fail is not surprising; after all, we have so far only used the natural LP which has a bad integrality gap. To resolve this issue, we strengthen the LP by *explicitly requiring y^T to be in the supply polyhedra*. Since we do not know T before solving the LP (after all our LP rounding generated it), we enforce this for *all* collections of complete-neighborhood sets. More precisely, for $T := (T_1, \ldots, T_L)$ of L disjoint complete neighborhood sets, let $\mathcal{I}_T$ denote the associated $Q|f_i|C_{min}$ demands.

$$\forall T := (T_1, \ldots, T_L) \text{ disjoint neighborhood subsets,} \quad y^T \in \mathcal{P}(\mathcal{I}_T) \qquad \text{(L7)}$$

Note that this is a feasible constraint to add to LP (L1)–(L6). In the OPT solution, for any T there must be enough supply dedicated for the clients responsible for these complete neighborhood sets. We don't know how (and don't expect) to check feasibility of (L7) for all collections T. However, we can still run ellipsoid method using the "round-and-cut" framework of [8,9,24,25]. To begin with, we start with the LP (L1)–(L6) and obtain feasible solution (x, y). Subsequently, we apply the decomposition Theorem 7 to obtain the collection $T = (T_1, \ldots, T_L)$. We then check if $y^T \in \mathcal{P}(\mathcal{I}_T)$ or not. Since we have a γ-approximate separation oracle for $\mathcal{P}(\mathcal{I}_T)$, we are either guaranteed that $y^T \in \mathcal{P}(\mathcal{I}_T')$ where the ℓth demand is now D_ℓ/γ; or we get a hyperplane separating y^T from $\mathcal{P}(\mathcal{I}_T)$ which also gives us a hyperplane separating y from LP (L1)–(L7). This can be fed to the ellipsoid algorithm to obtain a new (x, y) and the above process is repeated.

When this process stops, we will have a solution (x, y) such that the supply $\{y_p^T\}$ lies in the supply polyhedra $\mathcal{P}(\mathcal{I}_T)$. So our overall algorithm is to simply round the roundable sets (by rounding down), and solve the instance $\mathcal{I}_T$ with the supply vector $\{y_p^T\}$ using a suitable $Q|f_i|C_{min}$ algorithm.

We end the main body by noting that the configuration LP relaxation is in fact nearly the best possible supply polyhedra for $Q|f_i|C_{min}$.

Theorem 9. *For any instance $\mathcal{I}$ of $Q|f_i|C_{min}$, the natural configuration LP for $\mathcal{I}$ is an $O(\log D)$-approximate supply polyhedron with $(1 + \epsilon)$-approximate separation oracle for any $\varepsilon > 0$, where $D := D_{\max}/D_{\min}$. Moreover, there exists no supply polyhedra with approximation $o(\log D/\log\log D)$.*

4 Conclusion

In this paper we introduced and studied the Heterogeneous Cap-k-Center problem, and highlighted its connection to an interesting special case of the max-min allocation problems, namely $Q|f_i|C_{min}$. In our main result, we showed, using a decomposition theorem and the notion of supply polyhedra, a logarithmic approximation for $Q|f_i|C_{min}$, using which we showed a bicriteria $(O(1), O(\log n))$-approximation for Heterogeneous Cap-k-Center. We believe designing polynomial-time $O(1)$-approximations for $Q|f_i|C_{min}$ and bicriteria $(O(1), O(1))$ algorithms for Heterogeneous Cap-k-Center are very interesting open problems.

References

1. An, H.-C., Bhaskara, A., Chekuri, C., Gupta, S., Madan, V., Svensson, O.: Centrality of trees for capacitated k-center. In: Lee, J., Vygen, J. (eds.) IPCO 2014. LNCS, vol. 8494, pp. 52–63. Springer, Cham (2014). doi:10.1007/978-3-319-07557-0_5
2. An, H., Singh, M., Svensson, O.: Lp-based algorithms for capacitated facility location. In: 55th IEEE FOCS 2014, Philadelphia, PA, USA, 18–21 October 2014, pp. 256–265 (2014)
3. Asadpour, A., Feige, U., Saberi, A.: Santa claus meets hypergraph matchings. ACM Trans. Algorithms **8**(3), 24 (2012)
4. Babel, L., Kellerer, H., Kotov, V.: Thek-partitioning problem. Math. Meth. OR **47**(1), 59–82 (1998)
5. Bansal, M., Garg, N., Gupta, N.: A 5-approximation for capacitated facility location. In: Epstein, L., Ferragina, P. (eds.) ESA 2012. LNCS, vol. 7501, pp. 133–144. Springer, Heidelberg (2012). doi:10.1007/978-3-642-33090-2_13
6. Bansal, N., Sviridenko, M.: The santa claus problem. In: Proceedings of the 38th Annual ACM STOC 2006, pp. 31–40 (2006)
7. Bar-Ilan, J., Kortsarz, G., Peleg, D.: How to allocate network centers. J. Algorithms **15**(3), 385–415 (1993)
8. Carr, R.D., Fleischer, L., Leung, V.J., Phillips, C.A.: Strengthening integrality gaps for capacitated network design and covering problems. In: Proceedings of the Eleventh Annual ACM-SIAM SODA, San Francisco, CA, USA, 9–11 January 2000, pp. 106–115 (2000)

9. Chakrabarty, D., Chekuri, C., Khanna, S., Korula, N.: Approximability of capacitated network design. In: Günlük, O., Woeginger, G.J. (eds.) IPCO 2011. LNCS, vol. 6655, pp. 78–91. Springer, Heidelberg (2011). doi:10.1007/978-3-642-20807-2_7
10. Chakrabarty, D., Chuzhoy, J., Khanna, S.: On allocating goods to maximize fairness. In: 50th Annual IEEE FOCS 2009, Atlanta, Georgia, USA, 25–27 October 2009, pp. 107–116 (2009)
11. Chakrabarty, D., Goyal, P., Krishnaswamy, R.: The non-uniform k-center problem. In: 43rd ICALP 2016, pp. 67:1–67:15 (2016)
12. Chakrabarty, D., Krishnaswamy, R., Kumar, A.: The heterogeneous capacitated k-center problem. CoRR, abs/1611.07414 (2016)
13. Cygan, M., Hajiaghayi, M., Khuller, S.: LP rounding for k-centers with non-uniform hard capacities. In: 53rd Annual IEEE Symposium on Foundations of Computer Science, FOCS 2012, New Brunswick, NJ, USA, 20–23 October 2012, pp. 273–282 (2012)
14. Demirci, H.G., Li, S.: Constant approximation for capacitated k-median with (1+epsilon)-capacity violation. In: 43rd ICALP 2016, pp. 73:1–73:14 (2016)
15. Feige, U.: On allocations that maximize fairness. In: Proceedings of the Nineteenth Annual ACM-SIAM SODA 2008, pp. 287–293 (2008)
16. Garg, N., Vazirani, V.V., Yannakakis, M.: Approximate max-flow min-(multi)cut theorems and their applications. SIAM J. Comput. 25(2), 235–251 (1996)
17. Gørtz, I.L., Molinaro, M., Nagarajan, V., Ravi, R.: Capacitated vehicle routing with nonuniform speeds. Math. Oper. Res. 41(1), 318–331 (2016)
18. Guha, S., Rastogi, R., Shim, K.: Cure: an efficient clustering algorithm for large databases. Inf. Syst. 26(1), 35–58 (2001)
19. Hochbaum, D.S., Shmoys, D.B.: A best possible heuristic for the k-center problem. Math. Oper. Res. 10(2), 180–184 (1985)
20. Im, S., Moseley, B.: Scheduling in bandwidth constrained tree networks. In: Proceedings of the 27th ACM on Symposium on Parallelism in Algorithms and Architectures, SPAA 2015, Portland, OR, USA, 13–15 June 2015, pp. 171–180 (2015)
21. Kellerer, H., Kotov, V.: A 3/2-approximation algorithm for 3/2-partitioning. Oper. Res. Lett. 39(5), 359–362 (2011)
22. Khuller, S., Sussmann, Y.J.: The capacitated K-center problem. SIAM J. Discrete Math. 13(3), 403–418 (2000)
23. Leighton, F.T., Rao, S.: Multicommodity max-flow min-cut theorems and their use in designing approximation algorithms. J. ACM 46(6), 787–832 (1999)
24. Li, S.: On uniform capacitated k-median beyond the natural LP relaxation. In: Proceedings of the Twenty-Sixth Annual ACM-SIAM SODA 2015, pp. 696–707 (2015)
25. Li, S.: Approximating capacitated k-median with $(1 + \epsilon)k$ open facilities. In: Proceedings of the Twenty-Seventh Annual ACM-SIAM Symposium on Discrete Algorithms, SODA 2016, Arlington, VA, USA, 10–12 January 2016, pp. 786–796 (2016)
26. Lupton, R., Maley, F.M., Young, N.E.: Data collection for the sloan digital sky survey - a network-flow heuristic. J. Algorithms 27(2), 339–356 (1998)
27. Morgan, H.L., Levin, K.D.: Optimal program and data locations in computer networks. Commun. ACM 20(5), 315–322 (1977)
28. Murthy, K., Kam, J.B., Krishnamoorthy, M.S.: An approximation algorithm to the file allocation problem in computer networks. In: PODS (1983)
29. Polácek, L., Svensson, O.: Quasi-polynomial local search for restricted max-min fair allocation. ACM Trans. Algorithms 12(2), 13 (2016)

30. Qiu, Z., Stein, C., Zhong, Y.: Minimizing the total weighted completion time of coflows in datacenter networks. In: Proceedings of the 27th ACM SPAA 2015, pp. 294–303 (2015)
31. Saha, B., Srinivasan, A.: A new approximation technique for resource-allocation problems. In: Innovations in Computer Science, ICS 2010, pp. 342–357 (2010)
32. Sen, G., Krishnamoorthy, M., Rangaraj, N., Narayanan, V.: Exact approaches for static data segment allocation problem in an information network. Comput. Oper. Res. **62**, 282–295 (2015)
33. Shmoys, D.B., Tardos, É.: An approximation algorithm for the generalized assignment problem. Math. Program. **62**, 461–474 (1993)
34. Woeginger, G.J.: A comment on scheduling two parallel machines with capacity constraints. Discrete Optim. **2**, 269–272 (2005)

Local Guarantees in Graph Cuts and Clustering

Moses Charikar[1], Neha Gupta[1(✉)], and Roy Schwartz[2]

[1] Stanford University, Stanford, CA 94305, USA
{moses,nehagupta}@cs.stanford.edu
[2] Technion, 3200003 Haifa, Israel
schwartz@cs.technion.ac.il

Abstract. Correlation Clustering is an elegant model that captures fundamental graph cut problems such as Min $s - t$ Cut, Multiway Cut, and Multicut, extensively studied in combinatorial optimization. Here, we are given a graph with edges labeled $+$ or $-$ and the goal is to produce a clustering that agrees with the labels as much as possible: $+$ edges within clusters and $-$ edges across clusters. The classical approach towards Correlation Clustering (and other graph cut problems) is to optimize a global objective. We depart from this and study local objectives: minimizing the maximum number of disagreements for edges incident on a single node, and the analogous max min agreements objective. This naturally gives rise to a family of basic min-max graph cut problems. A prototypical representative is Min Max $s - t$ Cut: find an $s - t$ cut minimizing the largest number of cut edges incident on any node. We present the following results: (1) an $O(\sqrt{n})$-approximation for the problem of minimizing the maximum total weight of disagreement edges incident on any node (thus providing the first known approximation for the above family of min-max graph cut problems), (2) a remarkably simple 7-approximation for minimizing local disagreements in complete graphs (improving upon the previous best known approximation of 48), and (3) a $1/(2+\epsilon)$-approximation for maximizing the minimum total weight of agreement edges incident on any node, hence improving upon the $1/(4+\epsilon)$-approximation that follows from the study of approximate pure Nash equilibria in cut and party affiliation games.

Keywords: Approximation algorithms · Graph cuts · Correlation clustering · Linear programming

1 Introduction

Graph cuts are extensively studied in combinatorial optimization, including fundamental problems such as Min $s - t$ Cut, Multiway Cut, and Multicut. Typically, given an undirected graph $G = (V, E)$ equipped with non-negative edge weights

M. Charikar and N. Gupta—Supported by NSF grants CCF-1617577, CCF-1302518 and a Simons Investigator Award.
R. Schwartz—Supported by ISF grant 1336/16.

F. Eisenbrand and J. Koenemann (Eds.): IPCO 2017, LNCS 10328, pp. 136–147, 2017.
DOI: 10.1007/978-3-319-59250-3_12

$c : E \to \mathcal{R}_+$ the goal is to find a *constrained* partition $\mathcal{S} = \{S_1, \ldots, S_\ell\}$ of V minimizing the total weight of edges crossing between different clusters of $\mathcal{S}$. e.g., in Min $s - t$ Cut, $\mathcal{S}$ has two clusters, one containing s and the other containing t. Similarly, in Multiway Cut, $\mathcal{S}$ consists of k clusters each containing exactly one of k given special vertices $t_1, \ldots, t_k$. In Multicut, the clusters of $\mathcal{S}$ must separate k given pairs of special vertices $\{s_i, t_i\}_{i=1}^k$.

The elegant model of Correlation Clustering captures all of the above fundamental graph cut problems, and was first introduced by Bansal *et al.* [5] more than a decade ago. In Correlation Clustering, we are given an undirected graph $G = (V, E)$ equipped with non-negative edge weights $c : E \to \mathcal{R}_+$. Additionally, E is partitioned into E^+ and E^-, where edges in E^+ (E^-) are considered to be labeled as $+$ ($-$). The goal is to find a partition of V into an *arbitrary* number of clusters $\mathcal{S} = \{S_1, \ldots, S_\ell\}$ that agrees with the edges' labeling as much as possible: the endpoints of $+$ edges are supposed to be placed in the same cluster and endpoints of $-$ edges in different clusters. Typically, the objective is to find a clustering that minimizes the total weight of misclassified edges. This models, *e.g.*, Min $s - t$ Cut, since one can label all edges in G with $+$, and add (s, t) to E with a label of $-$ and set its weight to $c_{s,t} = \infty$ (Multiway Cut and Multicut are modeled in a similar manner).

Correlation Clustering has been studied extensively for more than a decade [1,2,9,10,13,26]. In addition to the simplicity and elegance of the model, its study is also motivated by a wide range of practical applications: image segmentation [26], clustering gene expression patterns [3,7], cross-lingual link detection [25], and the aggregation of inconsistent information [15], to name a few (refer to the survey [26] and the references therein for additional details).

Departing from the classical global objective approach towards Correlation Clustering, we consider a broader class of objectives that allow us to bound the number of misclassified edges incident on any node (or alternatively edges classified correctly). We refer to this class as Correlation Clustering with *local guarantees*. First introduced by Puleo and Milenkovic [20], Correlation Clustering with local guarantees naturally arises in settings such as community detection without antagonists, *i.e.*, objects that are inconsistent with large parts of their community, and has found applications in diverse areas, *e.g.*, recommender systems, bioinformatics, and social sciences [11,18,20,24].

Local Minimization of Disagreements and Graph Cuts. A prototypical example when considering minimization of disagreements with local guarantees is the Min Max Disagreements problem, whose goal is to find a clustering that minimizes the maximum total weight of misclassified edges incident on any node. Formally, given a partition $\mathcal{S} = \{S_1, \ldots, S_\ell\}$ of V, for $u \in S_i$, define:

$$\text{disagree}_\mathcal{S}(u) \triangleq \sum_{v \notin S_i : (u,v) \in E^+} c_{u,v} + \sum_{v \in S_i : (u,v) \in E^-} c_{u,v} .$$

The objective of Min Max Disagreements is: $\min_\mathcal{S} \max_{u \in V} \{\text{disagree}_\mathcal{S}(u)\}$. This is NP-hard even on complete unweighted graphs and approximations are known for only a few special cases [20]. No approximation is known for general graphs.

Just as minimization of total disagreements in Correlation Clustering models fundamental graph cut problems, Min Max Disagreements gives rise to a variety of basic min-max graph cut problems. A natural problem here is Min Max $s - t$ Cut: Its input is identical to that of Min $s - t$ Cut, however its objective is to find an $s - t$ cut $(S, \overline{S})$ minimizing the total weight of cut edges incident on any node: $\min_{S \subseteq V : s \in S, t \notin S} \max_{u \in V} \{\sum_{v:(u,v) \in \delta(S)} c_{u,v}\}$.[1] Despite the fact that Min Max $s - t$ Cut is a natural graph cut problem, no approximation is known for it. Min Max Disagreements also gives rise to Min Max Multiway Cut and Min Max Multicut, defined similarly; no approximation is known for these. One of our goals is to highlight this family of min-max graph cut problems which we believe deserve further study. Other graph cut problems were studied from the min-max perspective, e.g., [6,22]. However, the goal there is to find a constrained partition that minimizes the total weight of cut edges incident on any *cluster* (as opposed to incident on any *node*).

Min Max Disagreements is a special case of the more general Min Local Disagreements problem. Given a clustering $\mathcal{S}$, consider the vector of all disagreement values $\text{disagree}_{\mathcal{S}}(V) \in \mathcal{R}_+^V$, where $(\text{disagree}_{\mathcal{S}}(V))_u = \text{disagree}_{\mathcal{S}}(u) \; \forall u \in V$. The objective of Min Local Disagreements is to find a partition $\mathcal{S}$ that minimizes $f(\text{disagree}_{\mathcal{S}}(V))$ for a given function f. For example, if f is the max function Min Local Disagreements reduces to Min Max Disagreements, and if f is the summation function Min Local Disagreements reduces to the classic objective of minimizing total disagreements.

Local Maximization of Agreements. Another natural objective of Correlation Clustering is that of maximizing the total weight of edges correctly classified [5,23]. A prototypical example for local guarantees is Max Min Agreements, *i.e.* finding a clustering that maximizes the minimum total weight of correctly classified edges incident on any node. Formally, given a partition $\mathcal{S} = \{S_1, \ldots, S_\ell\}$ of V, for $u \in S_i$, define:

$$\text{agree}_{\mathcal{S}}(u) \triangleq \sum_{v \in S_i : (u,v) \in E^+} c_{u,v} + \sum_{v \notin S_i : (u,v) \in E^-} c_{u,v} .$$

The objective of Max Min Agreements is: $\max_{\mathcal{S}} \min_{u \in V} \{\text{agree}_{\mathcal{S}}(u)\}$.

This is a special case of the more general Max Local Agreements problem. Given a clustering $\mathcal{S}$, consider the vector of all agreement values $\text{agree}_{\mathcal{S}}(V) \in \mathcal{R}_+^V$, where $(\text{agree}_{\mathcal{S}}(V))_u = \text{agree}_{\mathcal{S}}(u) \; \forall u \in V$. The objective of Max Local Agreements is to find a partition $\mathcal{S}$ that maximizes $g(\text{agree}_{\mathcal{S}}(V))$ for a given function g, where, g is required to satisfy the following two conditions: (1) for any $\mathbf{x}, \mathbf{y} \in \mathcal{R}_+^V$ if $\mathbf{x} \leq \mathbf{y}$ then $g(\mathbf{x}) \leq \mathbf{g(y)}$ (monotonicity), and (2) $g(\alpha \mathbf{x}) \geq \alpha g(\mathbf{x})$ for any $\alpha \geq 0$ and $\mathbf{x} \in \mathcal{R}_+^V$ (reverse scaling). Note that g is not required to be concave. For example, if g is the min function Max Local Agreements reduces to Max Min Agreements, and if g is the summation function Max Local Agreements reduces to the classic objective of maximizing total agreements.

Max Local Agreements is closely related to the computation of local optima for Max Cut, and the computation of pure Nash equilibria in cut and party

---

[1] $\delta(S)$ denotes the collection of edges crossing the cut $(S, \overline{S})$.

affiliation games [4,8,12,14,21] (a well studied special class of potential games [19]). In the setting of party affiliation games, each node of G is a player that can choose one of two sides of a cut. The player's payoff is the total weight of edges incident on it that are classified correctly. It is well known that such games admit a pure Nash equilibria via the *best response dynamics* (also known as *Nash dynamics*), and that each such pure Nash equilibrium is a $(1/2)$-approximation for Max Local Agreements. Unfortunately, in general the computation of a pure Nash equilibria in cut and party affiliation games is PLS-complete [17], and thus it is widely believed no polynomial time algorithm exists for solving this task. Nonetheless, one can apply the algorithm of Bhalgat *et al.* [8] for finding an approximate pure Nash equilibrium and obtain a $1/(4+\varepsilon)$-approximation for Max Local Agreements (for any constant $\varepsilon > 0$). This approximation is also the best known for the special case of Max Min Agreements.

Our Results. Focusing first on Min Max Disagreements on general graphs we prove that both the natural LP and SDP relaxations admit a large integrality gap of $n/2$. Nonetheless, we present an $O(\sqrt{n})$-approximation for Min Max Disagreements, bypassing the above integrality gaps.

Theorem 1. *The natural LP and SDP relaxations for Min Max Disagreements have an integrality gap of $n/2$.*

Theorem 2. *Min Max Disagreements admits an $O(\sqrt{n})$-approximation for general weighted graphs.*

Since Min Max $s-t$ Cut, along with Min Max Multiway Cut and Min Max Multicut, are a special case of Min Max Disagreements, Theorem 2 applies to them as well, thus providing the first known approximation for this family of cut problems.[2]

When considering the more general Min Local Disagreements problem, we present a remarkably simple approach that achieves an improved approximation of 7 for both complete graphs and complete bipartite graphs (where disagreements are measured w.r.t one side only). This improves upon and simplifies [20] who presented an approximation of 48 for the former and 10 for the latter.

Theorem 3. *Min Local Disagreements admits a 7-approximation for complete graphs.*
where f is required to satisfy the following three conditions: (1) for any $\mathbf{x}, \mathbf{y} \in \mathcal{R}_+^V$ if $\mathbf{x} \leq \mathbf{y}$ then $f(\mathbf{x}) \leq f(\mathbf{y})$ (monotonicity), (2) $f(\alpha\mathbf{x}) \leq \alpha f(\mathbf{x})$ for any $\alpha \geq 0$ and $\mathbf{x} \in \mathcal{R}_+^V$ (scaling), and (3) f is convex.

Theorem 4. *Min Local Disagreements admits a 7-approximation for complete bipartite graphs where disagreements are measured w.r.t. one side of the graph. where f is required to satisfy the following three conditions: (1) for any $\mathbf{x}, \mathbf{y} \in \mathcal{R}_+^V$ if $\mathbf{x} \leq \mathbf{y}$ then $f(\mathbf{x}) \leq f(\mathbf{y})$ (monotonicity), (2) $f(\alpha\mathbf{x}) \leq \alpha f(\mathbf{x})$ for any $\alpha \geq 0$ and $\mathbf{x} \in \mathcal{R}_+^V$ (scaling), and (3) f is convex.*

---

[2] Theorem 1 can be easily adapted to apply also for Min Max $s-t$ Cut, Min Max Multiway Cut, and Min Max Multicut, resulting in a gap of $(n-1)/2$.

Focusing on local maximization of agreements, we present a $1/(2+\varepsilon)$ approximation for Max Min Agreements without any assumption on the edge weights. This improves upon the previous known $1/(4+\varepsilon)$-approximation that follows from the computation of approximate pure Nash equilibria in party affiliation games [8]. As before, we show that both the natural LP and SDP relaxations for Max Min Agreements have a large integrality gap of $\frac{n}{2(n-1)}$.

Theorem 5. *For any $\varepsilon > 0$, Max Min Agreements admits a $1/(2+\varepsilon)$-approximation for general weighted graphs, where the running time of the algorithm is $poly(n, 1/\varepsilon)$.*

Theorem 6. *The natural LP and SDP relaxations for Max Min Agreements have an integrality gap of $\frac{n}{2(n-1)}$.*

Our main algorithmic results are summarized in Table 1.

Table 1. Results for Correlation Clustering with local guarantees.

Problem	Input graph	Approximation	
		This work	Previous work
Min Local Disagreements	Complete	7	48 [20]
	Complete bipartite (one sided)	7	10 [20]
Min Max Disagreements	General weighted	$O(\sqrt{n})$	–
Min Max $s - t$ Cut	General weighted	$O(\sqrt{n})$	–
Min Max Multiway Cut			
Min Max Multicut			
Max Min Agreements	General weighted	$1/(2+\varepsilon)$	$1/(4+\varepsilon)$ [8]

Approach and Techniques. The non-linear nature of Correlation Clustering with local guarantees makes problems in this family much harder to approximate than Correlation Clustering with classic global objectives.

Firstly, LP and SDP relaxations are not always useful when considering local objectives. For example, the natural LP relaxation for the global objective of minimizing total disagreements on general graphs has a bounded integrality gap of $O(\log n)$ [9,13,16]. However, we prove that for its local objective counterpart, *i.e.*, Min Max Disagreements, both the natural LP and SDP relaxations have a huge integrality gap of $n/2$ (Theorem 1). To overcome this our algorithm for Min Max Disagreements on general weighted graphs uses a *combination* of the LP lower bound and a combinatorial bound. Even though each of these bounds on its own is bad, we prove that their combination suffices to obtain an approximation of $O(\sqrt{n})$, thus bypassing the huge integrality gaps of $n/2$.

Secondly, randomization is inherently difficult to use for local guarantees, while many of the algorithms for minimizing total disagreements, *e.g.*, [1,2,10], as well as maximizing total agreements, *e.g.*, [23], are all randomized in nature. The reason is that a bound on the expected weight of misclassified edges incident on any node does not translate to a bound on the maximum of this quantity over all nodes (similarly the expected weight of correctly classified edges incident on any node does not translate to a bound on the minimum of this quantity over all nodes). To overcome this difficulty, all the algorithms we present are deterministic, *e.g.*, for Min Local Disagreements we propose a new remarkably simple method of clustering that greedily chooses a center node s^* and cuts a sphere of a fixed and predefined radius around s^*, and for Max Min Agreements we present a new *non-oblivious* local search algorithm that runs on a graph with modified edge weights and circumvents the need to compute approximate pure Nash equilibria in party affiliation games.

Paper Organization. Section 2 contains the improved approximations for Min Max Disagreements on general weighted graphs and for Min Local Disagreements on complete and complete bipartite graphs (Theorems 2, 3, and 4), along with the integrality gaps of the natural LP and SDP relaxations (Theorem 1). Section 3 contains the improved approximation for Max Min Agreements as well as the integrality gaps of the natural LP and SDP relaxations (Theorems 5 and 6).

2   Local Minimization of Disagreements and Graph Cuts

We consider the natural convex programming relaxation for Min Local Disagreements. The relaxation imposes a metric d on the vertices of the graph. For each node $u \in V$ we have a variable $D(u)$ denoting the total fractional *disagreement* of edges incident on u. Additionally, we denote by $\mathbf{D} \in \mathcal{R}_+^V$ the vector of all $D(u)$ variables. Note that the relaxation is solvable in polynomial time since f is convex.[3]

$$\min \quad f(\mathbf{D}) \tag{1}$$

$$\sum_{v:(u,v)\in E^+} c_{u,v} d(u,v) + \sum_{v:(u,v)\in E^-} c_{u,v}(1 - d(u,v)) = D(u) \qquad \forall u \in V$$

$$d(u,v) + d(v,w) \geq d(u,w) \qquad \forall u,v,w \in V$$

$$D(u) \geq 0, \ 0 \leq d(u,v) \leq 1 \qquad \forall u,v \in V$$

For the special case of Min Max Disagreements, *i.e.*, f is the max function, (1) can be written as an LP. The proof of Theorem 1, which states that even for the special case of Min Max Disagreements the above natural LP and in addition the natural SDP both have a large integrality gap of $n/2$, appears in the full version of this paper. We note that Theorem 1 also applies to Min Max $s - t$ Cut, a further special case of Min Max Disagreements.

2.1   Min Max Disagreements on General Weighted Graphs

Our algorithm for Min Max Disagreements on general weighted graphs cannot rely solely on the the lower bound of the LP relaxation, since it admits an integrality gap of $n/2$ (Theorem 1). Thus, a different lower bound must be used.

---

[3] The convexity of f is used only to show that relaxation (1) can be solved, and it is not required in the rounding process.

Let $c_{\max}$ be the maximum weight of an edge that is misclassified in some optimal solution $\mathcal{S}^*$. Clearly, $c_{\max}$ also serves as a lower bound on the value of an optimal solution. Hence, we can mix these two lower bounds and choose $\max\{\max_{u \in V}\{D(u)\}, c_{\max}\}$ to be the lower bound we use. Note that we can assume w.l.o.g. that $c_{\max}$ is known to the algorithm, as one can simply execute the algorithm for every possible value of $c_{\max}$ and return the best solution.

Our algorithm consists of two main phases. In the first we compute the LP metric d but require additional constraints that ensure no *heavy* edge, *i.e.*, an edge e having $c_e > c_{\max}$, is (fractionally) misclassified by d. In the second phase, we perform a careful *layered clustering* of an auxiliary graph consisting of all $+$ edges whose length in the metric d is short. At the heart of the analysis lies a distinction between $+$ edges whose length in the metric d is short and all other edges. The contribution of the former is bounded using the combinatorial lower bound, *i.e.*, $c_{\max}$, whereas the contribution of the latter is bounded using the LP. Our algorithm also ensures that in the final clustering no heavy edge is misclassified. Let us now elaborate on the two phases, before providing an exact description of the algorithm (Algorithm 1).

Phase 1 (constrained metric computation). Denote by,

$$E_{\text{heavy}}^+ \triangleq \{e \in E^+ : c_e > c_{\max}\} \quad \text{and} \quad E_{\text{heavy}}^- \triangleq \{e \in E^- : c_e > c_{\max}\}$$

the collection of all heavy $+$ and $-$ edges, respectively. We solve the LP relaxation (1) (recall that f is the max function) while adding the following additional constraints that ensure d does not (fractionally) misclassify heavy edges:

$$d(u,v) = 0 \qquad \forall e = (u,v) \in E_{\text{heavy}}^+ \tag{2}$$

$$d(u,v) = 1 \qquad \forall e = (u,v) \in E_{\text{heavy}}^- \tag{3}$$

If no feasible solution exists then our current guess for $c_{\max}$ is incorrect.

Phase 2 (layered clustering). Denote the collections of $+$ and $-$ edges which are *almost* classified correctly by d as $E_{\text{bad}}^+ \triangleq \{e = (u,v) \in E^+ : d(u,v) < 1/\sqrt{n}\}$ and $E_{\text{bad}}^- \triangleq \{e = (u,v) \in E^- : d(u,v) > 1 - 1/\sqrt{n}\}$, respectively. Intuitively, any edge $e \notin E_{\text{bad}}^+ \cup E_{\text{bad}}^-$ can use its length d to pay for its contribution to the cost, regardless of what the output is. This is not the case with edges in E_{bad}^+ and E_{bad}^-, therefore all such edges are considered *bad*. Additionally, denote by $E_0^+ \triangleq \{e = (u,v) \in E^+ : d(u,v) = 0\}$ the collection of $+$ edges for which d assigns a length of 0.[4]

We design the algorithm so it ensures that no mistakes are made for edges in E_0^+ and E_{bad}^-. However, the algorithm might make mistakes for edges in E_{bad}^+, thus a careful analysis is required. To this end we consider the auxiliary graph consisting of all edges in E_{bad}^+, *i.e.*, $G_{\text{bad}}^+ \triangleq (V, E_{\text{bad}}^+)$, and equip it with the distance function dist_ℓ defined as the shortest path metric with respect to the length function $\ell : E_{\text{bad}}^+ \to \{0,1\}$:

---

[4] Note that $E_{\text{heavy}}^+ \subseteq E_0^+ \subseteq E_{\text{bad}}^+$ and $E_{\text{heavy}}^- \subseteq E_{\text{bad}}^-$.

$$\ell(e) \triangleq \begin{cases} 0 & e \in E_0^+ \\ 1 & e \in E_{\text{bad}}^+ \setminus E_0^+ \end{cases}$$

Assume E_{bad}^- contains k edges and denote the endpoints of the ith edge by s_i and t_i. The algorithm partitions every connected component X of G_{bad}^+ into clusters as follows: as long as X contains s_i and t_i for some i, we examine the layers $\text{dist}_\ell(s_i, \cdot)$ defines and perform a carefully chosen level cut. This *layered clustering* suffices as we can prove that our choice of a level cut ensures (1) no mistakes are made for edges in E_0^+ and E_{bad}^-, and (2) the *number* of misclassified edges from $E_{\text{bad}}^+ \setminus E_0^+$ incident on any node is at most $O(\sqrt{n})$. This ends the description of the second phase.

Algorithm 1. Layered Clustering $(G = (V, E), c_{\max})$

1: $\mathcal{C} \leftarrow \emptyset$.
2: let d be a solution to LP (1) with the additional constraints (2) and (3)
3: **for** every connected component X in G_{bad}^+ **do**
4:     **while** X contains $\{s_i, t_i\}$ for some i **do**
5:         $r_i \leftarrow \text{dist}_\ell(s_i, t_i)$ and $L_j^i \leftarrow \{u : \text{dist}_\ell(s_i, u) = j\}$ for every $j = 0, 1, \ldots, r_i$.
6:         choose $j^* \leq (\sqrt{n}-1)/2$ s.t. $|L_{j^*}^i|, |L_{j^*+1}^i|, |L_{j^*+2}^i| \leq 16\sqrt{n}$.
7:         $S \leftarrow \cup_{j=0}^{j^*} L_j^i$.
8:         $X \leftarrow X \setminus S$ and $\mathcal{C} \leftarrow \mathcal{C} \cup \{S\}$.
9:     **end while**
10:    $\mathcal{C} \leftarrow \mathcal{C} \cup \{X\}$.
11: **end for**
12: Output $\mathcal{C}$.

Refer to Algorithm 1 for a precise description of the algorithm. The following Lemma states that the distance between any $\{s_i, t_i\}$ pair with respect to the metric dist_ℓ is large, its proof appears in the full version of this paper.

Lemma 1. *For every* $i = 1, \ldots, k$, $\text{dist}_\ell(s_i, t_i) > \sqrt{n} - 1$.

The following Lemma simply states that only a few layers could be too large, its proof appears in the full version of this paper. It implies Corollary 1, whose proof appears in the full version of this paper.

Lemma 2. *For every* $i = 1, \ldots, k$, *the number of layers* L_j^i *for which* $|L_j^i| > 16\sqrt{n}$ *is at most* $\sqrt{n}/16$.

Corollary 1. *Algorithm 1 can always find* j^* *as required.*

Lemma 3 proves that no mistakes are made for edges in E_0^+ and E_{bad}^-, whereas Lemma 4 bounds the *number* of misclassified edges from $E_{\text{bad}}^+ \setminus E_0^+$ incident on any node. Their proofs appear in the full version of this paper.

Lemma 3. *Algorithm 1 never misclassifies edges in* E_0^+ *and* E_{bad}^-.

Lemma 4. *Let $u \in V$ and S be the cluster in $\mathcal{C}$ Algorithm 1 assigned u to. Then,*
$$\left| \{ e \in E_{bad}^+ \setminus E_0^+ : e = (u,v), v \notin S \} \right| \leq 48\sqrt{n}.$$

We are now ready to prove the main result, Theorem 2.

Proof (of Theorem 2). We prove that Algorithm 1 achieves an approximation of $49\sqrt{n}$. The proof considers edges according to their type: (1) E_0^+ and E_{bad}^- edges, (2) $E_{bad}^+ \setminus E_0^+$ edges, and (3) all other edges. It is worth noting that the contribution of edges of type (2) is bounded using the combinatorial lower bound, *i.e.*, c_{max}, whereas the contribution of edges of type (3) is bounded using the LP, *i.e.*, $D(u)$ for every node $u \in V$ (as defined by the relaxation (1)).

First, consider edges of type (1). Lemma 3 implies Algorithm 1 does not make any mistakes with respect to these edges, thus their contribution to the value of the output $\mathcal{C}$ is always 0. Second, consider edges of type (2). Lemma 4 implies that every node u has at most $48\sqrt{n}$ edges of type (2) incident on it that are classified incorrectly. Additionally, the weight of every edge of type (2) is at most c_{max} since $E_{heavy}^+ \subseteq E_0^+$ and edges of type (2) do not contain any edge of E_0^+. Thus, we can conclude that for every node u the total weight of edges of type (2) that touch u and are misclassified is at most $48\sqrt{n} \cdot c_{max}$.

Finally, consider edges of type (3). Fix an arbitrary node u and let $D(u)$ be the fractional disagreement value the LP assigned to u (see (1)). Edge e of type (3) is either an edge $e \in E^+$ whose d length is at least $1/\sqrt{n}$, or an edge $e \in E^-$ whose d length is at most $1 - 1/\sqrt{n}$. Hence, in any case the fractional contribution of such an edge e to $D(u)$ is at least $c_e/\sqrt{n}$. Therefore, regardless of what the output is, the total weight of misclassified edges of type (3) incident on u is at most $\sqrt{n} \cdot D(u)$.

Summing over all types of edges, we can conclude that the total weight of misclassified edges incident on u in $\mathcal{C}$ (the output of Algorithm 1) is at most $48\sqrt{n}c_{max} + \sqrt{n} \cdot D(u)$. Since both c_{max} and $D(u)$ are lower bounds on the value of an optimal solution, the proof is concluded. $\qquad\square$

2.2 Min Local Disagreements on Complete Graphs

We consider a simple deterministic greedy clustering algorithm for complete graphs that iteratively partitions the graph. In every step it does the following: (1) greedily chooses a center node s^* that has many nodes *close* to it, and (2) removes from the graph a sphere around s^* which constitutes a new cluster. The greedy choice of s^* is similar to that of [20]. However, our algorithm departs from the approach of [20], as it *always* cuts a large sphere around s^*. The algorithm of [20], on the other hand, outputs either a singleton cluster containing s^* or some other large sphere around s^* (the average distance within the large sphere determines which of the two options is chosen), thus mimicking the approach of [9]. Surprisingly, restricting the algorithm's choice enables us not only to obtain a simpler algorithm, but also to improve upon the approximation guarantee from 48 to 7.

Algorithm 2 receives as input the metric d as computed by the relaxation (1), whereas the variables $D(u)$ are required only for the analysis. Additionally, we

---

Algorithm 2. Greedy Clustering ($\{d(u,v)\}_{u,v \in V}$)

1: $S \leftarrow V$ and $\mathcal{C} \leftarrow \emptyset$.
2: **while** $S \neq \emptyset$ **do**
3:     $s^* \leftarrow \text{argmax} \{|\text{Ball}_S(s, 1/7)| : s \in S\}$.
4:     $\mathcal{C} \leftarrow \mathcal{C} \cup \{\text{Ball}_S(s^*, 3/7)\}$.
5:     $S \leftarrow S \setminus \text{Ball}_S(s^*, 3/7)$.
6: **end while**
7: Output $\mathcal{C}$.

---

denote by $\text{Ball}_S(u, r) \triangleq \{v \in S : d(u,v) < r\}$ the sphere of radius r around u in subgraph S.

The following lemma summarizes the guarantee achieved by Algorithm 2 (its proof appears in the full version of this paper, which also contains an overview of our charging scheme).

Lemma 5. *Assuming the input is a complete graph, Algorithm 2 guarantees that* $\text{disagree}_{\mathcal{C}}(u) \leq 7D(u)$ *for every* $u \in V$.

Proof (of Theorem 3). Apply Algorithm 2 to the solution of the relaxation (1). Lemma 5 guarantees that for every node $u \in V$ we have that $\text{disagree}_{\mathcal{C}}(u) \leq 7D(u)$, *i.e.*, $\text{disagree}_{\mathcal{C}}(V) \leq 7\mathbf{D}$. The value of the output of the algorithm is $f(\text{disagree}_{\mathcal{C}}(V))$ and one can bound it as follows:

$$f(\text{disagree}_{\mathcal{C}}(V)) \overset{(1)}{\leq} f(7\mathbf{D}) \overset{(2)}{\leq} 7f(\mathbf{D}).$$

Inequality (1) follows from the monotonicity of f, whereas inequality (2) follows from the scaling property of f. This concludes the proof since $f(\mathbf{D})$ is a lower bound on the value of any optimal solution. $\qquad\square$

2.3 Min Local Disagreements on Complete Bipartite Graphs

Our algorithm for Min Local Disagreements on complete bipartite graphs (with one sided disagreements) is a natural extension of Algorithm 2. Similarly to the complete graph case, we are able to present a remarkably simple algorithm achieving an improved approximation of 7. The description of the algorithm and the proof of Theorem 4 appear in the full version of this paper.

3   Local Maximization of Agreements

As previously mentioned, Max Local Agreements is closely related to the computation of local optima for Max Cut and pure Nash equilibria in cut and party affiliation games, both of which are PLS-complete problems. We focus on the special case of Max Min Agreements.

The natural local search algorithm for Max Min Agreements can be defined similarly to that of Max Cut: it maintains a single cut $S \subseteq V$; a node u moves to the other side of the cut if the move increases the total weight of correctly classified edges incident on u. This algorithm terminates in a local optimum that is a $(1/2)$-approximation for Max Min Agreements. Unfortunately, it is known that such a local search algorithm can take exponential time, even for Max Cut.

When considering Max Cut, this can be remedied by altering the local search step as follows: a node u moves to the other side of the cut S if the move increases the total weight of edges crossing S by a multiplicative factor of at least $(1 + \varepsilon)$ (for some $\varepsilon > 0$). This approach *fails* for the computation of (approximate) pure Nash equilibria in party affiliation games, as well as for Max Min Agreements. The reason is that both of these problems have *local* requirements from nodes, as opposed to the *global* objective of Max Cut. Thus, not surprisingly, the current best known $1/(4+\varepsilon)$-approximation for Max Min Agreements follows from [8] who present the state of the art algorithm for finding approximate pure Nash equilibria in party affiliation games.

We propose a direct approach for approximating Max Min Agreements that circumvents the need to compute approximate pure Nash equilibria in party affiliation games. We improve upon the $1/(4+\varepsilon)$-approximation by considering a *non-oblivious* local search that is executed with altered edge weights. We are able to change the edges' weights in such a way that: (1) any local optimum is a $1/(2+\varepsilon)$-approximation, and (2) the local search performs at most $O(n/\varepsilon)$ iterations. The proof of Theorem 5 appears in the full version of this paper, along with some intuition for our non-oblivious local search algorithm. Additionally, we prove that the natural LP and SDP relaxations for Max Min Agreements on general graphs admit an integrality gap of $\frac{n}{2(n-1)}$ (Theorem 6). This appears in the full version of this paper.

References

1. Ailon, N., Avigdor-Elgrabli, N., Liberty, E., van Zuylen, A.: Improved approximation algorithms for bipartite correlation clustering. SIAM J. Comput. **41**(5), 1110–1121 (2012)
2. Ailon, N., Charikar, M., Newman, A.: Aggregating inconsistent information: ranking and clustering. J. ACM (JACM) **55**(5), 23 (2008)
3. Amit, N.: The bicluster graph editing problem. Ph.D. thesis, Tel Aviv University (2004)
4. Balcan, M.F., Blum, A., Mansour, Y.: Improved equilibria via public service advertising. In: SODA 2009, pp. 728–737 (2009)
5. Bansal, N., Blum, A., Chawla, S.: Correlation clustering. Mach. Learn. **56**(1–3), 89–113 (2004)
6. Bansal, N., Feige, U., Krauthgamer, R., Makarychev, K., Nagarajan, V., Naor, J., Schwartz, R.: Min-max graph partitioning and small set expansion. SIAM J. Comput. **43**(2), 872–904 (2014)
7. Ben-Dor, A., Shamir, R., Yakhini, Z.: Clustering gene expression patterns. J. Comput. Biol. **6**(3–4), 281–297 (1999)
8. Bhalgat, A., Chakraborty, T., Khanna, S.: Approximating pure nash equilibrium in cut, party affiliation, and satisfiability games. In: EC 2010, pp. 73–82 (2010)

9. Charikar, M., Guruswami, V., Wirth, A.: Clustering with qualitative information. In: FOCS 2003, pp. 524–533 (2003)
10. Chawla, S., Makarychev, K., Schramm, T., Yaroslavtsev, G.: Near optimal LP rounding algorithm for correlationclustering on complete and complete k-partite graphs. In: STOC 2015, pp. 219–228 (2015)
11. Cheng, Y., Church, G.M.: Biclustering of expression data. In: Ismb, vol. 8, pp. 93–103 (2000)
12. Christodoulou, G., Mirrokni, V.S., Sidiropoulos, A.: Convergence and approximation in potential games. In: Durand, B., Thomas, W. (eds.) STACS 2006. LNCS, vol. 3884, pp. 349–360. Springer, Heidelberg (2006). doi:10.1007/11672142_28
13. Demaine, E.D., Emanuel, D., Fiat, A., Immorlica, N.: Correlation clustering in general weighted graphs. Theor. Comput. Sci. **361**(2), 172–187 (2006)
14. Fabrikant, A., Papadimitriou, C., Talwar, K.: The complexity of pure Nash equilibria. In: Proceedings of the Thirty-Sixth Annual ACM Symposium on Theory of Computing, pp. 604–612. ACM (2004)
15. Filkov, V., Skiena, S.: Integrating microarray data by consensus clustering. Int. J. Artif. Intell. Tools **13**(04), 863–880 (2004)
16. Garg, N., Vazirani, V.V., Yannakakis, M.: Approximate max-flow min-(multi) cut theorems and their applications. In: STOC 1993, pp. 698–707 (1993)
17. Johnson, D.S., Papadimitriou, C.H., Yannakakis, M.: How easy is local search? J. Comput. Syst. Sci. **37**(1), 79–100 (1988)
18. Kriegel, H.P., Kröger, P., Zimek, A.: Clustering high-dimensional data: a survey on subspace clustering, pattern-based clustering, and correlation clustering. ACM Trans. Knowl. Discov. Data (TKDD) **3**(1), 1 (2009)
19. Monderer, D., Shapley, L.S.: Potential games. Games Econ. Behav. **14**(1), 124–143 (1996)
20. Puleo, G., Milenkovic, O.: Correlation clustering and biclustering with locally bounded errors. In: Proceedings of the 33rd International Conference on Machine Learning, pp. 869–877 (2016)
21. Schäffer, A.A., Yannakakis, M.: Simple local search problems that are hard to solve. SIAM J. Comput. **20**(1), 56–87 (1991)
22. Svitkina, Z., Tardos, É.: Min-max multiway cut. In: Jansen, K., Khanna, S., Rolim, J.D.P., Ron, D. (eds.) APPROX/RANDOM -2004. LNCS, vol. 3122, pp. 207–218. Springer, Heidelberg (2004). doi:10.1007/978-3-540-27821-4_19
23. Swamy, C.: Correlation clustering: maximizing agreements via semidefinite programming. In: SODA 2004, pp. 526–527 (2004)
24. Symeonidis, P., Nanopoulos, A., Papadopoulos, A., Manolopoulos, Y.: Nearest-biclusters collaborative filtering with constant values. In: Nasraoui, O., Spiliopoulou, M., Srivastava, J., Mobasher, B., Masand, B. (eds.) WebKDD 2006. LNCS, vol. 4811, pp. 36–55. Springer, Heidelberg (2007). doi:10.1007/978-3-540-77485-3_3
25. Van Gael, J., Zhu, X.: Correlation clustering for crosslingual link detection. In: IJCAI, pp. 1744–1749 (2007)
26. Wirth, A.: Correlation clustering. In: Sammut, C., Webb, G. (eds.) Encyclopedia of Machine Learning, pp. 227–231. Springer, Heidelberg (2010)

Verifying Integer Programming Results

Kevin K.H. Cheung[1], Ambros Gleixner[2], and Daniel E. Steffy[3(✉)]

[1] School of Mathematics and Statistics, Carleton University,
Ottawa, ON, Canada
kevin.cheung@carleton.ca
[2] Department of Mathematical Optimization, Zuse Institute Berlin,
Takustr. 7, 14195 Berlin, Germany
gleixner@zib.de
[3] Department of Mathematics and Statistics, Oakland University,
Rochester, MI, USA
steffy@oakland.edu

Abstract. Software for mixed-integer linear programming can return incorrect results for a number of reasons, one being the use of inexact floating-point arithmetic. Even solvers that employ exact arithmetic may suffer from programming or algorithmic errors, motivating the desire for a way to produce independently verifiable certificates of claimed results. Due to the complex nature of state-of-the-art MIP solution algorithms, the ideal form of such a certificate is not entirely clear. This paper proposes such a certificate format designed with simplicity in mind, which is composed of a list of statements that can be sequentially verified using a limited number of inference rules. We present a supplementary verification tool for compressing and checking these certificates independently of how they were created. We report computational results on a selection of MIP instances from the literature. To this end, we have extended the exact rational version of the MIP solver SCIP to produce such certificates.

Keywords: Correctness · Verification · Proof · Certificate · Optimality · Infeasibility · Mixed-integer linear programming

1 Introduction

The performance of algorithms for solving mixed-integer linear programs to optimality has improved significantly over the last decades [3,4]. As the complexity of the solvers increases, a question emerges: *How does one know if the computational results are correct?*

Although rarely, MIP solvers do occasionally return incorrect or dubious results [13]. Despite such errors, maintaining a skeptical attitude that borders on paranoia is arguably neither healthy nor practical. After all, machines do outperform humans on calculations by orders of magnitude and many tasks in life are now entrusted to automation. Hence, the motivation for asking how to verify

© Springer International Publishing AG 2017
F. Eisenbrand and J. Koenemann (Eds.): IPCO 2017, LNCS 10328, pp. 148–160, 2017.
DOI: 10.1007/978-3-319-59250-3_13

correctness of computational results is not necessarily because of an inherent distrust of solvers. Rather, it is the desire to seek ways to identify and reduce errors and to improve confidence in the computed results. Previous research on computing accurate solutions for MIP has utilized various techniques including interval arithmetic [37], exact rational arithmetic [6,13,17], and safely derived cuts [12]. Nevertheless, as stated in [13], "even with a very careful implementation and extensive testing, a certain risk of an implementation error remains".

One way to satisfy skeptics is formal code verification as is sometimes found in software for medical applications and avionics. For global optimization, progress in this direction has been made very recently [36,41]. For modern MIP solvers, which easily consist of several 100,000 lines of code, this may be an ambitious goal. An alternative is to build solvers that output extra information that facilitates independent checking. We shall use the word *certificate* to refer to such extra information for a given problem that has been solved. Ideally, the certificate should allow for checking the results using fewer resources than what are needed to solve the problem from scratch. Such a certificate could in principle be used in formal verification using a proof checker as done in the Flyspeck Project [19,39,42] for a formal proof of Kepler's Conjecture, or informal verification as done by Applegate *et al.* [5] for the Traveling Salesman Problem and by Carr *et al.* [11] in their unpublished work for MIP in general. Naturally, certificates should be as simple to verify as possible if they are to be convincing.

We highlight two specific applications where solution verification is desirable. First, Achterberg [1] presented MIP formulations for circuit design verification problems, for which solvers have been shown to return incorrect results [13]. Second, Pulaj [40] has recently used MIP to settle open questions related to Frankl's conjecture. Software developed in connection with this paper has been successfully used to generate and check certificates for MIP models coming from both of these applications.

For linear programming, duality theory tells us that an optimal primal solution and an optimal dual solution are sufficient to facilitate effective verification of optimality. In the case of checking infeasibility, a Farkas certificate will do. Therefore, verifying LP results, at least in the case when exact rational arithmetic is used, is rather straightforward. However, the situation with MIP is drastically different. From a theoretical perspective, even though some notions of duality for MIP have been formulated [24], small (i.e. polynomial size) certificates for infeasibility or optimality may not even exist. As a result, there are many forms that certificates could take: a branch-and-bound tree, a list of derived cutting planes, a superadditive dual function, or other possibilities for problems with special structures such as pure integer linear programming and binary programming [10,16,31,33]. Which format would be preferred for certificate verification is not entirely clear, and in this paper we provide reasoning behind our choice.

From a software perspective, MIP result certification is also considerably more complicated than LP certification. Even though most solvers adopt the branch-and-cut paradigm, they typically do not make the computed branch-and-bound tree or generated cuts readily available, and they may also utilize many other techniques including constraint propagation, conflict analysis, or reduced cost fixing. Thus, even if a solver did print out all information used to derive its solution, a verifier capable of interpreting such information would itself be highly complex, contradicting our desire for a simple verifier. As a result, other than accepting the results of an exact solver such as [13], the best that many people can do today to "verify" the results of a solver on a MIP instance is to solve the instance by several different solvers and check if the results match or minimally check that a returned solution is indeed feasible and has the objective function value claimed, as is done by the solution checker in [32].

The main contribution of this paper is the development of a certificate format for the verification of mixed-integer linear programs. Compared to the previous work of Applegate *et al.* [5] for the Traveling Salesman Problem and the unpublished work of Carr *et al.* [11] for general MIP, our certificate format has a significantly simpler structure. It consists of a sequence of statements that can be verified one by one using simple inference rules, facilitating verification in a manner akin to natural deduction. The approach is similar to that for verification of unsatisfiability proofs for SAT formulas. (See for example [27,43].) This simple certificate structure makes it easier for researchers to develop their own independent certificate verification programs, or check the code of existing verifiers, even without any expert knowledge of MIP solution algorithms.

To demonstrate the utility of the proposed certificate format, we have developed a reference checker in C++ and added the capability to produce such certificates to the exact version of the MIP solver SCIP [13,21]. We used these tools to verify results reported in [13]. To the best of our knowledge, this work also represents the first software for general MIP certificate verification that has been made available to the mathematical optimization community.

Organization of the paper. Even though the proposed format for the certificate is straightforward, some of the details are nevertheless technical. Therefore, in this paper, we discuss the certificate format at a conceptual level. The full technical specification is found in the accompanying computer files.[1] We begin with the necessary ingredients for the simple case of LP in Sect. 2. In Sect. 3, the ideas for dealing with LP are extended to pure integer linear programming. The full conceptual description of the format of the certificate is then given in Sect. 4. Computational experiments are reported in Sect. 5, and concluding remarks are given in Sect. 6. Throughout this paper, we assume that problems are specified and solved with exact rational arithmetic.

2   Certificates for Linear Programming

A certificate of optimality for an LP is a dual feasible solution whose objective function value matches the optimal value. However, there is no need to specify

---

[1] See https://github.com/ambros-gleixner/VIPR.

the dual when one views the task of certification as an inference procedure, see, e.g., [28]. Suppose we are given the system of linear constraints

$$Ax \geq b, A'x \leq b', A''x = b'', \tag{S}$$

where x is a vector of variables, $A \in \mathbb{R}^{m \times n}$, $A' \in \mathbb{R}^{m' \times n}$, $A'' \in \mathbb{R}^{m'' \times n}$, $b \in \mathbb{R}^m$, $b' \in \mathbb{R}^{m'}$, and $b'' \in \mathbb{R}^{m''}$ for some nonnegative integers n, m, m', and m''.

We say that $c^\mathsf{T} x \geq v$ is obtained by taking a *suitable linear combination* of the constraints in (S) if

$$c^\mathsf{T} = d^\mathsf{T} A + {d'}^\mathsf{T} A' + {d''}^\mathsf{T} A'', \quad v = d^\mathsf{T} b + {d'}^\mathsf{T} b' + {d''}^\mathsf{T} b''$$

for some $d \in \mathbb{R}^m$, $d' \in \mathbb{R}^{m'}$, and $d'' \in \mathbb{R}^{m''}$ with $d \geq 0$ and $d' \leq 0$. If x satisfies (S), then it necessarily satisfies $c^\mathsf{T} x \geq v$. We say that the inequality $c^\mathsf{T} x \geq v$ is *inferred* from (S). We will refer to this general inference procedure as *linear inequality inference*.

Remark 1. Together, d, d', d'' is simply a feasible solution to the linear programming dual of the linear program

$$\min\{c^\mathsf{T} x \mid Ax \geq b, A'x \leq b', A''x = b''\}. \tag{LP}$$

The inequality $c^\mathsf{T} x \geq v$ is sometimes called a *surrogate* of (S). (See [28].)

Suppose that an optimal solution to (LP) exists and the optimal value is v. Linear programming duality theory guarantees that $c^\mathsf{T} x \geq v$ can be inferred from (S). Therefore, linear inequality inference is sufficient to certify optimality for linear programming. Conceptually, the certificate that we propose is a listing of the constraints in (S) followed by the inequality $c^\mathsf{T} x \geq v$ with the associated multipliers used in the inference as illustrated in the following example.

Example 2. The following shows an LP problem and its associated certificate.

min $2x + y$
s.t.
$C1 : 5x - y \geq 2$
$C2 : 3x - 2y \leq 1$

Given	
$C1 : 5x - y \geq 2$	
$C2 : 3x - 2y \leq 1$	
Derived	**Reason**
obj : $2x + y \geq 1$	$\{1 \times C1 + (-1) \times C2\}$

Here, $C1$ and $C2$ are constraint labels. Taking the suitable linear combination $1 \times C1 + (-1) \times C2$ gives $2x + y \geq 1$, thus establishing that 1 is a lower bound for the optimal value.

Remark 3. This type of linear inference can also be used to derive $\leq$-inequalities or equality constraints. Assuming that all problem data is rational, rational multipliers are sufficient to certify infeasibility or optimality.

3   Handling Chvátal-Gomory Cutting Planes

Gomory [23] showed in theory that, for pure integer linear programming (IP), optimality or infeasibility can be established by a pure cutting-plane approach. Such an approach can also work in practice [8,44]. In addition to linear inequality inference, a rounding operation is needed.

Suppose that $c^\mathsf{T} x \geq v$ can be inferred from (S) by taking a suitable linear combination of the constraints. If $c_i \in \mathbb{Z}$ for $i \in I$ for some $I \subseteq \{1, \ldots, n\}$ and $c_i = 0$ for $i \notin I$, then any $x \in \mathbb{R}^n$ satisfying (S) with $x_i \in \mathbb{Z}$ for $i \in I$ must also satisfy $c^\mathsf{T} x \geq \lceil v \rceil$. We say that $c^\mathsf{T} x \geq \lceil v \rceil$ is obtained from $c^\mathsf{T} x \geq v$ by *rounding*. When $I = \{1, \ldots, n\}$, the inequality $c^\mathsf{T} x \geq \lceil v \rceil$ is known as a *Chvátal-Gomory cut* (CG-cut in short). It can then be added to the system and the process of obtaining another CG-cut can be repeated. Conceptually, a certificate for an IP instance solved using only CG-cuts can be given as a list of the original constraints followed by the derived constraints.

Example 4. The following shows an IP problem and its associated certificate.

<table>
<tr><td></td><td colspan="2">Given</td></tr>
<tr><td>min $x + y$</td><td colspan="2">$x, y \in \mathbb{Z}$</td></tr>
<tr><td>s.t.</td><td colspan="2">$C1 : 4x + y \geq 1$</td></tr>
<tr><td>$C1 : 4x + y \geq 1$</td><td colspan="2">$C2 : 4x - y \leq 2$</td></tr>
<tr><td>$C2 : 4x - y \leq 2$</td><td>Derived</td><td>Reason</td></tr>
<tr><td>$x, y \in \mathbb{Z}$</td><td>$C3 : y \geq -\frac{1}{2}$</td><td>$\{\frac{1}{2} \times C1 + (-\frac{1}{2}) \times C2\}$</td></tr>
<tr><td></td><td>$C4 : y \geq 0$</td><td>{round up $C3$}</td></tr>
<tr><td></td><td>$C5 : x + y \geq \frac{1}{4}$</td><td>$\{\frac{1}{4} \times C1 + \frac{3}{4} \times C4\}$</td></tr>
<tr><td></td><td>$C6 : x + y \geq 1$</td><td>{round up $C5$}</td></tr>
</table>

Note that the derived constraints in the certificate can be processed in a sequential manner. In the next section, we see how to deal with branching without sacrificing sequential processing.

4   Branch-and-Cut Certificates

In practice, most MIP instances are not solved by cutting planes alone. Thus, certificates as described in the previous section are of limited utility. We now propose a type of certificate for optimality or infeasibility established by a branch-and-cut procedure in which the generated cuts at any node can be derived as split cuts and branching is performed on a disjunction of the form $a^\mathsf{T} x \leq \delta \vee a^\mathsf{T} x \geq \delta + 1$ where $\delta \in \mathbb{Z}$ and $a^\mathsf{T} x$ is integral for all feasible x.

The use of split disjunctions allows us to consider branching and cutting under one umbrella. Many of the well-known cuts generated by MIP solvers can be derived as split cuts [14] and they are effective in closing the integrality gap in practice [18]. Branching typically uses only simple split disjunctions (where the a above is a unit vector), although some studies have considered the computational performance of branching on general disjunctions [15,20,30,38].

Recall that each branching splits the solution space into two subcases. At the end of a branch-and-bound (or branch-and-cut) procedure, each leaf of the branch-and-bound tree corresponds to one of the cases and the leaves together cover all the cases that need to be considered. Hence, if the branch-and-bound tree is valid, all one needs to look at are the LP results at the leaves.

Our proposal is to "flatten" the branch-and-bound tree into a list of statements that can be verified sequentially. Thus, our approach departs from the approaches in [5,11], which require explicit handling of the tree structure. The price we pay is that we can no longer simply examine the leaves of the tree. Instead, we process the nodes in a bottom-up fashion and discharge assumptions as we move up towards the root. We illustrate the ideas with an example.

Example 5. It is known that the following has no solution.

$$C1:\ 2x_1 + 3x_2 \geq 1$$
$$C2:\ 3x_1 - 4x_2 \leq 2$$
$$C3:\ -x_1 + 6x_2 \leq 3$$
$$x_1, x_2 \in \mathbb{Z}$$

Note that $(x_1, x_2) = (\frac{10}{17}, -\frac{1}{17})$ is an extreme point of the region defined by $C1$, $C2$, and $C3$. Branching on the integer variable x_1 leads to two cases:

- **Case 1.** $A1:\ x_1 \leq 0$
  Note that $(x_1, x_2) = (0, \frac{1}{3})$ satisfies $C1, C2, C3, A1$. We branch on x_2:
  Case 1a. $A3:\ x_2 \leq 0$
  Taking $C1 + (-2) \times A1 + (-3) \times A3$ gives the absurdity $C4:\ 0 \geq 1$.
  Case 1b. $A4:\ x_2 \geq 1$
  Taking $\left(-\frac{1}{3}\right) \times C3 + \left(-\frac{1}{3}\right) \times A1 + 2 \times A4$ gives the absurdity $C5:\ 0 \geq 1$.
- **Case 2.** $A2:\ x_1 \geq 1$
  Taking $\left(-\frac{1}{4}\right) \times C2 + \left(\frac{3}{4}\right) \times A2$ gives $C6:\ x_2 \geq \frac{1}{4}$. Rounding gives $C7:\ x_2 \geq 1$.
  Taking $\left(-\frac{1}{3}\right) \times C2 + (-1) \times C3 + \frac{14}{3} \times C7$ gives the absurdity $C8:\ 0 \geq 1$.

As all cases lead to $0 \geq 1$, we conclude that there is no solution. To issue a certificate as a list of derived constraints, we need a way to specify the different cases. To this end, we allow the introduction of constraints as assumptions.

Figure 1 shows a conceptual certificate for the instance. Notice how the constraints $A1$, $A2$, $A3$, and $A4$ are introduced to the certificate as assumptions. Since we want to end with $0 \geq 1$ without additional assumptions attached, we get there by gradually undoing the case-splitting operations. We call the undoing operation *unsplitting*. For example, $C4$ and $C5$ are both the absurdity $0 \geq 1$ with a common assumption $A1$. Since $A3 \vee A4$ is true for all feasible x, we can infer the absurdity $C9:\ 0 \geq 1$ assuming only $A1$ in addition to the original constraints. We say that $C9$ is obtained by *unsplitting* $C4, C5$ on $A3, A4$. Similarly, both $C8$ and $C9$ are the absurdity $0 \geq 1$ and $A2 \vee A1$ is true for all feasible x, we can therefore unsplit on $C8, C9$ on $A2, A1$ to obtain $C10:\ 0 \geq 1$ without any assumption in addition to the original constraints.

Given		
$x_1, y_1 \in \mathbb{Z}$		
$C1 : 2x_1 + 3x_2 \geq 1$		
$C2 : 3x_1 - 4x_2 \leq 2$		
$C3 : -x_1 + 6x_2 \leq 3$		
Derived	**Reason**	**Assumptions**
$A1 : x_1 \leq 0$	$\{$assume$\}$	
$A2 : x_1 \geq 1$	$\{$assume$\}$	
$A3 : x_2 \leq 0$	$\{$assume$\}$	
$C4 : 0 \geq 1$	$\{C1 + (-2) \times A1 + (-3) \times A3\}$	$A1, A3$
$A4 : x_2 \geq 1$	$\{$assume$\}$	
$C5 : 0 \geq 1$	$\left\{\left(-\frac{1}{3}\right) \times C3 + \left(-\frac{1}{3}\right) \times A1 + 2 \times A4\right\}$	$A1, A4$
$C6 : x_2 \geq \frac{1}{4}$	$\left\{\left(-\frac{1}{4}\right) \times C2 + \left(\frac{3}{4}\right) \times A2\right\}$	$A2$
$C7 : x_2 \geq 1$	$\{$round up $C6\}$	$A2$
$C8 : 0 \geq 1$	$\left\{\left(-\frac{1}{3}\right) \times C2 + (-1) \times C3 + \frac{14}{3} \times C7\right\}$	$A2$
$C9 : 0 \geq 1$	$\{$unsplit $C4, C5$ on $A3, A4\}$	$A1$
$C10 : 0 \geq 1$	$\{$unsplit $C8, C9$ on $A2, A1\}$	

Fig. 1. Certificate for Example 5

In practice, the list of assumptions associated with each derived constraint needs not be specified explicitly in the certificate, but can be deduced on the fly by a checker. For example, when processing $C4$, we see that it uses $A1$ and $A3$, both of which are assumptions. Hence, we associate $C4$ with the list of assumptions $A1, A3$. As any linear inequality can be introduced as an assumption, branching can be performed on general disjunctions.

Remark 6. Our proposed certificate can also be used to represent split cuts. Split cuts are inequalities that are valid for the defining inequalities taken together with each one of the inequalities in a split disjunction, $a^\mathsf{T} x \leq \delta \vee a^\mathsf{T} x \geq \delta + 1$, where δ is an integer and a is an integer vector that is nonzero only in components corresponding to integer variables. To derive a proof of a split cut's validity, the inequalities in the split disjunction can each be introduced as assumptions, the cut can be derived for each side of the split disjunction using linear inequality inference, and then unsplitting can be applied to discharge the assumptions.

5   Computational Experiments

In this section, we describe software developed to produce and check certificates for MIP results using the certificate format developed in this paper. It is freely available for download, along with a precise technical specification of the file format.[2] One of its features is that after each derived constraint an integer is printed to specify the largest index of any derived constraint that references it. This allows constraints to be freed from memory when they will no longer be needed. The following C++ programs are provided:

---

[2] See https://github.com/ambros-gleixner/VIPR.

- `viprchk` verifies MIP results provided in our specified file format. All computations are performed in exact rational arithmetic using the GMP library [22].
- `viprttn` performs simple modifications to "tighten" certificates. It removes unnecessary derived constraints to reduce the file size. In order to decrease peak memory usage during checking, it reorders the remaining ones using a depth-first topological sort and for each derived constraint that remains, it computes the largest index over constraints that references it.
- `vipr2html` converts a certificate file to a "human-readable" HTML file.

We again emphasize that the format was designed with simplicity in mind; the certificate verification program we have provided is merely a reference and others should be able to write their own verifiers without much difficulty.

In addition, we created a modified version of the exact rational MIP solver described in [13] and used it to compute certificates for several MIP instances from the literature. The exact rational MIP solver is based on SCIP [21] and uses a hybrid of floating-point and exact rational arithmetic to efficiently compute exact solutions using a pure branch-and-bound algorithm. In our experiments, the rational MIP solver uses CPLEX 12.6.0.0 [29] as its underlying floating-point LP solver and a modified version of QSopt_ex 2.5.10 [7] as its underlying exact LP solver. The exact MIP solver supports several methods for computing valid dual bounds and our certificate printing functionality is currently supported by the *Project-and-shift* method (for dual solutions only) and the *Exact LP* method (for both dual solutions and Farkas proofs), for details on these methods see [13]. This developmental version is currently available from the authors by request. We note that the certificate is printed concurrently with the solution process which leads to certificates that have potential for reduction and simplification by `viprttn`, or other routines. For example, as each node is processed its derived dual bound is printed to the certificate even though it may become redundant if branching is performed and new dual bounds are computed at the child nodes; also, discovery of a new primal solution might allow pruning of a large subtree, rendering many bound derivations redundant.

The program `viprttn` processes the list of derived constraints in two passes. In the first pass, it builds the dependency graph with nodes representing the derived constraints and arcs uv such that the derived constraint represented by u is referenced by the reason for deriving the constraint represented by v. In the second pass, it performs a topological sort using depth-first search on the component that contains the final constraint and writes out the reordered list of derived constraints with updated constraint indices.

In the following, we report some computational results on the time and memory required to produce and verify certificates. We considered the *easy* and *numerically difficult* (referred to here as '*hard*') test sets from [13]; these test sets consist of instances from well known libraries including [2,9,32,34,35]. Experiments were conducted on a cluster of Intel(R) Xeon(R) CPU E5-2660 v3 at 2.60 GHz; jobs were run exclusively to ensure accurate time measurement. Table 1 reports a number of aggregate statistics on these experiments. The columns under the heading SCIP report results from tests using the exact version of SCIP, using its default dual bounding strategy. The columns under SCIP+C report on tests involving the version of exact SCIP that generates certificates as it solves instances; it uses only the dual bounding methods *Project-and-shift* and *Exact LP* that support certificate printing, contributing to its slower speed. Columns under the heading VIPR report time and memory usage for certificate checking.

For each of the easy and hard test sets, we report information aggregated into four categories: 'all' reports statistics over all instances; 'solved' reports over instances solved by both SCIP and SCIP+C within a 1 h time limit and a 10 GB limit on certificate file size; 'memout' reports on instances where SCIP+C stopped because the certificate file size limit was reached; and 'timeout' reports on the remaining instances, where one of the solvers hit the time limit. All averages are reported as shifted geometric means with a shift of 10 s. for time and 1 MB for memory. The column N represents the number of instances in each category; N_{sol} represents the number in each category that were solved to optimality (or infeasibility) by a given solver; t_{MIP} represents the time (sec.) used to solve the instance and, when applicable, output a certificate; t_{ttn} is the time (sec.) required by the viprttn routine to tighten the certificate file; t_{chk} is the time (sec.) required to for viprchk to check the certificate file – on instances in the memout and timeout rows this represents the time to verify the primal and dual bounds present in the intermediate certificate printed before the solver was halted. The final three columns list the size of the certificate (in MB), before tightening, after tightening and then after being compressed to a gzipped file. Timings and memory usage for individual instances are available in a document hosted together with the accompanying software.

Table 1. Aggregated computational results over 107 instances from [13].

Test set	N	SCIP N_{sol}	t_{MIP}	SCIP+C N_{sol}	t_{MIP}	t_{ttn}	t_{chk}	VIPR $size_{raw}$	$size_{ttn}$	$size_{gz}$
easy-all	57	54	63.3	39	190.9	8.9	27.2	227	77	24
-solved	39	39	23.2	39	48.0	3.6	11.5	77	34	10
-memout	5	4	600.6	0	1760.4	47.8	138.3	10286	513	157
-timeout	13	11	338.3	0	3600.0	23.3	102.7	1309	434	129
hard-all	50	23	725.2	14	975.6	7.9	12.1	373	38	11
-solved	13	13	22.9	13	40.7	2.2	5.3	49	15	5
-memout	12	2	2476.4	0	1713.1	32.7	59.8	10266	235	67
-timeout	25	8	2052.1	1	3518.1	4.3	5.3	216	25	7

From this table, we can make a number of observations. First, there is a noticeable, but not prohibitive, cost to generate the certificates. The differences in t_{MIP} between SCIP and SCIP+C are due to both the difference in dual bounding strategies, and the overhead for writing the certificate files. In some additional experiments, we observed that on the 39 instances in the easy-solved category, the file I/O amounted to roughly 7% of the solution time, based on this we believe that future modifications to the code will allow us to solve and print certificates in times much closer to those in the SCIP column. Perhaps most importantly, we observe that the time to check the certificates is significantly less than the time to solve the instances.

Moreover, the certificate tightening program viprttn is able to make significant reductions in the certificate size, and the resulting certificate sizes are often surprisingly manageable. Most striking is the tightening in the memout categories, which significantly exceed the approximately 50% reduction that could be expected by removing the redundant linear inferences derived for internal nodes of the branch-and-bound tree. The most extreme tightening was achieved for the instance markshare1_1 in 'easy-memout', from 10 GB to 8 kB. This is explained by the fact that the root dual bound is already zero and the tree search is only performed for finding the optimal solution. Hence, the certificate is highly redundant and the derived constraints for all but the root node can be removed.

The average reductions in the other categories are smaller, but also strictly above 50%. This shows that viprttn performs more than just a removal of internal nodes. These results also show two aspects in which SCIP's certificate printing can be improved: by avoiding printing dual bound derivations for internal nodes using a buffering scheme, and by not generating dual bound derivations for nodes that do not improve upon the bound of the parent node.

6   Conclusion

This paper presented a certificate format for verifying integer programming results. We have demonstrated the practical feasibility of generating and checking such certificates on well-known MIP instances. We see this as the first step of many in verifying the results of integer programming solvers. We now discuss some future directions made possible by this work.

Even in the context of floating-point arithmetic, our certificate format could serve a number of purposes. Using methods described by [12,37], directed rounding and interval arithmetic may allow us to compute and represent valid certificates exclusively using floating-point data, allowing for faster computation and smaller certificate size. Additionally, generating approximate certificates with inexact data could be used for debugging solvers, or measuring the maximum or average numerical violation over all derivations. In a more rigorous direction, one could also convert our certificates to a form that could be formally verified by a proof assistant such as HOL Light [25].

Acknowledgements. We thank Kati Wolter for the exact version of SCIP [13] and Daniel Espinoza for QSopt_ex [7], which provided the basis for our experiments, and Gregor Hendel for his Ipet package [26], which was a big help in analyzing the experimental results. This work has been supported by the Research Campus MODAL *Mathematical Optimization and Data Analysis Laboratories* funded by the Federal Ministry of Education and Research (BMBF Grant 05M14ZAM). All responsibility for the content of this publication is assumed by the authors.

References

1. Achterberg, T.: Constraint Integer Programming. Ph.D. Thesis, TU Berlin (2007)
2. Achterberg, T., Koch, T., Martin, A.: The mixed integer programming library: MIPLIB 2003. Oper. Res. Lett. **34**(4), 361–372 (2006)
3. Achterberg, T., Wunderling, R.: Mixed integer programming: analyzing 12 years of progress. In: Jünger, M., Reinelt, G. (eds.) Facets of Combinatorial Optimization: Festschrift for Martin Grötschel, pp. 449–481. Springer, Heidelberg (2013)
4. Bixby, R.E., Fenelon, M., Gu, Z., Rothberg, E., Wunderling, R.: MIP: theory and practice - closing the gap. In: Powell, M.J.D., Scholtes, S. (eds.) System Modelling and Optimization, pp. 19–49 (2000)
5. Applegate, D.L., Bixby, R.E., Chvátal, V., Cook, W., Espinoza, D.G., Goycoolea, M., Helsgaun, K.: Certification of an optimal TSP tour through 85,900 cities. Oper. Res. Lett. **37**, 11–15 (2009)
6. Applegate, D.L., Cook, W.J., Dash, S., Espinoza, D.G.: Exact solutions to linear programming problems. Oper. Res. Lett. **35**(6), 693–699 (2007)
7. Applegate, D.L., Cook, W.J., Dash, S., Espinoza, D.G.: QSopt_ex: http://www.math.uwaterloo.ca/bico/qsopt/ex/. Last accessed 13 Nov 2016
8. Balas, E., Fischetti, M., Zanette, A.: A hard integer program made easy by lexicography. Math. Program. Ser. A **135**, 509–514 (2012)
9. Bixby, R.E., Ceria, S., McZeal, C.M., Savelsbergh, M.W.: An updated mixed integer programming library: MIPLIB 3.0. Optima **58**, 12–15 (1998)
10. Boland, N.L., Eberhard, A.C.: On the augmented Lagrangian dual for integer programming. Math. Program. Ser. A **150**(2), 491–509 (2015)
11. Carr, R., Greenberg, H., Parekh, O., Phillips, C.: Towards certificates for integer programming computations. Presentation, 2011 DOE Applied Mathematics PI meeting, October 2011. Slides www.csm.ornl.gov/workshops/applmath11/documents/talks/Phillips_talk.pdf. Last accessed 13 Nov 2016
12. Cook, W., Dash, S., Fukasawa, R., Goycoolea, M.: Numerically safe Gomory mixed-integer cuts. INFORMS J. Comput. **21**(4), 641–649 (2009)
13. Cook, W., Koch, T., Steffy, D., Wolter, K.: A hybrid branch-and-bound approach for exact rational mixed-integer programming. Math. Program. Comput. **3**, 305–344 (2013)
14. Cornuéjols, G.: Valid inequalities for mixed integer linear programs. Math. Program. Ser. B **112**, 3–44 (2008)
15. Cornuéjols, G., Liberti, L., Nannicini, G.: Improved strategies for branching on general disjunctions. Math. Program. Ser. A **130**, 225–247 (2011)
16. De Loera, J.A., Lee, J., Malkin, P.N., Margulies, S.: Computing infeasibility certificates for combinatorial problems through Hilberts Nullstellensatz. J. Symbolic Comp. **46**(11), 1260–1283 (2011)
17. Dhiflaoui, M., Funke, S., Kwappik, C., Mehlhorn, K., Seel, M., Schomer, E., Schulte, R., Weber, D.: Certifying and repairing solutions to large LPs, how good are LP-solvers? In: SODA 2003, pp. 255–256. ACM/SIAM, New York (2003)
18. Fukasawa, R., Goycoolea, M.: On the exact separation of mixed integer knapsack cuts. Math. Program. Ser. A **128**, 19–41 (2008)
19. The Flyspeck Project. https://code.google.com/archive/p/flyspeck/. Last accessed 13 Nov 2016

20. Gamrath, G., Melchiori, A., Berthold, T., Gleixner, A.M., Salvagnin, D.: Branching on multi-aggregated variables. In: Michel, L. (ed.) CPAIOR 2015. LNCS, vol. 9075, pp. 141–156. Springer, Cham (2015). doi:10.1007/978-3-319-18008-3_10

21. Gamrath, G., et al.: The SCIP Optimization Suite 3.2. ZIB-Report (15–60) (2016)

22. GNU MP: The GNU Multiple Precision Arithmetic Library version 6.1.1. http://gmplib.org. Last accessed 16 Nov 2016

23. Gomory, R.E.: Outline of an algorithm for integer solutions to linear programs. Bull. Amer. Math. Soc. **64**, 275–278 (1958)

24. Guzelsoy, M., Ralphs, T.K.: Duality for mixed-integer linear programs. Int. J. Oper. Res. **4**(3), 118–137 (2007)

25. Harrison, J.: HOL light: a tutorial introduction. In: Srivas, M., Camilleri, A. (eds.) FMCAD 1996. LNCS, vol. 1166, pp. 265–269. Springer, Heidelberg (1996). doi:10.1007/BFb0031814

26. Hendel, G.: Empirical analysis of solving phases in mixed integer programming. Master's thesis, Technische Universität Berlin (2014). urn:nbn:de:0297-zib-54270

27. Heule, M.J.H., Hunt, W.A., Wetzler, N.: Verifying refutations with extended resolution. In: Bonacina, M.P. (ed.) CADE 2013. LNCS (LNAI), vol. 7898, pp. 345–359. Springer, Heidelberg (2013). doi:10.1007/978-3-642-38574-2_24

28. Hooker, J.N.: Integrated Methods for Optimization, 2nd edn. Springer, New York (2012)

29. IBM ILOG. CPLEX. https://www-01.ibm.com/software/commerce/optimization/cplex-optimizer/. Last accessed 16 Nov 2016

30. Karamanov, M., Cornuéjols, G.: Branching on general disjunctions. Math. Program. Ser. A **128**, 403–436 (2011)

31. Klabjan, D.: Subadditive approaches in integer programming. Eur. J. Oper. Res. **183**, 525–545 (2007)

32. Koch, T., Achterberg, T., Andersen, E., Bastert, O., Berthold, T., Bixby, R.E., Danna, E., Gamrath, G., Gleixner, A.M., Heinz, S., Lodi, A., Mittelmann, H., Ralphs, T., Salvagnin, D., Steffy, D.E., Wolter, K.: MIPLIB 2010. Math. Program. Comp. **3**(2), 103–163 (2011)

33. Lasserre, J.B.: Generating functions and duality for integer programs. Discrete Optim. **1**(2), 167–187 (2004)

34. Lehigh University COR@L mixed integer programming collection. http://coral.ie.lehigh.edu/wiki/doku.php/info:datasets:mip. Last accessed 18 Nov 2016

35. Mittelmann, H.D.: Benchmarks for Optimization Software. http://plato.asu.edu/bench.html Last accessed 18 Nov 2016

36. Narkawicz, A., Muñoz, C.: A formally verified generic branching algorithm for global optimization. In: Cohen, E., Rybalchenko, A. (eds.) VSTTE 2013. LNCS, vol. 8164, pp. 326–343. Springer, Heidelberg (2014). doi:10.1007/978-3-642-54108-7_17

37. Neumaier, A., Shcherbina, O.: Safe bounds in linear and mixed-integer linear programming. Math. Program. **99**(2), 283–296 (2004)

38. Owen, J.H., Mehrotra, S.: Experimental results on using general disjunctions in branch-and-bound for general-integer linear programs. Comput. Optim. Appl. **20**, 159–170 (2001)

39. Obua, S., Nipkow, T.: Flyspeck II: the basic linear programs. Ann. Math. Artif. Intell **56**, 245–272 (2009)

40. Pulaj, J.: Cutting Planes for Families Implying Frankl's Conjecture. ZIB-Report (16–51) (2016). urn:nbn:de:0297-zib-60626

41. Smith, A.P., Muñoz, C.A., Narkawicz, A.J., Markevicius, M.: Kodiak: an Implementation Framework for Branch and Bound Algorithms. Technical report: NASA/TM-2015-218776 (2015)
42. Solovyev, A., Hales, T.C.: Efficient formal verification of bounds of linear programs. In: Davenport, J.H., Farmer, W.M., Urban, J., Rabe, F. (eds.) CICM 2011. LNCS, vol. 6824, pp. 123–132. Springer, Heidelberg (2011). doi:10.1007/978-3-642-22673-1_9
43. Wetzler, N., Heule, M.J.H., Hunt, W.A.: DRAT-trim: efficient checking and trimming using expressive clausal proofs. In: Sinz, C., Egly, U. (eds.) SAT 2014. LNCS, vol. 8561, pp. 422–429. Springer, Cham (2014). doi:10.1007/978-3-319-09284-3_31
44. Zanette, A., Fischetti, M., Balas, E.: Lexicography and degeneracy: can a pure cutting plane algorithm work? Math. Program. Ser. A **130**, 153–176 (2011)

Long Term Behavior of Dynamic Equilibria in Fluid Queuing Networks

Roberto Cominetti[1], José Correa[2], and Neil Olver[3,4(✉)]

[1] Facultad de Ingeniería y Ciencias, Universidad Adolfo Ibáñez, Santiago, Chile
[2] Facultad de Ingeniería Industrial, Universidad de Chile, Santiago, Chile
[3] Department of Econometrics and Operations Research,
Vrije Universiteit Amsterdam, Amsterdam, Netherlands
n.olver@vu.nl
[4] CWI, Amsterdam, Netherlands

Abstract. A fluid queuing network constitutes one of the simplest models in which to study flow dynamics over a network. In this model we have a single source-sink pair and each link has a per-time-unit capacity and a transit time. A dynamic equilibrium (or equilibrium flow over time) is a flow pattern over time such that no flow particle has incentives to unilaterally change its path. Although the model has been around for almost fifty years, only recently results regarding existence and characterization of equilibria have been obtained. In particular the long term behavior remains poorly understood. Our main result in this paper is to show that, under a natural (and obviously necessary) condition on the queuing capacity, a dynamic equilibrium reaches a steady state (after which queue lengths remain constant) in finite time. Previously, it was not even known that queue lengths would remain bounded. The proof is based on the analysis of a rather non-obvious potential function that turns out to be monotone along the evolution of the equilibrium. Furthermore, we show that the steady state is characterized as an optimal solution of a certain linear program. When this program has a unique solution, which occurs generically, the long term behavior is completely predictable. On the contrary, if the linear program has multiple solutions the steady state is more difficult to identify as it depends on the whole temporal evolution of the equilibrium.

1 Introduction

A *fluid queuing network* is a directed graph $G = (V, E)$ where each arc $e \in E$ consists of a fluid queue with capacity $\nu_e > 0$ followed by a link with constant

We warmly thank Schloss Dagstuhl for its hospitality during the Seminar 15412 *"Dynamic Traffic Models in Transportation Science"* at which this research started. We also express our sincere gratitude to Vincent Acary, Umang Bhaskar, and Martin Skutella for enlightening discussions. This work was partially supported by *Núcleo Milenio Información y Coordinación en Redes* (ICM-FIC RC130003), an NWO Veni grant, and an NWO TOP grant.

F. Eisenbrand and J. Koenemann (Eds.): IPCO 2017, LNCS 10328, pp. 161–172, 2017.
DOI: 10.1007/978-3-319-59250-3_14

delay $\tau_e \geq 0$ (see Fig. 1). A constant inflow rate $u_0 > 0$ enters the network at a fixed source $s \in V$ and travels towards a terminal node $t \in V$. A *dynamic equilibrium* models the temporal evolution of the flows in the network. Loosely speaking, it consists of a flow pattern in which every particle travels along a shortest path, accounting for the fact that travel times depend on the instant at which a particle enters the network as well as the state of the queues that will be encountered along its path by the time at which they are reached.

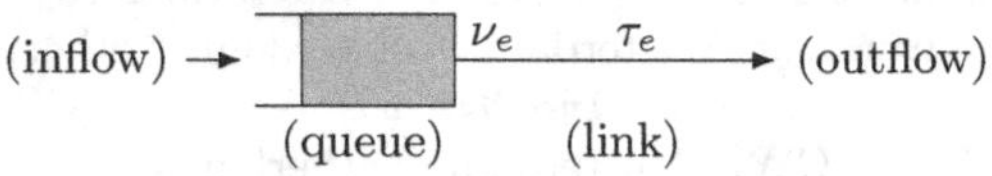

Fig. 1. An arc in the fluid queuing network.

Intuitively, if the queues are initially empty, the equilibrium should start by sending all the flow along shortest paths considering only the free-flow delays τ_e. These paths are likely to become overloaded so that queues will grow on some of its edges and at some point in time new paths will become competitive and will be incorporated into the equilibrium. These new paths may in turn build queues so that even longer paths may come into play. Hence one might expect that the equilibrium proceeds in phases in which the paths used by the equilibrium remain stable. However, it is unclear if the number of such phases is finite and whether the equilibrium will eventually reach a steady state in which the queues and travel times stabilize.

Although dynamic equilibria have been around for almost fifty years (see, e.g., [2–4,6,7,9–12]), their existence has only been proved recently by Zhu and Marcotte [13] though in a somewhat different setting, and by Meunier and Wagner [8] who gave the first existence result for a model that covers the case of fluid queuing networks. These proofs, however, rely heavily on functional analysis techniques and provide little intuition on the combinatorial structure of dynamic equilibria, their characterization, or feasible approaches to compute them. Substantial progress was recently achieved by Koch and Skutella [5] by introducing the concept of *thin flows with resetting* that characterize the time derivatives of a dynamic equilibrium, and which provide in turn a method to compute an equilibrium by integration. A slightly refined notion of *normalized* thin flows with resetting was considered by Cominetti et al. [1], who proved existence and uniqueness, and provided a constructive proof for the existence of a dynamic equilibrium.

In this paper we focus on the long term behavior of dynamic equilibria in fluid queuing networks. Clearly if the inflow u_0 is very large compared to the queuing capacities, the queues will grow without bound, and no steady state can be expected. More precisely, let $\delta(S)$ be an st-cut with minimum queuing capacity $\bar{\nu} = \sum_{e \in \delta(S)} \nu_e$; if there are multiple options, choose S (containing s) to be setwise minimal. If $u_0 > \bar{\nu}$ all the arcs in $\delta(S)$ will grow unbounded queues, whereas for $u_0 \leq \bar{\nu}$, it is natural to expect that the equilibrium should eventually reach a steady state, where queue lengths remain constant. This was not known—in fact, it was not even known that queue lengths remain bounded!

Our main goal in this paper is to show that both these properties do indeed hold: more precisely, when $u_0 \leq \bar{\nu}$, the dynamic equilibrium reaches a steady state in *finite* time. At first glance, these convergence properties might seem "obvious", and it might seem surprising that they are at all difficult to prove. We will present some examples that illustrate why this is not the case. For instance, it may occur that the flow across the cut $\delta(S)$ may temporarily exceed its capacity $\bar{\nu}$ by an arbitrarily large factor, forcing the queues to grow very large. This phenomenon may occur since the inflow u_0 entering the network at different points in time may experience different delays and eventually superpose at $\delta(S)$ which gets an inflow larger than u_0. In other cases some queues may grow during a period of time after which they reduce to zero and then grow again later on, so that no simple monotonicity arguments can be used to study the long term behavior.

Along the way to our main result, we provide a characterization of the steady state as an optimal solution of a certain linear programming problem and we discuss when this problem has a unique solution. Despite the fact that convergence to a steady state occurs in finite time, it remains as an open question whether this state is attained after finitely many phases or whether the dynamic equilibrium may exhibit Zeno-like oscillations in which queues alternate infinitely often over a finite time interval. In such a case the computation by integration would not yield a finite procedure. While this seems very unlikely, we have not been able to prove that it will never happen.

The paper is structured as follows. Section 2 reviews the model of fluid queuing networks, including the precise definition of dynamic equilibrium and the main results known so far. Then, in Sect. 3 we discuss the notion of steady state and provide a characterization in terms of a linear program. Inspired by the objective function of this linear program, in Sect. 4 we introduce a potential function and we prove that it is a Lyapunov function for the dynamics. This potential turns out to be piecewise linear in time with finitely many possible slopes. We then prove that the potential remains bounded so that there is a finite time at which its slope is zero, and we show that in that case the system has reached a steady state. Further, we provide an explicit pseudopolynomial bound on the convergence time. Finally, in Sect. 5 we discuss some counterexamples that rule out some natural properties that one might expect to hold in a dynamic equilibrium, and we state some related open questions.

2   Dynamic Equilibria in Fluid Queing Networks

In this section we recall the definition of dynamic equilibria in fluid queuing networks, and we briefly review the known results on their existence, characterization, and computation. The results are stated without proofs for which we refer to Koch and Skutella [5] and Cominetti et al. [1].

2.1   The Model

Consider a fluid queuing network $G = (V, E)$ with arc capacities ν_e and delays τ_e. The network dynamics are described in terms of the inflow rates $f_e^+(\theta)$ that enter each arc $e \in E$ at time θ, where $f_e^+ : [0, \infty) \to [0, \infty)$ is measurable.

Arc Dynamics. If the inflow $f_e^+(\theta)$ exceeds ν_e a queue $z_e(\theta)$ will grow at the entrance of the arc. The queues are assumed to operate at capacity, that is to say, when $z_e(\theta) > 0$ the flow is released at rate ν_e, whereas when the queue is empty the outflow is the minimum between $f_e^+(\theta)$ and the capacity ν_e. Hence the queue evolves from its initial state $z_e(0) = 0$ according to

$$\dot{z}_e(\theta) = \begin{cases} f_e^+(\theta) - \nu_e & \text{if } z_e(\theta) > 0 \\ [f_e^+(\theta) - \nu_e]_+ & \text{if } z_e(\theta) = 0. \end{cases} \tag{1}$$

These dynamics uniquely determine the queue lengths $z_e(\theta)$ as well as the arc outflows (Fig. 2)

$$f_e^-(\theta + \tau_e) = \begin{cases} \nu_e & \text{if } z_e(\theta) > 0 \\ \min\{f_e^+(\theta), \nu_e\} & \text{if } z_e(\theta) = 0. \end{cases} \tag{2}$$

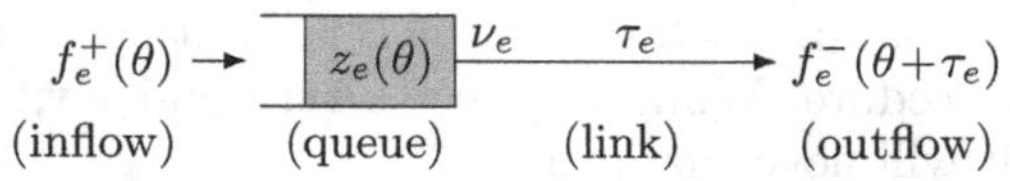

$$f_e^+(\theta) \to \boxed{z_e(\theta)} \xrightarrow{\nu_e \quad \tau_e} f_e^-(\theta + \tau_e)$$

$$\text{(inflow)} \qquad \text{(queue)} \qquad \text{(link)} \qquad \text{(outflow)}$$

Fig. 2. Dynamics of an arc in the queuing network.

Flow Conservation. A *flow over time* is a family $(f_e^+)_{e \in E}$ of arc inflows such that flow is conserved at every node $v \in V \setminus \{t\}$, namely for a.e. $\theta \geq 0$

$$\sum_{e \in \delta^+(v)} f_e^+(\theta) - \sum_{e \in \delta^-(v)} f_e^-(\theta) = \begin{cases} u_0 & \text{if } v = s \\ 0 & \text{if } v \neq s, t. \end{cases} \tag{3}$$

Dynamic Shortest Paths. A particle entering an arc e at time θ experiences a queuing delay $z_e(\theta)/\nu_e$ plus a free-flow delay τ_e to traverse the arc after leaving the queue, so that it will exit the arc at time

$$T_e(\theta) = \theta + \frac{z_e(\theta)}{\nu_e} + \tau_e. \tag{4}$$

Consider a particle entering the source node s at time θ. If this particle follows a path $p = e_1 e_2 \cdots e_k$, it will reach the end of the path at time

$$T_p(\theta) = T_{e_k} \circ \cdots \circ T_{e_2} \circ T_{e_1}(\theta). \tag{5}$$

Denoting $\mathcal{P}_v$ the set of all sv-paths, the minimal time at which node v can be reached is

$$\ell_v(\theta) = \min_{p \in \mathcal{P}_v} T_p(\theta). \tag{6}$$

The paths attaining these minima are called *dynamic shortest paths*. The arcs in these paths are said to be *active* at time θ and we denote them by E'_θ. Observe that $\ell_v(\theta)$ can also be defined through the dynamic Bellman's equations

$$\begin{cases} \ell_s(\theta) = \theta \\ \ell_w(\theta) = \min_{e=vw \in E} T_e(\ell_v(\theta)) \end{cases} \tag{7}$$

so that $e = vw$ is active iff $\ell_w(\theta) = T_e(\ell_v(\theta))$.

Dynamic Equilibrium. A dynamic equilibrium is a flow pattern that uses only dynamic shortest paths. More precisely, let $\Theta_e = \{\theta : e \in E'_\theta\}$ be the set of *entrance times* θ at which the arc e is active, and $\Xi_e = \ell_v(\Theta_e)$ the set of *local times* $\xi = \ell_v(\theta)$ at which e will be active. A flow over time $(f_e^+)_{e \in E}$ is called a *dynamic equilibrium* iff for a.e. $\xi \geq 0$ we have $f_e^+(\xi) > 0 \Rightarrow \xi \in \Xi_e$.

2.2   Characterization of Dynamic Equilibria

Since the inflows $f_e^+(\cdot)$ are measurable the same holds for $f_e^-(\cdot)$ and we may define the *cumulative inflows* and *cumulative outflows* as

$$F_e^+(\theta) = \int_0^\theta f_e^+(z)\,dz$$

$$F_e^-(\theta) = \int_0^\theta f_e^-(z)\,dz.$$

These cumulative flows allow to express the queues as $z_e(\theta) = F_e^+(\theta) - F_e^-(\theta + \tau_e)$. It turns out that a dynamic equilibrium can be equivalently characterized by the fact that for each arc $e = vw \in E$ we have

$$F_e^+(\ell_v(\theta)) = F_e^-(\ell_w(\theta)) \quad \forall\, \theta \geq 0. \tag{8}$$

In this case, the functions $x_e(\theta) \triangleq F_e^+(\ell_v(\theta))$ are static flows with

$$\sum_{e \in \delta^+(v)} x_e(\theta) - \sum_{e \in \delta^-(v)} x_e(\theta) = \begin{cases} u_0\theta & \text{if } v = s \\ -u_0\theta & \text{if } v = t \\ 0 & \text{if } v \neq s, t. \end{cases} \tag{9}$$

2.3   Derivatives of a Dynamic Equilibrium

The labels $\ell_v(\theta)$ and the static flows $x_e(\theta)$ are nondecreasing functions which are also absolutely continuous so that they can be reconstructed from their derivatives by integration.[1] Moreover, from these functions one can recover the equilibrium inflows $f_e^+(\cdot)$ using the relation $x'_e(\theta) = f_e^+(\ell_v(\theta))\ell'_v(\theta)$. Hence,

---

[1] These derivatives exist almost everywhere and are locally integrable.

finding a dynamic equilibrium reduces essentially to computing the derivatives $\ell'_v(\theta), x'_e(\theta)$.

Let θ be a point of differentiability and set $\ell'_v = \ell'_v(\theta) \geq 0$ and $x'_e = x'_e(\theta) \geq 0$. From (9) we see that x' is a static st-flow of size u_0, namely,

$$\sum_{e \in \delta^+(v)} x'_e - \sum_{e \in \delta^-(v)} x'_e = \begin{cases} u_0 & \text{if } v = s \\ -u_0 & \text{if } v = t \\ 0 & \text{if } v \neq s, t \end{cases} \tag{10}$$

while using (7), (4), (1) and the differentiation rule for a minimum we get

$$\begin{cases} \ell'_s = 1 \\ \ell'_w = \min_{e = vw \in E'_\theta} \rho_e(\ell'_v, x'_e) \end{cases} \tag{11}$$

where

$$\rho_e(\ell'_v, x'_e) = \begin{cases} x'_e/\nu_e & \text{if } e \in E^*_\theta \\ \max\{\ell'_v, x'_e/\nu_e\} & \text{if } e \notin E^*_\theta \end{cases} \tag{12}$$

with E^*_θ the set of arcs $e = vw$ with positive queue $z_e(\ell_v(\theta)) > 0$. In addition to this, the conditions for dynamic equilibria imply $E^*_\theta \subset E'_\theta$ as well as

$$\begin{aligned} (\forall e \in E'_\theta) \ x'_e > 0 &\Rightarrow \ell'_w = \rho_e(\ell'_v, x'_e) \\ (\forall e \notin E'_\theta) \ x'_e &= 0. \end{aligned} \tag{13}$$

These equations fully characterize the derivatives of a dynamic equilibrium. In fact, for all subsets $E^* \subseteq E' \subseteq E$ the system (10)–(13) admits at least one solution (ℓ', x') and moreover the ℓ' component is unique. These solutions are called *normalized thin flows with resetting* (NTFR) and can be used to reconstruct a dynamic equilibrium by integration, proving the existence of equilibria. We refer to [1] for the existence and uniqueness of NTFR's and to [5] for a description of the integration algorithm and how to find the equilibrium inflows $f_e^+(\cdot)$.

Observe that there are only finitely many options for E^* and E'. Since the corresponding ℓ' is unique, it follows that the functions $\ell_v(\theta)$ will be uniquely defined and piecewise linear with finitely many options for the derivatives. Although the static flows $x_e(\theta)$ are not unique in general, one can still find an equilibrium in which these functions are also piecewise linear by fixing a specific x' in the NTFR for each pair E^*, E'.

3   Steady States

We say that a dynamic equilibrium attains a *steady state* if for sufficiently large times all the queues are frozen to a constant $z_e(\theta) \equiv z_e^*$. This is clearly equivalent to the fact that the arc travel times become constant equal to $\tau_e^* = \tau_e + q_e^*$ with $q_e^* = z_e^*/\nu_e$ the corresponding queuing times.

Lemma 1. *A dynamic equilibrium attains a steady state iff there exists some $\theta^* \geq 0$ such that $\ell'_v(\theta) = 1$ for every node $v \in V$ and all $\theta \geq \theta^*$.*

Proof. In a steady state we clearly have $\ell_v(\theta) = \theta + d_v^*$ where d_v^* is the minimum travel time from s to v with arc times τ_e^*, so that $\ell_v'(\theta) = 1$. Conversely, if all these derivatives are equal to 1 then $\ell_v(\theta) = \theta + d_v^*$ for some constant d_v^* and $\theta \geq \theta^*$. Moreover, an arc $e = vw$ with nonempty queue must be active so that $\ell_w(\theta) = T_e(\ell_v(\theta))$ which yields

$$z_e(\theta + d_v^*) = z_e(\ell_v(\theta)) = \nu_e(\ell_w(\theta) - \ell_v(\theta) - \tau_e) = \nu_e(d_w^* - d_v^* - \tau_e)$$

which shows that all queues eventually become constant.   □

Theorem 2. *Consider a steady state with queues $z_e^* \geq 0$ and let d_v^* be the minimum travel times with arc times $\tau_e^* = \tau_e + q_e^*$ where $q_e^* = z_e^*/\nu_e$. Let (ℓ', x') with $\ell_v' = 1$ be a corresponding NTFR and denote by $\mathcal{F}_0$ the set of st-flows satisfying (10). Then x' and (d^*, q^*) are optimal solutions for the following pair of dual linear programs:*

$$\min_{x'} \quad \sum_{e \in E} \tau_e x_e'$$
$$s.t. \qquad x' \in \mathcal{F}_0 \tag{P}$$
$$\qquad 0 \leq x_e' \leq \nu_e \qquad \forall e \in E,$$

$$\max_{d,q} \quad u_0 d_t - \sum_{e \in E} \nu_e q_e \tag{D}$$
$$s.t. \qquad d_s = 0$$
$$\qquad d_w \leq d_v + \tau_e + q_e \qquad \forall e = vw \in E$$
$$\qquad q_e \geq 0 \qquad \forall e \in E.$$

Proof. Clearly (d^*, q^*) is feasible for (D). Also (10) gives $x' \in \mathcal{F}_0$, while (13) implies that if $x_e' > 0$ then $1 = \rho_e(1, x_e')$. This implies that $x_e' \leq \nu_e$, so x' is feasible for (P). If $x_e' > 0$ then by (13) the arc e is active, and hence $d_w^* = d_v^* + \tau_e + q_e^*$. And if $q_e^* > 0$, then (11) implies that $1 \leq \rho_e(1, x_e') = x_e'/\nu_e$, which yields $x_e' = \nu_e$. This proves that x' and (d^*, q^*) are complementary solutions, and hence are optimal for (P) and (D) respectively.   □

According to this result, if a dynamic equilibrium eventually settles to a steady state then the corresponding queue lengths must be optimal for (D). Generically (after perturbing capacities) this linear program has a unique solution in which case the steady state is fully characterized. Otherwise, if (D) has multiple solutions it is not evident which queue lengths will be obtained in steady state. Note that even if the min cost flow for (P) is unique, this does not mean that only one steady state situation is possible because there may be flexibility in the queue lengths. For instance, if $u_0 = 1$ and the network has a single link from s to t of unit capacity, if we create a queue of some length at time 0 this queue will remain in the steady state solution. This point will be further discussed in Example 3 in Sect. 5.

Remark. It is not difficult to show that when we start with initial conditions $z_e(0) = z_e^*$ where $z_e^* = \nu_e q_e^*$ with q^* optimal for (D), then the dynamic equilibrium is already at a steady state and the queues remain constant.

4   Convergence to a Steady State

In this section we prove that a steady state exists and that it is actually reached in finite time. To this end we introduce a Lyapunov potential function that increases along the evolution of the dynamic equilibrium. The potential function is inspired from the previous dual program and is given by

$$\Phi(\theta) := u_0(\ell_t(\theta) - \ell_s(\theta)) - \sum_{e=vw\in E} z_e(\ell_v(\theta)).$$

Theorem 3. $\Phi'(\theta)$ is nonnegative for all θ and strictly positive unless the dynamic equilibrium has reached a steady state.

Proof. The queues can be expressed as $z_e(\ell_v(\theta)) = \nu_e [\ell_w(\theta) - \ell_v(\theta) - \tau_e]_+$. Using the derivative of a max function and taking a NTFR (ℓ', x') at time θ, we thus obtain

$$\Phi'(\theta) = u_0(\ell_t' - \ell_s') - \sum_{e\in E_\theta'\setminus E_\theta^*} \nu_e[\ell_w' - \ell_v']_+ - \sum_{e\in E_\theta^*} \nu_e(\ell_w' - \ell_v').$$

Now, for $e \in E_\theta'\setminus E_\theta^*$ we have $\ell_w' \le \rho_e(\ell_v', x_e') = \ell_v'$ if $x_e' = 0$ and $\ell_w' = \rho_e(\ell_v', x_e') \ge \ell_v'$ if $x_e' > 0$, so that letting $E_\theta^+ = E_\theta^* \cup \{e \in E_\theta'\setminus E_\theta^* : x_e' > 0\}$ we may write

$$\Phi'(\theta) = u_0(\ell_t' - \ell_s') - \sum_{e\in E_\theta^+} \nu_e(\ell_w' - \ell_v').$$

Let us introduce a return arc ts with capacity $\nu_{ts} = u_0$ and flow $x_{ts}' = u_0$ so that x' is a circulation. Let $E_\theta^r = E_\theta^+ \cup \{ts\}$ and for each $e = vw \in E_\theta^r$ define the function

$$H_e(z) = \begin{cases} 1 & \text{if } \ell_v' \le z < \ell_w' \\ -1 & \text{if } \ell_w' \le z < \ell_v' \\ 0 & \text{otherwise.} \end{cases}$$

Then the derivative $\Phi'(\theta)$ can be expressed as

$$\Phi'(\theta) = -\int_0^\infty \sum_{e\in E_\theta^r} \nu_e H_e(z)\, dz.$$

For the remainder of the proof, let $\delta(S)$ denote the edges in E_θ^r crossing S (and similarly for $\delta^+(S)$ and $\delta^-(S)$). Let $V_z = \{v : \ell_v' \le z\}$ and consider an arc $e = vw \in E_\theta^+$. If $e \in \delta^+(V_z)$ then $\ell_v' \le z < \ell_w'$ and therefore $\ell_w' = x_e'/\nu_e$. Similarly, if $e \in \delta^-(V_z)$ then $\ell_w' \le z < \ell_v'$ which implies $e \in E_\theta^*$ and again

$\ell'_w = x'_e/\nu_e$. Hence $x'_e = \nu_e\ell'_w$ for all $e \in E_\theta^+ \cap \delta(V_z)$. This equality also holds for the return arc ts, while in the remaining arcs $x'_e = 0$. Hence

$$\sum_{e\in\delta^+(V_z)} \nu_e z \leq \sum_{e=vw\in\delta^+(V_z)} \nu_e\ell'_w = \sum_{e\in\delta^+(V_z)} x'_e = \sum_{e\in\delta^-(V_z)} x'_e = \sum_{e=vw\in\delta^-(V_z)} \nu_e\ell'_w \leq \sum_{e\in\delta^-(V_z)} \nu_e z$$

with strict inequality if $\delta^+(V_z)$ is nonempty. It follows that for all $z > 0$ we have

$$\sum_{e\in E_\theta^r} \nu_e H_e(z) = \sum_{e\in\delta^+(V_z)} \nu_e - \sum_{e\in\delta^-(V_z)} \nu_e \leq 0$$

and therefore $\Phi'(\theta) \geq 0$ with strict inequality unless $\delta^+(V_z)$ is empty for almost all $z \geq 0$. The latter occurs iff all ℓ'_v are equal. Since $\ell'_s = 1$ it follows that $\Phi'(\theta) = 0$ iff $\ell'_v = 1$ for all v which by Lemma 1 characterizes a steady state. $\square$

Theorem 4. Let $\bar{\nu} = \sum_{e\in C} \nu_e$ be the minimal queuing capacity among all st-cuts C. If $u_0 \leq \bar{\nu}$ then the dynamic equilibrium attains a steady state in finite time.

Proof. From Theorem 3 it follows that there is some $\kappa > 0$ such that $\Phi'(\theta) \geq \kappa$ for every phase other than the steady state. This is simply because the thin flow depends only on the current shortest path network E_θ and the set of queuing edges E_θ^*, and so there are only finitely many possible derivatives.

Thus, in order to prove that a steady state is reached in finite time it suffices to show that $\Phi(\theta)$ remains bounded. To this end we note that the condition $u_0 \leq \bar{\nu}$ implies that (P) is feasible and hence it has a finite optimal value α. The conclusion then follows by noting that the point (d, q) with $d_v = \ell_v(\theta) - \ell_s(\theta)$ and $q_e = z_e(\ell_v(\theta))/\nu_e$ is feasible for the dual (D) so that $\Phi(\theta) \leq \alpha$. $\square$

Given that convergence to a steady state does happen in finite time, it is natural to ask for explicit bounds. It is easy to see that a polynomial (in the input size encoding) is impossible; simply consider a network consisting of two parallel links, one with capacity $1 - 2^L$ and length zero, the other with capacity 1 and length 1. The first phase, where all traffic takes the shorter edge, lasts until time 2^L. However, we can give a *pseudopolynomial* bound on the convergence time (and hence, queue lengths).

Theorem 5. *Consider an instance for which $u_0 \in \mathbb{Z}_+$ and $\nu_e \in \mathbb{Z}_+$ for all $e \in E$. Let $M = \sum_{e\in E} \nu_e$ and $T = \sum_{e\in E} \tau_e$. Then assuming the dynamic equilibrium attains a steady state, it is reached by time $O(MT)$, and moreover, the waiting time in any queue never exceeds $O(M^3T)$.*

The argument to bound the convergence time involves showing that the difference between the smallest and largest label derivative is not too small. Combining this with an upper bound on the rate at which any queue can grow yields the second claim. We delay the proof to the full version of the paper.

5   Some Conjectures and Counterexamples

While we have settled the finite-time convergence to a steady state, there are a number of questions about dynamic equilibria that remain open. In this section we discuss some conjectures and provide counterexamples to some of them.

As mentioned in the introduction a first conjecture would be that, similarly to what happens for static flows, the flow across any cut is always bounded by the inflow. This would provide a way to estimate the queues and to prove their boundedness. Unfortunately the property fails in a dynamic equilibrium. The reason for this is that flow entering the network at different times may experience different delays in such a way that they later superpose across an intermediate cut. The following instance with unit inflow $u_0 = 1$ exhibits an outflow rate of 13/12 during a time interval.

Example 1. Consider the network consisting of the vertices $\{s, v, t\}$ with edges $e_1 = (s, t), e_2 = (s, v), e_3 = (v, t), e_4 = (v, t)$ and inflow $u_0 = u$. Capacities are $\nu_1 = u/3, \nu_2 = 3u/4, \nu_3 = u/3$, and $\nu_4 = u$, and delays are $\tau_1 = \tau_4 = \tau$, and $\tau_2 = \tau_3 = 0$. In this instance one can compute the derivative of the distance labels at node t as

$$
\ell'_t(\theta) = \begin{cases} 3 & \text{for } \theta \in [0, \tau/2) \\ 3/2 & \text{for } \theta \in [\tau/2, \tau/2 + \tau/5) \\ 12/13 & \text{for } \theta \in [\tau/2 + \tau/5, 2\tau) \\ 1 & \text{for } \theta \in [2\tau, \infty) \end{cases} .
$$

Thus the amount of flow arriving at t at time $\ell_t(\theta)$ can readily be computed as $u/\ell'_t(\theta)$. If we consider the local time at node t this flow is then

$$
f_1^-(\theta) + f_3^-(\theta) + f_4^-(\theta) = \begin{cases} u/3 & \text{for } \theta \in [0, 3\tau/2) \\ 2u/3 & \text{for } \theta \in [3\tau/2, 9\tau/5) \\ 13u/12 & \text{for } \theta \in [9\tau/5, 3\tau) \\ u & \text{for } \theta \in [3\tau, \infty). \end{cases}
$$

By chaining together slightly modified copies of this instance, one can blow up the maximum outflow to any desired quantity, even with unit inflow. Notice that the length of the "pulse" in the above construction can be made as large as required, by choosing τ appropriately. This pulse can be used to drive a second copy of the construction, with larger u. Figure 3 shows the construction with two copies; there is a phase with outflow $(13/12)^2$. The phases before the pulse of the left gadget only produce a queue on e', which has no impact on the behavior except for essentially shortening e' and h'. We delay the details to the full version of the paper.

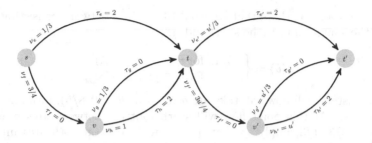

Fig. 3. Creating a larger pulse. Here, $u' = 13/12$.

Even though the previous example shows that intermediate flows can grow very large our main result states that a steady state is actually reached after finite time. This indeed implies that the queues remain bounded along the evolution of a dynamic equilibrium. However this also raises further questions. Indeed it is unclear whether the steady state is attained after finitely many phases of the Koch-Skutella algorithm. It is conceivable that in some situations the phases become shorter and shorter and that infinitely many of them occur in the finite time span before steady state is reached. The next example shows that there may actually be an exponential (in the input size) number of phases.

Example 2. Here we sketch the construction of an instance with an exponential number of phases; we defer the details to the full version of the paper. More precisely, for a given d we construct an instance with $\Omega(2^d)$ phases and $O(d^2)$ nodes. The main idea is to construct a "2-pulse" gadget, based on the "1-pulse" gadget described in Example 1. The outflow rate of this gadget has two, well-separated, periods where the outflow is large; outside of these two periods, the outflow is much smaller. Given a gadget with $\Omega(2^d)$ phases (call it H), we construct one with $\Omega(2^{d+1})$ phases roughly as follows. We begin with the 2-pulse gadget. To the output of this gadget, we attach both a single edge of small capacity and length 0 to the sink; and in parallel, we attach H. In between pulses, all flow uses the short low-capacity edge, and any queues in H decay. During each of the two pulses, flow enters H; inductively, this yields $\Omega(2^d)$ phases each time.

Knowing that the dynamic equilibrium always reaches a steady state, a natural question is whether steady state queues can be characterized without having to compute the full equilibrium evolution. While we already observe that this is the case when the dual problem (D) has a unique solution, which occurs generically, the following example suggests that this is likely not possible in general.

Example 3. Consider the network of Example 1, setting $\tau = 2$ and $u = 1$, with an extra node $\hat{t}$, which becomes the new sink, and two additional arcs, $a = (t, \hat{t})$ and $b = (t, \hat{t})$. Let $\nu_a = 2/3$, $\nu_b = 1/3$, $\tau_a = 0$, and $\tau_b = 1$. Clearly, up to time $3 + 3/5$ all flow will simply take arc a and will not queue at t. Therefore we can ignore this initial phase, and the queues that will form at equilibrium in arcs a

and b are the same as those that we would have in a network consisting of just nodes t (the source) and $\hat{t}$ (the sink) and inflow

$$u_0(\theta) = \begin{cases} 13/12 & \text{for } \theta \in [0, 2 + 2/5) \\ 1 & \text{for } \theta \in [2 + 2/5, \infty). \end{cases}$$

In this instance all flow will take arc a for time $\theta \in [0, 8/5)$, forming a queue $z_e(8/5) = 2/3$. At this point flow will start splitting between arcs a and b in proportions $2/3$, $1/3$, implying that queues will grow on both arcs until time $2 + 2/5$ where the steady state is achieved. The steady state queues will thus be $z_a^* = 32/45$ and $z_b^* = 1/45$. This example shows that the steady state queues are not minimal in any reasonable sense and that, furthermore, slightly changing the instance (e.g. τ_4) will change the steady state queues. Furthermore, if we slightly increase the capacity of arc b, say to $1/3 + \varepsilon$ the steady state queues jump to $z_a^* = 2/3$ and $z_b^* = 0$.

Additionally, one can observe from a slight variant of this instance, namely taking τ large and $\nu_b = 1/3 + \varepsilon$, that queues may grow very large in the transient and then go down to zero at steady state.

References

1. Cominetti, R., Correa, J., Larré, O.: Dynamic equilibria in fluid queueing networks. Oper. Res. **63**(1), 21–34 (2015)
2. Ford, L.R., Fulkerson, D.R.: Constructing maximal dynamic flows from static flows. Oper. Res. **6**, 419–433 (1958)
3. Friesz, T.L., Bernstein, D., Smith, T.E., Tobin, R.L., Wie, B.W.: A variational inequality formulation of the dynamic network user equilibrium problem. Ope. Res. **41**(1), 179–191 (1993)
4. Gale, D.: Transient flows in networks. Mich. Math. J. **6**, 59–63 (1959)
5. Koch, R., Skutella, M.: Nash equilibria and the price of anarchy for flows over time. Theor. Comput. Syst. **49**, 71–97 (2011)
6. Merchant, D.K., Nemhauser, G.L.: A model and an algorithm for the dynamic traffic assignment problems. Transp. Sci. **12**, 183–199 (1978)
7. Merchant, D.K., Nemhauser, G.L.: Optimality conditions for a dynamic traffic assignment model. Transp. Sci. **12**, 200–207 (1978)
8. Meunier, F., Wagner, N.: Equilibrium results for dynamic congestion games. Transp. Sci. **44**(4), 524–536 (2010)
9. Peeta, S., Ziliaskopoulos, A.K.: Foundations of dynamic traffic assignment: the past, the present and the future. Netw. Spat. Econ. **1**(3–4), 233–265 (2001)
10. Ran, B., Boyce, D.: Modeling Dynamic Transportation Networks. Springer, Berlin (1996)
11. Vickrey, W.S.: Congestion theory and transport investment. Am. Econ. Rev. **59**(2), 251–260 (1969)
12. Xu, Y.W., Wu, J.H., Florian, M., Marcotte, P., Zhu, D.L.: Advances in the continuous dynamic network loading problem. Transp. Sci. **33**(4), 341–353 (1999)
13. Zhu, D., Marcotte, P.: On the existence of solutions to the dynamic user equilibrium problem. Transp. Sci. **34**(4), 402–414 (2000)

A 4/5 - Approximation Algorithm
for the Maximum Traveling Salesman Problem

Szymon Dudycz, Jan Marcinkowski, Katarzyna Paluch$^{(\boxtimes)}$,
and Bartosz Rybicki

Institute of Computer Science, University of Wrocław, Wrocław, Poland
szymon.dudycz@gmail.com, jasiekmarc@cs.uni.wroc.pl, abraka@cs.uni.wroc.pl,
rybicki.bartek@gmail.com

Abstract. In the maximum traveling salesman problem (Max TSP) we
are given a complete undirected graph with nonnegative weights on the
edges and we wish to compute a traveling salesman tour of maximum
weight. We present a fast combinatorial $\frac{4}{5}$ – approximation algorithm
for Max TSP. The previous best approximation for this problem was
$\frac{7}{9}$. The new algorithm is based on a technique of eliminating difficult
subgraphs via gadgets with *half-edges*, a new method of edge coloring
and a technique of exchanging edges.

1 Introduction

The Maximum Traveling Salesman Problem (Max TSP) is a classical variant of
the famous Traveling Salesman Problem. In the problem we are given a complete
undirected graph $G = (V, E)$ with nonnegative weights on the edges and we
aim to compute a traveling salesman tour of maximum weight. Max TSP, also
informally known as the "taxicab ripoff problem", is both of theoretical and
practical interest.

Previous approximations of Max TSP have found applications in combinatorics and computational biology: the problem is useful in understanding RNA
interactions [27] and providing algorithms for compressing the results of DNA
sequencing [26]. It has also been applied to the problem of finding a maximum
weight triangle cover of the graph [14] and to a combinatorial problem called
bandpass-2 [7], where we are supposed to find the best permutation of rows in a
boolean-valued matrix, so that the weighted sum of structures called *bandpasses*
is maximised.

Previous Results. The first approximation algorithms for Max TSP were
devised by Fisher et al. [10]. They showed several algorithms having approxima-
tion ratio $\frac{1}{2}$ and one with a guarantee of $\frac{2}{3}$. In [16] Kosaraju, Park and Stein presented an improved algorithm giving a ratio of $\frac{19}{27}$ [4]. This was in turn improved

Partly supported by Polish National Science Center grant UMO-
2013/11/B/ST6/01748.

J. Marcinkowski—Partially supported by Polish NSC grant 2015/18/E/ST6/00456.

© Springer International Publishing AG 2017
F. Eisenbrand and J. Koenemann (Eds.): IPCO 2017, LNCS 10328, pp. 173–185, 2017.
DOI: 10.1007/978-3-319-59250-3_15

by Hassin and Rubinstein, who gave a $\frac{5}{7}$- approximation [12]. In the meantime
Serdyukov [25] presented (in Russian) a simple and elegant $\frac{3}{4}$-approximation algorithm. The algorithm is deterministic and runs in $O(n^3)$, where n denotes the number of vertices in the graph. Afterwards, Hassin and Rubinstein gave [13] a randomized algorithm with expected approximation ratio of at least $\frac{25(1-\epsilon)}{33-32\epsilon}$ and running
in $O(n^2(n + 2^{1/\epsilon}))$, where ϵ is an arbitrarily small constant. The first deterministic approximation algorithm with the ratio better than $\frac{3}{4}$ was given in [6] by Chen
et al. It is a $\frac{61}{81}$-approximation through a non-trivial derandomization of the algorithm from [13] that runs in $O(n^3)$. The currently best known approximation given
by Paluch et al. [22] achieves the ratio of $\frac{7}{9}$. Its running time is also $O(n^3)$.

Related Work. It is known that Max TSP is max-SNP-hard [3], so a constant
$\delta < 1$ exists, which is an upper bound on the approximation ratio of any algorithm for this problem. The geometric version of the problem, where all vertices
are in R^d and the weight of each edge is defined as the Euclidean distance of its
endpoints, was considered in [2] and shown to be solvable in polynomial time for
$d = 2$ and NP-hard for $d > 2$. Other metrics are also considered in that paper.

Regarding the path version of Max TSP – Max TSPP (the Maximum Traveling Salesman Path Problem), the approximation algorithms with ratios correspondingly $\frac{1}{2}$ and $\frac{2}{3}$ have been given in [19]. The first one for the case when
both endpoints of the path are specified and the other for the case when only
one endpoint is given.

Another related problem is called the maximum scatter TSP (see [1]), where
the goal is to find a TSP tour (or a path) maximizing the weight of the lightest
edge selected in the solution. The problem is motivated by medical imaging and
some manufacturing applications. In general there is no constant approximation
for this problem, but if the weights of the edges obey the triangle inequality, it
is possible to give a $\frac{1}{2}$-approximation algorithm. That paper also studies a more
general version of the maximum scatter TSP – the max-min-m-neighbour TSP.
The improved approximation results for the max-min-2-neighbour problem have
been given in [8].

The maximum metric symmetric traveling salesman problem, in which the
edge weights satisfy the triangle inequality - the best approximation factor is
$\frac{7}{8}$ [18]. For the maximum asymmetric traveling salesman problem with triangle
inequality the best approximation ratio currently equals $\frac{35}{44}$ [17].

In the Maximum Latency TSP problem we are given a complete undirected
graph with vertices $v_0, v_1, \ldots, v_n$. Our task is to find a Hamiltonian path starting
at a fixed vertex v_0, which maximizes the total latency of the vertices. If in a
given path P the weight of the i-th edge is w_i, then the latency of the j-th vertex
is $L_j = \sum_{i=1}^{j} w_i$ and the total latency is defined as $L(P) = \sum_{j=1}^{n} L_j$. A ratio
$\frac{1}{2}$-approximation algorithm for the metric version of the problem is presented
in [5]. Improved ratios for this and other versions (directed, nonmetric) of the
problem are shown in [11].

Our Approach and Results. We begin with computing a maximum weight *cycle cover* C_{max} of G. A cycle cover of a graph G is a collection of cycles such that each vertex belongs to exactly one of them. The weight of a maximum weight cycle cover C_{max} is an upper bound on OPT, where by OPT we denote the weight of a maximum weight traveling salesman tour. By computing a maximum weight perfect matching M we get another, even simpler than C_{max}, upper bound – on $OPT/2$. From C_{max} and M we build a multigraph G_1 which consists of two copies of C_{max} and one copy of M, (for each edge e of G the multigraph G_1 contains between zero and three copies of e). Thus the total weight of the edges of G_1 is at least $\frac{5}{2} OPT$. Next we would like to *path-3-color* G_1, that is to color the edges of G_1 with three colors, so that each color class contains only vertex-disjoint paths. The paths from the color class with maximum weight can then be patched in an arbitrary manner into a tour of weight at least $\frac{5}{6} OPT$.

Technique of Eliminating Difficult Subgraphs via Half-edges. Not every multigraph G_1 can, however, be path-3-colored. For example, a subgraph of G_1 obtained from a triangle T of C_{max} such that M contains one of the edges of T (such triangle is called a *3-kite of G_1*) cannot be path-3-colored as, clearly, it is impossible to color such seven edges with three colors and not create a monochromatic triangle. Similarly, a subgraph of G_1 obtained from a square S (i.e., a cycle of length four) of C_{max} such that M contains two edges connecting vertices of S (such square is called a *4-kite*) is not path-3-colorable. To find a way around this difficulty, we compute another cycle cover C_2 *improving C_{max} with respect to M*, which is a cycle cover that does not contain any 3-kite or 4-kite of G_1 and whose weight is also at least OPT. An important feature of C_2 is that it may contain *half-edges*. A half-edge of an edge e is, informally speaking, a half of the edge e that contains exactly one of its endpoints. Half-edges have already been introduced in [21]. Computing C_2 is done via a tailored reduction to a maximum weight perfect matching. It is, to some degree, similar to computing a directed cycle cover without length-two cycles in [21], but for Max TSP we need much more complex gadgets.

From one copy of C_2 and M we build another multigraph G_2 with weight at least $\frac{3}{2} OPT$. It turns out that G_2 can always be *path-2-colored*. The multigraph G_1 may be non-path-3-colorable – if it contains at least one kite. We notice, however, that if we remove one arbitrary edge from each kite, then G_1 becomes path-3-colorable. The edges removed from G_1 are added to G_2. As a result, the modified G_2 may cease to be path-2-colorable. To remedy this, we in turn remove some edges from G_2 and add them to G_1. In other words, we find two disjoint sets of edges – a set $F_1 \subseteq G_1$ and a set $F_2 \subseteq G_2$, called *exchange sets*, such that the multigraph $G'_1 = G_1 \backslash F_1 \cup F_2$ is path-3-colorable and the multigraph $G'_2 = G_2 \backslash F_2 \cup F_1$ is path-2-colorable. Since G_1 and G_2 have the total weight at least $4 OPT$, by path-3-coloring G'_1 and path-2-coloring G'_2 we obtain a $\frac{4}{5}$-approximate solution to Max TSP.

Edge Coloring. The presented algorithms for path-3-coloring and path-2-coloring are essentially based on a simple notion of a *safe edge* – an edge colored in such a way that it is guaranteed not to belong to any monochromatic cycle, used in an inductive way. The adopted approach may appear simple and straightforward. For comparison, let us point out that the method of path-3-coloring the multigraph obtained from two directed cycle covers described in [15] is rather convoluted.

Generally, the new techniques are somewhat similar to the ones used for the directed version of the problem – Max ATSP – in [20]. We are convinced that they will prove useful for other problems related with TSP, cycle covers or matchings.

The main result of the paper is

Theorem 1. *There exists a $\frac{4}{5}$-approximation algorithm for Max TSP. Its running time is $O(n^3)$ if the graph has an even number of vertices and $O(n^5)$ otherwise.*

Algorithm 1. A $\frac{4}{5}$-approximation for Max TSP

1: $C_{max} \leftarrow$ a maximum-weight cycle cover of G
2: $M \leftarrow$ a maximum-weight perfect matching in G
3: $G_1 \leftarrow C_{max} \uplus C_{max} \uplus M$
4: path-3-color G_1 with colors of $\mathcal{K}_3 = \{1, 2, 3\}$ leaving kites and edges of M incident to kites uncolored.   ▷ Section 2
5: $C_2 \leftarrow$ a maximum-weight *relaxed cycle cover improving C_{max} with respect to M*.   ▷ Section 3
6: $G_2 \leftarrow C_2 \uplus M$
7: $F_1 \subset C_{max}, F_2 \subset C_2 \leftarrow$ sets of edges such that the multigraph $G_1' = G_1 \backslash F_1 \cup F_2$ is path-3-colorable and $G_2' = G_2 \backslash F_2 \cup F_1$ is path-2-colorable.   ▷ Lemma 5
8: Path-2-color G_2' with colors of $\mathcal{K}_2 = \{4, 5\}$.   ▷ Full version of the paper
9: Extend the partial path-3-coloring of G_1 to the complete path-3-coloring of G_1'. ▷ Full version of the paper
10: Choose the heaviest color class $k \in \mathcal{K}_3 \cup \mathcal{K}_2$. Complete the disjoint paths of color k into a traveling salesman tour in an arbitrary way.

All missing proofs are contained in the full version of this paper [9].

2   Path-3-Coloring of G_1

We compute a maximum weight cycle cover C_{max} of a given complete undirected graph $G = (V, E)$ and a maximum weight perfect matching M of G. We are going to call cycles of length i, i.e., consisting of i edges i-**cycles**. Also sometimes 3-cycles will be called **triangles** and 4-cycles – **squares**. The multigraph G_1 consists of two copies of C_{max} and one copy of M. We want to color each edge of G_1 with one of three colors of $\mathcal{K}_3 = \{1, 2, 3\}$ so that each color class consists of vertex-disjoint paths. The *graph* G_1 is a subgraph of the *multigraph* G_1 that contains an edge (u, v) iff the multigraph G_1 contains an edge between u and v. The path-3-coloring of G_1 can be equivalently defined as coloring each edge of (the graph) G_1 with the number of colors equal to the number of copies contained in the multigraph G_1. From this time on, unless stated otherwise, G_1 denotes a graph and not a multigraph.

We say that a colored edge e of G_1 is **safe** if no matter how we color the so far uncolored edges of G_1 e is guaranteed not to belong to any monochromatic cycle of G_1. An edge e of M is said to be **external** if its two endpoints belong to two different cycles of C_{max}. Otherwise, e is **internal**. We say that an edge e is incident to a cycle c if it is incident to at least one vertex of c.

We prove the following useful lemma.

Lemma 1. *Consider a partial coloring of G_1. Let c be any cycle of C_{max} such that for each color $k \in K_3$ there exists an edge of M incident to c that is colored k. Then we can color c so that each edge of c and each edge incident to one of the edges of c is safe.*

Proof. The proposed procedure of coloring c is as follows.

If there exists an edge of c that also belongs to M, we color it with all three colors of K_3. For each uncolored edge of M incident to c, we color it with an arbitrary color of K_3. Next, we orient the edges of c (in any of the two ways) so that c becomes a directed cycle c. Let $e = (u, v)$ be any uncolored edge of c oriented from u to v. Then, there exists an edge e' of M incident to u. If e' is contained in c, then we color e with any two colors of K_3. Otherwise e' is colored with some color k of K_3. Then we color e with the two colors belonging to $K_3 \backslash k$. First, no vertex of c has three incident edges colored with the same color, as for each vertex its outgoing edge is colored with different colors than an incident matching edge. Second, as for each color $k \in K_3$ there is a matching edge incident to c colored with k, there exists an edge of c that is not colored k, thus c does not belong to any color class, i.e. there exists no color $k \in K_3$ such that each edge of c is colored with k. Let us consider now any edge $e = (u, v)$ of M incident to some edge of c and not belonging to c. The edge e is colored with some color k. Suppose also that vertex u belongs to c (v may or may not belong to c.) Let u' be any other vertex of c such that some edge of $M \backslash C_{max}$ colored k is incident to it (u' may be equal to v if e is internal). To show that e is safe, it suffices to show that there exists no path consisting of edges of $c \cup M$ that connects u and u' and whose every edge is colored k. However, by the way we color edges of c we know that the outgoing edges of u and u' are not colored with k because of the way we oriented the cycle, there is no path connecting u and u' contained in c that starts and ends with incoming edge.   □

For each cycle c of C_{max} we define its **degree of flexibility** denoted as $flex(c)$ and its **colorfulness**, denoted as $col(c)$. The degree of flexibility of a cycle c is the number of internal edges of M incident to c and the colorfulness of c is the number of colors of $\mathcal{K}_3$ that are used for coloring the external edges of M incident to c.

From Lemma 1 we can easily derive.

Lemma 2. *If a cycle c of C_{max} is such that $flex(c) + col(c) \geq 3$, then we can color c so that each edge of c and each edge incident to one of the edges of c is safe.*

Sometimes, even if a cycle c of C_{max} is such that $flex(c) + col(c) < 3$, we can color the edges of c so that each of them is safe. For example, suppose that c is a square consisting of edges $e_1, \ldots, e_4$ and there are four external edges of M incident to c, all colored 1. Suppose also that each external edge incident to c is already safe. Then we can color e_1 with 1 and 2, e_3 with 1 and 3 and both e_2 and e_4 with 2 and 3. We can notice that e_1 is guaranteed not to belong to a cycle colored 1 because external edges incident to e_1 are colored 1 and are safe. Analogously, we can easily check that each other edge of c is safe. However, for example, a triangle t of C_{max} that has three external edges of M incident to it, all colored with the same color of $\mathcal{K}_3$, cannot be colored in such a way that it does not contain a monochromatic cycle.

Consider a cycle c of C_{max} such that every external edge of M incident to c is colored. We say that c is **nice** if and only if (1) $flex(c) + col(c) \geq 3$ or (2) c contains at least $3 - flex(c) - col(c)$ vertex-disjoint edges, each of which has the property that it has exactly two incident external edges of M and the two external edges of M incident to it are colored with the same color of $\mathcal{K}_3$ or (3) c is a square such that $flex(c) = 1$.

Otherwise we say that c is **blocked**. We can see that a cycle c of C_{max} is blocked if and only if

- c is a triangle and all external edges of M incident to c are colored with the same color of $\mathcal{K}_3$,
- c is a square with two internal edges of M incident to it ($flex(c) = 2$),
- c is a cycle of even length, $flex(c) = 0$ and there exist two colors $k_1, k_2 \in \mathcal{K}_3$ such that external edges of M incident to c are colored alternately with k_1 and k_2.

Among blocked cycles we distinguish kites. We say that a cycle c is a **kite** if it is a triangle such that $flex(c) = 1$ and then we call it a 3-**kite** or it is a square, whose two edges belong to M (so $flex(c) = 2$) - called a 4-**kite**. We can assume that a square with two diagonals in M will not occur, as diagonals are heavier than any two opposite edges in this square (as they are in M), so they would be included in C_{max}. A cycle of C_{max} which is not a kite is said to be **non-kite**.

Now, we are ready to state the algorithm for path-3-coloring G_1. It is presented as Algorithm 2.

Lemma 3. *Let c be a non-kite cycle of C_{max} that at some step of Algorithm Color G_1 has the fewest uncolored external edges incident to it. Then, it is always*

Algorithm 2. Color G_1

1: **while** $\exists$ an uncolored non-kite cycle of C_{max} **do**
2:     $C \leftarrow$ a non-kite uncolored cycle of C_{max} with the fewest uncolored external
        edges incident to it.
3:     Color uncolored external edges incident to C so that no other
        cycle of C_{max} becomes blocked and either $flex(C) + col(C) \geq 3$ or
        its external matching edges are all safe.   ▷ Lemma 3
4:     Color C and internal edges incident to it in such a way, that each edge
        incident to C is safe.   ▷ Lemma 4
5: **end while**

---

possible to color all uncolored external edges incident to c so that no non-kite cycle of C_{max} becomes blocked. Moreover, if c has at least two uncolored external edges incident to c then, additionally, it is always possible to do it in such a way that $flex(c) + col(c) \geq 3$. If c has exactly one uncolored external edge e of M incident to it, then we can color e so that $flex(c) + col(c) \geq 3$ or so that e is safe.

From the above lemma we get

Corollary 1. *After all external edges are colored, each of them is incident to a cycle c of C_{max} such that $flex(c) + col(c) \geq 3$ or is safe.*

Lemma 4. *Let c be a nice cycle of C_{max} whose all incident external edges of M are already colored and safe. Then it is always possible to color c and internal edges incident to c in such a way that each edge incident to c is safe.*

3   A Cycle Cover Improving C_{max} with Respect to M

Since C_{max} may contain kites, we may not be able to path-3-color G_1. Therefore, our next aim is to compute another cycle cover C_2 of G such that it does not contain any kite of C_{max} and whose weight is an upper bound on OPT. Since computing such C_2 may be hard, we relax the notion of a cycle cover and allow C_2 to contain **half-edges**. A half-edge of the edge e is, informally speaking, a half of the edge e that contains exactly one of the endpoints of e. Let us also point out here that C_2 may contain kites which do not belong to C_{max}.

We say that an edge (u, v) is a **kite-edge** if u and v belong to the same kite (so it can be a side of a kite, but also a diagonal of a 4-kite). Every kite-edge $e = (u, v)$ is split into two half edges (u, x_e) and (x_e, v), each carrying half of the weight of e. The graph $\tilde{G} = (\tilde{V}, \tilde{E})$ will be G with kite-edges replaced with half-edges.

Definition 1. *A **relaxed cycle cover improving C_{max} with respect to M** is a subset $\tilde{C} \subseteq \tilde{E}$ such that*

(i) each vertex in V has exactly two incident edges in $\tilde{C}$;

(ii) for each 3-kite $\mathcal{T}$ of C_{max} the number of half-edges of kite-edges of $\mathcal{T}$ contained in $\tilde{C}$ is even and not greater than four;

(iii) for each 4-kite $\mathcal{S}$ of C_{max} the number of half-edges of kite-edges of $\mathcal{S}$ contained in $\tilde{C}$ is even and not greater than six.

To compute a *relaxed cycle cover* C_2 improving C_{max} with respect to M we construct the following graph $G' = (V', E')$ (by replacing kites with gadgets). The set of vertices V' is a superset of the set of verices $V(G)$. For each kite-edge (u, v) of G we add two vertices x_v^u, x_u^v to V' and edges $(u, x_v^u), (x_u^v, v)$ to E' (these represent the half-edges). For each kite-edge (u, v) which is not a diagonal of a 4-kite or one of the non-matching edges in 3-kite (for each 3-kite we choose arbitrarily one of them) we add also an edge (x_v^u, x_u^v). The edge (x_v^u, x_u^v) has weight 0 in G' and each of the edges $(u, x_v^u), (x_u^v, v)$ has weight equal to $\frac{1}{2}w(u, v)$. Each of the vertices x_v^u, x_u^v is called *a splitting vertex of the edge* (u, v).

For each 3-kite $\mathcal{T}$ on vertices u, v, w we add two vertices p^T, q^T to V'. Let's assume that u is incident to external edge of M and that (x_w^u, x_u^w) was the side not added to G'. The vertex p^T is connected to the splitting vertices of edges of $\mathcal{T}$ that are neighbors of u, i.e. to vertices x_v^u, x_w^u and to vertex x_w^v. The vertex q^T is connected to every other splitting vertex of $\mathcal{T}$, i.e. x_u^w, x_v^w, x_u^v. All edges incident to vertices p^T, q^T have weight 0 in G'.

For each 4-kite $\mathcal{S}$ of C_{max} on vertices u, v, w, z we add five vertices $p_u^S, p_v^S, p_w^S, p_z^S, q^S$ to V'. Vertex p_u^S is connected to the splitting vertices of edges of $\mathcal{S}$ that are neighbors of u, i.e. to vertices x_v^u, x_w^u, x_z^u. Vertices p_v^S, p_w^S, p_z^S are con-

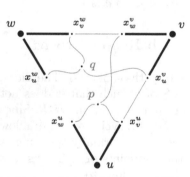

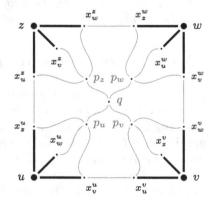

(a) $b(u) = b(v) = b(w) = 2, \forall k, l \in \{u, v, w\}$ $b(x_l^k) = 1, b(p) = b(q) = 1$

(b) $b(u) = b(v) = b(w) = b(z) = 2,$ $\forall k, l \in \{u, v, w, z\}$ $b(x_k^l) = 1, b(p_u) = b(p_v) = b(p_w) = b(p_z) = b(q) = 2$

Fig. 1. Gadgets for 3-kites **(a)** and 4-kites **(b)** of G_1 in graph G. Half-edges corresponding to the original edges are thickened, the auxiliary edges are thin. Original vertices (thick dots) are connected with all the other original vertices of graph G. The auxiliary vertices have no connections outside of the gadget. The figures are subtitled with the specifications of $b(v)$ values for different vertices. For a vertex t with $b(t) = i$, the resulting b-matching will contain exactly i edges ending in t.

nected analogously. Vertex q^S is connected to vertices $p_u^S, p_v^S, p_w^S, p_z^S$. All edges incident to vertices $p_u^S, p_v^S, p_w^S, p_z^S, q^S$ have weight 0.

For each edge (u, v) of G that is not a kite-edge we add it to E' with weight $w(u, v)$.

We reduce the problem of computing a relaxed cycle cover improving C_{max} with respect to M, to the problem of computing a perfect b-matching in the graph G'. We define the function $b : V' \to \mathbb{N}$ in the following way. For each vertex $v \in V$ we set $b(v) = 2$. For each splitting vertex v' of some problematic edge we set $b(v') = 1$. For all vertices p^T and q^T, where T denotes a 3-kite of C_{max} we have $b(p^T) = b(q^T) = 1$. For all vertices p_u^S and q^S, where S denotes a 4-kite of C_{max} and u one of its vertices we have $b(p_u^S) = b(q^S) = 2$ (Fig. 1).

Theorem 2. *Any perfect b-matching of G' yields a relaxed cycle cover C_2 improving C_{max} with respect to M. A maximum weight perfect b-matching of G' yields a relaxed cycle cover C_2 improving C_{max} with respect to M such that $w(C_2) \geq OPT$.*

4   Exchange Sets F_1, F_2 and Path-2-Coloring of G_2'

The multigraph G_2 is constructed from one copy of the relaxed cycle cover C_2 and one copy of the maximum weight perfect matching M. Since C_2 may contain half-edges and we want G_2 to contain only edges of G, for each half-edge of edge (u, v) contained in C_2, we will either include the whole edge (u, v) in G_2 or not include it at all. While doing so we have to ensure that the total weight of the constructed multigraph G_2 is at least $\frac{3}{2}OPT$.

The main idea behind deciding which half-edges are extended to full edges and included in G_2 is that we construct two sets Z_1 and Z_2 – for each kite in G_1 we distribute its edges corresponding to the half-edges so that half of them go into the set Z_1 and the other half to Z_2. (Note that by Definition 1 each kite in G_1 contains an even number of half-edges in C_2.) Let $I(C_2)$ denote the set consisting of whole edges of G contained in C_2. This way $w(C_2) = w(I(C_2)) + \frac{1}{2}(w(Z_1) + w(Z_2))$. Next, let Z denote the one of the sets Z_1 and Z_2 with larger weight. Then G_2 is defined as a multiset consisting of edges of M, edges of $I(C_2)$ and edges of Z. We reach the following

Fact 1. *The total weight of the constructed multigraph G_2 is at least $\frac{3}{2}OPT$.*

Proof. The weight of M is at least $\frac{1}{2}OPT$. The weight of $w(C_2) = w(I(C_2)) + \frac{1}{2}(w(Z_1) + w(Z_2))$ is at least OPT. Since $w(Z) = \max\{w(Z_1), w(Z_2)\}$, we conclude that $w(I(C_2)) + w(Z) \geq w(C_2)$. $\square$

If C_{max} contains at least one kite, G_1 is non-path-3-colorable. We can however notice, that if we remove one edge from each kite in the multigraph G_1, then the obtained multigraph is path-3-colorable.

If we manage to construct a set F_1 containing one edge from each kite, such that additionally the multigraph $G_2 \cup F_1$ is path-2-colorable, then we have a

$\frac{4}{5}$-approximation of Max TSP immediately. Since computing such F_1 may be difficult, we allow, in turn, certain edges of C_2 to be removed from G_2 and added to G_1. Thus, roughly, our goal is to compute such disjoint sets F_1, F_2 that:

1. $F_1 \subset C_{max}$ contains at least one edge of each kite;
2. $F_2 \subset I(C_2)$ contains one edge per each kite $C \in C_{max}$;
3. the multigraph $G'_1 = G_1 \backslash F_1 \cup F_2$ is path-3-colorable;
4. the multigraph $G'_2 = G_2 \backslash F_2 \cup F_1$ is path-2-colorable.

Let F_1 and F_2 be two sets of edges that satisfy properties 1. and 2. of the above. Then the set of edges $C'_2 = (I(C_2) \cup Z \cup F_1) \backslash F_2$ can be partitioned into *cycles and paths of* G'_2, where G'_2 denotes the resulting multigraph $G_2 \backslash F_2 \cup F_1$. The partition of C'_2 into cycles and paths is carried out in such a way that two incident edges of C'_2 belonging to a common path or cycle of C_2, belong also to a common path or cycle of C'_2 (and G'_2). Also, the partition is maximal, i.e., we can't add any edge e of C'_2 to any path $\mathcal{P}$ of G'_2 so that $\mathcal{P} \cup \{e\}$ is also a path or cycle of G'_2.

We say that e is a **double edge** of G'_2 if the multigraph G'_2 contains two copies of e. In any path-2-coloring of G'_2 every double edge must have both colors of $\mathcal{K}_2$ assigned to it.

We observe that in order for G'_2 to be path-2-colorable, we have to guarantee that there does not exist a cycle C of G'_2 of odd length l that has l incident double edges. When every two consecutive edges of C are incident to some double edge, they must be assigned different colors of $\mathcal{K}_2$ and if the length of C is odd, this is clearly impossible. The way to avoid this is to choose one edge of each such potential cycle and add it to F_2.

We say that a path $\mathcal{P}$ of G'_2 beginning at w and ending at v is **amenable** if

(i) neither v nor w has degree 4 in G'_2, or
(ii) v has degree 4, w has degree smaller than 4 and $\mathcal{P}$ ends with a double edge, the last-but-one edge of $\mathcal{P}$ is a double edge or the last-but-one and the last-but-three vertices in $\mathcal{P}$ are matched in M.

It turns out that G'_2 that does not contain odd cycles described above and whose every path is amenable is path-2-colorable — we show it in the full version of the paper. To facilitate the construction of G'_2, whose every path is amenable and to ensure that F_1 and F_2 have certain other useful properties we create two opposite orientations of $I(C_2)$: D_1 and D_2. In each of these orientations $I(C_2)$ contains directed cycles and paths and each kite has the same number of incoming and outgoing edges. This can be achieved by pairing the endpoints of paths ending at the same kite and combining them. For example, let us consider a 3-kite in Fig. 2. C_2 contains half-edges $h_1 = (w, x_{\{u,w\}})$ and $h_2 = (v, x_{\{v,w\}})$ of a certain 3-kite $\mathcal{T}$, so for the purpose of orientation we replace h_1 and h_2 with an edge (v, w). Then, if for example C_2 contains edges $e_1 = (w', w), e_2 = (v', v)$ in the orientation in which e_1 is directed from w' to w, the edge e_2 is directed from v to v' and vice versa.

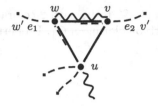   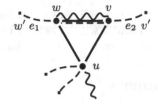

(a) Edges of C_2 incident to some 3-kite   (b) Graph used for purpose of orienting paths and cycles.

Fig. 2. Example of creating orientations D_1 and D_2

Apart from the whole edges C_2 also contains the half-edges. Let $H(C_2)$ denote the set of the edges of G such that C_2 contains exactly one half-edge of each of these edges. We would like to partition $H(C_2)$ into two sets Z_1, Z_2 so that for each kite c half of the edges of $H(C_2)$ is contained in Z_1 and the other half in Z_2. We associate Z_1 with orientation D_1 and Z_2 with orientation D_2. Thus, we assume that D_1 contains Z_1, with the edges of Z_1 being oriented in a consistent way with the edges of $I(C_2)$ under orientation D_1, and D_2 contains Z_2, with its edges being oriented accordingly. Depending on which of the sets Z_1, Z_2 has bigger weight, we either choose the orientation D_1 or D_2. Hence, from now on, we assume that the edges of $I(C_2) \cup Z$ are directed.

For example, for the triangle $\mathcal{T}$ described above (and presented in Fig. 2), the partition may be as follows. If e_1 is oriented from w to w' in D_1, then we assume that h_1 is in Z_1 and h_2 is in Z_2. Therefore, we can guarantee, that if h_1 is in Z, e_1 is oriented from v to v'.

The exact details of the construction of Z_1 and Z_2 are given in the proof of Lemma 5.

Lemma 5. *It is possible to compute the sets F_1, F_2 such that they, and the resulting G_2' satisfy:*

1. $F_1 \subset C_{max} \backslash ((Z \cup I(C_2)) \cap M)$;
2. $F_2 \subseteq I(C_2) \cup Z$;
3. *for each kite C, (i) the set F_1 contains exactly one edge of C and the set F_2 contains zero edges of C or (ii) (it can happen only for 4-kites) the set F_1 contains exactly two edges of C and the set F_2 contains one edge of $C \backslash M$;*
4. *for each kite C the set F_2 contains exactly one outgoing edge of C;*
5. *for each kite C and each vertex v of C the number of edges of F_2 incident to v is at most greater by one than the number of edges of F_1 incident to v;*
6. *there exists no cycle of G_2' of odd length l that has l double edges incident to it;*
7. *each path of G_2' is amenable.*

The property 1 of this lemma guarantees that G_2' does not contain more than two copies of any edge. It is shown in the full version of the paper that properties

6. and 7. are essentially sufficient for the multigraph G_2' to be path-2-colorable. Properties 4 and 5 will be helpful in finding a path-3-coloring of G_1'. Property 5 ensures that no vertex v has six incident edges in G_1'.

5 Summary

After the construction and path-2-coloring of G_2' we are presented with the task of extending the partial path-3-coloring of G_1 to the complete path-3-coloring of G_1'. In particular, we have to color the edges of kites and edges of F_2 that have been added during the construction of G_2'. This part of the algorithm is described in the full version of the paper.

The presented algorithm works for graphs with an even number of vertices. If the number of vertices of a given graph is odd, we proceed as follows. We select a vertex $v \in V$ arbitrarily. Then we guess its predecessor u and successor t in the optimal solution ($O(n^2)$ guesses). For each guess we replace the vertex v with two new vertices v_1, v_2 (so we have an even number of vertices). The edge (u, v_1) has weight $w(u, v)$, the edge (t, v_2) has weight $w(t, v)$ and all remaining edges incident to v_1 or v_2 have weight equal to 0. Then we run our Algorithm 1 on these instances. The approximation ratio of $\frac{4}{5}$ holds, because the computed solution can be always transformed into a tour in the original graph of at least the same weight, and the optimal tour is certainly present among the guesses.

References

1. Arkin, E.M., Chiang, Y., Mitchell, J.S.B., Skiena, S., Yang, T.: On the maximum scatter TSP (extended abstract). In: Proceedings of the Eighth Annual ACM-SIAM Symposium on Discrete Algorithms, pp. 211–220 (1997)
2. Barvinok, A.I., Fekete, S.P., Johnson, D.S., Tamir, A., Woeginger, G.J., Woodroofe, R.: The geometric maximum traveling salesman problem. J. ACM 50(5), 641–664 (2003)
3. Barvinok, A., Johnson, D.S., Woeginger, G.J., Woodroofe, R.: The maximum traveling salesman problem under polyhedral norms. In: Bixby, R.E., Boyd, E.A., Ríos-Mercado, R.Z. (eds.) IPCO 1998. LNCS, vol. 1412, pp. 195–201. Springer, Heidelberg (1998). doi:10.1007/3-540-69346-7_15
4. Bhatia, R.: Private communication
5. Chalasani, P., Motwani, R.: Approximating capacitated routing and delivery problems. SIAM J. Comput. 28(6), 2133–2149 (1999)
6. Chen, Z.Z., Okamoto, Y., Wang, L.: Improved deterministic approximation algorithms for max TSP. Inf. Process. Lett. 95(2), 333–342 (2005)
7. Chen, Z.-Z., Wang, L.: An improved approximation algorithm for the bandpass-2 problem. In: Lin, G. (ed.) COCOA 2012. LNCS, vol. 7402, pp. 188–199. Springer, Heidelberg (2012). doi:10.1007/978-3-642-31770-5_17
8. Chiang, Y.J.: New approximation results for the maximum scatter tsp. Algorithmica 41(4), 309–341 (2005)
9. Dudycz, S., Marcinkowski, J., Paluch, K.E., Rybicki, B.: A 4/5 - approximation algorithm for the maximum traveling salesman problem. CoRR abs/1512.09236 (2015). http://arxiv.org/abs/1512.09236

10. Fisher, M.L., Nemhauser, G.L., Wolsey, L.A.: An analysis of approximations for finding a maximum weight hamiltonian circuit. Oper. Res. **27**(4), 799–809 (1979)
11. Hassin, R., Levin, A., Rubinstein, S.: Approximation algorithms for maximum latency and partial cycle cover. Discrete Optim. **6**(2), 197–205 (2009)
12. Hassin, R., Rubinstein, S.: An approximation algorithm for the maximum traveling salesman problem. Inf. Process. Lett. **67**(3), 125–130 (1998)
13. Hassin, R., Rubinstein, S.: Better approximations for max TSP. Inf. Process. Lett. **75**(4), 181–186 (2000)
14. Hassin, R., Rubinstein, S.: An approximation algorithm for maximum triangle packing. In: Albers, S., Radzik, T. (eds.) ESA 2004. LNCS, vol. 3221, pp. 403–413. Springer, Heidelberg (2004). doi:10.1007/978-3-540-30140-0_37
15. Kaplan, H., Lewenstein, M., Shafrir, N., Sviridenko, M.: Approximation algorithms for asymmetric tsp by decomposing directed regular multigraphs. In: 44th Symposium on Foundations of Computer Science (FOCS 2003) (2003)
16. Kosaraju, S.R., Park, J.K., Stein, C.: Long tours and short superstrings. In: 35th Annual IEEE Symposium on Foundations of Computer Science (FOCS) (1994)
17. Kowalik, Ł., Mucha, M.: 35/44-approximation for asymmetric maximum TSP with triangle inequality. In: Dehne, F., Sack, J.-R., Zeh, N. (eds.) WADS 2007. LNCS, vol. 4619, pp. 589–600. Springer, Heidelberg (2007). doi:10.1007/978-3-540-73951-7_51
18. Kowalik, Ł., Mucha, M.: Deterministic 7/8-approximation for the metric maximum TSP. In: Goel, A., Jansen, K., Rolim, J.D.P., Rubinfeld, R. (eds.) APPROX/RANDOM -2008. LNCS, vol. 5171, pp. 132–145. Springer, Heidelberg (2008). doi:10.1007/978-3-540-85363-3_11
19. Monnot, J.: Approximation algorithms for the maximum hamiltonian path problem with specified endpoint(s). Eur. J. Oper. Res. **161**(3), 721–735 (2005)
20. Paluch, K.E.: Better approximation algorithms for maximum asymmetric traveling salesman and shortest superstring. CoRR (2014)
21. Paluch, K.E., Elbassioni, K.M., van Zuylen, A.: Simpler approximation of the maximum asymmetric traveling salesman problem. In: 29th International Symposium on Theoretical Aspects of Computer Science, STACS (2012)
22. Paluch, K., Mucha, M., Mądry, A.: A 7/9 - approximation algorithm for the maximum traveling salesman problem. In: Dinur, I., Jansen, K., Naor, J., Rolim, J. (eds.) APPROX/RANDOM -2009. LNCS, vol. 5687, pp. 298–311. Springer, Heidelberg (2009). doi:10.1007/978-3-642-03685-9_23
23. Papadimitriou, C.H., Yannakakis, M.: The traveling salesman problem with distances one and two. Math. Oper. Res. **18**(1), 1–11 (1993)
24. Schrijver, A.: Nonbipartite matching and covering. In: Combinatorial Optimization, vol. A, pp. 520–561. Springer (2003)
25. Serdyukov, A.I.: An algorithm with an estimate for the traveling salesman problem of maximum. Upravlyaemye Sistemy **25**, 80–86 (1984) (in Russian)
26. Sichen, Z., Zhao, L., Liang, Y., Zamani, M., Patro, R., Chowdhury, R., Arkin, E.M., Mitchell, J.S.B., Skiena, S.: Optimizing read reversals for sequence compression. In: Pop, M., Touzet, H. (eds.) WABI 2015. LNCS, vol. 9289, pp. 189–202. Springer, Heidelberg (2015). doi:10.1007/978-3-662-48221-6_14
27. Tong, W., Goebel, R., Liu, T., Lin, G.: Approximation algorithms for the maximum multiple RNA interaction problem. In: Widmayer, P., Xu, Y., Zhu, B. (eds.) COCOA 2013. LNCS, vol. 8287, pp. 49–59. Springer, Cham (2013). doi:10.1007/978-3-319-03780-6_5

Minimizing Multimodular Functions and Allocating Capacity in Bike-Sharing Systems

Daniel Freund$^{(\boxtimes)}$, Shane G. Henderson, and David B. Shmoys

Cornell University, Ithaca, NY 14853, USA
{df365,sgh9,dbs10}@cornell.edu

Abstract. The growing popularity of bike-sharing systems around the world has motivated recent attention to models and algorithms for the effective operation of these systems. Most of this literature focuses on their daily operation for managing asymmetric demand. In this work, we consider the more strategic question of how to allocate dock-capacity in such systems. Our main result is a practically fast polynomial-time allocation algorithm to compute optimal solutions for this problem, that can also handle a number of practically motivated constraints, such as a limit on the number of docks moved from a given allocation. Our work further develops connections between bike-sharing models and the literature on discrete convex analysis and optimization.

1 Introduction

As shared vehicle systems, such as bike-sharing and car-sharing, become an integral part of urban landscapes, novel lines of research seek to model and optimize the operations of these systems. In many systems, such as New York City's Citi Bike, users can rent and return bikes at any location throughout the city. This flexibility makes the system attractive for commuters and tourists alike. From an operational point of view, however, this flexibility leads to imbalances when demand is asymmetric as is commonly the case. The main contribution of this paper is to identify key questions in the design of operationally efficient bike-sharing systems, and to provide a polynomial algorithm for the associated discrete optimization problems.

Most bike-sharing systems are dock-based, meaning that they consist of stations, spread across the city, each of which has a number of docks in which bikes are locked. If a bike is present in a dock, users can rent it and return it at any other station with an open dock. However, system imbalance often causes some stations to have only empty docks and others to have only full docks. In the former case, users need to find alternate modes of transportation, whereas in the latter they might not be able to end their trip at the intended destination. In many bike-sharing systems, this has been found to be a leading cause of customer dissatisfaction (see e.g., [2]).

Work supported in part under NSF grants CCF-1526067, CMMI-1537394, CCF-1522054, and CMMI-1200315, and Army Research Office grant W911NF-17-1-0094.

F. Eisenbrand and J. Koenemann (Eds.): IPCO 2017, LNCS 10328, pp. 186–198, 2017.
DOI: 10.1007/978-3-319-59250-3_16

In order to meet demand in the face of asymmetric traffic, bike-sharing system operators seek to *rebalance* the system by moving bikes from locations with too few open docks to locations with too few bikes. To facilitate these operations, a burst of recent research has investigated models and algorithms to increase their efficiency and increase customer satisfaction. While similar in spirit to some of the literature on rebalancing, in this work we use a different control to increase customer satisfaction. Specifically, we answer the question *how should bike-sharing systems allocate dock capacity to stations within the system so as to minimize the number of dissatisfied customers?*

Related Work. Raviv and Kolka [17] defined a *user dissatisfaction function* that measures the expected number of out-of-stock events at an individual bike-share station. To do so, they define a continuous-time Markov chain on the possible number of bikes (between 0 and the capacity of the station). Bikes are rented with rate $\lambda(t)$ and returned with rate $\mu(t)$. Each arrival triggers a change in the state, either decreasing (rental) or increasing (return) the number of available bikes by one. When the number of bikes is 0 and a rental occurs, the customer experiences an out-of-stock event. Using a discrete Markov Chain, they approximate the expected number of out-of-stock events over a finite time-horizon. For fixed rates, the work of Schuijbroek et al. [19] and O'Mahony [15] give different techniques to compute the expected number of out-of-stock events exactly. A recursion suggested by Parikh and Ukkusuri [16] shows that these methods extend to settings in which rates are constant over intervals.

The definition of the user dissatisfaction function triggered a line of work around static rebalancing problems, in which a capacitated truck (or a fleet of trucks) is routed over a limited time horizon. The truck may pick up and drop off bikes at each station, so as to minimize the expected number of out-of-stock events that occur after the completion of the route. Variations of this setting include Raviv et al. [18], Forma et al. [4], Kaspi et al. [11], Ho et al. [8], and Freund et al. [6]. As in our work, all of these papers make the assumption that demand is exogeneous and independent among stations, i.e., reducing the number of bikes available for rentals upstream has no effect on the number of returns downstream.

In contrast to the other papers mentioned, O'Mahony [15] addressed the question of allocating both docks and bikes; he uses the user dissatisfaction function to design a mixed integer program over the possible allocations of bikes and docks. In essence, our work extends upon this by providing a fast polynomial-time algorithm for that same problem and an extension thereof. Optimal allocations of bikes and docks have also been studied by Jian and Henderson [10], Datner et al. [3], and by Jian et al. [9] who develop frameworks based on ideas from simulation optimization: while they also treat demand for bikes as being exogeneous, their framework captures the downstream effects of changes in supply upstream.

The work of Kaspi et al. [11] investigates the effects that broken bikes can have on the cost-function. Interestingly, some of their results can be viewed as analogous to ours. They prove that the cost-function at one station is M-convex (see the book by Murota [13] and the references therein); surprisingly, the existing literature on the minimization of naturally M convex functions does not seem to capture the optimization problems we consider. We discuss the similarities and differences below.

A broader overview of the related work can be found in the full version [5].

Our Contribution. We consider the problem of allocating dock capacity in bike-share systems in a setting in which we are given distributional knowledge about exogenous demand at each station that is independent of our solution. This allows us, using techniques from [15, 16, 19] to compute the expected number of out-of-stock events $c_i(d_i, b_i)$ at each station i for a given allocation of b_i bikes and d_i empty docks (i.e., $d_i + b_i$ docks in total) to station i.

Given this cost-function, we want to find the allocation of bikes and docks in the system that minimizes the total expected number of out-of-stock events within a system of n stations, i.e., $\sum_{i=1}^{n} c_i(d_i, b_i)$. However, due to the number of bikes and docks being limited, we need to accommodate a *budget constraint* B on the number of bikes in the system and another on the number of docks $D + B$ in the system. Other constraints are often important, such as lower and upper bounds on the allocation for a particular station; furthermore, one important issue that has arisen in our collaboration with Citi Bike in NYC is that we seek to optimize the allocation while limiting the number of docks moved from the current system configuration. Our methods are amenable to these *operational constraints*. Finally, one additional type of constraint is that the allocation given to disjoint neighborhoods must provide equitable access to the system; this can be modeled through a laminar family of set constraints, and our techniques can be extended to handle these by a standard dynamic programming approach, albeit with somewhat slower running times.

We design an algorithm that provably solves the minimization problem in $O(nT + (T + \log(n))(B + D))$ when given access to an oracle that computes $c_i(d, b)$ in $O(T)$ — [15] takes $O((d + b)^3)$. Our algorithm exploits the fact that the cost-function $c(\cdot, \cdot)$ is multimodular (cf. Definition 1) at each station.

Multimodularity provides an interesting connection to the literature on discrete convex analysis. Recent work [11] has shown independently that the number of out-of-stock events $F(b, U - d - b)$ at a bike-share station with fixed capacity U, b bikes and $U - d - b$ unusable bikes is M-natural convex in b and $U - d - b$. Unusable bikes effectively reduce the capacity at the station, since they are assumed to remain in the station over the entire time horizon. A station with capacity U, b bikes, and $U - b - d$ unusable bikes, must then have d empty docks; hence, $c(d, b) = F(b, U - d - b)$ for $d + b \le U$, which parallels our result that $c(\cdot, \cdot)$ is multimodular. Though this would suggest that algorithms to minimize M-convex functions could solve our problem optimally, we show in the full version that M-convexity is not preserved, even in the version with only budget constraints.[1] However, since multimodularity is preserved we believe that techniques by Murota [14], combined with the submodular function minimization algorithms of Lee et al. [12], give a $O(n^3 T + n^4(D + B))$ algorithm to solve the version with only budget constraints. By exploiting the separability of our objective function and the associated multimodularity of each station's cost function, we obtain algorithms with significantly stronger running-time guarantees that quickly yield solutions for instances at the scales that typically arise in practice.

2 Model

We denote by $X = (X_1, \ldots, X_s) \in \{\pm 1\}^s$ a sequence of s customers at a bikeshare station. The sign of X_t identifies whether customer t arrives to rent or to return a bike, i.e., if $X_t = 1$ customer t wants to return a bike and if $X_t = -1$ customer t wants to rent a bike. The truncated sequence $(X_1, \ldots, X_t)$ is written as $X(t)$. We denote throughout by d and b the number of open docks and available bikes at a station before any customer has arrived. Notice that a station with d open docks and b available bikes has $d+b$ docks in total. Whenever a customer arrives to return a bike at a station and there is an open dock, the customer returns the bike, the number of available bikes increases by 1 and the number of open docks decreases by 1. Similarly, a customer arriving to rent a bike when one is available decreases the number of available bikes by 1 and increases the number of open docks by 1. If however a customer arrives to rent (return) a bike when no bike (open dock) is available, then she disappears with an out-of-stock event. We assume that only customers affect the inventory-level at a station, i.e., no rebalancing occurs. It is useful then to write

$$\delta_{X(t)}(d, b) := \max\{0, \min\{d + b, \delta_{X(t-1)} - X_t\}\}, \ \delta_{X(0)}(d, b) = d$$
$$\beta_{X(t)}(d, b) := \max\{0, \min\{d + b, \beta_{X(t-1)} + X_t\}\}, \ \beta_{X(0)}(d, b) = b$$

as a shorthand for the number of open docks and available bikes after the first t customers.

Our cost function is based on the number of out-of-stock events. In accordance with the above-described model, customer t experiences an out-of-stock event if and only if $\delta_{X(t)}(d, b) = \delta_{X(t-1)}(d, b)$. Since $d+b = \delta_{X(t)}(d, b) + \beta_{X(t)}(d, b)$ for every t, this happens if and only if $\beta_{X(t)}(d, b) = \beta_{X(t-1)}(d, b)$. As we are interested in the number of out-of-stock events as a function of the initial number of open docks and available bikes, we can write our cost-function $c^{X(t)}(d, b)$

$$= |\{\tau : \tau \le t, X_\tau = 1, \delta_{X(\tau-1)}(d, b) = 0\}| + |\{\tau : \tau \le t, X_\tau = -1, \beta_{X(\tau-1)}(d, b) = 0\}|.$$

---

[1] Specifically, we (i) give an example in which a M-convex function restricted to a M-convex set is not M-convex, and (ii) show that this indeed means that Murota's algorithm for M-convex function minimization is not provably optimal in our setting.

It is then easy to see that with $c^{X(0)}(d,b) = 0$, $c^{X(t)}(d,b)$ fulfills the recursion

$$c^{X(t)}(d,b) = c^{X(t-1)}(d,b) + 1_{\{\beta_{X(t)}(d,b) = \beta_{X(t-1)}(d,b)\}}.$$

Given for each station $i \in [n]$ a distribution, which we call *demand-profile*, p_i over $\{(\pm 1)^s, s \in \mathbb{N}\}$, we can then write $c_i(d,b) = \mathbb{E}_{X \sim p_i}[c^X(d,b)]$ for the expected number of out-of-stock events at station i and $c(\boldsymbol{d}, \boldsymbol{b}) = \sum_i c_i(d_i, b_i)$. We then want to solve, for parameters D, B, $(\bar{\boldsymbol{d}}, \bar{\boldsymbol{b}})$, and z, as well as l_i, u_i for each $i \in [n]$, the following minimization problem

$$\begin{aligned}
\texttt{minimize}_{(\boldsymbol{d},\boldsymbol{b})} \quad & \sum_i c_i(d_i, b_i) \\
s.t. \quad & \sum_i d_i + b_i && \leq D + B, \\
& \sum_i b_i && \leq B, \\
& \sum_i |(\bar{d}_i + \bar{b}_i) - (d_i + b_i)| && \leq z, \\
\forall i \in [n]: \quad & l_i \leq d_i + b_i && \leq u_i.
\end{aligned}$$

Here, the first constraint corresponds to a budget on the number of docks, the second to a budget on the number of bikes, the third to the operational constraints and the fourth to the lower and upper bound on the number of docks at each station. We assume without loss of generality that there exists an optimal solution in which the second constraint holds with equality; to ensure that, we may add a dummy ("depot") station $\mathcal{D}$ that has $c_{\mathcal{D}}(\cdot, \cdot) = 0$, $l_{\mathcal{D}} = u_{\mathcal{D}} = B$, and run the algorithm with the budget on docks $(D + B)$ increased by B.

In Sect. 3 we prove that $c^X(\cdot, \cdot)$ fulfills a particular set of inequalities making it a so-called multimodular function.

Definition 1. *[1, 7] A function $f : \mathbb{N}_0^2 \to \mathbb{R}$ is called multimodular if*

$$f(d+1, b+1) - f(d+1, b) \geq f(d, b+1) - f(d, b); \tag{1}$$
$$f(d-1, b+1) - f(d-1, b) \geq f(d, b) - f(d, b-1); \tag{2}$$
$$f(d+1, b-1) - f(d, b-1) \geq f(d, b) - f(d-1, b); \tag{3}$$

for all d, b such that all terms are well-defined.

For future reference, we define the following additional inequalities, which are implied[2] by the above:

$$f(d+2, b) - f(d+1, b) \geq f(d+1, b) - f(d, b); \tag{4}$$
$$f(d, b+2) - f(d, b+1) \geq f(d, b+1) - f(d, b); \tag{5}$$
$$f(d+1, b+1) - f(d, b+1) \geq f(d+1, b) - f(d, b). \tag{6}$$

Even though we are motivated by the cost-functions defined in this section, our main results hold for arbitrary sums of such two-dimensional functions.

---

[2] (6) and (1) are equivalent, (1) and (2) imply (5), and (3) and (6) imply (4).

3   Multimodularity and an Allocation Algorithm

We first prove that the cost-functions defined in Sect. 2 are multimodular.

Lemma 2. $c^X(\cdot, \cdot)$ *is multimodular for all* X.

Proof. The proof of the lemma is straightforward by induction in $t = |X|$ and is left for the full version due to space constraints [5].

Corollary 3. $c_i(\cdot, \cdot)$ *is multimodular for any demand-profile* p_i.

Proof. The proof is immediate from Lemma 2 and linearity of expectation.   □

3.1   An Allocation Algorithm

In this section, we present our algorithm for settings without the operational constraints. Intuitively, in each iteration the algorithm picks one dock and at most one bike within the system and moves them from one station to another. It chooses the dock, and the bike, so as to maximize the reduction in objective value. To formalize this notion, we define the *movement of a dock* via the following transformations.

Definition 4. *We shall use the notation* $(\boldsymbol{v}_{-i}, \hat{v}_i) := (v_1 \ldots v_{i-1}, \hat{v}_i, v_{i+1} \ldots v_n)$. *Similarly,* $(\boldsymbol{v}_{-i,-j}, \hat{v}_i, \hat{v}_j) := (v_1 \ldots \hat{v}_i \ldots \hat{v}_j \ldots v_n)$. *Then a dock-move from* i *to* j *corresponds to one of the following transformations of feasible solutions:*

1. $o_{ij}(\boldsymbol{d}, \boldsymbol{b}) = \big((\boldsymbol{d}_{-i,-j}, d_i - 1, d_j + 1), \boldsymbol{b}\big)$ – *Moving one open dock from* i *to* j;
2. $e_{ij}(\boldsymbol{d}, \boldsymbol{b}) = \big(\boldsymbol{d}, (\boldsymbol{b}_{-i,-j}, b_i - 1, b_j + 1)\big)$ – *Moving a dock & a bike from* i *to* j;
3. $E_{ijh}(\boldsymbol{d}, \boldsymbol{b}) = \big((\boldsymbol{d}_{-i,-h}, d_i - 1, d_h + 1), (\boldsymbol{b}_{-j,-h}, b_j + 1, b_h - 1)\big)$ – *Moving one open dock from* i *to* j *and one bike from* h *to* j;
4. $O_{ijh}(\boldsymbol{d}, \boldsymbol{b}) = \big((\boldsymbol{d}_{-j,-h}, d_j + 1, d_h - 1), (\boldsymbol{b}_{-i,-h}, b_i - 1, b_h + 1)\big)$ – *Moving one bike from* i *to* h *and one open dock from* i *to* j.

Further, we define the neighborhood $N(\boldsymbol{d}, \boldsymbol{b})$ *of* $(\boldsymbol{d}, \boldsymbol{b})$ *as the set of allocations that are one dock-move away from* $(\boldsymbol{d}, \boldsymbol{b})$. *Formally,*

$$N(\boldsymbol{d}, \boldsymbol{b}) := \{o_{ij}(\boldsymbol{d}, \boldsymbol{b}), e_{ij}(\boldsymbol{d}, \boldsymbol{b}), E_{ijh}(\boldsymbol{d}, \boldsymbol{b}), O_{ijh}(\boldsymbol{d}, \boldsymbol{b}) : i, j, h \in [n]\}.$$

Finally, define the dock-move distance between $(\boldsymbol{d}, \boldsymbol{b})$ *and* $(\boldsymbol{d}', \boldsymbol{b}')$ *as*

$$\sum_i |(d_i + b_i) - (d_i' + b_i')|.$$

This gives rise to a very simple algorithm: we first find the optimal allocation of bikes for the current allocation of docks; the convexity of each c_i in the number of bikes, with fixed number of docks, implies that this can be done greedily by taking out all the bikes and then adding them one by one. Then, while there exists a dock-move that improves the objective, we find the best possible such dock-move and update the allocation accordingly (cf. Algorithm 1).

Remark: Each iteration of the algorithm can be implemented in (amortized) $O(T + \log(n))$ time by maintaining six binary heaps that contain for each change a dock-move could have at each station (i.e., add/take an open dock, add/take a bike, and add/take a bike and a dock) the change in objective this would yield. Instead of comparing all $O(n^2)$ possible moves, one can then find the argument of the minimum and update (d, b) in constant time, and then update the lists (for the stations involved in the dock-move) in $O(T + \log(n))$.

Algorithm 1. Greedy

1: Find optimal allocation of bikes for current dock allocation
2: **while** $c(d, b) > \min_{(d', b') \in N(d,b)} c(d', b')$ **do**
3:     $(d, b) \leftarrow \arg\min_{(d', b') \in N(d,b)} c(d', b')$
4: **end while**

3.2   Proof of Optimality

We prove that the algorithm returns an optimal solution by showing that the condition in the while-loop is false only if (d, b) globally minimizes the objective; else, the algorithm moves a dock to find a better solution. Thus, if the algorithm terminates, then the solution is optimal. Before we prove Lemma 7 to establish this, we first define an allocation of bikes and docks as bike-optimal if it minimizes the objective among allocations with the same number of docks at each station and prove that bike-optimality is an invariant of the while-loop.

Definition 5. *We call an allocation (d, b) bike-optimal if*

$$(d, b) \in \arg \min_{(\hat{d}, \hat{b}): \forall i, d_i + b_i = \hat{d}_i + \hat{b}_i, \sum_i \hat{b}_i = B} \{c(\hat{d}, \hat{b})\}.$$

Lemma 6. *Suppose (d, b) is bike-optimal. Given i and j, one of the possible dock-moves from i to j, i.e., $e_{ij}(d, b), o_{ij}(d, b), E_{ijh}(d, b),$ or $O_{ijh}(d, b)$, is bike-optimal, i.e., when moving a dock from i to j, one has to move at most one bike within the system to maintain bike-optimality.*

Proof. It is known that multimodular functions fulfill certain convexity properties (see e.g., [13,17]); in particular, for fixed d and b it is known that $c_i(k, d + b - k)$ is a convex function of $k \in \{0, \ldots, d + b\}$. Thus, if the best allocation out of $e_{ij}(d, b), o_{ij}(d, b), E_{ijh}(d, b),$ and $O_{ijh}(d, b)$, was not bike-optimal, there would have to be two stations such that moving a bike from one to the other improves the objective. By the bike-optimality of (d, b), at least one of these two stations must have been involved in the move. We prove that the result holds if e_{ij} was the best of the set of possible moves $\{e_{ij}, o_{ij}, E_{ijh}, O_{i,j,h}\}_{i,j,h \in [n]}$ – the other three cases are almost symmetric. Let ℓ denote a generic third station. Then a bike improving the objective could correspond to one being moved from ℓ to j, from i to j, from i to ℓ, from ℓ to i, from j to ℓ or from j to i. In this case, a move from ℓ to j, i to j and i to ℓ yield the allocations $E_{ij\ell}(\boldsymbol{d}, \boldsymbol{b})$, $o_{ij}(\boldsymbol{d}, \boldsymbol{b})$ and $O_{ij\ell}(\boldsymbol{d}, \boldsymbol{b})$, respectively. Since e_{ij} is assumed to be the minimizer among the possible dock-moves, none of these have objective smaller than that of $e_{ij}(\boldsymbol{d}, \boldsymbol{b})$. It remains to show that moving a bike from ℓ to i, j to ℓ or j to i yields no improvement. These all follow from bike-optimality of $(\boldsymbol{d}, \boldsymbol{b})$ and the multimodular inequalities. Specifically, an additional bike at i yields less improvement and a bike fewer at j has greater cost in $e_{ij}(\boldsymbol{d}, \boldsymbol{b})$ than in $(\boldsymbol{d}, \boldsymbol{b})$, since

$$c_i(d_i - 1, b_i) - c_i(d_i - 2, b_i + 1) \leq c_i(d_i, b_i) - c_i(d_i - 1, b_i + 1)$$
$$c_j(d_j + 2, b_j - 1) - c_j(d_j + 1, b_j) \geq c_j(d_j + 1, b_j - 1) - c_j(d_j, b_j).$$

Both of the above inequalities follow from inequality (3). $\qquad\square$

We are now ready to prove that when the algorithm terminates it must have found an optimal solution.

Lemma 7 (Neighborhood). *Suppose $(\boldsymbol{d}, \boldsymbol{b})$ is bike-optimal, but does not minimize $c(\cdot, \cdot)$ subject to budget constraints. Let $(\boldsymbol{d}^*, \boldsymbol{b}^*)$ denote a feasible solution with better objective at minimal dock-distance from $(\boldsymbol{d}, \boldsymbol{b})$. As $(\boldsymbol{d}, \boldsymbol{b})$ is bike-optimal, there exist j and k such that $b_j + d_j < b_j^* + d_j^*$ and $b_k + d_k > b_k^* + d_k^*$. Pick any such j and k; then there exists a dock-move to j or a dock-move from k that improves the objective of $(\boldsymbol{d}, \boldsymbol{b})$.*

Proof. The proof of the lemma follows a a case-by-case analysis, each of which resembles the same idea: $(\boldsymbol{d}^*, \boldsymbol{b}^*)$ minimizes the dock-move distance to $(\boldsymbol{d}, \boldsymbol{b})$ among solutions with lower function value than $(\boldsymbol{d}, \boldsymbol{b})$, i.e., among all $(\boldsymbol{d}^*, \boldsymbol{b}^*)$ such that $\sum_i d_i + b_i = \sum_i d_i^* + b_i^*$, $\sum_i b_i = \sum_i b_i^*$, and $c(\boldsymbol{d}^*, \boldsymbol{b}^*) < c(\boldsymbol{d}, \boldsymbol{b})$, $(\boldsymbol{d}^*, \boldsymbol{b}^*)$ has minimum dock-move distance to $(\boldsymbol{d}, \boldsymbol{b})$. We show that with j and k as in the statement of the lemma, either there exists a dock-move to j/from k that improves the objective or there exists a solution $(\boldsymbol{d}^{**}, \boldsymbol{b}^{**})$ with objective value lower than $(\boldsymbol{d}, \boldsymbol{b})$, $\sum_i d_i + b_i = \sum_i d_i^{**} + b_i^{**}$, $\sum_i b_i = \sum_i b_i^{**}$ and smaller dock-move distance to $(\boldsymbol{d}, \boldsymbol{b})$. Since the latter contradicts our choice of $(\boldsymbol{d}^*, \boldsymbol{b}^*)$, this proves, that in $(\boldsymbol{d}, \boldsymbol{b})$ there must be a dock-move to j/from k that yields a lower objective. We distinguish among the following cases:

1. $d_j < d_j^*$ and $d_k > d_k^*$;
2. $b_j < b_j^*$ and $b_k > b_k^*$;
3. $d_j < d_j^*$, $b_k > b_k^*$, and $b_j \geq b_j^*$
   (a) and there exists ℓ with $d_l + b_l \geq d_l^* + b_l^*$, $b_l < b_l^*$;
   (b) and there exists ℓ with $d_l + b_l < d_l^* + b_l^*$, $b_l < b_l^*$;
   (c) for all $\ell \notin \{j, k\}$, we have $b_l \geq b_l^*$, so $\sum_i b_i > \sum_i b_i^*$;
4. $b_j < b_j^*$, $d_j \geq d_j^*$, $b_k \leq b_k^*$ and $d_k > d_k^*$,
   (a) and there exists ℓ with $d_\ell + b_\ell > d_\ell^* + b_\ell^*$ and $b_\ell > b_\ell^*$;
   (b) and there exists ℓ with $d_\ell + b_\ell \leq d_\ell^* + b_\ell^*$ and $b_\ell > b_\ell^*$;
   (c) for all $\ell \notin \{j, k\}$, we have $b_\ell \leq b_\ell^*$, so $\sum_i b_i < \sum_i b_i^*$.

In the full version, we show that in case (1) a move from k to j yields improvement [5]. The proof for case (2) is symmetric. Thus, in cases (3a) and (4a) there exists a move from k to ℓ, respectively from ℓ to j, that yields improvement. The proofs for cases (3b) and (4b) are also symmetric and we present the proof for (3b) in the full version. Cases (3c) and (4c) contradict our assumption that $\sum_i b_i = \sum_i b_i^*$ and can thus be excluded.

4   Operational Constraints and Running Time

In this section, we show that the allocation algorithm is optimal for the operational constraints introduced in Sect. 2 and thereby also provide an analysis of the running-time of the algorithm. To do so, we first define the set of feasible solutions with respect to those constraints.

Definition 8. *Define the z-dock ball $S_z(d, b)$ around (d, b) as the set of allocations with dock-move distance at most $2z$, i.e., $S_0(d, b) = \{(d, b)\}$ and*

$$S_z(d, b) = S_{z-1}(d, b) \cup \left(\bigcup_{(d', b') \in S_{z-1}(d, b)} N(d', b') \right).$$

We now want to prove that Lemma 7 continues to hold in the constrained setting; in particular, we show that even with the operational constraints, local optima are global optima.

Lemma 9 (z-step neighborhood). *If $(\hat{d}, \hat{b}) \in S_z(d, b) \setminus S_{z-1}(d, b)$ is bike-optimal and $c(d^*, b^*) < c(\hat{d}, \hat{b})$ for some $(d^*, b^*) \in S_z(d, b) \setminus S_{z-1}(d, b)$, then there exists $(d', b') \in S_z(d, b) \cap N(\hat{d}, \hat{b})$ such that $c(d', b') < c(\hat{d}, \hat{b})$.*

Proof. Notice that this lemma closely resembles Lemma 7: the sole difference lies in Lemma 7 not enforcing the dock-move to maintain a bound on the distance to some allocation (d, b).

Define (d^*, b^*) as in Lemma 7 with the additional restriction that (d^*, b^*) be in $S_z(d, b)$, i.e., pick a solution in $S_z(d, b)$ that minimizes the dock-move distance to $(\hat{d}, \hat{b})$ among solutions with strictly smaller objective value. We argue again that bike-optimality of $(\hat{d}, \hat{b})$ implies that there exist j and k, such that $\hat{d}_j + \hat{b}_j < d_j^* + b_j^*$, and $\hat{d}_k + \hat{b}_k > d_k^* + b_k^*$. Further, for any such j and k, we can apply the proof of Lemma 7 to find a move involving at least one of the two that decreases both the objective value and the dock-move distance to (d^*, b^*).

We aim to find j and k such that the move identified, say from ℓ to m, is guaranteed to remain within $S_z(d, b)$. Notice that $|\{j\} \cap \{m\}| + |\{k\} \cap \{\ell\}| \geq 1$. We know that $d_m^* + b_m^* > \hat{d}_m + \hat{b}_m$ and $d_\ell^* + b_\ell^* < \hat{d}_\ell + \hat{b}_\ell$. Suppose the move from ℓ to m yields a solution outside of $S_z(d, b)$. It follows that $\hat{d}_m + \hat{b}_m \geq d_m + b_m$ and $\hat{d}_\ell + \hat{b}_\ell \leq d_\ell + b_\ell$, so in particular either $\hat{d}_j + \hat{b}_j \geq d_j + b_j$ or $\hat{d}_k + \hat{b}_k \leq d_k + b_k$. Thus, if we can identify j and k such that those two inequalities do not hold, we are guaranteed that the identified move remains within $S_z(d, b)$.

Define

$$k := \arg\max_i \{\hat{d}_i + \hat{b}_i - \max\{d_i + b_i, d_i^* + b_i^*\}\}$$

We can then write

$$\max_i \{\hat{d}_i + \hat{b}_i - \max\{d_i + b_i, d_i^* + b_i^*\}\} \geq$$
$$\min_i \{1, \max_{i:\hat{d}_i+\hat{b}_i > d_i+b_i} \{(\hat{d}_i + \hat{b}_i) - (d_i^* + b_i^*)\}\}$$

The latter is at least 1 unless it is the case for all i that if $\hat{d}_i + \hat{b}_i > d_i + b_i$ then $\hat{d}_i + \hat{b}_i \leq d_i^* + b_i^*$. Thus, unless the above condition fails, we have identified a k with the required properties. Suppose the condition does fail. Then

$$2z = \sum_i |(d_i + b_i) - (d_i^* + b_i^*)| = \sum_i |(d_i + b_i) - (\hat{d}_i + \hat{b}_i)|$$
$$\text{and } \sum_i d_i + b_i = \sum_i d_i^* + b_i^* = \sum_i \hat{d}_i + \hat{b}_i$$

imply that for all i with $\max\{\hat{d}_i + \hat{b}_i, d_i^* + b_i^*\} > d_i + b_i$, we have $\hat{d}_i + \hat{b}_i = d_i^* + b_i^*$. Thus, it must be the case that m fulfills $\hat{d}_m + \hat{b}_m < d_m + b_m$.

The argument for j is symmetric.   □

Theorem 10. *Starting with a bike-optimal allocation (d, b), in the z-th iteration, the greedy algorithm finds an optimal allocation among those in $S_z(d, b)$.*

Proof. We prove the theorem by induction in z. The base-case $z = 0$ holds trivially. Suppose in the zth iteration, the greedy algorithm has found the allocation $(d^z, b^z) \in \arg\min_{(d^*,b^*) \in S_z(d,b)} c(d^*, b^*)$. We need to show that

$$(d^{z+1}, b^{z+1}) := \arg\min_{(d^{z+1},b^{z+1}) \in N(d^z,b^z))} \{c(d^{z+1}, b^{z+1})\}$$

minimizes the cost function among solutions in $S_{z+1}(d, b)$.

We first observe that by Lemma 9, it suffices to show that there is no better solution in $S_{z+1}(d, b)$ that is just one dock-move away from (d^{z+1}, b^{z+1}). Further, by Lemma 6 and the choice of dock-moves in the greedy algorithm we know that (d^{z+1}, b^{z+1}) must be bike-optimal. Let i be the station from which a dock was moved and let j be the station to which it was moved in the $z + 1$st iteration. We denote a third station by h if the $z + 1$st move involved a third one (recall that a dock-move from i to j can take an additional bike from i to a third station h or take one from h to j). We can then immediately exclude the following cases:

1. Any dock-move in which i receives a dock from some station ℓ, including possibly $\ell = j$ or $\ell = h$, can be excluded since the greedy algorithm could have chosen to take a dock from ℓ instead of i and found a bike-optimal allocation (by Lemma 6).

2. The same holds for any dock-move in which a dock is taken from j.
3. A dock-move not involving either of i, j, and h yields the same improvement as it would have prior to the $z + 1$st iteration. Furthermore, if such a dock-move yields a solution within $S_{z+1}(d, b)$, then prior to the $z + 1$st iteration it would have yielded a solution within $S_z(d, b)$. Hence, by the induction assumption, it cannot yield any improvement.
4. A dock-move from station i (or to j), as is implied by the fourth, fifth, and sixth inequality in the definition of multimodularity increases the objective at i more (decreases the objective at j less) than it would have prior to the $z + 1$st iteration.

We are left with a dock-move from or to h as well as dock-moves that involve one of the three stations only via a bike being moved. Suppose that the dock-move in iteration $z + 1$ was E_{ijh}; the case of O_{ijh} is symmetric. In this case, a subsequent move of a dock and a bike from h, i.e., $o_{h\ell}$ or $O_{h\ell m}$ for some m, increases the objective at h by at least as much as it did before (by inequality (2)) and can thus be excluded. The same holds for the move of an empty dock to h (by inequality (3)).

However, subsequent moves of an empty dock from h (or a full dock to h) have a lower cost (greater improvement) and require a more careful argument. Suppose $e_{h\ell}$ yielded an improvement – the cases for $E_{h\ell m}$, $o_{\ell h}$, and $E_{\ell h m}$ are similar. Notice first that if it were the case that $d_h^z + b_h^z > d_h + b_h$ and $d_\ell^z + b_\ell^z < d_\ell + b_\ell$, then $e_{h\ell}(E_{ijh}(d^z, b^z)) \in S_z(d, b)$ and has a lower objective than (d^z, b^z) which contradicts the inductive assumption. Furthermore, since it must be the case that $e_{h\ell}(E_{ijh}(d^z, b^z)) \in S_{z+1}(d, b) \setminus S_z(d, b)$, it must also follow that either

1. $d_h^z + b_h^z > d_h + b_h$ and $d_\ell^z + b_\ell^z \geq d_\ell + b_\ell$ or
2. $d_h^z + b_h^z \leq d_h + b_h$ and $d_\ell^z + b_\ell^z < d_\ell + b_\ell$,

since otherwise a dock-move from h to ℓ would either yield a solution in S_z or one not in S_{z+1}. Notice further that the inductive assumption implies that $(d^{z+1}, b^{z+1}) \notin S_z(d, b)$. Thus, it must be the case that $d_i^{z+1} + b_i^{z+1} < d_i + b_i$ and $d_j^{z+1} + b_j^{z+1} < d_j + b_j$. We can thus argue in the following way about

$$c(e_{h\ell}(d^{z+1}, b^{z+1})) - c(d^{z+1}, b^{z+1}) =$$
$$c_h(d_h^z, b_h^z - 1) - c_h(d_h^z + 1, b_h^z - 1) + c_l(d_\ell^z + 1, b_\ell^z) - c_l(d_\ell^z, b_\ell^z).$$

In the first case, since $o_{h\ell}(d^z, b^z) \in S_z(d, b)$, the inductive assumption implies that $c_h(d_h^z, b_h^z - 1) + c_j(d_j^z, b_j^z + 1) \geq c_h(d_h^z, b_h^z) + c_j(d_j^z, b_j^z)$. Further, by the choice of the greedy algorithm, an additional empty dock at ℓ has no more improvement than an additional dock and an additional bike at j minus the cost of taking the bike from h; otherwise, the greedy algorithm would have moved an empty dock from h to ℓ in the $z + 1$st iteration. Thus,

$$c_\ell(d_\ell^z + 1, b_\ell^z) - c_\ell(d_\ell^z, b_\ell^z)$$
$$\leq c_j(d_j^z, b_j^z) - c_j(d_j^z, b_j^z + 1) - c_h(d_h^z + 1, b_h^z - 1) + c_h(d_h^z, b_h^z)$$
$$\leq c_h(d_h^z, b_h^z - 1) - c_h(d_h^z, b_h^z) - c_h(d_h^z + 1, b_h^z - 1) + c_h(d_h^z, b_h^z)$$
$$\leq \qquad c_h(d_h^z, b_h^z - 1) - c_h(d_h^z + 1, b_h^z - 1),$$

implying that $c(e_{h\ell}(\boldsymbol{d}^{z+1}, \boldsymbol{b}^{z+1})) - c(\boldsymbol{d}^{z+1}, \boldsymbol{b}^{z+1}) \geq 0$.

In the second case, since we know that $e_{i\ell}(\boldsymbol{d}^z, \boldsymbol{b}^z) \in S_z(\boldsymbol{d}, \boldsymbol{b})$, the inductive assumption implies $c_\ell(d_\ell^z + 1, b_\ell^z) + c_i(d_i^z - 1, b_i^z) \geq c_\ell(d_\ell^z, b_\ell^z) + c_i(d_i^z, b_i^z)$. Further, the choice of the greedy algorithm to take the dock from i, not h, implies that $c_i(d_i^z, b_i^z) - c_i(d_i^z - 1, b_i^z) \leq c_h(d_h^z, b_h^z - 1) - c_h(d_h^z + 1, b_h^z - 1)$. Combining these two inequalities again implies that $e_{h\ell}$ does not yield an improvement.

The remaining cases, in which a move only involves i, j, or h as the third station that a bike is taken from/added to, can be found in the full version [5].

An immediate corollary of the above result yields a bound on the number of iterations the greedy algorithm may run for.

Corollary 11. *The greedy algorithm terminates in at most $\sum_i d_i + b_i$ iterations.*

Technically, we might view the size of the input as $\log(B + D)$; however, in our application the physical entities of docks and bikes are truly given in an unary encoding.

Conclusion. Our work provides a fast and provably efficient algorithm for the problem of minimizing the sum of two-dimensional multimodular functions under constraints. It has strong connections to the literature on discrete convex analysis as well as more novel work on the optimization of bike-sharing systems.

References

1. Altman, E., Gaujal, B., Hordijk, A.: Multimodularity, convexity, and optimization properties. Math. Oper. Res. **25**(2), 324–347 (2000)
2. Capital Bikeshare: Bikeshare member survey report (2014)
3. Datner, S., Raviv, T., Tzur, M., Chemla, D.: Setting inventory levels in a bike sharing network (2015)
4. Forma, I.A., Raviv, T., Tzur, M.: A 3-step math heuristic for the static repositioning problem in bike-sharing systems. Transp. Res. Part B: Methodological **71**, 230–247 (2015)
5. Freund, D., Henderson, S.G., Shmoys, D.B.: Minimizing multimodular functions and allocating capacity in bike-sharing systems. arXiv preprint arXiv:1611.09304 (2016)
6. Freund, D., Norouzi-Fard, A., Paul, A., Henderson, S.G., Shmoys, D.B.: Data-driven rebalancing methods for bike-share systems, working paper (2016)
7. Hajek, B.: Extremal splittings of point processes. Math. Oper. Res. **10**(4), 543–556 (1985)

8. Ho, S.C., Szeto, W.: Solving a static repositioning problem in bike-sharing systems using iterated tabu search. Transp. Res. Part E: Logistics Transp. Rev. **69**, 180–198 (2014)
9. Jian, N., Freund, D., Wiberg, H.M., Henderson, S.G.: Simulation optimization for a large-scale bike-sharing system. In: Proceedings of the 2016 Winter Simulation Conference, pp. 602–613. IEEE Press (2016)
10. Jian, N., Henderson, S.G.: An introduction to simulation optimization. In: Proceedings of the 2015 Winter Simulation Conference, pp. 1780–1794. IEEE Press (2015)
11. Kaspi, M., Raviv, T., Tzur, M.: Bike-sharing systems: user dissatisfaction in the presence of unusable bicycles. IISE Trans. **49**(2), 144–158 (2017). http://dx.doi.org/10.1080/0740817X.2016.1224960
12. Lee, Y.T., Sidford, A., Wong, S.C.W.: A faster cutting plane method and its implications for combinatorial and convex optimization. In: 2015 IEEE 56th Annual Symposium on Foundations of Computer Science (FOCS), pp. 1049–1065. IEEE (2015)
13. Murota, K.: Discrete Convex Analysis: Monographs on Discrete Mathematics and Applications 10. Society for Industrial and Applied Mathematics, Philadelphia (2003)
14. Murota, K.: On steepest descent algorithms for discrete convex functions. SIAM J. Optim. **14**(3), 699–707 (2004)
15. O'Mahony, E.: Smarter Tools For (Citi) Bike Sharing. Ph.D. thesis, Cornell University (2015)
16. Parikh, P., Ukkusuri, S.V.: Estimation of optimal inventory levels at stations of a bicycle sharing system (2014)
17. Raviv, T., Kolka, O.: Optimal inventory management of a bike-sharing station. IIE Trans. **45**(10), 1077–1093 (2013)
18. Raviv, T., Tzur, M., Forma, I.A.: Static repositioning in a bike-sharing system: models and solution approaches. EURO J. Transp. Logistics **2**(3), 187–229 (2013)
19. Schuijbroek, J., Hampshire, R., van Hoeve, W.J.: Inventory rebalancing and vehicle routing in bike sharing systems. Eur. J. Oper. Res. **257**, 992–1004 (2017)

Compact, Provably-Good LPs for Orienteering and Regret-Bounded Vehicle Routing

Zachary Friggstad[1] and Chaitanya Swamy[2(✉)]

[1] Department of Computing Science, University of Alberta, Edmonton, Canada
zacharyf@ualberta.ca
[2] Combinatorics and Optimization, University of Waterloo, Waterloo, Canada
cswamy@uwaterloo.ca

Abstract. We develop polynomial-size LP-relaxations for *orienteering* and the *regret-bounded vehicle routing problem* (RVRP) and devise suitable LP-rounding algorithms that lead to various new insights and approximation results for these problems. In orienteering, the goal is to find a maximum-reward r-rooted path, possibly ending at a specified node, of length at most some given budget B. In RVRP, the goal is to find the minimum number of r-rooted paths of *regret* at most a given bound R that cover all nodes, where the regret of an r-v path is its length $- c_{rv}$. For *rooted orienteering*, we introduce a natural bidirected LP-relaxation and obtain a simple 3-approximation algorithm via LP-rounding. This is the *first LP-based* guarantee for this problem. We also show that *point-to-point* (P2P) *orienteering* can be reduced to a regret-version of rooted orienteering at the expense of a factor-2 loss in approximation. For RVRP, we propose two compact LPs that lead to significant improvements, in both approximation ratio and running time, over the approach in [10]. One is a natural modification of the LP for rooted orienteering; the other is an unconventional formulation motivated by certain structural properties of an RVRP-solution, which leads to a 15-approximation for RVRP.

1 Introduction

Vehicle-routing problems (VRPs) constitute a broad class of optimization problems that find a wide range of applications and have been widely studied in the Operations Research and Computer Science literature (see, e.g. [2,4,8,14,18]). Despite this extensive study, we have rather limited understanding of LP-relaxations for VRPs (with TSP and the minimum-latency problem, to a lesser extent, being exceptions), and this has been an impediment in the design of approximation algorithms for these problems.

A full version of the paper is available on the CS arXiv.

Z. Friggstad—Research supported by Canada Research Chairs program and NSERC Discovery Grant.

C. Swamy—Research supported by NSERC grant 327620-09 and an NSERC DAS Award.

F. Eisenbrand and J. Koenemann (Eds.): IPCO 2017, LNCS 10328, pp. 199–211, 2017.
DOI: 10.1007/978-3-319-59250-3_17

Motivated by this gap in our understanding, we investigate whether one can develop polynomial-size (i.e., compact) LP-relaxations with good integrality gaps for VRPs, focusing on the fundamental *orienteering* problem [4,8,13] and the related *regret-bounded vehicle routing problem* (RVRP) [5,10]. In *orienteering*, we are given rewards associated with clients located in a metric space, a length bound B, a start, and possibly end, location for the vehicle, and we seek a route of length at most B that gathers maximum reward. This problem frequently arises as a subroutine when solving VRPs, both in approximation algorithms—e.g., for minimum-latency problems (MLPs) [3,6,9,16], TSP with time windows [2], RVRP [5,10]—as well as in computational methods where orienteering corresponds to the "pricing" problem encountered in solving set covering/partitioning LPs (a.k.a configuration LPs) for VRPs via a column-generation or branch-cut-and-price method. In RVRP, we have a metric space $\{c_{uv}\}$ on client locations, a start location r, and a *regret* bound R. The regret of a path P starting at r and ending at location v is $c(P) - c_{rv}$. The goal in RVRP is to find a minimum number of r-rooted paths of regret at most R that visit all clients.

Our contributions. We develop polynomial-size LP-relaxations for orienteering and RVRP and devise suitable rounding algorithms for these LPs, which lead to various new insights and approximation results for these problems.

In Sect. 3, we introduce a natural, compact LP-relaxation for *rooted orienteering*, wherein only the vehicle start node is specified, and design a simple rounding algorithm to convert an LP-solution to an integer solution losing a factor of at most 3 in the objective value. This is the *first LP-based* approximation guarantee for orienteering. In contrast, all other approaches for orienteering utilize dynamic programming (DP) to stitch together suitable subpaths.

In Sect. 4, we consider the more-general *point-to-point* (P2P) *orienteering* problem, where both the start and end nodes of the vehicle are specified. We present a novel reduction showing that P2P-orienteering can be *reduced* to a *regret-version* of rooted orienteering, wherein the length bound is replaced by a *regret bound*, incurring a factor-2 loss (Theorem 6). No such reduction to a rooted problem was known previously, and all known algorithms for P2P-orienteering rely on approximations to suitable P2P-path problems. Typically, constraining a VRP by requiring that routes include a fixed node t causes an increase in the route lengths of the unconstrained problem (as we need to attach t to the routes); this would violate the length bound in orienteering, but, notably, we devise a way to avoid this in our reduction. We believe that the insights gained from our reduction may find further application. Our results for rooted orienteering translate to the regret-version of orienteering, and combined with the above reduction, give a compact LP for P2P-orienteering having integrality gap at most 6.

Although we do not improve the current-best approximation factor of $(2 + \epsilon)$ for orienteering [8], we believe that our LP-based approach is nevertheless appealing for various reasons. First, our LP-rounding algorithms are quite simple, and arguably, simpler than the DP-based approaches in [4,8]. Second, our LP-based approach offers the promising possibility that, by leveraging the key underlying ideas, one can obtain strong, compact LP-relaxations for other problems that utilize orienteering. Indeed, we already present evidence of such benefits by showing in Sect. 5.1 that our LP-insights for rooted orienteering yield a compact, provably-good LP for RVRP. (We remark that various configuration LPs considered for VRPs give rise to P2P-orienteering as the dual-separation problem, and utilizing our compact orienteering-LP in the dual could yield another way of obtaining a compact LP.) Finally, LP-based insights often tend to be powerful and have the potential to result in both improved guarantees, and algorithms for variants of the problem. In fact, we suspect that our orienteering LPs are better than what we have accounted for, and believe that they are a promising means of improving the state-of-the-art for orienteering.

Section 5 considers RVRP, and proposes two compact LP-relaxations for RVRP and corresponding rounding algorithms. Our LP-based algorithms not only yield improvements over the current-best 28.86-approximation for RVRP [10], but also result in substantial savings in running time compared to the algorithm in [10], which involves solving a configuration LP (with an exponential number of path variables) using the $\Omega(n^{1/\epsilon})$-time $(2 + \epsilon)$-approximation algorithm for orienteering in [8] as a subroutine. The first LP for RVRP is a natural modification of our LP for rooted orienteering, which we show has integrality gap at most 27 (Theorem 7). In Sect. 5.2, we formulate a rather atypical LP-relaxation (R2) for RVRP by exploiting certain key structural insights for RVRP. We observe that an RVRP-solution can be regarded as a collection of distance-increasing rooted paths covering some *sentinel* nodes S and a low-cost way of connecting the remaining nodes to S, and our LP aims to find the best such solution. We design a rounding algorithm for this LP that leads to a 15-approximation algorithm for RVRP, which is a significant improvement over the guarantee obtained in [10].

Finally, in Sect. 6, we observe that our techniques imply that the integrality gap of a Held-Karp style LP for the *asymmetric-TSP* (ATSP) *path* problem is 2 for the class of asymmetric metrics induced by the regret objective.

To give an overview of our techniques, a key tool that we use in our rounding algorithms, which also motivates our LP-relaxations, is an arborescence-packing result of [1] showing that an r-preflow $x \in \mathbb{R}_+^A$ in a digraph $D = (N, A)$ (i.e., $x(\delta^{in}(v)) \geq x(\delta^{out}(v)) \ \forall v \neq r$) dominates a weighted collection of r-rooted (non-spanning) out-arborescences (Theorem 3). An r-preflow x in the bidirected version of our metric, D, is a natural relaxation of an r-rooted path, and the $r \rightsquigarrow u$ connectivity under x abstracts whether u lies on this path. This leads to our LP (R-O) for (rooted) orienteering. The idea behind the rounding is that if we know the node v on the optimum path with maximum c_{rv} value, then we can enforce that our the LP-preflow x is consistent with v. Hence, we can decompose x into arborescences containing v of average length at most B, which yield r-v paths of average *regret* at most $2(B - c_{rv})$. These in turn can be converted (see Lemma 1) into a weighted collection of paths of total weight at most 3, where each path has regret at most $B - c_{rv}$ and ends at some node u with $c_{ru} \leq c_{rv}$; returning the maximum-reward path in this collection yields a 3-approximation.

Related work. The orienteering problem seems to have been first defined in [13]. Blum et al. [4] gave the first $O(1)$-factor approximation for rooted orienteering. They obtained an approximation ratio of 4, which was generalized to P2P-orienteering, and improved to 3 [2] and then to $2 + \epsilon$ [8].

Orienteering is closely related to the k-{*stroll*, MST, TSP} problems, which seek a minimum-cost rooted {path,tree,tour} respectively spanning at least k nodes (so the roles of objective and constraint are interchanged). k-MST has a rich history of study that culminated in a factor-2 approximation for both k-MST and k-TSP [11]. Chaudhuri et al. [7] obtained a $(2 + \epsilon)$-approximation algorithm for k-stroll. They also showed that for certain values of k, one can obtain a tree spanning k nodes and containing two specified nodes r, t, of cost at most the cheapest r-t path spanning k nodes. In particular, this holds for $k = n$, and yields an alternative way of obtaining a 2-approximation algorithm for the *minimum-regret TSP-path* problem considered in Sect. 6. The orienteering algorithms in [2, 4, 8] are all based on first obtaining suitable subpaths by approximating the *min-excess path* problem using a k-stroll algorithm as a subroutine, and then stitching together these subpaths via a DP. (For a rooted path, the notions of excess and regret coincide; we use the term regret as it is more in line with the terminology used in the vehicle-routing literature [15, 17].)

The use of regret as a vehicle-routing objective seems to have been first considered in [17], who present various heuristics, and RVRP is sometimes referred to as the *schoolbus problem* in the literature [5, 15, 17]. Bock et al. [5] were the first to consider RVRP from an approximation-algorithms perspective. They obtain approximation factors of $O(\log n)$ for general metrics and 3 for tree metrics. Subsequently, Friggstad and Swamy [10] gave the first constant-factor approximation algorithm for RVRP, obtaining a 28.86-approximation via an LP-rounding procedure for a configuration LP.

2   Preliminaries and Notation

Both orienteering and RVRP involve a complete undirected graph $G = (\{r\} \cup V, E)$, where r is a distinguished root (or depot) node, and metric edge costs $\{c_{uv}\}$. Let $n = |V| + 1$. We call a path P in G rooted if it begins at r. We always think of the nodes on P as being ordered in increasing order of their distance along P from r, and directing P away from r means that we direct each edge $uv \in P$ from u to v if u precedes v (under this ordering). We use D_v to denote c_{rv} for all $v \in V \cup \{r\}$. Let $\mathcal{T}$ denote the collection of all r-rooted trees in G. For a vector $d \in \mathbb{R}^E$, and a subset $F \subseteq E$, we use $d(F)$ to denote $\sum_{e \in F} d_e$. Similarly, for a vector $d \in \mathbb{R}^V$ and $S \subseteq V$, we use $d(S)$ to denote $\sum_{v \in S} d_v$.

Regret metric and RVRP. For every ordered pair $u, v \in V \cup \{r\}$, define the *regret distance* (with respect to r) to be $c_{uv}^{\text{reg}} := D_u + c_{uv} - D_v$. The regret distances $\{c_{uv}^{\text{reg}}\}$ form an asymmetric metric that we call the *regret metric*. The regret of a node v lying on a rooted path P is given by $c_P^{\text{reg}}(v) := c_P(v) - D_v = (c^{\text{reg}}$-length of the r-v portion of P), where $c_P(v)$ is the length of the r-v subpath of P.

Define the regret of P to be $c^{\text{reg}}(P)$, which is also the regret of the end-node of P. Observe that $c^{\text{reg}}(Z) = c(Z)$ for any cycle Z. We utilize the following results from [10].

Lemma 1 [10]. *Let $R \geq 0$. Given rooted paths $P_1, \ldots, P_k$ with total regret αR, we can efficiently find at most $k + \alpha$ rooted paths, each having regret at most R, that cover $\bigcup_{i=1}^{k} P_i$.*

Theorem 2 [10]. *Let $x = (x_P)_{P \in \mathcal{P}}$ be a weighted collection of rooted paths such that $\sum_{P \in \mathcal{P}: v \in P} x_P \geq 1$ for all $v \in V$. Let $R \geq 0$ be some given parameter. Let $k = \sum_{P \in \mathcal{P}} x_P$ and $\sum_{P \in \mathcal{P}} c^{\text{reg}}(P) x_P = \alpha R$. Then, for any $\theta \in (0, 1)$, we can round x to obtain a collection of at most $\left(\frac{6}{1-\theta} + \frac{1}{\theta}\right)\alpha + \lceil \frac{k}{\theta} \rceil$ rooted paths each of regret at most R that cover all nodes in V.*

Preflows and arborescence packing. Let $D = (\{r\} \cup V, A)$ be a digraph. We say that a vector $x \in \mathbb{R}_+^A$ is an *r-preflow* if $x(\delta^{\text{in}}(v)) \geq x(\delta^{\text{out}}(v))$ for all $v \in V$. When r is clear from the context, we simply say preflow. A key tool that we exploit is an arborescence-packing result of Bang-Jensen et al. [1] showing that we can decompose a preflow into out-arborescences rooted at r, and this can be done in polytime [16]. By an *out-arborescence rooted at r*, we mean a subgraph B whose undirected version is a tree containing r, and where every node spanned by B except r has exactly one incoming arc in B.

Theorem 3 [1,16]. *Let $D = (\{r\} \cup V, A)$ be a digraph and $x \in \mathbb{R}_+^A$ be a preflow. Let $\lambda_v := \min_{\{v\} \subseteq S \subseteq V} x(\delta^{\text{in}}(S))$ be the $r \rightsquigarrow v$ "connectivity" in D under capacities $\{x_a\}_{a \in A}$. Let $K > 0$ be rational. We can obtain out-arborescences $B_1, \ldots, B_q$ rooted at r, and rational weights $\gamma_1, \ldots, \gamma_q \geq 0$ such that $\sum_{i=1}^{q} \gamma_i = K$, $\sum_{i: a \in B_i} \gamma_i \leq x_a$ for all $a \in A$, and $\sum_{i: v \in B_i} \gamma_i = \min\{K, \lambda_v\}$ for all $v \in V$. Moreover, such a decomposition can be computed in time $\text{poly}(|V|, \text{ size of } K)$.*

3   Rooted Orienteering

In the *rooted orienteering* problem, we have a complete undirected graph $G = (\{r\} \cup V, E)$, metric edge costs $\{c_{uv}\}$, a distance bound $B \geq 0$, and nonnegative node rewards $\{\rho(v)\}_{v \in V}$. The goal is to find a rooted path with cost at most B that collects the maximum reward. Whereas all current approaches for orienteering rely on a dynamic program to stitch together suitable subpaths, we present a simple LP-rounding-based 3-approximation algorithm for rooted orienteering.

Let $D = (\{r\} \cup V, A)$ denote the bidirected version of G, where both (u, v) and (v, u) get cost c_{uv}. To introduce our LP and our rounding algorithm, first suppose that we know a node v on the optimum path that has maximum distance D_v among all nodes on the optimum path. In our relaxation, we model the path as one unit of flow $x \in \mathbb{R}_+^A$ that exits r, visits only nodes u with $D_u \leq D_v$ and v to an extent of 1, and has cost at most B. Since we do not know the endpoint of our path, we relax x to be a preflow. Letting z_u^v denote the $r \rightsquigarrow u$ connectivity (under capacities $\{x_a\}$), the reward earned by x is $\text{rewd}(x) := \sum_{u \in V} \rho(u) z_u^v$.

Our rounding procedure is based on the insight that Theorem 3 allows us to view x as a convex combination of arborescences, which we regard as r-rooted trees in G. Converting each tree into an r-v path (by standard doubling and shortcutting), we get a convex combination of rooted paths of average reward $\mathsf{rewd}(x)$, and average cost at most $2B - D_v$, and hence average c^{reg}-cost at most $2(B - D_v)$. Applying Lemma 1 to this collection, we then obtain a weighted collection of rooted paths of total weight at most 3 earning the same total reward, where each path has regret at most $B - D_v$, and hence, cost at most B (since it ends at some node u with $D_u \leq D_v$). Thus, the maximum-reward path in this collection yields a feasible solution with reward at least $\mathsf{rewd}(x)/3$.

Finally, we circumvent the need for "guessing" v by using variables z_v^v to indicate if v is the maximum-distance node on the optimum path. We impose that we have a preflow x^v of value z_v^v that visits v to an extent of z_v^v, and only visits nodes u with $D_u \leq D_v$, and z_u^v is now the $r \rightsquigarrow u$ connectivity under capacities x^v. (Note that $r \notin V$).

$$\max \quad \sum_{u,v \in V} \rho(u) z_u^v \tag{R-O}$$

$$\text{s.t.} \quad x^v\big(\delta^{\mathsf{in}}(u)\big) \geq x^v\big(\delta^{\mathsf{out}}(u)\big) \qquad \forall u, v \in V \tag{1}$$

$$x^v\big(\delta^{\mathsf{in}}(u)\big) = 0 \qquad \forall u, v \in V : D_u > D_v \tag{2}$$

$$x^v\big(\delta^{\mathsf{in}}(S)\big) \geq z_u^v \qquad \forall v \in V, S \subseteq V, u \in S \tag{3}$$

$$\sum_{a \in A} c_a x_a^v \leq B z_v^v \qquad \forall v \in V \tag{4}$$

$$x^v\big(\delta^{\mathsf{out}}(r)\big) = z_v^v \quad \forall v \in V, \qquad \sum_v z_v^v = 1, \quad x, z \geq 0.$$

This formulation can be converted to a compact LP by introducing flow variables $f^{u,v} = \{f_a^{u,v}\}_{a \in A}$, and encoding the cut constraints (3) by imposing that $f^{u,v} \leq x^v$, and that $f^{u,v}$ sends z_u^v units of flow from r to u. Observe that: (a) if $D_u > D_v$ then $z_u^v \leq x^v\big(\delta^{\mathsf{in}}(u)\big) = 0$; (b) we have $z_u^v \leq x^v\big(\delta^{\mathsf{in}}(V)\big) = x^v\big(\delta^{\mathsf{out}}(r)\big) = z_v^v$ for all u, v. Let (x^*, z^*) be an optimal solution to (R-O), of value OPT.

Theorem 4. *We can round (x^*, z^*) to a rooted-orienteering solution of value at least $OPT/3$.*

Proof. For each v with $z_v^{*v} > 0$ we apply Theorem 3 with $K = z_v^{*v}$ to obtain r-rooted out-arborescences, which we view as rooted trees in G, and associated nonnegative weights $\{\gamma_T^v\}_{T \in \mathcal{T}}$; recall that $\mathcal{T}$ is the collection of all r-rooted trees. So we have $\sum_T \gamma_T^v = z_v^{*v}$, $\sum_T \gamma_T^v c(T) \leq \sum_a c_a x_a^{*v} \leq B z_v^{*v}$, and $\sum_{T : u \in T} \gamma_T^v \geq z_u^{*v}$ for all $u \in V$. Note that for every T with $\gamma_T^v > 0$, we have $v \in T$, and $D_u \leq D_v$ for all $u \in T$ (as otherwise, we have $x^{*v}\big(\delta^{\mathsf{in}}(u)\big) = 0$). For every v and every tree T with $\gamma_T^v > 0$, we do the following. First, we double the edges not lying on the r-v path of T and shortcut to obtain a simple r-v path P_T^v. So

$$\sum_T \gamma_T^v c^{\mathsf{reg}}(P_T^v) \leq 2 \sum_T \gamma_T^v \big(c(T) - D_v\big) = 2 z_v^{*v}(B - D_v). \tag{5}$$

Next, we use Lemma 1 with regret-bound $B - D_v$ to break P_T^v into a collection $\mathcal{P}_T^v$ of at most $1 + \frac{c^{\text{reg}}(P_T^v)}{B - D_v}$ rooted paths, each having c^{reg}-cost at most $B - D_v$. Note that if $B = D_v$, then $c^{\text{reg}}(P_T^v) = 0$, and we use the convention that $0/0 = 0$, so $|\mathcal{P}_T^v| = 1$ in this case. Each path in $\mathcal{P}_T^v$ ends at a vertex u with $D_u \leq D_v$, so its c-cost is at most B. Now, for all $v \in V$, we have

$$\sum_T \gamma_T^v \sum_{P \in \mathcal{P}_T^v} \rho(P) = \sum_T \gamma_T^v \rho(P_T^v) \geq \sum_u \rho(u) z_u^{*v} \tag{6}$$

$$\sum_T \gamma_T^v |\mathcal{P}_T^v| \leq \sum_T \gamma_T^v \left(1 + \frac{c^{\text{reg}}(P_T^v)}{B - D_v}\right) \leq z_v^{*v} + 2z_v^{*v} = 3z_v^{*v} \tag{7}$$

where the last inequality in (7) follows from (5). Therefore, the maximum-reward path in $\bigcup_{v, T : \gamma_T^v > 0} \mathcal{P}_T^v$ earns reward at least

$$\left(\sum_{v, T} \gamma_T^v \sum_{P \in \mathcal{P}_T^v} \rho(P)\right) \Big/ \left(\sum_{v, T} \gamma_T^v |\mathcal{P}_T^v|\right) \geq \frac{\sum_{v, u} \rho_u z_u^{*v}}{3 \sum_v z_v^{*v}} = OPT/3.$$

$\square$

The following variant of rooted orienteering, which we call *regret orienteering*, will be useful in Sect. 4. In regret orienteering, instead of a cost bound B, we are given a *regret bound* R, and we seek a rooted path of *regret* at most R that collects the maximum reward. The LP-relaxation for regret-orienteering is very similar to (R-O); the only changes are that z_v^v now indicates if v is the *end node* of the optimum path, and so we drop (2) and replace (4) with $\sum_{a \in A} c_a x_a^v \leq (D_v + R) z_v^v$. The rounding algorithm is essentially unchanged: we convert the trees obtained from x^v into r-v paths, which are then split into paths of regret at most R. Theorem 4 yields the following corollary.

Corollary 5. *There is an LP-based 3-approximation for regret orienteering.*

4   Point-to-Point Orienteering

We now consider the generalization of rooted orienteering, where we have a start node r *and* an end node t, and we seek an r-t path with cost at most B that collects the maximum reward. We may assume that r and t have 0 reward, i.e., $\rho(r) = \rho(t) = 0$. The main result of this section is a novel reduction showing that *point-to-point* (P2P) *orienteering* problem can be *reduced* to regret orienteering losing a factor of at most 2 (Theorem 6). Combining this with our LP-approach for regret orienteering and Corollary 5, we obtain an LP-relaxation for P2P-orienteering having integrality gap at most 6 (described in the full version). We believe that the insights gained from this reduction may find further application.

Theorem 6. *An α-approximation algorithm for regret orienteering (where $\alpha \geq 1$) can be used to obtain a 2α-approximation algorithm for P2P-orienteering.*

Proof. Let $\big(G = (\{r, t\} \cup V, E), \{c_{uv}\}, \{\rho(u)\}, B\big)$ be a P2P-orienteering instance. Our reduction is simple. Let P^* be an optimal solution. We "guess" a node $v \in P^*$ (which could be r or t) such that $D_v + c_{vt} = \max_{u \in P^*}(D_u + c_{ut})$. (That is, we enumerate over all choices for v.) Let $S = \{u \in \{r,t\} \cup V : D_u + c_{ut} \leq D_v + c_{vt}\}$. We then consider two regret orienteering problems, both of which have regret bound $R = B - D_v - c_{vt}$ and involve only nodes in S (i.e., we equivalently set $\rho(u) = 0$ for all $u \notin S$); the first problem has root r, and the second has root t. Let P_1 and P_2 be the solutions obtained for these two problems respectively by our α-approximation algorithm. So for some $u_1, u_2 \in S$, P_1 is an r-u_1 path, and P_2 may be viewed as a u_2-t path. Notice that P_1 appended with the edge $u_1 t$ yields an r-t path of *cost* at most $D_{u_1} + c^{\text{reg}}(P_1) + c_{u_1 t} \leq D_{u_1} + c_{u_1 t} + B - D_v - c_{vt} \leq B$, since $u_1 \in S$. Similarly P_2 appended with the edge ru_2 yields an r-t path of cost at most B. We return $P_1 + u_1 t$ or $ru_2 + P_2$, whichever has higher reward.

To analyze this, we observe that the r-v portion of P^* is a feasible solution to the regret-orienteering instance with root r, since its cost is at most $B - c_{vt}$, and hence, its regret is at most R. Similarly, the v-t portion of P^* (viewed in reverse) is a feasible solution to the regret-orienteering instance with root t. Therefore, $\max\{\rho(P_1 + u_1 t), \rho(ru_2 + P_2)\} \geq \rho(P^*)/2\alpha$. $\square$

5   Compact LPs and Improved Guarantees for RVRP

Recall that in the *regret-bounded vehicle routing problem* (RVRP), we are given an undirected complete graph $G = (\{r\} \cup V, E)$ on n nodes with a distinguished root (depot) node r, metric edge costs or distances $\{c_{uv}\}$, and a regret-bound R. The goal is to find the minimum number of rooted paths that cover all nodes so that the regret of each node with respect to the path covering it is at most R. Throughout, let O^* denote the optimal value of the RVRP instance. We describe two compact LP-relaxations for RVRP and corresponding rounding algorithms that yield improvements, in both approximation ratio and running time, over the RVRP-algorithm in [10]. In Sect. 5.1, we observe that the compact LP for orienteering (R-O) yields a natural LP for RVRP; by combining the rounding ideas used for orienteering and Theorem 2, we obtain a 27-approximation algorithm for RVRP. In Sect. 5.2, we formulate an unorthodox, stronger LP-relaxation (R2) for RVRP by leveraging some key structural insights in [10]. We devise a rounding algorithm for this LP that leads to a 15-approximation algorithm for RVRP, which is a significant improvement over the guarantee obtained in [10].

5.1   Extending the Orienteering LP to RVRP

The LP-relaxation below can be viewed as a natural variant of the orienteering LP adapted to RVRP. As before, let $D = (\{r\} \cup V, A)$ be the bidirected version of G. For each node v, x^v is a preflow (constraint (8)) of value z_v^v such that the $r \rightsquigarrow u$ connectivity under capacities $\{x_a^v\}$ is at least z_u^v for all u, v (constraint (9)). As before, we can obtain a compact formulation by replacing the cut constraints (9) with constraints involving suitable flow variables.

$$\min \quad \sum_v z_v^v \tag{R1}$$

$$
\begin{aligned}
\text{s.t.} \quad & x^v(\delta^{\text{in}}(u)) \geq x^v(\delta^{\text{out}}(u)) && \forall u, v \in V && (8)\\
& x^v(\delta^{\text{in}}(S)) \geq z_u^v && \forall v \in V, S \subseteq V, u \in S && (9)\\
& \sum_{a \in A} c_a x_a^v \leq (D_v + R) z_v^v && \forall v \in V
\end{aligned}
$$

$$x^v(\delta^{\text{out}}(r)) = z_v^v \quad \forall v \in V, \quad \sum_{v \in V} z_u^v \geq 1 \quad \forall u \in V, \quad x, z \geq 0.$$

Theorem 7. *We can round an optimal solution to* (R1) *to obtain a 27-approximation for* RVRP.

5.2   A New Compact LP for RVRP Leading to a 15-Approximation

We now propose a different LP for RVRP, which leads to a much-improved 15-approximation for RVRP. To motivate this LP, we first collect some facts from [4,10] pertaining to the regret objective. By merging all nodes at distance 0 from each other, we may assume that $c_{uv} > 0$ for all $u, v \in V \cup \{r\}$, and hence $D_v > 0$ for all $v \in V$.

Definition 8 [10]. Let P be a rooted path ending at w. Consider an edge (u, v) of P, where u precedes v on P. We call this a *red* edge of P if there exist nodes x and y on the r-u portion and v-w portion of P respectively such that $D_x \geq D_y$; otherwise, we call this a *blue* edge of P. (Note that the first edge of P is always a blue edge).

For a node $x \in P$, let $\mathsf{red}(x, P)$ denote the maximal subpath Q of P containing x consisting of only red edges (which might be the trivial path $\{x\}$). Call the collection $\{\mathsf{red}(x, P) : x \in P\}$ of subpaths, the red intervals of P.

Lemma 9 [4]. *For any rooted path P, we have $\sum_{e \text{ red on } P} c_e \leq \frac{3}{2} c^{\text{reg}}(P)$.*

Lemma 10 [10]. *(i) Let u, v be nodes on a rooted path P such that u precedes v on P and $\mathsf{red}(u, P) \neq \mathsf{red}(v, P)$; then $D_u < D_v$. (ii) If P' is obtained by shortcutting P so that it contains at most one node from each red interval of P, then for every edge (x, y) of P' with x preceding y on P', we have $D_x < D_y$.*

We say that a node u on a rooted path of P is a *sentinel* of P if u is the first node of $\mathsf{red}(u, P)$. Part (ii) above shows that if we shortcut each path P of an optimal RVRP-solution past the non-sentinel nodes of P, then we obtain a distance-increasing collection of paths. Moreover, part (i) implies that if x and y are sentinels on P with x appearing before y, then $\max_{u \in \mathsf{red}(x,P)} D_u < \min_{u \in \mathsf{red}(y,P)} D_u$. Finally, every non-sentinel node is connected to the sentinel corresponding to its red interval via red edges, and Lemma 9 shows that the total (c-) cost of these edges at most $1.5R$(optimal value).

Thus, we can view an RVRP-solution as a collection of distance-increasing rooted paths covering some sentinel nodes S, and a low-cost way of connecting the nodes in $V \setminus S$ to S. Our LP-relaxation searches for the best such solution. Let $\mathcal{D} := \{D_v : v \in V\}$. For every $u \in V$, define $\mathcal{D}_u$ to be the collection $\{[d_1, d_2] : d_1, d_2 \in \mathcal{D}, \ d_1 \leq D_u \leq d_2\}$ of (closed) intervals. We have variables $x_{u,I,u}$ for every node $u \in V$ and interval $I = [d_1, d_2] \in \mathcal{D}_u$ to indicate if u is a sentinel and d_1, d_2 are the minimum and maximum distances (from r) respectively of nodes in the red interval corresponding to u; we say that I is u's distance interval. We also have variables $x_{u,I,v}$ for $v \neq u$ to indicate that v is connected to sentinel u with distance interval I, and edge variables $\{z_e\}_{e \in E}$ that encode these connections. Finally, we have flow variables $f_{r,u,I}, f_{u,I,v,J}, f_{u,I,t}$ for all $u, v \in V$ and $I \in \mathcal{D}_u$, $J \in \mathcal{D}_v$ that encode the distance-increasing rooted paths on the sentinels, with t representing a fictitious sink. We include constraints that encode that the distance intervals of sentinels lying on the same path are disjoint, and a non-sentinel v can be connected to (u, I) only if $D_v \in I$. We obtain the following LP.

$$\min \quad \sum_{u \in V, I \in \mathcal{D}_u} f_{r,u,I} \tag{R2}$$

$$\text{s.t.} \quad \sum_{u \in V, I \in \mathcal{D}_u} x_{u,I,v} \geq 1 \quad \forall v \in V \tag{10}$$

$$x_{u,I,v} \leq x_{u,I,u}, \quad x_{u,I,v} = 0 \quad \text{if } D_v \notin I \quad \forall u, v \in V, I \in \mathcal{D}_u \tag{11}$$

$$z\big(\delta(S)\big) \geq \sum_{u \notin S, I \in \mathcal{D}_u} x_{u,I,v} \quad \forall v \in V, \{v\} \subseteq S \subseteq V \tag{12}$$

$$f_{r,u,I} + \sum_{v \in V, J \in \mathcal{D}_v} f_{v,J,u,I} = x_{u,I,u} \quad \forall u \in V, I \in \mathcal{D}_u \tag{13}$$

$$\sum_{v \in V, J \in \mathcal{D}_v} f_{u,I,v,J} + f_{u,I,t} = x_{u,I,u} \quad \forall u \in V, I \in \mathcal{D}_u \tag{14}$$

$$f_{u,I,v,J} = 0 \quad \forall u, v \in V, I \in \mathcal{D}_u, J \in \mathcal{D}_v : I \cap J \neq \emptyset \text{ or } D_v \leq D_u \tag{15}$$

$$\sum_{u,v \in V, I \in \mathcal{D}_u, J \in \mathcal{D}_v} c_{uv}^{\text{reg}} f_{u,I,v,J} \leq R \cdot \sum_{u \in V, I \in \mathcal{D}_u} f_{r,u,I} \tag{16}$$

$$\sum_{e \in E} c_e z_e \leq 1.5R \cdot \sum_{u \in V, I \in \mathcal{D}_u} f_{r,u,I} \tag{17}$$

$$x, z, f \geq 0.$$

Constraint (10) encodes that every node v is either a sentinel or is connected to a sentinel; (11) ensures that if v is assigned to (u, I), then u is indeed a sentinel with distance interval I and that $D_v \in I$. Constraints (12) ensure that the z_es (fractionally) connect each non-sentinel v to the sentinel specified by the $x_{u,I,v}$ variables. Constraints (13), (14) encode that each sentinel (u, I) lies on rooted paths, and (15) ensures that these paths are distance increasing and moreover the distance intervals of the sentinels on the paths are disjoint. Finally, letting k denote the number of paths used (16), (17) encode that the total regret of the distance-increasing paths is at most kR (note that $c_{ru}^{reg} = 0$ for all u), and the total cost of the edges used to connect non-sentinels to sentinels is at most $1.5kR$. As before, the cut constraints (12) can be equivalently stated using flows to obtain a polynomial-size LP. Let (x^*, z^*, f^*) be an optimal solution to (R2) and OPT denote its objective value. We have already argued that an optimal RVRP-solution yields an integer solution to (R2), so we obtain that $\lceil OPT \rceil$ is at most the optimal value, O^*, of the RVRP instance.

Our rounding algorithm proceeds in a similar fashion as the RVRP-algorithm in [10]; yet, we obtain an improved guarantee since one can solve (R2) exactly whereas one can only obtain a $(2 + \epsilon)$-approximate solution to the configuration LP in [10]. Let $\theta \in (0, 1)$ be a parameter that we will set later. We first obtain a forest F of c-cost at most $\frac{3R}{1-\theta} \cdot OPT$ such that every component Z contains a witness node v that is assigned to an extent of at least θ to sentinels in Z. We argue that if we contract the components of F, then the distance-increasing

---

A1. For $S \subseteq V$, define $h(S) = 1$ if $\sum_{u \in S, I \in \mathcal{D}_u} x_{u,I,v}^* < \theta$ for all $v \in S$, and 0 otherwise. h is a *downwards-monotone* cut-requirement function: if $\emptyset \neq A \subseteq B$, then $h(A) \geq h(B)$. Use the LP-relative 2-approximation algorithm in [12] for $\{0,1\}$ downwards-monotone functions to obtain a forest F such that $|\delta(S) \cap F| \geq h(S)$ for all $S \subseteq V$.

A2. For every component Z of F with $r \notin Z$, pick a *witness node* $w \in Z$ such that $\sum_{u \in Z, I \in \mathcal{D}_u} x_{u,I,w}^* \geq \theta$. Let $\sigma(w) = \{(u, I) : u \in Z, x_{u,I,w}^* > 0\}$. Let $W \subseteq V$ be the set of all such witness nodes.

A3. f^* is an $r \rightsquigarrow t$ flow in an auxiliary graph having nodes r, t, and (u, I) for all $u \in V, I \in \mathcal{D}_u$, edges $(r, (u, I))$, $((u, I), t)$ for all $u \in V, I \in \mathcal{D}_u$, and edges $((u, I), (v, J))$ for all $u, v \in V, I \in \mathcal{D}_u, J \in \mathcal{D}_v$ such that $D_u < D_v$ and $I \cap J = \emptyset$. Let $\{f_P^*\}_{P \in \mathcal{P}}$ be a path-decomposition of this flow. Modify each flow path $P \in \mathcal{P}$ as follows. First, drop t from P. Shortcut P past the nodes in P that are not in $\{r\} \cup \bigcup_{w \in W} \sigma(w)$. The resulting path maps naturally to a rooted path in G (obtained by simply dropping the distance intervals), which we denote by $\pi(P)$. Clearly, $c^{reg}(\pi(P)) \leq \sum_{((u,I),(v,J)) \in P} c_{uv}^{reg}$ since shortcutting does not increase the regret cost.

A4. Let $\mathcal{Q}$ be the collection of rooted paths obtained by taking the paths $\{\pi(P) : P \in \mathcal{P}\}$ and contracting the components of F. Let H be the directed graph (which we prove is acyclic) obtained by directing the paths in $\mathcal{Q}$ away from r. To avoid notational clutter, for a component Z of F, we use Z to also denote the corresponding contracted node in H. For each $Q \in \mathcal{Q}$, define $y_Q = \sum_{P \in \mathcal{P} : \pi(P) \text{ maps to } Q} f_P^*$.

A5. Use the integrality property of flows to round the flow $\{\frac{y_Q}{\theta}\}_{Q \in \mathcal{Q}}$ to an integer flow of value $k \leq \lceil \frac{OPT}{\theta} \rceil$ and regret-cost at most $\frac{R}{\theta} \cdot OPT$. Since H is acyclic, this yields rooted paths $\hat{P}_1, \ldots, \hat{P}_k$ so that every component Z of F lies on exactly one $\hat{P}_i$ path.

A6. We map the $\hat{P}_i$s to rooted paths in G that cover V as follows. Consider a path $\hat{P}_i$. Let Z be a component lying on $\hat{P}_i$, and $u, v \in Z$ be the nodes where $\hat{P}_i$ enters and leaves Z respectively. We add to $\hat{P}_i$ a u-v path that covers all nodes of Z obtained by doubling all edges of Z except those on the u-v path in Z and shortcutting. Let $\tilde{P}_i$ be the rooted path in G obtained by doing this for all components lying on $\hat{P}_i$.

A7. Finally, we use Lemma 1 to convert $\tilde{P}_1, \ldots, \tilde{P}_k$ to an RVRP-solution.

sentinel flow paths yield an acyclic flow that covers every contracted component to an extent of at least θ. Hence, using the integrality property of flows, we obtain an integral flow, and hence a collection of at most $\lceil \frac{OPT}{\theta} \rceil$ rooted paths, that covers every component and has cost at most $\frac{R}{\theta} \cdot OPT$. Next, we show that we can uncontract the components and attach the component-nodes to these rooted paths incurring an additional cost of at most $\frac{6R}{1-\theta} \cdot OPT$. Finally, by applying Lemma 1, we obtain an RVRP solution with at most $\left(\frac{6}{1-\theta} + \frac{1}{\theta} \right) OPT + \lceil \frac{OPT}{\theta} \rceil$ rooted paths.

Theorem 11. *The above algorithm returns an* RVRP-*solution with at most* $\left(\frac{6}{1-\theta} + \frac{1}{\theta} \right) OPT + \lceil \frac{OPT}{\theta} \rceil$ *paths. Thus, taking* $\theta = \frac{1}{3}$, *we obtain at most* $15 \cdot O^*$ *paths.*

6   Minimum-Regret TSP-path

We now consider the *minimum-regret TSP-path* problem, wherein we have (as before), a complete graph $G = (V', E)$, $r, t \in V'$, metric edge costs $\{c_{uv}\}$, and we seek a minimum-regret r-t path that visits all nodes. Observe that this is precisely the ATSP-*path* problem under the asymmetric regret metric c^{reg}. We establish a tight bound of 2 on the integrality gap of the standard ATSP-path LP for the class of regret-metrics (induced by a symmetric metric). We consider the following LP for min-regret TSP path. Let $D = (V', A)$ be the bidirected version of G. Let $b_t = 1 = -b_r$ and $b_v = 0$ for all $v \in V' \setminus \{r, t\}$.

$$\min \sum_{a \in A} c_a^{\text{reg}} x_a \quad \text{s.t.} \quad x\left(\delta^{\text{in}}(v)\right) - x\left(\delta^{\text{out}}(v)\right) = b_v \forall v \in V', \quad x \geq 0 \quad \text{(R-TSP)}$$

$$x\left(\delta^{\text{in}}(S)\right) \geq 1 \quad \forall \emptyset \neq S \subseteq V \setminus \{r\}.$$

Theorem 12. *The integrality gap of* (R-TSP) *is 2 for regret metrics, and we can obtain in polytime a Hamiltonian* r-t *path* P *with* $c^{\text{reg}}(P) \leq 2 \cdot OPT_{R-TSP}$.

References

1. Bang-Jensen, J., Frank, A., Jackson, B.: Preserving and increasing local edge-connectivity in mixed graphs. SIAM J. Discrete Math. **8**(2), 155–178 (1995)
2. Bansal, N., Blum, A., Chawla, S., Meyerson, A.: Approximation algorithms for deadline-TSP and vehicle routing with time windows. In: 36th STOC (2004)
3. Blum, A., Chalasani, P., Coppersmith, D., Pulleyblank, B., Raghavan, P., Sudan, M.: The minimum latency problem. In: 26th STOC, pp. 163–171 (1994)
4. Blum, A., Chawla, S., Karger, D.R., Lane, T., Meyerson, A.: Approximation algorithms for orienteering and discount-reward TSP. SICOMP **37**, 653–670 (2007)
5. Bock, A., Grant, E., Könemann, J., Sanità, L.: The school bus problem on trees. In: Asano, T., Nakano, S., Okamoto, Y., Watanabe, O. (eds.) ISAAC 2011. LNCS, vol. 7074, pp. 10–19. Springer, Heidelberg (2011). doi:10.1007/978-3-642-25591-5_3
6. Chakrabarty, D., Swamy, C.: Facility location with client latencies: linear-programming based techniques for minimum-latency problems. Math. Oper. Res. **41**(3), 865–883 (2016)

7. Chaudhuri, K., Godfrey, P.B., Rao, S., Talwar, K.: Paths, trees and minimum latency tours. In: Proceedings of 44th FOCS, pp. 36–45 (2003)
8. Chekuri, C., Korula, N., Pál, M.: Improved algorithms for orienteering and related problems. ACM Trans. Algorithms $8(3)$, 661–670 (2012)
9. Fakcharoenphol, J., Harrelson, C., Rao, S.: The k-traveling repairman problem. ACM Trans. Alg. $3(4)$, Article 40 (2007)
10. Friggstad, Z., Swamy, C.: Approximation algorithms for regret-bounded vehicle routing and applications to distance-constrained vehicle routing. In: Proceedings of STOC, pp. 744–753 (2014). Detailed version posted on CS arXiv, November 2013
11. Garg, N.: Saving an epsilon: a 2-approximation for the k-MST problem in graphs. In: Proceedings of the 37th STOC, pp. 396–402 (2005)
12. Goemans, M.X., Williamson, D.P.: Approximating minimum-cost graph problems with spanning tree edges. Oper. Res. Lett. **16**, 183–189 (1994)
13. Golden, B.L., Levy, L., Vohra, R.: The orienteering problem. Naval Res. Logist. **34**, 307–318 (1987)
14. Haimovich, M., Kan, A.: Bounds and heuristics for capacitated routing problems. Math. Oper. Res. **10**, 527–542 (1985)
15. Park, J., Kim, B.: The school bus routing problem: a review. Eur. J. Oper. Res. **202**(2), 311–319 (2010)
16. Post, I., Swamy, C.: Linear-programming based techniques for multi-vehicle minimum latency problems. In: Proceedings of 26th SODA, pp. 512–531 (2015)
17. Spada, M., Bierlaire, M., Liebling, T.: Decision-aiding methodology for the school bus routing and scheduling problem. Transp. Sci. **39**, 477–490 (2005)
18. Toth, P., Vigo, D. (eds.): The Vehicle Routing Problem. SIAM Monographs on Discrete Mathematics and Applications, Philadelphia (2002)

Discrete Newton's Algorithm for Parametric Submodular Function Minimization

Michel X. Goemans[✉], Swati Gupta, and Patrick Jaillet

Massachusetts Institute of Technology, Cambridge, USA
goemans@math.mit.edu, {swatig,jaillet}@mit.edu

Abstract. We consider the line search problem in a submodular polyhedron $P(f) \subseteq \mathbb{R}^n$: Given an arbitrary $a \in \mathbb{R}^n$ and $x_0 \in P(f)$, compute $\max\{\delta : x_0 + \delta a \in P(f)\}$. The use of the discrete Newton's algorithm for this line search problem is very natural, but no strongly polynomial bound on its number of iterations was known (Iwata 2008). We solve this open problem by providing a quadratic bound of $n^2 + O(n \log^2 n)$ on its number of iterations. Our result considerably improves upon the only other known strongly polynomial time algorithm, which is based on Megiddo's parametric search framework and which requires $\tilde{O}(n^8)$ submodular function minimizations (Nagano 2007). As a by-product of our study, we prove (tight) bounds on the length of chains of ring families and geometrically increasing sequences of sets, which might be of independent interest.

Keywords: Discrete Newton's algorithm · Submodular functions · Line search · Ring families · Geometrically increasing sequence of sets · Fractional combinatorial optimization

1 Introduction

Let f be a submodular function on V, where $|V| = n$. We often assume that $V = [n] := \{1, 2, \cdots, n\}$. Let $P(f) = \{x \in \mathbb{R}^n \mid x(S) \leq f(S) \text{ for all } S \subseteq V\}$. The only assumption we make on f is that $f(\emptyset) \geq 0$ (otherwise $P(f)$ is empty). Given $x_0 \in P(f)$ (this condition can be verified by performing a single submodular function minimization) and $a \in \mathbb{R}^n$, we would like to find the largest δ such that $x_0 + \delta a \in P(f)$. For any vector $b \in \mathbb{R}^n$ and any set $S \subseteq V$, it is convenient to use the notation $b(S) := \sum_{e \in S} b_e$. By considering the submodular function f' taking the value $f'(S) = f(S) - x_0(S)$ for any set S, we can equivalently find the largest δ such that $\delta a \in P(f')$. Since $x_0 \in P(f)$ we know that $0 \in P(f')$ and thus f' is nonnegative. Thus, without loss of generality, we consider the problem

$$\delta^* = \max \left\{ \delta : \min_{S \subseteq V} f(S) - \delta a(S) \geq 0 \right\}, \tag{1}$$

for a nonnegative submodular function f.

© Springer International Publishing AG 2017
F. Eisenbrand and J. Koenemann (Eds.): IPCO 2017, LNCS 10328, pp. 212–227, 2017.
DOI: 10.1007/978-3-319-59250-3_18

Since $x_0 = 0 \in P(f)$ we know that $\delta^* \geq 0$ and that the minimum could be taken only over the sets S with $a(S) > 0$, although we will not be using this fact. To make this problem nontrivial, we assume that there exists some i with $a_i > 0$. Geometrically, the problem of finding δ^* is a line search problem. As we go along the line segment $\ell(\delta) = x_0 + \delta a$ (or just δa if we assume $x_0 = 0$), when do we exit the submodular polyhedron $P(f)$? This is a basic subproblem needed in many algorithmic applications. For example, for the algorithmic version of Carathéodory's theorem (over any polytope), one typically performs a line search from a vertex of the face being considered in a direction within the same face. This is, for example, also the case for variants of the Frank-Wolfe algorithm (see for instance (Freund et al. 2015)).

A natural way to solve this line search problem is to use a cutting plane approach. Start with any upper bound $\delta_1 \geq \delta^*$ and define the point $x^{(1)} = \delta_1 a$. One can then generate a most violated inequality for $x^{(1)}$, where most violated means the one minimizing $f(S) - \delta_1 a(S)$ over all sets S. The hyperplane corresponding to a minimizing set S_1 intersects the line in $x^{(2)} = \delta_2 a$. Proceeding analogously, we obtain a sequence of points and eventually will reach the optimum δ.

This cutting-plane approach is equivalent to Dinkelbach's algorithm or the discrete Newton's algorithm for solving (1). At the risk of repeating ourselves, we let $\delta_1 \geq \delta^*$. For example we could set $\delta_1 = \min_{i:a_i>0} f(\{i\})/a_i$. At iteration $i \geq 1$ of Newton's algorithm, we consider the submodular function $k_i(S) = f(S) - \delta_i a(S)$, and compute

$$h_i = \min_S k_i(S),$$

and define S_i to be *any* minimizer of $k_i(S)$. Now, let $f_i = f(S_i)$ and $g_i = a(S_i)$. As long as $h_i < 0$, we proceed and set

$$\delta_{i+1} = \frac{f_i}{g_i}.$$

As soon as $h_i = 0$, Newton's algorithm terminates and we have that $\delta^* = \delta_i$. We give the full description of the discrete Newton's algorithm in Algorithm 1.

Algorithm 1. DISCRETE NEWTON'S ALGORITHM

---

   input : submodular $f : 2^V \to \mathbb{R}$, f nonnegative, $a \in \mathbb{R}^n$
   output: $\delta^* = \max\{\delta : \min_S f(S) - \delta a(S) \geq 0\}$
   $i = 0, \delta_1 = \min_{i \in V, a(\{i\})>0} f(\{i\})/a(\{i\})$;
   repeat
      | $i = i + 1$;
      | $h_i = \min_{S \subseteq V} f(S) - \delta_i a(S)$;
      | $S_i \in \arg\min_{S \subseteq V} f(S) - \delta_i a(S)$;
      | $\delta_{i+1} = \frac{f(S_i)}{a(S_i)}$;
   until $h_i = 0$;
   Return $\delta^* = \delta_i$.

When $a \geq 0$, it is known that Newton's algorithm terminates in at most n iterations (for e.g. (Topkis 1978)). Even more, the function $g(\delta) := \min_S f(S) - \delta a(S)$ is a concave, piecewise affine function with at most n breakpoints (and $n + 1$ affine segments) since for any set $\{\delta_i\}_{i \in I}$ of δ values, the submodular functions $f(S) - \delta_i a(S)$ for $i \in I$ form a sequence of strong quotients (ordered by the δ_i's), and therefore the minimizers form a chain of sets. See (Iwata et al. 1997) for definitions of strong quotients and details.

When a is arbitrary (not necessarily nonnegative), little is known about the number of iterations of the discrete Newton's algorithm. The number of iterations can easily be bounded by the number of possible distinct positive values of $a(S)$, but this is usually very weak (unless, for example, the support of a is small as is the case in the calculation of exchange capacities). A weakly polynomial bound involving the sizes of the submodular function values is easy to obtain, but no strongly polynomial bound was known, as mentioned as an open question in (Nagano 2007, Iwata 2008). In this paper, we show that the number of iterations is quadratic. This is the first strongly polynomial bound in the case of an arbitrary a.

Theorem 1. *For any submodular function $f : 2^{[n]} \to \mathbb{R}_+$ and an arbitrary direction a, the discrete Newton's algorithm takes at most $n^2 + O(n \log^2(n))$ iterations.*

Previously, the only strongly polynomial algorithm to solve the line search problem in the case of an arbitrary $a \in \mathbb{R}^n$ was an algorithm of Nagano et al. (Nagano 2007) relying on Megiddo's parametric search framework. This requires $\tilde{O}(n^8)$ submodular function minimizations, where $\tilde{O}(n^8)$ corresponds to the current best running time known for *fully combinatorial* submodular function minimization (Iwata and Orlin 2009). On the other hand, our main result in Theorem 1 shows that the discrete Newton's algorithm takes $O(n^2)$ iterations, i.e. $O(n^2)$ submodular function minimizations, and we can use any submodular function minimization algorithm. Each submodular function minimization can be computed, for example, in $\tilde{O}(n^4 + \gamma n^3)$ time using a result of (Lee et al. 2015), where γ is the time for an evaluation of the submodular function.

Radzik (Radzik 1998) provides an analysis of the discrete Newton's algorithm for the related problem of $\max \delta : \min_{S \in \mathcal{S}} b(S) - \delta a(S) \geq 0$ where both a and b are modular functions and $\mathcal{S}$ is an arbitrary collection of sets. He shows that the number of iterations of the discrete Newton's algorithm is at most $O(n^2 \log^2(n))$. Our analysis does not handle an arbitrary collection of sets, but generalizes his setting as it applies to the more general case of submodular functions f. Note that considering submodular functions (as opposed to modular functions) makes the problem considerably harder since the number of input parameters for modular functions is only $2n$, whereas in the case of submodular functions the input is exponential (we assume oracle access for function evaluation).

Apart from the main result of bounding the number of iterations of the discrete Newton's algorithm for solving $\max \delta : \min_S f(S) - \delta a(S) \geq 0$ in Sect. 3, we prove results on ring families (set families closed under taking intersections

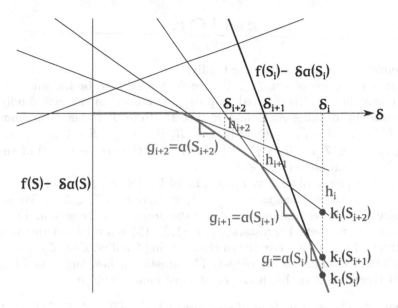

Fig. 1. Illustration of Newton's iterations and notation in Lemma 1.

and unions) and geometrically increasing sequences of sets, which may be of independent interest. As part of the proof of Theorem 1, we first show a tight (quadratic) bound on the length of a sequence $T_1, \cdots, T_k$ of sets such that no set in the sequence belongs to the smallest ring family generated by the previous sets (Sect. 2). Further, one of the key ideas in the proof of Theorem 1 is to consider a sequence of sets (each set corresponds to an iteration in the discrete Newton's algorithm) such that the value of a submodular function on these sets increases geometrically. We show a quadratic bound on the length of such sequences for any submodular function and construct two (related) examples to show that this bound is tight, in Sect. 4. Interestingly, one of these examples is a construction of intervals and the other example is a weighted directed graph where the cut function already gives such a sequence of sets.

2 Ring Families

A *ring family* $\mathcal{R} \subset 2^V$ is a family of sets closed under taking unions and intersections. From Birkhoff's representation theorem, we can associate to a ring family a directed graph $D = (V, E)$ in the following way. Let $A = \bigcap_{R \in \mathcal{R}} R$ and $B = \bigcup_{R \in \mathcal{R}} R$. Let $E = \{(i, j) \mid R \in \mathcal{R}, i \in R \Rightarrow j \in R\}$. Then for any $R \in \mathcal{R}$, we have that (i) $A \subseteq R$, (ii) $R \subseteq B$ and (iii) $\delta^+(R) = \{(i, j) \in E \mid i \in R, j \notin R\} = \emptyset$. But, conversely, any set R satisfying (i), (ii) and (iii) must be in $\mathcal{R}$. Indeed, for any $i \neq j$ with $(i, j) \notin E$, there must be a set $U_{ij} \in \mathcal{R}$ with $i \in U_{ij}$ and $j \notin U_{ij}$. To show that a set R satisfying (i), (ii) and (iii) is in $\mathcal{R}$, it suffices to observe that

$$R = \bigcup_{i \in R} \bigcap_{j \notin R} U_{ij}, \tag{2}$$

and therefore R belongs to the ring family.

Given a collection of sets $T \subseteq 2^V$, we define $\mathcal{R}(T)$ to be the smallest ring family containing T. The directed graph representation of this ring family can be obtained by defining A, B and E directly from T rather than from the larger $\mathcal{R}(T)$, i.e. $A = \bigcap_{R \in T} R = \bigcap_{R \in \mathcal{R}(T)} R$, $B = \bigcup_{R \in T} R = \bigcup_{R \in \mathcal{R}(T)} R$, and $E = \{(i,j) \mid R \in T, i \in R \Rightarrow j \in R\}$. Further, in the expression (2) of any set $R \in \mathcal{R}(T)$, we can use sets $U_{ij} \in T$.

Given a sequence of subsets $T_1, \cdots, T_k$ of V, define $\mathcal{L}_i := \mathcal{R}(\{T_1, \cdots, T_i\})$ for $1 \leq i \leq k$. Assume that for each $i > 1$, we have that $T_i \notin \mathcal{L}_{i-1}$. We should emphasize that this condition depends on the *ordering* of the sets, and not just on this collection of sets. For instance, $\{1\}, \{1,2\}, \{2\}$ is a valid ordering whereas $\{1\}, \{2\}, \{1,2\}$ is not. We have thus a chain of ring families: $\mathcal{L}_1 \subset \mathcal{L}_2 \subset \cdots \subset \mathcal{L}_k$ where all the containments are proper. The question is how large can k be, and the next theorem shows that it can be at most quadratic in n.

Theorem 2. *Consider a chain of ring families,* $\mathcal{L}_0 = \emptyset \neq \mathcal{L}_1 \subsetneq \mathcal{L}_2 \subsetneq \cdots \subsetneq \mathcal{L}_k$ *within* 2^V *with* $n = |V|$. *Then*

$$k \leq \binom{n+1}{2} + 1.$$

Before proving this theorem, we show that the bound on the number of sets is tight.

Example 1. Let $V = \{1, \cdots, n\}$. For each $1 \leq i \leq j \leq n$, consider intervals $[i,j] = \{k \in V \mid i \leq k \leq j\}$. Add also the empty set $\emptyset$ as the trivial interval $[0,0]$ (as $0 \notin V$). We have just defined $k = \binom{n+1}{2} + 1$ sets. Define a complete order on these intervals in the following way: $(i,j) \prec (s,t)$ if $j < t$ or $(j = t$ and $i < s)$. We claim that if we consider these intervals in the order given by $\prec$, we satisfy the main assumption of the theorem that $[s,t] \notin \mathcal{R}(T_{st})$ where $T_{st} = \{[i,j] \mid (i,j) \prec (s,t)\}$. Indeed, for $s = 1$ and any t, we have that $[1,t] \notin \mathcal{R}(T_{1t})$ since $\bigcup_{I \in T_{1t}} I = [1, t-1] \not\supseteq [1,t]$. On the other hand, for $s > 1$ and any t, we have that $[s,t] \notin \mathcal{R}(T_{st})$ since for all $I \in T_{st}$ we have $(t \in I \Rightarrow s - 1 \in I)$ while this is not the case for $[s,t]$.

Proof. For each $1 \leq i \leq k$, let $T_i \in \mathcal{L}_i \setminus \mathcal{L}_{i-1}$. We can assume that $\mathcal{L}_i = \mathcal{R}(\{T_1, \cdots T_i\})$ (otherwise a longer chain of ring families can be constructed). If none of the T_i's is the empty set, we can increase the length of the chain by considering (the ring families generated by) the sequence $\emptyset, T_1, T_2, \cdots, T_k$. Similarly if V is not among the T_i's, we can add V either in first or second position in the sequence. So we can assume that the sequence has $T_1 = \emptyset$ and $T_2 = V$, i.e. $\mathcal{L}_1 = \{\emptyset\}$ and $\mathcal{L}_2 = \{\emptyset, V\}$.

When considering $\mathcal{L}_2$, its digraph representation has $A = \emptyset$, $B = V$ and the directed graph $D = (V, E)$ is the bidirected complete graph on V. To show a weaker bound of $k \leq 2 + n(n-1)$ is easy: every T_i we consider in the sequence will remove at least one arc of this digraph and no arc will get added.

To show the stronger bound in the statement of the theorem, consider the digraph D' obtained from D by contracting every strongly connected component of D and discarding all but one copy of (possibly) multiple arcs between two vertices of D'. We keep track of two parameters of D': s is its number of vertices and a is its the number of arcs. Initially, when considering $\mathcal{L}_2$, we have $s = 1$ strongly connected component and D' has no arc: $a = 0$. Every T_i we consider will either keep the same strongly connected components in D (i.e. same vertices in D') and remove (at least) one arc from D', or will break up at least one strongly connected component in D (i.e. increases vertices in D'). In the latter case, we can assume that only one strongly connected component is broken up into two strongly connected components and the number of arcs added is at most s since this newly formed connected component may have a single arc to every other strongly connected component. Thus, in the worst case, we move either from a digraph D' with parameters (s, a) to one with $(s, a-1)$ or from (s, a) to $(s+1, a+s)$. By induction, we claim that if the original one has parameters (s, a) then the number of steps before reaching the digraph on V with no arcs with parameters $(n, 0)$ is at most

$$a + \binom{n+1}{2} - \binom{s+1}{2}.$$

Indeed, this trivially holds by induction for any step $(s, a) \rightarrow (s, a-1)$ and it also holds for any step $(s, a) \rightarrow (s+1, a+s)$ since:

$$(a + s) + \binom{n+1}{2} - \binom{s+2}{2} + 1 = a + \binom{n+1}{2} - \binom{s+1}{2}.$$

As the digraph corresponding to $\mathcal{L}_2$ has parameters $(1, 0)$, we obtain that $k \leq 2 + \binom{n+1}{2} - 1 = \binom{n+1}{2} + 1$. $\qquad \square$

3   Analysis of the Discrete Newton's Algorithm

To prove Theorem 1, we start by recalling Radzik's analysis of Newton's algorithm for the case of modular functions (Radzik 1998). First of all, the discrete Newton's algorithm, as stated in Algorithm 1 for solving $\max \delta$: $\min_{S \subseteq V} f(S) - \delta a(S) \geq 0$ terminates. Recall that $h_i = \min_{S \subseteq V} f(S) - \delta_i a(S)$, $S_i \in \arg \min_S f(S) - \delta_i a(S)$, $g_i = a(S_i)$ and $\delta_{i+1} = \frac{f(S_i)}{a(S_i)}$. Let $f_i = f(S_i)$ and $g_i = a(S_i)$. Figure 1 illustrates the discrete Newton's algorithm and the notation.

Lemma 1. *Newton's algorithm as described in Algorithm 1 terminates in a finite number of steps t and generate sequences:*

(i) $h_1 < h_2 < \cdots < h_{t-1} < h_t = 0,$
(ii) $\delta_1 > \delta_2 > \cdots > \delta_{t-1} > \delta_t = \delta^* \geq 0,$

(iii) $g_1 > g_2 > \cdots > g_{t-1} > g_t \geq 0$.

Furthermore, if $g_t > 0$ then $\delta^* = 0$.

The first proof of the above lemma is often attributed to (McCormick and Ervolina 1994) and is omitted for conciseness. As in Radzik's analysis, we use the following lemma, illustrated in Fig. 2, and we reproduce here its proof.

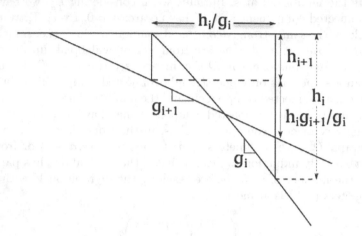

Fig. 2. Illustration for showing that $h_{i+1} + h_i \frac{g_{i+1}}{g_i} \leq h_i$, as in Lemma 2.

Lemma 2. *For any $i < t$, we have $\frac{h_{i+1}}{h_i} + \frac{g_{i+1}}{g_i} \leq 1$.*

Proof. By definition of S_i, we have that

$$h_i = f(S_i) - \delta_i a(S_i) = f_i - \delta_i g_i \leq f(S_{i+1}) - \delta_i a(S_{i+1}) = f_{i+1} - \delta_i g_{i+1}$$

$$= h_{i+1} + \frac{f_i}{g_i} g_{i+1} - \frac{f_i - h_i}{g_i} g_{i+1} = h_{i+1} + h_i \frac{g_{i+1}}{g_i}.$$

Since $h_i < 0$, dividing by h_i gives the statement.   □

Thus, in every iteration, either g_i or h_i decreases by a constant factor smaller than 1. We can thus partition the iterations into two types, for example as

$$J_g = \left\{ i \mid \frac{g_{i+1}}{g_i} \leq \frac{2}{3} \right\}$$

and $J_h = \{i \notin J_g\}$. Observe that $i \in J_h$ implies $\frac{h_{i+1}}{h_i} < \frac{1}{3}$. We first bound $|J_g|$ as was done in (Radzik 1998).

Lemma 3. $|J_g| = O(n \log n)$.

Proof sketch. Let $J_g = \{i_1, i_2, \cdots, i_k\}$ and let $T_j = S_{i_j}$. From the monotonicity of g, these sets T_j are such that $a(T_{j+1}) \leq \frac{2}{3} a(T_j)$. These can be viewed as linear inequalities with small coefficients involving the a_i's, and by normalizing and taking an extreme point of this polytope, Goemans (see (Radzik 1998)) has shown that the number k of such sets is $O(n \log n)$.

Although we do not need this for the analysis, the bound of $O(n \log n)$ on the number of geometrically decreasing sets defined on n numbers is tight, as was shown by Mikael Goldmann in 1993 by an intricate construction based on a Fourier-analytic approach of Håstad (Håstad 1994). As this was never published, we include (a variant of) this construction here. The reader is welcome to skip directly to Sect. 3.1 without break in continuity.

Theorem 3. *Let n be a power of 2. Then there exists $a \in \mathbb{R}^n$ and a sequence of sets $\{S_i\}_{i \in [k]}$ with $a(S_1) > 0$ and $a(S_i) \geq 2a(S_{i-1})$ for $i > 1$ where $k = \frac{1}{2}n \log_2 n - O(n \log \log n)$.*

Proof. Let m be such that $n = 2^m$. Consider all 2^m subsets of $[m]$ and order them as $\alpha_1, \alpha_2, \cdots, \alpha_n$ such that $|\alpha_i| \leq |\alpha_j|$ for $i < j$. Thus, $\alpha_1 = \emptyset$ and $\alpha_n = [m]$. We say $\alpha_i \prec \alpha_j$ if $i < j$. Consider the $n \times n$ Hadamard matrix Q in which the ith row and column are indexed by subset α_i of $[m]$ and

$$q_{ij} = (-1)^{|\alpha_i \cap \alpha_j|}.$$

Q is invertible and $Q^{-1} = \frac{1}{n}Q$. Set $b_1 = 0$ and $b_i = 2^{mi}$ for $i > 1$. Now, let $a \in \mathbb{R}^n$ be the solution to $Qa = b$. We claim that there is a sequence of sets of length $\frac{1}{2}nm + O(n \log \log n) = \frac{1}{2}n \log n + O(n \log \log n)$ whose $a(\cdot)$ values increase geometrically by a factor of 2.

First, observe that $q_{1j} = 1$ for all j and thus $a([n]) = 0$. This means that if we have a $r \in \{-1, 1\}^n$ such that $\langle r, a \rangle = p$ then $a(S) = \frac{p}{2}$ where $S = \{i | r_i = 1\} \subseteq [n]$. Thus we focus on constructing a sequence of vectors $r \in \{-1, 1\}^n$ whose inner product with a increases geometrically. We already have $n - 1$ such vectors, namely the rows $q_i \in \{-1, 1\}^n$ of Q for $i > 1$: $\langle q_i, a \rangle = 2^{mi}$.

Now, for each $i > 1$, we show how to construct ± 1 vectors v such that $2^{m(i-1)} < \langle v, a \rangle < 2^{mi}$ and whose a values increase geometrically. We will be able to construct one such set for almost all values between 1 and $|\alpha_i|$. Fix $i > 1$ and let $k = |\alpha_i|$. For any ℓ with $1 \leq \ell \leq k - 2$, consider a set $\alpha_{h_\ell} \subset \alpha_i$ of cardinality ℓ. Define the vector

$$w^{(\ell)} = \sum_{u : \alpha_{h_\ell} \subseteq \alpha_u \subseteq \alpha_i} q_u.$$

Its jth component is:

$$w_j^{(\ell)} = \sum_{u : \alpha_{h_\ell} \subseteq \alpha_u \subseteq \alpha_i} q_{uj} = \sum_{u : \alpha_{h_\ell} \subseteq \alpha_u \subseteq \alpha_i} (-1)^{|\alpha_u \cap \alpha_j|} = (-1)^{|\alpha_{h_\ell} \cap \alpha_j|} \sum_{u : \alpha_{h_\ell} \subseteq \alpha_u \subseteq \alpha_i} (-1)^{|(\alpha_u \setminus \alpha_{h_\ell}) \cap \alpha_j|}$$

$$= \begin{cases} 0 & \text{if } (\alpha_i \setminus \alpha_{h_\ell}) \cap \alpha_j = \emptyset \\ 2^{k-\ell}(-1)^{|\alpha_{h_\ell} \cap \alpha_j|} & \text{otherwise} \end{cases}$$

Now consider $v^{(\ell)} = 2^{1-(k-\ell)}w^{(\ell)} - q_h$. We claim this is a ± 1 vector. Its jth component is equal to $-q_{hj} \in \{-1, 1\}$ if $(\alpha_i \setminus \alpha_{h_\ell}) \cap \alpha_j = \emptyset$ and, otherwise, is equal to

$$2(-1)^{|\alpha_{h_\ell} \cap \alpha_j|} - (-1)^{|\alpha_{h_\ell} \cap \alpha_j|} = (-1)^{|\alpha_{h_\ell} \cap \alpha_j|} \in \{-1, 1\}.$$

Now for this vector $v^{(\ell)}$ (corresponding to a given pair $\alpha_{h_\ell} \subset \alpha_i$), we have:

$$\langle v^{(\ell)}, a \rangle = 2^{1-(k-\ell)} \langle w^{(\ell)}, a \rangle - \langle q_{h_\ell}, a \rangle = 2^{1-(k-\ell)} \left(\sum_{u:\alpha_{h_\ell} \subseteq \alpha_u \subseteq \alpha_i} b_u \right) - b_{h_\ell}. \quad (3)$$

Now the b_j's increase geometrically with j. In the summation (with $2^{k-\ell}$ terms), the dominant one will be $b_i = 2^{mi}$, and as a first approximation, we have that $\langle v^{(\ell)}, a \rangle$ is roughly $2^{mi+1-k+\ell}$, and therefore they appear to be between b_{i-1} and b_i, and increase appropriately by a factor 2. Unfortunately, lower terms matter and, therefore, we need to select carefully the indices h_ℓ's.

A simple construction of these sets $\{\alpha_{h_\ell}\}$ is as follows. Let $f = \lceil \log_2 k \rceil$. For any $f \leq \ell \leq k-f$, let α_{h_ℓ} be such that (i) $\alpha_{h_\ell} \cap [f]$ has as characteristic vector the f-bit representation of $k-f-\ell$ and (ii) the elements in $\alpha_{h_\ell} \cap ([k] \setminus [f])$ are chosen arbitrarily so that $|\alpha_{h_\ell}| = \ell$. Observe that (i) is possible for all $f \leq \ell \leq k-f$ since $k-f-\ell \leq k-2f \leq 2^f - 1$ and therefore $k-f-\ell$ can be represented by f bits. And (ii) is feasible as well by our choice of ℓ. We have just constructed $k - 2f + 1 \geq k - 2\log_2(k) - 1$ sets.

One can show (proof omitted for space considerations) that, for such a choice of $\{\alpha_{h_\ell}\}$, we have for $f \leq \ell < k - f$: $\langle v^{(\ell+1)}, a \rangle \geq 2\langle v^{(\ell)}, a \rangle$.

The number of vectors/sets we have constructed this way is therefore at least:

$$\sum_{k=0}^{m} \binom{m}{k} (k - 2\log_2(k) - 1) \geq \frac{m+1}{2} 2^m - 2\log_2(m)2^m - 2^m = \frac{1}{2} n \log_2(n) - O(n \log \log(n)),$$

and this completes the proof. □

3.1   Weaker Upper Bound

Before deriving the bound of $O(n^2)$ on $|J_g| + |J_h|$ for Theorem 1, we show how to derive a weaker bound of $O(n^3 \log n)$. For showing the $O(n^3 \log n)$ bound, first consider a block of *consecutive* iterations $[u, v] := \{u, u+1, \cdots, v\}$ within J_h.

Theorem 4. *Let $[u, v] \subseteq J_h$. Then $|[u, v]| \leq n^2 + n + 1$.*

The strategy of the proof is to show (i) that, for the submodular function $k_v(S) = f(S) - \delta_v a(S)$, the values of $k_v(S_i)$ for $i \in [u, v-1]$ form a geometrically decreasing series (Lemma 4), (ii) that each S_i cannot be in the ring family generated by $S_{i+1}, \ldots, S_{v-1}$ (Lemma 5 and Theorem 5), and (iii) then conclude using our Theorem 2 on the length of a chain of ring families.

Lemma 4. *Let $[u, v] \subseteq J_h$. Then for $k_v(S) = f(S) - \delta_v a(S)$, we have (i) $k_v(S_v) = \min_S k_v(S) = h_v$, (ii) $k_v(S_{v-1}) = 0$, (iii) $k_v(S_{v-2}) > 2|h_v|$ and (iv) $k_v(S_{i-1}) > 2k_v(S_i)$ for $i \in [u+1, v-1]$.*

Proof. Since $\frac{g_{i+1}}{g_i} > \frac{2}{3}$ for all $i \in [u, v]$, Lemma 2 implies that $\frac{h_{i+1}}{h_i} \leq \frac{1}{3}$, and thus

$$\frac{|h_{i+1}|}{g_{i+1}} \leq \frac{1}{2} \frac{|h_i|}{g_i}.$$

Since $\delta_{i+1} - \delta_i = \frac{f_i}{g_i} - \frac{f_i - h_i}{g_i} = \frac{h_i}{g_i}$. We deduce that

$$\delta_{i+1} - \delta_{i+2} = -\frac{h_{i+1}}{g_{i+1}} \leq \frac{1}{2}(\delta_i - \delta_{i+1}),\tag{4}$$

for all $i \in [u, v]$. Now, observe that for any $i \in [u, v-2]$, we have

$$\delta_{i+1} - \delta_v = \sum_{k=i+1}^{v-1} \delta_k - \delta_{k+1} \leq \frac{1}{2}\sum_{k=i+1}^{v-1}(\delta_{k-1} - \delta_k) = \frac{1}{2}(\delta_i - \delta_{v-1}) < \frac{1}{2}(\delta_i - \delta_v).$$

Thus

$$\delta_{i+1} - \delta_v < \frac{1}{2}(\delta_i - \delta_v),\tag{5}$$

and we can even extend the range of validity to $i \in [u, v]$ since for $i = v - 1$ or $i = v$, this follows from Lemma 1.

Consider the submodular function $k_v(S) = f(S) - \delta_v a(S)$. We have denoted its minimum value by $h_v < 0$ and S_v is one of its minimizers. For each $i \in [u, v-1]$ we have

$$k_v(S_i) = f_i - \delta_v g_i = g_i(\delta_{i+1} - \delta_v),$$

and therefore $k_v(S_{v-1}) = 0$ while $k_v(S_i) > 0$ for $i \in [u, v-2]$. Furthermore, (5) implies that

$$k_v(S_i) = g_i(\delta_{i+1} - \delta_v) < \frac{1}{2}\frac{g_i}{g_{i-1}}g_{i-1}(\delta_i - \delta_v) < \frac{1}{2}g_{i-1}(\delta_i - \delta_v) = \frac{1}{2}k_v(S_{i-1}),$$

and this is valid for $i \in [u, v-1]$. Thus the $k_v(S_i)$'s decrease geometrically with increasing i. In addition, we have $k_v(S_{v-2}) = g_{v-2}(\delta_{v-1} - \delta_v)$ while (by (4) and Lemma 1)

$$-k_v(S_v) = |h_v| = -h_v = g_v(\delta_v - \delta_{v+1}) < \frac{1}{2}g_{v-2}(\delta_{v-1} - \delta_v) = \frac{1}{2}k_v(S_{v-2}).$$

Summarizing, we have $k_v(S_v) = \min_S k_v(S) = h_v$, $k_v(S_{v-1}) = 0$, $k_v(S_{v-2}) > 2|h_v|$ and $k_v(S_{i-1}) > 2k_v(S_i)$ for $i \in [u, v-1]$. □

We now show that for any submodular function and any ring family on the same ground set, the values attained by the submodular function cannot increase much when the ring family is increased to the smallest ring family including a single additional set. This lemma follows from the submodularity of f and Birkhoff's representation theorem for subsets contained in a ring family.

Lemma 5. *Let* $f : 2^V \rightarrow \mathbb{R}$ *be a submodular function with* $f_{min} = \min_{S \subseteq V} f(S) \leq 0$. *Let* $\mathcal{L}$ *be any ring family over* V *and* $T \notin \mathcal{L}$. *Define* $\mathcal{L}' := \mathcal{R}(\mathcal{L} \cup \{T\})$, $m = \max_{S \in \mathcal{L}} f(S)$ *and* $m' = \max_{S \in \mathcal{L}'} f(S)$. *Then*

$$m' \leq 2(m - f_{min}) + f(T).$$

Proof. Consider $S \in \mathcal{L}'$. Using (2), we can express S as $S = \bigcup_{i \in S} S_i$ where S_i can be either (i) T, or (ii) R for some $R \in \mathcal{L}$, or (iii) $R \cap T$ for some $R \in \mathcal{L}$. Taking the union of the sets R of type (ii), resp. (iii), into P, resp. Q, we can express S as $S = P \cup T$ or as $S = P \cup (Q \cap T)$ where $P, Q \in \mathcal{L}$ (since the existence of any case (i) annihilates the need for case (iii)).

Now using submodularity, we obtain that

$$f(P \cup T) \leq f(P) + f(T) - f(P \cap T) \leq m + f(T) - f_{min},$$

in the first case and

$$
\begin{aligned}
f(P \cup (Q \cap T)) &\leq f(P) + f(Q \cap T) - f(P \cap Q \cap T) \\
&\leq f(P) + f(Q) + f(T) - f(Q \cup T) - f(P \cap Q \cap T) \\
&\leq 2m + f(T) - 2f_{min}.
\end{aligned}
$$

In either case, we get the desired bound on $f(S)$ for any $S \in \mathcal{R}'$.   □

We will now use the bound in Lemma 5 to show that if a sequence of sets increases in their submodular function value by a factor of 4, then any set in the sequence is not contained in the ring family generated by the previous sets.

Theorem 5. *Let* $f : 2^V \rightarrow \mathbb{R}$ *be a submodular function with* $f_{min} = \min_{S \subseteq V} f(S) \leq 0$. *Consider a sequence of distinct sets* $T_1, T_2, \cdots, T_q$ *such that* $f(T_1) = f_{min}$, $f(T_2) > -2f_{min}$, *and* $f(T_i) \geq 4f(T_{i-1})$ *for* $3 \leq i \leq q$. *Then* $T_i \notin \mathcal{R}(\{T_1, \cdots, T_{i-1}\})$ *for all* $1 < i \leq q$.

Proof. This is certainly true for $i = 2$. For any $i \geq 1$, define $\mathcal{L}_i = \mathcal{R}(\{T_1, \cdots, T_i\})$ and $m_i = \max_{S \in \mathcal{L}_i} f(S)$. We know that $m_1 = f_{min} \leq 0$ and $m_2 = f(T_2)$ since $T_1 \cap T_2$ and $T_1 \cup T_2$ cannot have larger f values than T_2 by submodularity of f and minimality of T_1.

We claim by induction that $m_k \leq 2f(T_k) + 2f_{min}$ for any $k \geq 2$. This is true for $k = 2$ since $m_2 = f(T_2) \leq 2f(T_2) + 2f_{min}$. Assume the induction claim to be true for $k - 1$.

We get that $m_{k-1} \leq 2f(T_{k-1}) + 2f_{min} < 4f(T_{k-1})$. Since $f(T_k) > m_{k-1}$, $T_k \notin \mathcal{L}_{k-1} = \mathcal{R}(T_1, \cdots, T_{k-1})$. Using Lemma 5, we get that

$$
\begin{aligned}
m_k &\leq 2(m_{k-1} - f_{min}) + f(T_k) \\
&\leq 2(2f(T_{k-1}) + 2f_{min} - f_{min}) + f(T_k) \\
&\leq 2f(T_k) + 2f_{min}.
\end{aligned}
$$

Thus proving the induction step for k, and hence the statement of the theorem.   □

We are now ready to prove Theorem 4.

Proof. (of Theorem 4) Apply Theorem 5 to the submodular function k_v given in Lemma 4. Let $T_1 = S_v$ and skip every other set to define $T_i = S_{v-2(i-1)}$ for $v - 2(i - 1) \geq u$ i.e. $i \leq q := 1 + (v - u)/2$. Then the conditions of Theorem 5 are satisfied (thanks to Lemma 4), and we obtain a sequence of sets $T_1, \cdots, T_q$ such that $T_i \notin \mathcal{R}(T_1, \cdots, T_{i-1})$. Therefore, Theorem 2 on the length of a chain of ring families implies that $q \leq \binom{n+1}{2} + 1$, or $v - u \leq (n+1)n$. This means $|[u,v]| \leq n^2 + n + 1$. $\qquad\square$

Since Lemma 3 shows that $|J_g| = O(n \log n)$ and we know from Theorem 4 that the intervals between two indices of J_g have length $O(n^2)$, this implies that $|J_g| + |J_h| = O(n \log n) \cdot O(n^2) = O(n^3 \log n)$.

3.2   Main Result of Theorem 1

The analysis of Theorem 4 can be improved by showing that we can extract a chain of ring families not just from one interval of J_h but from all of J_h. Instead of discarding every other set in J_h, we also need to discard the first $O(\log n)$ sets in every interval of J_h. This helps prove the main result of the paper that bounds the number of iterations in the discrete Newton's algorithm by at most $n^2 + O(n \log^2 n)$.

Theorem 6. *We have* $|J_h| = n^2 + O(n \log^2 n)$.

Before proving this, we need a variant of Lemma 5. The proof of the lemma again follows from the submodularity of f and Birkhoff's representation theorem for subsets contained in a ring family.

Lemma 6. *Let* $T \subseteq 2^V$ *and assume that* $f(S) \leq M$ *for all* $S \in T$. *Then for all* $S \in \mathcal{R}(T)$

$$f(S) \leq \frac{n^2}{4}(M - f_{min}).$$

Proof. Consider any $S \in \mathcal{R}(T)$. We know that $S = \bigcup_{i \in S} \bigcap_{j \notin S} U_{ij}$, for some $U_{ij} \in T$. Define $S_i = \bigcap_{j \notin S} U_{ij}$; thus $S = \bigcup_{i \in S} S_i$.

We first claim that, for any k sets $T_1, T_2, \cdots, T_k \in T$, we have that

$$f(\bigcap_{i=1}^{k} T_i) \leq kM - (k-1)f_{min}.$$

This is proved by induction on k, the base case of $k = 1$ being true by our assumption on f. Indeed, applying submodularity to $P = \bigcap_{i=1}^{k-1} T_i$ and T_k (and the inductive hypothesis), we get

$$f(\bigcap_{i=1}^{k} T_i) = f(P \cap T_k) \leq f(P) + f(T_k) - f(P \cup T_k) \leq (k-1)M - (k-2)f_{min}$$

$$+ M - f_{min} = kM - (k-1)f_{min}.$$

Using this claim, we get that for any $i \in S$, we have

$$f(S_i) = f(\bigcap_{j \notin S} U_{ij}) \leq |V \setminus S|M - (|V \setminus S| - 1)f_{min} \leq |V \setminus S|(M - f_{min}).$$

By a similar argument on the union of the S_i's, we derive that

$$f(S) \leq |S| \left(|V \setminus S| M - (|V \setminus S| - 1) f_{min} \right) - (|S| - 1) f_{min}$$
$$\leq |S||V \setminus S| M - (|S||V \setminus S| - 1) f_{min}$$
$$\leq \frac{n^2}{4} (M - f_{min}).$$

$\square$

We are now ready to prove Theorem 6.

Proof. (of Theorem 6) Let $J_h = \bigcup_{i=1}^{\ell} [u_i, v_i]$ where $u_{i-1} > v_i + 1$ for $1 < i \leq \ell$. Notice that these intervals are ordered in a *reverse* order (compared to the natural ordering). We construct a sequence of sets $T_1, \cdots$ such that each set in the sequence is not in the ring closure of the previous ones. The first sets are just every other set S_i from $[u_1, v_1]$ obtained as before by using Theorem 5 and Lemma 4 with the submodular function k_{v_1}. Let T_1 denote this sequence of sets.

Suppose now we have already considered the intervals $[u_j, v_j]$ for $j < i$ and have extracted a (long) sequence of sets T_{i-1} such that each set in the sequence is not in the ring closure of the previous ones. Consider now the submodular function $f := k_{v_i}$, and let $f_{min} \leq 0$ be its minimum value. Notice that from the order of iterations in the discrete Newton's algorithm we have that $f(T) < 0$ for $T \in T_{i-1}$. Therefore by Lemma 6 with $M = 0$ we have that $f(S) \leq -\frac{n^2}{4} f_{min}$ for all $S \in \mathcal{R}(T_{i-1})$. Using Lemma 4 with $f = k_{v_i}$, we have that only sets S_k with $k > v_i - \log(n^2/4)$ could possibly be in $\mathcal{R}(T_{i-1})$, and therefore we can safely add to T_{i-1} every other set in $[u_i, v_{i-O(\log n)}]$ while maintaining the property that every set is not in the ring closure of the previous ones. Over all i, we have thus constructed a chain of ring families of length $\frac{1}{2}|J_h| - O(\log n)\ell = \frac{1}{2}|J_h| - O(\log n)|J_g|$. The theorem now follows from Lemma 3 and Theorem 2. $\square$

Finally, combining Theorem 6 and Lemma 3 proves Theorem 1.

Proof. (of Theorem 1) In every iteration of discrete Newton's algorithm, either g_i or h_i decreases by a constant factor smaller than 1. Thus, the iterations can be partitioned into two types $J_g = \left\{ i \mid \frac{g_{i+1}}{g_i} \leq \frac{2}{3} \right\}$ and $J_h = \{i \notin J_g\}$. Lemma 3 shows that $|J_g| = O(n \log n)$ and Theorem 6 shows that $|J_h| = n^2 + O(n \log^2 n)$. Thus, the total number of iterations is $n^2 + O(n \log^2 n)$. $\square$

4   Geometrically Increasing Sequences

In the proof for Theorem 1, we considered a sequence of sets $S_1, \cdots, S_k$ such that $f(S_i) \geq 4f(S_{i-1})$ for all $i \leq k$ for a submodular function f. In the special case when f is modular, we know that the maximum length of such a sequence is at most $O(n \log n)$ (Lemma 3). When f is submodular, we show that the maximum length is at most $\binom{n+1}{2} + 1$ by applying Theorems 2 to 5. In this section, we show that the bound for the submodular case is tight by constructing two related examples: one that uses interval sets of the ground set $\{1, \cdots, n\}$, and the other that assigns weights to arcs in a directed graph such that the cut function already gives such a sequence of quadratic (in the number of vertices) number of sets.

4.1   Interval Submodular Functions

In this section, we show that the bound for the submodular case is tight by constructing a sequence of $\binom{n+1}{2} + 1$ sets $\emptyset, S_1, \cdots, S_{\binom{n+1}{2}}$ for a nonnegative submodular function f, such that $f(S_i) = 4f(S_{i-1})$ for all $i \leq \binom{n+1}{2}$.

For each $1 \leq i \leq j \leq n$, consider intervals $[i, j] = \{k \mid i \leq k \leq j\}$ and let the set of all intervals be $\mathcal{I} = \bigcup_{i,j}\{[i, j]\}$. Let $[i, j] = \emptyset$ whenever $i > j$. Consider a set function $f : \mathcal{I} \to \mathbb{R}_+$ such that $f(\emptyset) = 0$. We say f is *submodular on intervals* if for any $S, T \in \mathcal{I}$ such that $S \cup T \in \mathcal{I}$ and $S \cap T \in \mathcal{I}$, we have

$$f(S) + f(T) \geq f(S \cup T) + f(S \cap T).$$

Lemma 7. *Let τ and κ be monotonically increasing, nonnegative functions on the set $[n]$, then f defined by $f([i, j]) = \tau(i)\kappa(j)$ is submodular on intervals.*

Proof. Consider two intervals S and T. The statement follows trivially if $S \subseteq T$, so consider this is not the case. Let $S = [s_i, s_j]$ and $T = [t_i, t_j]$ and assume w.l.o.g that $s_j \geq t_j$.

i.   Case $S \cap T \neq \emptyset$. This implies $t_i < s_i$ and $s_i \leq t_j \leq s_j$. In this case, $f(S) + f(T) - f(S \cap T) - f(S \cup T) = \tau(s_i)\kappa(s_j) + \tau(t_i)\kappa(t_j) - \tau(s_i)\kappa(t_j) - \tau(t_i)\kappa(s_j) = (\tau(s_i) - \tau(t_i))(\kappa(s_j) - \kappa(t_j)) \geq 0.$
ii. Case $S \cap T = \emptyset, S \cup T = [t_i, s_j]$. In this case, $f(S) + f(T) - f(S \cup T) = \tau(s_i)\kappa(s_j) + \tau(t_i)\kappa(t_j) - \tau(t_i)\kappa(s_j) \geq \kappa(s_j)(\tau(s_i) - \tau(t_i)) \geq 0.$ □

We show that one can *extend* any function that is submodular on intervals to a submodular function (defined over the ground set). This construction is general, and might be of independent interest. For any set $S \subseteq V$, define $\mathcal{I}(S)$ to be the set of maximum intervals contained in S. For example, for $S = \{1, 2, 3, 6, 9, 10\}$, $\mathcal{I}(S) = \{[1, 3], [6, 6], [9, 10]\}$.

Lemma 8. *Consider a set function f defined over intervals such that (i) $f(\emptyset) = 0$, (ii) $f([i, j]) \geq 0$ for interval $[i, j]$, (iii) for any $S, T \in \mathcal{I}$ such that $S \cap T, S \cup T \in \mathcal{I}$, $f(S) + f(T) \geq f(S \cup T) + f(S \cap T)$. Then, $g(S) = \sum_{I \in \mathcal{I}(S)} f(I)$ is submodular over the ground set $\{1, \ldots, n\}$.*

Proof. We will show that g is submodular by proving that for any $T \subseteq S$ and any $k \notin S$, $g(S \cup \{k\}) - g(S) \leq g(T \cup \{k\}) - g(T)$. Let the marginal gain obtained by adding k to S be $g_k(S) = g(S \cup \{k\}) - g(S)$.

Note that $\mathcal{I}(S \cup k) \setminus \mathcal{I}(S)$ can either contain (i) $[s,k]$, for some $s \le k$, or (ii) $[k,u]$, for some $u > k$, or (iii) $[s,u]$ for $s \le k \le u$. In case (i), $g_k(S) = f([s,k]) - f([s,k-1])$; in case (ii), $g_k(S) = f([k,u]) - f([k+1,u])$; and in case (iii), $g_k(S) = f([s,u]) - f([s,k-1]) - f([k+1,u])$. Thus, when comparing the values of $g_k(S)$ and $g_k(T)$, we are only concerned with intervals that are modified due to the addition of k.

Let $S \cup \{k\}$ contain the interval $[s,k-1] \cup \{k\} \cup [k+1,u]$ and $T \cup \{k\}$ contain the interval $[t,k-1] \cup \{k\} \cup [k+1,v]$ where $s \le t$, $v \le u$ (as $T \subseteq S$) and $s \le k \le u$ ($s = k$ implies $[s,k-1] = \emptyset$ and $u = k$ implies that $[k+1,u] = \emptyset$) and $t \le k \le v$ ($t = k$ implies $[t,k-1] = \emptyset$ and $v = k$ implies that $[k+1,v] = \emptyset$).

$$
\begin{aligned}
&g(S \cup \{k\}) - g(S) - g(T \cup \{k\}) + g(T) \\
&= f([s,u]) - f([s,k-1]) - f([k+1,u]) - (f([t,v]) - f([t,k-1]) - f([k+1,v])) \\
&= f([s,u]) - f([s,k-1]) - f([k+1,u]) - f([t,v]) + f([t,k-1]) + f([k+1,v]) \\
&\le f([s,u]) - f([s,k-1]) - f([k+1,v]) - f([t,u]) + f([t,k-1]) + f([k+1,v]) \quad (6) \\
&= f([s,u]) - f([s,k-1]) - f([t,u]) + f([t,k-1]) \le 0. \quad\quad (7)
\end{aligned}
$$

where (6) follows from submodularity of f on intervals $[k+1,u]$ and $[t,v]$, i.e., $f([k+1,u]) + f([t,v]) \ge f([t,u]) + f([k+1,v])$, and (7) follows from submodularity of f on intervals $[s,k-1]$ and $[t,u]$.   □

Construction. Consider the function $f([i,j]) = 4^{\frac{j(j-1)}{2}} 4^i$ for $[i,j] \in \mathcal{I}$, obtained by setting $\tau(i) = 4^i$ and $\kappa(j) = 4^{\frac{j(j-1)}{2}}$. This is submodular on intervals from Lemma 7. This function defined on intervals can be extended to a submodular function g by Lemma 8. Consider the total order $\prec$ defined on intervals $[i,j]$ specified in Example 1 (Sect. 2). By our choice of τ and κ we have that $S \prec T$ implies $4g(S) \le g(T)$. The submodular function g thus contains a sequence of length $\binom{n+1}{2} + 1$ of sets that increase geometrically in their function values.

4.2 Cut Functions

The example from the previous section and Birkhoff's representation theorem motivates a construction of a complete directed graph $G = (V, A)$ ($|V| = n$) and a weight vector $w \in \mathbb{R}_+^{|A|}$ such that there exists a sequence of $m = \binom{n}{2}$ sets $\emptyset, S_1, \cdots, S_m \subseteq V$ that has $w(\delta^+(S_k)) \ge 4w(\delta^+(S_{k-1}))$ for all $k \ge 2$.

Construction. The sets S_i are all intervals of $[n-1]$, and are ordered by the complete order $\prec$ as defined previously. One can verify that the kth set S_k in the sequence is $S_k = [i,j]$ where $k = i + j(j-1)/2$.

Note that, if $i > 1$, for each interval $[i,j]$, arc $e_{i,j} := (j, i-1) \in \delta^+([i,j])$ and $(j, i-1) \notin \delta^+([s,t])$ for any $(s,t) \prec (i,j)$. For any interval $[1,j]$, arc $e_{1,j} := (j, j+1) \in \delta^+([1,j])$ and $(j, j+1) \notin \delta^+([s,t])$ for any $(s,t) \prec (1,j)$. Define arc weights w by $w(e_{i,j}) = 5^{i+j(j-1)/2}$. Thus, the arcs $e_{i,j}$ corresponding to the intervals $[i,j]$ increase in weight by a factor of 5. We claim that $w(\delta^+(S_k)) \ge 4w(\delta^+(S_{k-1}))$. This is true because $4\sum_{e_{s,t}:(s,t)\prec(i,j)} w(e_{s,t}) \le w(e_{i,j})$.

5   Open Question

In this paper, we showed an $O(n^2)$ bound on the number of iterations of the discrete Newton's algorithm for the problem of finding $\max \delta : \min_S f(S) - \delta a(S) \geq 0$ for an arbitrary direction $a \in \mathbb{R}^n$. Even though we showed that certain parts of our analysis were tight, we do not know whether this bound is tight. More fundamentally, we know little about the number of breakpoints of the piecewise linear function $g(\delta) = \min_S f(S) - \delta a(S)$ in the case of an arbitrary direction a. Our results do not imply anything on this number of breakpoints, and this number could still be quadratic, exponential or even linear. In the simpler, nonnegative setting $a \in \mathbb{R}^n_+$, it is not just that the discrete Newton's algorithm takes at most n iterations, but it is also the case that the number of breakpoints of the lower envelope is at most n (by the property of strong quotients). On the other hand, there exist instances of parametric minimum $s - t$ cut problems where the minimum cut value has an exponential number of breakpoints (Mulmuley 1999). However, this corresponds to the more general problem $\min_S f(S) - \delta a(S)$ where $f(\cdot)$ is submodular but the function $a(\cdot)$ is not modular (and not even supermodular or submodular as the slopes of the parametric capacities can be positive or negative).

References

Freund, R.M., Grigas, P., Mazumder, R.: An extended Frank-Wolfe method with "In-Face" directions, its application to low-rank matrix completion (2015). arXiv:1511.02204

Håstad, J.: On the size of weights for threshold gates. SIAM J. Discrete Math. **7**(3), 484–492 (1994)

Iwata, S.: Submodular function minimization. Math. Program. **112**(1), 45–64 (2008)

Iwata, S., Murota, K., Shigeno, M.: A fast parametric submodular intersection algorithm for strong map sequences. Math. Oper. Res. **22**(4), 803–813 (1997)

Iwata, S., Orlin, J.B.: A simple combinatorial algorithm for submodular function minimization. In: Proceedings of the Twentieth Annual ACM-SIAM Symposium on Discrete Algorithms, pp. 1230–1237. Society for Industrial and Applied Mathematics (2009)

Lee, Y.T., Sidford, A., Wong, S.C.: A faster cutting plane method and its implications for combinatorial and convex optimization. In: Foundations of Computer Science (FOCS), pp. 1049–1065. IEEE (2015)

McCormick, S.T., Ervolina, T.R.: Computing maximum mean cuts. Discrete Appl. Math. **52**(1), 53–70 (1994)

Mulmuley, K.: Lower bounds in a parallel model without bit operations. SIAM J. Comput. **28**(4), 1460–1509 (1999)

Nagano, K.: A strongly polynomial algorithm for line search in submodular polyhedra. Discrete Optim. **4**(3), 349–359 (2007)

Radzik, T.: Fractional combinatorial optimization. In: Du, D.Z., Pardalos, P.M. (eds.) Handbook of Combinatorial Optimization, pp. 429–478. Springer, Heidelberg (1998)

Topkis, D.M.: Minimizing a submodular function on a lattice. Oper. Res. **26**(2), 305–321 (1978)

Stochastic Online Scheduling on Unrelated Machines

Varun Gupta[1], Benjamin Moseley[2], Marc Uetz[3(✉)], and Qiaomin Xie[4]

[1] University of Chicago, Chicago, USA
varun.gupta@chicagobooth.edu
[2] Washington University in St. Louis, St. Louis, USA
bmoseley@wustl.edu
[3] University of Twente, Enschede, Netherlands
m.uetz@utwente.nl
[4] University of Illinois at Urbana-Champaign, Champaign, USA
qxie3@illinois.edu

Abstract. We derive the first performance guarantees for a combinatorial online algorithm that schedules stochastic, nonpreemptive jobs on unrelated machines to minimize the expectation of the total weighted completion time. Prior work on unrelated machine scheduling with stochastic jobs was restricted to the offline case, and required sophisticated linear or convex programming relaxations for the assignment of jobs to machines. Our algorithm is purely combinatorial, and therefore it also works for the online setting. As to the techniques applied, this paper shows how the dual fitting technique can be put to work for stochastic and nonpreemptive scheduling problems.

1 Introduction

The scheduling of jobs on multiple, parallel machines is a fundamental problem both in combinatorial optimization and systems theory. There is a vast amount of different model variants as well as applications, which is testified by the existence of the handbook [19]. A well studied class of problems is scheduling a set of n nonpreemptive jobs that arrive over time on m unrelated machines with the objective of minimizing the total weighted completion time. Here, unrelated machines refers to the fact that the matrix that describes the processing times of all jobs on all machines can have any rank larger than 1. The offline version of that problem is denoted $R|r_j|\sum w_j C_j$ in the three-field notation of Graham et al. [8], and it has always been a cornerstone problem for the development of new techniques in the design of (approximation) algorithms, e.g. [4,12,18,30].

We here address the online version of that problem with stochastic jobs. Online means that jobs arrive over time, and the set of jobs is unknown a priori. With respect to online models in scheduling, we refer to [14,27] for pointers to

B. Moseley—Supported in part by a Google Research Award, a Yahoo Research Award and NSF Grant CCF-1617724.

F. Eisenbrand and J. Koenemann (Eds.): IPCO 2017, LNCS 10328, pp. 228–240, 2017.
DOI: 10.1007/978-3-319-59250-3_19

relevant work. In many systems, the scheduler may not know the exact processing times of jobs upon arrival. Different approaches have been introduced to cope with this uncertainty. If jobs can be preempted, then non-clairvoyant schedulers have been studied that do not know the processing time of the jobs until the job is completed [5,9,13,16,26]. Unfortunately, if preemption is not allowed then any algorithm has poor performance in the non-clairvoyant model, as the lower bound for approximability is $\Omega(n)$.

That suggests that the non-clairvoyant model is perhaps too pessimistic. Even though exact processing times may be unknown, it is not unrealistic to assume that at least an estimate of the true processing times is available. For such systems, a model that is used is *stochastic scheduling*. In the stochastic scheduling model the job's processing times are given by random variables. A *non-anticipatory* scheduler only knows this random variable P_j that encodes the possible realizations of job j's processing time. If the scheduler starts a job on a machine, then that job must be run to completion *non-preemptively*, and it is only when the job completes that the scheduler learns the actual processing time. Both the scheduler and the optimal solution are non-anticipatory, which roughly means that the future is uncertain for both, the scheduler and the adversary. Stochastic scheduling has been well-studied, including fundamental work such as [23,24] and approximation algorithms, e.g. [21,25,29,31,32].

This paper considers online scheduling of non-preemptive, stochastic jobs in an unrelated machine environment to minimize the total weighted completion time. We address the same problem as Megow et al. [21], however for the most general, *unrelated* machines model. In the stochastic unrelated machine setting, that means that the scheduler is given a probability distribution of a job's processing time which is machine-dependent, and there need not be any correlation between the jobs' processing time distributions on different machines.

Restricting attention to non-preemptive policies, when all machines are identical, perhaps the most natural algorithm is Weighted Shortest Expected Processing Time (WSEPT) first, which always assigns a job with the maximum ratio of weight over expected size when a machine is free. With unit weights, this boils down to greedily scheduling jobs according to smallest expected size, or SEPT. When there is a single machine, WSEPT is optimal [28]. Further, in the case where job sizes are deterministic and arrive at the same time, SEPT is optimal [11]. In the identical machines setting, SEPT is optimal if job sizes are exponentially distributed [6,35], or more generally, are stochastically comparable in pairs [34]. Some extensions of these optimality results to the problem with weights exist as well [17]. However for more general distributions, simple solutions fail [33], and our knowledge of optimal scheduling policies is limited.

For this reason, approximation algorithms have been studied. With the notable exception of [15], all approximation algorithms have performance guarantees that depend on an upper bound Δ on the squared coefficient of variation of the underlying random variables. Möhring, Schulz and Uetz [25] established the first approximation algorithms for the problem via a linear programming relaxation for stochastic scheduling. Their work gave a $(3 + \Delta)$-approximation when jobs are released over time (yet offline), and they additionally showed that WSEPT is a $\frac{(3+\Delta)}{2}$-approximation when jobs arrive together[1]. These results have been built on and generalized in several settings [21,22,29,31–33], notably in [21] for the online setting. The currently best known result when jobs are released over time (yet offline) is a $(2 + \Delta)$-approximation by Schulz [29]. In the online setting Schulz gives a $(2.309 + 1.309\Delta)$-competitive algorithm [29]. These results build on an idea from [7] to use a preemptive, fast single machine relaxation, next to the relaxation of [25]. The work of Im, Moseley and Pruhs [15] gave the first results independent of Δ showing that there exist poly-logarithmic approximation algorithms under some assumptions. All these papers address problems with identical machines.

For some 15 years after the results of [25] for the identical machines case, no non-trivial results were known for the *unrelated* machines case despite being a major target in the area. Recently Skutella et al. [31] gave a $\frac{3+\Delta}{2}$-approximation algorithm when jobs arrive at the same time, and a $(2 + \Delta)$-approximation when jobs are released over time (yet offline). Central to unlocking an efficient approximation algorithm for the unrelated machines case was the introduction of a time-indexed linear program that lower bounds the objective value of the optimal non-anticipatory scheduling policy. It is this LP that allows the authors to overcome the complexities of the unrelated machines setting.

The present paper targets the more realistic *online* setting for the scheduling of stochastic jobs on unrelated machines. A priori, it is not clear that there should exist an algorithm with small competitive ratio at all. Prior work for the offline problem requires sophisticated linear [31] or convex [3] programming relaxations. Good candidates for online algorithms are simple and combinatorial, but even discovering an offline approximation algorithm that is simple and combinatorial remains a target.

Results. This paper shows that there exists an online, $O(\Delta)$-competitive, *combinatorial* algorithm for stochastic scheduling on unrelated machines. We thereby (1) develop the first combinatorial algorithm for stochastic scheduling on unrelated machines, (2) give the first simple and combinatorial online algorithm for unrelated machines that is competitive (even for the deterministic setting), and (3) introduce new techniques for bounding the performance of stochastic scheduling algorithms.

We address (1) and (2) by giving a simple greedy online algorithm for stochastic scheduling on unrelated machines. The algorithm rests on the straightforward idea to assign jobs to those machines where the expected increase of the objective is minimal, an idea that was used also before, e.g. in [2,20,21]. In the online-list model, where jobs arrive online (at time 0) and must be assigned to a machine immediately upon arrival, we establish a competitive ratio of $(8 + 4\Delta)$. In the online-time model, where jobs arrive over time, we derive a $(144 + 72\Delta)$-competitive algorithm. The $\Omega(\Delta)$ lower bound for fixed assignment policies in [31] yields that both these results are asymptotically tight in Δ.

---

[1] The ratio is slightly better, but for simplicity we ignore the additive $\Theta(1/m)$ term.

As to (3), we develop how to use dual fitting techniques for stochastic and non-preemptive algorithm analysis. The technique has been used in [1] for deterministic and preemptive problems. This paper establishes that dual fitting is a useful technique for bounding the performance of algorithms in stochastic settings. This is the first use of dual fitting for stochastic scheduling.

Due to space limitations, most proofs have been removed from this paper. The proofs can be found in the full version [10].

2   Notation and Preliminaries

We are given a set of unrelated parallel machines M of cardinality m. We consider two online models. In the first model, known as *online-list*, we are presented a sequence of jobs $j \in J$, which are presented to us one after the other, and whenever a job is presented we have to assign it to one of the machines. In this model, the machine assignment is decided when a job arrives, but the time when the job is processed can be deferred. It is unknown how many jobs will arrive, but once all jobs in J have arrived, the jobs assigned to any one of the machines must be scheduled on that machine. In the other model, known as *online-time*, time progresses and jobs appear over time at their individual release times r_j. At the moment of arrival a job must be assigned to a machine, but can possibly wait on that machine until it is finally processed. Each job j needs to be executed on any one of the machines $i \in M$, and each machine can process at most one job at a time.

The jobs are nonpreemptive. That means that a job, once started, must not be interrupted until its completion. Moreover, the jobs are stochastic, meaning that each job j's processing time is only revealed in the form of a random variable P_{ij} for every machine $i \in M$. If job j is assigned to machine i, its processing time will be random according to P_{ij}. It is allowed that certain jobs $j \in J$ cannot be processed on certain machines $i \in M$, in which case $\mathbb{E}[P_{ij}] = \infty$.

In the stochastic scheduling model, the actual realization of the processing time of a job j becomes only known at the moment that the job completes. We are looking for a non-anticipatory scheduling policy Π which minimizes the expected total weighted completion time $\mathbb{E}[\sum_j w_j C_j]$, where C_j denotes the completion time of job j.

We will assume for simplicity that the random variables P_{ij} are discrete and integer valued. This assumption comes at the cost of a multiplicative factor $(1+\varepsilon)$ in the final approximation ratio, for any $\varepsilon > 0$ [31]. We will subsequently make use of the following facts about first and second moments of discrete random variables; they also appear in [31].

Lemma 1. *Let X be an integer-valued, nonnegative random variable. Then,*

$$\sum_{r \in \mathbb{Z}_{\geq 0}} \mathbb{P}[X > r] = \mathbb{E}[X] \quad and \quad \sum_{r \in \mathbb{Z}_{\geq 0}} (r + \tfrac{1}{2}) \mathbb{P}[X > r] = \frac{1}{2}\mathbb{E}[X^2].$$

Definition 1. *Let X be a nonnegative random variable. The* squared coefficient of variation *is defined as the scaled variance of X, that is,*

$$\mathbb{CV}[X]^2 := \mathbb{V}\mathrm{ar}[X]/\mathbb{E}[X]^2,$$

where $\mathbb{V}\mathrm{ar}[X] = \mathbb{E}[X^2] - E[X]^2$.

2.1   Stochastic Online Scheduling and Policies

The setting that we consider in this paper is that of stochastic online scheduling as defined also in [21]. That means that (the existence of) a jobs j is unknown before it arrives, and upon arrival, only the random variables P_{ij} for the possible processing times on machine $i = 1, \ldots, m$ are known. At any given time t, a non-anticipatory online scheduling policy is allowed to use only that information that is available at time t. In particular, it may anticipate the (so far) realized processing times of jobs up to time t. For example, a job that has possible sizes 1, 3 or 4 with probabilities $1/3$ each, and has been running for 2 time units, will have processing times 3 or 4, each with probability $1/2$. That adaptivity over time may be relevant in order to minimize the expectation of the total weighted completion times is well known even in the offline setting, e.g. [33]. We refer to [21] for a more thorough discussion of the stochastic online model.

For simplicity of notation, we denote by OPT the expected total weighted completion time of an optimal, non-anticipatory online scheduling policy for the problem. That is, OPT is our benchmark, and we seek to find a non-anticipatory online scheduling policy (an algorithm) with expected performance ALG close to OPT. Note that, for convenience we use the same notation for both algorithm and its expected performance.

We remark that OPT is not restricted to assigning jobs to machine at the time of their arrival. The only restrictions on OPT is that it must schedule jobs nonpreemptively, and that it is non-anticipatory. In fact, our approximation guarantees hold against an even stronger adversary OPT which knows all jobs and their release times r_j and processing time distributions, but not the actual realizations of P_{ij}.

Finally, we may assume w.l.o.g. that no pair of job and machine exists with $\mathbb{E}[P_{ij}] = 0$, as then we can always schedule such job j at machine i (whenever released) at minimum possible cost. That said, we may further assume that $\mathbb{E}[P_{ij}] \geq 1$ for all machines i and jobs j, by scaling.

3   Linear Programming Relaxations

As previously discussed also in [31, Sect. 8], we are going to use variables y_{ijs} that denote the probability that job j is being processed on machine i within time interval $[s, s+1]$, under some given and fixed scheduling policy. It is well known that y_{ijs} can be linearly expressed in terms of the variables x_{ijt}, which denote the probability that job j is started at time t on machine i, as follows

$$y_{ijs} = \sum_{t=0}^{s} x_{ijt}\, \mathbb{P}[P_{ij} > s - t]. \tag{1}$$

The fact that any machine can process at most one job at a time can be written as

$$\sum_{j \in J} y_{ijs} \leq 1 \qquad \text{for all } i \in M, s \in \mathbb{Z}_{\geq 0}. \tag{2}$$

Moreover, making use of (1) and the first part of Lemma 1, the fact that each job needs to be completely processed translates into the constraints

$$\sum_{i \in M} \sum_{s \in \mathbb{Z}_{\geq 0}} \frac{y_{ijs}}{\mathbb{E}[P_{ij}]} = 1 \qquad \text{for all } j \in J. \tag{3}$$

Finally, with the help of (1) and the second part of Lemma 1, the expected completion time of a job j can be expressed in y_{ijs} variables as

$$C_j^S := \sum_{i \in M} \sum_{s \in \mathbb{Z}_{\geq 0}} \left(\frac{y_{ijs}}{\mathbb{E}[P_{ij}]} \left(s + \tfrac{1}{2}\right) + \frac{1 - \mathbb{CV}[P_{ij}]^2}{2} y_{ijs} \right) \qquad \text{for all } j \in J, \quad (4)$$

where we labeled the expected completion time variables with a superscript S for "stochastic", for reasons that will become clear shortly.

For the analysis to follow, we also need to express the fact that the expected completion time of a job cannot be smaller than its expected processing time

$$C_j^S \geq \sum_{i \in M} \sum_{s \in \mathbb{Z}_{\geq 0}} y_{ijs} \qquad \text{for all } j \in J. \tag{5}$$

That said, we can write down the following LP relaxation for the unrelated machine scheduling problem, which extends the one given in [31] by the additional constraints (5).

$$\min \quad z^S = \sum_{j \in J} w_j C_j^S$$

$$\text{s.t.} \quad (2), (3), (4), (5) \tag{S}$$

$$y_{ijs} \geq 0 \qquad \text{for all } j \in J, i \in M, s \in \mathbb{Z}_{\geq 0}.$$

Subsequently, we want to work with the dual of this relaxation. However the term $-\mathbb{CV}[P_{ij}]^2$ in the primal objective would appear in the dual constrains. As we do not know how to deal with this negative term in the analysis that is to follow, we are going to factor it out.

To that end, we first define a simpler, i.e., deterministic version for the expected completion times (4), labeled with "P" to distinguish it from the previous formulation, by letting

$$C_j^P = \sum_{i \in M} \sum_{s \in \mathbb{Z}_{\geq 0}} \left(\frac{y_{ijs}}{\mathbb{E}[P_{ij}]} \left(s + \tfrac{1}{2}\right) + \frac{y_{ijs}}{2} \qquad \text{for all } j \in J. \right) \tag{6}$$

Now consider the following linear programming problem

$$\min \quad z^P = \sum_{j \in J} w_j\, C_j^P$$

$$\text{s.t.} \quad (2), (3), (6) \tag{P}$$

$$y_{ijs} \geq 0 \qquad \text{for all } j \in J, i \in M, s \in \mathbb{Z}_{\geq 0}.$$

This corresponds to a time-indexed linear programming relaxation for a purely deterministic, unrelated machine scheduling problem where the random processing times are fixed at their expected values $\mathbb{E}[P_{ij}]$.

We are now going to establish a relation between these two relaxations. To do that, let us define an upper bound on the squared coefficient of variation by

$$\Delta := \max_{i,j} \mathbb{CV}[P_{ij}]^2.$$

Next, for any given solution y of (S) or (P), we define

$$H(y) := \sum_{j \in J} w_j \sum_{i \in M} \sum_{s \in \mathbb{Z}_{\geq 0}} y_{ijs}.$$

Now let y^S denote an optimal solution to (S) and recall that OPT is the expected total weighted completion time of an optimal non-anticipatory algorithm. By constraints (5),

$$H(y^S) = \sum_{j \in J} w_j \sum_{i \in M} \sum_{s \in \mathbb{Z}_{\geq 0}} y_{ijs}^S \leq \sum_{j \in J} w_j C_j^S = z^S(y^S) \leq \mathsf{OPT}.$$

The next lemma is crucial for our analysis and establishes the relation between the two relaxations.

Lemma 2. *The optimal solution values z^P and z^S of the linear programming relaxations (P) and (S) fulfill*

$$z^P \leq \left(1 + \frac{\Delta}{2}\right) z^S.$$

Recalling that (S) is a relaxation for the stochastic scheduling problem, we conclude the following.

Corollary 1. *The optimal solution value z^P of the linear programming relaxation (P) is bounded by the expected performance of an optimal scheduling policy by*

$$z^P \leq \left(1 + \frac{\Delta}{2}\right)\mathsf{OPT}.$$

Just like [1], we now consider the dual of (P), which will have unconstrained variables α_j for all $j \in J$ and nonnegative variables β_{is} for all $i \in M$ and $s \in \mathbb{Z}_{\geq 0}$.

The dual is

$$\max \quad z^D = \sum_{j \in J} \alpha_j - \sum_{i \in M} \sum_{s \in \mathbb{Z}_{\geq 0}} \beta_{is}$$

$$\text{s.t.} \quad \frac{\alpha_j}{\mathbb{E}[P_{ij}]} - \beta_{is} \leq w_j \left(\frac{s + \frac{1}{2}}{\mathbb{E}[P_{ij}]} + \frac{1}{2} \right) \quad \text{for all } i \in M, j \in J, s \in \mathbb{Z}_{\geq 0}, \tag{D}$$

$$\beta_{is} \geq 0 \qquad \qquad \text{for all } i \in M, s \in \mathbb{Z}_{\geq 0}.$$

We are going to define a feasible solution for the dual (D) by means of a simple online algorithm for the original scheduling problem. The same type of greedy algorithm has been used before, both in deterministic and stochastic scheduling on parallel machines, e. g. in [2,20,21].

4   Greedy Algorithm and Analysis

In this section the online-list model is considered. Let us assume w.l.o.g. that the jobs are presented in the order $1, 2 \ldots, |J|$. On any machine i, denote by $H(j, i)$ the set of all jobs that have higher priority according to their order in non-increasing order of ratios $w_j / \mathbb{E}[P_{ij}]$, breaking ties by index. That is, $H(j, i) := \{ k \in J \mid w_k / \mathbb{E}[P_{ik}] > w_j / \mathbb{E}[P_{ij}] \} \cup \{ k \in J \mid k \leq j, w_k / \mathbb{E}[P_{ik}] = w_j / \mathbb{E}[P_{ij}] \}$. Also, let $L(j, i) := J \setminus H(j, i)$. Further, denote by $k \to i$ the fact that a job k has been assigned to a machine i.

Greedy Algorithm. Whenever a new job $j \in J$ is presented to the algorithm, we compute for each of the machines $i \in M$ the *instantaneous expected increase* if the jobs already present on each machine were to be scheduled in non-increasing order of the ratios weight over expected processing time,

$$\text{EI}(j \to i) := w_j \left(\mathbb{E}[P_{ij}] + \sum_{k \to i, k < j, k \in H(j, i)} \mathbb{E}[P_{ik}] \right) + \mathbb{E}[p_{ij}] \sum_{k \to i, k < j, k \in L(j, i)} w_k.$$

We assign the job to one of the machines where this quantity is minimal, that is, a job is assigned to machine $i(j) := \text{argmin}_{i \in M} \{ \text{EI} j \to i \}$; ties broken arbitrarily. Once all jobs have arrived and are assigned, they will be sequenced in non-increasing order of ratios weight over expected processing time, which is optimal conditioned on the given assignment [28].

Now we define the dual solution (α, β) in a similar same way as it has been done in [1]. We let

$$\alpha_j := \text{EI}(j \to i(j)) \quad \text{for all } j \in J.$$

That is, α_j is defined as the instantaneous expected increase on the machine to which it is assigned by the greedy algorithm. Moreover, let

$$\beta_{is} := \sum_{j \in A_i(s)} w_j,$$

where $A_i(s)$ is defined as the total set of jobs assigned to machine i by the greedy algorithm, but restricted to those that have not yet been completed by time s if the jobs' processing times were their expected values $\mathbb{E}[P_{ij}]$. In other words, β_{is} is exactly the expected total weight of yet unfinished jobs on machine i at time s, given the assignment (and sequencing) of the greedy algorithm.

Fact 1. *The solution* $(\alpha/2, \beta/2)$ *is feasible for* (D).

Moreover, we have the following observations which follow more or less directly from the definition of the dual variables (α, β). Let us denote by ALG the total expected value achieved by the greedy algorithm.

Lemma 3. *The total expected value of the greedy algorithm is*

$$\mathsf{ALG} = \sum_{j \in J} \alpha_j = \sum_{i \in M} \sum_{s \in \mathbb{Z}_{\geq 0}} \beta_{is}.$$

5   Speed Augmentation and Analysis

The previous analysis of the dual feasible solution $(\alpha/2, \beta/2)$ yields a dual objective value 0 by Lemma 3, which is of little help. However following [1], we can define another dual solution which has an interpretation in the model where all machines run at faster speed $f \geq 1$, meaning that all (expected) processing times get scaled down by a factor f^{-1}. This will yield something useful.

So let us define ALG^f as the expected solution value obtained by the same greedy algorithm, only when all the machine run at speed f. Note that $\mathsf{ALG} = f\mathsf{ALG}^f$, by definition. We denote by (α^f, β^f) the exact same dual solution that was defined before, only for the new instance with faster machines. We now claim the following.

Lemma 4. *The solution* $(\frac{1}{2}\alpha^f, \frac{1}{2f}\beta^f)$ *is a feasible solution for the dual* (D) *in the* original *(unscaled) problem instance.*

Now we conclude with the first main theorem of the paper.

Theorem 1. *The greedy algorithm is a* $(8 + 4\Delta)$-*competitive algorithm for online scheduling of stochastic jobs to minimize the expectation of the total weighted completion times* $\mathbb{E}[\sum_j w_j C_j]$.

Proof. We know from Corollary 1 that $z^D(\frac{1}{2}\alpha^f, \frac{1}{2f}\beta^f) \leq z^D = z^P \leq \left(1 + \frac{\Delta}{2}\right)\mathsf{OPT}$.

Next, recall that $\mathsf{ALG}^f = \sum_{j \in J} \alpha_j^f = \sum_{i \in M} \sum_{s \in \mathbb{Z}_{\geq 0}} \beta_{is}^f$ by Lemma 3, and $\mathsf{ALG} = f\mathsf{ALG}^f$. The theorem now follows from evaluating the objective value of the specifically chosen dual solution $(\frac{1}{2}\alpha^f, \frac{1}{2f}\beta^f)$ for (D), as

$$z^D\left(\frac{1}{2}\alpha^f, \frac{1}{2f}\beta^f\right) = \frac{1}{2}\sum_{j \in J}\alpha_j^f - \frac{1}{2f}\sum_{i \in M}\sum_{s \in \mathbb{Z}_{\geq 0}}\beta_{is}^f = \frac{f-1}{2f}\mathsf{ALG}^f = \frac{f-1}{2f^2}\mathsf{ALG}.$$

Putting together this equality with the previous inequality yields a performance bound of $\frac{2f^2}{f-1}(1 + \frac{\Delta}{2})$, which is minimal for $f = 2$.   □

6 The Online Time Model

We now consider the online time model where jobs arrive over time. A job j arrives at time r_j. We can assume w.l.o.g. that $r_j \leq r_k$ for $j < k$. In the algorithm, which is the analogue of the one used in [21] for the parallel machine setting, each job will be irrevocably assigned to a machine upon arrival.

Modified Greedy Algorithm. *1. Assignment of jobs to machines:* At time r_j, we compute for each of the machines $\mathrm{EI}(j \to i)$ exactly in the same way as it has been done for the case without release times, and assign job j to one of the machines that minimizes $\mathrm{EI}j \to i$. *2. Scheduling:* For the case with release dates, it is well known that (long) jobs must be delayed in order to achieve competitive algorithms [20,21]. We do the same here, but we insert a little more forced idleness than these papers. For any job j assigned to machine i at time r_j, we modify its release date to $r'_j = \max\{2r_j, \mathbb{E}[P_{ij}]\}$. Now if a machine i falls idle at a time t, among all unfinished jobs assigned to i and with $r'_j \leq t$, we schedule the job with the highest ratio $w_j/\mathbb{E}[P_{ij}]$, by first forcing the machine to remain idle for another $\mathbb{E}[P_{ij}]$ units of time, and then beginning the actual processing of job j.

The main result of this section is:

Theorem 2. *For the stochastic online scheduling problem on unrelated parallel machines with release dates, if $\max_{i,j} \mathbb{CV}[P_{ij}]^2 \leq \Delta$, then the Modified Greedy Algorithm is $(144 + 72\Delta)$-competitive.*

Proof Sketch. The complete proof is a bit intricate and presented in the full version [10]. Here we sketch the main steps in the analysis. Defining the expected cost of the modified greedy algorithm as ALG_S and of the optimal non-anticipative policy as OPT, our goal is to prove $\mathsf{ALG}_S \leq (144 + 72\Delta)\mathsf{OPT}$. Step 1: As in the online-list model, the core of the argument proceeds via an instance with augmented machine speeds. Given instance $\{r_j, \{P_{ij}\}_{i \in M}\}_{j \in J}$, we define a family of instances parameterized by speed-up f with release times $r_j^f = r_j$ and processing times $P_{ij}^f = P_{ij}/f$. Denote by ALG_S^f the expected cost of a variant of the modified greedy algorithm where the scheduling rule is changed to use the modified release times as $R_j^f = \max\{r_j, \mathbb{E}[P_{ij}^f]\}$. Then a time scaling argument shows the following for $f = 2$

$$\mathsf{ALG}_S = 2 \cdot \mathsf{ALG}_S^2.$$

In fact, the equality is even in distribution and not just for expectations.
Step 2: For the stochastic instance $\{r_j^f, \{P_{ij}^f\}\}$, we define the deterministic instance where processing time of job j on machine i is non-stochastic and equals $\mathbb{E}[P_{ij}^f]$. Further, we begin processing the jobs as soon as they are scheduled, without the idleness. Let ALG_D^f denote the cost of our algorithm on this instance. We show,

$$\mathsf{ALG}_S^f \leq 6 \cdot \mathsf{ALG}_D^f.$$

Step 3: As in Sect. 3, we define the LP relaxation of the online stochastic machine scheduling problem (with optimal solution z^{S_o}). The only difference is that y_{ijs}

are forced to be 0 for $s \leq r_j$. Analogously, z^{P_o} denotes the optimal solution value of the corresponding deterministic version as in Sect. 3, giving us:

$$z^{P_o} \leq \left(1 + \frac{\Delta}{2}\right) z^{S_o} \leq \left(1 + \frac{\Delta}{2}\right) \mathsf{OPT}.$$

Finally, we use a dual fitting argument to prove, for any $f > 1$:

$$\mathsf{ALG}_D^f \leq \frac{6f}{f-1} z^{P_o}.$$

Now substituting $f = 2$, $\mathsf{ALG}_S = 2\mathsf{ALG}_S^2 \leq 2 \cdot 6 \cdot \mathsf{ALG}_D^2 \leq 2 \cdot 6 \cdot 12 \cdot z^{P_o} \leq 144 \left(1 + \frac{\Delta}{2}\right) \mathsf{OPT}$.

7   Conclusions

The main result of this paper is to show that simple, combinatorial online algorithms *can* be worst-case analyzed even for the most general of all machine scheduling models and uncertain job sizes. Further, note that the performance bounds are $O(\Delta)$, asymptotically the same as the identical machines setting.

Acknowledgements. This work was done while all four authors were with the Simons Institute for the Theory of Computing at UC Berkeley. The authors wish to thank the institute for the financial support and the organizers of the semester on "Algorithms & Uncertainty" for providing a very stimulating atmosphere.

References

1. Anand, S., Garg, N., Kumar, A.: Resource augmentation for weighted flow-time explained by dual fitting. In: Proceedings of the Twenty-Third Annual ACM-SIAM Symposium on Discrete Algorithms, SODA 2012, pp. 1228–1241 (2012)
2. Avrahami, N., Azar, Y.: Minimizing total flow time and total completion time with immediate dispatching. In: Proceedings of the 15th Symposium on Parallelism in Algorithms and Architectures (SPAA 2003), pp. 11–18. ACM (2003)
3. Balseiro, S., Brown, D., Chen, C.: Static routing in stochastic scheduling: Performance guarantees and asymptotic optimality. Technical report (2016)
4. Bansal, N., Srinivasan, A., Svensson, O.: Lift-and-round to improve weighted completion time on unrelated machines. In: Proceedings of 48th Annual ACM Symposium Theory Computing (STOC), pp. 156–167. ACM (2016)
5. Becchetti, L., Leonardi, S.: Non-clairvoyant scheduling to minimize the average flow time on single and parallel machines. In: STOC, pp. 94–103 (2001)
6. Bruno, J., Downey, P.J., Frederickson, G.: Sequencing tasks with exponential service times to minimize the expected flowtime or makespan. J. ACM **28**, 100–113 (1981)
7. Correa, J., Wagner, M.: LP-based online scheduling: from single to parallel machines. Math. Program. **119**, 109–136 (2008)
8. Graham, R.L., Lawler, E.L., Lenstra, J.K., Kan, A.H.G.R.: Optimization and approximation in deterministic sequencing and scheduling: a survey. Ann. Discrete Math. **5**, 287–326 (1979)

9. Gupta, A., Im, S., Krishnaswamy, R., Moseley, B., Pruhs, K.: Scheduling hetero-geneous processors isn't as easy as you think. In: Proceedings of the Twenty-Third Annual ACM-SIAM Symposium on Discrete Algorithms, SODA 2012, pp. 1242–1253 (2012)

10. Gupta, V., Moseley, B., Uetz, M., Xie, Q.: Stochastic online scheduling on unrelated machines. arXiv preprint arXiv:1703.01634 (2017)

11. Horn, W.: Minimizing average flowtime with parallel machines. Oper. Res. **21**, 846–847 (1973)

12. Horowitz, E., Sahni, S.: Exact and approximate algorithms for scheduling noniden-tical processors. J. ACM **23**(2), 317–327 (1976)

13. Im, S., Kulkarni, J., Munagala, K., Pruhs, K.: Selfishmigrate: a scalable algorithm for non-clairvoyantly scheduling heterogeneous processors. In: 55th IEEE Annual Symposium on Foundations of Computer Science, FOCS, pp. 531–540 (2014)

14. Im, S., Moseley, B., Pruhs, K.: A tutorial on amortized local competitiveness in online scheduling. SIGACT News **42**(2), 83–97 (2011)

15. Im, S., Moseley, B., Pruhs, K.: Stochastic scheduling of heavy-tailed jobs. In: STACS (2015)

16. Kalyanasundaram, B., Pruhs, K.: Speed is as powerful as clairvoyance. J. ACM **47**(4), 617–643 (2000)

17. Kämpke, T.: On the optimality of static priority policies in stochastic scheduling on parallel machines. J. Appl. Probab. **24**, 430–448 (1987)

18. Lenstra, J., Shmoys, D.B., Tardos, É.: Approximation algorithms for scheduling unrelated parallel machines. Math. Program. **46**, 259–271 (1990)

19. Leung, J.Y.-T. (ed.): Handbook of Scheduling: Algorithms, Models, and Perfor-mance Analysis. Chapman & Hall/CRC, Boca Raton (2004)

20. Megow, N., Schulz, A.: On-line scheduling to minimize average completion time revisited. Oper. Res. Lett. **32**, 485–490 (2004)

21. Megow, N., Uetz, M., Vredeveld, T.: Models and algorithms for stochastic online scheduling. Math. Oper. Res. **31**(3), 513–525 (2006)

22. Megow, N., Vredeveld, T.: A tight 2-approximation for preemptive stochastic scheduling. Math. Oper. Res. **39**, 1297–1310 (2011)

23. Möhring, R.H., Radermacher, F.J., Weiss, G.: Stochastic scheduling problems I: general strategies. ZOR - Z. Oper. Res. **28**, 193–260 (1984)

24. Möhring, R.H., Radermacher, F.J., Weiss, G.: Stochastic scheduling problems II: set strategies. ZOR - Z. Oper. Res. **29**, 65–104 (1985)

25. Möhring, R.H., Schulz, A.S., Uetz, M.: Approximation in stochastic scheduling: the power of LP-based priority policies. J. ACM **46**, 924–942 (1999)

26. Motwani, R., Phillips, S., Torng, E.: Non-clairvoyant scheduling. Theor. Comput. Sci. **130**(1), 17–47 (1994)

27. Pruhs, K., Sgall, J., Torng, E.: Online scheduling. In: Leung, J. (ed.) Handbook of Scheduling: Algorithms, Models and Performance Analysis. CRC Press, Boca Raton (2004)

28. Rothkopf, M.H.: Scheduling with random service times. Manage. Sci. **12**, 703–713 (1966)

29. Schulz, A.S.: Stochastic online scheduling revisited. In: Yang, B., Du, D.-Z., Wang, C.A. (eds.) COCOA 2008. LNCS, vol. 5165, pp. 448–457. Springer, Heidelberg (2008). doi:10.1007/978-3-540-85097-7_42

30. Skutella, M.: Convex quadratic and semidefinite programming relaxations in scheduling. J. ACM **48**, 206–242 (2001)

31. Skutella, M., Sviridenko, M., Uetz, M.: Unrelated machine scheduling with sto-chastic processing times. Math. Oper. Res. **41**(3), 851–864 (2016)

32. Skutella, M., Uetz, M.: Stochastic machine scheduling with precedence constraints. SIAM J. Comput. **34**, 788–802 (2005)
33. Uetz, M.: When greediness fails: examples from stochastic scheduling. Oper. Res. Lett. **31**, 413–419 (2003)
34. Weber, R., Varaiya, P., Walrand, J.: Scheduling jobs with stochastically ordered processing times on parallel machines to minimize expected owtime. J. Appl. Probab. **23**, 841–847 (1986)
35. Weiss, G., Pinedo, M.: Scheduling tasks with exponential service times on non-identical processors to minimize various cost functions. J. Appl. Probab. **17**, 187–202 (1980)

Online Matroid Intersection: Beating Half for Random Arrival

Guru Prashanth Guruganesh and Sahil Singla$^{(\boxtimes)}$

Computer Science Department, Carnegie Mellon University, Pittsburgh, USA
{ggurugan,ssingla}@cs.cmu.com

Abstract. For two matroids $\mathcal{M}_1$ and $\mathcal{M}_2$ defined on the same ground set E, the online matroid intersection problem is to design an algorithm that constructs a large common independent set in an online fashion. The algorithm is presented with the ground set elements one-by-one in a uniformly random order. At each step, the algorithm must irrevocably decide whether to pick the element, while always maintaining a common independent set. While the natural greedy algorithm—pick an element whenever possible—is half competitive, nothing better was previously known; even for the special case of online bipartite matching in the edge arrival model. We present the first randomized online algorithm that has a $\frac{1}{2} + \delta$ competitive ratio in expectation, where $\delta > 0$ is a constant. The expectation is over the random order and the coin tosses of the algorithm. As a corollary, we also obtain the first linear time algorithm that beats half competitiveness for offline matroid intersection.

Keywords: Online algorithms · Matroid intersection · Randomized algorithms · Competitive analysis · Linear-time algorithms

1 Introduction

The *online matroid intersection problem* in the random arrival model (OMI) consists of two matroids $\mathcal{M}_1 = (E, \mathcal{I}_1)$ and $\mathcal{M}_2 = (E, \mathcal{I}_2)$, where the elements in E are presented one-by-one to an online algorithm whose goal is to construct a large common independent set. As an element arrives, the algorithm must immediately and irrevocably decide whether to *pick* it, while ensuring that the picked elements always form a common independent set. We assume that the algorithm knows the size of E and has access to independence oracles for the already arrived elements. The greedy algorithm, which picks an element whenever possible, is half-competitive. The following is the main result of this paper.

Theorem 1. *The online matroid intersection problem in the random arrival model has a $(\frac{1}{2} + \delta)$-competitive randomized algorithm, where $\delta > 0$ is a constant.*

Supported in part by NSF awards CCF-1319811, CCF-1536002, and CCF-1617790.

F. Eisenbrand and J. Koenemann (Eds.): IPCO 2017, LNCS 10328, pp. 241–253, 2017.
DOI: 10.1007/978-3-319-59250-3_20

Our OMI algorithm makes only a linear number of calls to the independence oracles of both the matroids. Given recent interest in finding fast approximation algorithms for fundamental polynomial-time problems, this result is of independent interest even in the offline setting. Previously known algorithms that perform better than the greedy algorithm construct an "auxiliary graph", which already takes quadratic time [2,7].

Corollary 1. *The matroid intersection problem has a linear time $(\frac{1}{2}+\delta)$ approximation algorithm, where $\delta > 0$ is a constant.*

A special case of OMI where both the matroids are partition matroids already captures the *online bipartite matching problem* in the random edge arrival (OBME) model. Here, edges of a fixed (but adversarially chosen) bipartite graph G arrive in a uniformly random order and the algorithm must irrevocably decide whether to pick them into a matching. Despite tremendous progress made in the online vertex arrival model [5,9,12,16,20], nothing non-trivial was known in the edge arrival model where the edges arrive one-by-one. We present the first algorithm that performs better than greedy in the random arrival model. Besides being a natural theoretical question, it captures various online content systems such as online libraries where the participants are known to the matching agencies but the requests arrive in an online fashion.

Corollary 2. *The online bipartite matching problem in the random edge arrival model has a $(\frac{1}{2}+\delta)$-competitive randomized algorithm, where $\delta > 0$ is a constant.*

Finally, the simplicity of our OMI algorithm allows us to extend our results to the much more general problems of online matching in general graphs and to online k-matroid intersection; the latter problem being **NP**-Hard (proofs in full version).

Theorem 2. *The online matching problem for general graphs in the random edge arrival model has a $(\frac{1}{2}+\delta')$-competitive randomized algorithm, where $\delta' > 0$ is a constant.*

Theorem 3. *The online k-matroid intersection problem in the random arrival model has a $\left(\frac{1}{k} + \frac{\delta''}{k^4}\right)$-competitive randomized algorithm, where $\delta'' > 0$ is a constant.*

1.1   Comparison to Previous Work

Our main OMI result is interesting in two different aspects: It gives the first linear time algorithm that beats greedy for the classical offline matroid intersection problem; also, it is the first non-trivial algorithm for the general problem of online matroid intersection, where previously nothing better than half was known even for online bipartite matching. Since offline matroid intersection problem is a fundamental problem in the field of combinatorial optimization [19, Chap. 41] and online matching occupies a central position in the field of online algorithms [15], there is a long list of work in both these areas. We state the most relevant works here and refer readers to further related work in the full version.

Offline matroid intersection was brought to prominence in the groundbreaking work of Edmonds [3]. To illustrate the difficulty in moving from bipartite matching to matroid intersection, we note that while the first linear time algorithms that beat half for bipartite matching were designed more than 20 years ago [1,6], the fastest known matroid intersection algorithms till today that beat half make $\Omega(rm)$ calls to the independence oracles, where r is the rank of the optimal solution [2,7]. The quadratic term appears because matroid intersection algorithms rely on constructing auxiliary graphs that needs $\Omega(rm)$ calls [11, Chap. 13]. Until our work, achieving a competitive ratio better than half with linear number of independence oracle calls was not known. The key ingredient that allows us to circumvent these difficulties is the *Sampling Lemma* for matroid intersection. We do not construct an auxiliary graph and instead show that any maximal common independent is either already a $(\frac{1}{2} + \delta)$ approximation, or we can improve it to a $(\frac{1}{2} + \delta)$ approximation in a single pass over all the elements.

Online bipartite matching has been studied extensively in the vertex arrival model (see a nice survey by Mehta [15]). Since adversarial arrival order often becomes too pessimistic, the random arrival model (similar to the *secretary problem*) for online matching was first studied by Goel and Mehta [5]. Since then, this modeling assumption has become standard [8,12–14]. The only progress when edges arrive one-by-one has been in showing lower bounds: no algorithm can achieve a competitive ratio better than 0.57 (see [4]), even when the algorithm is allowed to drop edges.

While nothing was previously known for online matching in the random edge arrival model, similar problems have been studied in the streaming model, most notably by Konrad et al. [10]. They gave the first algorithm that beats half for bipartite matching in the random arrival streaming model. In this work we generalize their *Hastiness Lemma* to matroids. However, prior works on online matching are not useful as they are tailored to graphs—for instance their reliance on notion of "vertices" cannot be easily extended to the framework of matroids.

The simplicity of our OMI algorithm and flexibility of our analysis allows us to tackle problems of much greater generality, such as general graphs and k-matroid intersection, when previously even special cases like bipartite matching had been considered difficult in the online regime [17]. While our results are a qualitative advance, the quantitative improvement is small ($\delta > 10^{-4}$). It remains an interesting challenge to improve the approximation factor δ. Perhaps a more interesting challenge is to relax the random order requirement.

1.2 Our Techniques

In this section, we present an overview of our techniques to prove Theorem 1. Our analysis relies on two observations about the greedy algorithm that are encompassed in the Sampling Lemma and the Hastiness Lemma; former being the major contribution of this paper and the latter being useful to extend our linear time offline matroid intersection result to the online setting. Informally, the Sampling Lemma states that the greedy algorithm cannot perform poorly on a randomly generated OMI instance, and the Hastiness Lemma states that if the greedy algorithm performs poorly, then it picks most of its elements quickly.

Let OPT denote a fixed maximum independent set in the intersection of matroids $\mathcal{M}_1$ and $\mathcal{M}_2$. WLOG, we assume that the greedy algorithm is *bad*— returns a common independent set T of size $\approx \frac{1}{2}|\text{OPT}|$. For offline matroid intersection, by running the greedy algorithm once, one can assume that T is known. For online matroid intersection, we use the Hastiness Lemma to construct T. It states that even if we run the greedy algorithm for a small fraction f (say $< 1\%$) of elements, it already picks a set T of elements of size $\approx \frac{1}{2}|\text{OPT}|$. This lemma was first observed by Konrad et al. [10] for bipartite matching and is generalized to matroid intersection in this work. By running the greedy algorithm for this small fraction f, the lemma lets us assume that we start with an approximately maximal common independent set T with most of the elements $(1 - f > 99\%)$ still to arrive.

The above discussion reduces the problem to improving a common independent set T of size $\approx \frac{1}{2}|\text{OPT}|$ to a common independent set of size $\geq (\frac{1}{2}+\delta)|\text{OPT}|$ in a single pass over all the elements. (This is true for both linear-time offline and OMI problems.) Since T is approximately maximal, we know that picking most elements in T eliminates the possibility of picking two OPT elements (one for each matroid). Hence, to beat half-competitiveness, we drop a uniformly random p fraction of these "bad" elements in T to obtain a set S, and try to pick $(1 + \gamma)$ OPT elements (for constant $\gamma > 0$) per dropped element. Our main challenge is to construct an online algorithm that can get on average γ gain per dropped element of T in a single pass. The Sampling Lemma for matroid intersection, which is our main technical contribution, comes to rescue.

Sampling Lemma (Informal): *Suppose T is a common independent set in matroids $\mathcal{M}_1$ and $\mathcal{M}_2$, and define $\tilde{E} = \text{span}_1(T)$. Let S denote a random set containing each element of T independently with probability $(1 - p)$. Then,*

$$\mathbb{E}_S[|Greedy(\mathcal{M}_1/S, \mathcal{M}_2/T, \tilde{E})|] \geq \left(\frac{1}{1+p}\right) \cdot \mathbb{E}_S[|\text{OPT}(\mathcal{M}_1/S, \mathcal{M}_2/T, \tilde{E})|].$$

Intuitively, it says that if we restrict our attention to elements in $\text{span}_1(T)$ then dropping random elements from T allows us to pick more than $1/(1 + p) \geq 1/2$ fraction of the optimal intersection. The advantage over half yields the γ gain per dropped element. Applying the lemma requires care as we apply it twice, once for $(\mathcal{M}_1/S, \mathcal{M}_2/T)$ and once for $(\mathcal{M}_1/T, \mathcal{M}_2/S)$, while ensuring that the resulting solutions have few "conflicts" with each other. We overcome this by only considering elements that are in the span of T for exactly one of the matroids.

The proof of the Sampling Lemma involves giving an alternate view of the greedy algorithm for the random OMI instance. Using a carefully constructed invariant and the method of deferred decisions, we show that the expected greedy solution is not too small.

2   Warmup: Online Bipartite Matching

In this section, we consider a special case of online matroid intersection, namely online bipartite matching in the random edge arrival model. Although, this is a special case of the general Theorem 1, we present it because nothing non-trivial was known before and several of our ideas greatly simplify in this case (in particular the Sampling Lemma), allowing us to lay the framework of our ideas.

2.1   Definitions and Notation

An instance of the online bipartite matching problem (G, E, π, m) consists of a bipartite graph $G = (U \cup V, E)$ with $m = |E|$, and where the edges in E arrive according to the order defined by π. We assume that the algorithm knows m but does not know E or π. For $1 \leq i \leq j \leq m$, let $E^\pi[i, j]$ denote the set of edges that arrive in between positions i through j according to π^1. When permutation π is implicit, we abbreviate this to $E[i, j]$.

GREEDY denotes the algorithm that picks an edge into the matching whenever possible. Let OPT denote a fixed maximum offline matching of graph G. For $f \in [0, 1]$, let T_f^π denote the matching produced by GREEDY after seeing the first f-fraction of the edges according to order π. For a uniformly random chosen order π,

$$\mathcal{G}(f) := \frac{\mathbb{E}_\pi[|T_f^\pi|]}{|\text{OPT}|}.$$

Hence, $\mathcal{G}(1)|\text{OPT}|$ is the expected output size of GREEDY and $\mathcal{G}(\frac{1}{2})|\text{OPT}|$ is the expected output size of GREEDY after seeing half of the edges. We observe that GREEDY has a competitive ratio of at-least half and in the full version we show that this ratio is tight for worst case input graphs.[2]

2.2   Beating Half

Lemma 1 shows that we can restrict our attention to the case when the expected GREEDY size is small (proof in the full version). Theorem 4 gives an algorithm that beats half for this restricted case.

Lemma 1. *Suppose there exists an Algorithm $\mathcal{A}$ that achieves a competitive ratio of $\frac{1}{2} + \gamma$ when $\mathcal{G}(1) \leq (\frac{1}{2} + \epsilon)$ for some $\epsilon, \gamma > 0$. Then there exists an algorithm with competitive ratio at least $\frac{1}{2} + \delta$, where $\delta = \frac{\epsilon\gamma}{\frac{1}{2} + \epsilon + \gamma}$.*

Theorem 4. *If $\mathcal{G}(1) \leq (\frac{1}{2} + \epsilon)$ for some constant $\epsilon > 0$ then the MARKING-GREEDY algorithm outputs a matching of size at least $(\frac{1}{2} + \gamma)|\text{OPT}|$ in expectation, where $\gamma > 0$ is a constant.*

---

[1] We emphasize that our definition also works when i and j are non-integral.

[2] We also show that for regular graphs GREEDY is at least $(1 - \frac{1}{e})$ competitive, and that no online algorithm for OBME can be better than $\frac{69}{84} \approx 0.821$ competitive.

Before describing MARKING-GREEDY, we need the following property about the performance of GREEDY in the random arrival model — if GREEDY is bad then it makes most of its decisions quickly and incorrectly. We will be interested in the regime where $0 < \epsilon \ll f \ll 1/2$.

Lemma 2 (Hastiness property: Lemma 2 in [10]). *For any graph G if* $\mathcal{G}(1) \leq (\frac{1}{2} + \epsilon)$ *for some* $0 < \epsilon < \frac{1}{2}$, *then for any* $0 < f < 1/2$

$$\mathcal{G}(f) \geq \frac{1}{2} - \left(\frac{1}{f} - 2\right)\epsilon.$$

MARKING-GREEDY for Bipartite Matching

MARKING-GREEDY consists of two phases (see the pseudocode). In Phase (a), it runs GREEDY for the first f-fraction of the edges, but *picks* each edge *selected* by GREEDY into the final matching only with probability $(1 - p)$, where $p > 0$ is a constant. With the remaining probability p, it *marks* the edge e and its vertices, and behaves as if it had been picked. In Phase (b), which is for the remaining $1 - f$ fraction of edges, the algorithm runs GREEDY to pick edges on two restricted disjoint subgraphs G_1 and G_2, where it only considers edges incident to exactly one marked vertex in Phase (a).

Phase (a) is equivalent to running GREEDY to select elements, but then randomly dropping p fraction of the selected edges. The idea of marking some vertices (by marking an incident edge) is to "protect" them for augmentation in Phase (b). To distinguish if an edge is marked or picked, the algorithm uses auxiliary random bits Ψ that are unknown to the adversary. We assume that $\Psi(e) \sim \text{Bern}(1 - p)$ i.i.d. for all $e \in E$.

Algorithm 1. MARKING-GREEDY(G, E, π, m, Ψ)

    Phase (a)
1: Initialize S, T, N_1, N_2 to $\emptyset$
2: **for** each element $e \in E^\pi[1, fm]$ **do**   ▷ GREEDY *while picking and marking*
3:    **if** $T \cup e$ is a matching in G **then**
4:       $T \leftarrow T \cup e$ ▷ *Elements selected by* GREEDY
5:       **if** $\Psi(e) = 1$ **then**   ▷ *Auxiliary random bits* Ψ
6:          $S \leftarrow S \cup e$ ▷ *Elements picked into final solution*
    Phase (b)
7: Initialize set T_f to T. Let sets X_1, X_2 be vertices of U, V matched in T_f respectively.
8: Let G_1 be the subgraph of G induced on X_1 and $V \backslash X_2$.
9: Let G_2 be the subgraph of G induced on $U \backslash X_1$ and X_2.
10: **for** each edge $e \in (E^\pi[fm, m])$ **do**   ▷ GREEDY *on two disjoint subgraphs*
11:    **for** $i \in \{1, 2\}$ **do**
12:       **if** $e \in G_i$ and $S \cup N_i \cup e$ is a matching **then**   ▷ GREEDY *step*
13:          $N_i \leftarrow N_i \cup e$ ▷ *New edges picked*
14: **return** $S \cup N_1 \cup N_2$

Comparison to Konrad et al. [10] For the special case of bipartite matching, we can consider MARKING-GREEDY to be a variant of the streaming algorithm of [10]. For graphs where GREEDY is bad, both algorithms use the first phase to pick an approximately maximal matching T using the Hastiness Lemma. Konrad et al. [10] divides the remaining stream into two portions and uses each portion to find greedy matchings, say F_1 and F_2. Since decisions in the streaming setting are revocable, at the end of the stream they use edges in $F_1 \cup F_2$ to find sufficient number of three-augmenting paths w.r.t. T. Their algorithm is not online because it keeps all the matchings till the end. One can view the current algorithm as turning their algorithm into an online one by flipping a coin for each edge in T. In the second phase, it runs GREEDY on two random disjoint subgraphs and use the Sampling Lemma to argue that in expectation the algorithm picks sufficient number of augmenting paths.

While our online matching algorithm is simple and succinct, the main difficulty lies in extending it to OMI as the notions of marking and protecting vertices do not exist. This is also the reason why obtaining a linear time algorithm for offline matroid intersection problem, where Hastiness Lemma is not needed, had been open. Defining and proving the correct form of Sampling Lemma forms the core of our OMI analysis in Sect. 3.

Proof that MARKING-GREEDY Works for Bipartite Matching

Let G_i denote graphs G_1 or G_2 for $i \in \{1, 2\}$. For a fixed order π of the edges, graphs G_i in MARKING-GREEDY are independent of the randomness Ψ. Since the algorithm uses Ψ to pick a random subset of the GREEDY solution, this can be viewed as independently sampling each vertex matched by GREEDY in G_i. Lemma 3 shows that this suffices to pick in expectation more than the number of marked edges. In essence, we use the randomness Ψ to limit the power of an adversary deciding the order of the edges in Phase (b). While the proof follows from the more general Lemma 8, we include a simple self-contained proof for this case in the full version.

Lemma 3 (Sampling Lemma). *Consider a bipartite graph $H = (X \cup Y, \tilde{E})$ containing a matching $\tilde{I}$. Let $\Psi(x) \sim Bern(1 - p)$ i.i.d. for all $x \in X$, and define $X' = \{x \mid x \in X \text{ and } \Psi(x) = 0\}$. I.e., the vertices of X' are obtained by independently sampling each vertex in X with probability p. Let H' denote the subgraph induced on X' and Y. Then for any arrival order of the edges in H',*

$$\mathbb{E}_\Psi[\text{GREEDY}(H', \tilde{E})] \geq \frac{1}{1 + p} \left(p | \tilde{I} | \right).$$

We next prove the main lemma needed to prove Theorem 4. Setting $f = 0.07$, $p = 0.36$, and $\epsilon = 0.001$ in Lemma 4, the theorem follows by taking $\gamma > 0.05$.

Lemma 4. *For any $0 < f < 1/2$ and bipartite graph G, MARKING-GREEDY outputs a matching of expected size at least*

$$\left[(1 - p) \left(\frac{1}{2} - \left(\frac{1}{f} - 2 \right) \epsilon \right) + \frac{p}{1 + p} \left(1 - \frac{2\epsilon}{f} - f \right) \right] |\text{OPT}|.$$

Proof. We remind the reader that for any $f \in [0,1]$ and any permutation π of the edges, T_f^π denotes the matching that GREEDY produces on $E^\pi[1, fm]$. For $i \in \{1, 2\}$, let H_i denote the subgraph of G_i containing all its edges that appear in Phase (b). Let I_i denote the set of edges of OPT that appear in graph G_i. We use the following claim proved in the full version.

Claim 5.
$$\mathbb{E}_\pi\left[|I_1| + |I_2|\right] \geq \left(1 - \frac{2\epsilon}{f}\right)|\text{OPT}|.$$

For $i \in \{1, 2\}$, let $\tilde{I}_i \subseteq I_i$ denote the set of edges of OPT that appear in Phase (b) of MARKING-GREEDY, i.e., they appear in graph H_i. In expectation over uniform permutation π, at most $f|\text{OPT}|$ elements of OPT can appear in Phase (a). Hence,

$$\mathbb{E}_\pi\left[|\tilde{I}_1| + |\tilde{I}_2|\right] \geq \mathbb{E}_\pi\left[|I_1| + |I_2|\right] - f|\text{OPT}| \geq \left(1 - \frac{2\epsilon}{f} - f\right)|\text{OPT}|.$$

Marking a random subset of T_f^π independently is equivalent to marking a random subset of vertices independently. Thus, we can apply Lemma 3 to both H_1 and H_2. The expected number of edges in $N_1 \cup N_2$ is at least $\frac{p}{1+p}(|\tilde{I}_1| + |\tilde{I}_2|)$, where the expectation is over the auxilary bits Ψ that distinguishes the random set of edges marked. Taking expectations over π and noting that Phase (a) picks $(1 - p)\mathcal{G}(f)|\text{OPT}|$ edges, we have

$$\mathbb{E}_{\Psi,\pi}[|S \cup N_1 \cup N_2|] \quad = \quad \mathbb{E}_{\Psi,\pi}[|S|] + \mathbb{E}_{\Psi,\pi}[|N_1| + |N_2|]$$

$$\geq \quad \mathcal{G}(f)(1 - p)|\text{OPT}| + \frac{p}{1+p}\mathbb{E}_\pi\left[|\tilde{I}_1| + |\tilde{I}_2|\right]$$

$$\geq \left[(1 - p)\left(\frac{1}{2} - \left(\frac{1}{f} - 2\right)\epsilon\right) + \frac{p}{1+p}\left(1 - \frac{2\epsilon}{f} - f\right)\right]|\text{OPT}| \quad \text{(by Lemma 2)}.$$

3 Online Matroid Intersection

3.1 Definitions and Notation

An instance of the online matroid intersection problem $(\mathcal{M}_1, \mathcal{M}_2, E, \pi, m)$ consists of matroids $\mathcal{M}_1$ and $\mathcal{M}_2$ defined on ground set E of size m, and where the elements in E arrive according to the order defined by π. For any $1 \leq i \leq j \leq m$, let $E^\pi[i, j]$ denote the ordered set of elements of E that arrive in positions i through j according to π. For any matroid $\mathcal{M}$ on ground set E, we use $T \in \mathcal{M}$ to denote $T \subseteq E$ is an independent set in matroid $\mathcal{M}$. We use the terminology of matroid restriction and matroid contraction as defined in Oxley [18]. To avoid clutter, for any $e \in E$ we abbreviate $A \cup \{e\}$ to $A \cup e$ and $A\backslash\{e\}$ to $A\backslash e$.

We note that GREEDY is well defined even when matroids $\mathcal{M}_1$ and $\mathcal{M}_2$ are defined on larger ground sets as long as they contain E. This notation will be useful when we run GREEDY on matroids after contracting different sets in the two matroids. Since GREEDY always produces a maximal independent set, its

Algorithm 2. GREEDY $(\mathcal{M}_1, \mathcal{M}_2, E, \pi)$

1: Initialize set T to $\emptyset$
2: **for** each element $e \in E^\pi[1, |E|]$ **do**
3:      **if** $T \cup e \in \mathcal{M}_1 \cap \mathcal{M}_2$ **then**
4:          $T \leftarrow T \cup e$
5: **return** T

competitive ratio is at least half (see Theorem 13.8 in [11]). This is because an "incorrect" element creates at most two circuits in OPT, one for each matroid.

Let OPT denote a fixed maximum offline independent set in the intersection of both the matroids. For $f \in [0,1]$, let T_f^π denote the independent set that GREEDY produces after seeing the first f fraction of the edges according to order π. When clear from context, we will often abbreviate T_f^π with T_f. Let $\mathcal{G}(f) := \frac{\mathbb{E}_\pi[|T_f|]}{|\text{OPT}|}$, where π is a uniformly random chosen order.

For $i \in \{1, 2\}$, let $\text{span}_i(T) := \{e \mid (e \in E) \wedge (\text{rank}_{\mathcal{M}_i}(T \cup e) = \text{rank}_{\mathcal{M}_i}(T))\}$ denote the span of set $T \subseteq E$ in matroid $\mathcal{M}_i$. Suppose we have $T \in \mathcal{M}_i$ and $e \in \text{span}_i(T)$, then we denote the unique circuit of $T \cup e$ in matroid $\mathcal{M}_i$ by $C_i(T \cup e)$. If $i = 1$, we use $\bar{\imath}$ to denote 2, and vice versa.

3.2 Hastiness Property

Before describing our algorithm MARKING-GREEDY, we need an important hastiness property of GREEDY in the random arrival model. Intuitively, it states that if GREEDY's performance is bad then it makes most of its decisions quickly and incorrectly. This observation was first made by Konrad et al. [10] in the special case of bipartite matching. We extend this property to matroids in Lemma 5 (proof in the full version). We are interested in the regime where $0 < \epsilon \ll f \ll 1$.

Lemma 5 (Hastiness Lemma). *For any two matroids $\mathcal{M}_1$ and $\mathcal{M}_2$ on the same ground set E, let T_f^π denote the set selected by GREEDY after running for the first f fraction of elements E appearing in order π. Also, for $i \in \{1, 2\}$, let $\Phi_i(T_f^\pi) := \text{span}_i(T_f^\pi) \cap \text{OPT}$. Now for any $0 < f, \epsilon \le \frac{1}{2}$, if $\mathbb{E}_\pi[|T_1^\pi|] \le (\frac{1}{2} + \epsilon)|\text{OPT}|$ then*

$$\mathbb{E}_\pi\left[|\Phi_1(T_f^\pi) \cap \Phi_2(T_f^\pi)|\right] \le 2\epsilon |\text{OPT}| \qquad and$$

$$\mathbb{E}_\pi\left[|\Phi_1(T_f^\pi) \cup \Phi_2(T_f^\pi)|\right] \ge \left(1 - \frac{2\epsilon}{f} + 2\epsilon\right) |\text{OPT}|.$$

This implies $\mathcal{G}(f) := \frac{\mathbb{E}_\pi[|T_f^\pi|]}{|\text{OPT}|} \ge \left(\frac{1}{2} - \left(\frac{1}{f} - 2\right)\epsilon\right)$.

3.3 Beating Half for Online Matroid Intersection

Once again, we use Lemma 1 to restrict our attention to the case when the expected size of GREEDY is small. In Theorem 6, we give an algorithm that beats half for this restricted case, which when combined with Lemma 1 finishes the proof of Theorem 1.

Theorem 6. *For any two matroids $\mathcal{M}_1$ and $\mathcal{M}_2$ on the same ground set E, there exist constants $\epsilon, \gamma > 0$ and a randomized online algorithm* MARKING-GREEDY *such that if $\mathcal{G}(1) \leq \left(\frac{1}{2} + \epsilon\right)$ then* MARKING-GREEDY *outputs an independent set in the intersection of both the matroids of expected size at least $\left(\frac{1}{2} + \gamma\right) |\text{OPT}|$.*

MARKING-GREEDY **for OMI:**

Algorithm 3. MARKING-GREEDY $(\mathcal{M}_1, \mathcal{M}_2, E, \pi, m, \Psi)$

    Phase (a)
1: Initialize S, T to $\emptyset$
2: **for** each element $e \in E^\pi[1, fm]$ **do**   ▷ GREEDY *while picking and marking*
3:     **if** $T \cup e \in \mathcal{M}_1 \cap \mathcal{M}_2$ **then**
4:         $T \leftarrow T \cup e$ ▷ *Elements selected by* GREEDY
5:         **if** $\psi(e) = 1$ **then**   ▷ *Auxiliary random bits* Ψ
6:             $S \leftarrow S \cup e$ ▷ *Elements picked into the final solution*
    Phase (b)
7: Fix T_f to T and initialize sets N_1, N_2 to $\emptyset$
8: **for** each element $e \in E^\pi[fm, m]$ **do**   ▷ GREEDY *on two disjoint problems*
9:     **for** $i \in \{1, 2\}$ **do**
10:         **if** $e \in \text{span}_i(T_f)$ and $e \notin \text{span}_{\bar{i}}(T_f)$ **then**   ▷ *To ensure disjointness*
11:             **if** $(S \cup N_i \cup e \in \mathcal{M}_i)$ and $(T_f \cup N_i \cup e \in \mathcal{M}_{\bar{i}})$ **then** ▷ GREEDY *step*
12:                 $N_i \leftarrow N_i \cup e$ ▷ *Newly picked elements*
13: **return** $(S \cup N_1 \cup N_2)$

MARKING-GREEDY consists of two phases. In Phase (a), it runs GREEDY for the first f fraction of the elements, but *picks* each element *selected* by GREEDY into the final solution only with probability $(1 - p)$, where $p > 0$ is a constant. With the remaining probability p, it *marks* the element e, and behaves as if it had been selected. The idea of marking some elements in Phase (a) is that we hope to "augment" them in Phase (b). To distinguish if an element is marked or picked, the algorithm uses auxiliary random bits Ψ that are unknown to the adversary. We assume that $\Psi(e) \sim \text{Bern}(1 - p)$ i.i.d. for all $e \in E$.

In Phase (b), one needs to ensure that the augmentations of the marked elements do not conflict with each other. The crucial idea is to use the span of the elements selected by GREEDY in Phase (a) as a proxy to find two random disjoint OMI subproblems. The following Fact 7 underlies this intuition. It states that given any independent set S, we can substitute it by any other independent set contained in the span of S. In Lemma 6 we use it to prove the correctness of MARKING-GREEDY. Both Fact 7 and Lemma 6 are proved in the full version.

Fact 7. *Consider any matroid $\mathcal{M}$ and independent sets $A, B, C \in \mathcal{M}$ such that $A \subseteq \text{span}_\mathcal{M}(B)$ and $B \cup C \in \mathcal{M}$. Then, $A \cup C \in \mathcal{M}$.*

Lemma 6. MARKING-GREEDY *outputs sets* $S, N_1,$ *and* N_2 *such that*

$$(S \cup N_1 \cup N_2) \in \mathcal{M}_1 \cap \mathcal{M}_2.$$

Proof that MARKING-GREEDY **Works for OMI**

We know from Lemma 5 that $\mathcal{G}(f)$ is close to half for $\epsilon \ll f \ll 1$. In the following Lemma 7, we show that MARKING-GREEDY (which returns $S \cup N_1 \cup N_2$ by Lemma 6) gets an improvement over GREEDY. This completes the proof of Theorem 6 to give $\gamma \geq 0.03$ for $\epsilon = 0.001$, $f = 0.05$, and $p = 0.33$. The rest of the section is devoted to proving the following lemma.

Lemma 7. MARKING-GREEDY *outputs sets* $S, N_1,$ *and* N_2 *such that*

$$\mathbb{E}_{\pi,\Psi}[|S \cup N_1 \cup N_2|] \geq (1-p)\,\mathcal{G}(f)\,|\text{OPT}| + \frac{2p}{1+p}\left(1 - \frac{2\epsilon}{f} - 2\epsilon - f - \mathcal{G}(f)\right)|\text{OPT}|.$$

Proof (Lemma 7). We treat the sets $S \subseteq T_f, N_1,$ and N_2 as random sets depending on π and Ψ. Since MARKING-GREEDY ensures the sets are disjoint,

$$\mathbb{E}_{\pi,\Psi}[|S \cup N_1 \cup N_2|] = \mathbb{E}_{\pi,\Psi}[|S|] + \mathbb{E}_{\pi,\Psi}[|N_1| + |N_2|]$$
$$\geq (1-p)\,\mathcal{G}(f)\,|\text{OPT}| + \mathbb{E}_{\pi,\Psi}[|N_1| + |N_2|]. \tag{1}$$

Next, we lower bound $\mathbb{E}_{\pi,\Psi}[|N_1| + |N_2|]$ by observing that for $i \in \{1,2\}$, N_i is the result of running GREEDY on the following restricted set of elements.

Definition 1 (Sets $\tilde{E}_i$). *For* $i \in \{1,2\}$, *we define* $\tilde{E}_i$ *to be the set of elements* e *that arrive in Phase (b) and satisfy* $e \in \mathsf{span}_i(T_f)$ *and* $e \notin \mathsf{span}_{\bar{i}}(T_f)$.

It's easy to see that N_i is obtained by running GREEDY on the matroid $\mathcal{M}_i/S$ and $\mathcal{M}_{\bar{i}}/T_f$ with respect to elements in $\tilde{E}_i$, i.e. $N_i = \text{GREEDY}(\mathcal{M}_i/S, \mathcal{M}_{\bar{i}}/T_f, \tilde{E}_i)$. To lower bound $\mathbb{E}_{\pi,\Psi}[|N_1| + |N_2|]$, we use the following Sampling Lemma (see full version) that forms the core of our technical analysis. Intuitively, it says that if S is a random subset of T_f then for the obtained random OMI instance, with optimal solution of expected size $p|\tilde{I}|$, GREEDY performs better than half-competitiveness even for adversarial arrival order of ground elements.

Lemma 8 (Sampling Lemma). *Given matroids* $\mathcal{M}_1, \mathcal{M}_2$ *on ground set* E, *a set* $T \in \mathcal{M}_1 \cap \mathcal{M}_2$, *and* $\Psi(e) \sim Bern(1-p)$ *i.i.d. for all* $e \in T$, *we define set* $S := \{e \mid e \in T \text{ and } \Psi(e) = 1\}$. *I.e.,* S *is a set achieved by dropping each element in* T *independently with probability* p. *For* $i \in \{1,2\}$, *consider a set* $\tilde{E} \subseteq \mathsf{span}_i(T)$ *and a set* $\tilde{I} \subseteq \tilde{E}$ *satisfying* $\tilde{I} \in \mathcal{M}_i \cap (\mathcal{M}_{\bar{i}}/T)$. *Then for any arrival order of the elements of* $\tilde{E}$, *we have*

$$\mathbb{E}_{\Psi}[\text{GREEDY}(\mathcal{M}_i/S, \mathcal{M}_{\bar{i}}/T, \tilde{E})] \geq \frac{1}{1+p}\left(p|\tilde{I}|\right).$$

To use the Sampling Lemma, in Claim 8 we argue that in expectation there exist disjoint sets $\tilde{I}_i \subseteq \tilde{E}_i$ of "large" size that satisfy the preconditions of the Sampling Lemma (proof uses Hastiness Lemma and is deferred to full version).

Claim 8. If $\mathcal{G}(1) \leq \left(\frac{1}{2} + \epsilon\right)$ then for $i \in \{1, 2\}$ $\exists$ disjoint sets $\tilde{I}_i \subseteq \tilde{E}_i$ s.t.

(i) $\mathbb{E}_\pi\left[|\tilde{I}_1| + |\tilde{I}_2|\right] \geq 2\left(1 - \frac{2\epsilon}{f} - f - \mathcal{G}(f)\right)|\text{OPT}|.$

(ii) $\tilde{I}_i \in \mathcal{M}_i \cap (\mathcal{M}_{\bar{i}}/T_f).$

Finally, to finish the proof of Lemma 7, we use the sets $\tilde{I}_i$ from the above Claim 8 as $\tilde{I}$ and sets $\tilde{E}_i$ as $\tilde{E}$ in the Sampling Lemma 8. From Eq. (1) and Claim 8, we get

$$\mathbb{E}_{\pi,\Psi}[|S \cup N_1 \cup N_2|] \geq (1 - p)\,\mathcal{G}(f)\,|\text{OPT}| + \frac{p}{1 + p}\,\mathbb{E}_\pi\left[|\tilde{I}_1| + |\tilde{I}_2|\right]$$

$$\geq (1 - p)\,\mathcal{G}(f)\,|\text{OPT}| + \frac{2p}{1 + p}\left(1 - \frac{2\epsilon}{f} - f - \mathcal{G}(f)\right)|\text{OPT}|.$$

References

1. Aronson, J., Dyer, M., Frieze, A., Suen, S.: Randomized greedy matching. II. Random Struct. Algorithms **6**(1), 55–73 (1995)
2. Chekuri, C., Quanrud, K.: Fast approximations for matroid intersection. In: Proceedings of the Twenty-Seventh Annual ACM-SIAM Symposium on Discrete Algorithms (2016)
3. Edmonds, J.: Submodular functions, matroids, and certain polyhedra. In: Combinatorial Structures and Their Applications, pp. 69–87 (1970)
4. Epstein, L., Levin, A., Mestre, J., Segev, D.: Improved approximation guarantees for weighted matching in the semi-streaming model. SIAM J. Discrete Math. **25**(3), 1251–1265 (2011)
5. Goel, G., Mehta, A.: Online budgeted matching in random input models with applications to adwords. In: Proceedings of the Nineteenth Annual ACM-SIAM Symposium on Discrete Algorithms, pp. 982–991 (2008)
6. Hopcroft, J.E., Karp, R.M.: An $n^{5/2}$ algorithm for maximum matchings in bipartite graphs. SIAM J. Comput. **2**(4), 225–231 (1973)
7. Huang, C.-C., Kakimura, N., Kamiyama, N.: Exact and approximation algorithms for weighted matroid intersection. In: Proceedings of the Twenty-Seventh Annual ACM-SIAM Symposium on Discrete Algorithms. SIAM (2016)
8. Karande, C., Mehta, A., Tripathi, P.: Online bipartite matching with unknown distributions. In: Proceedings of the Forty-Third Annual ACM Symposium on Theory of Computing, pp. 587–596. ACM (2011)
9. Karp, R.M., Vazirani, U.V., Vazirani, V.V.: An optimal algorithm for on-line bipartite matching. In: Proceedings of the Twenty-Second Annual ACM Symposium on Theory of Computing, pp. 352–358 (1990)
10. Konrad, C., Magniez, F., Mathieu, C.: Maximum matching in semi-streaming with few passes. In: Gupta, A., Jansen, K., Rolim, J., Servedio, R. (eds.) APPROX/RANDOM -2012. LNCS, vol. 7408, pp. 231–242. Springer, Heidelberg (2012). doi:10.1007/978-3-642-32512-0_20
11. Korte, B., Vygen, J.: Combinatorial Optimization. Algorithms and Combinatorics, vol. 21. Springer, Berlin (2008)
12. Korula, N., Mirrokni, V., Zadimoghaddam, M.: Online submodular welfare maximization: greedy beats 1/2 in random order. In: Proceedings of the Forty-Seventh Annual ACM Symposium on Theory of Computing, pp. 889–898 (2015)

13. Korula, N., Pál, M.: Algorithms for secretary problems on graphs and hypergraphs. In: Albers, S., Marchetti-Spaccamela, A., Matias, Y., Nikoletseas, S., Thomas, W. (eds.) ICALP 2009. LNCS, vol. 5556, pp. 508–520. Springer, Heidelberg (2009). doi:10.1007/978-3-642-02930-1_42

14. Mahdian, M., Yan, Q.: Online bipartite matching with random arrivals: an approach based on strongly factor-revealing LPs. In: Proceedings of the Forty-Third Annual ACM Symposium on Theory of Computing, pp. 597–606 (2011)

15. Mehta, A.: Online matching and ad allocation. Theor. Comput. Sci. **8**(4), 265–368 (2012)

16. Mehta, A., Saberi, A., Vazirani, U., Vazirani, V.: Adwords and generalized online matching. J. ACM (JACM) **54**(5), 22 (2007)

17. Mehta, A., Vazirani, V.: Personal communication (2015)

18. Oxley, J.G.: Matroid Theory, vol. 3. Oxford University Press, Oxford (2006)

19. Schrijver, A.: Combinatorial Optimization: Polyhedra and Efficiency, vol. 24. Springer Science & Business Media, Heidelberg (2002)

20. Wang, Y., Wong, S.C.: Two-sided online bipartite matching and vertex cover: beating the greedy algorithm. In: Halldórsson, M.M., Iwama, K., Kobayashi, N., Speckmann, B. (eds.) ICALP 2015. LNCS, vol. 9134, pp. 1070–1081. Springer, Heidelberg (2015). doi:10.1007/978-3-662-47672-7_87

Number Balancing is as Hard as Minkowski's Theorem and Shortest Vector

Rebecca Hoberg[✉], Harishchandra Ramadas, Thomas Rothvoss,
and Xin Yang

University of Washington, Seattle, WA 98105, USA
{rahoberg,ramadas,rothvoss,yx1992}@uw.edu

Abstract. The number balancing (NBP) problem is the following: given
real numbers $a_1, \ldots, a_n \in [0, 1]$, find two disjoint subsets $I_1, I_2 \subseteq [n]$ so
that the difference $|\sum_{i \in I_1} a_i - \sum_{i \in I_2} a_i|$ of their sums is minimized.
An application of the pigeonhole principle shows that there is always a
solution where the difference is at most $O(\frac{\sqrt{n}}{2^n})$. Finding the minimum,
however, is NP-hard. In polynomial time, the *differencing algorithm* by
Karmarkar and Karp from 1982 can produce a solution with difference at
most $n^{-\Theta(\log n)}$, but no further improvement has been made since then.

In this paper, we show a relationship between NBP and Minkowski's
Theorem. First we show that an approximate oracle for Minkowski's
Theorem gives an approximate NBP oracle. Perhaps more surprisingly, we show that an approximate NBP oracle gives an approximate
Minkowski oracle. In particular, we prove that any polynomial time algorithm that guarantees a solution of difference at most $2^{\sqrt{n}}/2^n$ would give
a polynomial approximation for Minkowski as well as a polynomial factor
approximation algorithm for the Shortest Vector Problem.

1 Introduction

One of *six basic NP-complete problems* of Garey and Johnson [GJ97] is the *partition problem* that for a list of numbers $a_1, \ldots, a_n$ asks whether there is a partition of the indices so that the sums of the numbers in both partitions coincide.
Partition and related problems like knapsack, subset sum and bin packing are
some of the fundamental classical problems in theoretical computer science with
numerous practical applications; see for example the textbooks [MT90, KPP04]
and the article of Mertens [Mer06]. In this paper, we study a variant called the
number balancing problem (NBP), where the goal is to find two disjoint subsets
$I_1, I_2 \subseteq \{1, \ldots, n\}$ so that the difference $|\sum_{i \in I_1} a_i - \sum_{i \in I_2} a_i|$ is minimized.
Equivalently, given a vector of numbers $a = (a_1, \ldots, a_n) \in [0, 1]^n$, we want
to find a vector of *signs* $x \in \{-1, 0, 1\}^n \setminus \{0\}$ so that $|\langle a, x \rangle| = |\sum_{i=1}^n x_i a_i|$ is
minimized. Woeginger and Yu [WY92] studied this problem under the name

T. Rothvoss—Supported by NSF grant 1420180 with title *"Limitations of convex
relaxations in combinatorial optimization"*, an Alfred P. Sloan Research Fellowship
and a David and Lucile Packard Foundation Fellowship.

© Springer International Publishing AG 2017
F. Eisenbrand and J. Koenemann (Eds.): IPCO 2017, LNCS 10328, pp. 254–266, 2017.
DOI: 10.1007/978-3-319-59250-3_21

"equal-subset-sum" and showed that it is NP-hard to decide whether there are two disjoint subsets that sum up to the exact same value. This version has also been extensively studied in combinatorics [Lun88,Boh96,LY11].

On the positive side, it is not hard to prove that there is always a solution with exponentially small error. Suppose that $a_1, \ldots, a_n \in [0,1]$. Consider the list of 2^n many numbers $\sum_{i=1}^{n} a_i x_i$ for all $x \in \{0,1\}^n$. All these numbers fall into the interval $[0,n]$, hence by the *pigeonhole principle*, we can find two distinct vectors $x, x' \in \{0,1\}^n$ with $\left| \sum_{i=1}^{n} a_i x_i - \sum_{i=1}^{n} a_i x_i' \right| \leq \frac{n}{2^n - 1}$. Then $x - x'$ gives the desired solution. Note that the bound can be slightly improved to $O(\frac{\sqrt{n}}{2^n})$ by using the fact that due to *concentration of measure* effects, for a constant fraction of vectors $x \in \{0,1\}^n$, the sums $\sum_{i=1}^{n} a_i x_i$ fall into an interval of length $\sqrt{n}$ (instead of n).

However, since these arguments rely on the pigeonhole principle, they are non-constructive. Restricting the non-constructive argument to polynomially many "pigeons" provides a simple polynomial time algorithm to find at least an $x \in \{-1,0,1\}^n \backslash \{0\}$ with $|\langle a, x \rangle| \leq \frac{1}{\mathrm{poly}(n)}$ for an arbitrarily small polynomial. Interestingly, the only known polynomial time algorithm that gives a better guarantee is Karmarkar and Karp's *differencing algorithm* [KK82] which provides the bound $|\langle a, x \rangle| \leq n^{-c \log(n)}$ for some constant $c > 0$. Their algorithm uses a recursive scheme; find $\Theta(n)$ pairs of numbers a_i of distance at most $\Theta(\frac{1}{n})$ and create an instance consisting of their differences, then recurse.

This leads to the natural question: *Given $a_1, \ldots, a_n \in [0,1]$, what upper bound on $|\sum_{i=1}^{n} a_i x_i|$ can be guaranteed if $x \in \{-1,0,1\}^n \backslash \{0\}$ is to be chosen in polynomial time?* While answering this question directly seems out of reach, we note that NBP falls into the class PPP [Pap94], where good solutions are known to exist due to the pigeonhole principle. It is reasonable therefore to study the relationship between this problem and other problems in PPP.

Recall that given linearly independent vectors $b_1, \ldots, b_n \in \mathbb{R}^n$, a *(full rank) lattice* is the set $\Lambda := \{\sum_{i=1}^{n} \lambda_i b_i : \lambda_i \in \mathbb{Z} \; \forall i = 1, \ldots, n\}$. The set $\{b_1, \ldots, b_n\}$ is called a *basis* for Λ and we define $\det(\Lambda) := |\det(B)|$. For any lattice $\Lambda \subseteq \mathbb{R}^n$ with $\det(\Lambda) \geq 1$, *Minkowski's Theorem* tells us that any symmetric convex body $K \subseteq \mathbb{R}^n$ of volume at least 2^n must intersect $\Lambda \backslash \{0\}$, see for example [Mat02]. This theorem is proven by placing translates of $\frac{1}{2}K$ at any lattice point and then inferring an overlap due to the pigeonhole principle. Again, one can consider the algorithmic question: *given a symmetric convex body K with volume at least 2^n, for what factor ρ can one be guaranteed to find an $x \in (\rho K) \cap \mathbb{Z}^n \backslash \{0\}$ in polynomial time?*

We would like to point out that this factor ρ is within a polynomial factor of the approximability of the Shortest Vector Problem (SVP), the problem of finding the shortest[1] nonzero vector in a lattice. One direction follows from the fact K can be sandwiched between two ellipsoids that differ by a factor of $\sqrt{n}$ [Joh48]. In the other direction, a reduction of Lenstra and Schnorr shows that given a polynomial-time oracle to find a vector of length at most $f(n)$ in

---

[1] SVP can be defined for any norm, but anywhere the norm is not specified we consider the Euclidean norm.

a lattice with $\det(\Lambda) \leq 1$, there is a polynomial-time algorithm to find a vector within $f(n)^2$ of the shortest vector [Lov86]. This is a nontrivial reduction that uses the assumed oracle both on the original lattice and on its dual. Note that Minkowski's theorem guarantees the existence of a lattice vector of length at most $O(\sqrt{n})$.

The complexity of SVP is of great theoretical and practical interest. As a rarity in theoretical computer science, SVP admits (NP ∩ coNP)-certificates for a value that is at most a factor $O(\sqrt{n})$ away from the optimum [AR05], while the best known hardness under reasonable complexity assumptions lies at a sub-polynomial bound of $n^{\Theta(1/\log\log n)}$ [HR07]. The famous LLL-algorithm [LLL82] can find a $2^{n/2}$-approximation in polynomial time (the generalized *block reduction method* of Schnorr [Sch87] brings the factor down to $2^{n\log\log(n)/\log(n)}$). On the other hand, a polynomial factor approximation of SVP would be enough to break lattice-based cryptosystems [Ajt96, MR09].

It is not hard to use an *exact* oracle for Minkowski's Theorem to find a good number balancing solution, since the body $K := \{x \in (-2,2) : |\sum_{i=1}^{n} x_i a_i| \leq \Theta(\frac{n}{2^n})\}$ has volume at least 2^n. However, it is not clear how we could use an *approximate* oracle. For example, it is known that the LLL-algorithm can be used to find a nonzero integer vector $x \in \rho K$ for a factor of $\rho = \text{poly}(n) \cdot 2^{n/2}$. While the error guarantee of $|\sum_{i=1}^{n} a_i x_i| \leq \text{poly}(n) \cdot 2^{-n/2}$ outperforms the Karmarkar-Karp algorithm, we only know that $\|x\|_{\infty} < 2\rho$, which means that x will not be a valid solution if $\rho > 1$.[2] This leads us to the next question: *what factor ρ is needed for Minkowski's Theorem to improve over Karmarkar-Karp's bound?*

We have seen that in a certain sense NBP can be reduced to an oracle for Minkowski's Theorem, and in fact we will show that there is also a direct reduction to SVP in the ℓ^{∞} norm. This brings us to the question about the reverse: *given an oracle that solves NBP within an exponentially small error, can this give a non-trivial oracle for the Shortest Vector Problem or Minkowski's Theorem?*

Contribution. In this work, we provide some answers to the questions raised above, by relating the complexity of the number balancing problem to Minkowski's Theorem and the Shortest Vector Problem. First we give the precise definitions of the problems we will consider.

Definition 1. *Suppose $p \in [1, \infty]$ with $B^p(\mathbf{0}, 1)$ the closed unit ball in the ℓ^p norm. For $\delta \geq \frac{\Omega(\sqrt{n})}{2^n}$ and $\rho \geq 1$, we define the following problems.*

- *δ-**NBP**: Given $a \in [0,1]^n$, find $x \in \{-1,0,1\}^n$ with $|\langle a, x \rangle| \leq \delta$.*
- *ρ-**Minkowski Problem**: Given a lattice Λ and a symmetric convex body[3] $K \subseteq \mathbb{R}^n$ with $\text{vol}_n(K) \geq 2^n \det(\Lambda)$, find a vector $x \in (\rho K) \cap \Lambda \backslash \{\mathbf{0}\}$.*
- *ρ-**PromiseSVP**$_p$: Given a lattice $\Lambda \subset \mathbb{R}^n$ with $\text{vol}(B^p(\mathbf{0}, 1)) \geq 2^n \det(\Lambda)$, find a vector $x \in \Lambda$ with $\|x\|_p \leq \rho$.*

---

[2] If $\rho \leq 2 - \varepsilon$, then one can still obtain an error of $|\sum_{i=1}^{n} a_i x_i| \leq 2^{-\Theta(\varepsilon n)}$, but this breaks down if $\rho \geq 2$.

[3] We assume K is given to us by a separation oracle.

As already discussed, δ-NBP and the ρ-Minkowski Problem will always have a solution by nonconstructive arguments. Moreover, we notice that ρ-PromiseSVP$_p$ is just the ρ-Minkowski Problem on an ℓ^p ball, and so it is also guaranteed to have a solution. Notice also that ρ-PromiseSVP$_p$ would also follow immediately from a ρ-approximation to SVP in the ℓ^p norm, since a short enough vector is guaranteed to exist. We would like to stress that polynomial-time algorithms for any of the three problems are not known to be inconsistent with P $\neq$ NP. The hardness results of SVP do not apply to ρ-PromiseSVP$_p$ since we do not require it to give a shortest vector of the lattice.

We provide the following reduction:

Theorem 1. *Suppose there is a polynomial-time algorithm for the ρ-Minkowski problem for polytopes K with $O(n)$ facets. Then there is a polynomial-time algorithm for δ-NBP where $\delta := 2^{-n^{\Theta(1/\rho)}}$.*

In fact, to obtain an algorithm for δ-NBP, it suffices to have an approximate Minkowski oracle for the linear transformation of a cube, which is equivalent to an oracle for ρ-PromiseSVP$_\infty$.

Theorem 2. *Suppose that there is a polynomial-time algorithm for ρ-PromiseSVP$_\infty$. Then there is a polynomial-time algorithm for δ-NBP, where $\delta := 2^{-n^{\Theta(1/\rho)}}$.*

In particular an oracle for $\rho \leq c' \log(n)/\log\log(n)$ would imply an improvement over Karmarkar-Karp's algorithm, where $c' > 0$ is a small enough constant.

Finally, we can also prove that an oracle with exponentially small error for number balancing would provide an approximation for Minkowski's problem.

Theorem 3. *Suppose that there is a polynomial-time algorithm for δ-NBP with $\delta \leq 2^{\sqrt{n}}/2^n$. Then there is a polynomial-time algorithm for the ρ-Minkowski problem for $\rho = O(n^5)$. Here it suffices to have a separation oracle for the convex body $K \subseteq \mathbb{R}^n$.*

In fact, we will show that we can get within a $O(n^{4.5})$ factor of an ellipsoid of the volume of a unit ball, and so using the reduction of Lovasz we can guarantee a $O(n^9)$ approximation for SVP.

2   Reducing Number Balancing to Minkowski's Theorem

In this section we will show how to solve NBP with an oracle for Minkowski's problem. The idea is to consider a hypercube intersected with the constraint $|\langle a, x \rangle| \leq \delta$, and to show that this set has large enough volume. If we have an exact Minkowski oracle, this gives us $x \in \{-1, 0, 1\}^n \setminus \{0\}$ as desired. Here we state a more general version which uses only a ρ-approximate Minkowski oracle, and then show how we can use this more general version to solve NBP with a weaker bound. We present the proof in the full version of this paper.

Theorem 4. *Suppose we have a polynomial-time algorithm for the ρ-Minkowski problem, and let $k > 0$ be any positive integer. Then, for any $a \in [0,1]^n$, there is a polynomial-time algorithm to find $x \in \mathbb{Z}^n \backslash \{0\}$ with $\|x\|_\infty \leq k$ and so that*
$$|\langle a, x \rangle| \leq n \left(\frac{\rho}{k+1} \right)^{n-1}.$$

Suppose, for instance, that $\rho = (2 - \epsilon)$ for some $\epsilon \in (0,1]$. Then we can pick $k = 1$ and get $|\langle a, x \rangle| \leq 2^{-\Theta(\epsilon n)}$ with $\|x\|_\infty \leq 1$ and $x \in \mathbb{Z}^n \backslash \{0\}$. However, this line of arguments breaks down for $\rho \geq 2$ as these would in general not produce feasible solutions for number balancing.

Instead of using an oracle for the ρ-Minkowski Problem one can directly use an oracle for ρ-PromiseSVP$_\infty$. We need the following theorem, the proof of which we present in the full version of this paper:

Theorem 5. *Suppose that there is a polynomial-time algorithm for ρ-PromiseSVP$_\infty$. Then for any $a \in [0,1]^n$ there is a polynomial-time algorithm to find $x \in \mathbb{Z}^n \backslash \{0\}$ with $\|x\|_\infty \leq k$ and so that $|\langle a, x \rangle| \leq 2nk\rho(\frac{\rho}{k})^n$.*

However, we still face a problem in the case $\rho \geq 2$. It turns out that we can design a *recursive self-reduction* to allow us to use larger ρ. The main technical argument is to transform an algorithm that finds $x \in \mathbb{Z}^n \backslash \{0\}$ with $\|x\|_\infty \leq k$ for $k \geq 2$ into an algorithm that finds vectors $x \in \mathbb{Z}^n \backslash \{0\}$ with $\|x\|_\infty \leq \frac{k}{2}$, with a bounded decay in the error $|\langle a, x \rangle|$. Applying this recursively gives the following lemma.

Lemma 1. *Suppose that there is a polynomial-time algorithm that for any $a' \in [0,1]^n$ finds a vector $x' \in \{-k, \ldots, k\}^n \backslash \{0\}$ with $|\langle a', x' \rangle| \leq 2^{-n}$. If $k \leq \frac{\log n}{6 \log \log n}$, then there is also a polynomial-time algorithm that for any $a \in [0,1]^n$ finds a vector $x \in \mathbb{Z}^n$ with $|\langle a, x \rangle| \leq 2^{-n^{\frac{1}{3k}}}$ and $x \in \{-1,0,1\}^n \backslash \{0\}$.*

Before we go through the self reduction, we show how Lemma 1 gives Theorems 1 and 2.

Proof (Theorems 1 and 2). If $\rho \geq \frac{\log n}{48 \log \log n}$, then $2^{-n^{\Theta(1/\rho)}} = 2^{-\log^{O(1)} n}$. By choosing a proper constant on the exponent, this can be achieved with the Karmarkar-Karp algorithm. So we only need to work with $\rho < \frac{\log n}{48 \log \log n}$.

Now suppose we have a polynomial-time algorithm for the ρ-Minkowski Problem (resp. ρ-PromiseSVP$_\infty$). If we take $k = 3\rho$, Theorem 4 (resp. Theorem 5) gives a polynomial-time algorithm to find x with $\|x\|_\infty \leq k$ and $|\langle a, x \rangle| \leq 2^{-n}$.

Moreover, $k = 3\rho \leq \frac{\log n}{16 \log \log n}$, and hence the condition of Lemma 1 is satisfied. Then the bound given by Lemma 1 is $2^{-n^{\frac{1}{3k}}} \leq 2^{-n^{\Theta(1/\rho)}}$. □

We now prove Lemma 1. The way we do the self-reduction is the following. We partition our set of n numbers into subsets of size $\sqrt{n}$. First, for each subset ℓ, we find a number $b_\ell \neq 0$ for which we can (approximately) express $b_\ell, 2b_\ell, \ldots, kb_\ell$ as linear combinations of elements of that subset using only coefficients in $\{-\lfloor \frac{k}{2} \rfloor, \ldots \lfloor \frac{k}{2} \rfloor\}$. We then run our assumed algorithm on $b_1, \ldots, b_{\sqrt{n}}$

to obtain $\boldsymbol{y} \in \{-k, \ldots, k\}^n$ with $\langle \boldsymbol{b}, \boldsymbol{y} \rangle = \sum_{\ell=1}^{\sqrt{n}} y_\ell b_\ell$ being small. Since each of the summands can be expressed more efficiently in terms of our original set of numbers, we obtain a good solution $\boldsymbol{x}$ with coefficients in $\{-\lfloor \frac{k}{2} \rfloor, \ldots \lfloor \frac{k}{2} \rfloor\}$.

The following two lemmas go through this argument more precisely. Note that the interesting parameter choice is $r := \lceil k/2 \rceil$, so that the size of the coefficients is halved.

Lemma 2. *Let* $r, k \in \mathbb{N}$ *be parameters with* $0 < r < k$ *and let* $\delta \geq 0$. *Let* $\alpha_1, \ldots, \alpha_k \in \mathbb{R}$ *so that* $|\sum_{i=1}^{k} i \cdot \alpha_i| \leq \delta$ *and abbreviate* $\beta := \alpha_r + \ldots + \alpha_k$. *Then for any* $j \in \{0, \ldots, k\}$ *one can find coefficients* $\lambda_{i,j} \in \mathbb{Z}$ *with* $|\lambda_{i,j}| \leq \max\{r-1, k-r\}$ *and* $|j \cdot \beta - \sum_{i=1}^{k} \lambda_{i,j} \alpha_i| \leq \delta$.

Proof. By symmetry it suffices to consider $j \geq 0$. For $j \in \{0, \ldots, r-1\}$, we can obviously write

$$j \cdot \beta = j \cdot \alpha_r + \ldots + j \cdot \alpha_k.$$

Now consider $j \in \{r, \ldots, k\}$. The trick is to use that

$$j \cdot \beta - \sum_{i=1}^{k} i\alpha_i = \sum_{i=r}^{k} j \cdot \alpha_i - \sum_{i=1}^{k} i\alpha_i = \sum_{i=r}^{k} (j-i)\alpha_i + \sum_{i=1}^{r-1} (-i)\alpha_i.$$

If we inspect the size of the used coefficients, then for $i \in \{1, \ldots, r-1\}$ we have $|-i| \leq r-1$ and for $i \in \{r, \ldots, k\}$ we have $|j-i| \leq k-r$.   □

Lemma 3. *Let* $k, r \in \mathbb{N}$ *be parameters with* $0 < r < k$. *Let* $f : \mathbb{N} \to \mathbb{R}$ *be a nonnegative function such that* $f(n) \geq 4 \log n$. *Suppose that there is a polynomial-time algorithm that for any* $\boldsymbol{a}' \in [0,1]^n$ *finds a vector* $\boldsymbol{x}' \in \{-k, \ldots, k\}^n \backslash \{\boldsymbol{0}\}$ *with* $|\langle \boldsymbol{a}', \boldsymbol{x}' \rangle| \leq 2^{-f(n)}$. *Then there is also a polynomial-time algorithm that for any* $\boldsymbol{a} \in [0,1]^n$ *finds a vector* $\boldsymbol{x} \in \mathbb{Z}^n \backslash \{\boldsymbol{0}\}$ *with* $|\langle \boldsymbol{a}, \boldsymbol{x} \rangle| \leq 2^{-2f(\lfloor \sqrt{n} \rfloor)/3}$ *and* $\|\boldsymbol{x}\|_\infty \leq \max\{r-1, k-r\}$.

Proof. Let $\boldsymbol{a} \in [0,1]^n$ be the given vector of numbers. To keep notation simpler, we will assume n is a perfect square. If it is not, we can replace $\sqrt{n}$ by $\lfloor \sqrt{n} \rfloor$. Split $[n]$ into blocks $I_1, \ldots, I_{\sqrt{n}}$ each of size $|I_\ell| = \sqrt{n}$. For each block I_ℓ we use the oracle to find a vector $\boldsymbol{x}_\ell \in \{-k, \ldots, k\}^n \backslash \{\boldsymbol{0}\}$ with $\mathrm{supp}(\boldsymbol{x}_\ell) \subseteq I_\ell$ so that $|\langle \boldsymbol{a}, \boldsymbol{x}_\ell \rangle| \leq 2^{-f(\sqrt{n})}$. If for any ℓ one has $\|\boldsymbol{x}_\ell\|_\infty \leq r-1$, then we simply return $\boldsymbol{x} := \boldsymbol{x}_\ell$ and are done. Otherwise, we write the vector as $\boldsymbol{x}_\ell = \sum_{i=1}^{k} i \cdot \boldsymbol{x}_{\ell,i}$ with vectors $\boldsymbol{x}_{\ell,1}, \ldots, \boldsymbol{x}_{\ell,k} \in \{-1, 0, 1\}^n$. Note that these vectors will have disjoint support and $\mathrm{supp}(\boldsymbol{x}_{\ell,1}), \ldots, \mathrm{supp}(\boldsymbol{x}_{\ell,k}) \subseteq I_\ell$. Moreover we know that for every ℓ there is at least one index $i \in \{r, \ldots, k\}$ with $\boldsymbol{x}_{\ell,i} \neq \boldsymbol{0}$.

Now define a vector $\boldsymbol{b} \in \mathbb{R}^{\sqrt{n}}$ with $b_\ell := \sum_{i=r}^{k} \langle \boldsymbol{a}, \boldsymbol{x}_{\ell,i} \rangle$. Note that if for any ℓ we have $|b_\ell| \leq 2^{-f(\sqrt{n})}$, then we can set $\boldsymbol{x} = \sum_{i=r}^{k} \boldsymbol{x}_{\ell,i}$ and we are done. Therefore we may assume that $|b_\ell| > 2^{-f(\sqrt{n})}$ for all ℓ. Also note that since the $\boldsymbol{x}_{\ell,i}$ have disjoint support, we have $\|\boldsymbol{b}\|_\infty \leq \sqrt{n}$. We run the oracle again to find a vector $\boldsymbol{y} \in \{-k, \ldots, k\}^{\sqrt{n}} \backslash \{\boldsymbol{0}\}$ so that $|\langle \boldsymbol{b}, \boldsymbol{y} \rangle| \leq \sqrt{n} \cdot 2^{-f(\sqrt{n})}$. For each block $\ell \in [\sqrt{n}]$ we can use Lemma 2 to find integer coefficients $\lambda_{\ell,i}$ with $|\lambda_{\ell,i}| \leq$ $\max\{r-1, k-r\}$ so that

$$\left| y_\ell \cdot b_\ell - \sum_{i=1}^{k} \lambda_{\ell,i} \cdot \langle a, x_{\ell,i} \rangle \right| \leq 2^{-f(\sqrt{n})}.$$

We define

$$x := \sum_{\ell=1}^{\sqrt{n}} \sum_{i=1}^{k} \lambda_{\ell,i} x_{\ell,i}.$$

Then $\|x\|_\infty \leq \max\{r-1, k-r\}$ since the $x_{\ell,i}$'s have disjoint support and $\|x_{\ell,i}\|_\infty \leq 1$ for all ℓ, i. Moreover, since there is some $y_\ell \neq 0$ and $|b_\ell| > 2^{-f(\sqrt{n})}$, we have $x \neq 0$.

Finally we inspect that

$$|\langle a, x \rangle| \leq |\langle y, b \rangle| + \sum_{\ell=1}^{\sqrt{n}} \left| y_\ell b_\ell - \sum_{i=1}^{k} \lambda_{\ell,i} \langle a, x_{\ell,i} \rangle \right| \leq 2\sqrt{n} \cdot 2^{-f(\sqrt{n})} \leq 2^{-2f(\sqrt{n})/3}.$$

The last line comes from the fact that when $f(n) \geq 4\log n$, we have $2\sqrt{n} \leq 2^{\frac{2}{3}\log n} = 2^{\frac{4}{3}\log \sqrt{n}} \leq 2^{\frac{1}{3}f(\sqrt{n})}$. □

Now we can apply Lemma 3 recursively to prove Lemma 1.

Proof (Lemma 1). Suppose $k \leq \frac{\log n}{6\log\log n}$ and set $r = \lceil k/2 \rceil$. Consider the function $f_t(n) = 2^{-t}n^{2^{-t}}$ for $t = 0, \ldots, \lceil \log k \rceil$, and notice that $2^{-t} \geq \frac{1}{2k} \geq 2\frac{\log\log n}{\log n}$. Therefore we have

$$f_t(n) = 2^{-t}n^{2^{-t}} \geq \frac{1}{2k}n^{1/2k} \geq (2\frac{\log\log n}{\log n}) \cdot \log^2 n \geq 4\log n.$$

Finally, notice that $\frac{2}{3}f_t(\lfloor\sqrt{n}\rfloor) \geq \frac{1}{2}f_t(\sqrt{n}) = f_{t+1}(n)$. We are now able to apply Lemma 3. In particular, suppose we have an oracle to find x with $\|x\|_\infty \leq 2^{-t}k$ and $|\langle a, x \rangle| \leq 2^{-f_t(n)}$. Then Lemma 3 gives us an oracle to find x with $\|x\|_\infty \leq 2^{-(t+1)}k$ and $|\langle a, x \rangle| \leq 2^{-f_{t+1}(n)}$.[4] Running this $\lceil \log k \rceil \leq \log 2k$ times gives a bound of $2^{-f_{\log 2k}(n)} = 2^{-n^{1/2k}/2k} \leq 2^{-n^{1/3k}}$. Here we use the fact that when $k \leq \frac{1}{6}\frac{\log n}{\log\log n}$, we have $\frac{n^{\frac{1}{2k}}}{2k} \geq n^{\frac{1}{3k}}$. □

3 Reducing Minkowski's Theorem to Number Balancing

In this section we show that for small enough δ, an oracle for δ-NBP can be used to design an algorithm for ρ-Minkowski's Problem, where ρ is polynomial in n. The first helpful insight is that any symmetric convex body can be approximated

---

[4] Here we ignore the dependence of t on n - notice that t is nondecreasing in n, so replacing $t(n)$ by $t(\sqrt{n})$ only increases $f_t(n)$.

within a factor of $\sqrt{n}$ using an *ellipsoid* [Lov90, GLS12]. Recall that an ellipsoid is a set of the form

$$\mathcal{E} = \left\{ x \in \mathbb{R}^n \mid \sum_{i=1}^{n} \frac{1}{\lambda_i^2} \cdot \langle x, a_i \rangle^2 \leq 1 \right\} \tag{1}$$

with an *orthonormal basis* $a_1, \ldots, a_n \in \mathbb{R}^n$ defining the *axes* and positive coefficients $\lambda_1, \ldots, \lambda_n > 0$ that describe the *lengths of the axes*[5]. Overall, our reduction will operate in two steps:

(i) By combining *John's Theorem* with *lattice basis reduction*, we can show that it suffices to find integer points in an ellipsoid that is *well-rounded*, meaning that the lengths of the axes are bounded.
(ii) We show that a number balancing oracle allows a self-reduction to a generalized form where inner products with n vectors have to be minimized and additionally the solution space is $\mathbb{Z}^n$ instead of $\{-1, 0, 1\}^n$.

We begin by proving (*ii*) and postpone (*i*) until the end of this section.

3.1   A Self-reduction to a Generalized Form of Number Balancing

Recall that for $\delta > 0$, we defined δ-NBP as follows: given $a \in [0, 1]$, find a vector $x \in \{-1, 0, 1\}^n \setminus \{0\}$ with $|\langle a, x \rangle| \leq \delta$. Notice that we may allow for vectors $a \in [-1, 1]^n$ without changing the problem, as one can flip the signs of x as needed to accommodate for the changes in sign of a. The main technical result of this section is the following reduction:

Theorem 6. *Suppose there is a polynomial-time algorithm for δ-NBP with $\delta = \frac{g(n)}{2^n}$ and $g(n) \leq 2^{n/2}$. Then there is a polynomial-time algorithm that on input $a_1, \ldots, a_n \in [-1, 1]^n$ and $0 < \lambda_1 \leq \ldots \leq \lambda_n \leq 2^n$ with $\prod_{i=1}^{n} \lambda_i \geq 1$, finds a vector $x \in \mathbb{Z}^n \setminus \{0\}$ with*

$$|\langle x, a_i \rangle| \leq O(n^4) \cdot \lambda_i \cdot g(4n^2)^{1/n} \quad \forall i = 1, \ldots, n.$$

In particular if $g(n) \leq 2^{\sqrt{n}}$, then the right hand side in Theorem 6 simplifies to just $O(n^4) \cdot \lambda_i$. We will show this by introducing two extensions of the number balancing oracle. The first extension gives a weaker bound in terms of the error parameter, but allows for multiple vectors in $[-1, 1]^n$. In the second extension, we extend the range of coefficients from $\{-1, 0, 1\}$ to $\{-Q, \ldots, Q\}$ which leads to a much stronger error bound.

Lemma 4. *Suppose there is a polynomial-time algorithm for δ-NBP. Then there is a polynomial-time algorithm that given an input $a_1, \ldots, a_k \in [-1, 1]^n$ and parameters $\delta_1, \ldots, \delta_k \leq \frac{1}{2}$ with $\prod_{i=1}^{k} \delta_i \geq \delta$ finds a vector $x \in \{-1, 0, 1\}^n \setminus \{0\}$ with $|\langle a_i, x \rangle| \leq 2n^2 \delta_i$ for all $i = 1, \ldots, k$.*

---

[5] Strictly speaking, the length of axis i is $2\lambda_i$, but we will continue calling λ_i the "axis length".

Proof. The idea is that we will discretize all of the vectors and then run our oracle on their sum. The vector that we obtain will then have small inner product with all of the $\boldsymbol{a}_i$. To define the discretization $\tilde{\boldsymbol{a}}_i$, round elements of $\boldsymbol{a}_i$ down to the nearest multiple of $2n\delta_i$, and then multiply by $\prod_{j<i}\delta_j$. Defining $\tilde{\boldsymbol{a}}_i$ this way, notice that for all i we have $|\langle \tilde{\boldsymbol{a}}_i, \boldsymbol{x} \rangle| \leq n \prod_{j<i}\delta_j$ for any $\mathbf{x}$ satisfying $\|\mathbf{x}\|_\infty \leq 1$.

Now let $\boldsymbol{c} = \tilde{\boldsymbol{a}}_1 + \ldots + \tilde{\boldsymbol{a}}_k$. By our oracle, we can find $\boldsymbol{x} \in \{-1, 0, 1\}^n \backslash \{\boldsymbol{0}\}$ with $|\langle \boldsymbol{c}, \boldsymbol{x} \rangle| \leq \delta$. Recall that $\delta \leq \prod_{i=1}^k \delta_i \leq n \prod_{j\leq k}\delta_j$. We have

$$|\langle \tilde{\boldsymbol{a}}_1, \boldsymbol{x} \rangle| \leq |\langle \tilde{\boldsymbol{a}}_2, \boldsymbol{x} \rangle| + \ldots + |\langle \tilde{\boldsymbol{a}}_k, \boldsymbol{x} \rangle| + |\langle \boldsymbol{c}, \boldsymbol{x} \rangle| \leq n\delta_1 + n\prod_{j\leq 2}\delta_j + \ldots + n\prod_{j\leq k}\delta_j$$

$$\leq n\delta_1 \cdot \left(1 + \frac{1}{2} + \ldots + \frac{1}{2^k}\right) < 2n\delta_1.$$

Therefore $|\langle \tilde{\boldsymbol{a}}_1, \boldsymbol{x} \rangle| = 0$. Similarly, for $1 < i \leq k$, if $|\langle \tilde{\boldsymbol{a}}_1, \boldsymbol{x} \rangle|, \ldots, |\langle \tilde{\boldsymbol{a}}_{i-1}, \boldsymbol{x} \rangle| = 0$, then we have

$$|\langle \tilde{\boldsymbol{a}}_i, \boldsymbol{x} \rangle| \leq |\langle \tilde{\boldsymbol{a}}_{i+1}, \boldsymbol{x} \rangle| + \ldots + |\langle \tilde{\boldsymbol{a}}_k, \boldsymbol{x} \rangle| + |\langle \boldsymbol{c}, \boldsymbol{x} \rangle| < 2n\prod_{j\leq i}\delta_j,$$

and hence $|\langle \tilde{\boldsymbol{a}}_i, \boldsymbol{x} \rangle| = 0$ for all i. Notice that by definition of $\tilde{\boldsymbol{a}}_i$ we have $\|\prod_{j<i}\delta_j \boldsymbol{a}_i - \tilde{\boldsymbol{a}}_i\|_\infty \leq 2n \prod_{j\leq i}\delta_j$. Therefore $|\langle \prod_{j<i}\delta_j \boldsymbol{a}_i, \boldsymbol{x} \rangle| \leq 2n^2 \prod_{j\leq i}\delta_j$, and so we can conclude that $|\langle \boldsymbol{a}_i, \boldsymbol{x} \rangle| \leq 2n^2\delta_i$. $\qquad\square$

Now we come to a second reduction that takes the oracle constructed in Lemma 4 as a starting point:

Lemma 5. *Assume there exists a polynomial-time algorithm for δ-NBP where $\delta = f(n)$. Let $\boldsymbol{a}_1, \ldots, \boldsymbol{a}_k \in [-1, 1]^n$ be given with parameters $\delta_1, \ldots, \delta_k \leq \frac{1}{2}$ and a number Q that is a power of 2 and satisfies $\prod_{i=1}^k \delta_i \geq f(n \log Q)$. Then in polynomial time we can find a vector $\boldsymbol{x} \in \{-Q, \ldots, Q\}^n \backslash \{\boldsymbol{0}\}$ with $|\langle \boldsymbol{a}_i, \boldsymbol{x} \rangle| \leq \delta_i Q \cdot 2(n \log Q)^2$ for all $i = 1, \ldots, k$.*

Proof. For each i, we define $\boldsymbol{b}_i \in [-1, 1]^{n \log Q}$ by $\boldsymbol{b}_i(j, \ell) = \boldsymbol{a}_i(j)2^{-\ell}$ for $j = 1, \ldots, n$ and $\ell = 1, \ldots, \log Q$. Since $\prod_{i=1}^k \delta_i \geq f(n \log Q)$, we can apply Lemma 4 to find $\boldsymbol{y} \in \{-1, 0, 1\}^{n \log Q} \backslash \{\boldsymbol{0}\}$ with $|\langle \boldsymbol{b}_i, \boldsymbol{y} \rangle| \leq \delta_i \cdot 2(n \log Q)^2$.

Now define $\boldsymbol{x} \in \{-Q, \ldots, Q\}^n \backslash \{\boldsymbol{0}\}$ by $x_j := Q \sum_{\ell=1}^{\log Q} 2^{-\ell} y_{j\ell}$. Then for $i = 1, \ldots, k$ we have

$$\delta_i \cdot (2n \log Q)^2 \geq |\langle \boldsymbol{b}_i, \boldsymbol{y} \rangle| = \left| \sum_{j=1}^n \underbrace{\sum_{\ell=1}^{\log Q} y_{j\ell} 2^{-\ell}}_{=x_j/Q} a_i(j) \right| = \frac{1}{Q} \cdot |\langle \boldsymbol{a}_i, \boldsymbol{x} \rangle|$$

and rearranging gives the claim. $\qquad\square$

Finally we come to the proof of Theorem 6.

Proof (Theorem 6). Suppose that the oracle has parameter $f(n) = \rho(n)/2^n$. Suppose that $\boldsymbol{a}_1, \ldots, \boldsymbol{a}_n \in [-1,1]^n$ and $\lambda_1, \ldots, \lambda_n > 0$ with $\prod_{i=1}^n \lambda_i \geq 1$. We choose $Q := 2^{4n}$, which is a power of 2. Define $\delta_i = \lambda_i \cdot f(n \log Q)^{1/n}$. Note that $\delta_i \leq 2^{3n/2} \cdot f(4n^2)^{1/n} \leq \frac{1}{2}$ since $f(n) \leq 2^{-n/2}$. Then $\prod_{i=1}^n \delta_i \geq f(n \log Q)$, and so by Lemma 5 we can find $\boldsymbol{y} \in \{-Q, \ldots, Q\}^n \backslash \{\boldsymbol{0}\}$ with

$$|\langle \boldsymbol{a}_i, \boldsymbol{y} \rangle| \leq Q\delta_i \cdot 2(n \log Q)^2 \leq Q\lambda_i \cdot f(n \log Q)^{1/n} \cdot 2(n \log Q)^2$$
$$= \lambda_i \cdot \rho(4n^2)^{1/n} \cdot 2 \cdot (4n^2)^2.$$

$\square$

3.2 A Reduction to Well-Rounded Ellipsoids

Using John's Theorem [Joh48], the convex body K in Theorem 3 can be approximated by an ellipsoid $\mathcal{E}$ as defined in Eq. (1). The natural approach will then be to apply Theorem 6 to the axes of the ellipsoid. However, it will be crucial that the lengths of the axes of the ellipsoid are bounded by $2^{O(n)}$. We will now argue how to make an arbitrary ellipsoid well rounded.

Let us denote $\lambda_{\max}(\mathcal{E}) := \max\{\lambda_i : i = 1, \ldots, n\}$ as the maximum length of an axis. Recall that a matrix $U \in \mathbb{R}^{n \times n}$ is *unimodular* if $U \in \mathbb{Z}^{n \times n}$ and $|\det(U)| = 1$. In particular, the linear map $T : \mathbb{R}^n \to \mathbb{R}^n$ with $T(\boldsymbol{x}) = U\boldsymbol{x}$ is a bijection on the integer lattice, meaning that $T(\mathbb{Z}^n) = \mathbb{Z}^n$. It turns out that one can use the *lattice basis reduction* method by Lenstra et al. [LLL82] to find a unimodular linear transformation that "regularizes" any given ellipsoid. Note that it suffices to work with the regularized ellipsoid $T(\mathcal{E})$ since $\mathrm{vol}_n(T(\mathcal{E})) = \mathrm{vol}_n(\mathcal{E})$ and if we find a point $\boldsymbol{x} \in (\rho T(\mathcal{E})) \cap \mathbb{Z}^n$, then by linearity $T^{-1}(\boldsymbol{x}) \in \rho\mathcal{E}$ and $T^{-1}(\boldsymbol{x}) \in \mathbb{Z}^n$.

Given $\boldsymbol{b}_1, \ldots, \boldsymbol{b}_n \in \mathbb{R}^n$ we define the *Gram-Schmidt orthogonalization* iteratively as $\hat{\boldsymbol{b}}_j = \boldsymbol{b}_j - \sum_{i<j} \mu_{ij}\hat{\boldsymbol{b}}_i$, where $\mu_{ij} = \frac{\langle \boldsymbol{b}_j, \hat{\boldsymbol{b}}_i \rangle}{\|\hat{\boldsymbol{b}}_i\|_2^2}$. Notice that we can then write $\boldsymbol{b}_j = \hat{\boldsymbol{b}}_j + \sum_{i<j} \mu_{ij}\hat{\boldsymbol{b}}_i$. In particular, suppose B is the matrix with columns $\boldsymbol{b}_1, \ldots, \boldsymbol{b}_n$ and $\hat{B}$ is the matrix with columns $\hat{\boldsymbol{b}}_1, \ldots, \hat{\boldsymbol{b}}_n$. Then $B = \hat{B}V$ for an upper triangular matrix V with ones along the diagonal and $V_{ij} = \mu_{ij}$ for $i < j$.

Definition 2. *Let $B \in \mathbb{R}^{n \times n}$ be a lattice basis and let μ_{ij} be the coefficients from Gram-Schmidt orthogonalization. The basis is called LLL reduced if*

- *(Coefficient-reduced): $|\mu_{ij}| \leq \frac{1}{2}$ for all $1 \leq i < j \leq n$.*
- *(Lovász condition): $\|\hat{\boldsymbol{b}}_i\|_2^2 \leq 2\|\hat{\boldsymbol{b}}_{i+1}\|_2^2$ for $i = 1, \ldots, n-1$.*

LLL reduction has been widely used in diverse fields such as integer programming and cryptography [NV10]. One property of the LLL reduced basis is that the eigenvalues of the corresponding matrix B are bounded away from 0:

Lemma 6. *Let B denote the matrix with columns $\boldsymbol{b}_1, \ldots, \boldsymbol{b}_n$. If $\boldsymbol{b}_1, \ldots, \boldsymbol{b}_n$ is an LLL-reduced basis with $\|\boldsymbol{b}_i\|_2 \geq 1$ for all i, then $\|B\boldsymbol{x}\|_2 \geq 2^{-3n/2} \cdot \|\boldsymbol{x}\|_2$ for all $\boldsymbol{x} \in \mathbb{R}^n$.*

Proof. Let $\hat{B}$ denote the Gram-Schmidt orthogonalization of B, with columns $\hat{b}_1, \ldots, \hat{b}_n$. Now, for any k, we can use the properties of LLL reduction to gain the following bound (Proposition (1.7) in [LLL82]).

$$1 \le \|b_k\|_2^2 = \|\hat{b}_k\|_2 + \sum_{i<k} \mu_{ik}^2 \|\hat{b}_i\|_2^2 \le \left(1 + \frac{1}{4}\sum_{i<k} 2^{k-i}\right) \cdot \|\hat{b}_k\|_2^2 \le 2^k \cdot \|\hat{b}_k\|_2^2.$$

In particular, $\|\hat{b}_k\|^2 \ge 2^{-n}$ for all $k = 1, \ldots, n$. Now let V be the matrix so that $B = \hat{B}V$, and let $x \in \mathbb{R}^n$ with $\|x\|_2 = 1$. Let k denote the largest index with $|x_k| \ge 2^{-k}$. Then

$$|(Vx)_k| = \left|x_k + \sum_{j>k} \mu_{kj}x_j\right| \ge |x_k| - \frac{1}{2}\sum_{j>k}|x_j| \ge 2^{-k} - \frac{1}{2}\sum_{j>k} 2^{-j} \ge 2^{-n}.$$

Now, by the orthogonality of $\hat{b}_1, \ldots, \hat{b}_n$, we have

$$\|Bx\|_2^2 = \|\hat{B}Vx\|_2^2 = \sum_{i=1}^{n}(Vx)_i^2\|\hat{b}_i\|_2^2 \ge |(Vx)_k|^2 \cdot 2^{-n} \ge 2^{-3n}.$$

Taking square roots gives the claim. □

Lemma 7. *Let $\mathcal{E} = \{x \in \mathbb{R}^n : \|Ax\|_2^2 \le 1\}$ be an ellipsoid. Then in polynomial time, we can find*

(1) either a vector $x \in \mathcal{E} \cap \mathbb{Z}^n$
(2) or a linear transformation T so that $T(x) = Ux$ for a unimodular matrix U and $\lambda_{max}(T(\mathcal{E})) \le 2^{3n/2}$.

Proof. Use the algorithm of [LLL82] to find a unimodular matrix U such that $B = AU$ is LLL reduced. Let $b_1, \ldots, b_n$ denote the columns of B. Notice that if $\|b_i\|_2 \le 1$, then $A^{-1}b_i \in \mathcal{E} \cap \mathbb{Z}^n$, and so we are done. So assume now that $\|b_i\|_2 \ge 1$ for all i.

Define $T(x) = U^{-1}x$, and notice that $T(\mathcal{E}) = \{x \in \mathbb{R}^n : \|Bx\|_2^2 \le 1\}$. We then have

$$\lambda_{\max}(T(\mathcal{E})) = \max_{x \in f(\mathcal{E})} \|x\|_2 = \max_{\|Bx\|_2 \le 1} \|x\|_2 = \max_{x \ne 0} \frac{\|x\|_2}{\|Bx\|_2} \le 2^{3n/2},$$

where the last inequality follows from Lemma 6. □

Finally we can prove one of our main results, Theorem 3.

Proof (Theorem 3). Let $K \subseteq \mathbb{R}^n$ be a convex body with $\mathrm{vol}_n(K) \ge 2^n$. We compute an ellipsoid[6] $\mathcal{E} = \{x \in \mathbb{R}^n \mid \sum_{i=1}^{n} \frac{1}{\lambda_i^2}\langle x, a_i\rangle^2 \le 1\}$ so that $\frac{1}{5\sqrt{n}}\mathcal{E} \subseteq K \subseteq \frac{1}{5}\sqrt{n}\mathcal{E}$. Then

---

[6] Note that there *exists* an ellipsoid that approximates K within a factor of $\sqrt{n}$ and if K is a polytope with m facets, then this ellipsoid can be found in time polynomial in n and m. However, if one only has a separation oracle for K, then the best factor achievable in polynomial time is n [GLS12].

$$2^n \cdot 5^n \cdot n^{-n/2} \le \text{vol}_n(K) \cdot 5^n \cdot n^{-n/2} \le \text{vol}_n(\mathcal{E}) = \underbrace{\text{vol}(B(\mathbf{0}, 1))}_{\le 5^n n^{-n/2}} \cdot \prod_{i=1}^{n} \lambda_i.$$

and hence $\prod_{i=1}^{n} \lambda_i \ge 1$. We apply Lemma 7 to either find an integer point in $\mathcal{E}$ and we are done, or we find a unimodular transformation T so that the ellipsoid $\tilde{\mathcal{E}} := T(\mathcal{E})$ has all axes of length at most $2^{O(n)}$. Suppose the latter case happens. We write $\tilde{\mathcal{E}} = \{ \boldsymbol{x} \in \mathbb{R}^n \mid \sum_{i=1}^{n} \frac{1}{\tilde{\lambda}_i^2} \langle \boldsymbol{x}, \tilde{\boldsymbol{a}}_i \rangle^2 \le 1 \}$ and observe that still $\prod_{i=1}^{n} \tilde{\lambda}_i \ge 1$ as the volume of the ellipsoid has not changed. We make use of the δ-approximation for the number balancing problem to apply Theorem 6 to the vectors $\tilde{\boldsymbol{a}}_1, \ldots, \tilde{\boldsymbol{a}}_n$ and parameters $\tilde{\lambda}_1, \ldots, \tilde{\lambda}_n$ and obtain a vector $\boldsymbol{x} \in \mathbb{Z}^n \setminus \{0\}$ with $|\langle \tilde{\boldsymbol{a}}_i, \boldsymbol{x} \rangle| \le \tilde{\lambda}_i \cdot O(n^4)$. Then $\sum_{i=1}^{n} \frac{1}{\tilde{\lambda}_i^2} \langle \boldsymbol{a}_i, \boldsymbol{x} \rangle^2 \le O(n^9)$ and hence $\boldsymbol{x} \in O(n^{4.5}) \cdot \tilde{\mathcal{E}}$. Then $T^{-1}(\boldsymbol{x}) \in (O(n^5) \cdot K) \cap (\mathbb{Z}^n \setminus \{0\})$. $\qquad \square$

References

[Ajt96] Ajtai, M.: Generating hard instances of lattice problems. In: Proceedings of the 28th STOC, pp. 99–108. ACM (1996)

[AR05] Aharonov, D., Regev, O.: Lattice problems in NP cap conp. J. ACM **52**(5), 749–765 (2005)

[Boh96] Bohman, T.: A sum packing problem of erdös and the conway-guy sequence. Proc. AMS **124**(12), 3627–3636 (1996)

[GJ97] Garey, M.R., Johnson, D.S.: Computers and Intractability: A Guide to the Theory of NP-Completeness. W.H. Freeman, New York (1997)

[GLS12] Grötschel, M., Lovász, L., Schrijver, A.: Geometric Algorithms and Combinatorial Optimization, vol. 2, pp. 122–125. Springer, Heidelberg (2012)

[HR07] Haviv, I., Regev, O.: Tensor-based hardness of the shortest vector problem to within almost polynomial factors, pp. 469–477 (2007)

[Joh48] John, F.: Extremum problems with inequalities as subsidiary conditions. In: Studies and Essays Presented to R. Courant on his 60th Birthday, 8 January 1948, pp. 187–204. Interscience Publishers Inc., New York (1948)

[KK82] Karmarkar, N., Karp, R.: The differencing method of set partitioning. Technical report, CS Division, UC Berkeley (1982). http://digitalassets.lib.berkeley.edu/techreports/ucb/text/CSD-83-113.pdf

[KPP04] Kellerer, H., Pferschy, U., Pisinger, D.: Knapsack Problems. Springer, Heidelberg (2004)

[LLL82] Lenstra, A., Lenstra, H., Lovász, L.: Factoring polynomials with rational coefficients. Mathematische Annalen **261**(4), 515–534 (1982)

[Lov86] Lovász, L.: An Algorithmic Theory of Numbers, Graphs and Convexity. SIAM (1986)

[Lov90] Lovász, L.: Geometric Algorithms and Algorithmic Geometry. American Mathematical Society (1990)

[Lun88] Lunnon, W.: Integer sets with distinct subset-sums. Math. Comput. **50**(181), 297–320 (1988)

[LY11] Lev, V., Yuster, R.: On the size of dissociated bases. Electr. J. Comb. **18**(1), P117 (2011)

[Mat02] Matousek, J.: Lectures on Discrete Geometry. Springer, New York (2002)

[Mer06] Mertens, S.: The easiest hard problem: number partitioning. Comput. Complex. Stat. Phys. **125**(2), 125–139 (2006)

[MR09] Micciancio, D., Regev, O.: Lattice-based cryptography. In: Post-quantum Cryptography, pp. 147–191. Springer (2009)

[MT90] Martello, S., Toth, P.: Knapsack Problems: Algorithms and Computer Implementations. Wiley Inc., New York (1990)

[NV10] Nguyen, P., Vallée, B.: The lll algorithm. In: Nguyen, P., Vallée, B. (eds.) Information Security and Cryptography. Springer, Heidelberg (2010)

[Pap94] Papadimitriou, C.H.: On the complexity of the parity argument and other inefficient proofs of existence. J. Comput. Syst. Sci. **48**(3), 498–532 (1994)

[Sch87] Schnorr, C.: A hierarchy of polynomial time lattice basis reduction algorithms. Theor. Comput. Sci. **53**, 201–224 (1987)

[WY92] Woeginger, G., Yu, Z.: On the equal-subset-sum problem. Inf. Process. Lett. **42**(6), 299–302 (1992)

An Improved Deterministic Rescaling
for Linear Programming Algorithms

Rebecca Hoberg[(✉)] and Thomas Rothvoss

University of Washington, Seattle, WA 98105, USA
{rahoberg,rothvoss}@uw.edu

Abstract. The *perceptron algorithm* for linear programming, arising from machine learning, has been around since the 1950s. While not a polynomial-time algorithm, it is useful in practice due to its simplicity and robustness. In 2004, Dunagan and Vempala showed that a *randomized rescaling* turns the perceptron method into a polynomial time algorithm, and later Peña and Soheili gave a *deterministic rescaling*. In this paper, we give a deterministic rescaling for the perceptron algorithm that improves upon the previous rescaling methods by making it possible to rescale much earlier. This results in a faster running time for the rescaled perceptron algorithm. We will also demonstrate that the same rescaling methods yield a polynomial time algorithm based on the *multiplicative weights update* method. This draws a connection to an area that has received a lot of recent attention in theoretical computer science.

1 Introduction

One of the central algorithmic problems in theoretical computer science as well as in more practical areas like operations research is finding the solution to a *linear program*

$$\max\{c^T x \mid Ax \geq b\} \tag{1}$$

where $A \in \mathbb{R}^{m \times n}, c \in \mathbb{R}^n$ and $b \in \mathbb{R}^m$. On the theoretical side, linear programming relaxations are the backbone for many approximation algorithms [WS11, Vaz01]. On the practical side, many real-world problems can either be modeled as linear programs or as integer linear programs; the latter ones are then solved using Branch & Bound or Branch & Cut methods, both of which rely on repeatedly computing solutions to linear programs [CCZ14].

The first algorithm for solving linear programs was the *simplex method* due to Dantzig [Dan51]. While the method performs well in practice — and is still the method of choice today — for almost any popular pivoting rule one can construct instances where the algorithm takes exponential time [KM72]. In 1979, Khachiyan [Hač79, Sch86] developed the first polynomial-time algorithm. However, despite the desirable theoretical properties, Khachiyan's *ellipsoid method* turned out to be too slow for practical applications.

T. Rothvoss—Supported by an Alfred P. Sloan Research Fellowship. Both authors supported by NSF grant 1420180 with title *"Limitations of convex relaxations in combinatorial optimization"*.

© Springer International Publishing AG 2017
F. Eisenbrand and J. Koenemann (Eds.): IPCO 2017, LNCS 10328, pp. 267–278, 2017.
DOI: 10.1007/978-3-319-59250-3_22

In the 1980s, *interior point methods* were developed which were efficient in theory and in practice. Karmarkar's algorithm has a running time of $O(n^{3.5}L)$, where L is the number of bits in the input [Kar84]. As recently as 2015, it was shown that there is an interior-point method using only $\tilde{O}(\sqrt{\text{rank}(A)} \cdot L)^1$ many iterations; this upper bound essentially matches known lower bound barriers [LS15].

A common way to find a polynomial-time linear programming algorithm is with a greedy type procedure along with periodic rescaling. One famous example of this is the *perceptron algorithm* [Agm54], which we will focus on in this paper. Instead of solving (1) directly, this method finds a feasible point in the *open polyhedral cone*

$$P = \{x \in \mathbb{R}^n \mid Ax > \mathbf{0}\} \tag{2}$$

where $A \in \mathbb{R}^{m \times n}$ – using standard reductions one can interchange the representations (1) and (2) with at most a linear overhead. The classical perceptron algorithm starts at the origin and iteratively walks in the direction of any violated constraint. In the worst case this method is not polynomial time, but it is still useful due to its simplicity and robustness [Agm54]. In 2004, Dunagan and Vempala [DV06] showed that using a randomized rescaling procedure, the algorithm can be modified to find a point in (2) in polynomial time. Explicitly, their algorithm runs in time $\tilde{O}(mn^4 \log \frac{1}{\rho})$, where $\rho > 0$ is the radius of the largest ball in the intersection of P with the unit ball $B := B(0, 1)$. A deterministic rescaling procedure was provided by Peña and Soheili in [PS16] using techniques developed by Betke and Chubanov [Chu15, Chu12, Bet04]. Their algorithm uses an improved convergence of the perceptron algorithm based on *Nesterov's smoothing technique* [Nes05, PS12]. Overall, their algorithm takes time $\tilde{O}(m^2 n^{2.5} \log \frac{1}{\rho})$.

Another classical LP algorithm that we will discuss in this paper is based on a very general algorithmic framework called the *multiplicative weights update* (MWU) method. In its general form one imagines having m *experts* who each incur some *cost* in a sequence of iterations. In each iteration we have to select a convex combination of experts so that the expected cost is minimized, where we only have information on the past costs. The MWU method initially gives all experts the same weight and in each iteration the weight of expert i is multiplied by $\exp(-\varepsilon \cdot$ cost incurred by expert $i)$ where ε is some parameter. Then on average, the convex combination given by the weights will be nearly as good as the cost incurred by the best expert. MWU is an *online algorithm* that does not need to know the costs in advance, and it has numerous applications in machine learning, economics and theoretical computer science. In fact, MWU has been reinvented many times under different names in the literature. Recent applications in theoretical computer science include finding fast approximations to maximum flows [CKM+11], multicommodity flows [GK07, Mad10], solving LPs [PST95], and solving semidefinite programs [AHK05]. We refer to the survey of Arora, Hazan and Kale [AHK12] for a detailed overview.

When we apply the MWU framework to linear programming, the experts correspond to the linear constraints. Suppose we use this method to find a valid

---

[1] The $\tilde{O}$-notation suppresses any polylog(m, n) terms.

point in $P = \{x : Ax > 0\}$ where $\|A_i\|_2 = 1$ for every row A_i. At iteration t, the cost associated with expert i will be $\langle A_i, p^{(t)} \rangle$ for some vector $p^{(t)}$. Therefore the weight of expert i at time T will be $e^{-\langle A_i, x \rangle}$ where $x = \sum_{t=1}^{T} \varepsilon^{(t)} p^{(t)}$. The analysis of MWU consists of bounding the sum of the weights, which in this case is given by the *potential function* $\Phi(x) = \sum_{i=1}^{m} e^{-\langle A_i, x \rangle}$. If we choose the update vector $p^{(t)}$ to be a weighted sum of constraints at every iteration, notice that the resulting walk in $\mathbb{R}^n$ corresponds to gradient descent on Φ – in this case MWU terminates in $\tilde{O}(\frac{1}{\rho^2})$ iterations. However, ρ need not be polynomial in the input size, and in fact this method is not polynomial time in the worst case.

1.1   Our Contribution

For reference, the general form for the rescaled LP algorithms we will present in this paper is given in Algorithm 1. Throughout this paper we will assume that $P \cap B$ contains a ball of radius $\rho > 0$.

Algorithm 1

FOR $\tilde{O}(n \log \frac{1}{\rho})$ phases DO:

(1) **Initial phase:** Either find $x \in P$ or provide a $\lambda \in \mathbb{R}_{\geq 0}^{m}$, $\|\lambda\|_1 = 1$ with $\|\lambda A\|_2 \leq \Delta$.
(2) **Rescaling phase:** Find an invertible linear transformation F so that $\mathrm{vol}(F(P) \cap B)$ is a constant fraction larger than $\mathrm{vol}(P \cap B)$. Replace P by $F(P)$.

Our technical and conceptual contributions are as follows:

(1) *Improved rescaling:* We design a rescaling method that applies for a parameter of $\Delta = \Theta(\frac{1}{n})$, which improves over the threshold $\Delta = \Theta(\frac{1}{m\sqrt{n}})$ required by [PS16]. This results in a smaller number of iterations that are needed per phase until one can rescale the system.
(2) *Rescaling the MWU method:* We show that in $\tilde{O}(1/\Delta^2)$ iterations the MWU method can be made to implement the initial phase of Algorithm 1. The idea is that if gradient descent is making insufficient progress then the gradient must have small norm, and from this we can extract an appropriate λ. In particular, combining this with our rescaling method, we obtain a polynomial time LP algorithm based on MWU.
(3) *Faster gradient descent:* The standard gradient descent approach terminates in at most $\tilde{O}(1/\Delta^2)$ iterations, which matches the first approach in [PS16]. The more recent work of Peña and Soheili [PS12] uses *Nesterov smoothing* to bring the number of iterations down to $\tilde{O}(1/\Delta)$. We prove that essentially the same speedup can be obtained without modifying the objective function by projecting the gradient on a significant eigenspace of the Hessian.
(4) *Computing an approximate John ellipsoid:* For a general convex body K, computing a John ellipsoid is equivalent to finding a linear transformation so that $F(K)$ is well rounded. For our unbounded region P, our rescaling algorithm gives a linear transformation F so that $F(P) \cap B$ is well-rounded.

2   Rescaling of the Perceptron Algorithm

In this section we fix an initial phase for Algorithm 1 – in particular, the paper of Peña and Soheili gives a smooth variant of the perceptron algorithm that implements the initial phase of Algorithm 1 in time $\tilde{O}(\frac{mn}{\Delta})$ [PS16].

We then focus on the rescaling phase of the algorithm. Our main result is that we are able to rescale with $\Delta = O(\frac{1}{n})$.

Lemma 1. *Suppose* $\lambda \in \mathbb{R}_{\geq 0}^m$ *with* $\|\lambda\|_1 = 1$ *and* $\|\lambda A\|_2 \leq O(\frac{1}{n})$. *Then in time* $O(mn^2)$ *we can rescale* P *so that* $\mathrm{vol}(P \cap B)$ *increases by a constant factor.*

We introduce two new rescaling methods that achieve the guarantee of Lemma 1. First we show that we can extract a thin direction by sampling rows of A using a *random hyperplane*. The linear transformation that scales P in that direction, corresponding to a *rank-1 update*, will increase $\mathrm{vol}(P \cap B)$ by a constant factor.

Next we give an alternate rescaling which is no longer a rank-1 update but which has the potential to increase $\mathrm{vol}(P \cap B)$ by up to an exponential factor under certain conditions. In addition, if we take an alternate view where the cone P is left invariant and instead update the underlying *norm*, we see that this rescaling consists of adding a scalar multiple of a particular Hessian matrix to the matrix defining the norm. We also believe that this view is the right one to make potential use of the *sparsity* of the underlying matrix A, which would be a necessity for any practically relevant LP optimization method.

Combining the guarantees for the initial phase and rescaling phase gives us the following theorem:

Theorem 1. *There is an algorithm based on the perceptron algorithm that finds a point in* P *in time* $\tilde{O}(mn^3 \log(\frac{1}{\rho}))$.

2.1   Rescaling using a Thin Direction

In this section we will show how we can rescale by finding a direction in which the cone is *thin*. First we give the formal definition of width.

Definition 1. *Define the width of the cone* P *in the direction* $c \in \mathbb{R}^n \setminus \{0\}$ *as*

$$\mathrm{WIDTH}(P, c) = \frac{1}{\|c\|_2} \max_{x \in P \cap B} |\langle c, x \rangle|.$$

In [PS16], Peña and Soheili show that stretching P in a thin enough direction increases the volume of $P \cap B$ by a constant factor.

Lemma 2 [PS16]. *Suppose that there is a direction* $c \in \mathbb{R}^n \setminus \{0\}$ *with* WIDTH $(P, c) \leq \frac{1}{3\sqrt{n}}$. *Define* $F : \mathbb{R}^n \to \mathbb{R}^n$ *as the linear map with* $F(c) = 2c$ *and* $F(x) = x$ *for all* $x \perp c$. *Then*

$$vol(F(P) \cap B) \geq \frac{3}{2} \cdot vol(P \cap B).$$

Explicitly, assuming $\|c\|_2 = 1$, Lemma 2 updates our constraint matrix to $A(I - \frac{1}{2}cc^T)$. In particular, we apply a *rank-1 update* to the constraint matrix. Given a solution x to these new constraints, a solution to the original problem can be easily recovered as $(I - \frac{1}{2}cc^T)x$.

It remains to argue how one can extract a thin direction for P, given a convex combination λ so that $\|\lambda A\|_2$ is small. Here we will significantly improve over the bounds of [PS16] which require $\|\lambda A\|_2 \leq O(\frac{1}{m\sqrt{n}})$. We begin by a new generic argument to obtain a thin direction:

Lemma 3. *For any non-empty subset $J \subseteq [m]$ of constraints one has*

$$\text{WIDTH}\left(P, \sum_{i \in J} \lambda_i A_i\right) \leq \frac{\|\sum_{i=1}^m \lambda_i A_i\|_2}{\|\sum_{i \in J} \lambda_i A_i\|_2}.$$

Proof. First, note that by the full-dimensionality of P, we always have $\|\sum_{i \in J} \lambda_i A_i\|_2 > 0$. By definition of width, we can write

$$\text{WIDTH}\left(P, \sum_{i \in J} \lambda_i A_i\right) = \frac{1}{\|\sum_{i \in J} \lambda_i A_i\|_2} \max_{x \in P \cap B} \left\langle \sum_{i \in J} \lambda_i A_i, x \right\rangle.$$

Now, we know that $\langle A_i, x \rangle \geq 0$ for all $x \in P$ and so

$$\max_{x \in P \cap B} \left\langle \sum_{i \in J} \lambda_i A_i, x \right\rangle \leq \max_{x \in B} \left\langle \sum_{i=1}^m \lambda_i A_i, x \right\rangle = \|\lambda A\|_2$$

and the claim is proven.   □

So in order to find a direction of small width, it suffices to find a subset $J \subseteq [m]$ with $\|\sum_{i \in J} \lambda_i A_i\|_2$ large. Implicitly, the choice that Peña and Soheili [PS16] make is to select $J = \{i_0\}$ for $i_0 \in [m]$ maximizing λ_{i_0}. This approach gives a bound of $\|\sum_{i \in J} \lambda_i A_i\|_2 \geq \frac{1}{m}$. We will now prove the asymptotically optimal bound[2] using a random hyperplane:

Lemma 4. *Let $\lambda \in \mathbb{R}_{\geq 0}^m$ be any convex combination and $A \in \mathbb{R}^{m \times n}$ with $\|A_i\|_2 = 1$ for all i. Take a random Gaussian g and set $J := \{i \in [m] \mid \langle A_i, g \rangle \geq 0\}$. Then with constant probability $\|\sum_{i \in J} \lambda_i A_i\|_2 \geq \frac{1}{4\sqrt{\pi n}}$.*

Proof. We set $v := \frac{g}{\|g\|_2}$. Since v is unit vector we can lower bound the length of $\|\sum_{i \in J} \lambda_i A_i\|_2$ by measuring the projection on v and obtain $\|\sum_{i \in J} \lambda_i A_i\|_2 \geq \sum_{j \in J} \lambda_i \langle A_i, v \rangle$. By symmetry of the Gaussian it then suffices to argue that $\sum_{i=1}^m \lambda_i |\langle A_i, v \rangle| \geq \frac{1}{2\sqrt{\pi n}}$. First we will show that for an appropriate constant $\alpha \in (0, 1)$,

---

[2] It suffices here to consider the trivial example with $\lambda_1 = \ldots = \lambda_n = \frac{1}{n}$ and $A_i = e_i$ being the standard basis. Then $\|\sum_{i \in J} \lambda_i A_i\|_2 \leq \frac{1}{\sqrt{n}}$ for any subset J. The optimality of our rescaling can also be seen since the cone in the last iteration is $\tilde{O}(n)$-well rounded, which is optimal up to $\tilde{O}$-terms.

(1) $\Pr(\|g\|_2 \geq \sqrt{2n}) \leq \frac{2}{n}$

(2) $\Pr(\sum_{i=1}^{n} \lambda_i |\langle A_i, g \rangle| < \sqrt{\frac{1}{2\pi}}) \leq \alpha$.

Then, with probability at least $\gamma = \frac{1-\alpha}{2}$, we have $\sum_{i=1}^{n} \lambda_i |\langle A_i, v \rangle| \geq \frac{1}{2\sqrt{\pi n}}$.

For (1), notice that $\|g\|_2^2$ is just the chi-squared distribution with n degrees of freedom, and so it has variance $2n$ and mean n. Therefore Chebyshev's inequality tells us that $\Pr\left[\|g\|_2^2 \geq 2n\right] \leq \frac{2}{n}$. Now, for all i, $\langle A_i, g \rangle$ is a normal random variable with mean 0 and variance 1, and so the expectation of its absolute value is $\sqrt{\frac{2}{\pi}}$. Summing these up gives $\mathbb{E}\left[\sum_{i=1}^{m} |\langle \lambda_i A_i, g \rangle|\right] = \sqrt{\frac{2}{\pi}}$. Moreover, $\sum_{i=1}^{m} |\langle \lambda_i A_i, g \rangle|$ is Lipschitz in g with Lipschitz constant 1, and so[3]

$$\Pr\left(\sum_{i=1}^{m} |\langle \lambda_i A_i, g \rangle| < \sqrt{\frac{2}{\pi}} - t\right) \leq e^{-t^2/\pi^2}.$$

Letting $t = \sqrt{\frac{1}{2\pi}}$ gives (2). By a union bound, the probability either of these events happens is at most $\alpha + \frac{2}{n}$, and so with probability at least $\frac{1-\alpha}{2}$ neither occurs, which gives us the claim.   □

While the proof is probabilistic, one can use the method of *conditional expectation* to derandomize the sampling [AS04]. More concretely, consider the function $F(g) := \sum_{i=1}^{m} \lambda_i |\langle A_i, g \rangle| - \frac{1}{10\sqrt{n}}\|g\|_2$. The proof of Lemma 4 implies that the expectation of this function is at least $\Omega(1)$. Then we can find a desired vector $g = (g_1, \ldots, g_n)$ by choosing the coordinates one after the other so that the conditional expectation does not decrease. We are now ready to prove Lemma 1, which we restate here with explicit constants.

Lemma 5. *Suppose $\lambda \in \mathbb{R}_{\geq 0}^m$ with $\|\lambda\|_1 = 1$ and $\|\lambda A\|_2 \leq \frac{1}{12n\sqrt{\pi}}$. Then in time $O(mn^2)$ we can rescale P so that $\text{vol}(P \cap B)$ increases by a constant factor.*

Proof. Computing a random Gaussian and checking if it satisfies the conditions of Lemma 4 takes time $O(mn)$. Since the conditions will be satisfied with constant probability, the expected number of times we must do this is constant. Once the conditions are satisfied, finding a thin direction and rescaling can be done in time $O(n^3)$. Lemmas 2 and 3 guarantee we get a constant increase in the volume.   □

2.2   Deterministic Multi-rank Rescaling

We now introduce an alternate linear transformation we can use to rescale. This is no longer a rank-1 update, but it is inherently deterministic along with other

---

[3] Recall that a function $F : \mathbb{R}^n \to \mathbb{R}$ is *Lipschitz* with Lipschitz constant 1 if $|F(x) - F(y)| \leq \|x - y\|_2$ for all $x, y \in \mathbb{R}^n$. A famous concentration inequality by Sudakov, Tsirelson, Borell states that $\Pr[|F(g) - \mu| \geq t] \leq e^{-t^2/\pi^2}$, where g is a random Gaussian and μ is the mean of F under g.

nice properties. For one thing, although we only guarantee constant improvement in the volume, under certain circumstances the rescaling can improve the volume by an exponential factor. This transformation will also take a nice form when we change the view to consider rescaling the unit ball rather than the feasible region.

Lemma 6. *Suppose* $\lambda \in \mathbb{R}^m_{\geq 0}$, $\|\lambda\|_1 = 1$ *and* $\|\lambda A\|_2 \leq \frac{1}{10n}$. *Let* M *denote the matrix* $\sum_{i=1}^m \lambda_i A_i A_i^T$ *and suppose* $0 \leq \alpha \leq \frac{1}{\delta_{max}}$, *where* $\delta_{max} = \|M\|_{op}$ *denotes the maximal eigenvalue of* M. *Define* $F(x) = (I + \alpha M)^{1/2} x$. *Then* $\mathrm{vol}(F(P) \cap B) \geq e^{\alpha/5} \mathrm{vol}(P \cap B)$.

Proof. First notice that M is symmetric positive semi-definite with trace 1. Therefore the eigenvalues of $I + \alpha M$ take the form $1 + \alpha \delta_i$ where $0 \leq \delta_i \leq \alpha$ and $\sum_{i=1}^n \delta_i = 1$. Note that since $\alpha \delta_i \leq 1$, we can lower bound the eigenvalues by $1 + \alpha \delta_i \geq e^{\alpha \delta_i / 2}$. Therefore

$$\det(I + \alpha M) \geq \prod_{i=1}^n e^{\alpha \delta_i / 2} = \exp\left(\frac{\alpha}{2} \sum_{i=1}^n \delta_i\right) = e^{\alpha/2}.$$

In particular, $\det(F) \geq e^{\alpha/4}$.

So far we have shown that $\mathrm{vol}(F(P \cap B))$ is significantly larger than $\mathrm{vol}(P \cap B)$. However, the desired bound is on $\mathrm{vol}(F(P) \cap B)$, and so we need to ensure that we do not lose too much of the volume when we intersect with the unit ball. It turns out the bound on $\|\lambda A\|_2$ will allow us to do precisely this.

For any $x \in P \cap B$, we get the bound

$$\|F(x)\|_2^2 = x^T x + \alpha \sum_{i=1}^m \lambda_i \langle A_i, x\rangle^2 \leq 1 + \alpha \sum_{i=1}^m \lambda_i \langle A_i, x\rangle \leq 1 + \alpha\|\lambda A\|_2 \leq 1 + \frac{\alpha}{10n}.$$

The point is that every element of $F(P \cap B)$ has length at most $1 + \frac{\alpha}{20n}$, and so intersecting with the unit ball will not lose more volume than shrinking by a factor of $1 + \frac{\alpha}{20n}$. In particular, the volume decreases by at most $(1 + \frac{\alpha}{20n})^{-n} \geq e^{-\alpha/20}$, and so we have

$$\mathrm{vol}(F(P) \cap B) \geq e^{-\alpha/20} \mathrm{vol}(F(P \cap B))$$
$$\geq e^{-\alpha/20} \cdot e^{\alpha/4} \mathrm{vol}(P \cap B)$$
$$\geq e^{\alpha/5} \cdot \mathrm{vol}(P \cap B).$$

$\square$

Note that one always has $\delta_{max} \leq 1$ and hence in any case one can choose $\alpha \geq 1$. Therefore if $\|\lambda A\|_2 \leq \frac{1}{10n}$, we get constant improvement in $\mathrm{vol}(P \cap B)$. In fact, if the eigenvalues of M happen to be small, we could get up to exponential improvement. This computation can be carried out in time $O(mn^2)$ and so Lemma 6 proves Lemma 1 and hence Theorem 1.

2.3   An Alternate View of Rescaling

Instead of applying a linear transformation to the cone P, there is an equivalent view where instead one applies a linear transformation to the unit ball. We will now switch the view in the sense that we fix the cone P, but we update the norm in each rescaling step so that the unit ball becomes more representative of P.

Recall that a symmetric positive definite matrix $H \in \mathbb{R}^{n \times n}$ induces a *norm* $\|x\|_H := \sqrt{x^T H x}$. Note that also H^{-1} is a symmetric positive definite matrix and $\| \cdot \|_{H^{-1}}$ is the *dual norm* of $\| \cdot \|_H$. In this view we assume the rows A_i of A are normalized so that $\|A_i\|_{H^{-1}} = 1$.

Let $B_H := \{x \in \mathbb{R}^n \mid \|x\|_H \leq 1\}$ be the unit ball for the norm $\| \cdot \|_H$. Note that B_H is always an *ellipsoid*. We will measure progress in terms of the fraction of the ellipsoid B_H that lies in the cone P, namely $\mu(H) := \frac{\text{vol}(B_H \cap P)}{\text{vol}(B_H)}$. The goal of the rescaling step will then be to increase $\mu(H)$ by a constant factor. Note that we initially have $\mu(H) = \mu(I) \geq \rho^n$, and at any time $0 \leq \mu(H) \leq 1$, so we can rescale at most $O(n \log \frac{1}{\rho})$ times.

In this view, Lemma 6 takes the following form:

Lemma 7. *Let $H \in \mathbb{R}^{n \times n}$ be symmetric with $H \succ 0$. Suppose $\lambda \in \mathbb{R}_{\geq 0}^m$ with $\|\lambda\|_1 = 1$ and $\|\lambda A\|_{H^{-1}} \leq \frac{1}{10n}$ and let $M := \sum_{i=1}^m \lambda_i A_i A_i^T$. Let $0 \leq \alpha \leq \frac{1}{\delta_{\max}}$, where $\delta_{\max} := \|H^{-1}M\|_{op}$. Then for $\tilde{H} := H + \alpha M$ one has $\mu(\tilde{H}) \geq e^{\alpha/5} \cdot \mu(H)$.*

Algorithm 2 illustrates the multi-rank rescaling under the alternate view. Notice that the algorithm updates the norm matrix by adding a scalar multiple of the Hessian matrix of the MWU potential function discussed in Sect. 3. Moreover, throughout the algorithm our matrix H will have the form $I + \sum_{i=1}^m h_i A_i A_i^T$ for some $h_i \geq 0$. Note that this allows fairly compact representation as we only need $O(m)$ space to encode the coefficients h_i that define the norm matrix.

Algorithm 2

FOR $\tilde{O}(n \log \frac{1}{\rho})$ phases DO:

(1) **Initial phase:** Either find $x \in P$ or give $\lambda \geq 0$, $\|\lambda\|_1 = 1$ with $\|\lambda A\|_{H^{-1}} \leq O(\frac{1}{n})$.
(2) **Rescaling phase:** Update $H := H + \alpha M$, where $M = \sum_{i=1}^m \lambda_i A_i A_i^T$.

---

3   Rescaling for the MWU Algorithm

In this section we show that the same rescaling methods can be used to make the MWU method into a polynomial time LP algorithm. In particular, we show that a *MWU phase* can implement the initial phase of Algorithm 1 or 2. For ease of notation, we will assume $H = I$, but we can recover the general case by replacing A_i by $H^{-1/2} A_i$ and replacing the update direction p by $H^{1/2}p$.

Recall that the MWU algorithm corresponds to gradient descent on a particular potential function. First we show how the standard gradient descent approach implements the initial phase. We then introduce a modified gradient descent, which speeds up the MWU phase to $\tilde{O}(1/\Delta)$ iterations. Combining this with our rescaling results gives the following:

Theorem 2. *There is an algorithm based on the MWU algorithm that finds a point in P in time $\tilde{O}(mn^{\omega+1}\log(\frac{1}{\rho}))$, where $\omega \approx 2.373$ is the exponent of matrix multiplication.*

3.1   Standard Gradient Descent

Consider the potential function $\Phi(x) = \sum_{i=1}^{m} e^{-\langle A_i, x\rangle}$, where $\|A_i\|_2 = 1$ for all rows A_i. Notice that $\Phi(0) = m$ and that if $\Phi(x) < 1$ then $\langle A_i, x\rangle > 0$ for all i, and hence $x \in P$. In this section we analyze standard gradient descent on Φ, starting at the origin. Notice that the gradient takes the form

$$\nabla\Phi(x) = -\sum_{i=1}^{m} e^{-\langle A_i, x\rangle} A_i.$$

We begin by establishing some useful notation. Given $x \in \mathbb{R}^n$, define $\lambda_i = \frac{1}{\Phi(x)}e^{-\langle A_i, x\rangle}$, $y = -\frac{\nabla\Phi(x)}{\Phi(x)} = \sum_{i=1}^{m} \lambda_i A_i$ and $M = \frac{\nabla^2\Phi(x)}{\Phi(x)} = \sum_{i=1}^{m} \lambda_i A_i A_i^T$. Even though all three depend on x, we will not denote that here to keep the notation clean. In particular, since $\|\lambda\|_1 = 1$, if at any iteration $\|y\|_2$ is small enough, then we will be able to rescale. It remains to show, therefore, that if $\|y\|_2$ is large enough, then we get sufficient decrease in the potential function.

Lemma 8. *Suppose $x \in \mathbb{R}^n$ and abbreviate $y = -\frac{\nabla\Phi(x)}{\Phi(x)}$. Then*

$$\Phi(x + \frac{1}{2}y) \le \Phi(x) \cdot e^{-\|y\|_2^2/4}.$$

Proof. First note that since $\|\lambda\|_1 = 1$ and $\|A_i\|_2 = 1$, we know that $|\langle A_i, y\rangle| \le 1$ for all i. In our analysis we will also use the fact that for any $z \in \mathbb{R}$ with $|z| \le 1$ one has $e^z \le 1 + z + z^2$. We obtain the following.

$$\Phi(x + \frac{1}{2}y) = \sum_{i=1}^{m} e^{-\langle A_i, x + \frac{1}{2}y\rangle} = \sum_{i=1}^{m} e^{-\langle A_i, x\rangle} e^{-\frac{1}{2}\langle A_i, y\rangle}$$

$$\le \Phi(x) \cdot \sum_{i=1}^{m} \lambda_i(1 - \frac{1}{2}\langle A_i, y\rangle + \frac{1}{4}\langle A_i, y\rangle^2)$$

$$\le \Phi(x) \cdot (1 - \frac{1}{4}\|y\|_2^2). \qquad \square$$

Thus as long as $\|y\|_2 \ge \Omega(\frac{1}{n})$, gradient descent will decrease the potential function by a factor of $e^{-\Theta(1/n^2)}$ in each iteration, and so in at most $O(n^2 \ln(m))$ iterations we arrive at a point x with $\Phi(x) < 1$.

3.2 Modified Gradient Descent

With $\Delta = \Theta(\frac{1}{n})$, the standard gradient descent approach implements the initial phase of Algorithm 1 in $\tilde{O}(n^2)$ iterations. It turns out we can get the same guarantee in $\tilde{O}(n)$ iterations by choosing a more sophisticated update direction. While we do not know how to guarantee an update direction that decreases $\Phi(x)$ by factor of more than $e^{-\Theta(\|y\|^2)}$, we are able to decrease the *product* of $\Phi(x)$ and $\|\nabla\Phi(x)\|_2$ a lot faster. First we give general bounds for the change in Φ and $\|\nabla\Phi\|_2$ under an arbitrary update step.

Lemma 9. *For any $0 < \varepsilon \leq 1$ and $p \in \mathbb{R}^n$ with $\|p\|_2 \leq 1$, we have*

1. $\Phi(x + \varepsilon p) \leq \Phi(x) \cdot (1 - \varepsilon\langle y, p\rangle + \varepsilon^2 p^T M p)$
2. $\|\nabla\Phi(x + \varepsilon p)\|_2 \leq \|\nabla\Phi(x)\|_2 \cdot \left(1 - \varepsilon\frac{\langle y, Mp\rangle}{\|y\|_2^2} + \varepsilon^2\left(\frac{\langle p, Mp\rangle}{\|y\|_2} + \frac{\|Mp\|_2^2}{\|y\|_2^2}\right)\right)$

To get a sense of what Lemma 9 is saying, consider the case when y is an eigenvector of M with eigenvector α and $\varepsilon < \frac{1}{2}$. If we let $p = \frac{y}{\|y\|_2}$, we obtain

1. $\Phi(x + \varepsilon p) \leq \Phi(x) \cdot (1 - \varepsilon\|y\|_2 + \varepsilon^2\alpha)$
2. $\|\nabla\Phi(x + \varepsilon p)\|_2 \leq \|\nabla\Phi(x)\|_2 \cdot \left(1 - \frac{\varepsilon}{2}\frac{\alpha}{\|y\|_2} + \varepsilon^2\frac{\alpha^2}{\|y\|_2}\right)$

In particular, if $\varepsilon = \min\{\frac{1}{2}, \frac{\|y\|_2}{4\alpha}\}$ the product decreases by a factor of $e^{-\Omega(\|y\|_2)}$. While y may not be an eigenvector of M, it turns out that it suffices to find a vector p so that both p and Mp are close in angle with y.

Lemma 10. *Suppose $p \in \mathbb{R}^n$ with $\|p\|_2 \leq 1$ and constant $a > 0$ is such that either*

1. $\langle y, p\rangle \geq \frac{\|y\|_2}{(\log n)^a}$ *and* $\langle y, Mp\rangle \geq \frac{\|Mp\|_{H^{-1}} \cdot \|y\|_2}{(\log n)^a}$ *or*
2. $\langle y, p\rangle \geq \frac{\|y\|_2}{(\log n)^a}$ *and* $\|Mp\|_2 \leq O\left(\frac{1}{poly(n)}\right)$.

Then as long as $\|y\|_2 \geq \frac{\beta}{n}$, choosing $\varepsilon = \min\left\{\frac{\|y\|_2}{4(\log n)^{2a}\|Mp\|_2}, \frac{1}{2(\log n)^a}\right\}$ gives

$$\|\nabla\Phi(x + \varepsilon p)\|_2 \cdot \Phi(x + \varepsilon p) \leq \|\nabla\Phi(x)\|_2 \cdot \Phi(x)e^{-\tilde{\Theta}(1/n)}.$$

The idea for computing a direction p to satisfy Lemma 10 is to project y onto an appropriate eigenspace of M. More formally, suppose $M = \sum_{j=1}^n \alpha_j v_j v_j^T$ is the eigendecomposition of M. Notice that $\alpha_j \in [0, 1]$ for all j.

Given $K = \text{polylog}(n)$, define $S_k = \text{span}\{v_j \mid 2^{-k} \leq \alpha_j \leq 2^{-k+1}\}$ for $1 \leq k < K$ and define $S_K = \text{span}\{v_j \mid \alpha_j \leq 2^{-K+1}\}$. If $y_k = \text{proj}_{S_k} y$, there must be some $\|y_k\|_2 \geq \frac{1}{K}$. If $k < K$, we see that y_k satisfies case 1, and if $k = K$ it satisfies case 2 of Lemma 10. To speed up the computation of this vector, we do the following.

Lemma 11. *Suppose y and M are as above, and define $z = \frac{y}{\|y\|_2}$. For $k = 1, ..., K := \text{polylog}(n)$, define $z_k = (I - \frac{1}{2}M)^{2^k} z$. Then for some $k \leq K$, $p = z_k$ satisfies the conditions of Lemma 10.*

Such an update direction can be computed in time $\tilde{O}(mn^{\omega-1})$, and after $\tilde{O}(n)$ updates we obtain x with $\|\lambda A\|_2 \cdot \Phi(x)^2 \leq O(\frac{1}{n})$, which implies that either $\Phi(x) < 1$, in which case $x \in P$, or $\|\lambda A\|_2 \leq O(\frac{1}{n})$, in which case we can rescale.

4   Computing an Approximate John Ellipsoid

It turns out that our algorithm implicitly computes an approximate John ellipsoid for the considered cone P. Recall that a classical theorem of John [Joh48] shows that for any closed, convex set $Q \subseteq \mathbb{R}^n$, there is an ellipsoid E and a center z so that $z + E \subseteq Q \subseteq z + nE$. The bound of n is tight in general — for example for a simplex — but it can be improved to $\sqrt{n}$ for symmetric sets. This is equivalent to saying that for each convex body, there is a linear transformation that makes it n-*well rounded*. Here, a body Q is α-well rounded if $z + r \cdot B \subseteq Q \subseteq z + \alpha \cdot r \cdot B$ for some center $z \in \mathbb{R}$ and radius $r > 0$ [Bal97].

Suppose first that we use the modified gradient descent version of MWU phase for the initial phase of Algorithm 1, but we only terminate if $\Phi(x) < \frac{1}{e}$. Note that the final phase will terminate in at most $T = \tilde{O}(n)$ steps, and the step size is always bounded by $\frac{1}{2}$. Therefore if the final MWU phase outputs x we have $\|x\|_2 \leq \frac{T}{2}$ and moreover $\langle A_i, x \rangle \geq 1$ for all i. In particular, we have $B(\frac{1}{T}x, \frac{1}{T}) \subseteq P \cap B \subseteq B(\frac{1}{T}x, 2)$, which shows that $P \cap B$ is $\tilde{O}(n)$-well rounded.

Note that running the algorithm until $\Phi(x) < \frac{1}{e}$ only increases the worst case running times by a constant factor. Alternatively one can run the algorithm with standard gradient descent and a fixed step size of $\varepsilon := \Theta(\frac{1}{n})$ and only terminate when $\Phi(x) < \frac{1}{m}$. In the final phase we then obtain that $\|x\|_2 \leq O(n \ln m)$, and $\langle A_i, x \rangle \geq \ln m$ for all i. Therefore the final set $P \cap B$ will be $O(n)$-well rounded, thus removing the logarithmic terms suppressed by the $\tilde{O}$ notation.

Independent publication. The multi-rank rescaling was also discovered in a parallel and independent work by Dadush, Vegh and Zambelli [DVZ16] (see their Algorithm 5).

References

[Agm54]   Agmon, S.: The relaxation method for linear inequalities. Can. J. Math. **6**, 382–392 (1954)

[AHK05]   Arora, S., Hazan, E., Kale, S.: Fast algorithms for approximate semidefinite programming using the multiplicative weights update method. In: 46th IEEE FOCS, pp. 339–348 (2005)

[AHK12]   Arora, S., Hazan, E., Kale, S.: The multiplicative weights update method: a meta-algorithm and applications. Theor. Comp. **8**, 121–164 (2012)

[AS04]   Alon, N., Spencer, J.H.: The Probabilistic Method. Wiley Series in Discrete Mathematics and Optimization. Wiley, Hoboken (2004)

[Bal97]   Ball, K.: An elementary introduction to modern convex geometry. In: Silvio, L. (ed.) Flavors of Geometry, pp. 1–58. University Press, Cambridge (1997)

[Bet04]   Betke, U.: Relaxation, new combinatorial and polynomial algorithms for the linear feasibility problem. Discrete Comput. Geom. **32**(3), 317–338 (2004)

[CCZ14]   Conforti, M., Cornuejols, G., Zambelli, G.: Integer Programming. Springer Publishing Company Inc., Heidelberg (2014)

[Chu12] Chubanov, S.: A strongly polynomial algorithm for linear systems having a binary solution. Math. Program. **134**(2), 533–570 (2012)

[Chu15] Chubanov, S.: A polynomial projection algorithm for linear feasibility problems. Math. Program. **153**(2), 687–713 (2015)

[CKM+11] Christiano, P., Kelner, J.A., Madry, A., Spielman, D.A., Teng, S.: Electrical flows, laplacian systems, and faster approximation of maximum flow in undirected graphs. In: Proceedings of the 43rd ACM Symposium on Theory of Computing, New York, NY, USA, pp. 273–282 (2011)

[Dan51] Dantzig, G.B.: Maximization of a linear function of variables subject to linear inequalities. In: Activity Analysis of Production and Allocation, Cowles Commission Monograph, vol. 13, pp. 339–347. John Wiley & Sons Inc., Chapman & Hall Ltd., New York (1951)

[DV06] Dunagan, J., Vempala, S.: A simple polynomial-time rescaling algorithm for solving linear programs. Math. Program. **114**(1), 101–114 (2006)

[DVZ16] Dadush, D., Végh, L.A., Zambelli, G.: Rescaling algorithms for linear programming - part I: conic feasibility. CoRR, abs/1611.06427 (2016)

[GK07] Garg, N., Könemann, J.: Faster and simpler algorithms for multicommodity flow and other fractional packing problems. SIAM J. Comput. **37**(2), 630–652 (2007)

[Hač79] Hačijan, L.G.: A polynomial algorithm in linear programming. Dokl. Akad. Nauk SSSR **244**(5), 1093–1096 (1979)

[Joh48] John, F.: Extremum problems with inequalities as subsidiary conditions. In: Friedrichs, K.O., Neugebauer, O.E., Stoker, J.J. (eds.) Studies and Essays presented to R. Courant on his 60th Birthday, pp. 187–204. Interscience Publishers, New York (1948)

[Kar84] Karmarkar, N.: A new polynomial-time algorithm for linear programming. Combinatorica **4**(4), 373–395 (1984)

[KM72] Klee, V., Minty, G.: How good is the simplex algorithm? In: Inequalities, III (Proceedings Third Symposium, UCLA, 1969; Dedicated to the Memory of Theodore S. Motzkin), pp. 159–175. Academic Press, New York (1972)

[LS15] Lee, Y., Sinford, A.: A new polynomial-time algorithm for linear programming (2015). https://arxiv.org/abs/1312.6677

[Mad10] Madry, A.: Faster approximation schemes for fractional multicommodity flow problems via dynamic graph algorithms. In: Proceedings of the 42nd ACM Symposium on Theory of Computing, New York, NY, pp. 121–130 (2010)

[Nes05] Nesterov, Y.: Excessive gap technique in nonsmooth convex minimization. SIAM J. Optim. **16**(1), 235–249 (2005)

[PS12] Peña, J., Soheili, N.: A smooth perceptron algorithm. SIAM J. Optim. **22**(2), 728–737 (2012)

[PS16] Peña, J., Soheili, N.: A deterministic rescaled perceptron algorithm. Math. Program. **155**(1–2), 497–510 (2016)

[PST95] Plotkin, S.A., Shmoys, D.B., Tardos, E.: Fast approximation algorithms for fractional packing and covering problems. Math. Oper. Res. **20**(2), 257–301 (1995)

[Sch86] Schrijver, A.: Theory of linear and integer programming. Wiley-Interscience Series in Discrete Mathematics. John Wiley and Sons, Inc., New York (1986)

[Vaz01] Vazirani, V.: Approximation Algorithms. Springer, Heidelberg (2001)

[WS11] Williamson, D.P., Shmoys, D.B.: The Design of Approximation Algorithms. University Press, Cambridge (2011)

Min-Max Theorems for Packing and Covering Odd (u, v)-trails

Sharat Ibrahimpur and Chaitanya Swamy$^{(\boxtimes)}$

Combinatorics and Optimization, University of Waterloo,
Waterloo, ON N2L 3G1, Canada
{sharat.ibrahimpur,cswamy}@uwaterloo.ca

Abstract. We investigate the problem of packing and covering odd (u, v)-trails in a graph. A (u, v)-trail is a (u, v)-walk that is allowed to have repeated vertices but no repeated edges. We call a trail *odd* if the number of edges in the trail is odd. Let $\nu(u, v)$ denote the maximum number of edge-disjoint odd (u, v)-trails, and $\tau(u, v)$ denote the minimum size of an edge-set that intersects every odd (u, v)-trail.

We prove that $\tau(u, v) \leq 2\nu(u, v) + 1$. Our result is *tight*—there are examples showing that $\tau(u, v) = 2\nu(u, v) + 1$—and substantially improves upon the bound of 8 obtained in [5] for $\tau(u, v)/\nu(u, v)$. Our proof also yields a polynomial-time algorithm for finding a cover and a collection of trails satisfying the above bounds.

Our proof is simple and has two main ingredients. We show that (loosely speaking) the problem can be reduced to the problem of packing and covering odd $(\{u, v\}, \{u, v\})$-trails losing a factor of 2 (either in the number of trails found, or the size of the cover). Complementing this, we show that the odd-$(\{u, v\}, \{u, v\})$-trail packing and covering problems can be tackled by exploiting a powerful min-max result of [2] for packing vertex-disjoint nonzero A-paths in group-labeled graphs.

1 Introduction

Min-max theorems are a classical and central theme in combinatorics and combinatorial optimization, with many such results arising from the study of packing and covering problems. For instance, *Menger's theorem* [10] gives a *tight* min-max relationship for packing and covering edge-disjoint (or internally vertex-disjoint) (u, v)-paths: the maximum number of edge-disjoint (or internally vertex-disjoint) (u, v)-paths (i.e., *packing number*) is equal to the minimum number of edges (or vertices) needed to cover all u-v paths (i.e., *covering number*); the celebrated *max-flow min-cut theorem* generalizes this result to arbitrary edge-capacitated graphs. Another well-known example is the *Lucchesi-Younger theorem* [8], which shows that the maximum number of edge-disjoint directed cuts equals the minimum-size of an arc-set that intersects every directed cut.

A full version of the paper is available on the CS arXiv.
Research supported by NSERC grant 327620-09 and an NSERC DAS Award.

F. Eisenbrand and J. Koenemann (Eds.): IPCO 2017, LNCS 10328, pp. 279–291, 2017.
DOI: 10.1007/978-3-319-59250-3_23

Motivated by Menger's theorem, it is natural to ask whether similar (tight or approximate) min-max theorems hold for other variants of path-packing and path-covering problems. Questions of this flavor have attracted a great deal of attention. Perhaps the most prominent results known of this type are Mader's min-max theorems for packing vertex-disjoint S-paths [9,12], which generalize both the *Tutte-Berge formula* and Menger's theorem, and a further far-reaching generalization of this due to Chudnovsky et al. [2] regarding packing vertex-disjoint non-zero A-paths in group-labeled graphs.

We consider a different variant of the (u,v)-path packing and covering problems, wherein we impose *parity constraints* on the paths. Such constraints naturally arise in the study of multicommodity-flow problem, which can be phrased in terms of packing odd circuits in a signed graph, and consequently, such odd-circuit packing and covering problems have been widely investigated (see, e.g., [13], Chap. 75). Focusing on (u,v)-paths, a natural variant that arises involves packing and covering *odd* (u,v)-paths, where a (u,v)-path is odd if it contains an odd number of edges. However, there are simple examples [5] showing an unbounded gap between the packing and covering numbers in this setting.

In light of this, following [5], we investigate the min-max relationship for packing and covering odd (u,v)-trails. An *odd* $(u,v) - trail$ is a (u,v)-walk with *no repeated edges* and an odd number of edges. Churchley et al. [5] seem to have been the first to consider this problem. They showed that the (worst-case) ratio between the covering and packing numbers for odd (u,v)-trails is at most 8—which is in stark contrast with the setting of odd (u,v) paths, where the ratio is unbounded—and at least 2, so there is no tight min-max theorem like Menger's theorem. They also motivate the study of odd (u,v)-trails from the perspective of studying *totally-odd immersions*. In particular, determining if a graph G has k edge-disjoint odd (u,v)-trails is equivalent to deciding if the 2-vertex graph with k parallel edges has a totally-odd immersion into G.

Our results. We prove a *tight bound* on the ratio of the covering and packing numbers for odd (u,v)-trails, which also substantially improves the bound of 8 shown in [5] for this covering-vs-packing ratio.[1] Let $\nu(u,v)$ and $\tau(u,v)$ denote respectively the packing and covering numbers for odd (u,v)-trails. Our main result (Theorem 3.1) establishes that $\tau(u,v) \le 2\nu(u,v) + 1$. Furthermore, we obtain in polynomial time a certificate establishing that $\tau(u,v) \le 2\nu(u,v) + 1$. This is because we show that, for any integer $k \ge 0$, we can compute in polynomial time, a collection of k edge-disjoint odd (u,v)-trails, or an odd-(u,v)-trail cover of size at most $2k-1$. As mentioned earlier, there are examples showing $\tau(u,v) = 2\nu(u,v) + 1$ (see Fig. 1), so our result *settles* the question of obtaining worst-case bounds for the $\tau(u,v)/\nu(u,v)$ ratio.

Notably, our proof is also simple, and noticeably simpler than (and different from) the one in [5]. We remark that the proof in [5] constructs covers of a certain

---

[1] This bound was later improved to 5 [3,4,7]. We build upon some of the ideas in [7].

form; in the full version, we prove a *lower bound* showing that such covers cannot yield a bound better than 3 on the covering-vs-packing ratio.

Our techniques. We focus on showing that for any k, we can obtain either k edge-disjoint odd (u, v)-trails or a cover of size at most $2k - 1$. This follows from two other auxiliary results which are potentially of independent interest.

Our key insight is that one can *decouple* the requirements of parity and u-v connectivity when constructing odd (u, v)-trails. More precisely, we show that if we have a collection of k edge-disjoint odd $(\{u, v\}, \{u, v\})$-trails, that is, odd trails that start and end at a vertex of $\{u, v\}$, and the u-v edge connectivity, denoted $\lambda(u, v)$, is at least $2k$, then we can obtain k edge-disjoint odd (u, v)-trails (Theorem 3.3). Notice that if $\lambda(u, v) < 2k$, then a min u-v cut yields a cover of the desired size. So the upshot of Theorem 3.3 is that it reduces our task to the *relaxed* problem of finding k edge-disjoint odd $(\{u, v\}, \{u, v\})$-trails. The proof of Theorem 3.3 relies on elementary arguments (see Sect. 4). We show that given a fixed collection of $2k$ edge-disjoint (u, v)-paths, we can always modify our collection of edge-disjoint trails so as to make progress by decreasing the number of contacts that the paths make with the trails and/or by increasing the number of odd (u, v) trails in the collection. Repeating this process a small number of times thus yields the k edge-disjoint odd (u, v)-trails.

Complementing Theorem 3.3 we prove that we can either obtain k edge-disjoint $(\{u, v\}, \{u, v\})$-trails, or find an odd-$(\{u, v\}, \{u, v\})$-trail cover (which is also an odd-(u, v)-trail cover) of size at most $2k - 2$ (Theorem 3.2). This proof relies on a powerful result of [2] about *packing and covering nonzero A-paths in group-labeled graphs* (see Sect. 5, which defines these concepts precisely). The idea here is that [2] show that one can obtain either k vertex-disjoint nonzero A-paths or a set of at most $2k - 2$ vertices intersecting all nonzero A-paths, and this can be done in polytime [1,6]. This is the same type of result that we seek, except that we care about edge-disjoint trails, as opposed to vertex-disjoint paths. However, by moving to a suitable gadget graph where we replace each vertex by a clique, we can encode trails as paths, and edge-disjointness is captured by vertex-disjointness. Applying the result in [2] then yields Theorem 3.2.

Related work. Churchley et al. [5] initiated the study of min-max theorems for packing and covering odd (u, v)-trails. They cite the question of totally-odd immersions as motivation for their work. We say that a graph H has an *immersion* [11] into another graph G, if one can map V_H bijectively to some $U \subseteq V(G)$, and E_H to edge-disjoint trails connecting the corresponding vertices in U. (As noted by [5], trails are more natural objects than paths in the context of reversing an edge-splitting-off operation, as this, in general, creates trails.) An immersion is *strong* if the trails do not internally meet U, and *weak* otherwise. An immersion is called *totally odd* if all trails are of odd length.

In an interesting contrast to the unbounded gap between the covering and packing numbers for odd (u, v)-paths, [14] showed that the covering number is at most twice the *fractional packing number* (which is the optimal value of the natural odd-(u, v)-path-packing LP).

The notions of odd paths and trails can be generalized and abstracted in two ways. The first involves *signed graphs* [15], and there are various results on packing odd *circuits* in signed graphs, which are closely related to multicommodity flows (see [13], Chap. 75). The second involves *group-labeled graphs*, for which [1,2] present strong min-max theorems for packing and covering vertex-disjoint nonzero A-paths.

2   Preliminaries and Notation

Let $G = (V, E)$ be an undirected graph. For $X \subseteq V$, we use $E(X)$ to denote the set of edges having both endpoints in X and $\delta(X)$ to denote set of edges with exactly one endpoint in X. For disjoint $X, Y \subseteq V$, we use $E(X, Y)$ to denote the set of edges with one end in X and one end in Y.

A (p, q)-*walk* is a sequence $(x_0, e_1, x_1, e_2, x_2, \ldots, e_r, x_r)$, where $x_0, \ldots, x_r \in V$ with $x_0 = p$, $x_r = q$, and e_i is an edge with ends x_{i-1}, x_i for all $i = 1, \ldots, r$. The vertices $x_1, \ldots, x_{r-1}$ are called the *internal vertices* of this walk. We say that such a (p, q)-walk is a:

- (p, q)-*path*, if either $r > 0$ and all the x_is are distinct (so $p \neq q$), or $r = 0$, which we call a *trivial path*;
- (p, q)-*trail* if all the e_is are distinct (we could have $p = q$).

Thus, a (p, q)-trail is a (p, q)-walk that is allowed to have repeated vertices but *no repeated edges*. Given vertex-sets $A, B \subseteq V$, we say that a trail is an (A, B)-trail to denote that it is a (p, q)-trail for some $p \in A, q \in B$. A (p, q)-trail is called *odd* (respectively, *even*) if it has an odd (respectively, even) number of edges.

Definition 2.1. Let $G = (V, E)$ be a graph, and $u, v \in V$ (we could have $u = v$).

(a) The *packing number for odd* (u, v)-*trails*, denoted $\nu(u, v; G)$, is the maximum number of edge-disjoint odd (u, v)-trails in G.
(b) We call a subset of edges C an *odd* (u, v)-*trail cover* of G if it intersects every odd (u, v)-trail in G. The *covering number for odd* (u, v)-*trails*, denoted $\tau(u, v; G)$, is the minimum size of an odd (u, v)-trail cover of G.

We drop the argument G when it is clear from the context.

For any two distinct vertices x, y of G, we denote the size of a minimum (x, y)-cut in G by $\lambda(x, y; G)$, and drop G when it is clear from the context. By the max-flow min-cut (or Menger's) theorem, $\lambda(x, y; G)$ is also the maximum number of edge-disjoint (x, y)-paths in G.

3   Main Results and Proof Overview

Our main result is the following *tight* approximate min-max theorem relating the packing and covering numbers for odd (u, v) trails.

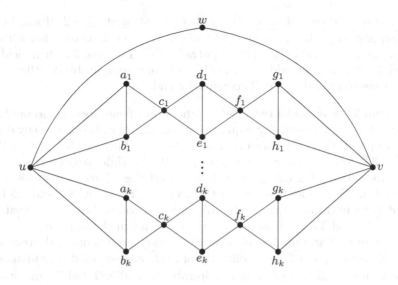

Fig. 1. Graph with $\nu(u,v) = k$, $\tau(u,v) = 2k+1$.

Theorem 3.1. *Let $G = (V,E)$ be an undirected graph, and $u,v \in V$. For any nonnegative integer k, we can obtain in polynomial time, either:*

1. *k edge-disjoint odd (u,v)-trails in G, or*
2. *an odd (u,v)-trail cover of G of size at most $2k-1$.*

Hence, we have $\tau(u,v;G) \le 2 \cdot \nu(u,v;G) + 1$.

Theorem 3.1 is tight (this was communicated to us by [3]), as can be seen from Fig. 1. The theorem follows readily from the following two results.

Theorem 3.2. *Let $G = (V,E)$ be an undirected graph and $s \in V$. For any nonnegative integer k, we can obtain in polynomial time:*

1. *k edge-disjoint odd (s,s)-trails in G, or*
2. *an odd (s,s)-trail cover of G of size at most $2k-2$.*

Theorem 3.3. *Let $G = (V,E)$ be an undirected graph, and $u,v \in V$ with $u \neq v$. Let $\widehat{T}$ be a collection of edge-disjoint odd $(\{u,v\},\{u,v\})$-trails in G. If $\lambda(u,v) \ge 2 \cdot |\widehat{T}|$, then we can obtain in polytime $|\widehat{T}|$ edge-disjoint odd (u,v)-trails in G.*

Proof of Theorem 3.1. If $u = v$, then Theorem 3.2 yields the desired statement. So suppose $u \neq v$. We may assume that $\lambda(u,v) \ge 2k$, since otherwise a minimum (u,v)-cut in G is an odd (u,v)-trail cover of the required size. Let E_{uv} be the uv edge(s) in G (which could be $\emptyset$). Let $\widehat{G}$ be obtained from $G - E_{uv}$ by identifying u and v into a new vertex s. (Note that $\widehat{G}$ has no loops.) Any odd (u,v)-trail in $G - E_{uv}$ maps to an odd (s,s)-trail in $\widehat{G}$. We apply Theorem 3.2 to $\widehat{G}, s, k' = k - |E_{uv}|$. If this returns an odd-(s,s)-trail cover C of size at most $2k'-2$, then

$C \cup E_{uv}$ is an odd-(u, v)-trail cover for G of size at most $2k - 2$. If we obtain a collection of k' edge-disjoint odd (s, s)-trails in $\widehat{G}$, then these together with E_{uv} yield k edge-disjoint odd $(\{u, v\}, \{u, v\})$-trails in G. Theorem 3.3 then yields the required k edge-disjoint odd (u, v)-trails. Polytime computability follows from the polytime computability in Theorems 3.2 and 3.3.   □

Theorem 3.3 is our chief technical insight, which facilitates the decoupling of the parity and u-v connectivity requirements of odd (u, v)-trails, thereby driving the entire proof. (It can be seen as a refinement of Theorem 5.1 in [7].) While Theorem 3.2 returns $(\{u, v\}, \{u, v\})$-trails with the right *parity*, Theorem 3.3 supplies the missing ingredient needed to convert these into (u, v)-trails (of the same parity). We give an overview of the proofs of Theorems 3.2 and 3.3 below before delving into the details in the subsequent sections. We remark that both Theorems 3.2 and 3.3 are *tight* as well; we show this in the full version.

The proof of Theorem 3.3 relies on elementary arguments and proceeds as follows (see Sect. 4). Let $\mathcal{P}$ be a collection of $2 \cdot |\widehat{T}|$ edge-disjoint (u, v)-paths. We provide a simple, efficient procedure to iteratively modify $\widehat{T}$ (whilst maintaining $|\widehat{T}|$ edge-disjoint odd $(\{u, v\}, \{u, v\})$-trails) and eventually obtain $|\widehat{T}|$ odd (u, v)-trails. Let $\mathcal{P}_0 \subseteq \mathcal{P}$ be the collection of paths of $\mathcal{P}$ that are edge-disjoint from trails in $\widehat{T}$. First, we identify the trivial case where $|\mathcal{P}_0|$ is sufficiently large. If so, these paths and $\widehat{T}$ directly yield odd (u, v)-trails as follows: odd-length paths in $\mathcal{P}_0$ are already odd (u, v)-trails, and even-length paths in $\mathcal{P}_0$ can be combined with odd (u, u)- and odd (v, v)- trails to obtain odd (u, v)-trails.

The paths in $\mathcal{P} \setminus \mathcal{P}_0$, all share at least one edge with some trail in $\widehat{T}$. Each path is a sequence of edges from u to v. If the first edge that a path $P \in \mathcal{P}$ shares with a trail in $\widehat{T}$ lies on a (v, v)-trail T, then it is easy to use parts of P and T to obtain an odd (u, v)-trail that is edge-disjoint from all other trails in $\widehat{T}$, and thereby make progress by increasing the number of odd (u, v)-trails in the collection. A similar conclusion holds if the last edge that a path shares with a trail in $\widehat{T}$ lies on a (u, u)-trail. If neither of the above cases apply, then the paths in $\mathcal{P} \setminus \mathcal{P}_0$ are in a sense *highly tangled* (which we formalize later) with trails in $\widehat{T}$. We then infer that $\mathcal{P} \setminus \mathcal{P}_0$ and $\widehat{T}$ must satisfy some simple structural properties, and leverage this to carefully modify the collection $\widehat{T}$ (while preserving edge-disjointness) so that the new set of trails are "less tangled" with $\mathcal{P}$ than $\widehat{T}$, and thereby make progress. Continuing this procedure a polynomial number of times yields the desired collection of $|\widehat{T}|$ edge-disjoint odd (u, v)-trails.

The proof of Theorem 3.2 relies on the key observation that we can cast our problem as the problem of packing and covering nonzero A-paths in a group-labeled graph (H, Γ) [2] for a suitable choice of $A, H,$ and Γ (see Sect. 5). In the latter problem, (1) H denotes an oriented graph whose arcs are labeled with elements of a group Γ, and (2) a non-zero A-path is a path in the undirected version of H whose ends lie in A, whose Γ-length, which is the sum of $\pm\gamma_e$s (suitably defined) for arcs in P, is non-zero. Chudnovsky et al. [2] show that either there are k *vertex-disjoint* non-zero A-paths, or there is a vertex-set of size at most $2k - 2$ intersecting every non-zero A-path (Theorem 1.1 in [2]).

We show that applying their result to a suitable "gadget graph" H (essentially the line graph of G), yields Theorem 3.2 (see Sect. 5). Polytime computability follows because a subsequent paper [1] gave a polytime algorithm for finding a maximum-size collection of vertex-disjoint non-zero A-paths, and it is implicit in their proof that this also yields a suitable vertex-covering of non-zero A-paths [6].

We remark that while the use of the packing-covering result in [2] yields quite a compact proof of Theorem 3.2, it also makes the resulting proof somewhat opaque since we apply the result in [2] to the gadget graph. However, it is possible to translate the min-max theorem for packing vertex-disjoint nonzero A-paths proved in [2] to our setting and obtain the following more-accessible min-max theorem for packing edge-disjoint odd (s, s)-trails (stated in terms of G and not the gadget graph). In the full version, we prove that

$$\nu(s, s; G) = \min\left(|E(S) \setminus F| + \sum_{H \in \mathrm{comp}(G-S)} \left\lfloor \frac{|E(S, H)|}{2} \right\rfloor \right)$$

where the minimum is taken over all bipartite subgraphs (S, F) of G such that $s \in S$. (Notice that Theorem 3.2 follows easily from this min-max formula.)

4   Proof of Theorem 3.3: Converting Edge-Disjoint Odd $(\{u, v\}, \{u, v\})$-trails to Edge-Disjoint Odd (u, v)-trails

Recall that $\widehat{T}$ is a collection of edge-disjoint odd $(\{u, v\}, \{u, v\})$-trails in G. We denote the subset of odd (u, u)-trails, odd (v, v)-trails, and odd (u, v)-trails in $\widehat{T}$ by $\widehat{T}_{uu}$, $\widehat{T}_{vv}$, and $\widehat{T}_{uv}$, respectively. Let $k_{uu}(\widehat{T}) = |\widehat{T}_{uu}|$, $k_{vv}(\widehat{T}) = |\widehat{T}_{vv}|$, and $k_{uv}(\widehat{T}) = |\widehat{T}_{uv}|$. To keep notation simple, we will drop the argument $\widehat{T}$ when its clear from the context. Since we are given that $\lambda(u, v) \geq 2 \cdot |\widehat{T}|$, we can obtain a collection $\mathcal{P}$ of $2 \cdot |\widehat{T}|$ edge-disjoint (u, v)-paths in G. In the sequel, while we will modify our collection of odd $(\{u, v\}, \{u, v\})$-trails, $\mathcal{P}$ stays fixed.

We now introduce the key notion of a *contact* between a trail T and a (u, v)-path P. Suppose that $P = (x_0, e_1, x_1, \ldots, e_r, x_r)$ for some $r \geq 1$.

Definition 4.1. A *contact* between P and T is a *maximal subpath* S of P containing at least one edge such that S is also a subtrail of T i.e., for $0 \leq i < j \leq r$, we say that $(x_i, e_{i+1}, x_{i+1}, \ldots, e_j, x_j)$ is a contact between P and T if $(x_i, e_{i+1}, x_{i+1}, \ldots, e_j, x_j)$ is a subtrail of T, but neither $(x_{i-1}, e_i, x_i, \ldots e_j, x_j)$ (if $i > 0$) nor $(x_i, e_{i+1}, x_{i+1}, \ldots, x_j, e_{j+1}, x_{j+1})$ (if $j < r$) is a subtrail of T.

Define $\mathcal{C}(P, T) = \left| \{ (i, j) : 0 \leq i < j \leq r, \quad (x_i, e_{i+1}, x_{i+1}, \ldots, e_j, x_j) \right.$

$\left. \text{is a contact between } P \text{ and } T \} \right|$

By definition, contacts between P and T are edge disjoint. For an edge-disjoint collection $\mathcal{T}$ of trails, we use $\mathcal{C}(P, \mathcal{T})$ to denote $\sum_{T \in \mathcal{T}} \mathcal{C}(P, T)$. So if $\mathcal{C}(P, \mathcal{T}) = 0$, then P is edge-disjoint from every trail in $\mathcal{T}$. Otherwise, we use the term *first contact* of P to refer to the contact arising from the first edge that P shares with some trail in $\mathcal{T}$ (note that P is a (u,v)-walk so is a sequence from u to v). Similarly, the *last contact* of P is the contact arising from the last edge that P shares with some trail in $\mathcal{T}$. If $\mathcal{C}(P,\mathcal{T}) = 1$, then the first and last contacts of P are the same. We further overload notation and use $\mathcal{C}(\mathcal{P},\mathcal{T})$ to denote $\sum_{P\in\mathcal{P}} \mathcal{C}(P,\mathcal{T}) = \sum_{P\in\mathcal{P}, T\in\mathcal{T}} \mathcal{C}(P,T)$. We use $\mathcal{C}(\mathcal{P},\mathcal{T})$ as a measure of how "tangled" $\mathcal{T}$ is with $\mathcal{P}$. The following lemma classifies five different cases that arise for any pair of edge-disjoint collections of odd $(\{u,v\},\{u,v\})$-trails and (u,v)-paths.

Lemma 4.2. *Let $\mathcal{T}$ be a collection of edge-disjoint odd $(\{u,v\},\{u,v\})$-trails in G. If $|\mathcal{P}| \geq 2 \cdot |\mathcal{T}|$, then one of the following conditions holds.*

(a) *There are at least $k_{uu}(\mathcal{T}) + k_{vv}(\mathcal{T})$ paths in $\mathcal{P}$ that make no contact with any trail in $\mathcal{T}$.*
(b) *There exists a path $P \in \mathcal{P}$ that makes its first contact with a trail $T \in \mathcal{T}_{vv}$.*
(c) *There exists a path $P \in \mathcal{P}$ that makes its last contact with a trail $T \in \mathcal{T}_{uu}$.*
(d) *There exist three distinct paths $P_1, P_2, P_3 \in \mathcal{P}$ which make their first contact with a trail $T \in \mathcal{T}_{uu} \cup \mathcal{T}_{uv}$.*
(e) *There exist three distinct paths $P_1, P_2, P_3 \in \mathcal{P}$ which make their last contact with a trail $T \in \mathcal{T}_{uv} \cup \mathcal{T}_{vv}$.*

Proof. To keep notation simple, we drop the argument $\mathcal{T}$ in the proof. Suppose that conclusion (a) does not hold. Then there are at at least $2 \cdot |\mathcal{T}| - (k_{uu} + k_{vv} - 1) = 2k_{uv} + k_{uu} + k_{vv} + 1$ paths in $\mathcal{P}$ that make at least one contact with some trail in $\mathcal{T}$. Let $\mathcal{P}' \subseteq \mathcal{P}$ be this collection of paths. If either conclusions (b) or (c) hold (for some $P \in \mathcal{P}'$), then we are done, so assume that this is not the case. Then, every path $P \in \mathcal{P}'$ makes its first contact with a trail in $\mathcal{T}_{uu} \cup \mathcal{T}_{uv}$ and its last contact with a trail in $\mathcal{T}_{uv} \cup \mathcal{T}_{vv}$. Note that the number of first and last contacts are both at least $2k_{uv} + k_{uu} + k_{vv} + 1 > 2 \cdot \min(k_{uv} + k_{uu}, k_{uv} + k_{vv})$. So if $k_{uu} \leq k_{vv}$, then by the Pigeonhole principle, there are at least 3 paths that make their first contact with some $T \in \mathcal{T}_{uu} \cup \mathcal{T}_{uv}$, i.e., conclusion (d) holds. Similarly, if $k_{vv} \leq k_{uu}$, then conclusion (e) holds. □

We now leverage the above classification and show that in each of the above five cases, we can make progress by "untangling" the trails (i.e., decreasing $\mathcal{C}(\mathcal{P},\mathcal{T})$) and/or increasing the number of odd (u,v)-trails in our collection.

Lemma 4.3. *Let $\mathcal{T}$ be a collection of edge-disjoint odd $(\{u,v\},\{u,v\})$-trails. If $|\mathcal{P}| \geq 2 \cdot |\mathcal{T}|$, we can obtain another collection $\mathcal{T}'$ of edge-disjoint odd $(\{u,v\},\{u,v\})$-trails such that at least one of the following holds.*

(i) $k_{uv}(\mathcal{T}') = |\mathcal{T}|$.
(ii) $\mathcal{C}(\mathcal{P},\mathcal{T}') \leq \mathcal{C}(\mathcal{P},\mathcal{T})$ *and* $k_{uv}(\mathcal{T}') = k_{uv}(\mathcal{T}) + 1$.
(iii) $\mathcal{C}(\mathcal{P},\mathcal{T}') \leq \mathcal{C}(\mathcal{P},\mathcal{T}) - 1$ *and* $k_{uv}(\mathcal{T}') \geq k_{uv}(\mathcal{T}) - 1$.

Proof. If $k_{uv}(\mathcal{T}) = |\mathcal{T}|$, then (i) holds trivially by taking $\mathcal{T}' = \mathcal{T}$. So we may assume that $\mathcal{T}$ contains some odd (u,u)- or odd (v,v)-trail. Observe that $\mathcal{T}$ and $\mathcal{P}$ satisfy the conditions of Lemma 4.2, so at least one of the five conclusions of Lemma 4.2 applies. We handle each case separately.

(a) At least $k_{uu}(\mathcal{T}) + k_{vv}(\mathcal{T})$ paths in $\mathcal{P}$ have zero contacts with $\mathcal{T}$. Let $\mathcal{P}_0 = \{P \in \mathcal{P} : \mathcal{C}(P, \mathcal{T}) = 0\}$. Consider some $P \in \mathcal{P}_0$. If P is odd, we can replace an odd (u, u)- or odd (v, v)- trail in $\mathcal{T}$ with P. If P is even, then P can be combined with an odd (u, u)- or odd (v, v)- trail to obtain an odd (u, v)-trail. Since $|\mathcal{P}_0| \geq k_{uu}(\mathcal{T}) + k_{vv}(\mathcal{T})$, we create $k_{uu}(\mathcal{T}) + k_{uv}(\mathcal{T})$ odd (u, v)-trails this way, and this new collection $\mathcal{T}'$ satisfies (i).

(b) Some $P \in \mathcal{P}$ makes its first contact with an odd (v, v)-trail $T \in \mathcal{T}$. Let the first vertex in the first contact between P and T be x. Observe that x partitions the trail T into two subtrails S_1 and S_2. Since T is an odd trail, exactly one of S_1 and S_2 is odd. We can now obtain an odd (u, v)-trail T' by traversing P from u to x, and then traversing S_1 or S_2, whichever yields odd parity (see Fig. 2). Since P already made a contact with T, we have $\mathcal{C}(P, T') \leq \mathcal{C}(P, T)$, and $\mathcal{C}(Q, T') \leq \mathcal{C}(Q, T)$ for any other path $Q \in \mathcal{P}$. Thus, taking $T' = (\mathcal{T} \cup \{T'\}) \setminus \{T\}$, we have $\mathcal{C}(\mathcal{P}, T') \leq \mathcal{C}(\mathcal{P}, \mathcal{T})$, and (ii) holds.

(c) Some $P \in \mathcal{P}$ makes its last contact with an odd (u, u)-trail $T \in \mathcal{T}$. This is completely symmetric to (b), so a similar strategy works and we satisfy (ii).

(d) Paths $P_1, P_2, P_3 \in \mathcal{P}$ that make their first contact with an odd $(u, \{u, v\})$-trail $T \in \mathcal{T}$. Note that all contacts between paths in $\mathcal{P}$ and trails in $\mathcal{T}$ are edge disjoint, since the paths in $\mathcal{P}$ are edge disjoint and the trails in $\mathcal{T}$ are edge disjoint. For $i = 1, 2, 3$, let the first vertex in the first contact of P_i (with T) be x_i. Let Q_i denote the subpath of P_i between u and x_i. Note that T is a sequence of edges from u to some vertex in $\{u, v\}$. Without loss of generality, assume that in T, the first contact of P_1 appears before the first contact of P_2, which appears before the first contact of P_3. The vertices x_1, x_2, x_3 partition the trail T into four subtrails S_0, S_1, S_2, S_3 (see Fig. 3). For a trail X, we denote the reverse sequence of X by $\overline{X}$. Now consider the following trails (where $+$ denotes concatenation):

$$T_1 = S_0 + \overline{Q}_1, \quad T_2 = Q_1 + S_1 + \overline{Q}_2, \quad T_3 = Q_2 + S_2 + \overline{Q}_3, \quad T_4 = Q_3 + S_3.$$

Observe that the disjoint union of edges in T_1, T_2, T_3, and T_4 has the same parity as that of T, and hence at least one of the T_is is an odd trail; call this trail T'. Let $\mathcal{T}' = \mathcal{T} \cup \{T'\} \setminus \{T\}$. By construction, every T_i avoids at least one of the (first) contacts made by P_1, P_2, or P_3 (with T). Also, for any other path $Q \in \mathcal{P} \setminus \{P_1, P_2, P_3\}$, we have $\mathcal{C}(Q, T') \leq \mathcal{C}(Q, T)$. Therefore, $\mathcal{C}(\mathcal{P}, T') \leq \mathcal{C}(\mathcal{P}, \mathcal{T}) - 1$. It could be that T was an odd (u, v)-trail, which is now replaced by an odd (u, u)-trail, so $k_{uv}(\mathcal{T}') \geq k_{uv}(\mathcal{T}) - 1$. So we satisfy (iii).

(e) Paths $P_1, P_2, P_3 \in \mathcal{P}$ make their last contact with an odd $(\{u, v\}, v)$-trail in $\mathcal{T}$. This is symmetric to (d); the same approach works, so (iii) holds.   □

Theorem 3.3 now follows by simply applying Lemma 4.3 starting with the initial collection $\mathcal{T}^0 := \widehat{\mathcal{T}}$ until conclusion (i) of Lemma 4.3 applies. The $\mathcal{T}'$ returned by this final application of Lemma 4.3 then satisfies the theorem statement.

We now argue that this process terminates in at most $2 \cdot |E(G)| + |\widehat{\mathcal{T}}|$ steps, which will conclude the proof. Let $k = |\widehat{\mathcal{T}}|$. Consider the following potential

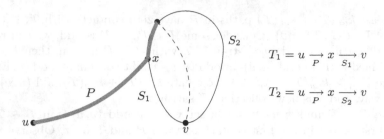

Fig. 2. Path P makes its first contact with an odd (v, v)-trail.

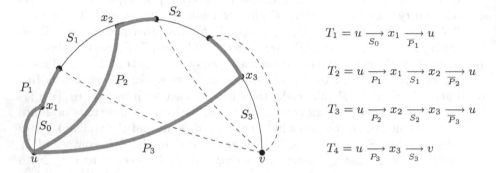

Fig. 3. Paths P_1, P_2, P_3 make their first contact with an odd (u, v)-trail.

function defined on a collection $\mathcal{T}$ of k edge-disjoint odd $(\{u, v\}, \{u, v\})$-trails: $\phi(\mathcal{T}) := 2 \cdot \mathcal{C}(\mathcal{P}, \mathcal{T}) - k_{uv}(\mathcal{T})$. Consider any iteration where we invoke Lemma 4.3 and move from a collection $\mathcal{T}$ to another collection $\mathcal{T}'$ with $k_{uv}(\mathcal{T}') < k$. Then, either conclusion (ii) or (iii) of Lemma 4.3 applies, and it is easy to see that $\Phi(\mathcal{T}') \leq \phi(\mathcal{T}) - 1$. Finally, we have $-k \leq \Phi(\mathcal{T}) \leq 2 \cdot |E(G)|$ for all $\mathcal{T}$ since $0 \leq \mathcal{C}(\mathcal{P}, \mathcal{T}) \leq |E(G)|$ as the contacts between paths in $\mathcal{P}$ and trails in $\mathcal{T}$ are edge-disjoint, so the process terminates in at most $2|E(G)| + k$ steps.

5   Proof of Theorem 3.2

Our proof relies on two reductions both involving non-zero A-paths in a group-labeled graph, which we now formally define. A *group-labeled graph* is a pair (H, Γ), where Γ is a group, and $H = (N, E')$ is an oriented graph (i.e., for any $u, v \in N$, if $(u, v) \in E'$ then $(v, u) \notin E'$) whose arcs are labeled with elements of Γ. All addition (and subtraction) operations below are always with respect to the group Γ. A path P in H is a sequence $(x_0, e_1, x_1, \ldots, e_r, x_r)$, where the x_is are distinct, and each e_i has ends x_i, x_{i+1} but could be oriented either way (i.e., as (x_i, x_{i+1}) or (x_{i+1}, x_i)). (So upon removing arc directions, P yields a path in the undirected version of H.) We say that P traverses e_i in the direction (x_i, x_{i+1}). The Γ-*length* (or simply length) of P, denoted $\gamma(P)$, is the sum of $\pm\gamma_e$s for arcs in P, where we count $+\gamma_e$ for e if P's traversal of e matches e's orientation and $-\gamma_e$ otherwise. Given $A \subseteq N$, an A-path is a path $(x_0, e_1, \ldots, e_r, x_r)$ where $r \geq 1$, and $x_0, x_r \in A$; finally, call an A-path P a nonzero A-path if $\gamma(P) \neq 0$ (where 0 denotes the identity element for Γ).

Chudnovsky et al. [2] proved the following theorem as a consequence of a min-max formula they obtain for the maximum number of nonzero vertex-disjoint A-paths. Subsequently, [1] devised a polytime algorithm to compute the maximum number of vertex-disjoint A-paths. Their algorithm also implicitly computes the quantities needed in (the minimization portion of) their min-max formula to show the optimality of the collection of A-paths they return [6]; this in turn easily yields the vertex-set mentioned in Theorem 5.1.

Theorem 5.1 [1,2]. *Let $\big(H = (N, E'), \Gamma\big)$ be a group-labeled graph, and $A \subseteq V$. Then, for any integer k, one can obtain in polynomial time, either:*

1. *k vertex-disjoint nonzero A-paths, or*
2. *a set of at most $2k - 2$ vertices that intersects every nonzero A-path.*

Recall that G is the undirected graph in the theorem statement, and $s \in V$. For a suitable choice of a group-labeled graph (H, Γ), and a vertex-set A, we show that: (a) vertex-disjoint nonzero A-paths in (H, Γ) yield edge-disjoint odd (s, s)-trails; and (b) a *vertex-set* covering all nonzero A-paths in (H, Γ) yields an odd (s, s)-trail cover of G. Combining this with Theorem 5.1 finishes the proof.

Since we are dealing with parity, it is natural to choose $\Gamma = \mathbb{Z}_2$ (so the orientation of edges in H will not matter). To translate vertex-disjointness (and vertex-cover) to edge-disjointness (and edge-cover), we essentially work with the line graph of G, but slightly modify it to incorporate edge labels. We replace each vertex $x \in V$ with a clique of size $\deg_G(x)$, with each clique-node corresponding to a distinct edge of G incident to x; we use $[x]$ to denote this clique, both its set of nodes and edges; the meaning will be clear from the context. For every edge $e = xy \in E$, we create an edge between the clique nodes of $[x]$ and $[y]$ corresponding to e. We arbitrarily orient the edges to obtain H. We give each clique edge a label of 0, and give every other edge a label of 1. Finally, we let $A = [s]$. The proof of the following lemma is straightforward.

Lemma 5.2. *The following properties hold.*

(a) *Every A-path P in H maps to an (s, s)-trail $T = \pi(P)$ in G such that $\gamma(P) = 1$ iff T is an odd trail.*
(b) *If two A-paths P, Q are vertex disjoint then the (s, s)-trails $\pi(P)$ and $\pi(Q)$ are edge disjoint.*
(c) *Every (s, s)-trail T in G with at least one edge maps to an A-path $P = \sigma(T)$ in G such that: T is an odd trail iff $\gamma(P) = 1$, and P contains a vertex x iff T contains the corresponding edge of G.*

To complete the proof of Theorem 3.2, we apply Theorem 5.1 to the nonzero A-paths instance constructed above. If we obtain k vertex-disjoint nonzero A-paths in H, then parts (a) and (b) of Lemma 5.2 imply that we can map these to k edge-disjoint odd (s, s)-trails. Alternatively, if we obtain a set C of at most $2k - 2$ vertices of H that intersect every nonzero A-path, then we obtain a cover F for odd (s, s)-trails in G by taking the set of edges in G corresponding to the vertices in C. To see why F is a cover, suppose that the graph $G - F$ has an odd (s, s)-trail. This then maps to a nonzero A-path P in H such that $P \cap C = \emptyset$ by part (c) of Lemma 5.2, which yields a contradiction.

6   Extensions

Odd trails in signed graphs. A *signed graph* is a tuple $(G = (V, E), \Sigma)$, where G is undirected and $\Sigma \subseteq E$. A set F of edges is now called odd if $|F \cap \Sigma|$ is odd. Our results extend to the more-general setting of packing and covering odd (u, v)-trails in a signed graph. In particular, Theorems 3.1, 3.2 and 3.3 *hold without any changes*. Theorem 3.2 follows simply because it utilizes Theorem 5.1, which applies to the even more-general setting of group-labeled graphs. Theorem 3.3 holds because it uses basic parity arguments: if we simply replace parity with parity with respect to Σ (i.e., instead of parity of F, we now consider parity of $|F \cap \Sigma|$), then everything goes through. Finally, as before, combining the above two results yields (the extension of) Theorem 3.1.

Odd (C,D)-trails. This is the generalization of the odd (u, v)-trails setting, where we have disjoint sets $C, D \subseteq V$. Our results yield a factor-2 gap between the the minimum number of edges needed to cover all odd (C, D)-trails and the maximum number of edge-disjoint odd (C, D)-trails. First, we utilize Theorem 5.1 to prove a generalization of Theorem 3.2 showing that for any integer $k \geq 0$, we can either obtain k edge-disjoint odd $(C \cup D, C \cup D)$-trails, or an odd-$(C \cup D, C \cup D)$-trail cover of size at most $2k - 2$. Next, we observe that Theorem 3.3 can still be applied in this more-general setting to show that if we have a collection $\widehat{T}$ of k edge-disjoint odd $(C \cup D, C \cup D)$-trails, and (at least) $2k$ edge-disjoint (C, D)-paths, then we can obtain k edge-disjoint odd (C, D)-trails.

References

1. Chudnovsky, M., Cunningham, W.H., Geelen, J.: An algorithm for packing non-zero A-paths in group-labelled graphs. Combinatorica **28**(2), 145–161 (2008)
2. Chudnovsky, M., Geelen, J., Gerards, B., Goddyn, L.A., Lohman, M., Seymour, P.D.: Packing non-zero A-paths in group-labelled graphs. Combinatorica **26**(5), 521–532 (2006)
3. Churchley, R.: Personal communication (2016)
4. Churchley, R.: Odd disjoint trails and totally odd graph immersions. Ph.D. thesis, Simon Fraser University (2017). https://theses.lib.sfu.ca/sites/all/files/public_copies/etd9973-thesis.pdf
5. Churchley, R., Mohar, B., Wu, H.: Weak duality for packing edge-disjoint odd (u, v)-trails. In: Proceedings of the 27th SODA, pp. 2086–2094 (2016)
6. Geelen, J.: Personal communication (2016)

7. Ibrahimpur, S.: Packing and covering odd (u,v)-trails in a graph. Master's thesis, University of Waterloo (2016). https://uwspace.uwaterloo.ca/handle/10012/10939
8. Lucchesi, C.L., Younger, D.H.: A minimax theorem for directed graphs. J. Lond. Math. Soc. **17**, 369–374 (1978)
9. Mader, W.: Über die maximalzahl kreuzungsfreier H-Wege. Archiv der Math. **31**(1), 387–402 (1978)
10. Menger, K.: Zur allgemeinen kurventheorie. Fund. Math. **10**, 96–115 (1927)
11. Robertson, N., Seymour, P.: Graph minors XXIII. Nash-Williams' immersion conjecture. J. Combin. Theory Ser. B **100**(2), 181–205 (2010)
12. Schrijver, A.: A short proof of Mader's S-paths theorem. JCTB **82**, 319–321 (2001)
13. Schrijver, A.: Combinatorial Optimization: Polyhedra and Efficiency. Algorithms and Combinatorics. Springer, Heidelberg (2003)
14. Schrijver, A., Seymour, P.D.: Packing odd paths. JCTB **62**, 280–288 (1994)
15. Zaslavsky, T., Glossary of signed, gain graphs, allied areas. Electron. J. Combin., Dynamic Survey 9 **5**, 41 pp. (electronic) (1998)

Breaking $1 - 1/e$ Barrier for Non-preemptive Throughput Maximization

Sungjin Im[1(✉)], Shi Li[2], and Benjamin Moseley[3]

[1] University of California, Merced, CA 95344, USA
sim3@ucmerced.edu
[2] University at Buffalo, Buffalo, NY 14260, USA
shil@buffalo.edu
[3] Washington University in St. Louis, St. Louis, MO 63130, USA
bmoseley@wustl.edu

Abstract. In this paper we consider one of the most basic scheduling problems where jobs have their respective arrival times and deadlines. The goal is to schedule as many jobs as possible *non-preemptively* by their respective deadlines on m identical parallel machines. For the last decade, the best approximation ratio known for the single machine case ($m = 1$) has been $1 - 1/e - \epsilon \approx 0.632$ due to [Chuzhoy-Ostrovsky-Rabani, FOCS 2001 and MOR 2006]. We break this barrier and give an improved 0.644-approximation. For the multiple machine case, we give an algorithm whose approximation guarantee becomes arbitrarily close to 1 as the number of machines increases. This improves upon the previous best $1 - 1/(1 + 1/m)^m$ approximation due to [Bar-Noy et al., STOC 1999 and SICOMP 2009], which converges to $1 - 1/e$ as m goes to infinity. Our result for the multiple-machine case extends to the weighted throughput objective where jobs have different weights, and the goal is to schedule jobs with the maximum total weight. Our results show that the $1 - 1/e$ approximation factor widely observed in various coverage problems is not tight for the non-preemptive maximum throughput scheduling problem.

1 Introduction

Scheduling jobs with arrival times and deadlines is a fundamental problem in numerous areas of computer science and other fields. Due to this, there has been a large amount of research focusing on the topic. However, relatively little is known when jobs must be scheduled *non-preemptively*. Nonetheless, non-preemptive job scheduling occurs frequently in practice for a variety of reasons including because jobs cannot be stopped during execution due to practical constraints or because overhead costs are prohibitively large.

S. Im—Supported in part by NSF grants CCF-1409130 and CCF-1617653.
S. Li—Supported in part by NSF grant CCF-1566356.
B. Moseley—Supported in part by a Google Research Award, a Yahoo Research Award and NSF Grant CCF-1617724.

© Springer International Publishing AG 2017
F. Eisenbrand and J. Koenemann (Eds.): IPCO 2017, LNCS 10328, pp. 292–304, 2017.
DOI: 10.1007/978-3-319-59250-3_24

A central problem in the scheduling literature is determining how to schedule jobs by their deadline. In many cases when jobs have deadlines, not all jobs can be scheduled by their deadline. In these situations an alternative goal is to complete as many jobs as possible by their deadline. In this paper, we consider this problem a.k.a. *throughput maximization*. There are m identical machines and n jobs. Each job j has size p_j, arrival/release time r_j, and deadline d_j; all these quantities are assumed to be integers in $[0, T]$. The goal is to schedule as many jobs as possible by their deadlines non-preemptively on the m machines. Non-preemptive scheduling means that once a job starts being processed at time s_j on a machine, then the job must be scheduled until time $s_j + p_j$ on the machine. A machine can process at most one job at a time. To highlight the non-preemptive aspect of the problem, we will call this problem the Job Interval Scheduling (JIS). Not surprisingly, JIS has various applications in practice. For examples, see [6, 8, 10, 13].

It was shown by Garey and Johnson that this problem is NP-Hard [9]. Bar-Noy et al. [4] showed that there is an algorithm achieving an approximation ratio $1 - 1/(1 + \frac{1}{m})^m$. The approximation ratio gets better when m becomes larger. In particular, the ratio is $1/2$ if $m = 1$ and converges to $1 - 1/e$ as m tends to infinity.

Later Chuzhoy et al. [7] gave a $(1 - 1/e - \epsilon)$-approximation algorithm for a discrete version of this problem. In this version, we are explicitly given a set of intervals $\mathcal{I}_j$ in $(0, T]$ (which may have different lengths) for each job j. To schedule the job, we need to select an interval from the set $\mathcal{I}_j$. A schedule is valid if the intervals selected for all the scheduled jobs are disjoint. In this problem, adding more machines does not add more generality to this problem[1]. We will refer to the problem we consider as the *continuous* variant to distinguish it from this work. It seems that the discrete version generalizes the continuous version of the problem we consider: for each job j with arrival time r_j, deadline d_j and processing time p_j, the set of intervals for j is all sub-intervals of $(r_j, d_j]$ of length p_j with integer end-points. However, there is a small caveat: the number of intervals can be exponential in n. It was not known how to handle this tricky issue using the algorithm of [7]. Thus, when T is not polynomially bounded by n, the $\left(1 - 1/(1 + \frac{1}{m})^m\right)$-approximation due to [4] remains the state-of-art for this problem; in particular, a $1/2$-approximation is the best known when $m = 1$ [1, 4, 16].

Our Results: In this paper, we improve upon the state-of-art approximations for JIS for both the single-machine and multiple-machine cases. First, we show that for constant m, there is a $0.6448 > (1 - 1/e)$-approximation for JIS.

Theorem 1. *For some $\alpha_0 > 0.6448 > 1 - 1/e$ and any $\epsilon > 0$, there exists an $(\alpha_0 - \epsilon)$-approximation algorithm for the (unweighted) Job Interval Scheduling (JIS) problem with running time $n^{O(m/\epsilon^5)}$.*

---

[1] Suppose there are m machines. Then, in our new instance, the time horizon is $(0, mT]$, which can be viewed as the concatenation of m horizons of length T. If a job can be scheduled in $(A, B] \subseteq (0, T]$ in the original instance, it can be scheduled in $(iT + A, iT + B]$ for every $i = 0, 1, \cdots, m - 1$ in the new instance.

To complement the result, we give a second algorithm whose approximation ratio approaches 1 as the number m of machines goes to infinity, improving upon the previous $1 - 1/e$ limit. Thus, we can make our approximation ratio better than $1 - 1/e$ for any m: we run the first algorithm if m is at most a small constant; we run the second algorithm if m is large. Indeed, our second algorithm works for the more general weighted version of the problem, provided that T is polynomially bounded by n. In this version, each job i has some positive weight w_i and the goal is to maximize the total weight of the jobs completed by their deadline. We remark that for the unweighted version, we do not require T to be polynomially bounded.

Theorem 2. *For any $\epsilon > 0$, there exists a $\left(1 - O\left(\sqrt{(\log m)/m}\right) - \epsilon\right)$-approximation for unweighted JIS on m machines. If $T = \mathrm{poly}(n)$, there exists a $\left(1 - O\left(\sqrt{(\log m)/m}\right)\right)$-approximation for weighted JIS on m machines.*

Our Techniques: Our result in Theorem 2 will follow from a simple rounding procedure based on the naive LP relaxation for the problem. We scale down a naive LP solution by $(1 - \epsilon)$, apply a standard rounding technique to obtain a tentative schedule. Then we convert the tentative schedule to one that is feasible by removing jobs in a greedy manner. We show that the probability that a job is removed from the tentative schedule is exponentially small in m. Another technical contribution from this result is a method to solve the naive LP for unweighted JIS when T is not bounded, that only sacrifices a $(1 - \epsilon)$-factor in the LP value. This was not known previously.

Our main technical contribution is in obtaining an $\alpha_0 - \epsilon \approx 0.6448$-approximation stated in Theorem 1. The algorithm is based on a slightly different variation of the configuration LP used in [7]. We highlight our algorithmic ideas as follows assuming $m = 1$.

Chuzhoy et al. considered a configuration LP to obtain an approximation ratio $1 - 1/e - \epsilon \approx 0.632$ [7]; it is known that a naive LP has an integrality gap of 2 when $m = 1$ [4,16]. The configuration LP considered in [7] is fairly natural and builds on "blocks" of jobs: a block is a window (a time interval) together with k jobs scheduled in it, for some fixed k. It is straightforward to construct the set of blocks from an integral schedule: take the window in which the first k jobs are scheduled, take the window in which the second k jobs are scheduled, and so on. [7] used an involved preprocessing step to guess the windows where a block of k jobs are scheduled in some optimum solution. Then for each window corresponding to where a block of k jobs are scheduled in an optimum solution, there are variables encoding which k jobs are scheduled inside it. In our configuration LP, we do not guess where blocks are scheduled in an optimum solution. Rather, we have variables for blocks of k jobs that are scheduled and, allowing the blocks to be scheduled fractionally, we ensure that at most one fractional block covers every time point. Thus, instead partitioning time based on guessing where blocks are in an optimum solution, we partition the time horizon $(0, T]$ using the fractional blocks obtained from the configuration LP. We remark that this novelty is not essential in obtaining the improved approximation ratio; the improved approximation ratio could be obtained using the involved preprocessing step and the configuration LP in [7]. However, our configuration LP yields the following byproducts: (1) our configuration LP can handle the case when T is super polynomial; (2) we can reduce the dependence of running time on ϵ from double exponential to single exponential; (3) we can obtain the improved $(\alpha_0 - \epsilon)$-approximation for any constant m^2.

Now we state some intuition on how the configuration LP can help give a better approximation. Suppose we know the time steps when the optimal solution schedules k additional jobs, $t_1, t_2, t_3, \cdots$. (Since we do not know, we need to lose $1 - \epsilon$ factor in the approximation ratio and k needs to be a large constant.) Then, we only need to consider windows $\mathcal{W} = \{(t_1, t_2], (t_2, t_3], (t_3, t_4], \cdots\}$; let's call this the optimal partition. The configuration LP gives a distribution over sets of k jobs scheduled in each window. By randomly choosing one set of k jobs for each window, we can easily show a $(1 - 1/e)$-approximation following a standard analysis for the maximum coverage problem.

To improve the $(1-1/e)$-approximation, we use a second rounding procedure, which works only for the continuous version of JIS. Suppose each $(r_j, d_j]$ is exactly the union of some windows in $\mathcal{W}$; in other words, $r_j = t_i$ and $d_j = t_{i'}$ for some $i < i'$. The rounding procedure is based on individual jobs as opposed to individual windows as in the first rounding procedure. We assign each job to one of the windows according to how much the job is assigned to each individual window in the LP solution. Here a crucial observation is that the job can be scheduled anywhere in such windows – the only constraint we have to ensure is that we do not assign too much volume of jobs to the same window. With an additional preprocessing step of removing "big" jobs, we can show that such a bad overflow event rarely occurs, and this leads to a $(1 - \epsilon)$-approximation. Since each $(r_j, d_j]$ may not be aligned with the partition $\mathcal{W}$, we do not get this $(1 - \epsilon)$-approximation in general. Among all the windows in $\mathcal{W}$ that intersect $(r_j, d_j]$, the first one and the last one are special, since we can not schedule j anywhere inside these two windows. However, if the fraction of the job j assigned to these two windows is large, then we observe that, in fact, the first rounding algorithm can give better than a $1 - 1/e$ factor for the probability we schedule job j. Thus, taking the best solution given by these two rounding procedures will lead to an approximation ratio better than $1 - 1/e$.

Removing Dependency on T: As mentioned above, in our problem, the continuous version of JIS, the $\frac{1}{2}$-approximation was the best known polynomial time algorithm for the single machine case [4]. Interestingly, we also use the configuration LP to remove the dependency on T. This is somewhat counter-intuitive since the configuration LP is more complicated than the standard LP which is a special case of the configuration LP where each block has only one job. Thus it will seem that using the configuration LP is in the opposite direction to reduce

---

[2] As mentioned before, [7] focuses on the discrete version of JIS while our work does on the continuous version. The approach in [7] does not seem to easily extend to give a better than $1 - 1/e$-approximation for multiple machines.

the number of LP variables to obtain a true polynomial time algorithm. One of our key observations is that if a set of k jobs are very flexible, that is, can be scheduled seamlessly in "many" places, then such a block can be added later. Then we show one job can be kicked out to schedule k additional jobs. A similar configuration LP is used in the $\left(1 - O(\sqrt{\frac{\log m}{m}})\right)$-approximation to reduce the dependence on T, although only the naive LP is needed when $T = \text{poly}(n)$.

Related Work: A simple greedy algorithm that schedules a job with the earliest deadline is known to be a $\frac{1}{2}$-approximation for the single machine case [1,16]. There are $\frac{1}{2}$-approximations known for the weighted throughput objective in the multiple machines setting [4,5]. [2] considered JIS when the algorithm is given resource augmentation and gave an $O(1)$-speed 1-approximation. If preemption is allowed, it is known that if $m - 1$ then there exists a polynomial time optimal algorithm [3]. When $m \geq 2$ then the problem becomes NP-Hard [11]. To see why the problem is hard, note that if all jobs have the same release time and deadline then finding the minimum number of machines to schedule the jobs on is effectively the bin packing problem. The problem has also been considered in the online setting [12,14].

Organization: Due to the space constraints, in this paper we only prove our theorems under certain simplifying assumptions. Specifically, we show Theorem 1 when the number of machines, m is a constant and $T = \text{poly}(n)$; recall that $(0, T]$ is the time horizon we are considering. We continue to show Theorem 2 in Sect. 3 under the assumption that T is polynomially bounded. The proof for the general cases will be included in the full version of this paper.

2   Proof of Theorem 1 when $m = O(1)$ and $T = \text{poly}(n)$

Our algorithm is based on a configuration LP relaxation for the problem. We will then use this relaxation to partition the time horizon into disjoint windows. We remark that this step can replace the involved preprocessing step of [7]. With the definition of windows in place, we can run the rounding procedure of [7]; this will give us $(1 - 1/e - \epsilon)$-approximation for the unweighted case. To obtain the improved $(\alpha_0 - \epsilon)$ approximation ratio, we run a different rounding procedure and choose the better solution from the two procedures. This will give us Theorem 1 when $T = \text{poly}(n)$.

2.1   Linear Programming

We define a block as a triple $B = (L_B, R_B, \mathcal{J}_B)$ where L_B and R_B are two integer time points such that $0 \leq L_B < R_B \leq T$, and $\mathcal{J}_B$ is a subset of jobs that can be scheduled in the interval $(L_B, R_B]$ non-preemptively on m machines. We assume that B is associated with a specific schedule where jobs in $\mathcal{J}_B$ are scheduled in $(L_B, R_B]$ on m machines. The size of block B is defined as the number of jobs in $\mathcal{J}_B$, which is denoted as w_B. We say that B has *block window* $(L_B, R_B]$.

Let $k = \lceil 3/\epsilon \rceil$ and $\Delta = 2mk^5$; recall that ϵ is a parameter that stands for the proximity to the desired approximation factor. The integer programming for JIS is defined as follows. We only consider the blocks B with either $w_B = \Delta$, or $R_B = T$ and $w_B < \Delta$ in the IP; for simplicity we omit this constraint.

$$\max \quad \sum_B w_B \cdot x_B \qquad (\text{LP}_{\text{conf}})$$

$$\sum_{B:L_B < t \leq R_B} x_B \leq 1 \qquad \forall t \in [T] \qquad (1)$$

$$\sum_{B:j \in J_B} x_B \leq 1 \qquad \forall j \in \mathcal{J} \qquad (2)$$

$$x_B \in \{0,1\} \quad \forall B$$

In the above IP, Constraint (1) ensures that block windows are disjoint, and Constraint (2) requires each job to be scheduled at most once. Note that the number of constraints in (1) is polynomially bounded when $T = \text{poly}(n)$ – as mentioned earlier, we discuss how to handle non-polynomially bounded T in the full version of the paper.

It is easy to see that any solution to the IP gives a valid schedule. On the other hand, not every schedule can be converted to a feasible IP solution when $m > 1$. Thus the IP may not give the optimum throughput. However we show that the loss is small. To see this, fix an optimal schedule. Given the optimum schedule, we sort all the jobs according to their completion time. Let $L = 0$ initially. In each iteration, we take the first Δ jobs $\mathcal{J}'$ from the sequence and let R be the completion time of the Δ-th job; if there are less than Δ jobs in the sequence, we let $\mathcal{J}'$ be all the jobs in the sequence and let $R = T$. We create a block $B = (L, R, \mathcal{J}')$ and set $x_B = 1$. Then, we remove all jobs whose starting time is before R from the sequence. Then let $L = R$ and start a new iteration. The process ends when the sequence becomes empty. It is easy to see that the blocks we created have disjoint windows. Moreover, if $|\mathcal{J}'| = \Delta$, we remove at most $\Delta + m - 1$ jobs from the sequence: other than the Δ jobs in $\mathcal{J}'$, we may remove at most $m - 1$ extra jobs who are scheduled intersecting the interval $(R - 1, R]$. If $|\mathcal{J}'| < \Delta$ in the last iteration, we only remove $|\mathcal{J}'|$ jobs from the sequence. Thus, the value of the IP is at least $\frac{\Delta}{\Delta + m - 1}$ times the optimum throughput.[3]

The LP is obtained by relaxing the constraints $x_B \in \{0,1\}$ to $x_B \geq 0$. Note that the running time of solving the LP is $n^{O(\Delta)} = n^{O(m/\epsilon^5)}$. Let $\{x_B^*\}$ denote the optimal solution to the above LP. Let $\text{OPT}_{\text{LP}} = \sum_B w_B x_B^*$ denote the optimal LP objective.

---

[3] It is worth noting that this is where we crucially use the assumption that jobs have uniform weights.

2.2   Preprocessing

In the preprocessing step, we break the time horizon $(0, T]$ into a set $\mathcal{W}$ of disjoint intervals, which we call *base* windows to distinguish them from job windows and block windows. We also construct a new solution $\{x'_B\}$. The main goal is two-fold: (i) to preserve most of the LP objective and (ii) to make each block window completely contained in a base window; this makes the rounding procedures more applicable.

We now formally show how to break $(0, T]$ into base windows and obtain $\{x'_B\}$ from $\{x^*_B\}$. Note that the blocks in the support of $\{x^*_B\}$ can overlap with one another. To create $\{x'_B\}$, we iteratively cut at the time point when an additional $1/k$ fraction of blocks end in $\{x^*_B\}$. Formally, we associate each integer time-point $t \in (0, T]$ with a weight $e_t = \sum_{B:R_B=t} x^*_B$, which is the sum of x^*_B over all blocks B ending at time t. Let $L = 0$ and $\mathcal{W} = \emptyset$ initially. Each iteration works as follows. Let R be the first time point such that $\sum_{t=L+1}^{R} e_t \geq 1/k$, or let $R = T$ if no such time point exists – note that the sum is counted from time $L+1$. Create a base window $(L, R]$ and add it to $\mathcal{W}$. Let $L = R$ and start a new iteration. The procedure terminates when $L = T$.

Once we defined the base windows $\mathcal{W}$, for every B with $x^*_B > 0$, we cut B into multiple blocks at the boundaries of the base windows. Formally, for every base window $(L, R]$ that intersect $(L_B, R_B]$, we create a block $B' = (\max\{L, L_B\}, \min\{R, R_B\}, \mathcal{J}')$ where $\mathcal{J}'$ is the set of jobs in $\mathcal{J}_B$ whose scheduling intervals are contained in $(L, R]$. For all these created blocks B', we let $x'_{B'} = x^*_B$. Notice that the jobs across the boundaries of base windows are deleted in this process. After that, we delete big jobs from each created block B': a job j in $\mathcal{J}_{B'}$ is said to be big compared to B' if $p_j \geq (R_{B'} - L_{B'})/k^3$.[4]

We have constructed a set $\mathcal{W}$ of disjoint base windows and derived a new fractional solution $\{x'_B\}$ from $\{x^*_B\}$ that satisfies the following properties. All the blocks B in the description are restricted to the ones with $x'_B > 0$.

Properties of $\{x'_B\}$:

1. For every block B, the block window $(L_B, R_B]$ is fully contained in some base window in $\mathcal{W}$.
2. No job j in $\mathcal{J}_B$ is big compared to B. That is, for all $k \in \mathcal{J}_B$ it is the case that $p_j < (R_{B'} - L_{B'})/k^3$.
3. If $\mathrm{OPT}_{\mathrm{LP}} \geq \Delta$, then $\sum_{B'} w_B x'_B \geq (1 - \epsilon/3)\mathrm{OPT}_{\mathrm{LP}}$.
4. For all windows $(L, R] \in \mathcal{W}$, $\sum_{B:(L_B,R_B]\subseteq(L,R]} x'_B \leq 1 + 1/k$.
5. For all jobs j, $\sum_{B:j\in\mathcal{J}_B} x'_B \leq 1$.

Properties (1), (2) and (5) are very easy to check. To see Property (4) holds, consider a base window $(L, R] \in \mathcal{W}$. We know that at most 1 fractional block intersects R due to the Constraints (1). Due to the way we defined base windows, at most $1/k$ fractional block can end during $(L, R - 1]$.

---

[4] This is another place where we rely on the assumption that jobs have uniform weights.

Property (3) is the most non-trivial one. Observe that there are at most m fractional jobs across the boundary of two adjacent base windows. Each time (except for the last one) we build a base window, we collected at least $1/k$ fractional blocks, and thus Δ/k fractional jobs. So the total number of boundaries is at most $\text{OPT}_{\text{LP}}/(\Delta/k) = k\text{OPT}_{\text{LP}}/\Delta$. Thus the total jobs we discarded due to "boundary crossing" is at most $km\text{OPT}_{\text{LP}}/\Delta$. Also at most mk^3 fractional big jobs are discarded from each base window. Thus at most $mk^3 \times (k\text{OPT}_{\text{LP}}/\Delta+1)$ fractional big jobs are removed. If $\mathsf{opt} \geq \Delta$ and $k \geq 3$, the total number of fractional big jobs removed is at most $2mk^4\text{OPT}_{\text{LP}}/\Delta = \text{OPT}_{\text{LP}}/k \leq \epsilon\text{OPT}_{\text{LP}}/3$. Hence Property (3) follows. If the condition $\text{OPT}_{\text{LP}} \geq \Delta$ is not satisfied, we can simply guess the optimal solution using enumeration.

2.3   The First Rounding Procedure

In this subsection, we show how to round $\{x'_B\}$ to obtain an improved approximation. As mentioned before, we have two rounding procedures. The first rounding is an independent rounding that samples a block from the set of blocks contained in each base window $\mathcal{W}$. Formally, for each base window $(L, R] \in \mathcal{W}$, we sample a block B with $(L_B, R_B] \subseteq (L, R]$ with probability $\frac{x'_B}{1+1/k}$. This is well defined due to Property (4) – it says that there are only $(1 + 1/k)$ fractional blocks to be considered for each base window. If a job is scheduled more than once, we keep only one scheduling of the job. This completes the description of the first rounding.

A standard analysis for independent rounding can only show that each job can be scheduled with probability at least $(1 - 1/e)/(1 + 1/k)$ times the fraction by which the job is scheduled. To derive an improved approximation better than $1 - 1/e$, we need to do a more careful analysis. Let's focus on each job j. Consider the set of base windows that intersect $(r_j, d_j]$. We call the first and the last of these base windows the boundary base windows. All these base windows except the boundary ones are completely contained in $(r_j, d_j]$. Job j may appear in multiple base windows, more precisely in blocks contained in multiple base windows. Let a_j be the fraction by which job j is scheduled in boundary base windows, scaled down by $1 + 1/k$. There may be only one boundary window for j, but it only helps the approximation ratio, hence for simplicity, let's proceed with our analysis assuming that there are two boundary base windows for every job. Now we turn our attention to non-boundary windows. Observe that in every non-boundary base window in which j is scheduled, we can schedule j anywhere inside it. Let b_j be the fraction by which job j is scheduled in non-boundary base windows, scaled down by $1 + 1/k$.

We show that the first rounding schedules j with probability at least:

$$1 - (1 - a_j/2)^2 e^{-b_j} \tag{3}$$

Let's take a close look at why this is the case. Let a, a' be the fractions by which job j is scheduled on the two boundary base windows, scaled down by $1 + 1/k$. Likewise, let $b(u)$ be the fraction by which job j is scheduled on a non-boundary base window u, scaled down by $1 + 1/k$. Note that $\sum_u b(u) = b_j$. Then, j is scheduled with probability at least $1 - (1 - a)(1 - a') \prod_u (1 - b(u)) \geq 1 - (1 - a)(1 - a') \prod_u e^{-b(u)} = 1 - (1 - a)(1 - a')e^{-b_j} \geq 1 - (1 - a_j/2)^2 e^{-b_j}$ where we use the well-known inequality $e^x \geq 1 + x$.

2.4   The Second Rounding Procedure

The second rounding makes use of the flexibility of non-boundary base windows. This rounding procedure completely ignores the boundary base windows, and assigns jobs individually. Consider each job j together with its non-boundary base windows. The fractional solution x'_B tells us how much job j can be scheduled in each of its base windows, and we randomly assign the job to one of them exactly as the fractional solution suggests. Then what is the probability that job j cannot be scheduled since a lot of jobs are assigned to the same base window? We can show such a probability is tiny by scaling down the assignment probability slightly and using the fact that all jobs are small compared to base windows. We show that each job j is successfully scheduled with probability at least

$$(1 - \epsilon/3)b_j \tag{4}$$

Formally, the second rounding algorithm is as follows. Let $f_{j,W}$ be the amount by which job j is assigned to a base window W, i.e. $f_{j,W} := \sum_{B:(L_B,R_B] \subseteq W, j \in \mathcal{J}_B} x'_B$. Then b_j is $\sum_W f_{j,W}/(1 + 1/k)$, where W is over all non-boundary base windows of j. Consider each job j. We assign job j to one of its non-boundary base windows W with probability $f_{j,W}/(1 + 1/k)$. Let $\mathcal{J}(W)$ be the set of jobs selected to be scheduled in the base window W. We schedule these jobs greedily on m machines within W. Since each job j in $\mathcal{J}(W)$ can be scheduled anywhere within the window, and all jobs in $\mathcal{J}(W)$ are small compared to W, the greedy packing is pretty good.

Lemma 1. *Consider a base window $W = (L, R]$. If the total size of jobs in $\mathcal{J}(W)$ is no greater than $(1 - 1/k^3)m(R - L)$, then all jobs in $\mathcal{J}(W)$ can be scheduled on m machines within the window W.*

Proof. The proof immediately follows from the fact that all jobs in $\mathcal{J}(W)$ have sizes no greater than $(R - L)/k^3$ (Property (2)), and all jobs in $\mathcal{J}(W)$ can be scheduled everywhere inside W.   □

If the total size of jobs in $\mathcal{J}(W)$ is greater than $(1 - k^3)m$, we simply discard all jobs in $\mathcal{J}(W)$. Our goal is to show that this bad event happens with low probability. The following observation is immediate.

Lemma 2. *The total size of jobs in $\mathcal{J}(W)$ is at most $m(R - L)/(1 + 1/k)$ in expectation.*

Proof. Notice that $\sum_{B:t \in (L_B,R_B]} x'_B \leq 1$ for every $t \in (L, R]$. Thus, $\sum_{B:(L_B,R_B] \subseteq (L,R]} (R_B - L_B)x'_B = \sum_{t \in (L,R]} \sum_{B:t \in (L_B,R_B]} x'_B \leq (R - L)$. Since we can schedule at most $m(R_B - L_B)x'_B$ volume of jobs in B, the total volume of jobs scheduled in B is at most $m(R - L)$; here the volume of a job j refers to p_j times the fraction by which the job is scheduled. The claim follows since we scaled down the assignment probability down by a factor of $1 + 1/k$.   □

This claim, together with the fact that all jobs are small compared to the base window, will allow us to show that the bad event happens with a low probability. To show this, fix a base window $W = (L, R)$. The upper bound in the following lemma easily follows from a well known concentration inequality.

Lemma 3. *For any window $W = (L, R)$, the total size of jobs in $\mathcal{J}(W)$ is at most $(1 - 1/2k)(R - L)$ with probability at least $1 - \epsilon/3$.*

Proof. Let X_j be p_j if job j is in $\mathcal{J}(W)$, and otherwise 0. Note that $X_j \leq (R - L)/k^3$. Let $Z = \sum_j X_j$. By Lemma 2, we know that $\mu := \mathbf{E}[Z] \leq m(R - L)/(1 + 1/k) \leq (1 - 0.9/k)m(R - L)$ when k is large enough. By adding enough dummy random variables, we may assume $\mu = (1 - 0.9/k)m(R - L)$; this only increases $\Pr[Z \geq (1 - 1/2k)m(R - L)]$. Using the following concentration inequality (see Theorem 2.3 in [15]).

Theorem 3. *Let Z be the sum of n independent random variables where each random variable takes value in $[0, K]$. Let $\mu = E[Z]$. Then for any $\lambda \in [0, 1]$, we have*

$$\Pr\left[Z \geq (1 + \lambda)\mu\right] \leq e^{-\lambda^2 \mu/3K}.$$

we have

$$\Pr[Z \geq (1 - 1/2k)m(R - L)]$$
$$\leq \exp\left(-\frac{(0.4/k)^2/(1 - 0.9/k)^2 \times (1 - 0.9/k)m(R - L)}{3(R - L)/k^3}\right)$$
$$= \exp\left(-\frac{0.16km}{3(1 - 0.9/k)}\right) \leq \exp(-k/20),$$

which is at most $\epsilon/3$ when ϵ is small enough.   □

If the total size of jobs assigned to the base window $(L, R]$ is at most $(1 - 1/2k)m(R - L) \leq (1 - 1/k^3)m(R - L)$, then all these jobs can be scheduled in $(L, R]$ on m machines. Further, this happens with probability at least $(1 - \epsilon/3)$ due to Lemma 3. Since a job is assigned to one of a non-boundary window with probability b_j, the probability that job j is scheduled due to the second rounding is at least $(1 - \epsilon/3)b_j$. This shows the probability claimed in (4).

2.5   Combining the Two Rounding Procedures

Finally, we take the better between the two rounding solutions. That is, we take the maximum of the two lower bounds, (3) and (4). The following lemma lower bounds (3) by a linear combination of a_j and b_j, which follows by approximating e^x by a piecewise linear function and performing some case analysis. The proof is deferred to the full version of this paper.

Lemma 4. *For all a_j, b_j such that $0 \leq a_j + b_j \leq 1$, $\left(1 - (1 - a_j/2)^2 e^{-b_j}\right) \geq \lambda_1 a_j + \lambda_2 b_j$ where $\lambda_1 = 0.69$ and $\lambda_2 = 0.62$.*

Then the expected number of jobs we schedule is at least

$$(1 - \epsilon/3) \max \left\{ \sum_j \left(1 - (1 - a_j/2)^2 e^{-b_j}\right), \sum_j b_j \right\}$$

$$\geq (1 - \epsilon/3) \max \left\{ \lambda_1 \sum_j a_j + \lambda_2 \sum_j b_j, \sum_j b_j \right\} \geq (1 - \epsilon/3) \frac{\lambda_1}{\lambda_1 - \lambda_2 + 1} \left(\sum_j a_j + \sum_j b_j \right).$$

Let $\alpha_0 = \frac{\lambda_1}{\lambda_1 - \lambda_2 + 1} \geq 0.6448$. Notice that $\sum_j a_j + \sum_j b_j$ is the total number of jobs scheduled by the solution $\{x'_B\}$, scaled down by $1 + 1/k$, which is at least $(1 - 1/k)(1 - \epsilon/3) \mathrm{OPT}_{\mathrm{LP}}$, due to the Property (3). Noticing that $\mathrm{OPT}_{\mathrm{LP}}$ is at least $\frac{\Delta}{\Delta + m - 1} \geq (1 - \epsilon/3)$ times the optimum throughput, our approximation ratio is at least $(1 - \epsilon/3)(1 - 1/k)(1 - \epsilon/3)(1 - \epsilon/3)\alpha_0 \geq (1 - 4\epsilon/3)\alpha_0 \geq \alpha_0 - \epsilon$. This proves Theorem 1.

3 $1 - O\left(\sqrt{(1/m)\ln m}\right)$-Approximation for JIS

In this section our goal is to prove Theorem 2 assuming that $T = \mathrm{poly}(n)$.

We start by describing our algorithm which works by rounding the naive LP relaxation for the problem. The relaxation is the following. Let $x_{j,t}$ denote whether job j is started at time t. This variable is defined if $r_j \leq t \leq d_j - p_j$.

$$\max \quad \sum_j \sum_t w_j x_{j,t} \qquad (\mathrm{LP_{naive}})$$

$$\sum_j \sum_{t'=\max\{r_j, t - p_j\}}^{\min\{d_j - p_j, t - 1\}} x_{j,t'} \leq m \quad \forall t \in [T]; \qquad \sum_t x_{j,t} \leq 1 \quad \forall j \in \mathcal{J}$$

where $x_{j,t} \geq 0$ for all $j \in \mathcal{J}, t \in [r_j, d_j - p_j]$. The first constraint ensures that at most m jobs are scheduled at any point in time. The second constraints ensure that each job is scheduled at most once.

Our algorithm works as follows. After solving the LP, we round the solution. For each job, we do randomized rounding. Each job j selects a starting time t to be scheduled with probability $(1 - \epsilon)x_{j,t}$, and j is not scheduled with probability $1 - (1 - \epsilon)\sum_t x_{j,t}$ where $\epsilon < 1$ is a parameter depending on m which will be fixed later. This is the tentative schedule, which could be infeasible. Then we order the jobs by their *starting* times in the tentative schedule. Consider a job j, whose starting time is t in the tentative schedule. Then we schedule job j at time t whenever we can. It is easy to see that we can schedule j if and only if the time slot $(t, t + 1]$ is covered by less than m already-assigned jobs.

To bound the quality of the solution, our goal is to bound the probability that a job is scheduled. The proof idea is quite simple. We consider any fixed job j. We condition on the event that we chose to tentatively schedule j at time t; this happens with probability $(1 - \epsilon)x_{j,t}$. We then bound the probability that j is removed from the tentative schedule. Here we can apply concentration inequalities since each job is rounded independently of other jobs.

Lemma 5. *Each job* j *is scheduled with probability at least* $(1 - \epsilon)$ $\left(1 - \exp\left(-m\frac{\epsilon^2}{3(1-\epsilon)}\right)\right)\sum_t x_{j,t}$.

We set $\epsilon = \sqrt{\frac{2\ln m}{m}}$. Then $\exp\left(-\frac{\epsilon^2 m}{3(1-\epsilon)}\right) \leq m^{-2/3}$ and $(1 - \epsilon)$ $\left(1 - \exp\left(-\frac{\epsilon^2 m}{3(1-\epsilon)}\right)\right) \geq 1 - \sqrt{\frac{2\ln m}{m}} - m^{-2/3} = 1 - O\left(\sqrt{\frac{\log m}{m}}\right)$. This implies the second half of Theorem 2.

References

1. Adler, M., Rosenberg, A.L., Sitaraman, R.K., Unger, W.: Scheduling time-constrained communication in linear networks. Theor. Comput. Syst. **35**(6), 599–623 (2002)
2. Bansal, N., Chan, H.L., Khandekar, R., Pruhs, K., Stein, C., Schieber, B.: Non-preemptive min-sum scheduling with resource augmentation. In: FOCS, pp. 614–624 (2007)
3. Baptiste, P.: An $O(n^4)$ algorithm for preemptive scheduling of a single machine to minimize the number of late jobs. Oper. Res. Lett. **24**(4), 175–180 (1999)
4. Bar-Noy, A., Guha, S., Naor, J., Schieber, B.: Approximating the throughput of multiple machines in real-time scheduling. SIAM J. Comput. **31**(2), 331–352 (2001)
5. Berman, P., DasGupta, B.: Improvements in throughout maximization for real-time scheduling. In: Proceedings of the Thirty-Second Annual ACM Symposium on Theory of Computing, pp. 680–687. ACM (2000)
6. Błażewicz, J., Ecker, K.H., Pesch, E., Schmidt, G., Weglarz, J.: Scheduling Computer and Manufacturing Processes. Springer Science & Business Media, Heidelberg (2013)
7. Chuzhoy, J., Ostrovsky, R., Rabani, Y.: Approximation algorithms for the job interval selection problem and related scheduling problems. Math. Oper. Res. **31**(4), 730–738 (2006)
8. Fischetti, M., Martello, S., Toth, P.: The fixed job schedule problem with spread-time constraints. Oper. Res. **35**(6), 849–858 (1987)
9. Garey, M.R., Johnson, D.S.: Two-processor scheduling with start-times and deadlines. SIAM J. Comput. **6**(3), 416–426 (1977)
10. Hall, N.G., Magazine, M.J.: Maximizing the value of a space mission. Eur. J. Oper. Res. **78**(2), 224–241 (1994)
11. Hong, K.S., Leung, J.Y.T.: Preemptive scheduling with release times and deadlines. Real Time Syst. **1**(3), 265–281 (1989)
12. Koren, G., Shasha, D.: D^{over}: an optimal on-line scheduling algorithm for overloaded uniprocessor real-time systems. SIAM J. Comput. **24**(2), 318–339 (1995)
13. Lawler, E.L., Lenstra, J.K., Rinnooy Kan, A., Shmoys, D.: Sequencing and scheduling: algorithms and complexity. Hanbooks Oper. Res. Manage. Sci. **4**, 445–522 (1993)
14. Lipton, R.J., Tomkins, A.: Online interval scheduling. In: SODA, pp. 302–311 (1994)

15. McDiarmid, C.: Concentration. In: Habib, M., McDiarmid, C., Ramirez-Alfonsin, J., Reed, B. (eds.) Probabilistic Methods for Algorithmic Discrete Mathematics. Algorithms and Combinatorics, vol. 16, pp. 195–248. Springer, Heidelberg (1998)
16. Spieksma, F.C.: On the approximability of an interval scheduling problem. J. Sched. **2**(5), 215–227 (1999)

A Quasi-Polynomial Approximation
for the Restricted Assignment Problem

Klaus Jansen and Lars Rohwedder[(✉)]

University of Kiel, 24118 Kiel, Germany
{kj,lro}@informatik.uni-kiel.de

Abstract. Scheduling jobs on unrelated machines and minimizing the makespan is a classical problem in combinatorial optimization. A job j has a processing time p_{ij} for every machine i. The best polynomial algorithm known for this problem goes back to Lenstra et al. and has an approximation ratio of 2. In this paper we study the RESTRICTED ASSIGNMENT problem, which is the special case where $p_{ij} \in \{p_j, \infty\}$. We present an algorithm for this problem with an approximation ratio of $11/6 + \epsilon$ and quasi-polynomial running time $n^{\mathcal{O}(1/\epsilon \log(n))}$ for every $\epsilon > 0$. This closes the gap to the best estimation algorithm known for the problem with regard to quasi-polynomial running time.

Keywords: Approximation · Scheduling · Unrelated machines · Local search

1 Introduction

In the problem we consider, which is known as SCHEDULING ON UNRELATED MACHINES, a schedule $\sigma : \mathcal{J} \to \mathcal{M}$ of the jobs $\mathcal{J}$ to the machines $\mathcal{M}$ has to be computed. On machine i the job j has a processing time of p_{ij}. We want to minimize the makespan, i.e., $\max_{i \in \mathcal{M}} \sum_{j \in \sigma^{-1}(i)} p_{ij}$. The classical 2-approximation by Lenstra et al. [8] is still the algorithm of choice for this problem.

Recently a special case, namely the RESTRICTED ASSIGNMENT problem, has drawn much attention in the scientific community. Here each job j has a processing time p_j, which is independent from the machines, and a set of machines $\Gamma(j)$. A job j can only be assigned to $\Gamma(j)$. This is equivalent to the former problem when $p_{ij} \in \{p_j, \infty\}$. For both the general and the restricted variant there cannot be a polynomial algorithm with an approximation ratio better than $3/2$, unless P = NP [8]. If the exponential time hypothesis (ETH) holds, such an algorithm does not even exist with sub-exponential (in particular, quasi-polynomial) running time [5].

In a recent breakthrough, Svensson has proved that the configuration-LP, a natural linear programming relaxation, has an integrality gap of at most $33/17$ [10]. We have later improved this bound to $11/6$ [7]. By approximating

Research supported by German Research Foundation (DFG) project JA 612/15-1.

F. Eisenbrand and J. Koenemann (Eds.): IPCO 2017, LNCS 10328, pp. 305–316, 2017.
DOI: 10.1007/978-3-319-59250-3_25

the configuration-LP this yields an $(11/6 + \epsilon)$-estimation algorithm. However, no polynomial algorithm is known that can produce a solution of this value.

For instances with only two processing times additional progress has been made. Chakrabarty et al. gave a polynomial $(2 - \delta)$-approximation for a very small δ [4]. Later Annamalai surpassed this with a $(17/9 + \epsilon)$-approximation for every $\epsilon > 0$ [1]. For this special case it was also shown that the integrality gap is at most $5/3$ [6].

In [6, 7, 10] the critical idea is to design a local search algorithm, which is then shown to produce good solutions. However, the algorithm has a potentially high running time; so it was only used to prove the existence of such a solution. A similar algorithm was used in the RESTRICTED MAX-MIN FAIR ALLOCATION problem. Here a quasi-polynomial variant by Polácek et al. [9] and a polynomial variant by Annamalai et al. [2] were later discovered.

In this paper, we present a variant of the local search algorithm, that admits a quasi-polynomial running time. The algorithm is purely combinatorial and uses the configuration-LP only in the analysis.

Theorem 1. *For every $\epsilon > 0$ there is an $(11/6 + \epsilon)$-approximation algorithm for the* RESTRICTED ASSIGNMENT *problem with running time* $\exp(\mathcal{O}(1/\epsilon \cdot \log^2(n)))$, *where* $n = |\mathcal{J}| + |\mathcal{M}|$.

The main idea is the concept of layers. The central data structure in the local search algorithm is a tree of so-called blockers and we partition this tree into layers, that are closely related to the distance of a blocker from the root. Roughly speaking, we prevent the tree from growing arbitrarily high. A similar approach was taken in [9].

1.1   The Configuration-LP

A well known relaxation for the problem of SCHEDULING ON UNRELATED MACHINES is the configuration-LP (see Fig. 1). The set of configurations with respect to a makespan T are defined as $\mathcal{C}_i(T) = \{C \subseteq \mathcal{J} : \sum_{j \in C} p_{ij} \leq T\}$. We refer to the minimal T for which this LP is feasible as the optimum or OPT*. In the RESTRICTED ASSIGNMENT problem a job j can only be used in configurations of machines in $\Gamma(j)$ given T is finite. We can find a solution for the LP with a value of at most $(1 + \epsilon)$OPT* in polynomial time for every $\epsilon > 0$ [3].

1.2   Preliminaries

In this section we simplify the problem we need to solve. The approximation ratio we will aim for is $1 + R$, where $R = 5/6 + 2\epsilon$. We assume that $\epsilon < 1/12$ for our algorithm, since otherwise the 2-approximation in [8] can be used.

We will use a binary search to obtain a guess T for the value of OPT*. In each iteration, our algorithm either returns a schedule with makespan at most $(1+R)T$ or proves that T is smaller than OPT*. After polynomially many iterations, we will have a solution with makespan at most $(1 + R)$OPT*. To shorten notation,

$$\sum_{C\in\mathcal{C}_i(T)} x_{i,C} \leq 1 \quad \forall i \in \mathcal{M}$$

$$\sum_{i\in\mathcal{M}}\sum_{C\in\mathcal{C}_i(T):j\in C} x_{i,C} \geq 1 \quad \forall j \in \mathcal{J}$$

$$x_{i,C} \geq 0$$

$$\min \sum_{i\in\mathcal{M}} y_i - \sum_{j\in\mathcal{J}} z_j$$

$$s.t.$$

$$y_i \geq \sum_{j\in C} z_j \quad \forall i \in \mathcal{M}, C \in \mathcal{C}_i(T)$$

$$y_i, z_j \geq 0$$

Fig. 1. Primal (left) and dual (right) of the configuration-LP w.r.t. makespan T

we scale each size by $1/T$ within an iteration, that is to say our algorithm has to find a schedule of makespan $1 + R$ or show that OPT$^* > 1$. Unless otherwise stated we will assume that $T = 1$ when speaking about configurations or feasibility of the configuration-LP.

Definition 1 (Small, big, medium, huge jobs). A job j is small if $p_j \leq 1/2$ and big otherwise; A big job is medium if $p_j \leq 5/6$ and huge if $p_j > 5/6$.

The sets of small (big, medium, huge) jobs are denoted by $\mathcal{J}_S$ (respectively, $\mathcal{J}_B$, $\mathcal{J}_M$, $\mathcal{J}_H$). Note that at most one big job can be in a configuration (w.r.t. makespan 1).

Definition 2 (Valid partial schedule). We call $\sigma : \mathcal{J} \to \mathcal{M} \cup \{\bot\}$ a valid partial schedule if (1) for each job j we have $\sigma(j) \in \Gamma(j) \cup \{\bot\}$, (2) for each machine $i \in \mathcal{M}$ we have $p(\sigma^{-1}(i)) \leq 1 + R$, and (3) each machine is assigned at most one huge job.

$\sigma(j) = \bot$ means that job j has not been assigned. In each iteration of the binary search, we will first find a valid partial schedule for all medium and small jobs and then extend the schedule one huge job at a time. We can find a schedule for all small and medium jobs with makespan at most $11/6$ by applying the algorithm by Lenstra, Shmoys, and Tardos [8]. This algorithm outputs a solution with makespan at most OPT$^* + p_{max}$, where p_{max} is the biggest processing time (in our case at most $5/6$). The problem that remains to be solved is given in below.

Input: An instance of RESTRICTED ASSIGNMENT, a valid partial schedule σ, a huge job j_{new} with $\sigma(j_{new}) = \bot$.
Output: Either: (1) A valid partial schedule σ' with $\sigma'(j_{new}) \neq \bot$ and $\sigma(j) \neq \bot \Rightarrow \sigma'(j) \neq \bot$ for all $j \in \mathcal{J}$, or (2) 'error' (indicating that OPT$^* > 1$).

Without loss of generality let us assume that the jobs are identified by natural numbers, that is $\mathcal{J} = \{1, 2, \ldots, |\mathcal{J}|\}$, and $p_1 \leq p_2 \leq \cdots \leq p_{|\mathcal{J}|}$. This gives us a total order on the jobs that will simplify the algorithm.

2   Algorithm

Throughout the paper, we make use of modified processing times $\overline{P}_j$ and $\overline{p}_j$, which we obtain by rounding the sizes of huge jobs up or down, that is

$$\overline{P}_j = \begin{cases} 1 & \text{if } p_j > 5/6, \\ p_j & \text{if } p_j \leq 5/6; \end{cases} \quad \text{and} \quad \overline{p}_j = \begin{cases} 5/6 & \text{if } p_j > 5/6, \\ p_j & \text{if } p_j \leq 5/6. \end{cases}$$

Definition 3 (Moves, valid moves). A pair (j, i) of a job j and a machine i is a *move*, if $i \in \Gamma(j) \backslash \{\sigma(j)\}$. A move (j, i) is *valid*, if (1) $\overline{P}(\sigma^{-1}(i)) + p_j \leq 1 + R$ and (2) j is not huge or no huge job is already on i.

We note that by performing a valid move (j, i) the properties of a valid partial schedule are not compromised.

Definition 4 (Blockers). A blocker is a tuple (j, i, Θ), where (j, i) is a move and Θ is the type of the blocker. There are 6 types with the following abbreviations: (SA) *small-to-any blockers*, (HA) *huge-to-any blockers*, (MA) *medium-to-any blockers*, (BH) *huge-/medium-to-huge blockers*, (HM) *huge-to-medium blockers*, and (HL) *huge-to-least blockers*.

The algorithm maintains a set of blockers called the blocker tree $\mathcal{T}$. We will discuss the tree analogy later. The blockers wrap moves that the algorithm would like to execute. By abuse of notation, we write that a move (j, i) is in $\mathcal{T}$, if there is a blocker (j, i, Θ) in $\mathcal{T}$ for some Θ. The type Θ determines how the algorithm treats the machine i as we will elaborate below.

The first part of a type's name refers to the size of the blocker's job, e.g., small-to-any blockers are only used with small jobs, huge-to-any blockers only with huge jobs, etc. The latter part of the type's name describes the undesirable jobs on the machine: The algorithm will try to remove jobs from this machine if they are undesirable; at the same time it does not attempt to add such jobs to the machine. On machines of small-/medium-/huge-to-any blockers all jobs are undesirable; on machines of huge-/medium-to-huge blockers huge jobs are undesirable; on machines of huge-to-medium blockers medium jobs are undesirable and finally on machines of huge-to-least blockers only those medium jobs of index smaller or equal to the smallest medium job on i are undesirable.

The same machine can appear more than once in the blocker tree. In that case, the undesirable jobs are the union of the undesirable jobs from all types. Also, the same job can appear multiple times in different blockers.

The blockers corresponding to specific types are written as $\mathcal{T}_{SA}$, $\mathcal{T}_{HA}$, etc. From the blocker tree, we derive the machine set $\mathcal{M}(\mathcal{T})$ which consists of all machines corresponding to moves in $\mathcal{T}$. This notation is also used with subsets of $\mathcal{T}$, e.g., $\mathcal{M}(\mathcal{T}_{HA})$.

Definition 5 (Blocked small jobs, active jobs). A small job j is *blocked*, if it is undesirable on all other machines it allowed on, that is $\Gamma(j) \backslash \{\sigma(j)\} \subseteq$

$\mathcal{M}(\mathcal{T}_{SA} \cup \mathcal{T}_{MA} \cup \mathcal{T}_{HA})$. We denote the set of blocked small jobs by $S(\mathcal{T})$. The set of active jobs $\mathcal{A}$ includes j_{new}, $S(\mathcal{T})$ as well as all those jobs, that are undesirable on the machine, they are currently assigned to.

We define for all machines i the job sets $S_i(\mathcal{T}) = S(\mathcal{T}) \cap \sigma^{-1}(i)$, $\mathcal{A}_i(\mathcal{T}) = \mathcal{A}(\mathcal{T}) \cap \sigma^{-1}(i)$, $M_i = \sigma^{-1}(i) \cap \mathcal{J}_M$ and $H_i = \sigma^{-1}(i) \cap \mathcal{J}_H$. Moreover, set $M_i^{\min} = \{\min M_i\}$ if $M_i \neq \emptyset$ and $M_i^{\min} = \emptyset$ otherwise.

2.1   Tree and Layers

The blockers in $\mathcal{T}$ and an additional root can be imagined as a tree. The parent of each blocker $\mathcal{B} = (j, i, \Theta)$ is only determined by j. If $j = j_{\mathrm{new}}$ it is the root node; otherwise it is a blocker $\mathcal{B}' \in \mathcal{T}$ for machine $\sigma(j)$ with a type for which j is regarded undesirable. If this applies to several blockers, we use the one that was added to the blocker tree first. We say that $\mathcal{B}'$ *activates* j.

Let us now introduce the notion of a layer. Each blocker is assigned to exactly one layer. The layer roughly correlates with the distance of the blocker to the root node. In this sense, the children of a blocker are usually in the next layer. There are some exceptions, however, in which a child is in the same layer as its parent. We now define the layer of the children of a blocker $\mathcal{B}$ in layer k.

1. If $\mathcal{B}$ is a huge-/medium-to-huge blocker, all its children are in layer k as well;
2. if $\mathcal{B}$ is a huge-to-any blocker, children regarding small jobs are in layer k as well;
3. in every other case, the children are in layer $k + 1$.

We note that by this definition for an active job j all blockers $(j, i, \Theta) \in \mathcal{T}$ must be in the same layer; in other words, it is unambiguous in which layer blockers for it would be placed in. We say j is k-*headed*, if blockers for j would be placed in layer k. The blockers in layer k are denoted by $\mathcal{T}^{(k)}$. The set of blockers in layer k and below is referred to by $\mathcal{T}^{(\leq k)}$. We use this notation in combination with qualifiers for the type of blocker, e.g., $\mathcal{T}_{HA}^{(k)}$.

We establish an order between the types of blockers within a layer and refer to this order as the sublayer number. The huge-/medium-to-huge blockers form the first sublayer of each layer, huge-to-any and medium-to-any blockers the second, small-to-any blockers the third, huge-to-least the fourth and huge-to-medium blockers the fifth sublayer (see also Table 1 and Fig. 2). By saying a sublayer is after (before) another sublayer we mean that either its layer is higher (lower) or both layers are the same and its sublayer number is higher (lower).

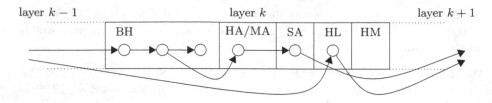

layer $k-1$ layer k layer $k+1$

Fig. 2. Example layer

Algorithm 1: Quasi-polynomial local search

```
1   initialize empty blocker tree T;
2   loop
3     if a move in T is valid then
4       choose a blocker (j,i,Θ) in the lowest sublayer,
5           where (j,i) is valid;
6       let B be the blocker that activated j;
7       // Update the schedule
8       σ(j) ← i;
9       remove all sublayers after B from T;
10      if j = j_new then
11          return σ;
12      end
13      if not conditions*(B) then
14          remove the sublayer of B from T;
15      end
16    else
17      let ℓ be the minimum layer to which we can
18          add a potential move;
19      if ℓ > K or no such ℓ exists then
20          return 'error';
21      end
22      add potential move (j,i) of highest priority to layer ℓ;
23      remove all sublayers after (j,i) from T;
24    end
25  end
```

In the final algorithm whenever we remove one blocker, we also remove all blockers in its sublayer and all later sublayers (in particular, all descendants). Also, when we add a blocker to a sublayer, we remove all later sublayers. Among other properties, this guarantees that the connectivity of the tree is never compromised. It also means that, if j is undesirable regarding several blockers for $\sigma(j)$, then the parent is in the lowest sublayer among these blockers, since a blocker in a lower sublayer cannot have been added after one in a higher sublayer.

The running time will be exponential in the number of layers; hence this should be fairly small. We introduce an upper bound $K = 2/\epsilon\lceil\ln(|\mathcal{M}|) + 1\rceil = \mathcal{O}(1/\epsilon \cdot \log(|\mathcal{M}|))$ and will not add any blockers to a layer higher than K.

2.2 Detailed Description of the Algorithm

The algorithm (see Algorithm 1) contains a loop that terminates once j_{new} is assigned. In each iteration the algorithm performs a valid move in the blocker tree if possible and otherwise adds a new blocker.

Adding blockers. We only add a move to $\mathcal{T}$, if it meets certain requirements. A move that does is called a potential move. For each type of blocker we also define a type of potential move: Potential small-to-any moves, potential huge-to-any moves, etc. When a potential move is added to the blocker tree, its type will then be used for the blocker. Let k be a layer and let $j \in \mathcal{A}(\mathcal{T})$ be k-headed. For a move (j, i) to be a potential move of a certain type, it has to meet the following requirements.

1. (j, i) is not already in $\mathcal{T}$;
2. the size of j corresponds to the type, for instance, if j is big, (j, i) cannot be a small-to-any move;
3. j is not undesirable on i w.r.t. $\mathcal{T}^{(\leq k)}$, i.e., (a) $i \notin \mathcal{M}(\mathcal{T}_{SA}^{(\leq k)} \cup \mathcal{T}_{MA}^{(\leq k)} \cup \mathcal{T}_{HA}^{(\leq k)})$ and (b) if j is huge, then $i \notin \mathcal{M}(\mathcal{T}_{BH}^{(\leq k)})$; (c) if j is medium, then $i \notin \mathcal{M}(\mathcal{T}_{HM}^{(\leq k)})$ and either $i \notin \mathcal{M}(\mathcal{T}_{HL}^{(\leq k)})$ or $\min M_i < j$.
4. The load of the target machine has to meet certain conditions (see Table 1).

Comparing the conditions in the table we notice that for moves of small and medium jobs there is always exactly one type that applies. For huge jobs

Table 1. Types of blockers/potential moves

Type	Conditions	S	P	Undesirable
Huge-/Medium-to-huge (BH)	$p(\sigma^{-1}(i)\backslash H_i) + p_j \leq 1 + R$	1	5	Huge jobs
Small-to-any (SA)	None	3	4	All jobs
Medium-to-any (MA)	$* p(\sigma^{-1}(i)\backslash H_i) + p_j > 1 + R$	2	3	All jobs
Huge-to-any (HA)	$* p(\sigma^{-1}(i)\backslash H_i) + p_j > 1 + R$ $p(S_i(\mathcal{T}^{(\leq k)}) \cup M_i) + p_j \leq 1 + R$	2	3	All jobs
Huge-to-least (HL)	$* p(S_i(\mathcal{T}^{(\leq k)}) \cup M_i^{\min}) + p_j > 1 + R$ $p(S_i(\mathcal{T}^{(\leq k)})) + p_j \leq 1 + R$	4	2	Medium jobs j_M with $j_M \leq \min M_i$
Huge-to-medium (HM)	$* p(S_i(\mathcal{T}^{(\leq k)}) \cup M_i) + p_j > 1 + R$ $* p(S_i(\mathcal{T}^{(\leq k)}) \cup M_i^{\min}) + p_j \leq 1 + R$	5	1	Medium jobs

The conditions are meant in respect to a move (j, i) where j is k-headed. Column S stands for the sublayer and P for the priority of a blocker type. Conditions marked with a star ($*$) are additionally checked whenever a job activated by this blocker is moved.

there is exactly one type if $p(S_i(T^{(\leq k)})) + p_j \leq 1 + R$ and no type applies, if $p(S_i(T^{(\leq k)})) + p_j > 1 + R$. The table also lists a priority for each type of move. It is worth mentioning that the priority does not directly correlate with the sublayer. The algorithm will choose the move that can be added to the lowest layer and among those has the highest priority. After adding a blocker, all higher sublayers are deleted.

Performing valid moves. The algorithm performs a valid move in T if there is one. It chooses a blocker (j, i, Θ) in T, where the blocker's sublayer is minimal and (j, i) is valid. Besides assigning j to i, T has to be updated as well.

Let $\mathcal{B}$ be the blocker that activated j. When certain conditions for $\mathcal{B}$ are no longer met, we will delete $\mathcal{B}$ and its sublayer. The conditions that need to be checked depend on the type of $\mathcal{B}$ and are marked in Table 1 with a star ($*$). In any case, the algorithm will discards all blockers in higher sublayers than $\mathcal{B}$ is.

3   Analysis

The analysis of the algorithm has two critical parts. First, we show that it does not get stuck, i.e., there is always a blocker that can be added to the blocker tree or a move that can be executed. Then we show that the number of iterations is bounded by $\exp(\mathcal{O}(1/\epsilon \log^2(n)))$.

Theorem 2. *If the algorithm returns 'error', then* $\mathrm{OPT}^* > 1$.

The proof consists in the construction of a solution (z^*, y^*) for the dual of the configuration-LP. The value z_j^* is composed of $\overline{p}_j$ and a scaling coefficient (a power of $\delta := 1 - \epsilon$). The idea of the scaling coefficient is that values for jobs activated in higher layers are supposed to get smaller and smaller. We set $z_j^* = 0$ if $j \notin \mathcal{A}(T)$ and $z_j^* = \delta^k \cdot \overline{p}_j$, if $j \in \mathcal{A}(T)$ and k is the smallest layer such that j is k-headed or $j \in S(T^{(\leq k)})$. For all $i \in \mathcal{M}$ let

$$w_i = \begin{cases} z^*(\mathcal{A}_i(T)) + \delta^k \frac{1}{6} & \text{if } i \in \mathcal{M}(T_{HA}^{(k)}), \\ z^*(\mathcal{A}_i(T)) - \delta^k \frac{1}{6} & \text{if } i \in \mathcal{M}(T_{SA}^{(k)}), \\ z^*(\mathcal{A}_i(T)) & \text{otherwise.} \end{cases}$$

Finally set $y_i^* = \delta^K + w_i$. Note that w is well-defined, since a machine i can be in at most one of the sets $\mathcal{M}(T_{HA}^{(1)}), \mathcal{M}(T_{SA}^{(1)}), \mathcal{M}(T_{HA}^{(2)}), \mathcal{M}(T_{SA}^{(2)}), \ldots$ On a small-/huge-to-any blocker all jobs are undesirable, that is to say as long as one of such blockers remains in the blocker tree, the algorithm will not add another blocker with the same machine. Also note that $z^*(\mathcal{A}_i(T))$ and $z^*(\sigma^{-1}(i))$ are interchangeable.

Lemma 1. *If there is no valid move in T and no potential move of a k-headed job for a $k \leq K$, the value of the solution is negative, i.e.,* $\sum_{j \in \mathcal{J}} z_j^* > \sum_{i \in \mathcal{M}} y_i^*$.

Proof. Using the Taylor series and $\epsilon < 1/12$ it is easy to check $\ln(1 - \epsilon) \geq -\epsilon/2$. This gives

$$K \geq \frac{2}{\epsilon}(\ln(|\mathcal{M}|) + 1) \geq \frac{\ln(2|\mathcal{M}|)}{\epsilon/2} \geq -\frac{\ln(2|\mathcal{M}|)}{\ln(1 - \epsilon)} = \log_\delta\left(\frac{1}{2|\mathcal{M}|}\right).$$

Claim 1 *(Proof is omitted to conserve space). For all $k \leq K$ we have $|\mathcal{M}(T_{HA}^{(k)})| \leq |\mathcal{M}(T_{SA}^{(k)})|$.*

Using this claim we find that

$$\sum_{j \in \mathcal{J}} z_j^* \geq z_{j_{new}}^* + \sum_{i \in \mathcal{M}} z^*(\sigma^{-1}(i))$$

$$\geq \delta 1\frac{5}{6} + \sum_{i \in \mathcal{M}} y_i^* - \delta^K |\mathcal{M}| + \sum_{k=1}^{K}[\delta^k \frac{1}{6}|\mathcal{M}(T_{SA}^{(k)})| - \delta^k \frac{1}{6}|\mathcal{M}(T_{HA}^{(k)})|]$$

$$\geq \delta 1\frac{5}{6} + \sum_{i \in \mathcal{M}} y_i^* - \frac{1}{2} + 0 > \sum_{i \in \mathcal{M}} y_i^*.$$

$\qquad\square$

Lemma 2. *If there is no valid move in T and no potential move of a k-headed job for a $k \leq K$, the solution is feasible, i.e., $z^*(C) \leq y_i^*$ for all $i \in \mathcal{M}$, $C \in \mathcal{C}_i$.*

Proof. We will make the following assumptions, that can be shown with an exhaustive case analysis.

Claim 2 *(Proof is omitted to conserve space). Let $k \leq K$, $i \notin \mathcal{M}(T_{SA}^{(\leq k)} \cup T_{MA}^{(\leq k)} \cup T_{HA}^{(\leq k)})$, $C \in \mathcal{C}_i$, $j \in C$ k-headed and big with $\sigma(j) \neq i$. Then $z_j^* \leq z^*(\mathcal{A}_i(T^{(\leq k)})\backslash C)$.*

Claim 3 *(Proof is omitted to conserve space). Let $k \leq K$ and $i \in \mathcal{M}(T_{SA}^{(k)} \cup T_{MA}^{(k)} \cup T_{HA}^{(k)})$. Then*

$$w_i \geq z^*(\mathcal{A}_i(T)) + \delta^k \cdot (1 - \delta \overline{p}(\mathcal{A}_i(T))).$$

Let $C_0 \in \mathcal{C}_i$ and $C \subseteq C_0$ denote the set of jobs j with $z_j^* \geq \delta^K \overline{p}_j$. In particular, C does not contain jobs that have potential moves. It is sufficient to show that $z^*(C) \leq w_i$, as this would imply

$$z^*(C_0) = z^*(C) + z^*(C_0\backslash C) \leq w_i + \delta^K \overline{p}(C_0) \leq y_i^*.$$

Loosely speaking, the purpose of δ^K in the definition of y^* is to compensate for ignoring all $(K + 1)$-headed jobs.

First, consider the case where $i \notin \mathcal{M}(T_{SA} \cup T_{MA} \cup T_{HA})$. There cannot be a small and activated job $j_S \in C$ with $\sigma(j_S) \neq i$, because then (j_S, i) would be a potential move; hence $C \cap \mathcal{J}_S \cap \mathcal{A}(T) \subseteq C \cap \mathcal{A}_i(T)$. If there is a big job $j_B \in C$ with $\sigma(j_B) \neq i$, then

$$z^*(C) = z_{j_B}^* + z^*(C \cap \mathcal{J}_S) \leq z^*(\mathcal{A}_i(T)\backslash C) + z^*(C \cap \mathcal{A}_i(T)) = z^*(\mathcal{A}_i(T)) = w_i.$$

If there is no such job, then $C \cap \mathcal{A}(T) \subseteq \mathcal{A}_i(T)$ and in particular $z^*(C) \le w_i$.

In the remainder of this proof we assume that $i \in \mathcal{M}(T_{SA}^{(\ell+1)} \cup T_{MA}^{(\ell+1)} \cup T_{HA}^{(\ell+1)})$. Note that for any $k \ne \ell + 1$ we have $i \notin \mathcal{M}(T_{SA}^{(k)} \cup T_{MA}^{(k)} \cup T_{HA}^{(k)})$. Also, since all jobs on i are active we have that $z_j^* \ge \delta^{\ell+2}\overline{p}_j$ for all $j \in \sigma^{-1}(i)$. Because there is no potential move (j_S, i) for a small job j_S with $z_{j_S}^* \ge \delta^\ell \overline{p}_{j_S}$, we have for all small jobs $j_S \in C \backslash \mathcal{A}_i(T)$: $z_{j_S}^* \le \delta^{\ell+1}\overline{p}_{j_S}$.

Case 1. For every big job $j \in C$ with $\sigma(j) \ne i$ we have $z_j^* \le \delta^{\ell+1}\overline{p}_j$.
This implies

$$z^*(C \backslash \mathcal{A}_i(T)) \le \delta^{\ell+1}\overline{p}(C \backslash \mathcal{A}_i(T)) = \delta^{\ell+1}(\overline{p}(C) - \overline{p}(\mathcal{A}_i(T) \cap C))$$
$$\le \delta^{\ell+1}(1 - \delta\overline{p}(\mathcal{A}_i(T) \cap C)).$$

Therefore

$$z^*(C) = z^*(\mathcal{A}_i(T) \cap C) + z^*(C \backslash \mathcal{A}_i(T))$$
$$\le z^*(\mathcal{A}_i(T) \cap C) + \delta^{\ell+1}(1 - \delta\overline{p}(\mathcal{A}_i(T) \cap C))$$
$$\le z^*(\mathcal{A}_i(T)) + \delta^{\ell+1}(1 - \delta\overline{p}(\mathcal{A}_i(T))) \le w_i.$$

Case 2. There is a big job $j \in C$ with $\sigma(j) \ne i$ and $z_j^* \ge \delta^\ell \overline{p}_j$.
Let $k \le \ell$ with $z_j^* = \delta^k \overline{p}_j$, that is to say j is k-headed. Then

$$z_j^* - \delta^{\ell+1}\overline{p}_j = (1 - \delta^{\ell+1-k})z_j^* \le (1 - \delta^{\ell+1-k})z^*(\mathcal{A}_i(T^{(\le k)}) \backslash C)$$
$$\le z^*(\mathcal{A}_i(T^{(\le k)}) \backslash C) - \delta^{\ell+2}\overline{p}(\mathcal{A}_i(T^{(\le k)}) \backslash C)$$
$$\le z^*(\mathcal{A}_i(T) \backslash C) - \delta^{\ell+2}\overline{p}(\mathcal{A}_i(T) \backslash C).$$

In the second inequality we use that for every $j' \in \mathcal{A}_i(T^{(\le k)})$ we have $z_{j'}^* \ge \delta^{k+1} \cdot \overline{p}_{j'}$. This implies that

$$z^*(C) = z_j^* + z^*(C \backslash \{j\})$$
$$= z_j^* + z^*(\mathcal{A}_i(T) \cap C) + z^*(C \backslash (\{j\} \cup \mathcal{A}_i(T)))$$
$$\le z_j^* + z^*(\mathcal{A}_i(T) \cap C) + \delta^{\ell+1}(\overline{p}(C) - \overline{p}_j - \overline{p}(\mathcal{A}_i(T) \cap C))$$
$$\le z_j^* + z^*(\mathcal{A}_i(T) \cap C) + \delta^{\ell+1}(1 - \overline{p}_j - \delta\overline{p}(\mathcal{A}_i(T) \cap C))$$
$$\le z^*(\mathcal{A}_i(T)) + \delta^{\ell+1}(1 - \delta\overline{p}(\mathcal{A}_i(T))) \le w_i.$$

$\square$

We can now complete the proof of Theorem 2.

Proof (Theorem 2). Suppose toward contradiction there is no potential move of a k-headed job, where $k \le K$, and no move in the blocker tree is valid. It is obvious that since Lemmas 1 and 2 hold for (y^*, z^*), they also hold for a scaled solution $(\alpha \cdot y^*, \alpha \cdot z^*)$ with $\alpha > 0$. We can use this to obtain a solution with an arbitrarily low objective value; thereby proving that the dual is unbounded regarding makespan 1 and therefore $OPT^* > 1$. $\square$

Theorem 3. *The algorithm terminates in time* $\exp(\mathcal{O}(1/\epsilon \cdot \log^2(n)))$.

Proof. Let $\ell \leq K$ be the index of the last non-empty layer in $\mathcal{T}$. We will define the so-called signature vector as $s(\mathcal{T}, \sigma) = (s_1, s_2, \ldots, s_\ell)$, where s_k is given by

$$
s_k = \left(\sum_{(j,i,\Theta) \in T_{BH}^{(k)}} [|\mathcal{J}| - |H_i|], \quad \sum_{(j,i,\Theta) \in T_{MA}^{(k)} \cup T_{HA}^{(k)}} [|\mathcal{J}| - |\sigma^{-1}(i)|], \right.
$$

$$
\left. \sum_{(j,i,\Theta) \in T_{SA}^{(k)}} [|\mathcal{J}| - |\sigma^{-1}(i)|], \quad \sum_{(j,i,\Theta) \in T_{HL}^{(k)}} [\min M_i], \quad \sum_{(j,i,\Theta) \in T_{HM}^{(k)}} [|\mathcal{J}| - |M_i|] \right).
$$

Each component in s_k represents a sublayer within layer k and it is the sum over certain values associated with its blockers. Note that these values are all strictly positive, since j_{new} is not assigned and therefore $|\sigma^{-1}(i)| < |\mathcal{J}|$.

Claim 4 *(Proof is omitted to conserve space). The signature vector increases lexicographically after polynomially many iterations of the loop.*

This means that the number of possible vectors is an upper bound on the running time (except for a polynomial factor). Each sublayer has at most $|\mathcal{J}| \cdot |\mathcal{M}|$ many blockers (since there are at most this many moves) and the value for every blocker in each of the five cases is easily bounded by $\mathcal{O}(|\mathcal{J}|)$. This implies there are at most $(\mathcal{O}(n^3))^5 = n^{\mathcal{O}(1)}$ values for each s_k. Using $K = \mathcal{O}(1/\epsilon \log(n))$ we bound the number of different signature vectors by $n^{\mathcal{O}(K)} = \exp(\mathcal{O}(1/\epsilon \log^2(n)))$. $\qquad\square$

4   Conclusion

We have greatly improved the running time of the local search algorithm for the RESTRICTED ASSIGNMENT problem. At the same time we were able to maintain almost the same approximation ratio. We think there are two important directions for future research. The first is to improve the approximation ratio further. For this purpose, it makes sense to first find improvements for the much simpler variant of the algorithm given in [7].

The perhaps most important open question, however, is whether the running time can be brought down to a polynomial one. Recent developments in the RESTRICTED MAX-MIN FAIR ALLOCATION problem indicate that a layer structure similar to the one in this paper may also help in that regard [2]. In the mentioned paper moves are only performed in large groups. This concept is referred to as laziness. The asymptotic behavior of the partition function (the number of integer partitions of a natural number) is then used in the analysis for a better bound on the number of possible signature vectors. This approach appears to have a great potential for the RESTRICTED ASSIGNMENT problem as well. In [1] it was already adapted to the special case of two processing times.

References

1. Annamalai, C.: Lazy local search meets machine scheduling. CoRR abs/1611.07371 (2016). http://arxiv.org/abs/1611.07371
2. Annamalai, C., Kalaitzis, C., Svensson, O.: Combinatorial algorithm for restricted max-min fair allocation. In: Proceedings of the Twenty-Sixth Annual ACM-SIAM Symposium on Discrete Algorithms, SODA 2015, San Diego, CA, USA, 4–6 January 2015, pp. 1357–1372 (2015). doi:10.1137/1.9781611973730.90
3. Bansal, N., Sviridenko, M.: The santa claus problem. In: Proceedings of the 38th Annual ACM Symposium on Theory of Computing, Seattle, WA, USA, 21–23 May 2006, pp. 31–40 (2006). doi:10.1145/1132516.1132522
4. Chakrabarty, D., Khanna, S., Li, S.: On $(1, \epsilon)$-restricted assignment makespan minimization. In: Proceedings of the Twenty-Sixth Annual ACM-SIAM Symposium on Discrete Algorithms, SODA 2015, San Diego, CA, USA, 4–6 January 2015, pp. 1087–1101 (2015). doi:10.1137/1.9781611973730.73
5. Jansen, K., Land, F., Land, K.: Bounding the running time of algorithms for scheduling and packing problems. SIAM J. Discrete Math. **30**(1), 343–366 (2016). doi:10.1137/140952636
6. Jansen, K., Land, K., Maack, M.: Estimating the makespan of the two-valued restricted assignment problem. In: Proceedings of the 15th Scandinavian Symposium and Workshops on Algorithm Theory, SWAT 2016, Reykjavik, Iceland, 22–24 June 2016, pp. 24:1–24:13 (2016). doi:10.4230/LIPIcs.SWAT.2016.24
7. Jansen, K., Rohwedder, L.: On the Configuration-LP of the restricted assignment problem. In: Proceedings of the Twenty-Eighth Annual ACM-SIAM Symposium on Discrete Algorithms, SODA 2017, Barcelona, Spain, Hotel Porta Fira, 16–19 January, pp. 2670–2678 (2017). doi:10.1137/1.9781611974782.176
8. Lenstra, J.K., Shmoys, D.B., Tardos, E.: Approximation algorithms for scheduling unrelated parallel machines. Math. Program. **46**(3), 259–271 (1990). doi:10.1007/BF01585745
9. Polácek, L., Svensson, O.: Quasi-polynomial local search for restricted max-min fair allocation. ACM Trans. Algorithms **12**(2), 13 (2016). doi:10.1145/2818695
10. Svensson, O.: Santa claus schedules jobs on unrelated machines. SIAM J. Comput. **41**(5), 1318–1341 (2012). doi:10.1137/110851201

Adaptive Submodular Ranking

Prabhanjan Kambadur[1], Viswanath Nagarajan[2], and Fatemeh Navidi[2(✉)]

[1] Bloomberg LP, 731 Lexington Avenue, New York City, NY, USA
pkambadur@bloomberg.net
[2] Department of Industrial and Operations Engineering,
University of Michigan, Ann Arbor, MI, USA
{viswa,navidi}@umich.edu

Abstract. We study a general stochastic ranking problem where an algorithm needs to adaptively select a sequence of elements so as to "cover" a random scenario (drawn from a known distribution) at minimum expected cost. The coverage of each scenario is captured by an individual submodular function, where the scenario is said to be covered when its function value goes above some threshold. We obtain a logarithmic factor approximation algorithm for this adaptive ranking problem, which is the best possible (unless $P = NP$). This problem unifies and generalizes many previously studied problems with applications in search ranking and active learning. The approximation ratio of our algorithm either matches or improves the best result known in each of these special cases. Moreover, our algorithm is simple to state and implement. We also present preliminary experimental results on a real data set.

1 Introduction

Many stochastic optimization problems can be viewed as sequential decision processes of the following form. There is an *a priori* distribution $\mathcal{D}$ over a set of scenarios, and the goal is to cover the realized scenario $i^* \leftarrow \mathcal{D}$. In each step, an algorithm chooses an element with some given cost which partially covers i^* and also provides some feedback depending on i^*. This feedback is then used to refine the distribution of i^* which is used to select elements in the subsequent steps. So any solution in this setting is an adaptive sequence of elements. The objective is to minimize the expected cost to cover scenario i^*.

Furthermore, many different criteria to cover a scenario can be modeled as covering a suitable submodular function. Submodular functions are widely used to model utilities in game theory, influence maximization in social networks, diversity in search ranking etc.

In this paper, we study an abstract stochastic optimization problem in the setting described above which unifies and generalizes many previously-studied problems such as optimal decision trees [8,10,11,18,20,24], equivalence class determination [6,15], decision region determination [23] and submodular ranking [2,21]. We obtain an algorithm with the best-possible approximation guarantee in all these special cases. We also obtain the first approximation algorithms

© Springer International Publishing AG 2017
F. Eisenbrand and J. Koenemann (Eds.): IPCO 2017, LNCS 10328, pp. 317–329, 2017.
DOI: 10.1007/978-3-319-59250-3_26

for some other problems such as stochastic knapsack cover and matroid basis with correlated elements. Moreover, our algorithm is very simple to state and implement. We also present experimental results for two of these special cases (on a real data set), and our algorithm performs quite well.

For some stochastic optimization problems, one can come up with approximately optimal solutions using static (non-adaptive) solutions that are insensitive to the feedback obtained, eg. [5,12]. However, this is not the case for the adaptive submodular ranking problem. Even for the special cases above, there are instances where the optimal adaptive value is much less than the optimal non-adaptive value. Thus, it is important to come up with an adaptive algorithm.

Problem Statement. We start with some basics. A set function $f : 2^U \to \mathbb{R}_+$ on ground set U is said to be submodular if $f(A) + f(B) \geq f(A \cap B) + f(A \cup B)$ for all $A, B \subseteq U$. The function f is said to be monotone if $f(A) \leq f(B)$ for all $A \subseteq B \subseteq U$. We assume that set functions are given in the standard *value oracle* model, i.e. we can evaluate $f(S)$ for any $S \subseteq U$ in polynomial time.

In the adaptive submodular ranking problem (ASR) we have a ground set U of n *elements* with positive costs $\{c_e\}_{e \in U}$, that we can assume are integer without loss of generality. We also have m *scenarios* with a probability distribution $\mathcal{D}$ given by probabilities $\{p_i\}_{i=1}^m$ totaling to one. Each scenario $i \in [m] := \{1, \cdots, m\}$ is specified by:

(i) a *monotone submodular* function $f_i : 2^U \to [0,1]$ where $f_i(\emptyset) = 0$ and $f_i(U) = 1$ (any monotone submodular function can be expressed in this form by scaling), and

(ii) a *feedback* function $r_i : U \to G$ where G is a set of possible feedback values.

We note that f_i and r_i need not be related in any way: this flexibility allows us to capture many different applications. Scenario $i \in [m]$ is said to be *covered* by any subset $S \subseteq U$ of elements such that $f_i(S) = 1$. The goal in ASR is to find a sequence of elements in U that minimizes the expected cost to cover a *random scenario i^** drawn from $\mathcal{D}$. The identity of i^* is initially unknown to the algorithm. When the algorithm selects an element $e \in U$, it receives feedback value $g = r_{i^*}(e) \in G$ which can be used to update the probability distribution of i^* (by eliminating all scenarios i with $r_i(e) \neq g$), which in turn can be used to select subsequent elements. The sequence of selected elements is Z adaptive because it depends on the feedback received.

A solution to ASR is represented by a decision tree $\mathcal{T}$, where each node is labeled by an element $e \in U$ and the branches out of such a node are labeled by the feedback received after selecting e. Each node in $\mathcal{T}$ also corresponds to a *state* (E, H), where E is the set of elements shown until this node, and $H \subseteq [m]$ is the subset of scenarios compatible with all the feedback obtained until this node. Every scenario $i \in [m]$ traces a root-leaf path in the decision tree $\mathcal{T}$ which takes the $r_i(e)$-branch out of any node labeled by $e \in U$; let T_i denote the sequence of elements on this path. In a feasible decision tree $\mathcal{T}$, each scenario $i \in [m]$ must be covered, i.e. $f_i(T_i) = 1$. The cost incurred under scenario i in decision tree $\mathcal{T}$ is the total cost of the shortest prefix $\bar{T}_i$ of T_i such that $f_i(\bar{T}_i) = 1$. The objective in ASR is to minimize the expected cost $\sum_{i=1}^{m} p_i \cdot \left(\sum_{e \in \bar{T}_i} c_e \right)$. We emphasize that multiple scenarios may trace the same path in $\mathcal{T}$: in particular, it is *not* necessary to identify the realized scenario i^* in order to cover it.

An important parameter in the analysis of our algorithm is the following:

$$\epsilon \quad := \quad \min_{\substack{e \in U : f_i(S \cup e) > f_i(S) \\ i \in [m], S \subseteq U}} \quad f_i(S \cup e) - f_i(S). \tag{1}$$

It measures the minimum positive incremental value of any element. Such a parameter appears in all results on the submodular cover problem, eg. [2,27].

Results. Our main result is an $O(\log \frac{1}{\epsilon} + \log m)$-approximation algorithm for adaptive submodular ranking where $\epsilon > 0$ is as defined in (1) and m is the number of scenarios. Assuming $P \neq NP$, this result is the best possible (up to a constant factor) as the set cover problem [13] is a special case of ASR even when $m = 1$. Our algorithm is a simple adaptive greedy-style algorithm. At each step, we assign a score to each remaining element and select the element with maximum score. Such a simple algorithm was previously unknown even in the special case of optimal decision tree (under arbitrary costs/probabilities), despite a large number of papers [1,8,10,11,14,17,18,20,24] on this topic.

The following are direct applications of our framework.

- *Optimal decision tree:* This involves uniquely identifying the scenario i^* at minimum expected cost. We obtain an $O(\log m)$-approximation which matches the previous best approximation ratio [18] with a faster and much simpler algorithm. We can also handle a natural extension where the goal is to obtain a small subset of scenarios (of specified size) that contains i^*.
- *Adaptive multiple intent re-ranking:* This is an adaptive version of a problem studied in [3,4,26] with applications to search ranking. We obtain the first approximation algorithm for the adaptive problem.
- *Stochastic correlated knapsack cover:* This is a stochastic version of the covering knapsack problem, where elements have random (correlated) sizes and deterministic cost. We obtain the first approximation algorithm here.

More applications and details can be seen in the full version [25].

Techniques. Our algorithm involves repeatedly selecting an element that maximizes a combination of (i) the expected increase in function value relative to the target of one, and (ii) a measure of gain in identifying the realized scenario. See (2) for the formal selection criterion. Our analysis provides new ways of reasoning about adaptive decision trees. At a high level, our approach is similar to minimum-latency problems [7,9]: we upper bound the probability that the algorithm incurs a certain power-of-two cost 2^k in terms of the probability that the optimal solution incurs cost $2^k/\alpha$, which is then used to establish an $O(\alpha)$ approximation ratio. Our main technical contribution is in relating these completion probabilities in the algorithm and the optimal solution (see Lemma 2). In particular, a key step in our proof is a coupling of "bad" states in the algorithm (where the gain in terms of our selection criterion is small) with "bad" states in the optimum (where the cost incurred is high). This is reflected in the classification of the algorithm's states as good/ok/bad (Definition 1) and the proof that the *expected* gain of the algorithm is large (Lemma 4).

Related Works. The basic submodular cover problem (select a min-cost subset of elements that covers a given submodular function) was first considered by [27] who proved that the natural greedy algorithm is a $(1 + \ln \frac{1}{\epsilon})$-approximation algorithm. This result is tight because set cover is a special case. The submodular cover problem corresponds to the special case of ASR with $m = 1$.

The deterministic submodular ranking problem was introduced by [2] who obtained an $O(\log \frac{1}{\epsilon})$-approximation algorithm when all costs are unit. This is a special case of ASR when there is no feedback (i.e. $G = \emptyset$); note that the optimal ASR solution in this case is just a fixed sequence of elements. The result in [2] was based on an interesting "reweighted" greedy algorithm: the second term in our selection criterion (2) is similar to this. A different proof of the submodular ranking result (using a min-latency type analysis) was obtained in [21] which also implied an $O(\log \frac{1}{\epsilon})$-approximation algorithm with costs. We also use a min-latency type analysis for ASR.

The first $O(\log m)$-approximation algorithm for optimal decision tree was obtained in [18], which is known to be tight [8]. This result was extended to the equivalence class determination problem in [10], Previous results, eg. [1,8,11,17, 24], based on a simple greedy "splitting" algorithm, had a logarithmic dependence on either costs or probabilities which can be exponential in m. The algorithms in [10,18] were more complex than what we obtain here as a special case of ASR. In particular these algorithms proceeded in $O(\log m)$ phases, each of which required solving an auxiliary subproblem that reduced the number of possible scenarios by a constant factor. Using such a "phase based" algorithm and analysis for the general ASR problem only leads to an $O(\log m \cdot \log \frac{1}{\epsilon})$-approximation ratio because the subproblem to be solved in each phase is submodular ranking which only has an $O(\log \frac{1}{\epsilon})$-approximation ratio. Our work is based on a much simpler greedy-style approach.

A different stochastic version of submodular ranking was considered in [21] where (i) the feedback was independent across elements and (ii) all the submodular functions needed to be covered. In contrast, ASR involves a correlated scenario-based distribution and only the submodular function of the "realized" scenario i^* needs to be covered. Due to these differences, both the algorithm and analysis for ASR are different from [21]: our selection criterion (2) involves an additional "information gain" term, and our analysis requires a lot more work in order to handle correlations. We note that unlike ASR, the stochastic submodular ranking problem in [21] does not capture the optimal decision tree problem and its variants [10,23].

For some previously-studied special cases [6,15,23] of ASR, one could obtain approximation algorithms via the framework of "adaptive submodularity" [14]. However, this approach does not apply to the general ASR problem and the approximation ratio obtained is at least $\Omega(\log 1/p_{min})$ where $p_{min} = \min_{i=1}^{m} p_i$ can be exponentially small in m.

Recently, [16] considered the scenario submodular cover problem, defined in terms of adaptive submodularity. This involves a *single* integer-valued submodular function for all scenarios which is defined on an expanded groundset $U \times G$ (i.e. pairs of "element, feedback" values). This problem is a special case of ASR: our algorithm matches (in fact, improves slightly) the approximation ratio in [16] with a much simpler algorithm and analysis. We note that ASR is strictly more general than this problem: submodular ranking [2] is a special case of ASR but not of scenario submodular cover.

2 The Algorithm

The state of our algorithm (i.e. any node in its decision tree) is represented by (i) the set $E \subseteq U$ of previously displayed elements, and (ii) the set $H \subseteq [m]$ of scenarios that are compatible with feedback (on E) received so far and are still uncovered. At any state (E, H), our algorithm does the following. For each element $e \in U \setminus E$, let $B_e(H)$ denote the set with maximum cardinality amongst $\{i \in H : r_i(e) = t\}$ for $t \in G$, and we define $L_e(H) = H \setminus B_e(H)$. Then we select element $e \in U \setminus E$ that maximizes:

$$\frac{1}{c_e} \cdot \left(\sum_{j \in L_e(H)} p_j + \sum_{i \in H} p_i \cdot \frac{f_i(e \cup E) - f_i(E)}{1 - f_i(E)} \right). \tag{2}$$

Note that H only contains uncovered scenarios. So for all $i \in H, f_i(E) < 1$ and the denominator in the sum above is always positive. Next we analyze the performance of this algorithm. For any subset $T \subseteq [m]$ of scenarios, we use $\Pr(T) = \sum_{i \in T} p_i$.

Let OPT denote an optimal solution to the ASR instance and ALG be the solution found by the above algorithm. Set $L := 15(1 + \ln 1/\epsilon + \log_2 m)$ and its choice will be clear later. We refer to the total cost incurred at any point in a solution as the *time*. For any $k = 0, 1, \cdots$, we define the following quantities:

- A_k is the set of uncovered scenarios of ALG at time $L \cdot 2^k$, and $a_k = \Pr(A_k)$.
- x_k is the probability of uncovered scenarios of OPT at time 2^{k-1}.

Lemma 1. *The expected cost of ALG and OPT can be bounded as follows.*

$$C_{ALG} \leq L \sum_{k \geq 0} 2^k a_k + L \quad and \quad C_{OPT} \geq \frac{1}{2} \sum_{k \geq 1} 2^{k-1} x_k \tag{3}$$

Proof. Can be found in the full version [25]. □

Thus, if we upper bound each a_k by some multiple of x_k, it will be easy to obtain the approximation factor. However, this is not the case, instead we prove:

Lemma 2. *For all $k \geq 1$ we have $a_k \leq 0.2a_{k-1} + 3x_k$.*

This lemma implies our main result:

Theorem 1. *Our algorithm is an $\mathcal{O}(\log 1/\epsilon + \log m)$-approximation algorithm.*

Proof. Can be found in the full version [25]. $\square$

We now prove Lemma 2 for a fixed $k \geq 1$. Consider any time t between $L \cdot 2^{k-1}$ and $L \cdot 2^k$. Note that ALG's decision tree induces a partition of all the uncovered scenarios at time t, where each part H consists of all scenarios that are at a particular node (E, H) at time t. Let $R(t)$ denote the set of parts in this partition. Note that all scenarios in A_k appear in $R(t)$ as these scenarios are uncovered even at time $L \cdot 2^k \geq t$. Similarly, all scenarios in $R(t)$ are in A_{k-1}. See Fig. 2.

For any part $H \in R(t)$, let (E, H) denote the node in ALG's decision tree corresponding to H. We note that E consists of all elements that have been completely displayed by time t. The element that is selected at this node is *not* included in E. Let $T_H(k)$ denote the subtree of OPT that corresponds to paths traced by scenarios in H up to time 2^{k-1}; see Fig. 1. Note that each node (labeled by any element $e \in U$) in $T_H(k)$ has at most $|G|$ outgoing branches and one of them is labeled by the feedback corresponding to $B_e(H) = H \setminus L_e(H)$. We define $\mathsf{Stem}_k(H)$ to be the path in $T_H(k)$ that at each node (labeled e) follows the branch corresponding to $H \setminus L_e(H)$. Below we also use $\mathsf{Stem}_k(H)$ to denote the set of elements that are completely displayed on this path. We have the following observation:

Observation 1. *Consider any node (E, H). For each $e \in E$, (i) the feedback values $\{r_i(e) : i \in H\}$ are all identical, and (ii) $L_e(H) = \emptyset$. (See full version [25] for more details).*

Definition 1. *Each node (E, H) in ALG is exactly of one of the following types:*

- *"bad" if the probability of uncovered scenarios in H at the end of $\mathsf{Stem}_k(H)$ is at least $\frac{\Pr(H)}{3}$.*
- *"okay" if it is not bad and $\Pr(\cup_{e \in \mathsf{Stem}_k(H)} L_e(H))$ is at least $\frac{\Pr(H)}{3}$.*
- *"good" if it is neither bad nor okay and the probability of scenarios in H that get covered by $\mathsf{Stem}_k(H)$ is at least $\frac{\Pr(H)}{3}$.*

See Fig. 2. This is well defined, because by definition of $\mathsf{Stem}_k(H)$ each scenario in H is (i) uncovered at the end of $\mathsf{Stem}_k(H)$, or (ii) in $L_e(H)$ for some $e \in \mathsf{Stem}_k(H)$, or (iii) covered by some prefix of $\mathsf{Stem}_k(H)$, i.e. the function value reaches 1 on $\mathsf{Stem}_k(H)$. So the total probability of the scenarios in one of these 3 categories must be at least $\frac{\Pr(H)}{3}$. Therefore each node (E, H) is at least of one of these three types.

Lemma 3. *For any $t \in (L2^{k-1}, L2^k]$, we have $\sum_{\substack{H \in R(t) \\ H:bad}} \Pr(H) \leq 3x_k$.*

Proof. Can be found in [25]. $\square$

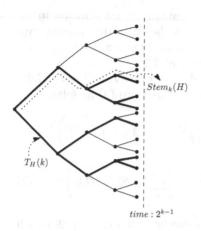

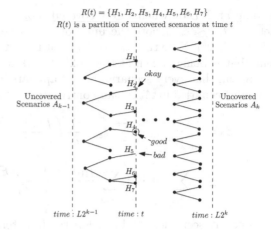

Fig. 1. $Stem_k(H)$ in OPT for $|G| = 2$

Fig. 2. Bad, good and okay nodes in ALG

The following quantity turns out to be useful in our proof of Lemma 2.

$$
Z := \sum_{t>L2^{k-1}}^{L2^k} \sum_{H \in R(t)} \max_{e \in U \setminus E} \frac{1}{c_e} \cdot \left(\Pr(L_e(H)) + \sum_{i \in H} p_i \cdot \frac{f_i(e \cup E) - f_i(E)}{1 - f_i(E)} \right)
\tag{4}
$$

Above, for any $H \in R(t)$ the set E of elements comes from the node (E, H) in ALG corresponding to H. Note that Z corresponds to the total "gain" according to our algorithm's selection criterion (2) accrued from time $L2^{k-1}$ to $L2^k$. Now, we obtain a lower and upper bound for Z and combine them to prove Lemma 2. The lower bound views Z as sum of terms over t, and uses the fact that the gain is "high" for good/ok nodes as well as the bound on probability of bad nodes (Lemma 3). The upper bound views Z as a sum of terms over scenarios and uses the fact that if the total gain for a scenario is "high" then it must be covered.

Lemma 4. *We have $Z \geq L \cdot (a_k - 3x_k)/3$.*

Proof Sketch. We will show that the term in (4) corresponding to each t is at least $\frac{a_k - 3x_k}{3 \cdot 2^{k-1}}$. We know that for each t, we can partition $R(t)$ into three groups $R_{ok}, R_{good}, R_{bad}$. In the proof we decompose the inner sum of (4) to two terms for okay nodes and good nodes. Then we use Lemma 3 and this fact that all scenarios in A_k appear in $R(t)$ to obtain the result. The complete proof can be found in the full version [25]. □

Lemma 5. *We have $Z \leq a_{k-1} \cdot (1 + \ln 1/\epsilon + \log m)$.*

Proof. For any scenario $i \in A_{k-1}$ (i.e. uncovered in ALG by time $L2^{k-1}$) let π_i be the path traced by i in ALG's decision tree, starting from time $2^{k-1}L$ and ending at $2^k L$ or when i gets covered. For each element e that appears in π_i, let $1 \leq t_{e,i} \leq c_e$ denote the units of time when e is shown during the interval $(L2^{k-1}, L2^k]$. Note that there can be at most two elements e in π_i with $t_{i,e} < c_e$: one that is being shown at time $L2^{k-1}$ and another at $L2^k$.

Recall that every scenario in $R(t)$ appears in A_{k-1} for any $L2^{k-1} < t \leq L2^k$. So only scenarios in A_{k-1} can contribute to Z and we rewrite (4) as follows:

$$Z = \sum_{i \in A_{k-1}} p_i \cdot \sum_{e \in \pi_i} t_{e,i} \cdot \frac{1}{c_e} \left(\frac{f_i(e \cup E) - f_i(E)}{1 - f_i(E)} + \mathbb{1}[i \in L_e(H)] \right)$$

$$\leq \sum_{i \in A_{k-1}} p_i \cdot \left(\sum_{e \in \pi_i} \frac{f_i(e \cup E) - f_i(E)}{1 - f_i(E)} + \sum_{e \in \pi_i} \mathbb{1}[i \in L_e(H)] \right) \qquad (5)$$

Fix any scenario $i \in A_{k-1}$. For the first term, we use Claim 2.1 in [2] which relies on the definition of ϵ in (1). This implies $\sum_{e \in \pi_i} \frac{f_i(e \cup E) - f_i(E)}{1 - f_i(E)} \leq 1 + \ln \frac{1}{\epsilon}$. To bound the second term, note that if scenario $i \in L_e(H)$ when ALG selects element e then number of possible scenarios decreases by at least a factor of *two* in path π_i. So such an event can happen at most $\log_2 m$ times along the path π_i. Thus we can write $\sum_{e \in \pi_i} \mathbb{1}[i \in L_e(H)] \leq \log m$. The lemma follows from (5). $\square$

We now complete the proof of Lemma 2. By Lemmas 4 and 5 we have:

$$L \cdot (a_k - 3x_k)/3 \leq Z \leq a_{k-1} \cdot (1 + \ln 1/\epsilon + \log m) = a_{k-1} \cdot \frac{L}{15}$$

Rearranging, we obtain $a_k \leq 0.2 \cdot a_{k-1} + 3x_k$ as needed. $\square$

3   Applications

In this section we discuss various applications of ASR. For problems that have been previously studied, our result matches the best-known approximation ratio with a simpler (and faster) algorithm. For the other problems, we provide the first approximation ratio. The results are summarized in Table 1. We explain some of these problems here and the rest can be found in [25].

Optimal Decision Tree. This problem captures many applications in active learning, medical diagnosis, databases etc., and has been studied extensively. There are m possible hypotheses (which will correspond to scenarios in ASR) with a distribution $\mathcal{D}$ given by probabilities $\{p_i\}_{i=1}^m$, from which an unknown hypothesis i^* is drawn. There are also a number of binary tests (elements in ASR); each test e costs c_e and returns a positive result if i^* lies in some subset T_e of hypotheses and a negative result if $i^* \in [m] \setminus T_e$. We assume that i^* can be uniquely identified by performing all tests. The goal is to identify the hypothesis i^* at the minimum expected cost. This can be cast as a special case of ASR as follows. For each test (element) e and hypothesis (scenario) $i \in [m]$ define $T_e(i) = [m] \setminus T_e$ if $i \in T_e$ and $T_e(i) = T_e$ if $i \notin T_e$. For each scenario $i \in [m]$,

Table 1. Applications of adaptive submodular ranking.

Problem	Previous best result	Our result
Deterministic submodular ranking	$\mathcal{O}(\log \frac{1}{\epsilon})$	$\mathcal{O}(\log \frac{1}{\epsilon})$
Adaptive multiple intent re-ranking	-	$\mathcal{O}(\log \max_{i \in [m]} K_i + \log m)$
Optimal decision tree	$\mathcal{O}(\log m)$	$\mathcal{O}(\log m)$
Generalized optimal decision tree	-	$\mathcal{O}(\log m)$
Equivalence class determination	$\mathcal{O}(\log m)$	$\mathcal{O}(\log m)$
Decision region determination	$\mathcal{O}(\min\{r, d\} \cdot \log m)$ in exp time in r	$\mathcal{O}(r \log m)$ in poly time, $\mathcal{O}(d \log m)$ in exp time in d
Stochastic Knapsack cover	-	$\mathcal{O}(\log m + \log W)$
MSF in random graph	-	$\mathcal{O}(\log m + \log q)$
Scenario submodular cover	$\mathcal{O}(\log m + \log Q)$	$\mathcal{O}(\log m + \log \frac{1}{\epsilon})$

define submodular function $f_i(S) = |\cup_{e \in S} T_e(i)| \cdot \frac{1}{m-1}$. Note that $\epsilon = \frac{1}{m}$; so by Theorem 1 we obtain an $\mathcal{O}(\log m)$-approximation algorithm which is best-possible [8]. We can also handle a natural extension where the goal is to obtain a subset of scenarios containing i^* with a specified size (not necessarily one).

Adaptive Multiple Intent Re-ranking. This is an adaptive version of the multiple intent re-ranking problem, introduced in [3] with applications to search ranking. There are n results (will correspond to elements in ASR) to a particular search query, and m different users (scenarios in ASR). Each user i is characterized by a subset S_i of the results that s/he is interested in and a threshold $K_i \leq |S_i|$: user i gets "covered" after seeing at least K_i results from the subset S_i. The goal is to find an ordering of the results that minimizes the expected number of results to cover a random user i^*, drawn from a known distribution $\mathcal{D}$. This can be modeled as ASR as follows. For each user $i \in [m]$, define submodular function $f_i(S) = \min(|S \cap S_i|, K_i)/K_i$. Note that $\epsilon = 1/\max_{i \in [m]} K_i$, so Theorem 1 implies an $\mathcal{O}(\log \max_{i \in [m]} K_i + \log m)$-approximation algorithm. This almost matches the $\mathcal{O}(\log \max_{i \in [m]} K_i)$ bound from [3] for the deterministic special case. We note however that there are better $\mathcal{O}(1)$-approximation algorithms for the deterministic problem [4,22,26] based on a different LP-based approach: extending this to the adaptive case is still an interesting open question.

Minimum Spanning Forest in Random Graph. We are given a graph $G(V, E)$ with $|V| = q$ and $|E| = n$. Every edge $e \in E$ fails with some probability and its status (active or failed) can be checked at cost c_e. (Any edge that does not fail is called active.) Edge failures are correlated, and we assume a scenario-based joint probability distribution. There are m scenarios, and each scenario $i \in [m]$ is specified by a probability p_i and subset $E_i \subseteq E$ of edges that are active in that scenario; we have $\sum_{i=1}^{m} p_i = 1$. So the set of active edges $E^* = E_i$ with probability p_i for $i \in [m]$. The "active subgraph" (V, E^*) need not be connected. The goal is to find a spanning forest in the active subgraph at minimum expected cost. We can model this as ASR by viewing the edges E as elements and defining for each scenario $i \in [m]$, the submodular function $f_i(S) = \frac{\mathsf{rank}_i(S \cap E_i)}{\mathsf{rank}_i(E_i)}$ where rank_i is the rank function of the graphic matroid on (V, E_i).

The f_is are monotone submodular functions due to the submodularity of matroid rank functions. Moreover, $\epsilon \geq \frac{1}{q}$ and so Theorem 1 implies an $\mathcal{O}(\log m + \log q)$-approximation algorithm. This result also holds for a general matroid: where a random (correlated) subset of elements is active and the goal is to find a basis on the active elements at minimum expected cost.

4   Experiments

In this section, we present experimental results for two special cases of ASR: (generalized) optimal decision tree (ODT) and adaptive multiple intent re-ranking (MIR). Further tests and explanations can be found in the full version [25]. We use a real-world dataset (called ML-100) from the MovieLens [19] repository: this involves the ratings of 943 users (scenarios) for 1682 movies (elements). We binarized this dataset by setting all ratings < 3 to 0 and ratings ≥ 3 to 1. User i is interested in movie e if the corresponding entry is 1. With this dataset, we use the power-law $(Pr[X = x; \alpha] = \alpha x^{\alpha-1})$ with $\alpha = 1, 2, 3$; note that when $\alpha = 1$, we get a uniform distribution. To get a better understanding of the performance results, we generate multiple permutations of scenario distributions for the same value of α. For the ODT problem, we also use a synthetic dataset — SYN-K — that is parameterized by k; this is based on a hard instance for the greedy algorithm [24]. Given k, we generate $m = 2k + 3$ sets, $n = k + 2$ elements, with $4k + 4$ non-zeros as follows: (a) elements $i \in [1, k]$ are contained in scenarios $2i - 1$ and $2i$, (b) element $k + 1$ is contained in all odd numbered scenarios, and (c) element $k + 2$ is contained in all even numbered scenarios and scenario $2k + 3$. The probabilities for the scenarios are as follows: $Pr[2i - 1] = Pr[2i] = 2^{-i-2}$ for $i \in [1, k]$, $Pr[2k + 1] = Pr[2k + 2] = 2^{-k-2} - \epsilon$, where $0 < \epsilon < 2^{-k-2}$, and $Pr[2k + 3] = 2^{-1} + 2\epsilon$.

There is not much hope for computing optimal values on these instances. Due to the adaptive nature of these problems, we do not know reasonable mathematical programming formulations even for small size instances. Therefore, we compare the performance of our algorithm against other natural heuristics.

Optimal Decision Tree. Given a distribution $\mathcal{D}$ on users, the goal is to identify a random user i^* (drawn from $\mathcal{D}$) by asking questions of the form "is user i^* interested in movie e?". We compare (a) ODT-adsub, the algorithm presented

Table 2. Expected cost of running ODT-adsub, ODT-greedy, and ODT-ml.

Dataset	ODT-adsub	ODT-greedy	ODT-ml
SYN-50	2.75	27.50	21.00
SYN-100	2.75	52.50	21.64
SYN-150	2.75	77.50	56.43
SYN-200	2.75	102.5	53.01
SYN-250	2.75	127.5	69.52

Table 3. Expected cost of ODT-adsub, ODT-greedy and ODT-ml.

α	ODT-adsub	ODT-greedy	ODT-ml
	9.841	9.778	18.283
$\alpha = 2$	9.827	9.759	18.359
	9.834	9.775	18.144
	9.651	9.473	18.455
$\alpha = 3$	9.642	9.484	18.486
	9.672	9.497	18.725

T	ODT-adsub	ODT-greedy	ODT-ml
1	9.967	9.967	18.219
$[1, 5)$	8.936	8.870	17.121
$[5, 10)$	7.524	7.516	15.466

in this paper, (b) ODT-greedy, the classic greedy algorithm [1,8,11,17,24], and (c) ODT-ml, a "machine learning" algorithm that tries to exploit similarities between users/movies by clustering (see the full version [25] for details). Table 3 (left) shows the performance of the three algorithms when $\mathcal{D}$ comes from a power law distribution with $\alpha = 2, 3$. For each α, we tried three different distributions $\mathcal{D}$ obtained by permuting users before assigning probabilities. We also tested our algorithm for the "generalized" ODT problem, where users are associated with bounds $\{t_i\}_{i=1}^m$ and the goal is to find a subset I containing the random user i^* with $|I| \leq t_{i^*}$. Table 3 (right) shows the performance of the three algorithms (for uniform $\mathcal{D}$) where t_is are drawn randomly from the specified intervals.

Note that although ODT-greedy performs the best, ODT-adsub is very competitive. Combined with the fact that ODT-greedy performs poorly on worst-case instances (Table 2), ODT-adsub is a good alternative in practice.

Adaptive Multiple Intent Re-ranking. Given distribution $\mathcal{D}$ on users and thresholds $\{K_i\}_{i=1}^m$, the goal is to present at least K_{i^*} movies that a random user i^* is interested in. We compare (a) MIR-adsub, our algorithm, (b) MIR-static, the algorithm from [3] which uses the same sequence for all users, (c) MIR-adstatic, which improves on MIR-static by using feedback to eliminate movies that only belong to incompatible users, and (d) MIR-ml, a "machine learning" algorithm based on clustering.

Table 4 (left) shows the performance when $\mathcal{D}$ is uniform; each row corresponds to an instance where the thresholds K_i are chosen randomly from the specified interval. Table 4 (right) shows the performance when the thresholds are $K_i = |S_i|$ (their maximum value) and $\mathcal{D}$ is a power law distribution with

Table 4. Expected cost of MIR-adsub, MIR-static, MIR-adstatic and MIR-ml.

K_i	adsub	static	adstatic	ml
$|S_i|$	92.50	932.42	93.70	101.78
$[|S_i|/2, |S_i|)$	70.38	386.86	71.58	79.70
$[|S_i|/4, |S_i|)$	60.98	304.92	62.19	69.67
$[1, |S_i|/2)$	26.21	84.29	27.31	35.62
$[1, |S_i|/4)$	14.48	38.60	15.54	22.93

α	adsub	static	adstatic	ml
	92.50	932.42	93.70	101.78
$\alpha = 2$	26.21	84.29	27.31	35.62
	14.48	38.60	15.45	22.93
	86.20	905.14	87.78	96.18
$\alpha = 3$	92.72	934.54	94.42	101.52
	91.23	927.03	92.98	100.67

$\alpha = 2, 3$. Again, we tried three different distributions $\mathcal{D}$ (for each α) by permuting users: the cost instability across permutations indicates an inherent skew in the dataset. We note that MIR-adsub consistently outperforms the other three algorithms. The poor performance of MIR-static (the only "non adaptive" algorithm) demonstrates the importance of *adaptive* algorithms.

Acknowledgements. Part of V. Nagarajan's work was done while visiting the Simons institute for theoretical computer science (UC Berkeley). The authors thank Lisa Hellerstein for a clarification on [16] regarding the OR construction of submodular functions.

References

1. Adler, M., Heeringa, B.: Approximating optimal binary decision trees. Algorithmica **62**(3–4), 1112–1121 (2012)
2. Azar, Y., Gamzu, I.: Ranking with submodular valuations. In: SODA, pp. 1070–1079 (2011)
3. Azar, Y., Gamzu, I., Yin, X.: Multiple intents re-ranking. In: STOC, pp. 669–678 (2009)
4. Bansal, N., Gupta, A., Krishnaswamy, R.: A constant factor approximation algorithm for generalized min-sum set cover. In: SODA, pp. 1539–1545 (2010)
5. Bansal, N., Gupta, A., Li, J., Mestre, J., Nagarajan, V., Rudra, A.: When LP is the cure for your matching woes: improved bounds for stochastic matchings. Algorithmica **63**(4), 733–762 (2012)
6. Bellala, G., Bhavnani, S.K., Scott, C.: Group-based active query selection for rapid diagnosis in time-critical situations. IEEE Trans. Inf. Theor. **58**(1), 459–478 (2012)
7. Blum, A., Chalasani, P., Coppersmith, D., Pulleyblank, W.R., Raghavan, P., Sudan, M.: The minimum latency problem. In: STOC, pp. 163–171 (1994)
8. Chakaravarthy, V.T., Pandit, V., Roy, S., Awasthi, P., Mohania, M.K.: Decision trees for entity identification: approximation algorithms and hardness results. ACM Trans. Algorithms **7**(2), 15 (2011)
9. Chaudhuri, K., Godfrey, B., Rao, S., Talwar, K.: Paths, trees, and minimum latency tours. In: FOCS, pp. 36–45 (2003)
10. Cicalese, F., Laber, E.S., Saettler, A.M.: Diagnosis determination: decision trees optimizing simultaneously worst and expected testing cost. In: ICML, pp. 414–422 (2014)
11. Dasgupta, S.: Analysis of a greedy active learning strategy. In: NIPS (2004)
12. Dean, B.C., Goemans, M.X., Vondrák, J.: Approximating the stochastic knapsack problem: the benefit of adaptivity. Math. Oper. Res. **33**(4), 945–964 (2008)
13. Feige, U.: A threshold of ln n for approximating set cover. J. ACM **45**(4), 634–652 (1998)
14. Golovin, D., Krause, A.: Adaptive submodularity: theory and applications in active learning and stochastic optimization. J. Artif. Intell. Res. **42**, 427–486 (2011)
15. Golovin, D., Krause, A., Ray, D.: Near-optimal Bayesian active learning with noisy observations. In: NIPS, pp. 766–774 (2010)
16. Grammel, N., Hellerstein, L., Kletenik, D., Lin, P.: Scenario submodular cover. CoRR abs/1603.03158 (2016). (to appear in WAOA 2016)
17. Guillory, A., Bilmes, J.: Average-case active learning with costs. In: Gavaldà, R., Lugosi, G., Zeugmann, T., Zilles, S. (eds.) ALT 2009. LNCS, vol. 5809, pp. 141–155. Springer, Heidelberg (2009). doi:10.1007/978-3-642-04414-4_15

18. Gupta, A., Nagarajan, V., Ravi, R.: Approximation algorithms for optimal decision trees and adaptive TSP problems. In: Abramsky, S., Gavoille, C., Kirchner, C., Meyer auf der Heide, F., Spirakis, P.G. (eds.) ICALP 2010. LNCS, vol. 6198, pp. 690–701. Springer, Heidelberg (2010). doi:10.1007/978-3-642-14165-2_58
19. Harper, F.M., Konstan, J.A.: The movielens datasets: history and context. ACM Trans. Interact. Intell. Syst. (TiiS) **5**(4), 19 (2015)
20. Hyafil, L., Rivest, R.L.: Constructing optimal binary decision trees is NP-complete. Inf. Process. Lett. **5**(1), 15–17 (1976/1977)
21. Im, S., Nagarajan, V., Zwaan, R.: Minimum latency submodular cover. In: Czumaj, A., Mehlhorn, K., Pitts, A., Wattenhofer, R. (eds.) ICALP 2012. LNCS, vol. 7391, pp. 485–497. Springer, Heidelberg (2012). doi:10.1007/978-3-642-31594-7_41
22. Im, S., Sviridenko, M., van der Zwaan, R.: Preemptive and non-preemptive generalized min sum set cover. Math. Program. **145**(1–2), 377–401 (2014)
23. Javdani, S., Chen, Y., Karbasi, A., Krause, A., Bagnell, D., Srinivasa, S.S.: Near optimal bayesian active learning for decision making. In: AISTATS, pp. 430–438 (2014)
24. Kosaraju, S.R., Przytycka, T.M., Borgstrom, R.: On an optimal split tree problem. In: Dehne, F., Sack, J.-R., Gupta, A., Tamassia, R. (eds.) WADS 1999. LNCS, vol. 1663, pp. 157–168. Springer, Heidelberg (1999). doi:10.1007/3-540-48447-7_17
25. Navidi, F., Kambadur, P., Nagarajan, V.: Adaptive submodular ranking. arXiv preprint arXiv:1606.01530 (2016)
26. Skutella, M., Williamson, D.P.: A note on the generalized min-sum set cover problem. Oper. Res. Lett. **39**(6), 433–436 (2011)
27. Wolsey, L.: An analysis of the greedy algorithm for the submodular set covering problem. Combinatorica **2**(4), 385–393 (1982)

On the Notions of Facets, Weak Facets, and Extreme Functions of the Gomory–Johnson Infinite Group Problem

Matthias Köppe[✉] and Yuan Zhou

Department of Mathematics, University of California, Davis, USA
{mkoeppe,yzh}@math.ucdavis.edu

Abstract. We investigate three competing notions that generalize the notion of a facet of finite-dimensional polyhedra to the infinite-dimensional Gomory–Johnson model. These notions were known to coincide for continuous piecewise linear functions with rational breakpoints. We show that two of the notions, extreme functions and facets, coincide for the case of continuous piecewise linear functions, removing the hypothesis regarding rational breakpoints. We then separate the three notions using discontinuous examples.

1 Introduction

1.1 Facets in the Finite-Dimensional Case

Let G be a finite index set. The space $\mathbb{R}^{(G)}$ of real-valued functions $y\colon G \to \mathbb{R}$ is isomorphic to and routinely identified with the Euclidean space $\mathbb{R}^{|G|}$. Let $\mathbb{R}^G$ denote its dual space. It is the space of functions $\alpha\colon G \to \mathbb{R}$, which we consider as linear functionals on $\mathbb{R}^{(G)}$ via the pairing $\langle \alpha, y \rangle = \sum_{r \in G} \alpha(r)y(r)$. Again it is routinely identified with the Euclidean space $\mathbb{R}^{|G|}$, and the dual pairing $\langle \alpha, y \rangle$ is the Euclidean inner product. A (closed, convex) rational polyhedron of $\mathbb{R}^{(G)}$ is the set of $y\colon G \to \mathbb{R}$ satisfying $\langle \alpha_i, y \rangle \geq \alpha_{i,0}$, where $\alpha_i \in \mathbb{Z}^G$ are integer linear functionals and $\alpha_{i,0} \in \mathbb{Z}$, for i ranging over another finite index set I.

Consider an integer linear optimization problem in $\mathbb{R}^{(G)}$, i.e., the problem of minimizing a linear functional $\eta \in \mathbb{R}^G$ over a feasible set $F \subseteq \{ y\colon G \to \mathbb{Z}_+ \} \subset \mathbb{R}^{(G)}_+$, or, equivalently, over the convex hull $R = \operatorname{conv} F \subset \mathbb{R}^{(G)}_+$. A *valid inequality* for R is an inequality of the form $\langle \pi, y \rangle \geq \pi_0$, where $\pi \in \mathbb{R}^G$, which holds for all $y \in R$ (equivalently, for all $y \in F$). If R is closed, it is exactly the set of all y that satisfy all valid inequalities. In the following we will restrict ourselves to the case that $R \subseteq \mathbb{R}^{(G)}_+$ is a polyhedron of blocking type, in which case we only need to consider normalized valid inequalities with $\pi \geq 0$ and $\pi_0 = 1$.

Let $P(\pi)$ denote the set of functions $y \in F$ for which the inequality $\langle \pi, y \rangle \geq 1$ is tight, i.e., $\langle \pi, y \rangle = 1$. If $P(\pi) \neq \emptyset$, then $\langle \pi, y \rangle \geq 1$ is a *tight valid inequality*.

The authors gratefully acknowledge partial support from the National Science Foundation through grant DMS-1320051, awarded to M. Köppe.

F. Eisenbrand and J. Koenemann (Eds.): IPCO 2017, LNCS 10328, pp. 330–342, 2017.
DOI: 10.1007/978-3-319-59250-3_27

Then R is exactly the set of all $y \geq 0$ that satisfy all tight valid inequalities. A valid inequality $\langle \pi, y \rangle \geq 1$ is called *minimal* if there is no other valid inequality $\pi' \neq \pi$ such that $\pi' \leq \pi$ pointwise. One can show that a minimal valid inequality is tight. A valid inequality $\langle \pi, y \rangle \geq 1$ is called *facet-defining* if

$$\text{for every valid inequality } \langle \pi', y \rangle \geq 1 \text{ such that } P(\pi) \subseteq P(\pi'), \tag{wF}$$
$$\text{we have } P(\pi) = P(\pi'),$$

or, in other words, if the face induced by $\langle \pi, y \rangle \geq 1$ is maximal. Because R is of blocking type, its recession cone is $\mathbb{R}_+^{(G)}$ and therefore R has full affine dimension. Thus, we get the following characterization of facet-defining inequalities:

$$\text{for every valid inequality } \langle \pi', y \rangle \geq 1 \text{ such that } P(\pi) \subseteq P(\pi'), \tag{F}$$
$$\text{we have } \pi = \pi'.$$

The theory of polyhedra gives another characterization of facets:

$$\text{If } \langle \pi^1, y \rangle \geq 1 \text{ and } \langle \pi^2, y \rangle \geq 1 \text{ are valid inequalities, and } \pi = \tfrac{1}{2}(\pi^1 + \pi^2) \tag{E}$$
$$\text{then } \pi = \pi^1 = \pi^2.$$

1.2 Facets in the Infinite-Dimensional Gomory–Johnson Model

It is perhaps not surprising that the three conditions (wF), (F), and (E) are no longer equivalent when R is a general convex set that is not polyhedral, and in particular when we change from the finite-dimensional to the infinite-dimensional setting. In the present paper, however, we consider a particular case of an infinite-dimensional model, in which this question has eluded researchers for a long time. Let $G = \mathbb{Q}$ or $G = \mathbb{R}$ and let $\mathbb{R}^{(G)}$ now denote the space of finite-support functions $y \colon G \to \mathbb{R}$. The so-called *infinite group problem* was introduced by Gomory and Johnson in their seminal papers [9,10]. Let $F = F_f(G, \mathbb{Z}) \subseteq \mathbb{R}_+^{(G)}$ be the set of all finite-support functions $y \colon G \to \mathbb{Z}_+$ satisfying the equation

$$\sum_{r \in G} r\, y(r) \equiv f \pmod{1} \tag{1}$$

where f is a given element of $G \setminus \mathbb{Z}$. We study its convex hull $R = R_f(G, \mathbb{Z}) \subseteq \mathbb{R}_+^{(G)}$, whose elements are understood as finite-support functions $y \colon G \to \mathbb{R}_+$.

Valid inequalities for R are of the form $\langle \pi, y \rangle \geq \pi_0$, where π comes from the dual space $\mathbb{R}^G$, which is the space of all real-valued functions (without the finite-support condition). When $G = \mathbb{Q}$, then R is again of "blocking type" (see, for example, [7, Sect. 5]), and so we again may assume $\pi \geq 0$ and $\pi_0 = 1$.

If $G = \mathbb{R}$ (the setting of the present paper), typical pathologies from the analysis of functions of a real variable come into play. For example, by [4, Proposition 2.4] there is an infinite-dimensional space of valid equations $\langle \pi^*, y \rangle = 0$, where π^* are constructed using a Hamel basis of $\mathbb{R}$ over $\mathbb{Q}$. Each of these functions π^* has a graph whose topological closure is $\mathbb{R}^2$. In order to tame these pathologies, it is common to make further assumptions. Gomory–Johnson [9,10] only considered continuous functions π. However, this rules out many interesting functions such as the Gomory fractional cut. Instead it has become common in the literature to build the assumption $\pi \geq 0$ into the definition; then we can again normalize $\pi_0 = 1$. We call such functions π *valid functions*.

(Minimal) valid functions π that satisfy the conditions (wF), (F), and (E), are called *weak facets*, *facets*, and *extreme functions*, respectively. The relation of these notions, in particular of facets and extreme functions, has remained unclear in the literature. For example, Basu et al. [1] wrote:

> The statement that extreme functions are facets appears to be quite non-trivial to prove, and to the best of our knowledge there is no proof in the literature. We therefore cautiously treat extreme functions and facets as distinct concepts, and leave their equivalence as an open question.

The survey [4, Sect. 2.2, Fig. 2] summarizes what was known about the relation of the three notions: Facets form a subset of the intersection of extreme functions and weak facets. In the case of continuous piecewise linear functions with rational breakpoints, [4, Proposition 2.8] and [5, Theorem 8.6] proved that (E) $\Leftrightarrow$ (F). We note that in this case, (wF) $\Rightarrow$ (F) can be shown by restriction with oversampling to finite group problems. Thus (E), (F), (wF) are equivalent when π is a continuous piecewise linear function with rational breakpoints.

1.3   Contribution of this Paper

A well known sufficient condition for facetness of a minimal valid function π is the Gomory–Johnson Facet Theorem. In its strong form, due to Basu–Hildebrand–Köppe–Molinaro [6], it reads:

Theorem 1.1 (Facet Theorem, strong form, [6, Lemma 34]; see also [4, Theorem 2.12]). *Suppose for every minimal valid function π', $E(\pi) \subseteq E(\pi')$ implies $\pi' = \pi$. Then π is a facet.*

(Here $E(\pi)$ is the *additivity domain* of π, defined in Sect. 2.) We show (Theorem 4.3 below) that, in fact, **this holds as an "if and only if" statement.**

As we mentioned above, for the case of continuous piecewise linear functions with rational breakpoints, Basu et al. [4, Proposition 2.8] showed that the notions of extreme functions and facets coincide. This was a consequence of Basu et al.'s finite oversampling theorem [2]. We **sharpen this result by removing the hypothesis regarding rational breakpoints.**

Theorem 1.2. *In the case of continuous piecewise linear functions (not necessarily with rational breakpoints), {extreme functions} = {facets}.*

Then we investigate the notions of facets and weak facets in the case of discontinuous functions. This appears to be a first in the published literature. All papers that consider discontinuous functions only used the notion of extreme functions. We give **three discontinuous functions that furnish the separation of the three notions** (Theorem 6.1): A function ψ that is extreme, but is neither a weak facet nor a facet; a function π that is not an extreme function (nor a facet), but is a weak facet; and a function π_{lifted} that is extreme and a weak facet but is not a facet.

It remains an open question whether this separation can also be done using continuous (non–piecewise linear) functions.

2   Minimal Valid Functions and Their Perturbations

Following [4], given a locally finite one-dimensional polyhedral complex $\mathcal{P}$, we call a function $\pi\colon \mathbb{R} \to \mathbb{R}$ *piecewise linear* over $\mathcal{P}$, if it is affine linear over the relative interior of each face of the complex. Under this definition, piecewise linear functions can be discontinuous. We say the function π is *continuous piecewise linear* over $\mathcal{P}$ if it is affine over each of the cells of $\mathcal{P}$ (thus automatically imposing continuity).

For a function $\pi\colon \mathbb{R} \to \mathbb{R}$, define the *subadditivity slack* of π as $\Delta\pi(x,y) := \pi(x) + \pi(y) - \pi(x+y)$; then π is subadditive if and only if $\Delta\pi(x,y) \geq 0$ for all $x, y \in \mathbb{R}$. Denote the *additivity domain* of π by

$$E(\pi) = \{\, (x,y) \mid \Delta\pi(x,y) = 0 \,\}.$$

By a theorem of Gomory and Johnson [9] (see [4, Theorem 2.6]), the minimal valid functions are exactly the subadditive functions $\pi\colon \mathbb{R} \to \mathbb{R}_+$ that are periodic modulo 1 and satisfy the *symmetry condition* $\pi(x) + \pi(f - x) = 1$ for all $x \in \mathbb{R}$. As a consequence, minimal valid functions are bounded between 0 and 1.

To combinatorialize the additivity domains of piecewise linear subadditive functions, we work with the two-dimensional polyhedral complex $\Delta\mathcal{P}$, whose faces are $F(I, J, K) = \{\, (x,y) \in \mathbb{R}\times\mathbb{R} \mid x \in I, y \in J, x+y \in K \,\}$ for $I, J, K \in \mathcal{P}$. Define the projections $p_1, p_2, p_3\colon \mathbb{R} \times \mathbb{R} \to \mathbb{R}$ as $p_1(x,y) = x$, $p_2(x,y) = y$, $p_3(x,y) = x+y$.

In the continuous case, since the function π is piecewise linear over $\mathcal{P}$, we have that $\Delta\pi$ is affine linear over each face $F \in \Delta\mathcal{P}$. Let π be a minimal valid function for $R_f(\mathbb{R}, \mathbb{Z})$ that is piecewise linear over $\mathcal{P}$. Following [4], we define the *space of perturbation functions with prescribed additivities $E = E(\pi)$*

$$\bar{\Pi}^E(\mathbb{R}, \mathbb{Z}) = \left\{ \bar{\pi}\colon \mathbb{R} \to \mathbb{R} \,\middle|\, \begin{array}{l} \bar{\pi}(0) = 0 \\ \bar{\pi}(f) = 0 \\ \bar{\pi}(x) + \bar{\pi}(y) = \bar{\pi}(x+y) \text{ for all } (x,y) \in E \\ \bar{\pi}(x+t) = \bar{\pi}(x) \qquad \text{for all } x \in \mathbb{R},\, t \in \mathbb{Z} \end{array} \right\}. \quad (2)$$

When π is discontinuous, one also needs to consider the limit points where the subadditivity slacks are approaching zero. Let F be a face of $\Delta\mathcal{P}$. For $(x,y) \in F$, we denote

$$\Delta\pi_F(x,y) := \lim_{\substack{(u,v) \to (x,y) \\ (u,v) \in \text{rel int}(F)}} \Delta\pi(u,v).$$

Define
$$E_F(\pi) = \{\, (x,y) \in F \mid \Delta\pi_F(x,y) \text{ exists, and } \Delta\pi_F(x,y) = 0 \,\}.$$

Notice that in the above definition of $E_F(\pi)$, we include the condition that the limit denoted by $\Delta\pi_F(x,y)$ exists, so that this definition can as well be applied to functions π (and $\bar{\pi}$) that are not piecewise linear over $\mathcal{P}$.

We denote by $E_\bullet(\pi, \mathcal{P})$ the family of sets $E_F(\pi)$, indexed by $F \in \Delta\mathcal{P}$. Define the *space of perturbation functions with prescribed additivities and limit-additivities* $E_\bullet = E_\bullet(\pi, \mathcal{P})$

$$\bar{\Pi}^{E_\bullet}(\mathbb{R}, \mathbb{Z}) = \left\{ \bar{\pi} \colon \mathbb{R} \to \mathbb{R} \;\middle|\; \begin{array}{l} \bar{\pi}(0) = 0 \\ \bar{\pi}(f) = 0 \\ \Delta\bar{\pi}_F(x,y) = 0 \quad \text{for } (x,y) \in E_F,\ F \in \Delta\mathcal{P} \\ \bar{\pi}(x+t) = \bar{\pi}(x) \ \text{for } x \in \mathbb{R},\ t \in \mathbb{Z} \end{array} \right\} . (3)$$

Remark 2.1. Let $\bar{\pi} \in \bar{\Pi}^{E}(\mathbb{R}, \mathbb{Z})$. The third condition of (2) is equivalent to $E(\pi) \subseteq E(\bar{\pi})$. Let $\bar{\pi} \in \bar{\Pi}^{E_\bullet}(\mathbb{R}, \mathbb{Z})$. The third condition of (3) is equivalent to $E_F(\pi) \subseteq E_F(\bar{\pi})$ for all faces $F \in \Delta\mathcal{P}$, which is stronger than $E(\pi) \subseteq E(\bar{\pi})$ in (2). Thus, in general, $\bar{\Pi}^{E_\bullet}(\mathbb{R}, \mathbb{Z}) \subseteq \bar{\Pi}^{E}(\mathbb{R}, \mathbb{Z})$. If π is continuous, then $E(\pi) \subseteq E(\bar{\pi})$ implies that $E_F(\pi) \subseteq E_F(\bar{\pi})$ for all faces $F \in \Delta\mathcal{P}$, hence $\bar{\Pi}^{E_\bullet}(\mathbb{R}, \mathbb{Z}) = \bar{\Pi}^{E}(\mathbb{R}, \mathbb{Z})$.

3   Effective Perturbation Functions

Following [13], we define the *space of effective perturbation functions*

$$\tilde{\Pi}^{\pi}(\mathbb{R}, \mathbb{Z}) = \{\, \tilde{\pi} \colon \mathbb{R} \to \mathbb{R} \mid \exists\, \epsilon > 0 \text{ s.t. } \pi^{\pm} = \pi \pm \epsilon\tilde{\pi} \text{ are minimal valid} \,\}. \quad (4)$$

Because of [4, Lemma 2.11(i)], a function π is extreme if and only if $\tilde{\Pi}^{\pi}(\mathbb{R}, \mathbb{Z}) = \{0\}$. Note that every function $\tilde{\pi} \in \tilde{\Pi}^{\pi}(\mathbb{R}, \mathbb{Z})$ is bounded.

It is clear that if $\tilde{\pi} \in \tilde{\Pi}^{\pi}(\mathbb{R}, \mathbb{Z})$, then $\tilde{\pi} \in \bar{\Pi}^{E_\bullet}(\mathbb{R}, \mathbb{Z})$, where $E_\bullet = E_\bullet(\pi, \mathcal{P})$; see [2, Lemma 2.7] or [13, Lemma 2.1].

The other direction does not hold in general, but requires additional hypotheses. Let $\bar{\pi} \in \bar{\Pi}^{E_\bullet}(\mathbb{R}, \mathbb{Z})$. In [3, Theorem 3.13] (see also [4, Theorem 3.13]), it is proved that if π and $\bar{\pi}$ are continuous and $\bar{\pi}$ is piecewise linear, we have $\bar{\pi} \in \tilde{\Pi}^{\pi}(\mathbb{R}, \mathbb{Z})$. (Similar arguments also appeared in the earlier literature, for example in the proof of [2, Theorem 3.2].)

We will need a more general version of this result. Consider the following definition. Given a locally finite one-dimensional polyhedral complex $\mathcal{P}$, we call a function $\bar{\pi} \colon \mathbb{R} \to \mathbb{R}$ *piecewise Lipschitz continuous* over $\mathcal{P}$, if it is Lipschitz continuous over the relative interior of each face of the complex. Under this definition, piecewise Lipschitz continuous functions can be discontinuous.

Theorem 3.1. *Let π be a minimal valid function that is piecewise linear over a polyhedral complex $\mathcal{P}$. Let $\bar{\pi} \in \bar{\Pi}^{E_\bullet}(\mathbb{R}, \mathbb{Z})$ be a perturbation function, where $E_\bullet = E_\bullet(\pi, \mathcal{P})$. Suppose that $\bar{\pi}$ is piecewise Lipschitz continuous over $\mathcal{P}$. Then $\bar{\pi}$ is an effective perturbation function, $\bar{\pi} \in \tilde{\Pi}^{\pi}(\mathbb{R}, \mathbb{Z})$.*

We omit the proof in this extended abstract.

4   Extreme Functions and Facets

In this section, we discuss the relations between the notions of extreme functions and facets. We first review the definition of a facet, following [4, Sect. 2.2.3]; cf. ibid. for a discussion of this notion in the earlier literature, in particular [8,11].

Let $P(\pi)$ denote the set of functions $y\colon \mathbb{R} \to \mathbb{Z}_+$ with finite support satisfying

$$\sum_{r\in\mathbb{R}} ry(r) \in f + \mathbb{Z} \quad \text{and} \quad \sum_{r\in\mathbb{R}} \pi(r)y(r) = 1.$$

A valid function π is called a *facet* if for every valid function π' such that $P(\pi) \subseteq P(\pi')$ we have that $\pi' = \pi$. Equivalently, a valid function π is a facet if this condition holds for all such *minimal* valid functions π' [6].

Remark 4.1. In the discontinuous case, the additivity in the limit plays a role in extreme functions, which are characterized by the non-existence of an effective perturbation function $\tilde{\pi} \not\equiv 0$. However facets (and weak facets, see the next section) are defined through $P(\pi)$, which does not capture the limiting additive behavior of π. The additivity domain $E(\pi)$, which appears in the Facet Theorem as discussed below, also does not account for additivity in the limit.

A well known sufficient condition for facetness of a minimal valid function π is the Gomory–Johnson Facet Theorem. We have stated its strong form, due to Basu–Hildebrand–Köppe–Molinaro [6], in the introduction as Theorem 1.1. In order to prove our "if and only if" version, we need the following lemma.

Lemma 4.2. *Let π and π' be minimal valid functions. Then $E(\pi) \subseteq E(\pi')$ if and only if $P(\pi) \subseteq P(\pi')$.*

Proof. The "if" direction is proven in [6, Theorem 20]; see also [4, Theorem 2.12]. We now show the "only if" direction, using the subadditivity of π. Assume that $E(\pi) \subseteq E(\pi')$. Let $y \in P(\pi)$. Let $\{r_1, r_2, \ldots, r_n\}$ denote the finite support of y. By definition, the function y satisfies that $y(r_i) \in \mathbb{Z}_+$, $\sum_{i=1}^n r_i y(r_i) \equiv f \pmod 1$, and $\sum_{i=1}^n \pi(r_i)y(r_i) = 1$. Since π is a minimal valid function, we have that $1 = \sum_{i=1}^n \pi(r_i)y(r_i) \geq \pi\left(\sum_{i=1}^n r_i y(r_i)\right) = \pi(f) = 1$. Thus, each subadditivity inequality here is tight for π, and is also tight for π' since $E(\pi) \subseteq E(\pi')$. We obtain $\sum_{i=1}^n \pi'(r_i)y(r_i) = \pi'\left(\sum_{i=1}^n r_i y(r_i)\right) = \pi'(f) = 1$, which implies that $y \in P(\pi')$. Therefore, $P(\pi) \subseteq P(\pi')$.   □

Theorem 4.3 (Facet Theorem, "if and only if" version). *A minimal valid function π is a facet if and only if for every minimal valid function π', $E(\pi) \subseteq E(\pi')$ implies $\pi' = \pi$.*

Proof. It follows from the Facet Theorem in the strong form (Theorem 1.1) and Lemma 4.2.   □

Now we come to the proof of a main theorem stated in the introduction.

Proof (of Theorem 1.2). Let π be a continuous piecewise linear minimal valid function. As mentioned in [4, Sect. 2.2.4], [6, Lemma 1.3] showed that if π is a facet, then π is extreme.

We now prove the other direction by contradiction. Suppose that π is extreme, but is not a facet. Then by Theorem 4.3, there exists a minimal valid function $\pi' \neq \pi$ such that $E(\pi) \subseteq E(\pi')$. Since π is continuous piecewise linear and $\pi(0) = \pi(1) = 0$, there exists $\delta > 0$ such that $\Delta\pi(x,y) = 0$ and $\Delta\pi(-x,-y) = 0$ for $0 \leq x, y \leq \delta$. The condition $E(\pi) \subseteq E(\pi')$ implies that $\Delta\pi'(x,y) = 0$ and $\Delta\pi'(-x,-y) = 0$ for $0 \leq x, y \leq \delta$ as well. As the function π' is bounded, it follows from the Interval Lemma (see [4, Lemma 4.1], for example) that π' is affine linear on $[0, \delta]$ and on $[-\delta, 0]$. We also know that $\pi'(0) = 0$ as π' is minimal valid. Using the subadditivity, we obtain that π' is Lipschitz continuous. Let $\bar{\pi} = \pi' - \pi$. Then $\bar{\pi} \not\equiv 0$, $\bar{\pi} \in \bar{\Pi}^E(\mathbb{R}, \mathbb{Z})$ where $E = E(\pi)$, and $\bar{\pi}$ is Lipschitz continuous. Since π is continuous, we have $\bar{\Pi}^E(\mathbb{R}, \mathbb{Z}) = \bar{\Pi}^{E\bullet}(\mathbb{R}, \mathbb{Z})$. By Theorem 3.1, there exists $\epsilon > 0$ such that $\pi^\pm = \pi \pm \epsilon\bar{\pi}$ are distinct minimal valid functions. This contradicts the assumption that π is an extreme function.

Therefore, {extreme functions} = {facets}. $\qquad\qquad\square$

5 Weak Facets

We first review the definition of a weak facet, following [4, Sect. 2.2.3]; cf. ibid. for a discussion of this notion in the earlier literature, in particular [8,11]. A valid function π is called a *weak facet* if for every valid function π' such that $P(\pi) \subseteq P(\pi')$ we have $P(\pi) = P(\pi')$.

As we mentioned above, to prove that π is an extreme function or is a facet, it suffices to consider π' that is minimal valid. The following lemma shows it is also the case in the definition of weak facets.

Lemma 5.1

(1) Let π be a valid function. If π is a weak facet, then π is minimal valid.

(2) Let π be a minimal valid function. Suppose that for every minimal valid function π', we have that $P(\pi) \subseteq P(\pi')$ implies $P(\pi) = P(\pi')$. Then π is a weak facet.

(3) A minimal valid function π is a weak facet if and only if for every minimal valid function π', we have that $E(\pi) \subseteq E(\pi')$ implies $E(\pi) = E(\pi')$.

Proof. (1) Suppose that π is not minimal valid. Then, by [6, Theorem 1], π is dominated by another minimal valid function π', with $\pi(x_0) > \pi'(x_0)$ at some x_0. Let $y \in P(\pi)$. We have

$$1 = \sum \pi(r_i)y(r_i) \geq \sum \pi'(r_i)y(r_i) \geq \pi'\left(\sum r_i y(r_i)\right) = \pi'(f) = 1.$$

Hence equality holds throughout, implying that $y \in P(\pi')$. Therefore, $P(\pi) \subseteq P(\pi')$. Now consider y with $y(x_0) = y(f - x_0) = 1$ and $y(x) = 0$ otherwise. It is easy to see that $y \in P(\pi')$, but $y \notin P(\pi)$ since $\pi(x_0) + \pi(f - x_0) >$

$\pi'(x_0) + \pi'(f - x_0) = 1$. Therefore, $P(\pi) \subsetneq P(\pi')$, a contradiction to the weak facet assumption on π.

(2) Consider any valid function π^* (not necessarily minimal) such that $P(\pi) \subseteq P(\pi^*)$. Let π' be a minimal function that dominates π^*: $\pi' \leq \pi^*$. From the proof of (1) we know that $P(\pi^*) \subseteq P(\pi')$. Thus, $P(\pi) \subseteq P(\pi')$. By hypothesis, we have that $P(\pi) = P(\pi^*) = P(\pi')$. Therefore, π is a weak facet.

(3) Direct consequence of (2) and Lemma 4.2. $\square$

Theorem 5.2. *Let $\mathcal{F}$ be a family of functions such that existence of an effective perturbation implies existence of a piecewise linear effective perturbation. Let π be a continuous piecewise linear function (not necessarily with rational breakpoints) such that $\pi \in \mathcal{F}$. The following are equivalent. (E) π is extreme, (F) π is a facet, (wF) π is a weak facet.*

Remark 5.3. As shown in [2] (for a stronger statement, see [5, Theorem 8.6]), the family of continuous piecewise linear functions with rational breakpoints is such a family $\mathcal{F}$ where existence of an effective perturbation implies existence of a piecewise linear effective perturbation. A forthcoming paper will investigate larger such families $\mathcal{F}$.

Proof (of Theorem 5.2). By Theorem 1.2 and the fact that {facets} $\subseteq$ {extreme functions} $\cap$ {weak facets}, it suffices to show that {weak facets} $\subseteq$ {extreme functions}.

Assume that π is a weak facet, thus π is minimal valid by Lemma 5.1. We show that π is extreme. For the sake of contradiction, suppose that π is not extreme. By the assumption $\pi \in \mathcal{F}$, there exists a piecewise linear perturbation function $\bar{\pi} \not\equiv 0$ such that $\pi \pm \bar{\pi}$ are minimal valid functions. Furthermore, by [4, Lemma 2.11], we know that $\bar{\pi}$ is continuous, and $E(\pi) \subseteq E(\bar{\pi})$. By taking the union of the breakpoints, we can define a common refinement, which will still be denoted by $\mathcal{P}$, of the complexes for π and for $\bar{\pi}$. In other words, we may assume that π and $\bar{\pi}$ are both continuous piecewise linear over $\mathcal{P}$. Since $\Delta\bar{\pi} \not\equiv 0$, we may assume without loss of generality that $\Delta\bar{\pi}(x, y) > 0$ for some $(x, y) \in \text{vert}(\Delta\mathcal{P})$. Define

$$\epsilon = \min\left\{ \frac{\Delta\pi(x, y)}{\Delta\bar{\pi}(x, y)} \;\middle|\; (x, y) \in \text{vert}(\Delta\mathcal{P}),\ \Delta\bar{\pi}(x, y) > 0 \right\}.$$

Notice that $\epsilon > 0$, since $\Delta\pi \geq 0$ and $E(\pi) \subseteq E(\bar{\pi})$. Let $\pi' = \pi - \epsilon\bar{\pi}$. Then π' is a bounded continuous function piecewise linear over $\mathcal{P}$, such that $\pi' \neq \pi$.

The function π' is subadditive, since $\Delta\pi'(x, y) \geq 0$ for each $(x, y) \in \text{vert}(\Delta\mathcal{P})$. As in the proof of Theorem 3.1, it can be shown that π' is non-negative, $\pi'(0) = 0$, $\pi'(f) = 1$, and that π' satisfies the symmetry condition. Therefore, π' is a minimal valid function. Let (u, v) be a vertex of $\Delta\mathcal{P}$ satisfying $\Delta\bar{\pi}(u, v) > 0$ and $\Delta\pi(u, v) = \epsilon\Delta\bar{\pi}(u, v)$. We know that $\Delta\pi'(u, v) = \Delta\pi(u, v) - \epsilon\Delta\bar{\pi}(u, v) = 0$, hence $(u, v) \in E(\pi')$. However, $(u, v) \notin E(\pi)$, since $\Delta\bar{\pi}(u, v) > 0$ implies that $\Delta\pi(u, v) \neq 0$. Therefore, $E(\pi) \subsetneq E(\pi')$. By Lemma 5.1(3), we have that π is not a weak facet, a contradiction. $\square$

6 Separation of the Notions in the Discontinuous Case

The definition of facets fails to account for additivities-in-the-limit, which are a crucial feature of the extremality test for discontinuous functions. This allows us to separate the two notions. Below we do this by observing that a discontinuous piecewise linear extreme function from the literature, `hildebrand_discont_3_slope_1()` (https://github.com/mkoeppe/infinite-group-relaxation-code/search?q=%22def+hildebrand_discont_3_slope_1(%22), constructed by Hildebrand (2013, unpublished; reported in [4]), works as a separating example.

The other separations appear to require more complicated constructions. Recently, the authors constructed a two-sided discontinuous piecewise linear minimal valid function, `kzh_minimal_has_only_crazy_perturbation_1`, which is not extreme, but which is not a convex combination of other piecewise linear minimal valid functions; see [13] for the definition. This function has two special "uncovered" pieces on the intervals (l, u) and $(f - u, f - l)$, where $f = \frac{4}{5}$, $l = \frac{219}{800}$, $u = \frac{269}{800}$, on which every nonzero perturbation is microperiodic (invariant under the action of the dense additive group $T = \langle t_1, t_2 \rangle_{\mathbb{Z}}$, where $t_1 = \frac{77}{7752}\sqrt{2}$, $t_2 = \frac{77}{2584}$). Below we prove that it furnishes another separation.

For the remaining separation, we construct an extreme function π_{lifted} as follows. Define π_{lifted} by perturbing the function $\pi = $ `kzh_minimal_has_only_crazy_perturbation_1()` on infinitely many cosets of the group T on the two uncovered intervals as follows.

$$\pi_{\text{lifted}}(x) = \begin{cases} \pi(x) & \text{if } x \notin (l, u) \cup (f - u, f - l), \text{ or} \\ & \text{if } x \in (l, u) \text{ such that } x + T \in C, \text{ or} \\ & \text{if } x \in (f - u, f - l) \text{ such that } f - x + T \in C; \\ \pi(x) + s & \text{if } x \in (l, u) \text{ such that } x + T \in C^+, \text{ or} \\ & \text{if } x \in (f - u, f - l) \text{ such that } f - x + T \in C^+; \\ \pi(x) - s & \text{otherwise,} \end{cases} \tag{5}$$

where $x_{39} = \frac{4899}{5000}$, $s = \pi(x_{39}^-) + \pi(1 + l - x_{39}) - \pi(l) = \frac{19}{23998}$,

$$C = \left\{ x \in \mathbb{R}/T \mid x = \tfrac{l+u}{2} + T \text{ or } \tfrac{l+u-t_1}{2} + T \text{ or } \tfrac{l+u-t_2}{2} + T \right\},$$
$$C^+ = \left\{ x \in \mathbb{R}/T \mid \text{arbitrary choice of one element of } \{x, \phi(x)\}, x \notin C \right\}$$

with $\phi \colon \mathbb{R}/T \ni x \mapsto l + u - x$.

Theorem 6.1

(1) The function $\psi = $ `hildebrand_discont_3_slope_1()` is extreme, but is neither a weak facet nor a facet.

(2) The function $\pi = $ `kzh_minimal_has_only_crazy_perturbation_1()` is not an extreme function (nor a facet), but is a weak facet.

(3) The function π_{lifted} is extreme; it is a weak facet but is not a facet.

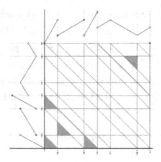
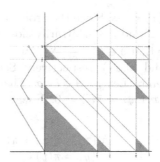

Fig. 1. Two diagrams of functions h (*blue graphs on the top and the left*) and polyhedral complexes $\Delta\mathcal{P}$ (*gray solid lines*) with additive domains $E(\mathbf{h})$ (*shaded in green*), as plotted by the command `plot_2d_diagram_additive_domain_sans_limits(h)`. (*Left*) h = `hildebrand_discont_3_slope_1()` $= \pi$. (*Right*) h $= \pi'$ from the proof of Theorem 6.1(2). (Color figure online)

Proof. (1) The function $\psi =$ `hildebrand_discont_3_slope_1()` is extreme (Hildebrand, 2013, unpublished, reported in [4]). This can be verified using the extremality test implemented in [12]. Consider the minimal valid function π' defined by

$$\pi'(x) = \begin{cases} 2x & \text{if } x \in [0, \tfrac{1}{2}]; \\ \pi(x) & \text{if } x \in (\tfrac{1}{2}, 1). \end{cases}$$

Observe that $E(\psi)$ is a strict subset of $E(\pi')$. See Fig. 1 for an illustration. Thus, by Lemma 5.1(3), the function ψ is not a weak facet (nor a facet).

(2) By [13, Theorem 4.1], the function $\pi =$ `kzh_minimal_has_only_crazy_perturbation_1()` is minimal valid, but is not extreme. Let π' be a minimal valid function such that $E(\pi) \subseteq E(\pi')$. We want to show that $E(\pi) = E(\pi')$. Consider $\bar{\pi} = \pi' - \pi$, which is a bounded $\mathbb{Z}$-periodic function satisfying that $E(\pi) \subseteq E(\bar{\pi})$. We apply the proof of [13, Theorem 4.1, Part (ii)] to the perturbation $\bar{\pi}$, and obtain that

(i) $\bar{\pi}(x) = 0$ for $x \notin (l, u) \cup (f - u, f - l)$;
(ii) $\bar{\pi}$ is constant on each coset in $\mathbb{R}/T$ on the pieces (l, u) and $(f - u, f - l)$.

Furthermore, it follows from the additivity relations of π and $E(\pi) \subseteq E(\bar{\pi})$ that

(iii) $\bar{\pi}(x) + \bar{\pi}(y) = 0$ for $x, y \in (l, u)$ such that $x + y \in \{l + u, l + u - t_1, l + u - t_2\}$;
(iv) $\bar{\pi}(x) + \bar{\pi}(y) = 0$ for $x \in (l, u)$, $y \in (f - u, f - l)$ such that $x + y = f$.

We now show that $\bar{\pi}$ also satisfies the following condition:

(v) $|\bar{\pi}(x)| \leq s$ for all $x \in (l, u) \cup (f - u, f - l)$, where s is the constant from the definition of π_{lifted}.

Indeed, by (iii) and (iv), it suffices to show that for any $x \in (l, u)$, we have $\bar{\pi}(x) \geq -s$. Suppose, for the sake of contradiction, that there is $\bar{x} \in (l, u)$ such that $\bar{\pi}(\bar{x}) < -s$. Since the group T is dense in $\mathbb{R}$, we can find $x \in (l, u)$ such that $x \in \bar{x}+T$ and x is arbitrarily close to $1+l-x_{39}$. Let $\delta = x-(1+l-x_{39})$. We may assume that $\delta \in (0, \frac{-s-\bar{\pi}(\bar{x})}{c_2-c_3})$, where c_2 and c_3 denote the slope of π on the pieces (l, u) and $(0, x_1)$, respectively. See [13, Table 1] for the concrete values of the parameters. Let $y = 1+l-x$. Then $y = x_{39}-\delta$. It follows from (i) that $\bar{\pi}(y) = 0$ and $\bar{\pi}(x + y) = \bar{\pi}(l) = 0$. Now consider $\Delta\pi'(x, y) = \pi'(x) + \pi'(y) - \pi'(x + y)$, where

$$
\begin{aligned}
\pi'(x) &= \bar{\pi}(x) + \pi(x) = \bar{\pi}(x) + \pi(1 + l - x_{39}) + \delta c_2; \\
\pi'(y) &= \pi(y) = \pi(x_{39}^-) - \delta c_3; \\
\pi'(x + y) &= \pi(x + y) = \pi(l).
\end{aligned}
$$

Since $x - \bar{x} \in T$, the condition (ii) implies that $\bar{\pi}(x) = \bar{\pi}(\bar{x})$. We have

$$
\begin{aligned}
\Delta\pi'(x, y) &= \bar{\pi}(\bar{x}) + [\pi(1 + l - x_{39}) + \pi(x_{39}^-) - \pi(l)] + \delta(c_2 - c_3) \\
&= \bar{\pi}(\bar{x}) + s + \delta(c_2 - c_3) < 0,
\end{aligned}
$$

a contradiction to the subadditivity of π'. Therefore, $\bar{\pi}$ satisfies condition (v).

Let F be a face of $\Delta\mathcal{P}$. Denote by $n_F \in \{0, 1, 2\}$ the number of projections $p_i(\mathrm{rel\,int}(F))$ for $i = 1, 2, 3$ that intersect with $(l, u) \cup (f - u, f - l)$. It follows from the conditions (i) and (v) that

$$
|\Delta\bar{\pi}(x, y)| \leq n_F \cdot s \quad \text{for any } (x, y) \in \mathrm{rel\,int}(F).
$$

It can be verified computationally that, if $F \in \Delta\mathcal{P}$ has $n_F \neq 0$, then either

(a) $\Delta\pi_F(u, v) = 0$ for all $(u, v) \in \mathrm{vert}(F)$, or
(b) $\Delta\pi_F(u, v) \geq n_F \cdot s$ for all $(u, v) \in \mathrm{vert}(F)$, and the inequality is strict for at least one vertex.

Let $(x, y) \in [0, 1)^2$ such that $(x, y) \notin E(\pi)$. Then $\Delta\pi(x, y) > 0$ since π is subadditive. Consider the (unique) face $F \in \Delta\mathcal{P}$ such that $(x, y) \in \mathrm{rel\,int}(F)$. We will show that $\Delta\pi'(x, y) > 0$. If $n_F = 0$, then $\Delta\bar{\pi}(x, y) = 0$, and hence $\Delta\pi'(x, y) = \Delta\pi(x, y) > 0$. Now assume that $n_F \neq 0$. Since $\Delta\pi_F$ is affine linear on F, $\Delta\pi(x, y)$ is a convex combination of $\{\Delta\pi_F(u, v) \mid (u, v) \in \mathrm{vert}(F)\}$. We have $\Delta\pi(x, y) > 0$ by assumption. Thus the above case (b) applies, which implies that $\Delta\pi(x, y) > n_F \cdot s$. Hence $\Delta\pi'(x, y) = \Delta\pi(x, y) + \Delta\bar{\pi}(x, y) > 0$ holds when $n_F \neq 0$ as well. Therefore, $(x, y) \notin E(\pi')$. We obtain that $E(\pi') \subseteq E(\pi)$. This, together with the assumption $E(\pi) \subseteq E(\pi')$, implies that $E(\pi) = E(\pi')$.

We conclude, by Lemma 5.1(3), that π is a weak facet.

Remark: Conversely, if a $\mathbb{Z}$-periodic function $\bar{\pi}$ satisfies the conditions (i) to (v), then $\pi^\pm = \pi \pm \bar{\pi}$ are minimal valid functions, and $E(\pi) = E(\pi^+) = E(\pi^-)$.

(3) Let $\bar{\pi} = \pi_{\mathrm{lifted}} - \pi$. Observe that $\bar{\pi}$ satisfies the conditions (i) to (v) in (2). Thus, the function π_{lifted} is minimal valid and $E(\pi_{\mathrm{lifted}}) = E(\pi)$. Let π' be a minimal valid function such that $E(\pi_{\mathrm{lifted}}) \subseteq E(\pi')$. Then, as shown in (2), we have $E(\pi_{\mathrm{lifted}}) = E(\pi')$. It follows from Lemma 5.1(3) that π_{lifted} is a weak facet. However, the function π_{lifted} is not a facet, since $E(\pi_{\mathrm{lifted}}) = E(\pi)$ but $\pi_{\mathrm{lifted}} \neq \pi$. Next, we show that π_{lifted} is an extreme function.

Suppose that π_{lifted} can be written as $\pi_{\text{lifted}} = \frac{1}{2}(\pi^1 + \pi^2)$, where π^1, π^2 are minimal valid functions. Then $E(\pi_{\text{lifted}}) \subseteq E(\pi^1)$ and $E(\pi_{\text{lifted}}) \subseteq E(\pi^2)$. Let $\bar{\pi}^1 = \pi^1 - \pi$ and $\bar{\pi}^2 = \pi^2 - \pi$. We have that $E(\pi) \subseteq E(\bar{\pi}^1)$ and $E(\pi) \subseteq E(\bar{\pi}^2)$. Hence, as shown in (2), $\bar{\pi}^1$ and $\bar{\pi}^2$ satisfy the conditions (i) to (v). We will show that $\bar{\pi}^1 = \bar{\pi}^2$.

For $x \notin (l, u) \cup (f - u, f - l)$, we have $\bar{\pi}^i(x) = 0$ $(i = 1, 2)$ by condition (i). It remains to prove that $\bar{\pi}^1(x) = \bar{\pi}^2(x)$ for $x \in (l, u) \cup (f - u, f - l)$. By the symmetry condition (iv), it suffices to consider $x \in (l, u)$. We distinguish three cases. If $x + T \in C$, then condition (iii) implies $\bar{\pi}^i(x) = 0$ $(i = 1, 2)$. If $x + T \in C^+$, then $\bar{\pi}(x) = s$ by definition. Notice that $\bar{\pi}^1 + \bar{\pi}^2 = \pi^1 + \pi^2 - 2\pi = 2\pi_{\text{lifted}} - 2\pi = 2\bar{\pi}$, and that $\bar{\pi}^i(x) \leq s$ $(i = 1, 2)$ by condition (v). We have $\bar{\pi}^i(x) = s$ $(i = 1, 2)$ in this case. If $x + T \notin C$ and $x + T \notin C^+$, then $\bar{\pi}(x) = -s$, and hence $\bar{\pi}^i(x) = -s$ $(i = 1, 2)$. Therefore, $\bar{\pi}^1 = \bar{\pi}^2$ and $\pi^1 = \pi^2$, which proves that the function π_{lifted} is extreme. $\qquad\square$

References

1. Basu, A., Conforti, M., Cornuéjols, G., Zambelli, G.: A counterexample to a conjecture of Gomory and Johnson. Math. Program. Ser. A **133**(1–2), 25–38 (2012). doi:10.1007/s10107-010-0407-1
2. Basu, A., Hildebrand, R., Köppe, M.: Equivariant perturbation in Gomory and Johnson's infinite group problem. I. The one-dimensional case. Math. Oper. Res. **40**(1), 105–129 (2014). doi:10.1287/moor.2014.0660
3. Basu, A., Hildebrand, R., Köppe, M.: Equivariant perturbation in Gomory and Johnson's infinite group problem-III: foundations for the k-dimensional case with applications to $k = 2$. Math. Program. **163**(1), 301–358 (2017). doi:10.1007/s10107-016-1064-9
4. Basu, A., Hildebrand, R., Köppe, M.: Light on the infinite group relaxation I: foundations and taxonomy. 4OR **14**(1), 1–40 (2016). doi:10.1007/s10288-015-0292-9
5. Basu, A., Hildebrand, R., Köppe, M.: Light on the infinite group relaxation II: sufficient conditions for extremality, sequences, and algorithms. 4OR **14**(2), 107–131 (2016). doi:10.1007/s10288-015-0293-8
6. Basu, A., Hildebrand, A., Köppe, M., Molinaro, M.: A $(k+1)$-slope theorem for the k-dimensional infinite group relaxation. SIAM J. Optim. **23**(2), 1021–1040 (2013). doi:10.1137/110848608
7. Conforti, M., Cornuéjols, G., Zambelli, G.: Corner polyhedra and intersection cuts. Surv. Oper. Res. Manage. Sci. **16**, 105–120 (2011)
8. Dey, S.S., Richard, J.-P.P.: Facets of two-dimensional infinite group problems. Math. Oper. Res. **33**(1), 140–166 (2008). doi:10.1287/moor.1070.0283
9. Gomory, R.E., Johnson, E.L.: Some continuous functions related to corner polyhedra, I. Math. Program. **3**, 23–85 (1972). doi:10.1007/BF01584976
10. Gomory, R.E., Johnson, E.L.: Some continuous functions related to corner polyhedra, II. Math. Program. **3**, 359–389 (1972). doi:10.1007/BF01585008
11. Gomory, R.E., Johnson, E.L.: T-space and cutting planes. Math. Program. **96**, 341–375 (2003). doi:10.1007/s10107-003-0389-3

12. Hong, C.Y., Köppe, M., Zhou, Y.: SageMath program for computation and experimentation with the 1-dimensional Gomory–Johnson infinite group problem (2014). https://github.com/mkoeppe/infinite-group-relaxation-code
13. Köppe, M., Zhou, Y.: Equivariant perturbation in Gomory and Johnson's infinite group problem. VI. The curious case of two-sided discontinuous functions. eprint arXiv:1605.03975 [math.OC] (2016)

Minimum Birkhoff-von Neumann Decomposition

Janardhan Kulkarni[1], Euiwoong Lee[2], and Mohit Singh[3(✉)]

[1] Microsoft Research, Redmond, USA
[2] Carnegie Mellon University, Pittsburgh, USA
[3] Georgia Institute of Technology, Atlanta, USA
mohitsinghr@gmail.com

Abstract. Motivated by the applications in routing in data centers, we study the problem of expressing an $n \times n$ doubly stochastic matrix as a linear combination using the smallest number of (sub)permutation matrices. The Birkhoff-von Neumann decomposition theorem proves that there exists such a decomposition, but does not give a representation with the smallest number of permutation matrices. In particular, we consider the case when the optimal decomposition uses a constant number of matrices. We show that the problem is not fixed parameter tractable, and design a logarithmic approximation to the problem.

1 Introduction

A non-negative $n \times n$ matrix A is called a doubly stochastic matrix if the sum of entries in every row and every column is equal to 1. A matrix P is called a permutation matrix if every row and every column has exactly one non-zero entry with value 1. A beautiful result of Birkhoff-von Neumann (BvN Theorem) states that any doubly stochastic matrix can be presented as a linear combination of permutation matrices [13]. Formally, there exist non-negative constants $\lambda_1, \lambda_2, \ldots, \lambda_k$ for some $k > 0$ such that

$$A = \lambda_1 P_1 + \lambda_2 P_2 + \ldots \lambda_k P_k, \quad \text{and} \quad \forall i, \lambda_i > 0 \tag{1}$$

In graph theoretic language, the BvN theorem states that given a non-negative edge weighted bipartite graph, where for every vertex the total weight of edges incident on it is equal to 1 (fractional perfect matching), then it can be represented a convex combination of integral matchings. A proof of the BvN theorem follows from the fact that vertices of the doubly stochastic matrix polytope, called Birkhoff polytope, correspond to permutation matrices [13]. Note that a BvN decomposition of a doubly stochastic matrix may not be unique - both in terms of permutation matrices used and also in the number of matrices used to produce such a decomposition. In this paper, motivated by its applications to routing in data centers, we are interested in BvN representations of doubly stochastic matrices with small number of matrices. However, in our setting, the matrices in the representation can be sub-permutation matrices instead of permutation matrices. A matrix is a sub-permutation matrix if every row and every

© Springer International Publishing AG 2017
F. Eisenbrand and J. Koenemann (Eds.): IPCO 2017, LNCS 10328, pp. 343–354, 2017.
DOI: 10.1007/978-3-319-59250-3_28

column has *at most* one non-zero entry with value 1. In graph theoretic language, we seek to represent a fractional perfect matching of a bipartite graph as a linear combination of integral matchings (not necessarily integral perfect matchings). That is,

$$A = \lambda_1 M_1 + \lambda_2 M_2 + \ldots \lambda_k M_k, \quad \text{and} \quad \forall i, \lambda_i > 0 \tag{2}$$

where each M_i is a sub-permutation matrix. We call such a representation *matching decomposition* of doubly stochastic matrix A. In this paper we study the problem of finding a minimum matching decomposition of a doubly stochastic matrix. From Carathodory's Theorem [15] we know that there is a representation with at most $n^2 + 1$ matrices, which can be tightened to $n^2 - 2n + 2$ matrices by applying the Marcus-Ree Theorem [14]. Yet, for a given doubly stochastic matrix A, the number of matrices needed in a matching decomposition of A can be much smaller than these bounds. Consider for example doubly stochastic matrices that lie on a line connecting two vertices of the Birkhoff polytope; such matrices can be represented as convex combination of only two permutation matrices. We say that a matching decomposition of a doubly stochastic matrix is *minimum* if there is no other matching decomposition with a smaller number of sub-permutation matrices. Our goal is to find a minimum matching decomposition of a doubly stochastic matrix.

Dufossé and Uçar [7] show that the problem of finding a minimum matching decomposition is NP-hard, building on the work of Brualdi and Gibson [3,4]. Hence, we focus our attention on the case when the optimal representation uses a constant number of sub-permutation matrices. We ask, *is minimum matching decomposition fixed parameter tractable in k?*, where k is number of matrices used in the optimal solution. In other words, is there an algorithm that finds a matching decomposition with the minimum number of sub-permutation matrices that runs in time that is polynomial in n but can depend arbitrarily in k.

Our main motivation to study the fixed parameter tractability of the problem comes from the application of the BvN theorem in traffic routing in reconfigurable data centers and more broadly in software defined networks. One of the emerging technologies to connect servers within a data center is to use light (laser). An advantage of such an approach is that as traffic between servers changes over time topology can be reconfigured. In such contexts, the Birkhoff-von Neumann decomposition theorem has been extensively used to route traffic among servers [2,4–6,9,10,18,20].

In routing applications, a doubly stochastic matrix represents traffic that needs to be routed among a set of n servers. (Although traffic matrices need not be doubly stochastic, in the applications of interest it is reduced to a doubly stochastic matrix by appropriate scaling. See [18] more details.) A routing decision at any time step is a matching between senders and receivers, a BvN decomposition of the traffic matrix gives a schedule to route traffic. Switching between matchings, however, involves reconfiguring hardware (moving laser pointers and receivers) that comes at a cost. Consequently, finding a decomposition with few permutation matrices improves performance [2,12,18]. Moreover, the empirical evidence shows that the number of servers that are active is small compared to the total number of servers in a data center, and hence it is observed that BvN decomposition has a small support [12]. There is a growing body of work in understanding BvN decompositions in the context of reconfigurable data center architectures and we refer the reader to [2,10,18] and references there in for more details. Apart from its applications in data centers, the BvN theorem has also been used in routing in wireless networks; we refer the readers to [4,5] more details.

In this paper we show that the minimum matching decomposition problem is not fixed parameter tractable.

Theorem 1. *There exists a universal constant $k \geq 4$ for which it is NP-hard to find a minimum matching decomposition of a doubly stochastic matrix that admits a decomposition into k matchings.*

Since the optimal value k is a constant, our result implies that the problem is APX-hard and does not admit a PTAS. In addition to fixed parameter tractable algorithms, it also rules out an algorithm that runs in time $n^{f(k)}$ for any function f. Interestingly, the problem is polynomial time solvable for $k = 2, 3$, and we believe that it becomes NP-hard for $k = 4$. On the positive side, we show that there exists a logarithmic approximation to the problem.

Theorem 2. *There is an algorithm that is $O(\log k)$ approximation to minimum matching decomposition problem, which runs in time polynomial in n and doubly exponential in k.*

In particular, our algorithm finds a representation of A using at most $O(\log k) \cdot k$ sub-permutation matrices if there is an optimal solution with at most k sub-permutation matrices.

There is an algorithmic proof of the BvN decomposition theorem [7], and it is natural to ask what is approximation factor of that algorithm. We show an exponential lower bound on the approximation factor of the BvN decomposition algorithm. We also show that our lower bound example extends to all known variants of BvN decomposition algorithms. Another related question, in a spirit similar to the approximate Carathéodory's Theorem, is if there is a small representation of ϵ-close matrix of A. For any $n \times n$ doubly stochastic matrix A, call a $n \times n$ matrix A' ϵ-close if $\forall a_{ij} \in A, a'_{ij} \in A', |a_{ij} - a'_{ij}| \leq \epsilon$. We show that for this problem there exists a tight representation using at most $1/\epsilon$ matrices. This is an improvement over the result of Barman [1] who showed a representation using at most $O(\log(n)/\epsilon^2)$ matrices.

Discussion. Our results leave open several important questions. The most interesting question is if there is a constant factor fixed parameter tractable approximation algorithm to the problem. Another interesting direction is to understand the approximability of the problem for the general case when k is not a constant.

2 Approximation Algorithm

In this section we prove Theorem 2. We design an algorithm that runs in time $f(n)g(k)$, where $f(n)$ is polynomial in n and $g(k)$ is doubly exponential.

Overview of the Algorithm. Our algorithm consists of three main steps. In the first step, our algorithm finds a set of values $\lambda_1, \lambda_2, \ldots \lambda_k$ such that every entry in A can be represented as sum of some subset of the values. We find such a set of λ values by a combination of a brute force search and solving a sequence of linear equations, and hence this step of our algorithm runs in time that is exponential in k. Given λ values, we reduce our problem to a combinatorial problem called *generalized bipartite edge coloring* (GBEC), which is a generalization of the bipartite edge coloring problem. Here, for each edge we are given a list S_e where each element $s \in S_e$ is a subset of $\{\lambda_1, \lambda_2, \ldots \lambda_k\}$ (we allow the same number to appear multiple times). Our goal is to assign each edge e an element $s_e \in S_e$ such that the maximum degree in the induced bipartite graph corresponding to each λ_i is as small as possible. An edge e is in the induced bipartite graph for λ_i if $\lambda_i \in s_e$ and e is assigned s_e. Note that if there is a matching decomposition of A using k matchings and our "guess" of λ values was correct, then the induced bipartite graph corresponding to each λ_i would be a matching. Finally, we give a LP rounding based algorithm to get a logarithmic approximation to the generalized bipartite edge coloring problem.

2.1   Computing λ Values

Let $w_1, w_2, \ldots w_\ell$ be the set of distinct elements of matrix A. We say that a set of k real numbers $S = \{\lambda_1, \lambda_2, \ldots \lambda_k\}$ is *feasible* if every entry w_i can be represented as a sum of subset of values from S. That is, for all w_i, $\exists s \subseteq S$, such that $w_i = \sum_{\lambda_i \in s} \lambda_i$. Our first observation is that one can find a feasible set for distinct elements of A in time that is polynomial in n and exponential in k.

Lemma 1. *A feasible set representing all distinct entries in A can be found in time $O(2^{k2^k})$.*

Proof (sketch). First we observe that there cannot be more than 2^k distinct entries in A. This is true since every distinct entry in A needs to be represented as some subset of $[k]$, and there are at most 2^k such subsets. Now we set up a sequence of linear equations to find a feasible set.

   The variables of our linear equations are $x_1, x_2, x_3 \ldots \ldots x_k$, which are intended to represent a feasible set of $\lambda_1, \lambda_2, \ldots \lambda_k$. For every distinct entry a_{ij}, we "guess" a subset $s \subset [k]$ such that $\sum_{i \in s} x_i = a_{ij}$ and add it as a constraint in our linear equations. Therefore, our systems of linear equations has k variables and a constraint for every distinct entry in A. We write a sequence of such linear equations for every possible guess of a_{ij} values. If A admits a matching decomposition using at most k distinct matrices, then at least one of 2^{k2^k} system of linear equations should have a feasible solution, which will find a feasible subset of λ values.   □

   We remark that although the above algorithm finds a feasible set of λ values, it is only a necessary condition for a matrix to have matching decomposition using at most k entries. For the sake of getting a good approximation algorithm this also turns out to be sufficient: In particular, if for a feasible set of λ values S the LP in Sect. 2.2 has a valid solution, then we show a representation using the values in S.

Given the $\lambda_1, \lambda_2, \ldots \lambda_k$ values, our next step is to find a set of sub-permutation matrices such that A can be represented as a linear combination of sub-permutation matrices. Our idea to get a good approximation algorithm is to first find an *intermediate representation* of A by a set of k incidence matrices corresponding to a set of bipartite graphs such that maximum degree of any vertex is minimized.

Formally, we first represent

$$A = \lambda_1 B_1 + \lambda_2 B_2 + \ldots \lambda_k B_k, \tag{3}$$

where B_i is 0–1 matrix; that is, all non-zero entries in B_i have value 1. Further, each B_i corresponds to a bipartite graph. We call this bipartite graph-decomposition of matrix A. Our first intermediate goal is to represent A using bipartite graphs $B_1, B_2, \ldots B_k$ such that the degree of every vertex in B_i for all $i \in [k]$ is as small as possible. Let $\Delta = \max_i \{\max_{v \in B_i} \delta_i(v)\}$, where $\delta_i(v)$ is the degree of vertex v in B_i.

Lemma 2. *If A admits a bipartite graph-decomposition (3) with maximum degree Δ, then there is a matching decomposition of A with at most $k\Delta$ sub-permutation matrices.*

Proof. To prove the lemma we make use of the Vizing's theorem [19], which states that any bipartite graph with maximum degree Δ can be decomposed into Δ matchings. Therefore, by decomposing every B_i into at most Δ matchings (and replicating the λ_i coefficient) we get a matching decomposition with at most $k\Delta$ matchings. $\square$

The rest of the section is devoted to finding a bipartite decomposition of A such that maximum degree is minimized. We formulate this problem as a linear program.

2.2   LP Formulation

For each entry $a_{ij} \in A$, let S_{ij} denote the set of all subsets of $[k]$ such that a_{ij} is equal to sum of the corresponding λ's. That is,

$$\forall s \in S_{ij}, s \subseteq [k], a_{ij} = \sum_{t \in s} \lambda_t.$$

Let $G(A) = (V, E)$ denote the bipartite graph represented by A, where the weight of an edge $e := (i, j)$ is equal to a_{ij}. For every subset $s \in S_{ij}$ and for every entry a_{ij}, we create a variable x_{es} where $e := (i, j)$. The LP relaxation we write has the following three simple constraints:

$$\sum_{s \in S_e} x_{es} = 1, \qquad\qquad \forall e \in E \qquad\qquad (4)$$

$$\sum_{e:e \to v} \sum_{s:i \in s, s \in S_e} x_{es} \le 1, \qquad\qquad v \in V, i \in [k] \qquad\qquad (5)$$

$$x_{es} \ge 0 \qquad\qquad\qquad\qquad (6)$$

The first set of constraints (4) ensure that every edge e is assigned some valid set $s \in S_e$. The second set of constraints (5) are matching constraints: Since A admits a matching decomposition, for every coefficient λ_i, there is a bipartite graph B_i such that the maximum degree is at most 1.

If the LP has no feasible solution, it implies that our guess of λ values was not valid. So, we iterate over all possible guesses of λ values (as outlined in Lemma (1)) till the LP (4–6) returns a feasible solution.

2.3 Rounding

Let $\mathbf{x}$ be a feasible solution to the above LP (4–6). Let x_{es}^* be the value of variable x_{es} in the solution $\mathbf{x}$. We do a randomized rounding of $\mathbf{x}$. That is, we assign edge e to subset $s \in S_e$ with probability x_{es}^*. The first set constraints of from LP (4) imply that every edge e gets a $s \in S_e$. Let B_i represent the bipartite graph corresponding to the set of edges e such that $e \to s$ and $i \in s$. We use $e \to s$ to denote that the edge e was assigned to subset s in our randomized rounding. Thus, we get an intermediate representation of A using bipartite graphs, and it remains to bound Δ of this representation.

Let $\delta_i(v)$ be the degree of vertex v in the bipartite graph i after the randomized rounding. (That is, bipartite graph corresponding to coefficient λ_i). Note that $\delta_i(v)$ is a random variable and from the second set of LP constraints we have $\mathbb{E}[\delta_i(v)] \le 1$. We now show that probability that $\delta_i(v) \ge O(\log k)$ is at most $O(1/k^2)$. Towards that we need the following version of Chernoff bound.

Theorem 3 *(Chernoff Bounds [17]). Let $X_1, X_2, \cdots X_n$ be n independent random variables with $X_i = 1$ with probability p_i and $X_i = 0$ with probability $1 - p_i$. Let $X = \sum_i X_i$. Then for any $\epsilon > 1$,*

$$\mathbb{P}(X \ge (1 + \epsilon)\mathbb{E}(X)) \le exp(-\epsilon/3 \cdot \mathbb{E}(X)).$$

A simple application of the above theorem gives the following result.

Lemma 3. $\mathbb{P}(\delta_i(v) \ge c \log(k)) \le \frac{1}{k^3}$, *for $c \ge 10$.*

Proof. First we note that $\delta_i(v)$ is the sum of independent random variables X_e. Formally,

$$\delta_i(v) = \sum_{e:e \to v} \sum_{s:i \in s} \mathbf{1}(e \to s)$$

The indicator function $\mathbf{1}(e \to s)$ denotes that edge e is assigned set s in our random coloring. From the constraints of LP

$$\mathbb{E}[\delta_i(v)] = \sum_{e:e \to v} \sum_{s:i \in s} \mathbb{P}(e \to s) = \sum_{e:e \to v} \sum_{s:i \in s} x_{es}^* \le 1$$

Now we apply the Chernoff bound by taking $\epsilon = 9\log(k)$ for the random variable $\delta_i(v)$ to complete the proof. □

However, the fact that $\mathbb{P}(\delta_i(v) \geq c\log(k)) \leq \frac{1}{k^3}$ for fixed v and i does not guarantee that $\mathbb{P}(\max_{i,v}\{\delta_i(v)\} \leq c\log(k))$ is non-zero. This is because there are nk events corresponding to every pair of (v, i) and probability of failure of each event is only polynomial in $1/k$. Since n can be much larger than k, union bound does not give a non-zero probability for $\mathbb{P}(\max_{i,v}\{\delta_i(v)\} \leq c\log(k))$. To overcome this, we apply Lovaśz Local Lemma (LLL).

Theorem 4 *(Lovaśz Local Lemma [8]). Let $T_1, T_2 \ldots T_m$ be events such that: (1) $\mathbb{P}[T_i] \leq p$, (2) each T_i depends on at most d other events, and (3) $4 \cdot p \cdot d \leq 1$. Then there is a nonzero probability that none of the events occurs; $\mathbb{P}(\bigcap_{i=1}^{m} \overline{T_i}) > 0$.*

We now apply LLL to show that when we do randomized rounding of LP solution, with non-zero probability no vertex gets a degree more than $O(\log k)$ in any bipartite graph B_i.

Lemma 4. *With non-zero probability, no vertex $v \in B_i, i \in [k]$, gets a degree more than $c\log k$ for some constant $c \geq 10$.*

Proof. We prove the lemma by applying LLL. We define a set of bad events $T_{i,v}$ corresponding to every vertex v in the bipartite graph B_i. An event $T_{i,v}$ is bad if $\delta_i(v)$ is greater than $10\log k$. From Lemma (3), we have $\mathbb{P}(T_{i,v}) \leq \frac{1}{k^3}$.

Next, we bound dependency degree of an event $T_{i,v}$. Towards this, we define a bipartite graph with a vertex for every bad event $T_{i,v}$ in the left-hand side. The right-hand side of this bipartite graph consists of random variables X_e, where X_e represents the random variable for an edge e in $G(A)$. Now, observe that each bad event $T_{i,v}$ depends on at most k random variables X_e. This follows from the observation that at most k edges can be incident at a vertex v in $G(A)$ since A admits a BvN decomposition of size k. Further, each random variable X_e can affect at most $2k$ bad events corresponding to (i, v), where $i \in [k]$ and $e \to v$. This means that each $A_{i,v}$ depends on at most $2k^2$ other events.

With these two facts, it is easy to verify the LLL condition that $4 \cdot p \cdot d \leq 1$ since $p \leq 1/k^3$ and $d \leq 2k^2$. Thus, with non-zero probability none of the bad events occur. By applying Moser-Tardos [16] framework, we can find such an outcome in polynomial time. □

Putting all the pieces together we have the following theorem.

Theorem 5. *The randomized rounding of LP (4-6) is an $O(\log(k))$ approximation to the problem of finding minimum matching decomposition of a doubly stochastic matrix.*

Proof. From Lemma 4, we conclude that there is a decomposition of A into bipartite graphs such that maximum degree (Δ) is at most $O(\log(k))$. Then it follows from Lemma 2 that there is a matching decomposition with at most $O(k\log k)$ matrices. This completes the proof. □

Now we show that our rounding of LP is almost optimal by showing an instance where LP has $\Omega(\log k / \log\log k)$ integrality gap.

2.4 LP Gap

Our LP in Sect. 2.2 fractionally assigns a subset s_e of $[k]$ to each edge e, where s_e must belong to the given collection S_e of subsets. For each integral solution $\{s_e\}_{e \in E}$ with $s_e \in S_e$, recall that $\delta_i(v) = |\{e \in E : e \to v \text{ and } i \in s_e\}|$. Each edge e satisfies $\sum_{i \in s} \lambda_i = \sum_{i \in s'} \lambda_i$ for every $s, s' \in S_e$, so S_e cannot be an arbitrary collection of subsets. We show that if we ignore restrictions given by $\{\lambda_i\}_{i \in [k]}$ and allow S_e to be an arbitrary collection of subsets, our LP in Sect. 2.2 has a gap of $\Omega(\frac{\log k}{\log \log k})$.

Lemma 5. *There is a bipartite graph $G = (V, E)$, $k \in \mathbb{N}$, and $\{S_e \subseteq 2^{[k]}\}_{e \in E}$ such that the LP (4–6) is feasible, but for any integral solution $\{s_e\}_{e \in E}$ with $s_e \in S_e$, there exist at least $k/2$ numbers $p_1, \ldots, p_{k/2} \in [k]$ such that for each p_i, $\max_v \delta_{p_i}(v) \geq \Omega(\frac{\log k}{\log \log k})$.*

Proof. We first present an instance that is feasible for the LP and each integral solution has one $p \in [k]$ with $\max_v \delta_p(v) \geq \Omega(\frac{\log k}{\log \log k})$. Then we show how to extend this construction to achieve $p_1, \ldots, p_{k/2}$. Our instance is parameterized by integers k and d, where d divides k. It has $d+1$ vertices $\{u, v_1, \ldots, v_d\}$ and d edges $\{(u, v_i)\}_{1 \leq i \leq d}$. Let $e_i := (u, v_i)$. For each i, e_i is associated with d subsets $s_{i,1}, \ldots, s_{i,d}$ in the following way: for each element $p \in [k]$, pick a random number $j \in [d]$ and put p into $s_{i,j}$. So for each i, $s_{i,1}, \ldots, s_{i,d}$ are disjoint and their union is $[k]$. Fractionally, if each e_i picks every $s_{i,j}$ with $\frac{1}{d}$, at vertex u, every $p \in [k]$ is picked exactly once.

We want to claim that if each e_i integrally picks one s_{i,j_i}, one $p \in [k]$ is picked many times. Fix one integral solution $(j_1, \ldots, j_d)$, which represents that e_i picks s_{i,j_i}. For each $p \in [k]$, the number of occurrences of p in $s_{1,j_1}, \ldots, s_{d,j_d}$ is a random variable drawn from $B(d, \frac{1}{d})$, the binomial distribution with d trials and probability $1/d$. Note that each $p \in [k]$ is independent. Let m be an integer fixed later, and let $P_m := \Pr[X \geq m]$ where X is drawn from $B(d, \frac{1}{d})$. The probability that every $p \in [k]$ occurs strictly less than m times is $(1 - P_m)^k \leq e^{-P_m k}$.

We can lower bound P_m by

$$P_m \geq \Pr[X = m] = \binom{d}{m}(\frac{1}{d})^m(1 - \frac{1}{d})^{d-m} \geq \frac{(d-m)^m}{m^m} \cdot \frac{1}{d^m} \cdot \frac{1}{e} \geq \frac{1}{(2m)^m},$$

for $d \geq 4m$. There are d^d tuples $(j_1, \ldots, j_d)$, so as long as

$$(1 - P_m)^k \cdot d^d < 1 \Leftarrow e^{-P_m k} \cdot e^{d \log d} < 1 \Leftarrow \frac{1}{(2m)^m} > \frac{d \log d}{k},$$

there is an instance where for each integral solution $\{s_e\}_{e \in E}$ there exists $p \in [k]$ with $\max_v \delta_p(v) \geq m$. It works when $d = \sqrt{k}$ and $m = \Omega(\frac{\log k}{\log \log k})$.

To extend the above strategy to find $p_1, \ldots, p_{k/2}$ simultaneously for each integral solution, create ℓ disjoint copies of the above instance independently. As the above argument, it will be feasible for the LP if each edge picks each of d subsets with $\frac{1}{d}$. Fix an integral solution (there are now $(d^d)^\ell = d^{d\ell}$ choices).

For each $p \in [k]$, the probability that the number occurrences of p is less than m in each copy is $(1 - P_m)^\ell$. The probability that there are $k/2$ numbers that occur less than m times in each copy is at most $\binom{k}{k/2} \cdot (1 - P_m)^{\frac{\ell k}{2}} \leq e^k \cdot e^{-P_m \frac{\ell k}{2}}$. Union bounding over all $d^{d\ell}$ choices, as long as (take $\ell \gg k$)

$$e^{d\ell \log d} \cdot e^k \cdot e^{-P_m \frac{\ell k}{2}} < 1$$

$$\Leftarrow P_m \frac{\ell k}{2} > d\ell \log d + k$$

$$\Leftarrow P_m k > 4d \log d$$

$$\Leftarrow \frac{1}{(2m)^m} > \frac{d \log d}{4k},$$

there is an instance where for every integral solution, there are at least $k/2$ numbers $p_1, \ldots, p_{k/2}$ that occur more than m times in a single copy. Since all edges in one copy is incident on a single vertex, for each p_i, $\max_v \delta(v)_{p_i} \geq m$. It works again with $d = \sqrt{k}$ and $m = \Omega(\frac{\log k}{\log \log k})$. $\qquad \square$

Remark: There is a natural configuration LP for the problem which overcomes our intergrality gap example. It would be interesting to see if it can be rounded to get a constant approximation to the problem.

3 Hardness Results

In this section we show that it is NP-hard to find a minimum matching decomposition of a doubly stochastic matrix even when $k = O(1)$. To obtain the hardness for minimum matching decomposition, we show NP-hardness *generalized bipartite edge coloring* (GBEC) introduced in Sect. 2. Recall that this problem was formally defined as follows.

- Input: A bipartite graph $G = (V, E)$, an integer $k \in \mathbb{N}$. For each edge $e \in E$, a collection of subsets $S_e \subseteq 2^{[k]}$.
- Output: For each edge e, $s_e \in S_e$.
- Goal: Minimize $\max_{i \in [k], v \in V} \delta_i(v)$, where $\delta_i(v) := |\{e \in E : e \to v \text{ and } i \in s_e\}|$.

Our algorithm in Sect. 2 gives an $O(\log k)$-approximation algorithm. We complement our algorithmic result by showing that GBEC is NP-hard even when $k = 28$ in Sect. 3.1.

Theorem 6. *Given an instance of* GBEC *with* $k = 28$, *it is NP-hard to distinguish whether the optimal value is* 1 *or higher.*

We also show that there is an efficient reduction from GBEC to minimum matching decomposition that keeps k to be constant. The proof is deferred to the full version of the paper. These two results show NP-hardness of minimum matching decomposition when k is a universal constant, proving Theorem 1.

Lemma 6. *For some function $k' : \mathbb{N} \to \mathbb{N}$, there is a polynomial time reduction ϕ from GBEC to minimum matching decomposition such that the instance $\mathcal{I}$ of GBEC with k has the optimal value at most 1 if and only if the instance $\phi(\mathcal{I})$ of minimum matching decomposition admits a decomposition into at most $k' = k'(k)$ matchings.*

3.1 Hardness of GBEC

This section proves Theorem 6, showing NP-hardness of GBEC even when k is a constant. We reduce from EDGE COLORING in 3-regular (general) graphs to GBEC. Vizing's theorem shows that every 3-regular graph can be 4-edge-colorable, but Holyer [11] shows that it is NP-hard to decide whether it is 3-edge-colorable or 4-edge-colorable.

Given a cubic graph $G = (V, E)$ for EDGE COLORING, let $c^1 : V \mapsto [4]$ be a 4-vertex coloring, and $c^2 : E \mapsto [4]$ be a 4-edge coloring (both are easily computable). The instance of GBEC is defined as follows.

- $G' = (V', E')$ where $V' = V \cup E$ and $E' = \{(v, e) : v \in V, e \in E, v \in e\}$. It is clearly bipartite.
- Each edge $e = (u, v) \in E$ is divided into two edges $(u, e), (v, e)$ in G'. Arbitrarily call one of them head and the other tail.
- $k = 28$.
- Define $\{T_{i,d}\}_{i,d}$ for each $i \in [3]$ and $d \in \{\text{head}, \text{tail}\}$ such that
  - $T_{i,d} \subseteq [4]$ and $|T_{i,d}| = 2$ for each i, d.
  - $T_{i,\text{head}} \cup T_{i,\text{tail}} = [4]$ for each i.
  - $T_{i,d} \cap T_{j,d'} \neq \emptyset$ for every $i \neq j$ and d, d'.
  - Simply, $T_{1,\text{head}} = \{1, 2\}, T_{1,\text{tail}} = \{3, 4\}, T_{2,\text{head}} = \{1, 3\}, T_{2,\text{tail}} = \{2, 4\}, T_{3,\text{head}} = \{1, 4\}, T_{3,\text{tail}} = \{2, 3\}$ works.
- Fix an edge $e' = (e, u) \in E'$ where $e = (u, v) \in E$ and $d \in \{\text{head}, \text{tail}\}$ be the type of e'. $S_{e'}$ has the following three subsets. When $T \subseteq [4]$ and $j \in \mathbb{N}$, let $(T + j)$ denote $\{t + j : t \in T\}$.
  - Let $s_{e',i} := \{i + 3(c_u^1 - 1)\} \cup (T_{i,d} + (4c_e^2 + 8))$.
  - $S_{e'} := \{s_{e',i} : i \in [3]\}$.

Intuition. For any of subset $s_{e',i} = \{i + 3(c_u^1 - 1)\} \cup (T_{i,d} + (4c_e^2 + 8))$, note that $\{i + 3(c_u^1 - 1)\} \subseteq \{1, \ldots, 12\}$ and $(T_{i,d} + (4c_e^2 + 8)) \subseteq \{13, \ldots, 28\}$. Out of 28 colors for GBEC, $\{13, \ldots, 28\}$ ensures that for each $e \in E$, its head and tail get the same color in $[3]$ for EDGE COLORING. The subset $\{1, \ldots, 12\}$ checks that for each vertex $v \in V$, all incident edges have different colors. We use many ($k = 28$) colors to perform these checks because we want these checks only between desired edges (decided by c^1 and c^2).

Lemma 7. *If G is 3-edge colorable, the optimum of GBEC for G' is 1.*

Proof. Let $c^* : E \mapsto [3]$ be an edge 3-coloring of G. For edge $e = (u, v) \in E$ with head (e, u) and tail (e, v), let

$$s_{(e,u)} \leftarrow s_{(e,u),c^*_e} = \{c^*_e + 3(c^1_u - 1)\} \cup (T_{c^*_e, \text{head}} + (4c^2_e + 8)),$$
$$s_{(e,v)} \leftarrow s_{(e,u),c^*_v} = \{c^*_e + 3(c^1_v - 1)\} \cup (T_{c^*_e, \text{tail}} + (4c^2_e + 8)).$$

Let $e'_1, e'_2 \in E'$ be adjacent edges. There are two cases.

- $e'_1 = (e, u)$ and $e'_2 = (e, v)$ for some edge $e = (u, v) \in E$: One of them is head, and the other is tail. In $\{13, \ldots, 28\}$, they are disjoint since $T_{c^*_e, \text{head}}$ and $T_{c^*_e, \text{tail}}$ are disjoint. In $\{1, \ldots, 12\}$, they are disjoint $c^1_u \neq c^1_v$.
- $e'_1 = (e, u)$ and $e'_2 = (f, u)$ for $e \neq f \in E$: In $\{13, \ldots, 28\}$, they are disjoint since $c^2_e \neq c^2_f$. In $\{1, \ldots, 12\}$, they are disjoint $c^*_e \neq c^*_f$.

Therefore, the optimum of GBEC for G' is 1. □

Lemma 8. *If the optimum of GBEC for G' is 1, G is 3-edge colorable.*

Proof. Let $c : E' \to [3]$ be a solution of GBEC where each $e' \in E'$ chooses $s_{e',c(e')}$ and the optimum is 1. The fact that the optimum is 1 means that for every adjacent $e', f' \in E'$, $s_{e',c(e')}$ and $s_{f',c(f')}$ are disjoint.

For each edge $e = (u, v) \in E$, let $e'_1 = (e, u)$ be its head and $e'_2 = (e, v)$ be its tail. We must have $c(e'_1) = c(e'_2)$ to have the optimum at most 1 since otherwise $(T_{c(e'_1), \text{head}} + (4c^2_e + 8)) \subseteq s_{e'_1, c(e'_1)}$, $(T_{c(e'_2), \text{tail}} + (4c^2_e + 8)) \subseteq s_{e'_2, c(e'_2)}$, and $T_{c(e'_1), \text{head}}$ and $T_{c(e'_2), \text{tail}}$ intersect if $c(e'_1) \neq c(e'_2)$.

For any $e = (u, v) \in E$ and $f = (u, w) \in E$ that meet at vertex u, let $e' = (u, e) \in E'$ and $f' = (u, f) \in E'$. We must have $c(e') \neq c(f')$ to have the optimum at most 1 since otherwise $c(e') + 3(c^1_u - 1)$ is contained in both $s_{e',c(e')}$ and $s_{f',c(f')}$.

Therefore, c gives the same color to head and tail of the same edge of E, and adjacent edges must have different colors. Therefore, it gives a proper 3-edge coloring of G. □

References

1. Barman, S.: Approximating nash equilibria and dense bipartite subgraphs via an approximate version of caratheodory's theorem. In: Proceedings of the Forty-Seventh Annual ACM on Symposium on Theory of Computing, pp. 361–369. ACM (2015)
2. Bojja, S., Mohammad Alizadeh, V., Viswanath, P.: Costly circuits, submodular schedules and approximate carathéodory theorems. In: Proceedings of the ACM SIGMETRICS International Conference on Measurement and Modeling of Computer Science, pp. 75–88. ACM (2016)
3. Brualdi, R.A.: Notes on the birkhoff algorithm for doubly stochastic matrices. Can. Math. Bull. **25**(2), 191–199 (1982)
4. Brualdi, R.A., Gibson, P.M.: Convex polyhedra of doubly stochastic matrices. I. Applications of the permanent function. J. Comb. Theor. Ser. A **22**(2), 194–230 (1977)

5. Chang, C.-S., Chen, W.-J., Huang, H.-Y.: On service guarantees for input-buffered crossbar switches: a capacity decomposition approach by Birkhoff and von Neumann. In: Seventh International Workshop on Quality of Service, IWQoS 1999, pp. 79–86. IEEE (1999)
6. Chen, K., Singla, A., Singh, A., Ramachandran, K., Lei, X., Zhang, Y., Wen, X., Chen, Y.: Osa: an optical switching architecture for data center networks with unprecedented flexibility. IEEE/ACM Trans. Netw. **22**(2), 498–511 (2014)
7. Dufossé, F., Uçar, B.: Notes on Birkhoff-von Neumann decomposition of doubly stochastic matrices. Linear Algebra Appl. **497**, 108–115 (2016)
8. Erdos, P., Lovász, L.: Problems and results on 3-chromatic hypergraphs and some related questions. Infinite Finite Sets **10**(2), 609–627 (1975)
9. Farrington, N., Porter, G., Radhakrishnan, S., Hajabdolali Bazzaz, H., Subramanya, V., Fainman, Y., Papen, G., Vahdat, A.: Helios: a hybrid electrical/optical switch architecture for modular data centers. ACM SIGCOMM Comput. Commun. Rev. **40**(4), 339–350 (2010)
10. Ghobadi, M., Mahajan, R., Phanishayee, A., Devanur, N., Kulkarni, J., Ranade, G., Blanche, P.-A., Rastegarfar, H., Glick, M., Kilper, D.: Projector: agile reconfigurable data center interconnect. In: Proceedings of the Conference on ACM SIGCOMM Conference, pp. 216–229. ACM (2016)
11. Holyer, I.: The NP-completeness of edge-coloring. SIAM J. Comput. **10**(4), 718–720 (1981)
12. Liu, H., Mukerjee, M.K., Li, C., Feltman, N., Papen, G., Savage, S., Seshan, S., Voelker, G.M., Andersen, D.G., Kaminsky, M., et al.: Scheduling techniques for hybrid circuit/packet networks. In: ACM CoNEXT (2015)
13. Lovász, L., Plummer, M.D.: Matching Theory, vol. 367. American Mathematical Soc., Providence (2009)
14. Marcus, M., Ree, R.: Diagonals of doubly stochastic matrices. Q. J. Math. **10**(1), 296–302 (1959)
15. Matoušek, J.: Lectures on Discrete Geometry, vol. 108. Springer, New York (2002)
16. Moser, R.A., Tardos, G.: A constructive proof of the general Lovász local lemma. J. ACM (JACM) **57**(2), 11 (2010)
17. Motwani, R., Raghavan, P.: Randomized Algorithms. Chapman & Hall/CRC, Boca Raton (2010)
18. Porter, G., Strong, R., Farrington, N., Forencich, A., Chen-Sun, P., Rosing, T., Fainman, Y., Papen, G., Vahdat, A.: Integrating microsecond circuit switching into the data center, vol. 43. ACM (2013)
19. Vizing, V.G.: On an estimate of the chromatic class of a p-graph. Diskret. Analiz **3**(7), 25–30 (1964)
20. Wang, G., Andersen, D.G., Kaminsky, M., Konstantina Papagiannaki, T.S., Ng, M.K., Ryan, M.: c-through: part-time optics in data centers. In: ACM SIGCOMM Computer Communication Review, vol. 40, pp. 327–338. ACM (2010)

Maximum Matching in the Online
Batch-Arrival Model

Euiwoong Lee and Sahil Singla[(✉)]

Computer Science Department, Carnegie Mellon University, Pittsburgh, USA
{euiwoonl,ssingla}@cs.cmu.com

Abstract. Consider a *two-stage matching* problem, where edges of an input graph are revealed in two stages (batches) and in each stage we have to immediately and irrevocably extend our matching using the edges from that stage. The natural greedy algorithm is half competitive. Even though there is a huge literature on online matching in adversarial *vertex arrival model*, no positive results were previously known in adversarial *edge arrival model*.

For two-stage bipartite matching problem, we show that the optimal competitive ratio is exactly 2/3 in both the fractional and the randomized-integral models. Furthermore, our algorithm for fractional bipartite matching is *instance optimal*—achieves the best competitive ratio for any *given* first stage graph. We also study natural extensions of this problem to general graphs and to s stages, and present randomized-integral algorithms with competitive ratio $\frac{1}{2} + 2^{-O(s)}$.

Our algorithms use a novel **LP** and combine graph decomposition techniques with online primal-dual analysis.

Keywords: Online algorithms · Matching · Primal-dual analysis · Edmonds-Gallai decomposition · Competitive ratio · Semi-streaming

1 Introduction

The field of online algorithms has had tremendous success in modeling optimization problems under uncertain input (see books [3,9,20]). The framework involves an underlying optimization problem (e.g., max matching), where the input arrives in stages (e.g., edges or vertices of a graph) and we have to make immediate and irrevocable decisions at the end of each stage (e.g., whether to match the vertex or edge). The goal is to design online algorithms with high *competitive ratio*—for the worst possible input instance the expected ratio of the online algorithm to the best algorithm in hindsight.

Most prior works on competitive analysis have only considered "single" element arrival in each stage. Since the amount of information revealed in each stage

E. Lee—This work was partially supported by Samsung Scholarship and Simons Award for Graduate Students in Theoretical Computer Science.

S. Singla—This work was supported by CMU Presidential Fellowship and NSF awards CCF-1319811, CCF-1536002, and CCF-1617790.

© Springer International Publishing AG 2017
F. Eisenbrand and J. Koenemann (Eds.): IPCO 2017, LNCS 10328, pp. 355–367, 2017.
DOI: 10.1007/978-3-319-59250-3_29

is small, the interesting regime is when the number of stages are *large* (linear in input size). Although powerful, this model often becomes too pessimistic and algorithms with good competitive ratios cannot be obtained. Here we consider the alternate *online batch arrival model*, where a "large" portion of the input (batch) arrives in each stage and the algorithm makes an irrevocable decision at the end of each stage. For a single stage this model captures the offline optimization problem and for a linear number of stages it captures the standard online model. Can we obtain competitive ratios better than the standard online model for a "small" number of stages, say even two stages?

The motivation to study online batch arrival model is based on the fact that in many scenarios that involve decision making under uncertainty, it is conceivable that instead of making an irrevocable decision for each arrival, the decision-maker prefers to gather some information for a certain amount of time and make a collective decision based on it. Indeed, multistage and especially two-stage robust/stochastic optimization problems have been actively studied in both computer science and operations research (see recent paper of Golovin et al. [12] and references therein, or a survey of Swamy and Shmoys [22]).

In this paper we study the online matching problem in the batch arrival model, where edges of a graph arrive in s stages/batches. For the basic online model where edges arrive one-by-one, its competitive ratio is not well understood. In particular, it is still open that the competitive ratio is strictly bigger than $1/2$ or not, which is achieved by a simple greedy algorithm. We prove that in our online batch arrival model, the competitive ratio is strictly bigger than $1/2$ for any fixed number of stages. In particular, when $s = 2$, the tight competitive ratio is exactly $2/3$. For $s = 2$, we also present a new LP relaxation that guarantees *instance optimality*, which means that our algorithm's decision in the first stage is optimal over the arbitrary choice of the second batch.

Our algorithms use classical tools from matching theory, such as Edmonds-Gallai decomposition and TDIness of the matching polytope, combined with carefully chosen parameters to prove large competitive ratios (some inequalities are computer-assisted). These positive results imply that our online batch arrival model allows interesting algorithmic ideas to work, and may not be as pessimistic as the original online model.

1.1   Our Model and Results

Online matching in the *vertex-arrival model* started with the seminal work of Karp et al. [15]. In this setting, vertices on one side of a bipartite graph are revealed one-by-one along with their edges to the other side, and the problem is to immediately and irrevocably match this vertex. Since this problem occupies a central position in online algorithms and has many applications in online advertisement, many of its variants have been studied in great depth (see survey [20]). This includes problems like AdWords [4,5,21], vertex-weighted [1,6], edge-weighted [13,17], stochastic matching [8,19], random vertex arrival [11,14,18], and vertex arrival on both sides [2,24].

Even though there is a long list of work in the vertex arrival model, no non-trivial algorithms are known in the equally natural *edge-arrival model*. Here edges of a (bipartite) graph are revealed one-by-one and the online problem is to immediately and irrevocably decide whether to *pick* the revealed edge into a matching. The best known algorithm is greedy, which picks an edge whenever possible and is half-competitive. Even when edges incident to a vertex are revealed together, it already captures online bipartite matching with vertex arrival on both sides, where nothing more than half is known [7,24].

The *two-stage (fractional) bipartite matching problem* is formally defined as follows. Edges of a bipartite graph $G = ((U_1, U_2), E)$ are revealed in two stages: Stage 1 reveals a subgraph $G^{(1)} = ((U_1, U_2), E^{(1)})$ of G and we need to immediately and irrevocably decide which of its edges to pick into a (fractional) matching $X^{(1)}$, without any knowledge of the remaining edges. Unmatched Stage 1 edges then disappear. Stage 2 reveals the remaining edges $E^{(2)} = E \backslash E^{(1)}$ of G as a subgraph $G^{(2)} = ((U_1, U_2), E^{(2)})$, and we (fractionally) pick a subset $X^{(2)}$ of them while ensuring that $X^{(1)} \cup X^{(2)}$ forms a (fractional) matching. This paper gives randomized algorithms that maximize the competitive ratio for this problem, i.e. ratio of the expected size of (fractional) matching obtained by the algorithm to the size of the maximum matching in G. In the integral version, the algorithm is allowed to use internal randomness, since otherwise the optimal competitive ratio is half. Note that the optimal competitive ratio for the fractional version is at least that of the integral version.

Our first main result is for the two-stage fractional bipartite matching problem. We say an algorithm is *instance optimal* if given the first stage graph $G^{(1)}$, it outputs a fractional matching $X^{(1)}$ that achieves the optimal competitive ratio over the adversarial choice of $G^{(2)}$.

Theorem 1. *There exists an instance optimal algorithm for the two-stage fractional bipartite matching problem.*

Although the above algorithm is instance optimal, it does not prove that for any $G^{(1)}$ the competitive ratio is more than half. Also, it does not work if one can only select an integral matching in $G^{(1)}$. We show that a 2/3-competitive algorithm is always possible and that this ratio cannot be improved. Indeed, our result is on a generalization to multiple stages. In the *s-stage general matching problem*, edges of a graph G are revealed in s stages. At the end of each stage, we immediately and irrevocably decide which of that stage's edges to pick into a matching, while the other edges disappear. We show that one can still beat half for a small number of stages.

Theorem 2. *There exists a $(\frac{1}{2} + \frac{1}{2^{s+1}-2})$-competitive algorithm for the s-stage integral bipartite matching problem. The competitive ratio $\frac{2}{3}$ for $s = 2$ is information-theoretically tight.*

We also prove similar results for general graphs.

Theorem 3. *There exists a* $(\frac{1}{2} + \frac{1}{2^{O(s)}})$-*competitive algorithm for the s-stage integral general matching problem. For the two-stage fractional general matching problem, there exists a 0.6-competitive algorithm.*

Our proofs for bipartite matching results extend to corresponding *multistage bipartite vertex cover* results[1]. We describe the problem and prove these corollaries in the full version.

1.2   Our Techniques

Consider a simple example where Stage 1 reveals a single edge (u, v). Should our algorithm pick this edge irrevocably? Suppose it does not, then no edge might appear in Stage 2, which makes the competitive ratio 0. On the other hand, if it picks (u, v) then Stage 2 might reveal two edges (u', u) and (v, v'), where $u' \neq v'$. Now the maximum matching has size two, but the algorithm only picks a single edge, which is half-competitive. This example already shows that no deterministic algorithm can be more than half competitive and that no randomized algorithm can be more than 2/3 competitive—picking (u, v) with probability 2/3 is optimal.

A natural extension of the above algorithm is to pick a (carefully chosen) maximum matching with probability 2/3. It turns out that picking a maximum matching with probability δ for any $\delta \in (0, 1)$ fails to achieve a 2/3-competitive ratio, regardless of how the maximum matching is chosen. This establishes the fact that different parts of the graph must use "different local distributions" to sample a matching. On the other hand, another important consideration is to avoid matching one vertex too much, since otherwise the adversary can ensure that the optimal edge indeed appears in Stage 2. Intuitively, the vertices should somehow be "uniformly matched". Our algorithms balance the above two seemingly contradictory objects by exploiting graph decomposition techniques and carefully chosen probability distributions.

We construct an online primal-dual algorithm, which finds online both a random matching $X^{(1)}$ and a dual solution $Y^{(1)}$ (vertex-cover) such that for any $G^{(2)}$ we can pick $X^{(2)} \subseteq E^{(2)}$ and a dual solution $Y^{(2)}$ (where $X^{(1)} \cup X^{(2)}$ is a matching) that satisfy the following two properties:

(i) ℓ_1 norm of $Y := Y^{(1)} + Y^{(2)}$ is the same as the cardinality of set $X := X^{(1)} \cup X^{(2)}$.
(ii) In expectation the dual solution $Y := Y^{(1)} + Y^{(2)}$ approximately satisfies every dual constraint, i.e., covers every edge of G by at least 2/3.

Relaxed complementary slackness conditions now imply that the algorithm is 2/3 competitive [23].

We view our technical contribution in two categories. The instance optimal algorithm for two-stage fractional bipartite matching uses a novel **LP** that computes both a matching and a vertex-cover (dual) simultaneously in the primal

---

[1] For general graphs, approximating vertex cover more than half is UGC hard even when the entire graph is given [16].

linear program. The **LP** allows both the algorithm and the adversary to interpret their optimal strategies. On the other hand, to prove concrete competitive ratios for various models, our technical contribution lies in the design of algorithms themselves, which are based on graph decomposition techniques and carefully chosen probability distributions to ensure Properties (i) and (ii). We believe these ideas to be useful for future research on online matching. In the following, we give brief overview of our results in more details.

Fractional Bipartite Matching: Instance Optimality using a New LP.
We write a linear program on $G^{(1)}$ to solve two-stage fractional bipartite matching problem. Our contribution in proving Theorem 1 is a new technique that strengthens the linear program for an online problem by moving the dual constraints (approximate edge-coverage) to the primal linear program. We believe that this technique might be of independent interest and will have other applications. Since our solutions are fractional, we use lower case letters x and y instead of X and Y. We maximize the competitive ratio α such that there exists a fractional matching $x^{(1)}$ in $E^{(1)}$ and a fractional vertex cover dual $y^{(1)}$ that satisfies: (a) ℓ_1 norms of $x^{(1)}$ and $y^{(1)}$ are equal and (b) every edge in $E^{(1)}$ is α-approximately covered by $y^{(1)}$. It turns out that these constraints are necessary but not sufficient by themselves. This means that for the optimal ratio α^* of the linear program, one can provide a Stage 2 graph $G^{(2)}$ where the algorithm cannot be more than α^* competitive; however, there might be graphs where α^* is not achievable by the algorithm. To further strengthen this linear program and prove our theorem, we add new constraints that force the dual y to cover "highly-matched" vertices in Stage 1.

Integral Bipartite Matching: Using Bipartite Matching Skeleton of [10]. For two-stage integral bipartite matching problem, the natural approach of rounding the instance optimal fractional matching solution fails. This has been also observed in previous online matching results [24]. Indeed, there is an example where going from fractional to integral setting strictly decreases the competitive ratio. Our first observation is that in the special case where $E^{(1)}$ contains a perfect matching, the algorithm that selects a perfect matching in $E^{(1)}$ w.p. 2/3, and no edge otherwise, is 2/3 competitive. To prove this we construct $Y^{(1)}$ by giving every matched vertex a value of 1/2 (in expectation 1/3). With simple case analysis, we show that for any $G^{(2)}$ we can always find $X^{(2)}$ and $Y^{(2)}$ that satisfy Properties (i) and (ii).

To obtain a 2/3-competitive algorithm for any bipartite graph, we use a decomposition into bipartite matching skeleton due to Goel et al. [10]. It partitions the vertices of $G^{(1)}$ into disjoint *expanding pairs* (S_j, T_j) that satisfy $\alpha_j \cdot |N(S) \cap T_j| \geq |S|$ for any $S \subseteq S_j$; here j is an integer, $0 < \alpha_j \leq 1$, and $N(S)$ denotes the set of neighbors of S in $G^{(1)}$ (see Sect. 2). For each expanding pair (S_j, T_j), our online primal-dual algorithm finds a probability δ_j with which it picks a random maximum matching in (S_j, T_j), with some correlation between different pairs, and no edge in (S_j, T_j) otherwise. Moreover, we find ϵ_j that tells us how to distribute the mass of any picked edge between its vertices in the dual solution $y^{(1)}$. Some careful case analysis allows us to show that for any $E^{(2)}$ one can obtain both $X^{(2)}$ and $y^{(2)}$ that satisfy Properties (i) and (ii).

Integral General Matching: Using a New General Matching Skeleton.
To beat half for two-stage general graph matching, we rely on Edmonds-Gallai decomposition. It gives us a characterization of any maximum matching in $G^{(1)}$ by partitioning the vertices of $G^{(1)}$ into three sets C, A, and D (see Sect. 2.2), where vertices in A form a "bridge" between vertices in D and C and the subgraph $G^{(1)}(C)$ of $G^{(1)}$ induced on C contains a perfect matching. For $G^{(1)}(C)$, as above, our algorithm again picks a perfect matching in C w.p. 2/3, and no edge otherwise, while distributing the duals equally to all the vertices in C. Most of our effort goes in designing an algorithm for the induced subgraph $G^{(1)}(A \cup D)$.

The crucial difference between bipartite and general matching is that D is no longer independent and any maximum matching contains edges inside D. We choose $D' \subseteq D$ and apply our bipartite matching algorithm to the bipartite graph induced by $A \cup D'$ (ignoring edges inside A). Finally, we match edges inside D and construct duals to satisfy Properties (i) and (ii). Special care is taken for vertices in D' since they may be matched by both procedures. Analysis involves more technical work since the number of dual variables for general graph matching is exponential.

2   Preliminaries and Notation

In the s-stage matching problem, for each Stage i ($1 \leq i \leq s$), the graph $G^{(i)} = (V, E^{(i)})$ is given. Let $(G^{(1)} \cup \cdots \cup G^{(i)})$ denote the graph $(V, E^{(1)} \cup \cdots \cup E^{(i)})$.

For the integral matching problem, in stage i, the algorithm is supposed to return $X^{(i)} \subseteq E^{(i)}$ such that $X^{(1)} \cup \cdots \cup X^{(i)}$ is a matching in $G^{(1)} \cup \cdots \cup G^{(i)}$. The algorithm is allowed to use internal randomness. For the fractional matching problem, in stage i, the algorithm is supposed to return $x^{(i)} \in [0,1]^{E^{(i)}}$ such that $x^{(1)} + \cdots + x^{(i)}$ is in the matching polytope of $G^{(1)} \cup \cdots \cup G^{(i)}$. By definition, the competitive ratio for the integral matching problem is at most that of the fractional matching problem.

2.1   Notation

Let $G = (V, E)$ be an arbitrary graph. For $S \subseteq V$, let $G(S)$ be the subgraph induced by S and let $N(S) := \{v \in V \backslash S : (u,v) \in E \text{ for some } u \in S\}$. Let $G \backslash S := G(V \backslash S)$. Let $E(S)$ be the set of edges of $G(S)$. Let $o(G)$ be the number of odd components in G. Call an odd component $S \subseteq V$ *factor-critical* if for any $s \in S$, the induced subgraph $G(S \backslash \{s\})$ has a perfect matching.

2.2   Graph Decompositions

For bipartite graphs, Goel et al. [10] showed the following bipartite matching skeleton. Since no such result was known for general graphs, in the full version we show a general matching skeleton using the classical Edmonds-Gallai decomposition.

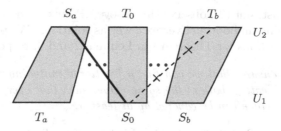

Fig. 1. Bipartite matching skeleton [10]. Solid edges can exist but not dashed edges.

Lemma 1 (Bipartite matching skeleton [10]). *Let $G = ((U_1, U_2), E)$ be a bipartite with no isolated vertex. There exists a partition of the vertices (see Fig. 1) into pairs of subsets $\{(S_j, T_j)\}_j$, where j is an integer in the interval $[a, b]$ for integers $a \leq 0 \leq b$, such that*

1. *$S_j \subseteq U_1$ and $T_j \subseteq U_2$ for $j \geq 0$, and vice versa for $j < 0$.*
2. *$|T_j| = \frac{1}{\alpha_j}|S_j|$ and for any $P \subseteq S_j$, one has $|N(P) \cap T_j| \geq \frac{1}{\alpha_j}|P|$.*
3. *$\alpha_a < \alpha_{a+1} \cdots < \alpha_0 = 1 > \cdots > \alpha_{b-1} > \alpha_b$.*
4. *There exists a fractional matching between S_j and T_j such that vertices in S_j are perfectly matched and vertices in T_j are exactly α_j matched. Call (S_j, T_j) an α_j-expanding pair.*
5. *There is no edge of G between vertices in S_j and T_k for j, k where $\alpha_j > \alpha_k$.*
6. *There is no edge of G between vertices in T_j and T_k for any j, k.*

3   Instance Optimal Two-stage Fractional Bipartite Matching

In this section, we present a polynomial time instance optimal algorithm for the two-stage fractional bipartite matching, proving Theorem 1. Recall that an algorithm is *instance optimal* if for every $G^{(1)}$, it is guaranteed to achieve the optimal competitive ratio *given $G^{(1)}$*.

Theorem 4. *For any bipartite $G^{(1)}$, the following **LP** computes the optimal competitive ratio given $G^{(1)}$.*

$$\max \quad \alpha$$

$$\text{s.t.} \quad f_u = \sum_{v \in N_1(u)} x_{u,v}^{(1)} \qquad \forall u \in V$$

$$f_u \leq 1, \qquad \forall u \in V$$

$$\sum_{(u,v) \in E^{(1)}} x_{u,v}^{(1)} \geq \sum_u y_u^{(1)},$$

$$y_u^{(1)} + y_v^{(1)} \geq \alpha, \qquad \forall (u,v) \in E^{(1)} \quad (1)$$

$$y_u^{(1)} \geq f_u - (1 - \alpha), \qquad \forall u \in V \quad (2)$$

$$x_{u,v}^{(1)}, y_u^{(1)} \geq 0, \qquad \forall u, v \in V$$

In absence of Constraint (2), observe that above LP outputs $\alpha = 1$ with optimal maximum matching and optimal vertex cover as solutions. We prove sufficiency and necessity directions for Theorem 4 in Lemmas 2 and 3, respectively.

Lemma 2. *LP ensures that we can find a fractional matching x and fractional vertex cover certificate y, both of the same value, such that every edge appearing in first or second stage can be covered by at least α.*

Proof. Consider any 2nd stage extension LP for a given first stage solution $x^{(1)}$. Also, consider its dual.

$$
\begin{array}{ll}
\max \quad \sum_{(u,v) \in E^{(2)}} x_{u,v}^{(2)} & \quad \min \quad \sum_u y'_u(1 - f_u) \\[2mm]
\text{s.t.} \quad \sum_{v \in N_2(u)} x_{u,v}^{(2)} \leq 1 - f_u, \quad \forall u \in V & \quad \text{s.t.} \quad y'_u + y'_v \geq 1, \quad \forall(u,v) \in E^{(2)} \qquad (3) \\[2mm]
x_{u,v}^{(2)} \geq 0, \quad \forall u, v \in V & \quad \phantom{\text{s.t.}} \quad y'_u \geq 0, \quad \forall u \in V
\end{array}
$$

For any vertex u, define second stage vertex cover $y_u^{(2)}$ to be $y'_u(1 - f_u)$. Hence the fractional vertex cover y is defined as $y_u := y_u^{(1)} + y_u^{(2)} = y_u^{(1)} + y'_u(1 - f_u)$. It can be easily verified that $\|y\|_1$ is the same as the obtained fraction matching. Equation (1) tells that any first stage edge is α covered by y. We next show that any second stage edge (u, v) is also α covered by y. This is because $y_u + y_v$

$$
\begin{aligned}
&= y_u^{(1)} + y_v^{(1)} + y'_u(1 - f_u) + y'_v(1 - f_v) \\
&\geq y_u^{(1)} + y_v^{(1)} + (y'_u + y'_v)(1 - \max\{f_u, f_v\}) \\
&\geq -(1 - \alpha) + 1 = \alpha \qquad\qquad\qquad\qquad \text{(using Eqs. (2) and (3))}.
\end{aligned}
$$

Lemma 3. *LP is tight, i.e. we can produce a Stage 2 graph s.t. no algorithm can be better than α competitive.*

Proof. We prove by contradiction and consider any decision x^* at the end of the first stage (this also fixes f_u^*) by an optimal algorithm with competitive ratio $\beta > \alpha$. We note that the optimal value of the following LP is greater than $\sum x_{u,v}^*$ as otherwise we get a feasible solution to **LP** with value β, and this is a contradiction that α is the optimal value of **LP**.

$$
\begin{array}{lll}
\min & \sum_u y'_u & \\[2mm]
\text{s.t.} & y'_u + y'_v \geq \alpha, & \forall(u,v) \in E^{(1)} \\[2mm]
& y'_u \geq f_u^* - (1 - \alpha), & \forall u \in V_1 \\[2mm]
& y'_u \geq 0, & \forall u \in V
\end{array}
$$

Also, consider its dual linear program.

$$\max \qquad \sum_{(u,v)\in E^{(1)}} \alpha\, Z_{u,v} + \sum_u (f_u^* - (1-\alpha))\, Y_u \qquad\qquad (4)$$

$$\text{s.t.} \qquad \sum_{v\in N_1(u)} Z_{u,v} + Y_u \le 1, \qquad\qquad \forall u \in V_1$$

$$Z_{u,v}, Y_u \ge 0, \qquad\qquad \forall u, v \in V_1.$$

Proposition 1. *The above dual linear program has an optimal integral solution.*

We prove the proposition in the full version. Given the solution to the above dual LP, the second stage graph consists $\sum Y_u$ disjoint edges, each adjacent to exactly one vertex with $Y_u = 1$. Note that due to Eq. 4, edges with $Z_{u,v} = 1$ can never be adjacent to a vertex with $Y_u = 1$. Hence the optimum matching for this two stage graph is at least $\sum_u Y_u + \sum_{(u,v)\in E^{(1)}} Z_{u,v}$. On the other hand, the two stage algorithm's value is $\sum_u x_{u,v}^* + \sum_u Y_u (1 - f_u^*)$. Combining,

$$\alpha\,\mathsf{OPT} - \mathsf{ALG} \ge \alpha \left(\sum_u Y_u + \sum_{(u,v)\in E^{(1)}} Z_{u,v}\right) - \left(\sum_{u,v} x_{u,v}^* + \sum_u Y_u (1 - f_u^*)\right)$$

$$= \left(\sum_{(u,v)\in E^{(1)}} \alpha\, Z_{u,v} + \sum_u (f_u^* - (1-\alpha))\, Y_u\right) - \sum_{u,v} x_{u,v}^* \quad > 0.$$

4   Two-stage Integral Bipartite Matching

Even though the previous section gave us an instance optimal two-stage fractional bipartite matching solution, it does not prove that for any $G^{(1)}$ the competitive ratio is more than half. Also, it does not work if one can only select an integral matching in $G^{(1)}$. In this section, we show that the optimal competitive ratio for two-stage integral bipartite matching is exactly 2/3, proving Theorem 2 for $s = 2$. We already know from Sect. 1 that no algorithm can be better than 2/3-competitive for two-stage integral bipartite matching problem. To prove the other direction, the idea is to find matching $X^{(1)}$ in a way that we have a corresponding fractional dual solution $Y^{(1)}$ such that for any Stage 2 graph $G^{(2)}$, we can find a matching $X^{(2)}$ and a dual $Y^{(2)}$ where in expectation $Y^{(1)} + Y^{(2)}$ covers every edge in G by 2/3.

4.1   Warmup: $G^{(1)}$ Contains a Perfect Matching

For intuition, we first analyze a simple case where $G^{(1)}$ contains a perfect matching. Consider the algorithm that picks the perfect matching into $X^{(1)}$ w.p. 2/3, and no edge otherwise. In Stage 2, the algorithm picks the maximum possible matching $X^{(2)} \in E^{(2)}$ such that $X^{(1)} \cup X^{(2)}$ is a matching. This is equivalent to finding a maximum matching in $G^{(2)}$ after ignoring the vertices matched in $X^{(1)}$.

To prove that the above algorithm is 2/3 competitive, for any vertex u we set $Y_u^{(1)} = 1/2$ whenever it's matched. For Stage 2, since bipartite maximum matching is equivalent to bipartite vertex cover, it also gives us integral vertex dual $Y^{(2)}$ such that $|Y^{(2)}| = |X^{(2)}|$ and every edge in $G^{(2)}$, with none of its vertices matched in $X^{(1)}$, is covered by $Y^{(2)}$. To prove that the algorithm is 2/3 competitive, we show that for any edge $(u, v) \in E^{(1)} \cup E^{(2)}$,

$$\mathsf{E}[Y_u] + \mathsf{E}[Y_v] = \mathsf{E}[Y_u^{(1)}] + \mathsf{E}[Y_v^{(1)}] + \mathsf{E}[Y_u^{(2)}] + \mathsf{E}[Y_v^{(2)}] \geq 2/3.$$

For $(u, v) \in E^{(1)}$, the above equation is simply true because $\mathsf{E}[Y_u^{(1)}] + \mathsf{E}[Y_v^{(1)}] = \Pr[\text{Perfect matching picked}] \cdot (\frac{1}{2} + \frac{1}{2}) = \frac{2}{3}(\frac{1}{2} + \frac{1}{2}) = \frac{2}{3}$.

Now, consider an edge $(u, v) \in E^{(2)}$. Consider first the case where both u and v have an edge incident to them in Stage 1. Then, similar to above, we have $\mathsf{E}[Y_u^{(1)}] + \mathsf{E}[Y_v^{(1)}] = \frac{2}{3}(\frac{1}{2} + \frac{1}{2}) = \frac{2}{3}$. So WLOG assume v has no edge incident to it in $E^{(1)}$. Now,

$$\mathsf{E}[Y_u] + \mathsf{E}[Y_v] = \Pr[u \text{ is matched in } X^{(1)}] \cdot \mathsf{E}[Y_u^{(1)} + Y_v^{(1)} \mid u \text{ is matched in } X^{(1)}] +$$
$$\Pr[u \text{ is not matched in } X^{(1)}] \cdot \mathsf{E}[Y_u^{(2)} + Y_v^{(2)} \mid u \text{ is not matched in } X^{(1)}]$$
$$= \Pr[u \text{ is matched in } X^{(1)}] \cdot \frac{1}{2} + \Pr[u \text{ is not matched in } X^{(1)}] \cdot 1$$
$$\geq \frac{2}{3} \qquad (\text{because } \Pr[u \text{ is matched in } X^{(1)}] \leq 2/3).$$

4.2 Any Bipartite Graph $G^{(1)}$

Algorithm and Construction of Duals. The algorithm starts by constructing a matching skeleton for $G^{(1)}$ as described in Sect. 2.2. For $j \in \{a, \ldots, -1, 0, 1, \ldots, b\}$, let (S_j, T_j) denote the obtained expanding pair with expansion α_j between 0 and 1. We define

$$\delta_j := \frac{3 - \alpha_j}{3} \quad \text{and} \quad \epsilon_j := \frac{2 - \alpha_j}{3 - \alpha_j}.$$

The algorithm chooses a uniformly random r between $[0, 1]$ and picks a random maximum matching between all (S_j, T_j) with $r < \delta_j$. Note that picking a matching in (S_j, T_j) implies every vertex in S_j is matched w.p. 1 and each vertex in T_j is matched w.p. probability α_j (not independently). The dual variables $Y^{(1)}$ are given values in a natural way: for any edge (u, v) in (S_j, T_j) picked into matching $X^{(1)}$, we assign $Y_u^{(1)} = \epsilon_j$ and $Y_v^{(1)} = 1 - \epsilon_j$. This clearly satisfies $\|Y^{(1)}\|_1 = |X^{(1)}|$.

In Stage 2, the algorithm picks the maximum possible matching $X^{(2)} \in E^{(2)}$ such that $X^{(1)} \cup X^{(2)}$ is a matching. This is equivalent to finding a maximum matching in $G^{(2)}$ after ignoring the vertices matched in $X^{(1)}$. Since bipartite maximum matching is equivalent to bipartite vertex cover, this step also gives us integral vertex dual $Y^{(2)}$ with $|Y^{(2)}| = |X^{(2)}|$ and every edge in $G^{(2)}$, with none of its vertices matched in $X^{(1)}$, is covered by $Y^{(2)}$. In fact, the next section shows that for any edge $(u, v) \in E^{(1)} \cup E^{(2)}$, we have

$$\mathsf{E}[Y_u] + \mathsf{E}[Y_v] = \mathsf{E}[Y_u^{(1)}] + \mathsf{E}[Y_v^{(1)}] + \mathsf{E}[Y_u^{(2)}] + \mathsf{E}[Y_v^{(2)}] \geq 2/3.$$

Algorithm 1. Two-stage Integeral Bipartite Matching Algorithm

    Stage 1

1: Use Lemma 1 to construct a matching skeleton of $G^{(1)}$ with expanding pairs (S_j, T_j) and expansion α_j.

2: Define $\delta_j := \frac{3 - \alpha_j}{3}$ and generate a uniformly random real $r \in [0, 1]$.

3: **for** each j **do**

4:     **if** $r < \delta_j$ **then**

5:         Using Property 4 in Lemma 1, pick a random maximum matching between (S_j, T_j) that matches all vertices of S_j w.p. 1 & each vertex of T_j w.p. exactly α_j.

6:     **end if**

7: **end for**

8: **Stage 2**   Pick the optimal matching extension in $G^{(2)}$.

Analysis of the Online Primal-Dual Algorithm. We show $\mathsf{E}[Y]$ covers every edge in $E^{(1)} \cup E^{(2)}$ by 2/3. First, consider any Stage 1 edge (u, v). The only possible cases due to Lemma 1 are: (a) $u \in S_j$ and $v \in T_k$ for $\alpha_j \le \alpha_k$, and (b) $u \in S_j$ and $v \in S_k$.

(a) $u \in S_j$ and $v \in T_k$ for $\alpha_j \le \alpha_k$: Using linearity of expectation and noting that v is matched w.p. $\delta_k \alpha_k$,

$$\mathsf{E}[Y_u^{(1)}] + \mathsf{E}[Y_v^{(1)}] = \delta_j \epsilon_j + \delta_k \alpha_k (1 - \epsilon_k) \ge \delta_j \epsilon_j + \delta_j \alpha_j (1 - \epsilon_j) = 2/3,$$

    where the inequality is because $\delta\alpha(1 - \epsilon) = \frac{\alpha}{3}$ decreases with decrease in α.

(b) $u \in S_j$ and $v \in S_k$: Using linearity of expectation,

$$\mathsf{E}[Y_u^{(1)}] + \mathsf{E}[Y_v^{(1)}] = \delta_j \epsilon_j + \delta_k \epsilon_k \ge 2\,\delta_0 \epsilon_0 = 2/3, \text{ as } \delta\epsilon \text{ is minimum for } \alpha = 1.$$

    Next, consider any Stage 2 edge (u, v). Since Lemma 1 does not apply to Stage 2 edges, we need to consider all the following cases: (a) $u \in S_j$ and $v \in T_k$, (b) $u \in S_j$ and $v \in S_k$, (c) $u \in T_j$ and $v \in T_k$, and (d) $u \in S_j$ and v new, (e) u new and $v \in T_k$. We only discuss Case (c), and defer others to the full version.

    Case (c) ($u \in T_j$ and $v \in T_k$): WLOG assume $\alpha_j \ge \alpha_k$. In Stage 1 vertex u is matched w.p. $\delta_j \alpha_j$ and vertex v w.p. $\delta_k \alpha_k$. Using linearity of expectation,

$$\mathsf{E}[Y_u^{(1)}] + \mathsf{E}[Y_v^{(1)}] = \delta_j \alpha_j (1 - \epsilon_j) + \delta_k \alpha_k (1 - \epsilon_k) = \alpha_j/3 + \alpha_k/3. \tag{5}$$

    Also, Stage 2 dual gets value 1 if $r > \delta_k$, or v is not matched for $r \in [\delta_j, \delta_k]$, or both u, v not matched for $r < \delta_j$.

$$\mathsf{E}[Y_u^{(2)}] + \mathsf{E}[Y_v^{(2)}] = (1 - \delta_k) + (\delta_k - \delta_j)(1 - \alpha_k) + \delta_j(1 - \alpha_j)(1 - \alpha_k). \tag{6}$$

Summing (5) and (6), and substituting for δ_j and δ_k gives,

$$\mathsf{E}[Y_u] + \mathsf{E}[Y_v] = \mathsf{E}[Y_u^{(1)} + Y_u^{(2)}] + \mathsf{E}[Y_v^{(1)} + Y_v^{(2)}]$$
$$= \frac{1}{3}\left(2 + (1 - \alpha_j - \alpha_k)^2 + \alpha_j \alpha_k (1 - \alpha_j)\right) \ge 2/3.$$

References

1. Aggarwal, G., Goel, G., Karande, C., Mehta, A.: Online vertex-weighted bipartite matching & single-bid budgeted allocations. In: Proceedings of the Twenty-Second Annual ACM-SIAM Symposium on Discrete Algorithms, pp. 1253–1264 (2011)
2. Blum, A., Sandholm, T., Zinkevich, M.: Online algorithms for market clearing. J. ACM (JACM) **53**(5), 845–879 (2006)
3. Borodin, A., El-Yaniv, R.: Online Computation and Competitive Analysis. Cambridge University Press, UK (2005)
4. Buchbinder, N., Jain, K., Naor, J.S.: Online primal-dual algorithms for maximizing ad-auctions revenue. In: Arge, L., Hoffmann, M., Welzl, E. (eds.) ESA 2007. LNCS, vol. 4698, pp. 253–264. Springer, Heidelberg (2007). doi:10.1007/978-3-540-75520-3_24
5. Devanur, N.R., Hayes, T.P.: The adwords problem: online keyword matching with budgeted bidders under random permutations. In: Proceedings of the 10th ACM Conference on Electronic Commerce, pp. 71–78. ACM (2009)
6. Devanur, N.R., Jain, K., Kleinberg, R.D.: Randomized primal-dual analysis of ranking for online bipartite matching. In: Proceedings of the Twenty-Fourth Annual ACM-SIAM Symposium on Discrete Algorithms, pp. 101–107 (2013)
7. Epstein, L., Levin, A., Segev, D., Weimann, O.: Improved bounds for online preemptive matching. In: 30th International Symposium on Theoretical Aspects of Computer Science, pp. 389–399 (2013)
8. Feldman, J., Mehta, A., Mirrokni, V., Muthukrishnan, S.: Online stochastic matching: beating 1−1/e. In: 50th Annual IEEE Symposium on Foundations of Computer Science, FOCS 2009, pp. 117–126. IEEE (2009)
9. Fiat, A., Algorithms, O.: The State of the Art (LNCS) (1998)
10. Goel, A., Kapralov, M., Khanna, S.: On the communication and streaming complexity of maximum bipartite matching. In: Proceedings of the Twenty-third Annual ACM-SIAM Symposium on Discrete Algorithms, pp. 468–485. SIAM (2012)
11. Goel, G., Mehta, A.: Online budgeted matching in random input models with applications to adwords. In: Proceedings of the Nineteenth Annual ACM-SIAM Symposium on Discrete Algorithms, pp. 982–991. SIAM (2008)
12. Golovin, D., Goyal, V., Polishchuk, V., Ravi, R., Sysikaski, M.: Improved approximations for two-stage min-cut and shortest path problems under uncertainty. Math. Program. **149**(1–2), 167–194 (2015)
13. Haeupler, B., Mirrokni, V.S., Zadimoghaddam, M.: Online stochastic weighted matching: improved approximation algorithms. In: Chen, N., Elkind, E., Koutsoupias, E. (eds.) WINE 2011. LNCS, vol. 7090, pp. 170–181. Springer, Heidelberg (2011). doi:10.1007/978-3-642-25510-6_15
14. Karande, C., Mehta, A., Tripathi, P.: Online bipartite matching with unknown distributions. In: Proceedings of the Forty-Third Annual ACM Symposium on Theory of Computing, pp. 587–596. ACM (2011)
15. Karp, R.M., Vazirani, U.V., Vazirani, V.V.: An optimal algorithm for on-line bipartite matching. In: Proceedings of the Twenty-Second Annual ACM Symposium on Theory of Computing, pp. 352–358 (1990)
16. Khot, S., Regev, O.: Vertex cover might be hard to approximate to within 2-ε. J. Comput. Syst. Sci. **74**(3), 335–349 (2008)

17. Korula, N., Pál, M.: Algorithms for secretary problems on graphs and hypergraphs. In: Albers, S., Marchetti-Spaccamela, A., Matias, Y., Nikoletseas, S., Thomas, W. (eds.) ICALP 2009. LNCS, vol. 5556, pp. 508–520. Springer, Heidelberg (2009). doi:10.1007/978-3-642-02930-1_42
18. Mahdian, M., Yan, Q.: Online bipartite matching with random arrivals: an approach based on strongly factor-revealing LPs. In: Proceedings of the Forty-Third Annual ACM Symposium on Theory of Computing, pp. 597–606 (2011)
19. Manshadi, V.H., Gharan, S.O., Saberi, A.: Online stochastic matching: online actions based on offline statistics. Math. Oper. Res. **37**(4), 559–573 (2012)
20. Mehta, A.: Online matching and ad allocation. Theor. Comput. Sci. **8**(4), 265–368 (2012)
21. Mehta, A., Saberi, A., Vazirani, U., Vazirani, V.: Adwords and generalized online matching. J. ACM (JACM) **54**(5), 22 (2007)
22. Swamy, C., Shmoys, D.B.: Approximation algorithms for 2-stage stochastic optimization problems. ACM SIGACT News **37**(1), 33–46 (2006)
23. Vazirani, V.V.: Approximation Algorithms. Springer Science & Business Media, New York (2013)
24. Wang, Y., Wong, S.C.: Two-sided online bipartite matching and vertex cover: beating the greedy algorithm. In: Halldórsson, M.M., Iwama, K., Kobayashi, N., Speckmann, B. (eds.) ICALP 2015. LNCS, vol. 9134, pp. 1070–1081. Springer, Heidelberg (2015). doi:10.1007/978-3-662-47672-7_87

Budget Feasible Mechanisms on Matroids

Stefano Leonardi[1], Gianpiero Monaco[2], Piotr Sankowski[3],
and Qiang Zhang[1(✉)]

[1] Sapienza University of Rome, Rome, Italy
`leonardi@dis.uniroma1.it, csqzhang@gmail.com`
[2] University of L'Aquila, L'Aquila, Italy
`gianpiero.monaco@univaq.it`
[3] University of Warsaw, Warsaw, Poland
`sank@mimuw.edu.pl`

Abstract. Motivated by many practical applications, in this paper we study *budget feasible mechanisms* where the goal is to procure independent sets from matroids. More specifically, we are given a matroid $\mathcal{M} = (E, \mathcal{I})$ where each ground (indivisible) element is a selfish agent. The cost of each element (i.e., for selling the item or performing a service) is only known to the element itself. There is a buyer with a budget having additive valuations over the set of elements E. The goal is to design an incentive compatible (truthful) budget feasible mechanism which procures an independent set of the matroid under the given budget that yields the largest value possible to the buyer. Our result is a deterministic, polynomial-time, individually rational, truthful and budget feasible mechanism with 4-approximation to the optimal independent set. Then, we extend our mechanism to the setting of matroid intersections in which the goal is to procure common independent sets from multiple matroids. We show that, given a polynomial time deterministic blackbox that returns α-approximation solutions to the matroid intersection problem, there exists a deterministic, polynomial time, individually rational, truthful and budget feasible mechanism with $(3\alpha + 1)$-approximation to the optimal common independent set.

1 Introduction

Procurement auctions (a.k.a. reverse auctions), often carried out by governments or private companies, deal with the scenarios where a buyer would like to purchase objects from a set of sellers. These objects are not limited to physical items. For instance they can be services provided by sellers. In this work we consider the problem where a buyer with a budget is interested in a set of indivisible objects for which he has additive valuations. We assume that each object is a selfish agent. More specifically, we assume agents have quasi-linear utilities and they

This work was partially supported by FET IP project MULTIPEX 317532, polish funds for years 2011–2015 for co-financed international projects, NCN grant UMO-2014/13/B/ST6/00770 and ERC project PAAL-POC 680912. It was also partly supported by the Google Focused Award on Algorithms for Large-scale Data Analysis.

F. Eisenbrand and J. Koenemann (Eds.): IPCO 2017, LNCS 10328, pp. 368–379, 2017.
DOI: 10.1007/978-3-319-59250-3_30

are rational (i.e., they aim to maximize the differences between the payments they receive and their true costs). We also restrict ourself to the case where the buyer is constrained to purchase a subset of objects that forms an independent set with respect to an underlying matroid structure. A wide variety of research studies have shown that matroids are linked to many interesting applications, for example, auctions [2,9,13], spectrum market [17], scheduling matroids [8] and house market [14].

One challenge in such procurement auctions involves providing incentives to sellers for declaring their true costs when those costs are their *private information*. A classical mechanism, namely Vickrey-Clark-Groves (VCG) mechanism [7,11,18], provides an intuitive solution to this problem. The VCG mechanism returns a procurement that maximizes the valuation of the buyer and the payments for sellers are their externalities to the procurement. The VCG mechanism is a *truthful* mechanism, i.e., no seller will improve its utility by manipulating its cost regardless the costs declared by others. However, the VCG mechanism also has its drawbacks. One of the drawbacks, which makes VCG mechanism impractical, is that the payments to sellers could be very high. To overcome this problem two different approaches have been proposed and investigated. The first one is studying the *frugality* of mechanisms [12], which studies the minimum payment the buyer needs to pay for a set of objects when sellers are rational utility maximizers. The other approach is developing *budget feasible mechanisms* [16], where the goal is to maximize the buyer's value for the procurement under a given budget when sellers are rational utility maximizers. Singer [16] showed that budget feasible mechanisms could approximate optimal procurements that "magically" know the costs of sellers, when buyers have nondecreasing submodular valuations.

Our Results. The goal of this study is to design budget feasible mechanisms for procuring objects that form an independent set in a given matroid structure. To the best of our knowledge it is the first time that matroid constraints are considered in the budget feasible mechanisms setting examined here. Previous work was mainly devoted to different types of valuations for the buyer (see the Related Work subsection). Our results are positive. In Sect. 3 we give a deterministic, polynomial time, individually rational, truthful and budget feasible mechanism with 4-approximation to the optimal independent set (i.e., the independent set with maximum value for the buyer under the given budget) within the budget of the buyer when the buyer has additive valuations. To generalize this result we also provide a similar mechanism to procure the intersection of independent sets in multiple matroids. In particular, given a deterministic polynomial time α-approximation algorithm for the matroid intersection problems as a blackbox, in Sect. 4 we present a deterministic, polynomial time, individually rational, truthful and budget feasible mechanism with $(3\alpha + 1)$-approximation to the optimal independent set within the budget of the buyer when the buyer has additive valuations. It is also good to know the limitations (e.g. lower bounds) of such budget feasible mechanisms. In particular the lower bound to any deterministic mechanism of $1 + \sqrt{2}$ for additive valuations with one buyer presented in [6] (it is worth noticing that such lower bound do not rely on any computational or complexity assumption), suggests that our mechanisms are not far away from the optimal ones.

Due to space limitations, some of the proofs are available in the full version.

Related Work. The study of budget feasible mechanisms was initiated in [16]. It essentially focuses on the procurement auctions when sellers have private costs for their objects and a buyer aims to maximize his valuation function on subsets of objects, conditioned on that the sum of the payments given to sellers *cannot* exceed a given budget of the buyer. In particular Singer [16] considered budget feasible mechanisms when the valuation function of the buyer is nondecreasing submodular. For general nondecreasing submodular functions, Singer [16] gave a lower bound of 2 for deterministic budget feasible mechanisms and a randomized budget feasible mechanism with 112-approximation. When the valuation function of the buyer is additive, a special class of nondecreasing submodular functions, Singer [16] gave a polynomial deterministic budget feasible mechanism with 6-approximation and a lower bound of 2 for any deterministic budget feasible mechanism. All results were improved in [6], for example, a deterministic budget feasible mechanism with $2 + \sqrt{2}$-approximation and an improved lower bound of $1 + \sqrt{2}$ for any deterministic budget feasible mechanism for additive valuations were given. Furthermore, Bei et al. [3] gave a 768-approximation mechanism for XOS valuations and extended their study to Bayesian settings. Chan and Chen [5] studied budget feasible mechanisms in the settings in which each seller processes multiple copies of the objects. They gave logarithmic mechanisms for concave additive valuations and sub-additive valuations.

Budget feasible mechanisms are attractive to many communities due to their various applications. In crowdsourcing the goal is to assign skilled workers to tasks when workers have private costs. By injecting some characteristics in crowdsourcing, budget feasible mechanisms have been further developed and improved. For example, Goel et al. [10] developed budget feasible mechanisms that achieve $\frac{2e-1}{e-1}$-approximation to the optimal social welfare by exploiting the assumption that one worker has limited contribution to the social welfare. Furthermore Anari et al. [1] gave a budget feasible mechanism that achieves a competitive ratio of $1 - 1/e \approx 0.63$ by using the assumption that the cost of any worker is relatively small compared to the budget of the buyer. Another work close to ours is [4], which studies the "dual" problem of maximizing the revenue by selling the maximum independent set of a matroid. They proposed a truthful ascending auction in which a seller is constrained to sell objects that forms a basis in a matroid.

2   Preliminaries

Matroids. A matroid $\mathcal{M}$ is a pair of $(E, \mathcal{I})$ where E is a ground set of finite elements and $\mathcal{I} \subseteq 2^E$ consists of subsets of the ground set satisfying the following properties:

- Hereditary property: If $I \in \mathcal{I}$, then $J \in \mathcal{I}$ for every $J \subset I$.
- Exchange property: For any pair of sets $I, J \in \mathcal{I}$, if $|I| < |J|$, then there exists an element $e \in J$ such that $I \cup \{e\} \in \mathcal{I}$.

The sets in $\mathcal{I}$ are called *independent sets*. Given a matroid $\mathcal{M} = (E, \mathcal{I})$ and $T \subseteq E$ is a subset of E, the *restriction* of $\mathcal{M}$ to T, denoted by $\mathcal{M}|T$, is the matroid in which the ground set is T and the independent sets are the independent sets of $\mathcal{M}$ that are contained in T. That is, $\mathcal{M}|T = (T, \mathcal{I}(\mathcal{M}|T))$ where $\mathcal{I}(\mathcal{M}|T) = \{I \subseteq T : I \in \mathcal{I}\}$. Similarly, the *deletion* of $\mathcal{M}$, denoted by $\mathcal{M} \backslash T$, is the matroid in which the ground set is $E - T$ and the independent sets are the independent sets of $\mathcal{M}$ that do not contain any element in T. That is, $\mathcal{M} \backslash T = (E - T, \mathcal{I}(\mathcal{M} \backslash T))$ where $\mathcal{I}(\mathcal{M} \backslash T) = \{I \subseteq E - T : I \in \mathcal{I}\}$.

Matroid Budget Feasible Mechanisms. In an instance of the matroid budget feasible mechanism design problem, we are given a matroid $\mathcal{M} = (E, \mathcal{I})$ consisting of n ground elements, each of whom is associated with a weight $w_e \in \mathbb{R}_+$. Each element $e \in E$ is also associated with a private cost $c_e \in \mathbb{R}_+$, which is only known to the element itself. Our goal is to design a truthful mechanism that gives incentives to elements for declaring their private costs truthfully and then selects an independent set conditioned on that the total payment given to the elements does not exceed a given budget b. Given an independent set $I \in \mathcal{I}$, the value of the independent set is defined by $w(I) = \sum_{e \in I} w(e)$. We compare the value of the independent set selected by the mechanism against the value of the maximum-value independent set in which the total true cost of elements does not exceed the budget.

We use $\mathbf{w} = \langle w_1, \ldots, w_n \rangle$ to denote the weight of the ground elements and use $\mathbf{d} = \langle d_1, \ldots, d_n \rangle$ to denote the costs declared by the ground elements. Let τ be the maximum-weight element (breaking ties arbitrarily), that is, $w_\tau = \max_{e \in E} w_e$. We assume that $d_e \in \mathbb{R}_+$ and $d_e \leq b$ for any $e \in E$ since elements with costs greater than b cannot be selected by any mechanism due to the budget constraint. This also implies that no element could improve its utility by declaring $d_i > b$. Given a subset of element T, we use $\mathbf{w}_{-T}$ and $\mathbf{d}_{-T}$ to denote the weight and cost vector excluding elements in T. Similarly, we use $\mathbf{w}_T$ and $\mathbf{d}_T$ to denote the weight and cost vector only including elements in T. For each element $e \in E$, $\mathsf{bb}(e) = \frac{d_e}{w_e}$ is called the *buck-per-bang* rate for element e.[1]

A deterministic mechanism $M = (f, p)$ consists of an allocation function $f : \mathcal{M}, \mathbf{w}, \mathbf{d}, b \to I \in \mathcal{I}$ and a payment function $p : \mathcal{M}, \mathbf{w}, \mathbf{d}, b \to \mathbb{R}_+^n$. Given the weights and declared costs of the ground elements, the allocation function returns an independent set in the matroid and the payment function indicates the payments for all elements. Let $\mathbf{f}^M(\mathcal{M}, \mathbf{w}, \mathbf{d}, b)$ and $\mathbf{p}^M(\mathcal{M}, \mathbf{w}, \mathbf{d}, b)$ be the independent set and payments returned by M, respectively. If element e is in the independent set obtained by M, then $f_e^M(\mathcal{M}, \mathbf{w}, \mathbf{d}, b) = 1$. Otherwise, $f_e^M(\mathcal{M}, \mathbf{w}, \mathbf{d}, b) = 0$. It is assumed that $p_e^M(\mathcal{M}, \mathbf{w}, \mathbf{d}, b) = 0$ if

---

[1] $\frac{w_e}{c_e}$ is usually known as the *bang-per-buck* rate. To simplify the presentation, we call $\frac{d_e}{w_e}$ the *buck-per-bang* rate.

$f_e^M(\mathcal{M}, \mathbf{w}, \mathbf{d}, b) = 0$. The utility of an element is the difference between the payment received from the mechanism and its true cost. More specifically, the utility of element e is given by $u_e^M(\mathcal{M}, \mathbf{w}, \mathbf{d}, b) = p_e^M(\mathcal{M}, \mathbf{w}, \mathbf{d}, b) - f_e^M(\mathcal{M}, \mathbf{w}, \mathbf{d}, b) \cdot c_e$.

Individual Rationality: A mechanism M is *individually rational* if $p_e^M(\mathcal{M}, \mathbf{w}, \mathbf{d}, b) - f_e^M(\mathcal{M}, \mathbf{w}, \mathbf{d}, b) \cdot d_e \geq 0$ for any $\mathcal{M}$, any $\mathbf{w} \in \mathbb{R}_+^n$, any $\mathbf{d} \in \mathbb{R}_+^n$, any $b \in \mathbb{R}_+$ and any element $e \in E$. That is, no element in the selected independent set is paid less than the cost it declared.

Truthfulness: A mechanism M is *truthful* if it holds $u_e^M(\mathcal{M}, \mathbf{w}, \mathbf{d}_{-e}, c_e, b) \geq u_e^M(\mathcal{M}, \mathbf{w}, \mathbf{d}_{-e}, d_e, b)$ for any $\mathcal{M}$, any $\mathbf{w} \in \mathbb{R}_+^n$, any $\mathbf{d}_{-e} \in \mathbb{R}_+^{n-1}$, any $d_e \in \mathbb{R}_+$, any $c_e \in \mathbb{R}_+$, $b \in \mathbb{R}_+$ and any $e \in E$, where $\mathbf{d}_{-e} = \langle d_1, \ldots, d_{e-1}, d_{e+1}, \ldots, d_n \rangle$. When the context is clear, we sometimes abuse some notations. For example, here we write $u_e^M(\mathcal{M}, \mathbf{w}, \mathbf{d}_{-e}, c_e, b)$ instead of $u_e^M(\mathcal{M}, \mathbf{w}, \langle \mathbf{d}_{-e}, c_e \rangle, b)$. A truthful mechanism prevents any element improving its utility by mis-declaring its cost regardless the costs declared by other elements.

Budget Feasibility: A mechanism M is *budget feasible* if $\sum_{e \in E} p_e^M(\mathcal{M}, \mathbf{w}, \mathbf{d}, b) \leq b$ for any $\mathcal{M}, \mathbf{w} \in \mathbb{R}_+^n$, any $\mathbf{d} \in \mathbb{R}_+^n$ and any $b \in \mathbb{R}_+$.

Competitiveness: A mechanism M is α-*competitive* if $w(f^M(\mathcal{M}, \mathbf{w}, \mathbf{d}, b)) \geq \frac{1}{\alpha} w(\mathsf{OPT}(\mathcal{M}, \mathbf{w}, \mathbf{d}, b))$ for any $\mathbf{w} \in \mathbb{R}_+^n, \mathbf{d} \in \mathbb{R}_+^n$ and $b \in \mathbb{R}_+$, where $\mathsf{OPT}(\mathcal{M}, \mathbf{w}, \mathbf{d}, b)$ is the maximum-value independent set in which the total cost of the elements is at most b. We often call $\mathsf{OPT}(\mathcal{M}, \mathbf{w}, \mathbf{d}, b)$ the optimal independent set and simplify it as $\mathsf{OPT}(\mathcal{M}, b)$ throughout the paper when the weights and the costs of elements are clear. Similarly we use $\mathsf{MAX}(\mathcal{M}, \mathbf{w})$, shorten by $\mathsf{MAX}(\mathcal{M})$, to denote the maximum-value independent set in $\mathcal{M}$ without the budget constraint.

Simplifying Notations: From now on to avoid heavy notations we sometimes simplify the notations. For example we will write f^M, f_e^M, p^M, p_e^M when the inputs of the mechanism are clear. And we will use $\mathsf{OPT}(\mathcal{M} \backslash T, b)$ instead of $\mathsf{OPT}(\mathcal{M} \backslash T, \mathbf{w}_{-T}, \mathbf{d}_{-T}, b)$ to denote the optimal independent set in matroid $\mathcal{M} \backslash T$. Similarly we will use $\mathsf{OPT}(\mathcal{M}|T, b)$ instead of $\mathsf{OPT}(\mathcal{M}|T, \mathbf{w}_T, \mathbf{d}_T, b)$ to denote the optimal independent set in matroid $M|T$. Furthermore we use $\mathsf{MAX}(\mathcal{M} \backslash T)$ instead of $\mathsf{MAX}(\mathcal{M} \backslash T, \mathbf{w}_{-T})$ to denote the maximum-value independent set in $\mathcal{M} \backslash T$ without considering the costs of the elements and the budget.

3   Mechanisms for Matroids

In this section we provide our main result. We give a deterministic, polynomial time, individually rational, truthful and budget feasible mechanism that is 4-approximating the optimal independent set. Before providing the mechanism we discuss some intuition that guides us in the design of Mechanism 1. First imagine that there exists an element with a very high weight, i.e., any independent set without this element results in a poor value compared to the optimal independent set. In this case that element may strategically declare a high cost in order to increase its utility as it knows that any competitive mechanism has to select it. To avoid that this happens we remove element τ (i.e., the element with the largest weight) from the matroid via matroid deletion operation, and compare it with the independent set computed later by the mechanism. Second we observe that most of the existing budget feasible mechanisms adopt proportional payment schemes, where elements (i.e., agents) are paid proportionally according to their contribution in the solution. In other words in a proportional payment scheme there is an uniform price such that the payments for elements in the solution are the products of their contribution and this price. In addition greedy algorithms are commonly used in matroid systems. Combining these two observations our plan is to start from a high price and compute the maximum-value independent set in the matroid at each iteration. If there is enough budget to pay this independent set at the current price then we proceed to the final step of the mechanism. Otherwise we reduce the price and remove an element from the matroid. The buck-per-bang rate of that element becomes an upper bound of the payment on each contribution in the next iteration. The mechanism performs the procedure described above until the payment of the maximum-value independent set is within budget b. As we will show next, if the value of the optimal independent set does not come from a single element, we are able to retain most of the value of the optimal independent set after removing those elements. Finally, we show that returning the better solution between the maximum-value independent set found and element τ approximates the value of the optimal independent set within a factor of 4.

Mechanism 1: A budget feasible mechanism for procuring independent sets in matroids

---

    Input: $\mathcal{M} = (E, \mathcal{I}), \mathbf{w}, \mathbf{d}, b$

    Output: $\mathbf{f}, \mathbf{p}$

1 Sort elements in $E - \tau$ in a non-increasing order of buck per bang, i.e. $\mathsf{bb}(i) \geq \mathsf{bb}(j)$ if $i < j$, break ties arbitrarily;

2 Let $\mathsf{bb}(0) = +\infty$, $i = 1$ and $T = \emptyset$;

3 Set $r = \mathsf{bb}(i)$;

4 **while** $w(\mathsf{MAX}(\mathcal{M} \backslash (T \cup \tau))) \cdot r > b$ **do**

5 $T = T \cup \{i\}$ and $i = i + 1$;

6 $r = \mathsf{bb}(i)$;

7 $r = \min\{\frac{b}{w(\mathsf{MAX}(\mathcal{M} \backslash (T \cup \tau)))}, \mathsf{bb}(i - 1)\}$;

8 **if** $w(\mathsf{MAX}(\mathcal{M} \backslash (T \cup \tau))) > w_\tau$ **then**

9   For each $e \in E$, if $e \in \mathsf{MAX}(\mathcal{M} \backslash (T \cup \tau))$, $f_e = 1$ and $p_e = r \cdot w_e$. Otherwise, $f_e = 0$ and $p_e = 0$;

10 **else**

11 $f_\tau = 1, p_\tau = b$. For edge $e \in E - \tau$, $f_e = 0, p_e = 0$;

12 **return** $\mathbf{f}, \mathbf{p}$;

Theorem 3.1. *Mechanism 1 is a deterministic, polynomial time, individually rational, truthful and budget feasible mechanism that is 4-competitive against the optimal independent set given a budget.*

3.1   Approximation

Recall that T is the set of elements removed from the matroid. $\mathsf{MAX}(\mathcal{M}\backslash(T\cup\tau))$ is the independent set found when Mechanism 1 stops, and it is also the maximal-value independent set in matroid $M\backslash(T\cup\tau)$. The roadmap of the proof is to first show that, the independent set $\mathsf{MAX}(\mathcal{M}\backslash(T\cup\tau))$ well approximates the optimal independent set in matroid $M\backslash\tau$. Next we show that returning the maximum between τ and $\mathsf{MAX}(\mathcal{M}\backslash(T\cup\tau))$ gives 4-approximation to the optimal independent set in matroid $\mathcal{M}$.

Lemma 3.1. *Given any* $\mathcal{M}, \mathbf{w}, \mathbf{d}, b$, *when Mechanism 1 stops, it holds*

$$w(\mathsf{OPT}(\mathcal{M}\backslash\tau, b)) \leq 2w(\mathsf{MAX}(\mathcal{M}\backslash(T\cup\tau))) + w_\tau$$

Proof. It is trivial to see that this lemma holds when τ is the only element in matroid $\mathcal{M}$. The rest of the proof uses a similar idea in [10] and is divided into two cases depending on whether the full budget b is spent or not. Consider $E - \{\tau\}$ is partitioned into two disjoint sets, $E - \{\tau\} - T$ and T. The value of maximum-value independent set $w(\mathsf{OPT}(\mathcal{M}\backslash\tau, b))$ is bounded by

$$w(\mathsf{OPT}(\mathcal{M}|T, b)) + w(\mathsf{OPT}(\mathcal{M}\backslash(T\cup\tau), b))$$

As the buck-per-bang is at least r for every element in T, the weight of the optimal independent set given a budget b in $\mathcal{M}|T$, i.e. $w(\mathsf{OPT}(\mathcal{M}|T, b))$, is at most b/r. When the full budget is spent, the weight of independent set f^M is b/r in Mechanism 1. On the other hand, f^M is the maximum-value independent set in $\mathcal{M}\backslash(T\cup\tau)$. It implies that $w(\mathsf{MAX}(\mathcal{M}\backslash(T\cup\tau))) \geq w(\mathsf{OPT}(\mathcal{M}\backslash(T\cup\tau), b))$. The above analysis concludes that

$$w(\mathsf{OPT}(\mathcal{M}\backslash\tau, b)) \leq 2w(\mathsf{MAX}(\mathcal{M}\backslash(T\cup\tau)))$$

Now we turn to the case that some budget is left in Mechanism 1. Note that it happens because $r = \mathsf{bb}(i-1)$ (see Line 7) during the execution of Mechanism 1. Since Mechanism 1 does not stop while considering $r = \mathsf{bb}(i-1)$ at previous iterations in the loop, it implies that the maximum-value independent set found was not budget feasible at previous iteration. After removing element $i-1$, the maximum-value independent set becomes budget feasible. These together imply

$$w(\mathsf{MAX}(\mathcal{M}\backslash(T'\cup\tau))) \cdot \mathsf{bb}(i-1) > b > w(\mathsf{MAX}(\mathcal{M}\backslash(T\cup\tau))) \cdot \mathsf{bb}(i-1)$$

where $T' = T - \{i-1\}$. This further implies that budget left is at most $\mathsf{bb}(i-1) \cdot w_{i-1}$. By the similar argument as in previous case, the optimal independent set in $\mathcal{M}|T$ is at most b/r, while the value of the independent set $\mathsf{MAX}(\mathcal{M}\backslash(T\cup\tau))$ is at least $(b - \mathsf{bb}(i-1) \cdot w_{i-1})/r$, which is at least $b/r - w_{i-1}$. Therefore, we have

$$w(\mathsf{OPT}(\mathcal{M}\backslash\tau, b)) \leq 2w(\mathsf{MAX}(\mathcal{M}\backslash(T\cup\tau))) + w_{i-1}$$

Substituting w_{i-1} with w_τ completes the proof.   □

Next, we show that returning the maximum between τ and $\mathsf{MAX}(\mathcal{M}\backslash(T\cup\tau))$ is 4-competitive against the optimal independent set in $\mathcal{M}$.

Lemma 3.2. *Given any* $\mathcal{M}, \mathbf{w}, \mathbf{d}, b$, *the independent set returned by Mechanism 1, i.e., the maximum between* τ *and* $\mathsf{MAX}(\mathcal{M}\backslash(T\cup\tau))$, *is 4-competitive against the optimal independent set.*

Proof. The optimal independent set in $\mathcal{M}$ is bounded by

$$w(\mathsf{OPT}(\mathcal{M}, b)) \leq w_\tau + w(\mathsf{OPT}(\mathcal{M}\backslash\tau, b))$$

By Lemma 3.1, we have

$$w(\mathsf{OPT}(\mathcal{M}, b)) \leq 2w_\tau + 2w(\mathsf{MAX}(\mathcal{M}\backslash(T\cup\tau)))$$

Therefore, the maximum between τ and $\mathsf{MAX}(\mathcal{M}\backslash(T\cup\tau))$ approximates the optimal independent set within a factor of 4. $\qquad\qquad\square$

3.2 Truthfulness

We show that Mechanism 1 is truthful by considering following different cases.

Lemma 3.3. *The element with the maximum weight, i.e., element* τ, *could not improve his utility by declaring cost* $d_\tau \neq c_\tau$.

Lemma 3.4. *Assume an element* k *is in* T *when it declares its cost truthfully. Then, element* k *could not improve his utility by declaring a cost* $d_k \neq c_k$.

Lemma 3.5. *Assume an element* k *is in* $E - \tau - T - \mathsf{MAX}(\mathcal{M}\backslash(T\cup\tau))$ *when it declares its cost truthfully. Then, element* k *could not improve his utility by declaring a cost* $d_k \neq c_k$.

Lemma 3.6. *Assume an element* k *is in* $\mathsf{MAX}(\mathcal{M}\backslash(T\cup\tau))$ *when it declares its cost truthfully. Then, element* k *could not improve his utility by declaring a cost* $d_k \neq c_k$.

3.3 Individual Rationality

When Mechanism 1 returns τ, the utility of τ is non-negative as $c_\tau \leq b$. The utilities for other edges are zero. When Mechanism 1 returns $\mathsf{MAX}(\mathcal{M}\backslash(T\cup\tau))$, for any element $e \in \mathsf{MAX}(\mathcal{M}\backslash(T\cup\tau))$, that is, $f_e = 1$, its utility is $r \cdot w_e - c_e$ which is non-negative since $r \geq \mathsf{bb}(e)$. For other edges, their utilities are zero.

3.4 Budget Feasibility

When Mechanism 1 returns τ, it only pays b to edge τ. Hence, it is budget feasible. On the other hand, when Mechanism 1 returns $\mathsf{MAX}(\mathcal{M}\backslash(T\cup\tau))$, r is used as payment per contribution. As $r = \min\{\frac{b}{w(\mathsf{MAX}(\mathcal{M}\backslash(T\cup\tau)))}, \mathsf{bb}(i-1)\}$, it guarantees the budget feasibility.

3.5   Remarks

In Mechanism 1, we iteratively compute the maximum-value independent set (e.g. Line 4). In the case that the maximum-value independent set is not unique, we assume there is a deterministic tie-breaking rule. Note that all the results still hold under this assumption. For example, the truthfulness of the mechanism will not be compromised since the the maximum-value independent set only consider the weights of the elements that is the public knowledge.

4   Mechanisms for Matroid Intersections

In this section we extend our mechanism to matroid intersections. The matroid intersection problem (i.e., finding the maximum-value common independent set) is NP-hard in general when more than three matroids are involved. Some interesting cases of matroid intersection problems can be solved efficiently (i.e., they can be formulated as the intersection of two matroids), for example, matchings in bipartite graphs, arborescences in directed graphs, spanning forests in undirected graphs, etc. Nevertheless we point out that a very similar mechanism to the one presented in last section achieves a 4 approximation for the case when, instead of a matroid, we are given an undirected weighted (general) graph where the selfish agents are the edges of the graph and the buyer wants to procure a matching under the given budget that yields the largest value possible to him.

For general matroid intersections, our main result is the following. Given a deterministic polynomial time blackbox APX that achieves an α-approximation to k-matroid intersection problems, we provide a polynomial time, individually rational, truthful and budget feasible deterministic mechanism that is $(3\alpha + 1)$-competitive against the maximum-value common independent set. The mechanism is similar to Mechanism 1 by changing MAX to APX. It is well-known that the VCG payment rule does not preserve the property of truthfulness in the presence of approximated solutions (i.e., non-optimal outcome). However unlike the VCG mechanism, we show that Mechanism 2 preserves its truthfulness when APX is used. This result will make our contribution more practical.

4.1   Matroid Intersections

Given k-matroid $\mathcal{M}_1, \ldots, \mathcal{M}_k$, let $\mathcal{M} = (E, \mathcal{I})$ be the "true matroid" where E is the common ground elements and $\mathcal{I} = \bigcap_j \mathcal{I}_j$ is the "true independent sets". Similar as the notations we used before, let $\mathsf{OPT}(\mathcal{M}\backslash\mathcal{T}, b)$ and $\mathsf{OPT}(\mathcal{M}|\mathcal{T}, b)$ denote the optimal independent set satisfying the budget constraint in matroid $\mathcal{M}\backslash T$ and $\mathcal{M}|T$, respectively. Let $\mathsf{APX}(\mathcal{M}\backslash\mathcal{T}, b)$ be the maximum-value independent set in matroid $\mathcal{M}\backslash T$ returned by the α-approximation algorithm.

4.2   Obtaining $O(\alpha)$ Approximation

We show the following key lemma, which is similar to Lemma 3.1 and implies the approximation of our mechanism for matroid intersections.

Mechanism 2: A budget feasible mechanism for procuring independent sets in matroid intersections

---

Input: $\mathcal{M} = (E, \mathcal{I}), \mathbf{w}, \mathbf{d}, b$

Output: $\mathbf{f}, \mathbf{p}$

1 Sort elements in $E - \tau$ in a non-increasing order of buck per bang, i.e. $\mathsf{bb}(i) \geq \mathsf{bb}(j)$ if $i < j$, break ties arbitrarily;

2 Let $\mathsf{bb}(0) = +\infty$, $i = 1$ and $T = \emptyset$;

3 Set $r = \mathsf{bb}(i)$;

4 **while** $w(\mathsf{APX}(\mathcal{M}\backslash T)) \cdot r > b$ **do**

5 $\quad\big|\quad T = T \cup \{i\}$ and $i = i + 1$;

6 $\quad\big\lfloor\quad r = \mathsf{bb}(i)$;

7 $r = \min\{\frac{b}{w(\mathsf{APX}(\mathcal{M}\backslash T))}, \mathsf{bb}(i-1)\}$;

8 **if** $w(\mathsf{APX}(\mathcal{M}\backslash T)) > w_\tau$ **then**

9 $\quad\big|\quad$ For each $e \in E$, if $e \in \mathsf{APX}(\mathcal{M}\backslash T)$, $f_e = 1$ and $p_e = r \cdot w_k$. Otherwise, $f_e = 0$ and $p_e = 0$;

10 **else**

11 $\quad\big\lfloor\quad f_\tau = 1, p_\tau = b$. For edge $e \in E - \tau, f_e = 0, p_e = 0$;

12 **return** $\mathbf{f}, \mathbf{p}$;

Lemma 4.1. *Given any* $\mathcal{M}, \mathbf{w}, \mathbf{d}, b$, *when Mechanism 2 stops, it holds*

$$w(\mathsf{OPT}(\mathcal{M}\backslash\tau, b)) \leq 2 \cdot \alpha \cdot w(\mathsf{APX}(\mathcal{M}\backslash(T \cup \tau))) + \alpha \cdot w_\tau$$

Proof. The proof has the same spirit as the proof of Lemma 3.1. We consider two cases depending on whether the full budget b is spent or not. Consider $E - \{\tau\}$ is partitioned into two disjoint sets, $E - \{\tau\} - T$ and T. Similar to Lemma 3.1, when the full budget is spent, we get

$$w(\mathsf{OPT}(\mathcal{M}\backslash\tau, b)) \leq w(\mathsf{OPT}(\mathcal{M}|T, b)) + w(\mathsf{OPT}(\mathcal{M}\backslash(T \cup \tau), b))$$
$$\leq \frac{b}{r} + \alpha \cdot w(\mathsf{APX}(\mathcal{M}\backslash(T \cup \tau)))$$
$$\leq (\alpha + 1) \cdot w(\mathsf{APX}(\mathcal{M}\backslash(T \cup \tau)))$$

When there is some budget left in Mechanism 2, the analysis involves one more step compared to Lemma 3.1 although the idea is still to bound the budget left. Since Mechanism 2 does not stop when $r = \mathsf{bb}(i - 1)$, it implies that the independent set returned by APX was not budget feasible at previous iteration. It further implies that the maximum-value independent set is not budget feasible either if the payment per weight is r. After removing element $i - 1$, the independent set returned by APX becomes budget feasible when $r = \mathsf{bb}(e_{i-1})$. These together imply

$$w(\mathsf{MAX}(\mathcal{M}\backslash(T' \cup \tau))) \cdot \mathsf{bb}(i-1) \geq w(\mathsf{APX}(\mathcal{M}\backslash(T' \cup \tau))) \cdot \mathsf{bb}(i-1)$$
$$> b$$
$$> w(\mathsf{APX}(\mathcal{M}\backslash(T \cup \tau))) \cdot \mathsf{bb}(i-1)$$

where $T' = T - \{i-1\}$. As the sum of $w(\mathsf{APX}(\mathcal{M}\backslash(T \cup \tau)))$ and $w(i-1)$ is at least $\frac{1}{\alpha}$ fraction of $w(\mathsf{MAX}(\mathcal{M}\backslash(T' \cup \tau)))$, we get

$$\left(w(\mathsf{APX}(\mathcal{M}\backslash(T \cup \tau)) + w_{i-1}\right) \cdot \mathsf{bb}(i-1) \geq \frac{1}{\alpha} \cdot w(\mathsf{MAX}(\mathcal{M}\backslash(T' \cup \tau))) \cdot \mathsf{bb}(i-1) > \frac{b}{\alpha}$$

Hence, we get $w(\mathsf{APX}(\mathcal{M}\backslash(T \cup \tau))) + w_{i-1} > \frac{b}{\alpha \cdot \mathsf{bb}(i-1)}$. Finally,

$$\mathsf{OPT}(\mathcal{M}\backslash\tau, b) \leq w(\mathsf{OPT}(\mathcal{M}|T, b)) + w(\mathsf{OPT}(\mathcal{M}\backslash(T \cup \tau), b))$$
$$\leq \frac{b}{\mathsf{bb}(i-1)} + \alpha \cdot w(\mathsf{APX}(\mathcal{M}\backslash(T \cup \tau))$$
$$\leq 2 \cdot \alpha \cdot w(\mathsf{APX}(\mathcal{M}\backslash(T \cup \tau) + \alpha \cdot w_{i-1}$$

Substituting w_{i-1} with w_τ completes the proof.   □

4.3 Preserving the Truthfulness

In this section, we will show that replacing MAX by APX preserve the truthfulness of the mechanism for matroid intersections. The reason behind is that the mechanism works in a greedy fashion and at each iteration the cost declared by elements *does not* affect the independent set computed in the mechanism. The property of the truthfulness replies on the greedy approach instead of the optimality of the independent set. The proofs are similar to the proofs in Sect. 3.2.

5 Applications

Uniform Matroid. Additive valuation has been studied in the design of budget feasible mechanisms, e.g. [6,16]. In such settings a buyer would like to maximize his valuation by procuring items under the constraint that his payment is at most his budget. Our result generalizes to the case where the buyer has not only the budget constraint but also has a limit on the number of items he can buy. For example hiring people in companies is not only constraint by budgets but also limited by the office space.

Scheduling Matroid. Our mechanism could be used to purchase processing time in the context of job scheduling. One special case is the following. Each job is associated with a deadline and a profit, and requires a unit of processing time. As jobs may conflict with each other, only one job can be scheduled at the same time. The buyer would like to maximize his profit by completing jobs under the constraint that he does not spend more than his budget in purchasing processing time.

Spectrum Market. Tse and Hanly [17] showed that the set of achievable rates in a Gaussian multiple-access, known as the Cover-Wyner capacity region, forms a polymotroid. It is known there is a pseudopolynomial reduction from integral polymatroids to matroids [15]. Therefore, our mechanism can be used to purchase transmission rates by tele-communication companies.

References

1. Anari, N., Goel, G., Nikzad, A.: Mechanism design for crowdsourcing: an optimal 1-1/e competitive budget-feasible mechanism for large markets. In: 55th Annual IEEE Symposium on Foundations of Computer Science (FOCS), pp. 266–275. IEEE (2014)
2. Ausubel, L.M.: An efficient ascending-bid auction for multiple objects. Am. Econ. Rev. **94**, 1452–1475 (2004)
3. Bei, X., Chen, N., Gravin, N., Lu, P.: Budget feasible mechanism design: from prior-free to bayesian. In: Proceedings of the Forty-fourth Annual ACM Symposium on Theory of Computing (STOC), pp. 449–458. ACM (2012)
4. Bikhchandani, S., de Vries, S., Schummer, J., Vohra, R.V.: An ascending vickrey auction for selling bases of a matroid. Oper. Res. **59**(2), 400–413 (2011)
5. Chan, H., Chen, J.: Truthful multi-unit procurements with budgets. In: Liu, T.-Y., Qi, Q., Ye, Y. (eds.) WINE 2014. LNCS, vol. 8877, pp. 89–105. Springer, Cham (2014). doi:10.1007/978-3-319-13129-0_7
6. Chen, N., Gravin, N., Lu, P.: On the approximability of budget feasible mechanisms. In: Proceedings of the Twenty-Second Annual ACM-SIAM Symposium on Discrete Algorithms (SODA), pp. 685–699. SIAM (2011)
7. Clarke, E.H.: Multipart pricing of public goods. Public choice **11**(1), 17–33 (1971)
8. Demange, G., Gale, D., Sotomayor, M.: Multi-item auctions. J. Polit. Econ. **94**, 863–872 (1986)
9. Goel, G., Mirrokni, V., Leme, R.P.: Polyhedral clinching auctions and the adwords polytope. J. ACM (JACM) **62**(3), 18 (2015)
10. Goel, G., Nikzad, A., Singla, A.: Allocating tasks to workers with matching constraints: truthful mechanisms for crowdsourcing markets. In: Proceedings of the Companion Publication of the 23rd International Conference on World Wide Web Companion, pp. 279–280 (2014)
11. Groves, T.: Incentives in teams. Econometrica: J. Econometric Soc. **41**, 617–631 (1973)
12. Karlin, A.R., Kempe, D.: Beyond VCG: frugality of truthful mechanisms. In: 46th Annual IEEE Symposium on Foundations of Computer Science (FOCS), pp. 615–624. IEEE (2005)
13. Kleinberg, R., Weinberg, S.M.: Matroid prophet inequalities. In: Proceedings of the Forty-Fourth Annual ACM Symposium on Theory of Computing, pp. 123–136. ACM (2012)
14. Krysta, P., Zhang, J.: House markets with matroid and knapsack constraints. In: Proceedings of The 43rd International Colloquium on Automata, Languages and Programming (ICALP) (2016)
15. Schrijver, A.: Combinatorial Optimization. Algorithms and Combinatorics, vol. 24. Springer, Berlin (2003)
16. Singer, Y.: Budget feasible mechanisms. In: 51st Annual IEEE Symposium on Foundations of Computer Science (FOCS), pp. 765–774 (2010)
17. Tse, D.N.C., Hanly, S.V.: Multiaccess fading channels. I. Polymatroid structure, optimal resource allocation and throughput capacities. IEEE Trans. Inf. Theor. **44**(7), 2796–2815 (1998)
18. Vickrey, W.: Counterspeculation, auctions, and competitive sealed tenders. J. Financ. **16**(1), 8–37 (1961)

Deterministic Discrepancy Minimization via the Multiplicative Weight Update Method

Avi Levy, Harishchandra Ramadas[(✉)], and Thomas Rothvoss

University of Washington, Seattle, USA
{avius,rothvoss}@uw.edu, ramadas@math.washington.edu

Abstract. A well-known theorem of Spencer shows that any set system with n sets over n elements admits a coloring of discrepancy $O(\sqrt{n})$. While the original proof was non-constructive, recent progress brought polynomial time algorithms by Bansal, Lovett and Meka, and Rothvoss. All those algorithms are randomized, even though Bansal's algorithm admitted a complicated derandomization.

We propose an elegant deterministic polynomial time algorithm that is inspired by Lovett-Meka as well as the Multiplicative Weight Update method. The algorithm iteratively updates a fractional coloring while controlling the exponential weights that are assigned to the set constraints.

A conjecture by Meka suggests that Spencer's bound can be generalized to symmetric matrices. We prove that $n \times n$ matrices that are block diagonal with block size q admit a coloring of discrepancy $O(\sqrt{n} \cdot \sqrt{\log(q)})$. Bansal, Dadush and Garg recently gave a randomized algorithm to find a vector x with entries in $\{-1, 1\}$ with $\|Ax\|_\infty \le O(\sqrt{\log n})$ in polynomial time, where A is any matrix whose columns have length at most 1. We show that our method can be used to deterministically obtain such a vector.

1 Introduction

The classical setting in (combinatorial) *discrepancy theory* is that a set system $S_1, \ldots, S_m \subseteq \{1, \ldots, n\}$ over a ground set of n elements is given and the goal is to find *bi-coloring* $\chi : \{1, \ldots, n\} \to \{\pm 1\}$ so that the worst imbalance $\max_{i=1,\ldots,m} |\chi(S_i)|$ of a set is minimized. Here we abbreviate $\chi(S_i) := \sum_{j \in S_i} \chi(j)$. A seminal result of Spencer [Spe85] says that there is always a coloring χ where the imbalance is at most $O(\sqrt{n \cdot \log(2m/n)})$ for $m \ge n$. The proof of Spencer is based on the *partial coloring method* that was first used by Beck in 1981 [Bec81]. The argument applies the *pigeonhole principle* to obtain that many of the 2^n many colorings χ, χ' must satisfy $|\chi(S_i) - \chi'(S_i)| \le O(\sqrt{n \cdot \log(2m/n)})$ for all sets S_i. Then one can take the *difference* between such a pair of colorings with $|\{j \mid \chi(j) \ne \chi'(j)\}| \ge \frac{n}{2}$ to obtain a *partial coloring* of low discrepancy.

T. Rothvoss—Supported by NSF grant 1420180 with title *"Limitations of convex relaxations in combinatorial optimization"*, an Alfred P. Sloan Research Fellowship and a David & Lucile Packard Foundation Fellowship.

F. Eisenbrand and J. Koenemann (Eds.): IPCO 2017, LNCS 10328, pp. 380–391, 2017.
DOI: 10.1007/978-3-319-59250-3_31

This partial coloring can be used to color half of the elements. Then one iterates the argument and again finds a partial coloring. As the remaining set system has only half the elements, the bound in the second iteration becomes better by a constant factor. This process is repeated until all elements are colored; the total discrepancy is then given by a convergent series with value $O(\sqrt{n} \cdot \log(2m/n))$. More general arguments based on convex geometry were given by Gluskin [Glu89] and by Giannopoulos [Gia97], but their arguments still relied on a pigeonhole principle with exponentially many pigeons and pigeonholes and did not lead to polynomial time algorithms.

In fact, Alon and Spencer [AS08] even conjectured that finding a coloring satisfying Spencer's theorem would by intractable. In a breakthrough, Bansal [Ban10] showed that one could set up a *semi-definite program* (SDP) to find at least a vector coloring, using Spencer's Theorem to argue that the SDP has to be feasible. He then argued that a random walk guided by updated solutions to that SDP would find a coloring of discrepancy $O(\sqrt{n})$ in the balanced case $m = n$. However, his approach needed a very careful choice of parameters.

A simpler and truly constructive approach that does not rely on Spencer's argument was provided by Lovett and Meka [LM12], who showed that for $x^{(0)} \in [-1, 1]^n$, any polytope of the form $P = \{x \in [-1, 1]^n : |\langle v_i, x - x^{(0)} \rangle| \leq \lambda_i \ \forall i \in [m]\}$ contains a point that has at least half of the coordinates in $\{-1, 1\}$. Here it is important that the polytope P is large enough; if the normal vectors v_i are scaled to unit length, then the argument requires that $\sum_{i=1}^{m} e^{-\lambda_i^2/16} \leq \frac{n}{16}$ holds. Their algorithm surprisingly simple: start a Brownian motion at $x^{(0)}$ and stay inside any face that is hit at any time. They showed that this random walk eventually reaches a point with the desired properties.

More recently, the third author provided another algorithm which simply consists of taking a random Gaussian vector x and then computing the nearest point to x in P. In contrast to both of the previous algorithms, this argument extends to the case that $P = Q \cap [-1, 1]^n$ where Q is any symmetric convex set with a large enough Gaussian measure.

However, all three algorithms described above are randomized, although Bansal and Spencer [BS13] could derandomize the original arguments by Bansal. They showed that the random walk already works if the directions are chosen from a 4-wise independent distribution, which then allows a polynomial time derandomization.

In our algorithm, we think of the process more as a *multiplicative weight update* procedure, where each constraint has a weight that increases if the current point moves in the direction of its normal vector. The potential function we consider is the sum of those weights. Then in each step we simply need to select an update direction in which the potential function does not increase.

The multiplicative weight update method is a meta-algorithm that originated in game theory but has found numerous recent applications in theoretical computer science and machine learning. In the general setting one imagines having a set of experts (in our case the set constraints) that are assigned an exponential weight that reflects the value of the gain/loss that expert's decisions had in previous rounds. Then in each iteration one selects an update, which can be a convex combination of experts, where the convex coefficient is proportional to the current weight of the expert[1]. We refer to the very readable survey of Arora et al. [AHK12] for a detailed discussion.

1.1 Related Work

If we have a set system $S_1, \ldots, S_m$ where each element lies in at most t sets, then the partial coloring technique described above can be used to find a coloring of discrepancy $O(\sqrt{t} \cdot \log n)$ [Sri97]. A linear programming approach of Beck and Fiala [BF81] showed that the discrepancy is bounded by $2t - 1$, independent of the size of the set system. On the other hand, there is a non-constructive approach of Banaszczyk [Ban98] that provides a bound of $O(\sqrt{t \log n})$ using convex geometry arguments. Only very recently, a corresponding algorithmic bound was found by Bansal et al. [BDG16]. A conjecture of Beck and Fiala says that the correct bound should be $O(\sqrt{t})$. This bound can be achieved for the vector coloring version, see Nikolov [Nik13].

More generally, the theorem of Banaszczyk [Ban98] shows that for any convex set K with Gaussian measure at least $\frac{1}{2}$ and any set of vectors $v_1, \ldots, v_m$ of length $\|v_i\|_2 \leq \frac{1}{5}$, there exist signs $\varepsilon_i \in \{\pm 1\}$ so that $\sum_{i=1}^{m} \varepsilon_i v_i \in K$.

A set of k permutations on n symbols induces a set system with kn sets given by the prefix intervals. One can use the partial coloring method to find a $O(\sqrt{k} \log n)$ discrepancy coloring [SST], while a linear programming approach gives a $O(k \log n)$ discrepancy [Boh90]. In fact, for any k one can always color half of the elements with a discrepancy of $O(\sqrt{k})$ — this even holds for each induced sub-system [SST]. Still, [NNN12] constructed 3 permutations requiring a discrepancy of $\Theta(\log n)$ to color all elements.

Also the recent proof of the Kadison-Singer conjecture by Marcus et al. [MSS13] can be seen as a discrepancy result. They show that a set of vectors $v_1, \ldots, v_m \in \mathbb{R}^n$ with $\sum_{i=1}^{m} v_i v_i^T = I$ can be partitioned into two halves S_1, S_2 so that $\sum_{i \in S_j} v_i v_i^T \preceq (\frac{1}{2} + O(\sqrt{\varepsilon}))I$ for $j \in \{1, 2\}$ where $\varepsilon = \max_{i=1,\ldots,m}\{\|v_i\|_2^2\}$ and I is the $n \times n$ identity matrix. Their method is based on interlacing polynomials; no polynomial time algorithm is known to find the desired partition.

For a symmetric matrix $A \in \mathbb{R}^{m \times m}$, let $\|A\|_{op}$ denote the largest singular value; in other words, the largest absolute value of any eigenvalue. The discrepancy question can be generalized from sets to symmetric matrices $A_1, \ldots, A_n \in \mathbb{R}^{m \times m}$ with $\|A_i\|_{op} \leq 1$ by defining $\mathrm{disc}(\{A_1, \ldots, A_n\}) := \min\{\|\sum_{i=1}^{n} x_i A_i\|_{op} : x \in \{-1, 1\}^n\}$. Note that picking 0/1 diagonal matrices A_i corresponding to the incidence vector of element i would exactly encode the set coloring setting. Again the interesting case is $m = n$; in contrast to the diagonal case it is only known that the discrepancy is bounded by $O(\sqrt{n \cdot \log(n)})$, which is already attained by a random coloring. Meka[2] conjectured that the discrepancy of n matrices can be bounded by $O(\sqrt{n})$.

---

[1] We should mention for the sake of completeness that our update choice is *not* a convex combination of the experts weighted by their exponential weights.

[2] See the blog post https://windowsontheory.org/2014/02/07/discrepancy-and-be ating-the-union-bound/.

For a very readable introduction into discrepancy theory, we recommend Chap. 4 in the book of Matoušek [Mat99] or the book of Chazelle [Cha01].

1.2 Our Contribution

Our main result is a deterministic version of the theorem of Lovett and Meka:

Theorem 1. *Let* $v_1, \ldots, v_m \in \mathbb{R}^n$ *unit vectors,* $x^{(0)} \in [-1, 1]^n$ *be a starting point and let* $\lambda_1 \geq \ldots \geq \lambda_m \geq 0$ *be parameters so that* $\sum_{i=1}^m \exp(-\lambda_i^2/16) \leq \frac{n}{32}$. *Then there is a deterministic algorithm that computes a vector* $x \in [-1, 1]^n$ *with* $\langle v_i, x - x^{(0)} \rangle \leq 8\lambda_i$ *for all* $i \in [m]$ *and* $|\{i : x_i = \pm 1\}| \geq \frac{n}{2}$, *in time* $O(\min\{n^4 m, n^3 m \lambda_1^2\})$.

By setting $\lambda_i = O(1)$ this yields a deterministic version of Spencer's theorem in the balanced case $m = n$:

Corollary 1. *Given* n *sets over* n *elements, there is a deterministic algorithm that finds a* $O(\sqrt{n})$-*discrepancy coloring in time* $O(n^4)$.

Furthermore, Spencer's *hyperbolic cosine algorithm* [Spe77] can also be interpreted as a multiplicative weight update argument. However, the techniques of [Spe77] are only enough for a $O(\sqrt{n \log(n)})$ discrepancy bound for the balanced case. Our hope is that similar arguments can be applied to solve open problems such as whether there is an extension of Spencer's result to balance matrices [Zou12] and to better discrepancy minimization techniques in the Beck-Fiala setting. To demonstrate the versatility of our arguments, we show an extension to the matrix discrepancy case.

We say that a symmetric matrix $A \in \mathbb{R}^{m \times m}$ is *q-block diagonal* if it can be written as $A = \text{diag}(B_1, \ldots, B_{m/q})$, where each B_j is a symmetric $q \times q$ matrix.

Theorem 2. *For given* q-*block diagonal matrices* $A_1, \ldots, A_n \in \mathbb{R}^{m \times m}$ *with* $\|A_i\|_{op} \leq 1$ *for* $i = 1, \ldots, n$ *one can compute a coloring* $x \in \{-1, 1\}^n$ *with* $\|\sum_{i=1}^n x_i A_i\|_{op} \leq O(\sqrt{n \log(\frac{2qm}{n})})$ *deterministically in time* $O(n^5 + n^4 m^3)$.

Finally, we can also give the first deterministic algorithm for the result of Bansal et al. [BDG16].

Theorem 3. *Let* $A \in \mathbb{R}^{m \times n}$ *be a matrix with* $\|A^j\|_2 \leq 1$ *for all columns* $j = 1, \ldots, n$. *Then there is a deterministic algorithm to find a coloring* $x \in \{-1, 1\}^n$ *with* $\|Ax\|_\infty \leq O(\sqrt{\log n})$ *in time* $O(n^3 \log(n) \cdot (m + n))$.

While [BDG16] need to solve a semidefinite program in each step of their random walk, our algorithm does not require solving any SDPs. Note that we do not optimize running times such as by using fast matrix multiplication.

In the Beck-Fiala setting, we are given a set system over n elements, where each element is contained in at most t subsets. Theorem 3 then provides the first polynomial-time deterministic algorithm that produces a coloring with discrepancy $O(\sqrt{t \log n})$; we simply choose the matrix A whose rows are the incidence vectors of members of the set system, scaled by $1/\sqrt{t}$.

For space reasons, we defer the proof of Theorem 2 to the full version of the paper.[3]

2   The Algorithm for Partial Coloring

We will now describe the algorithm proving Theorem 1. First note that for any $\lambda_i > 2\sqrt{n}$ we can remove the constraint $\langle v_i, x - x_0 \rangle \leq \lambda_i$, as it does not cut off any point in $[-1, 1]^n$. Thus we assume without loss of generality that $2\sqrt{n} \geq \lambda_1 \geq \cdots \geq \lambda_m \geq 0$. Let $\delta := \frac{1}{\lambda_1}$ denote the step size of our algorithm. The algorithm will run for $O(n/\delta^2)$ iterations, each of computational cost $O(n^2 m)$. Note that $\delta = O(1/\sqrt{n})$ so the algorithm terminates in $O(n^2)$ iterations. The total runtime is hence $O(n^2 m \cdot n/\delta^2) = O(n^3 m \lambda_1^2) \leq O(n^4 m)$.

For a symmetric matrix $M \in \mathbb{R}^{n \times n}$ we know that an *eigendecomposition* $M = \sum_{j=1}^n \mu_j u_j u_j^T$ can be computed in time $O(n^3)$. Here $\mu_j := \mu_j(M)$ is the *jth eigenvalue* of M and $u_j := u_j(M)$ is the corresponding *eigenvector* with $\|u_j\|_2 = 1$. We make the convention that the eigenvalues are sorted as $\mu_1 \geq \ldots \geq \mu_n$. The algorithm is as follows:

(1) Set weights $w_i^{(0)} = \exp(-\lambda_i^2)$ for all $i = 1, \ldots, m$.
(2) FOR $t = 0$ TO ∞ DO
    (3) Define the following subspaces
        - $U_1^{(t)} := \mathrm{span}\{e_j : -1 < x_j^{(t)} < 1\}$
        - $U_2^{(t)} := \{x \in \mathbb{R}^n \mid \langle x, x^{(t)} \rangle = 0\}$
        - $U_3^{(t)} := \{x \in \mathbb{R}^n \mid \langle v_i, x \rangle = 0 \ \forall i \in I^{(t)}\}$. Here $I^{(t)} \subseteq [m]$ are the $|I^{(t)}| = \frac{n}{16}$ indices with maximum weight $w_i^{(t)}$.
        - $U_4^{(t)} := \{x \in \mathbb{R}^n \mid \langle v_i, x \rangle = 0 \ \forall i \text{ with } \lambda_i \leq 1\}$
        - $U_5^{(t)} := \{x \in \mathbb{R}^n \mid \langle x, \sum_{i=1}^m \lambda_i w_i^{(t)} \cdot \exp\left(-\frac{4\delta^2 \lambda_i^2}{n}\right) v_i \rangle = 0\}$
        - $U_6^{(t)} := \mathrm{span}\{u_j(M^{(t)}) : \frac{1}{16}n \leq j \leq n\}$, for $M^{(t)} := \sum_{i=1}^m w_i^{(t)} \lambda_i^2 v_i v_i^T$.
        - $U^{(t)} := U_1^{(t)} \cap \ldots \cap U_6^{(t)}$
    (4) Let $z^{(t)}$ be any unit vector in $U^{(t)}$
    (5) Choose a maximal $\alpha^{(t)} \in (0, 1]$ so that $x^{(t+1)} := x^{(t)} + \delta \cdot y^{(t)} \in [-1, 1]^n$, with $y^{(t)} = \alpha^{(t)} z^{(t)}$.
    (6) Update $w_i^{(t+1)} := w_i^{(t)} \cdot \exp(\lambda_i \cdot \delta \cdot \langle v_i, y^{(t)} \rangle) \cdot \exp\left(-\frac{4\delta^2 \lambda_i^2}{n}\right)$.
    (7) Let $A^{(t)} := \{j \in [n] : -1 < x_j^{(t)} < 1\}$. If $|A^{(t)}| < \frac{n}{2}$, then set $T := t$ and stop.

The intuition is that we maintain weights $w_i^{(t)}$ for each constraint i that increase exponentially with the one-sided discrepancy $\langle v_i, x^{(t)} - x^{(0)} \rangle$. Those weights are discounted in each iteration by a factor that is slightly less than 1 — with a bigger discount for constraints with a larger parameter λ_i. The subspaces $U_1^{(t)}$ and $U_2^{(t)}$ ensure that the length of $x^{(t)}$ is monotonically increasing and fully colored elements remain fully colored.

---

[3] See https://arxiv.org/abs/1611.08752.

2.1 Bounding the Number of Iterations

First, note that if the algorithm terminates, then at least half of the variables in $x^{(T)}$ will be either -1 or $+1$. In particular, once a variable is set to ± 1, it is removed from the set $A^{(t)}$ of active variables and the subsequent updates will leave those coordinates invariant. To bound the number of iterations, we use the fact that the algorithm always makes a step of length δ orthogonal to the current position — except for the steps where it hits the boundary.

Lemma 1. *The algorithm terminates after $T = O(\frac{n}{\delta^2})$ iterations.*

Proof. First, we can analyze the length increase

$$\|x^{(t+1)}\|_2^2 = \|x^{(t)} + \delta \cdot y^{(t)}\|_2^2 = \|x^{(t)}\|_2^2 + 2\delta \underbrace{\langle x^{(t)}, y^{(t)} \rangle}_{=0} + \delta^2 \|y^{(t)}\|_2^2,$$

using that $y^{(t)} \in U_2^{(t)}$. Whenever $\alpha^{(t)} = 1$, we have $\|x^{(t+1)}\|_2^2 \geq \|x^{(t)}\|_2^2 + \delta^2$. It happens that $\alpha^{(t)} < 1$ at most n times, simply because in each such iteration $|A^{(t)}|$ must decrease by at least one. We know that $x^{(T)} \in [-1,1]^n$. Suppose for the sake of contradiction that $T > \frac{2n}{\delta^2}$, then $\|x^{(T)}\|_2^2 \geq (T-n) \cdot \delta^2 > n$, which is impossible. We can hence conclude that the algorithm will terminate in step (7) after at most $\frac{2n}{\delta^2}$ iterations.

2.2 Properties of the Subspace $U^{(t)}$

One obvious condition to make the algorithm work is to guarantee that the subspace $U^{(t)}$ satisfies $\dim(U^{(t)}) \geq 1$. In fact, its dimension will even be linear in n.

Lemma 2. *In any iteration t, one has $\dim(U^{(t)}) \geq \frac{n}{8}$.*

Proof. By accounting for all linear constraints that define $U^{(t)}$, we get

$$\dim(U^{(t)}) \geq |A^{(t)}| - |I^{(t)}| - |\{i : \lambda_i \leq 1\}| - \frac{n}{16} - 2 \geq \frac{n}{2} - \frac{n}{16} - \frac{n}{8} - \frac{n}{16} - 2 \geq \frac{n}{8}$$

assuming that $n \geq 16$.

Another crucial property will be that every vector in $U^{(t)}$ has a bounded *quadratic error term*:

Lemma 3. *For each unit vector $y \in U^{(t)}$ one has $y^T M^{(t)} y \leq \frac{16}{n} \sum_{i=1}^{m} w_i^{(t)} \lambda_i^2$.*

Proof. We have $\mathrm{Tr}[v_i v_i^T] = 1$ since each v_i is a unit vector, hence $\mathrm{Tr}[M^{(t)}] = \sum_{i=1}^{m} w_i^{(t)} \lambda_i^2 \mathrm{Tr}[v_i v_i^T] = \sum_{i=1}^{m} w_i^{(t)} \lambda_i^2$. Because $M^{(t)}$ is positive semidefinite, we know that $\mu_1, \ldots, \mu_n \geq 0$, where $\mu_j := \mu_j(M^{(t)})$ is the jth eigenvalue. Then by *Markov's inequality* at most a $\frac{1}{16}$ fraction of eigenvalues can be larger than $\frac{16}{n}$ · $\mathrm{Tr}[M^{(t)}]$. The claim follows as $U_6^{(t)}$ is spanned by the $\frac{15}{16}n$ eigenvectors $v_j(M^{(t)})$ belonging to the smallest eigenvalues, which means $\mu_j \leq \frac{16}{n}\mathrm{Tr}[M^{(t)}]$ for $j = \frac{1}{16}n, \ldots, n$.

2.3   The Potential Function

So far, we have defined the weights by iterative update steps, but it is not hard to verify that in each iteration t one has the explicit expression

$$w_i^{(t)} = \exp\left(\lambda_i\langle v_i, x^{(t)} - x^{(0)}\rangle - \lambda_i^2 \cdot \left(1 + t\cdot\frac{4\delta^2}{n}\right)\right). \tag{1}$$

Inspired by the multiplicative weight update method, we consider the *potential function* $\Phi^{(t)} := \sum_{i=1}^m w_i^{(t)}$ that is simply the sum of the individual weights. At the beginning of the algorithm we have $\Phi^{(0)} = \sum_{i=1}^m w_i^{(0)} = \sum_{i=1}^m \exp(-\lambda_i^2/16) \leq \frac{n}{32}$ using the assumption in Theorem 1. Next, we want to show that the potential function does not increase. Here the choice of the subspaces $U_5^{(t)}$ and $U_6^{(t)}$ will be crucial to control the error.

Lemma 4. *In each iteration t one has $\Phi^{(t+1)} \leq \Phi^{(t)}$.*

Proof. Let us abbreviate $\rho_i := \exp\left(-\frac{4\delta^2\lambda_i^2}{n}\right)$ as the *discount factor* for the ith constant. Note that in particular $0 < \rho_i \leq 1$ and $\rho_i \leq 1 - \frac{2\delta^2\lambda_i^2}{n}$. The change in one step can be analyzed as follows:

$$
\begin{aligned}
\Phi^{(t+1)} &= \sum_{i=1}^m w_i^{(t+1)} = \sum_{i=1}^m w_i^{(t)}\cdot\exp\left(\lambda_i\delta\langle v_i, y^{(t)}\rangle\right)\cdot\rho_i\\
&\overset{(*)}{\leq} \sum_{i=1}^m w_i^{(t)}\cdot\left(1 + \lambda_i\delta\langle v_i, y^{(t)}\rangle + \lambda_i^2\delta^2\langle v_i, y^{(t)}\rangle^2\right)\cdot\rho_i\\
&= \sum_{i=1}^m w_i^{(t)}\cdot\rho_i + \delta\underbrace{\left\langle\sum_{i=1}^m\lambda_i w_i^{(t)}\rho_i v_i, y^{(t)}\right\rangle}_{=0\text{ since }y^{(t)}\in U_5^{(t)}} + \delta^2\sum_{i=1}^m w_i^{(t)}\lambda_i^2\underbrace{\rho_i}_{\leq 1}\langle v_i, y^{(t)}\rangle^2\\
&\leq \sum_{i=1}^m w_i^{(t)}\cdot\rho_i + \delta^2\cdot(y^{(t)})^T M^{(t)}y^{(t)} \overset{(**)}{\leq} \sum_{i=1}^m w_i^{(t)}\cdot\rho_i + \delta^2\frac{16}{n}\sum_{i=1}^m w_i^{(t)}\lambda_i^2\\
&\overset{(***)}{\leq} \sum_{i=1}^m w_i^{(t)} = \Phi^{(t)}.
\end{aligned}
$$

In $(*)$, we use the inequality $e^x \leq 1 + x + x^2$ for $|x| \leq 1$ together with the fact that $\lambda_i\delta|\langle v_i, y^{(t)}\rangle| \leq \lambda_i\delta \leq 1$. In $(**)$ we bound $(y^{(t)})^T M^{(t)}y^{(t)}$ using Lemma 3. In $(***)$ we finally use the fact that $\rho_i + \frac{16}{n}\delta^2 \leq 1$.

Typically in the multiplicative weight update method one can only use the fact that $\max_{i\in[m]} w_i^{(t)} \leq \Phi^{(t)}$ which would lead to the loss of an additional $\sqrt{\log n}$ factor. The trick in our approach is that there is always a *linear* number of weights of order $\max_{i\in[m]} w_i^{(t)}$ since the updates are always chosen orthogonal to the $\frac{n}{16}$ constraints with highest weight.

Lemma 5. *At the end of the algorithm,* $\max\{w_i^{(T)} : i \in [m]\} \leq 2.$

Proof. Suppose, for contradiction, that $w_i^{(T)} > 2$ for some i. Let t^* be the last iteration when i was not among the $\frac{n}{16}$ constraints with highest weight. After iteration $t^* + 1$, $w_i^{(t)}$ only decreases in each iteration. Then

$$2 < w_i^{(T)} \leq w_i^{(t^*+1)} = w_i^{(t^*)} \cdot \underbrace{\exp(\lambda_i \cdot \delta \cdot \langle v_i, y^{(t)} \rangle)}_{\leq e} \cdot \underbrace{\rho_i}_{\leq 1} \leq w_i^{(t^*)} \cdot e,$$

hence $w_i^{(t^*)} > \frac{2}{e}$. This implies that $\Phi^{(t^*)} \geq \frac{n}{16} \cdot \frac{2}{e} > \frac{n}{32}$, contradicting Lemma 4.

Lemma 6. *If* $w_i^{(T)} \leq 2$, *then* $\langle v_i, x^{(T)} - x^{(0)} \rangle \leq 11\lambda_i.$

Proof. First note that the algorithm always walks orthogonal to all constraint vectors v_i if $\lambda_i \leq 1$ and in this case $\langle v_i, x^{(T)} - x^{(0)} \rangle = 0$. Now suppose that $\lambda_i > 1$. We know that $w_i^{(T)} \overset{(1)}{=} \exp\left(\lambda_i \cdot \langle v_i, x^{(T)} - x^{(0)} \rangle - \lambda_i^2 \cdot \left(1 + 4 \cdot T \cdot \frac{\delta^2}{n}\right)\right) \leq 2.$ Taking logarithms on both sides and dividing by λ_i then gives

$$\langle v_i, x^{(T)} - x^{(0)} \rangle \leq \underbrace{\frac{\log(2)}{\lambda_i}}_{\leq 2} + \lambda_i \left(1 + 4T\underbrace{\frac{\delta^2}{n}}_{\leq 2}\right) \leq 11\lambda_i.$$

This lemma concludes the proof of Theorem 1.

2.4   Application to Set Coloring

Now we come to the main application of the partial coloring argument from Theorem 1, which is to color set systems:

Lemma 7. *Given a set system* $S_1, \ldots, S_m \subseteq [n]$, *we can find a coloring* $x \in \{-1, 1\}^n$ *with* $|\sum_{j \in S_i} x_j| \leq O(\sqrt{n \log \frac{2m}{n}})$ *for every* i *deterministically in time* $O(n^3 m \log(\frac{2m}{n}))$.

Proof. For a fractional vector x, let us abbreviate $\mathrm{disc}(S, x) := |\sum_{j \in S} x_j|$ as the discrepancy with respect to set S. Set $x^{(0)} := \mathbf{0}$. For $s = 1, \ldots, \log_2(n)$ many phases we do the following. Let $A^{(s)} := \{i \in [n] : -1 < x_i^{(s-1)} < 1\}$ be the not yet fully colored elements. Define a vector $v_i := \frac{1}{\sqrt{|A^{(s)}|}} \mathbf{1}_{S_i \cap A^{(s)}}$ of length $\|v_i\|_2 \leq 1$ with parameters $\lambda_i := C\sqrt{\log(\frac{2m}{|A^{(s)}|})}$. Then apply Theorem 1 to find $x^{(s)} \in [-1, 1]^n$ with $\mathrm{disc}(S_i, x^{(s)} - x^{(s-1)}) \leq O(\sqrt{|A^{(s)}| \log(\frac{2m}{|A^{(s)}|})})$ such that $x_i^{(s)} = x_i^{(s-1)}$ for $i \notin A^{(s)}$. Since each time at least half of the elements get fully colored we have $|A^{(s)}| \leq 2^{-(s-1)}n$ for all s. Then $x := x^{(\log_2 n)} \in \{-1, 1\}^n$ and

$$\mathrm{disc}(S_i, x) \leq \sum_{s \geq 1} O\left(\sqrt{2^{-(s-1)}n \log\left(\frac{2m}{2^{-(s-1)}n)}\right)}\right) \leq O\left(\sqrt{n \log(\frac{2m}{n})}\right)$$

using that this convergent sequence is dominated by the first term.

In each application of Theorem 1 one has $\delta \geq \Omega(1/\sqrt{\log(\frac{2m}{n})})$. Thus phase s runs for $O(2^{-(s-1)}n/\delta^2) = O(2^{-(s-1)}n \log(\frac{2m}{n}))$ iterations, each of which takes $O((2^{-(s-1)}n)^2 m)$ time, for a total runtime of $O((2^{-(s-1)}n)^3 m \log(\frac{2m}{n}))$ in phase s. Summing the geometric series for $s = 1, \ldots, \log_2 n$ results in a total running time of $O(n^3 m \log(\frac{2m}{n}))$.

By setting $m = n$ in Lemma 7, we obtain Corollary 1.

3   Matrix Balancing

In this section we prove Theorem 2. We begin with some preliminaries. For matrices $A, B \in \mathbb{R}^{n \times n}$, let $A \bullet B := \sum_{i=1}^{n} \sum_{j=1}^{n} A_{ij} \cdot B_{ij}$ be the *Frobenius inner product*. Recall that any symmetric matrix $A \in \mathbb{R}^{n \times n}$ can be written as $A = \sum_{j=1}^{n} \mu_j u_j u_j^T$, where μ_j is the eigenvalue corresponding to eigenvector u_j. The *trace* of A is $\text{Tr}[A] = \sum_{i=1}^{n} A_{ii} = \sum_{j=1}^{n} \mu_j$ and for symmetric matrices A, B one has $\text{Tr}[AB] = A \bullet B$. If A has only nonnegative eigenvalues, we say that A is *positive semidefinite* and write $A \succeq 0$. Recall that $A \succeq 0$ if and only if $y^T A y \geq 0$ for all $y \in \mathbb{R}^n$. For a symmetric matrix A, we denote $\mu_{\max} := \max\{\mu_j : j = 1, \ldots, n\}$ as the largest Eigenvalue and $\|A\|_{op} := \max\{|\mu_j| : j = 1, \ldots, n\}$ as the largest singular value. Note that if $A \succeq 0$, then $|A \bullet B| \leq \text{Tr}[A] \cdot \|B\|_{op}$. If $A, B \succeq 0$, then $A \bullet B \geq 0$. Finally, note that for any symmetric matrix A one has $A^2 := AA \succeq 0$.

From the eigendecomposition $A = \sum_{j=1}^{n} \mu_j u_j u_j^T$, one can easily show that the maximum singular value also satisfies $\|A\|_{op} = \max\{\|Ay\|_2 : \|y\|_2 = 1\}$ and $\|A\|_{op} = \max\{|y^T A y| : \|y\|_2 = 1\}$. For any function $f : \mathbb{R} \to \mathbb{R}$ we define $f(A) := \sum_{j=1}^{n} f(\mu_j) u_j u_j^T$ to be the symmetric matrix that is obtained by applying f to all Eigenvalues. In particular we will be interested in the *matrix exponential* $\exp(A) := \sum_{j=1}^{n} e^{\mu_j} u_j u_j^T$. For any symmetric matrices $A, B \in \mathbb{R}^n$, the *Golden-Thompson inequality* says that $\text{Tr}[\exp(A + B)] \leq \text{Tr}[\exp(A) \exp(B)]$. (It is not hard to see that for diagonal matrices one has equality.) We refer to the textbook of Bhatia [Bha97] for more details.

Theorem 4. *Let $A_1, \ldots, A_n \in \mathbb{R}^{m \times m}$ be q-block diagonal matrices with $\|A_i\|_{op} \leq 1$ for $i = 1, \ldots, m$ and let $x^{(0)} \in [-1, 1]^n$ be a starting point. Then there is a deterministic algorithm that finds an $x \in [-1, 1]^n$ with*

$$\left\| \sum_{i=1}^{n} (x_i - x_i^{(0)}) \cdot A_i \right\|_{op} \leq O\left(\sqrt{n \log\left(\frac{2qm}{n}\right)} \right)$$

in time $O(n^5 + n^4 m^3)$. Moreover, at least $\frac{n}{2}$ coordinates of x will be in $\{-1, 1\}$.

Our algorithm computes a sequence of iterates $x^{(0)}, \ldots, x^{(T)}$ such that $x^{(T)}$ is the desired vector x with half of the coordinates being integral. In our algorithm the step size is $\delta = \frac{1}{\sqrt{n}}$ and we use a parameter $\varepsilon = \frac{1}{\sqrt{n}}$ to control the scaling of the following potential function:

$$\Phi^{(t)} := \mathrm{Tr}\left[\exp\left(\varepsilon \sum_{i=1}^{n}(x_i^{(t)} - x_i^{(0)}) \cdot A_i\right)\right].$$

Suppose $B_{i,k} \in \mathbb{R}^{q \times q}$ are symmetric matrices so that $A_i = \mathrm{diag}(B_{i,1}, \ldots, B_{i,m/q})$. Then we can decompose the weight function as $\Phi^{(t)} = \sum_{k=1}^{m/q} \Phi_k^{(t)}$ with $\quad \Phi_k^{(t)} :=$ $\mathrm{Tr}\left[\exp\left(\varepsilon \sum_{i=1}^{n}(x_i^{(t)} - x_i^{(0)})B_{i,k}\right)\right]$. In other words, the potential function is simply the sum of the potential function applied to each individual block. The algorithm is as follows:

(1) FOR $t = 0$ TO ∞ DO
  (2) Define weight matrix $W^{(t)} := \exp(\varepsilon \sum_{i=1}^{n}(x_i^{(t)} - x_i^{(0)})A_i)$
  (3) Define the following subspaces
    – $U_1^{(t)} := \mathrm{span}\{e_j : -1 < x_j^{(t)} < 1\}$
    – $U_2^{(t)} := \{x \in \mathbb{R}^n \mid \langle x, x^{(t)} \rangle = 0\}$
    – $U_3^{(t)} := \{x \in \mathbb{R}^n \mid \sum_{i=1}^{n} x_i B_{i,k} = \mathbf{0} \ \forall k \in I^{(t)}\}$. Here $I^{(t)} \subseteq [m]$ are the $|I^{(t)}| = \frac{1}{16} \cdot \frac{n}{q^2}$ indices k with maximum weight $\Phi_k^{(t)}$.
    – $U_4^{(t)} := \{x \in \mathbb{R}^n \mid \sum_{i=1}^{n} x_i \cdot (W^{(t)} \bullet A_i) = 0\}$
    – $U_5^{(t)}$ is the subspace defined in Lemma 9, with $k = 16$.
    – $U^{(t)} := U_1^{(t)} \cap \ldots \cap U_5^{(t)}$
  (4) Let $z^{(t)}$ be any unit vector in $U^{(t)}$.
  (5) Choose a maximal $\alpha^{(t)} \in (0, 1]$ so that $x^{(t+1)} := x^{(t)} + \delta \cdot y^{(t)} \in [-1, 1]^n$, where $y^{(t)} = \alpha^{(t)} z^{(t)}$.
  (6) Let $A^{(t)} := \{j \in [n] : -1 < x_j^{(t)} < 1\}$. If $|A^{(t)}| < \frac{n}{2}$, then set $T := t$ and stop.

The analysis of our algorithm follows a sequence of lemmas, the proofs of most of which we defer to the full version of the paper. By exactly the same arguments as in Lemma 1 we know that the algorithm terminates after $T \leq \frac{2n}{\delta^2}$ iterations. Each iteration can be done in time $O(n^2 m^3 + n^3)$ (c.f. Lemma 9).

Lemma 8. *In each iteration t one has $\dim(U^{(t)}) \geq \frac{n}{4}$.*

Proof. By accounting for all linear constraints that define $U^{(t)}$, we get

$$\dim(U^{(t)}) \geq \underbrace{|A^{(t)}|}_{U_1^{(t)}} - \underbrace{|I^{(t)}|}_{U_3^{(t)}} - \underbrace{\frac{n}{16}}_{} - \underbrace{2}_{U_2^{(t)}, U_4^{(t)}} \geq \frac{n}{2} - \frac{n}{16q^2} \cdot q^2 - \frac{n}{16} - 2 \geq \frac{n}{4}$$

$$\underbrace{\qquad}_{U_5^{(t)}}$$

assuming that $n \geq 16$.

To analyze the behavior of the potential function, we first prove the existence of a suitable subspace $U_5^{(t)}$ that will bound the quadratic error term.

Lemma 9. *Let $W \in \mathbb{R}^{m \times m}$ be a symmetric positive semidefinite matrix, let $A_1, \ldots, A_n \in \mathbb{R}^{m \times m}$ be symmetric matrices with $\|A_i\|_{op} \leq 1$ and let $k > 0$ be a parameter. Then in time $O(n^2 m^3 + n^3)$ one can compute a subspace $U \subseteq \mathbb{R}^n$ of dimension $\dim(U) \geq (1 - \frac{1}{k})n$ so that*

$$W \bullet \left(\sum_{i=1}^n y_i A_i \right)^2 \leq k \cdot Tr\,[W] \quad \forall y \in U \text{ with } \|y\|_2 = 1. \tag{2}$$

Proof. See the full version.

Again, we bound the increase in the potential function. This gives us a bound on the potential function at the end of the algorithm.

Lemma 10. *In each iteration t, one has $\Phi^{(t+1)} \leq (1 + 16\varepsilon^2 \delta^2) \cdot \Phi^{(t)}$.*

Proof. See the full version.

Lemma 11. *At the end of the algorithm, $\Phi^{(T)} \leq m \cdot \exp(32\varepsilon^2 n)$.*

Proof. Since $\Phi^{(0)} = \text{Tr}[\exp(\mathbf{0})] = \text{Tr}[I] = m$, we get that $\Phi^{(T)} \leq m \cdot (1 + 16\varepsilon^2 \delta^2)^T \leq m \cdot \exp(32\varepsilon^2 n)$, using the fact that $T \leq \frac{2n}{\delta^2}$.

Lemma 12. *We have $\mu_{\max}(\sum_{i=1}^n (x_i^{(T)} - x_i^{(0)}) \cdot A_i) = O(\sqrt{n \log(\frac{2qm}{n})})$.*

Proof. See the full version.

These lemmas put together give us Theorem 4: an algorithm that yields a partial coloring with the claimed properties. We run the algorithm in phases to obtain Theorem 2, by boosting the partial coloring to a full coloring using a similar technique as in Lemma 7. The interested reader may refer to the full version of the paper for details.

References

[AHK12] Arora, S., Hazan, E., Kale, S.: The multiplicative weights update method: a meta-algorithm and applications. Theor. Comput. **8**(6), 121–164 (2012)

[AS08] Alon, N., Spencer, J.H.: The Probabilistic Method. Wiley-Interscience Series in Discrete Mathematics and Optimization, 3rd edn. John Wiley & Sons Inc., Hoboken (2008). With an appendix on the life and work of Paul Erdős

[Ban98] Banaszczyk, W.: Balancing vectors and Gaussian measures of n-dimensional convex bodies. Random Struct. Algorithms **12**(4), 351–360 (1998)

[Ban10] Bansal, N.: Constructive algorithms for discrepancy minimization. In: FOCS, pp. 3–10 (2010)

[BDG16] Bansal, N., Dadush, D., Garg, S: An algorithm for komlós conjecture matching banaszczyk's bound. CoRR, abs/1605.02882 (2016)

[Bec81] Beck, J.: Roth's estimate of the discrepancy of integer sequences is nearly sharp. Combinatorica **1**(4), 319–325 (1981)

[BF81] Beck, J., Fiala, T.: "Integer-making" theorems. Discrete Appl. Math. **3**(1), 1–8 (1981)

[Bha97] Rajendra, B.: Matrix Analysis. Graduate Texts in Mathematics. Springer, New York (1997)

[Boh90] Bohus, G.: On the discrepancy of 3 permutations. Random Struct. Algorithms 1(2), 215–220 (1990)

[BS13] Bansal, N., Spencer, J.: Deterministic discrepancy minimization. Algorithmica 67(4), 451–471 (2013)

[Cha01] Chazelle, B.: The Discrepancy Method - Randomness and Complexity. University Press, Cambridge (2001)

[Gia97] Giannopoulos, A.: On some vector balancing problems. Stud. Math. 122(3), 225–234 (1997)

[Glu89] Gluskin, E.D.: Extremal properties of orthogonal parallelepipeds and their applications to the geometry of banach spaces. Math. USSR Sb. 64(1), 85 (1989)

[LM12] Lovett, S., Meka, R.: Constructive discrepancy minimization by walking on the edges. In: FOCS, pp. 61–67 (2012)

[Mat99] Matoušek, J.: Geometric Discrepancy. Algorithms and Combinatorics. Springer, Berlin (1999). An illustrated guide

[MSS13] Marcus, A., Spielman, D.A., Srivastava, N.: Interlacing families, I.I.: mixed characteristic polynomials and the Kadison-singer problem. arXiv e-prints, June 2013

[Nik13] Nikolov, A.: The komlos conjecture holds for vector colorings. arXiv e-prints, Jan 2013

[NNN12] Newman, A., Neiman, O., Nikolov, A.: Beck's three permutations conjecture: a counterexample and some consequences. In: FOCS, pp. 253–262 (2012)

[Spe77] Spencer, J.: Balancing games. J. Comb. Theor. Ser. B 23(1), 68–74 (1977)

[Spe85] Spencer, J.: Six standard deviations suffice. Trans. Am. Math. Soc. 289(2), 679–706 (1985)

[Sri97] Srinivasan, A.: Improving the discrepancy bound for sparse matrices: Better approximations for sparse lattice approximation problems. In: SODA 1997, ACM SIGACT, SIAM, Philadelphia, PA, pp. 692–701 (1997)

[SST] Spencer, J.H., Srinivasan, A., Tetali, P.: The discrepancy of permutation families. Unpublished manuscript

[Zou12] Zouzias, A.: A Matrix hyperbolic cosine algorithm and applications. In: Czumaj, A., Mehlhorn, K., Pitts, A., Wattenhofer, R. (eds.) ICALP 2012. LNCS, vol. 7391, pp. 846–858. Springer, Heidelberg (2012). doi:10.1007/978-3-642-31594-7_71

Mixed-Integer Convex Representability

Miles Lubin, Ilias Zadik, and Juan Pablo Vielma[✉]

Massachusetts Institute of Technology, Cambridge, MA, USA
mlubin@mit.edu,jvielma@mit.edu

Abstract. We consider the question of which nonconvex sets can be represented exactly as the feasible sets of mixed-integer convex optimization problems. We state the first complete characterization for the case when the number of possible integer assignments is finite. We develop a characterization for the more general case of unbounded integer variables together with a simple necessary condition for representability which we use to prove the first known negative results. Finally, we study representability of subsets of the natural numbers, developing insight towards a more complete understanding of what modeling power can be gained by using convex sets instead of polyhedral sets; the latter case has been completely characterized in the context of mixed-integer linear optimization.

1 Introduction

Early advances in solution techniques for mixed-integer linear programming (MILP) motivated studies by Jeroslow and Lowe [6] and others (recently reviewed in [9]) on understanding precisely which sets can be encoded as projections of points within a closed polyhedron satisfying integrality restrictions on a subset of the variables. These sets can serve as feasible sets in mixed-integer linear optimization problems and therefore potentially be optimized over in practice by using branch-and-bound techniques (ignoring issues of computational complexity). Jeroslow and Lowe, for example, proved that the epigraph of the piecewise linear function $f(x)$ which equals 1 if $x > 0$ and 0 if $x = 0$, is not representable over the domain $x \geq 0$. Such a function would naturally be used to model a fixed cost in production. It is now well known that an upper bound on x is required in order to encode such fixed costs in an MILP formulation.

Motivated by recent developments in methods for solving mixed-integer convex programming (MICP) problems [1,8], in this work we address the analogous question of which *nonconvex* sets may be represented as projections of points within a *convex* set satisfying integrality restrictions on a subset of the variables. To our knowledge, we are the first authors to consider this general case. Related but more specific analysis has been developed by Del Pia and Poskin [3] where

---

M. Lubin and I. Zadik—Contributed equally to this work.

J.P. Vielma—Supported by NSF under grant CMMI-1351619.

We acknowledge the anonymous referees for improving the presentation of this work.

F. Eisenbrand and J. Koenemann (Eds.): IPCO 2017, LNCS 10328, pp. 392–404, 2017.
DOI: 10.1007/978-3-319-59250-3_32

they characterized the case where the convex set is an intersection of a polyhedron with an ellipsoidal region and by Dey and Morán [4] where they studied the structure of integer points within convex sets but without allowing a mix of continuous and discrete variables.

After a brief study in Sect. 2.1 of restricted cases, e.g., when there is a finite number of possible integer assignments, we focus primarily on the more challenging general case where we seek to understand the structure of countably infinite unions of slices of convex sets induced by mixed-integer constraints. In Sect. 3 we develop a general, yet hard to verify, characterization of representable sets as families of convex sets with specific properties, and in Sect. 4 we prove a much simpler necessary condition for representability which enables us to state a number of nonrepresentability results. Using that condition, we prove, for example, that the set of $m \times n$ matrices with rank at most 1 is not representable when $m, n \geq 2$. In Sect. 5 we conclude with an in-depth study of the representability of subsets of the natural numbers. The special case of the natural numbers is a sufficiently challenging first step towards a general understanding of the structure of representable sets. We prove, for example, that the set of prime numbers is not representable, an interesting case that separates mixed-integer convex representability from mixed-integer polynomial representability [5]. By adding rationality restrictions to the convex set in the MICP formulation, we completely characterize representability of subsets of natural numbers, discovering that one can represent little beyond what can be represented by using rational polyhedra.

2   Preliminaries

We use the notation $[\![k]\!]$ to denote the set $\{1, 2, \ldots, k\}$. Also by $\mathbb{N}$ we will refer to the nonnegative integers $\{0, 1, 2, \ldots\}$. We will often work with projections of a set $M \subseteq \mathbb{R}^{n+p+d}$ for some $n, p, d \in \mathbb{N}$. We identify the variables in $\mathbb{R}^n$, $\mathbb{R}^p$ and $\mathbb{R}^d$ of this set as x, y and z and we let

$$\operatorname{proj}_x(M) = \left\{ x \in \mathbb{R}^n : \exists\, (y, z) \in \mathbb{R}^{p+d} \text{ s.t. } (x, y, z) \in M \right\}.$$

We similarly define $\operatorname{proj}_y(M)$ and $\operatorname{proj}_z(M)$.

Definition 1. *Let $M \subseteq \mathbb{R}^{n+p+d}$ be a closed, convex set and $S \subseteq \mathbb{R}^n$. We say M induces an MICP formulation of S if and only if*

$$S = \operatorname{proj}_x\left(M \cap \left(\mathbb{R}^{n+p} \times \mathbb{Z}^d \right) \right). \tag{1}$$

*A set $S \subseteq \mathbb{R}^n$ is **MICP representable** if and only if there exists an MICP formulation of S. If such formulation exists for a closed polyhedron M then we say S is (additionally) MILP representable.*

Definition 2. *A set S is **bounded MICP (MILP)** representable if there exists an MICP (MILP) formulation which satisfies $\left| \operatorname{proj}_z\left(M \cap \left(\mathbb{R}^{n+p} \times \mathbb{Z}^d \right) \right) \right| < \infty$. That is, there are only finitely many feasible assignments of the integer variables z.*

Definition 3. *For a set of integral vectors $z_1, z_2, \ldots, z_k \in \mathbb{Z}^d$ we define the integral cone $\operatorname{intcone}(z_1, z_2, \ldots, z_k) := \{ \sum_{i=1}^k \lambda_i z_i \,|\, \lambda_i \in \mathbb{N}, i \in [\![k]\!] \}$.*

2.1   Bounded and Other Restricted MICP Representability Results

It is easy to see that bounded MICP formulations can represent *at most* a finite union of projections of closed, convex sets. To date, however, there are no precise necessary conditions over these sets for the existence of a bounded MICP formulation. For instance, Ceria and Soares [2] provide an MICP formulation for the finite union of closed, convex sets under the condition that the sets have the same *recession cone* (set of unbounded directions). In the following proposition we close this gap and give a simple, explicit formulation for any finite union of projections of closed, convex sets without assumptions on recession directions.

Proposition 1. *$S \subseteq \mathbb{R}^n$ is bounded MICP representable if and only if there exist nonempty, closed, convex sets $T_1, T_2, \ldots, T_k \subset \mathbb{R}^{n+p}$ for some $p, k \in \mathbb{N}$ such that $S = \bigcup_{i=1}^k \mathrm{proj}_x T_i$. In particular a formulation for such S is given by*

$$x = \sum_{i=1}^k x_i, \quad (x_i, y_i, z_i) \in \hat{T}_i \; \forall i \in [\![k]\!], \quad \sum_{i=1}^k z_i = 1, \; z \in \{0,1\}^k, \quad (2a)$$

$$\|x_i\|_2^2 \leq z_i t, \quad \forall i \in [\![k]\!], t \geq 0 \quad (2b)$$

where $\hat{T}_i$ is the closed conic hull of T_i, i.e., the closure of $\{(x, y, z) : (x,y)/z \in T_i, z > 0\}$.

Known MICP representability results for unbounded integers are more limited. For the case in which M is a rational polyhedron Jerowslow and Lowe [6] showed that a set $S \subseteq \mathbb{R}^n$ is (unbounded) rational MILP representable if and only if there exist $r_1, r_2, \ldots, r_t \subseteq \mathbb{Z}^n$ and rational polytopes S^i for $i \in [\![k]\!]$ such that

$$S = \bigcup_{i=1}^k S^i + \mathrm{intcone}(r_1, r_2, \ldots, r_t). \quad (3)$$

Characterization (3) does not hold in general for non-polyhedral M. However, using results from [4] it is possible to show that it holds for some pure integer cases as well. For instance, Theorem 6 in [4] can be used to show that for any $\alpha > 0$, $S_\alpha := \{x \in \mathbb{Z}^2 : x_1 x_2 \geq \alpha\}$ satisfies (3) with S^i containing a single integer vector for each $i \in [\![k]\!]$.

The only mixed-integer and non-polyhedral result we are aware of is a characterization of the form (3) when M is the intersection of a rational polyhedron with an ellipsoidal cylinder having a rational recession cone [3]. An identical proof also holds when the recession cone of M is a rational subspace and M is contained in a rational polyhedron with the same recession cone as M. We can further extend this result to the following simple proposition whose proof is in the extended version of this paper [7].

Proposition 2. *If M induces an MICP-formulation of S and $M = C + K$ where C is a compact convex set and K is a rational polyhedral cone, then S satisfies representation (3) with S^i now being compact convex sets for each $i \in [\![k]\!]$.*

Unfortunately, MICP-representable sets in general may not have a representation of the form (3), even when S^i is allowed to be any convex set. We illustrate this with a simple variation on the pure-integer example above.

Example 1. Let $S := \{x \in \mathbb{N} \times \mathbb{R} : x_1 x_2 \geq 1\}$ be the set depicted in Fig. 1. For each $z \in \mathbb{N}, z \neq 0$ let $A_z := \{x \in \mathbb{R}^2 : x_1 = z, x_2 \geq 1/z\}$ so that $S = \bigcup_{z=1}^{\infty} A_z$. Suppose for contradiction that S satisfies (3) for convex sets S^i. By convexity of S^i and finiteness of k there exists $z_0 \in \mathbb{Z}$ such that $\bigcup_{i=1}^{k} S^i \subset \bigcup_{z=1}^{z_0-1} A_z$. Because $\min_{x \in A_{z_0}} x_2 < \min_{x \in A_z} x_2$ for all $z \in [\![z_0 - 1]\!]$ we have that there exists $j \in [\![k]\!]$ such that the second component of r_j is strictly negative. However, this implies that there exists $x \in S$ such that $x_2 < 0$ which is a contradiction with the definition of S.

3   A General Characterization of MICP Representability

The failure of characterizations of the form (3) to hold calls for a more general characterization of MICP-representable sets as projections of families of sets with particular structure. Example 1 hints at the union of a countable number of convex sets indexed by a set of integers. The following definition shows the precise conditions on this sets and indexes for the existence of a MICP formulation.

Definition 4. *Let $C \subseteq \mathbb{R}^d$ be a convex set and $(A_z)_{z \in C}$ be a family of convex sets in $\mathbb{R}^n$. We say that the family of sets is **convex** if for all $z, z' \in C$ and $\lambda \in [0,1]$ it holds $\lambda A_z + (1 - \lambda)A_{z'} \subseteq A_{\lambda z + (1-\lambda)z'}$.*

*We further say that the family is **closed** if A_z is closed for all $z \in C$ and for any convergent sequences $\{z_m\}_{m \in \mathbb{N}}, \{x_m\}_{m \in \mathbb{N}}$ with $z_m \in C$ and $x_m \in A_{z_m}$ we have $\lim_{m \to \infty} x_m \in A_{\lim_{m \to \infty} z_m}$.*

Lemma 1. *Let $(A_z)_{z \in C}$ be a convex family and $C' \subseteq C$ be a convex set. Then $(\text{proj}(A_z))_{z \in C'}$ is a convex family, where proj is any projection onto a subset of the variables.*

The proof of the above lemma is simple and it is omitted.

Theorem 1. *A set $S \subseteq \mathbb{R}^n$ is MICP representable if and only if there exists $d \in \mathbb{N}$, a convex set $C \subseteq \mathbb{R}^d$ and a closed convex family $(B_z)_{z \in C}$ in $\mathbb{R}^{n+p}$ such that $S = \bigcup_{z \in C \cap \mathbb{Z}^d} \text{proj}_x (B_z)$.*

Proof. Suppose that S is MICP representable. Then there exists $p, d \in \mathbb{N}$ and a closed and convex set $M \subseteq \mathbb{R}^{n+p+d}$ satisfying (1). Let $C = \text{proj}_z (M)$ and for any $z \in C$ let $B_z = \{(x, y) \in \mathbb{R}^{n+p} : (x, y, z) \in M\}$. The result follows by noting that $(B_z)_{z \in C}$ is a closed convex family because M is closed and convex.

For the converse, let $M := \overline{\text{conv}} \left(\bigcup_{z \in C \cap \mathbb{Z}^d} B_z \times \{z\} \right)$. Set M is closed and convex by construction and hence the only thing that remains to prove is that $B_z = \{(x, y) : (x, y, z) \in M\}$ for all $z \in C \cap \mathbb{Z}^d$. The left to right containment is direct. For the reverse containment let $M' := \text{conv} \left(\bigcup_{z \in C \cap \mathbb{Z}^d} B_z \times \{z\} \right)$ so that $M = \overline{M'}$ and $B'_z = \{(x, y) : (x, y, z) \in M'\}$ for all $z \in C$. Because

$(B_z)_{z \in C}$ is a convex family we have $B'_z \subseteq B_z$ for all $z \in C$. Let $z \in C \cap \mathbb{Z}^d$ and $\{(x_m, y_m, z_m)\}_{m \in \mathbb{N}} \subseteq M'$ be a convergent sequence such that $\lim_{m \to \infty} z_m = z$. We have for all m, $(x_m, y_m) \subseteq B'_{z_m} \subseteq B_{z_m}$ so $\lim_{m \to \infty}(x_m, y_m) \in B_z$ because $(B_z)_{z \in C}$ is a closed convex family.

Definition 5. *For an MICP representable set $S \subseteq \mathbb{R}^n$ we let its MICP-dimension be the smallest $d' \in \mathbb{N}$ such that the representation from Theorem 1 holds with $\dim(C) = d'$.*

Remark 1. If $S = \bigcup_{z \in C \cap \mathbb{Z}^d} \text{proj}_x(B_z)$ for a convex set $C \subseteq \mathbb{R}^d$ and a closed convex family $(B_z)_{z \in C}$ in $\mathbb{R}^{n+p}$ and $C' = \text{conv}(C \cap \mathbb{Z}^d)$, then $(B_z)_{z \in C'}$ is a closed convex family and $S = \bigcup_{z \in C' \cap \mathbb{Z}^d} \text{proj}_x(B_z)$.

Remark 2. Using the convex family characterization it can be proven that sets like the union of expanding circles with concave radii or the set described in Example 1 are MICP representable; see Fig. 1.

4   A Necessary Condition for MICP Representability

In this section we prove an easy to state, and usually also to check, necessary property for any MICP representable set. Intuitively, it is saying that despite the fact that MICP representable sets could be nonconvex, they will never be very nonconvex in an appropriately defined way.

Definition 6. *We say that a set $S \subseteq \mathbb{R}^n$ is **strongly nonconvex**, if there exists a subset $R \subseteq S$ with $|R| = \infty$ such that for all pairs $x, y \in R$,*

$$\frac{x + y}{2} \notin S, \tag{4}$$

that is, an infinitely large subset of points in S such that the midpoint between any pair is not in S.

Lemma 2 (The midpoint lemma). *Let $S \subseteq \mathbb{R}^n$. If S is strongly nonconvex, then S is not MICP representable.*

Proof. Suppose we have R as in the statement above and there exists an MICP formulation of S, that is, a closed convex set $M \subset \mathbb{R}^{n+p+d}$ such that $x \in S$ iff $\exists z \in \mathbb{Z}^d, y \in \mathbb{R}^p$ such that $(x, y, z) \in M$. Then for each point $x \in R$ we associate at least one integer point $z_x \in \mathbb{Z}^d$ and a $y_x \in \mathbb{R}^p$ such that $(x, y_x, z_x) \in M$. If there are multiple such pairs of points z_x, y_x then for the purposes of the argument we may choose one arbitrarily.

We will derive a contradiction by proving that there exist two points $x, x' \in R$ such that the associated integer points $z_x, z_{x'}$ satisfy

$$\frac{z_x + z_{x'}}{2} \in \mathbb{Z}^d. \tag{5}$$

Indeed, this property combined with convexity of M, i.e., $\left(\frac{x+x'}{2}, \frac{y_x+y_{x'}}{2}, \frac{z_x+z_{x'}}{2}\right) \in M$ would imply that $\frac{x+x'}{2} \in S$, which contradicts the definition of R.

Recall a basic property of integers that if $i, j \in \mathbb{Z}$ and $i \equiv j \pmod{2}$, i.e., i and j are both even or odd, then $\frac{i+j}{2} \in \mathbb{Z}$. We say that two integer vectors $\alpha, \beta \in \mathbb{Z}^d$ have the same *parity* if α_i and β_i are both even or odd for each component $i = 1, \ldots, d$. Trivially, if α and β have the same parity, then $\frac{\alpha+\beta}{2} \in \mathbb{Z}^d$. Given that we can categorize any integer vector according to the 2^d possible choices for whether its components are even or odd, and we notice that from any infinite collection of integer vectors we must have at least one pair that has the same parity. Therefore since $|R| = \infty$ we can find a pair $x, x' \in R$ such that their associated integer points $z_x, z_{x'}$ have the same parity and thus satisfy (5). $\square$

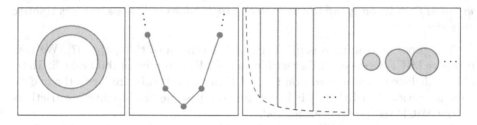

Fig. 1. From left to right, the annulus and the piece-wise linear function connecting the integer points on the parabola *are not* mixed-integer convex representable. The mixed-integer hyperbola and the collection of balls with increasing and concave radius *are* mixed-integer convex representable.

Proposition 3. *Fix $n, m \in \mathbb{N}$ with $m, n \geq 2$. The set of matrices of dimension $m \times n$ with rank at most 1, i.e., $C_1 := \{X \in \mathbb{R}^{m \times n} : \text{rank}(X) \leq 1\}$ is strongly nonconvex and therefore not MICP representable.*

Proof. We can assume $m = 2$. We set for all $k \in \mathbb{N}$ the matrix $A_k = \begin{bmatrix} 1 & k & O_{1 \times n-2} \\ k & k^2 & O_{1 \times n-2} \end{bmatrix} \in C_1$. We then set $R = \{A_k | k \in \mathbb{N}\}$. Clearly $|R| = \infty$. It is easy to verify that $\text{rank}(\frac{1}{2}(A_k + A_{k'})) = 2$ for $k \neq k'$. Therefore for any pair of distinct points in R, their midpoint is not in C_1. Therefore C_1 is strongly nonconvex and in particular not MICP representable.

One may use the midpoint lemma to verify that the epigraph of a twice differentiable function is MICP representable if and only if the function is convex and that the graph of a twice differentiable function is MICP representable if and only if f is linear. In Fig. 1, we illustrate two more sets whose nonrepresentability follows directly from the midpoint lemma: the annulus and the piecewise linear function connecting the points $\{(x, y) \in \mathbb{Z}^2 : y = x^2\}$.

5   MICP Representability of Subsets of Natural Numbers

Recall that we define $\mathbb{N} = \{0, 1, 2, \ldots\}$ to be the set of natural numbers. In this section we investigate the limitations of MICP for representing subsets of the natural numbers. We remind the reader that in the MILP case, it is known that a subset of the natural numbers is rational-MILP (the coefficients of the continuous relaxation polyhedron are rational numbers) representable if and only if the set is equal to the Minkowski summation of finitely many natural numbers plus the set of nonnegative integer combinations of a finite set of integer generators [9]. We simplify this characterization since we are dealing with subsets of the natural numbers.

We define an infinite arithmetic progression in the natural numbers to be a sequence of natural numbers of the form $am + b, m \in \mathbb{N}$ for some fixed $a, b \in \mathbb{N}$.

Lemma 3. *Let $S \subseteq \mathbb{N}$. S is rational-MILP representable if and only if S is the union of finitely many infinite arithmetic progressions with the same nonnegative step size.*

The proof of Lemma 3 is in the extended version of this paper [7]. We now compare MICP representability with rational-MILP representability on $\mathbb{N}$. To be able to deduce a characterization for MICP in the naturals similar to the one we have for rational-MILP in $\mathbb{N}$ it is natural to put some "rationality" restrictions on the MICP representations as well.

For instance we could require M to have a representation of the form $Ax - b \in K$ where A and b are an appropriately sized rational matrix and rational vector, and K is a specially structured convex cone (e.g. the semidefinite cone or a product of Lorentz cones defined as $\mathcal{L}_n := \{(t, x) \in \mathbb{R}^n : ||x||_2 \leq t\}$) or to have polynomial constraints with rational coefficients. Unfortunately these restrictions can still result in representable sets that do not contain any infinite arithmetic progression and hence are far from being rational-MILP representable. We present such an example below.

Example 2. For $x \in \mathbb{R}$ let $f(x) = x - \lfloor x \rfloor$. For $\varepsilon > 0$ consider the set

$$K_\varepsilon = \{x \in \mathbb{R}^2 : (x_2 + \varepsilon, x_1, x_1) \in \mathcal{L}_3, (2x_1 + 2\varepsilon, x_2, x_2) \in \mathcal{L}_3, x_1, x_2 \geq 0\} \quad (6)$$

$$= \{x \in \mathbb{R}^2 : \sqrt{2}x_1 - \varepsilon \leq x_2 \leq \sqrt{2}x_1 + \sqrt{2}\varepsilon, \quad x_1, x_2 \geq 0\} \quad (7)$$

and $S_\varepsilon = \{x_1 \in \mathbb{R} : \exists x_1 \text{ s.t. } (x_1, x_2) \in K_\varepsilon \cap \mathbb{Z}^2\} = \{x \in \mathbb{N} : f(\sqrt{2}x) \notin (\varepsilon, 1 - \sqrt{2}\varepsilon)\}$. Let $\varepsilon_0 < 1/(1 + \sqrt{2})$ be rational (e.g. $\varepsilon = 0.4$). Suppose that for some $a, b \in \mathbb{N}, a \geq 1$ it holds $ak + b \in S_{\varepsilon_0}$ for all $k \in \mathbb{N}$. $\emptyset \neq (\varepsilon_0, 1 - \sqrt{2}\varepsilon_0) \subseteq (0, 1)$, so by Kroneckers Approximation Theorem we have that there exist $k_0 \in \mathbb{N}$ such that $f(\sqrt{2}(ak_0 + b)) \in (\varepsilon_0, 1 - \sqrt{2}\varepsilon_0)$ which is a contradiction. Therefore the set S_ε does not contain an arithmetic progression and in particular it is not rational-MILP representable.

We follow now a different path to define what rational MICP-representability is and we characterize it completely. Quite surprisingly it becomes almost equivalent with rational-MILP representability.

We give the following definitions.

Definition 7. *We say that an unbounded convex set* $C \subseteq \mathbb{R}^d$ *is **rationally unbounded** if the image* C' *of any rational linear mapping of* C, *is either bounded or there exists* $r \in \mathbb{Z}^d \setminus \{0\}$ *such that* $x + \lambda r \in C'$ *from any* $x \in \mathbb{Z}^d \cap C'$ *and* $\lambda \geq 0$.

Let $A \subseteq \mathbb{Z}^d$ *be an infinite set of integer points. We say that* A *is **rationally unbounded** if there exists a finite subset* $I \subset A$ *such that the set* $\mathrm{conv}(A \setminus I)$ *is rationally unbounded.*

Finally, we say that a set S *is **rational-MICP representable** if there exists an MICP representation for* S *with convex family* $(A_x)_{x \in C}$ *with* $A_x \neq \emptyset, \forall x \in C$ *such that the set of integer points* $C \cap \mathbb{Z}^d$ *is either bounded or rationally unbounded.*

It is easy to see that the set K_ϵ from Example 2 is not rationally unbounded.

To completely characterize the rational-MICP representable subsets of the natural numbers we will use the following lemmata. We prove Lemma 4 in the extended version of this paper [7] and Lemma 5 in Sect. 6.

Lemma 4. *Any union of finitely many infinite arithmetic progressions is equal to a union of finitely many arithmetic progressions with the same step length.*

Lemma 5. *Suppose* $S \subseteq \mathbb{N}$ *is rational MICP representable with MICP-dimension* d'. *Then either* S *is a finite set or there exists* $k \in \mathbb{N}$ *such that* $S = T_0 \cup S_0 \cup \bigcup_{i=1}^{k} S_i$ *where* T_0 *is a finite set of natural numbers,* S_0 *is a finite union of infinite arithmetic progressions and for each* $i \in [\![k]\!]$, S_i *is rational-MICP representable with MICP-dimension at most* $d' - 1$.

Theorem 2. *Suppose that* $S \subseteq \mathbb{N}$. *Then following are equivalent.*

(a) S *is rational-MICP representable*
(b) *There exists* $k \in \mathbb{N}$ *such that* $S = A_0 \cup \left(\bigcup_{i=1}^{k} A_i \right)$, *where* $A_0 \subseteq \mathbb{N}$ *is a finite set and for each* $i = 1, \ldots, k$, $A_i \subseteq \mathbb{N}$ *is an infinite arithmetic progression.*
(c) *There exists a finite set* A_0 *and a rational-MILP representable set* T *such that* $S = A_0 \cup T$.

Proof. We start by proving that (c) implies (a). We will use Lemma 3. Say $A_0 = \{a_1, \ldots, a_m\}$ and $T = \{b_1, \ldots, b_n\} + \mathrm{intcone}(z)$. Then $x \in A_0 \cup T$ iff $\exists x_1, x_2, \beta, q, \alpha, \nu, \lambda, \eta, t$ such that

$$x = x_1 + x_2, \quad x_1 = \sum_{i=1}^{n} \beta_i + qz, \quad x_2 = \sum_{i=1}^{m} \alpha_i, \quad \beta_i = b_i \nu_i \; \forall i \in [\![n]\!],$$

$$\alpha_i = a_i \lambda_i \; \forall i \in [\![m]\!], \quad \sum_{i=1}^{n} \nu_i = \eta, \quad \sum_{i=i}^{m} \lambda_i = 1 - \eta, \quad q^2 \leq \eta t,$$

$$t \geq 0, \quad \lambda_i \in \{0, 1\} \; \forall i \in [\![m]\!], \quad \nu_i \in \{0, 1\} \; \forall i \in [\![n]\!], \quad \eta \in \{0, 1\}, q \in \mathbb{N}$$

We claim that this is a rational-MICP representation. Consider the integer variables $\lambda_i, \nu_i, \eta, q$. By excluding the finitely many integer extreme points with $\eta = 0$ we have that it is enough to consider only the integer points with $\eta = 1$ which imply all $\lambda_i = 0$ and hence we have to consider only the integer variables ν_i, q

that satisfy $\sum_{i=1}^{n} \nu_i = 1, \nu_i \in \{0,1\}, i = 1, 2, \ldots, n$ and $q \in \mathbb{N}$. The convex hull of these integer points is $\mathbb{R}_+ \times \{x \in \mathbb{R}_+^n | \sum_{i=1}^{n} x_i = 1\}$. But this is rationally unbounded, as it has exactly one rational recession direction $e_1 := (1, 0, 0, \ldots, 0)$ and any rational linear map t will either satisfy $t(e_1) \neq 0$, in which case $t(e_1)$ is a rational recession direction for the image, or $t(e_1) = 0$, in which case the image is bounded.

Now (b) implies (c) because of Lemma 4 and noticing that finite union of infinite arithmetic progressions with the same step length is immediately rational-MILP representable because of Lemma 3.

Finally we prove that (a) implies (b). Suppose S is rational-MICP representable. We will use Lemma 5. We first apply to S. If it is finite we are done. If not we apply it to each of the S_i, $i = 1, 2, \ldots, m$ produced by Lemma 5. Continuing like this with at most d iterations we prove our result.

Despite the similarity that the above result indicates, rational MICP-representable subsets of the natural numbers and rational MILP-representable subsets of the natural numbers are not identical as the example below illustrates.

Example 3. Consider the set $S = \{1\} \cup 2\mathbb{N}$. Then the set is rational-MICP representable from the above theorem. On the other hand, it cannot be written as the Minkowski summation of a finite set plus a finitely generated integral monoid and therefore it is not rational-MILP representable. To see the last, suppose it could be written like this by contradiction. Then consider one of the generators of the monoid z_1. Assume z_1 is odd. Then $2 + z_1$ should belong to S but it is odd and bigger than 1, a contradiction. Assume z_1 is even. Then $1 + z_1$ should belong to S but it is odd and bigger than 1, a contradiction. The proof is complete.

We end the section with a global limitation of MICP representability in the subsets of the integers which hold without any type of rationality restriction. Its proof is based on the midpoint lemma.

Theorem 3. *The set of prime numbers $\mathbb{P}$ is strongly nonconvex and therefore not MICP representable.*

Proof. We will inductively construct a subset of primes such that no midpoint of any two elements in the set is prime.

Let $\{p_1, \ldots, p_n\}$ be a set of such primes. We will find a prime p such that $\{p_1, \ldots, p_n, p\}$ has no prime midpoints. (We may start the induction with $p_1 = 3$, $p_2 = 5$.)

Set $M = \prod_{i=1}^{n} p_i$. Choose any prime p (not already in our set and not equal to 2) such that $p \equiv 1 \pmod{M!}$. By Dirichlet's theorem on arithmetic progressions there exist an infinite number of primes of the form $1 + kM!$ because 1 and $M!$ are coprime, so we can always find such p.

Suppose for some i we have $q := \frac{p+p_i}{2} \in \mathbb{P}$. By construction, we have $p + p_i \equiv 1 + p_i \pmod{M!}$, so $\exists k$ such that $p + p_i = k \cdot M! + 1 + p_i$. Note that M is larger than p_i so $M!$ will contain $(1+p_i)$ as a factor; in other words, $(1+p_i)$ divides $M!$, so it divides also $k \cdot M! + 1 + p_i = p + p_i$. In fact we can write $p + p_i = k'(1+p_i)$ for some $k' \in \mathbb{Z}_{\geq 0}$. We claim that $k' = 1$. Indeed $q = \frac{p+p_i}{2} = k' \frac{1+p_i}{2}$. Note $1 + p_i$ is even, so $\frac{1+p_i}{2}$ is an integer bigger than 1 as $p_i > 1$. But q is prime and therefore since it is written as the product of k' and $\frac{1+p_i}{2} > 1$ it must be the case that $k' = 1$ as claimed. But $k' = 1$ implies that $p + p_i = 1 + p_i$, i.e., $p = 1$ which is a contradiction.

6   Proof of Lemma 5

We first state Lemma 6 whose proof is given in [7].

Lemma 6. *Let $C \subseteq \mathbb{R}^d$ be a convex set, $h : C \to \mathbb{R}$ a nonpositive convex function and $\{x^i\}_{i=1}^k \subset C$ such that $h(x^1) = 0$ and $x^1 \in \operatorname{relint}\left(\operatorname{aff}\left(\{x^i\}_{i=1}^k\right) \cap C\right)$. Then $h(x) = 0$ for all $x \in \operatorname{aff}\left(\{x^i\}_{i=1}^k\right) \cap C$.*

Proof (of Lemma 5). Let $C \subseteq \mathbb{R}^d$ be the convex set such that $\dim(C) = d'$ and $\{B_z\}_{z \in C}$ be the closed convex family such that $S = \bigcup_{z \in C \cap \mathbb{Z}^d} \operatorname{proj}_x(B_z)$. Since S is rational-MICP representable, $C \cap \mathbb{Z}^d$ is either finite or rationally unbounded. In the first case S is finite so we can assume that $C \cap \mathbb{Z}^d$ is rationally unbounded.

Since $n = 1$ and convex subsets of the real line are intervals, we may define $f, g : C \to \mathbb{R}$ such that $f(z)$ and $g(z)$ represent the lower and upper endpoints of the intervals $\operatorname{proj}_x(B_z)$ for all $z \in C$. Because $\{\operatorname{proj}_x(B_z)\}_{z \in C}$ is a convex family we have that $h := f - g : C \to \mathbb{R}$ is a convex function and $h(z) \leq 0$ for all $z \in C$. Furthermore, since $S \subseteq \mathbb{N}$ we have $h(z) = 0$ for all $z \in C \cap \mathbb{Z}^d$.

Let $I \subseteq C \cap \mathbb{Z}^d$ be the finite set such that $C' = \operatorname{conv}\left((C \cap \mathbb{Z}^d) \setminus I\right)$ is rationally unbounded. By letting $T_0 := \{\operatorname{proj}_x(B_z)\}_{z \in I} \subset \mathbb{N}$ be the finite set in the statement of Lemma 5 and noting that $C' \subseteq C$ we may redefine C to be equal to C'. Let $r \in \mathbb{Z}^d$ be the direction from Definition 7 such that $l_z := \{z + \lambda r : \lambda \geq 0\} \subseteq C$. Because $|l_z \cap \mathbb{Z}| = \infty$ and $h(z') = 0$ for all $z' \in l_z \cap \mathbb{Z}$ we have that $h(z') = 0$ for all $z' \in l_z$ by Lemma 6. Hence for all $z' \in l_z$ we have $\operatorname{proj}_x(B_{z'}) = \{f(z')\} = \{g(z')\}$. Given that $\operatorname{proj}_x(B_{z'}) \in \mathbb{N}$ for $z' \in l_z$ and being a convex and a concave function is equivalent to being an affine function, we further have that $\{\operatorname{proj}_x(B_z)\}_{z \in l_z}$ being a convex family implies that there exist $\alpha_z \in \mathbb{Z}^d$ and $\beta_z \in \mathbb{Z}$ such that $f(z') = g(z') = \alpha_z \cdot z' + \beta_z$ for all $z' \in l_z$. Then $\{\operatorname{proj}_x(B_{z'})\}_{z' \in l_z \cap \mathbb{Z}^d} = \{a_z m + b_z : m \in \mathbb{N}\}$, where $a_z = (\alpha_z \cdot r)/\gcd(r_1, \ldots, r_d)$ and $b_z = \beta_z$. If $\alpha_z \cdot r > 0$ this corresponds to an infinite arithmetic progression and if $\alpha_z \cdot r = 0$ it corresponds to a single point.

Let $\{T_i\}_{i=1}^{2^d}$ be such that $C \cap \mathbb{Z}^d = \bigcup_{i=1}^{2^d} T_i$ and $z_j \equiv z'_j \mod 2$ for all $j \in [\![d]\!]$, $i \in [\![2^d]\!]$ and $z, z' \in T_i$. For fixed $i \in [\![2^d]\!]$ we have $\frac{z+z'}{2} \in C \cap \mathbb{Z}^d$ and $l_{\frac{z+z'}{2}} \subset C$ for any $z, z' \in T_i$. Then $P := \operatorname{conv}\left(\left\{l_z, l_{\frac{z+z'}{2}}, l_{z'}\right\}\right) \subset C$ and $h(\tilde{z}) = 0$ for all $\tilde{z} \in P \cap \mathbb{Z}^d$. Then, by Lemma 6 $h(\tilde{z}) = 0$ for all $\tilde{z} \in P$. By the same argument in the previous paragraph there exist $\alpha_P \in \mathbb{Z}^d$ and $\beta_P \in \mathbb{Z}$ such that $f(\tilde{z}) = g(\tilde{z}) = \alpha_P \cdot \tilde{z} + \beta_P$ for all $\tilde{z} \in P$. In particular, $\alpha_P \cdot \tilde{z} + \beta_P = \alpha_z \cdot \tilde{z} + \beta_z$ for all $\tilde{z} \in l_z$ and $\alpha_P \cdot \tilde{z} + \beta_P = \alpha_{z'} \cdot \tilde{z} + \beta_{z'}$ for all $\tilde{z} \in l_{z'}$. Hence $\alpha_z \cdot r = \alpha_{z'} \cdot r$

and then $s_i := a_z = a_{z'}$. Then $\{\text{proj}_x(B_{\bar{z}})\}_{\bar{z} \in l_z \cap \mathbb{Z}^d} = \{s_i m + b_z : m \in \mathbb{N}\}$ for all $z \in T_i$. Unfixing i we may define S_0 from the statement of Lemma 5 to be

$$S_0 := \bigcup_{i \in [\![2^d]\!]:s_i>0} \bigcup_{z \in T_i} \{s_i m + b_z : m \in \mathbb{N}\} \subseteq \bigcup_{i \in [\![2^d]\!]:s_i>0} \bigcup_{b=0}^{s_i-1} \{s_i m + b : m \in \mathbb{N}\}.$$

The last inclusion implies S_0 is a finite union of infinite arithmetic progressions. It then only remains to consider sets $\{\text{proj}_x(B_{\bar{z}})\}_{\bar{z} \in l_z \cap \mathbb{Z}^d}$ for $z \in T_i$ and $i \in [\![2^d]\!]$ such that $s_i = 0$ ($s_i \geq 0$ because $S \subseteq \mathbb{N}$). Say we have k such i's (WLOG $i = 1, \ldots, k$) and we will show that for every such i, $S_i := \bigcup_{z' \in T_i} \text{proj}_x(B_{z'}^i)$ is rational-MICP representable with MICP-dimension at most $d' - 1$. Because $S = T_0 \cup S_0 \cup \bigcup_{i=1}^k S_i$ the proof will be complete.

For a fixed $i \in [\![2^d]\!]$ such that $s_i = 0$, let $t^i \in T_i$ so that $T_i = (t^i + 2\mathbb{Z}^d) \cap C$. Let $\{v_i\}_{i=1}^d$ a rational orthogonal basis of $\mathbb{R}^d$ such that $r = v_d$ and $\{v_i\}_{i=d-d'+1}^d$ is an orthonormal basis of the linear subspace $L(C)$ parallel to $\text{aff}(C)$ (i.e. $L(C) := \text{aff}(C - z)$ for any $z \in C$). Let $A \in \mathbb{R}^{d \times d}$ such that for $i \leq d - 1$ the i-th row of A is v_i^T and the d-th row of A has all components equal to zero. Also, let $A_{1:d-1}$ be the restriction of A to the first $d - 1$ rows and let $H \in \mathbb{R}^{(d-1) \times (d-1)}$ and $U \in \mathbb{R}^{d \times d}$ be a unimodular matrix such that

$$A_{1:d-1} = [H|0]U \tag{8}$$

(e.g. Hermite normal form). Finally, let $l_z^{\pm} := \{z + \lambda r : \lambda \in \mathbb{R}\} \cap C$ and $C_i = U^{-1} \begin{bmatrix} H^{-1} & 0 \\ 0 & 1 \end{bmatrix} A(C - t^i)/2 = U^{-1} \begin{bmatrix} I & 0 \\ 0 & 0 \end{bmatrix} U(C - t^i)/2$.

We claim that $\bigcup_{w \in T_i} l_w^{\pm} = \bigcup_{w \in C_i \cap \mathbb{Z}^d} l_{2w+t^i}^{\pm}$. Indeed, $z' \in T_i$ if and only if there exists $z'' \in \mathbb{Z}^d \cap (C - t^i)/2$ such that $z' = t^i + 2z''$. Then $Az'' \in A(C-t^i)/2$ and $l_{z'}^{\pm} = l_{t^i+2z''}^{\pm}$. But we know $A = \begin{bmatrix} H & 0 \\ 0 & 0 \end{bmatrix} U$ which gives after some algebra $Uz'' \in \begin{bmatrix} I & 0 \\ 0 & 0 \end{bmatrix} U(C - t^i)/2 + \begin{bmatrix} 0 & 0 \\ 0 & 1 \end{bmatrix} \mathbb{R}^d$. However, because U is unimodular and $z'' \in \mathbb{Z}^d$ we have $Uz'' \in \mathbb{Z}^d$ we can replace $\mathbb{R}^d$ by $\mathbb{Z}^d$ and there exist $y \in \mathbb{Z}$ and $z \in C_i \cap \mathbb{Z}^d$ such that $z'' = z + U^{-1} \begin{bmatrix} 0 \\ y \end{bmatrix}$. From (8), orthogonality of $\{v_i\}_{i=1}^d$ and unimodularity of U we have $U^{-1}e_d = \alpha r$ for some $\alpha \in \mathbb{Z}$ and hence $z'' = z + y\alpha r$. Then $l_{z'}^{\pm} = l_{t^i+2z''}^{\pm} = l_{t^i+2z}^{\pm} \subseteq \bigcup_{w \in C_i \cap \mathbb{Z}^d} l_{2w+t^i}^{\pm}$.

For the other direction, if $z \in C_i \cap \mathbb{Z}^d$ then there exist $z'' \in (C - t^i)/2$ such that $Uz = \begin{bmatrix} I & 0 \\ 0 & 0 \end{bmatrix} Uz'' = Uz'' - e^d[Uz'']_d = U(z'' - \alpha r[Uz'']_d)$ for the $\alpha \in \mathbb{Z}$ such that $U^{-1}e_d = \alpha r$. Let $z''' = z'' - \alpha r[Uz'']_d$ and $\mu = \alpha[Uz'']_d$. Then $z''' \in \mathbb{Z}^d$ and $z''' + \mu r \in (C - t^i)/2$ and hence there exist $k \geq \mu$ such that $\bar{z} = z''' + kr \in (C - t^i)/2 \cap \mathbb{Z}^d$ (because without loss of generality we may replace C with $\text{conv}(C \cap \mathbb{Z}^d)$ so that r is a recession direction of C). Finally, $z = z'' - \alpha r[Uz'']_d = \bar{z} - kr$. Then $z' := t^i + 2(z + kr)$ is such that $z' \in T_i$ and $l_{t^i+2z}^{\pm} = l_{z'}^{\pm} \subseteq \bigcup_{w \in T_i} l_w^{\pm}$ as claimed.

Now let $l^{\pm}(z, \lambda) := z + \lambda r$ and $\Lambda := \{\lambda \in \mathbb{R} : l^{\pm}(z, \lambda) \in C\}$ so that $l_z^{\pm} = \{z + \lambda r : \lambda \in \mathbb{R}\} \cap C = \bigcup_{\lambda \in \Lambda} l^{\pm}(z, \lambda)$ and Λ is a convex set in $\mathbb{R}$. Furthermore, for each $z \in C_i$ let $\tilde{B}_z^i = \bigcup_{\lambda \in \Lambda} \left(B_{l^{\pm}(t^i + 2z, \lambda)} \times \{\lambda\} \right)$. We can check that $\left(\tilde{B}_z^i \right)_{z \in C_i}$ is a closed convex family. Both convexity of $\tilde{B}_z^i$ and the convex family property hold because $(B_z)_{z \in C}$ is a convex family, $l_z^{\pm}$ is convex and $l^{\pm}(z, \lambda)$ is affine. To see that the family is closed, suppose we have a sequence $(x^m, y^m, \lambda^m, z^m)$ converging to (x, y, λ, z) with $(x^m, y^m, \lambda^m) \in \tilde{B}_{z^m}^i$. Then $(x^m, y^m) \in B_{l^{\pm}(t^i + 2z^m, \lambda^m)}$ and hence by closedness of $(B_z)_{z \in C}$ and continuity of $l^{\pm}(z, \lambda)$ we have $(x, y) \in B_{l^{\pm}(t^i + 2z, \lambda)}$ and hence $(x, y, \lambda) \in \tilde{B}_z^i$. Finally, for each $z \in C_i \cap \mathbb{Z}^d$ we have

$$\text{proj}_x \left(\tilde{B}_z^i \right) = \bigcup_{\lambda \in \Lambda} \text{proj}_x \left(B_{l^{\pm}(t^i + 2z, \lambda)} \right) = \bigcup_{\tilde{z} \in l_{t^i + 2z}^{\pm}} \text{proj}_x (B_{\tilde{z}}) = \bigcup_{\tilde{z} \in l_{z'}^{\pm}} \text{proj}_x (B_{\tilde{z}})$$

for some $z' \in T_i$ where we set $B_y = \emptyset$ for all $y \notin C$ and we have used $\bigcup_{w \in C_i \cap \mathbb{Z}^d} l_{2w + t^i}^{\pm} \subset \bigcup_{w \in T_i} l_w^{\pm}$. However, because $s_i = 0$ we have that $\text{proj}_x (B_{\tilde{z}}) = \{b_{z'}\}$ for all $\tilde{z} \in l_{z'}^{\pm} \cap C$. Hence for all $z \in C_i \cap \mathbb{Z}^d$, $\text{proj}_x \left(\tilde{B}_z^i \right) = \{b_{z'}\} = \text{proj}_x (B_{z'})$ for some $z' \in T_i$. But for any $z' \in T_i$ since $\bigcup_{w \in T_i} l_w^{\pm} \subset \bigcup_{w \in C_i \cap \mathbb{Z}^d} l_{2w + t^i}^{\pm}$ it holds also $\text{proj}_x (B_{z'}) = \text{proj}_x \left(\tilde{B}_z^i \right)$ for some $z \in C_i \cap \mathbb{Z}^d$. Hence $S_i = \bigcup_{z' \in T_i} \text{proj}_x (B_{z'}) = \bigcup_{z \in C_i \cap \mathbb{Z}^d} \text{proj}_x \left(\tilde{B}_z^i \right)$. The result finally follows since $\tilde{B}_z^i$ is a closed convex family, by noting that $\dim(C_i) \leq d' - 1$, C_i is rationally unbounded by definition of C as a rational map of C and hence S_i is rational-MICP representable with MICP-dimension at most $d' - 1$.

References

1. Bonami, P., Kilinç, M., Linderoth, J.: Algorithms and software for convex mixed integer nonlinear programs. In: Lee, J., Leyffer, S. (eds.) Mixed Integer Nonlinear Programming. The IMA Volumes in Mathematics and its Applications, vol. 154, pp. 1–39. Springer, New York (2012)
2. Ceria, S., Soares, J.: Convex programming for disjunctive convex optimization. Math. Program. **86**(3), 595–614 (1999)
3. Del Pia, A., Poskin, J.: Ellipsoidal mixed-integer representability. Optimization Online (2016). http://www.optimization-online.org/DB_HTML/2016/05/5465.html
4. Dey, S.S., Morán, R., Diego, A.: Some properties of convex hulls of integer points contained in general convex sets. Math. Program. **141**, 507–526 (2013)
5. James, J.P., Sato, D., Wada, H., Wiens, D.: Diophantine representation of the set of prime numbers. Am. Math. Mon. **83**(6), 449–464 (1976)
6. Jeroslow, R.G., Lowe, J.K.: Modeling with integer variables. Math. Program. Stud. **22**, 167–184 (1984)
7. Lubin, M., Zadik, I., Vielma, J.P.: Mixed-integer convex representability. ArXiv e-prints arXiv:1611.07491, March 2017

8. Lubin, M., Yamangil, E., Bent, R., Vielma, J.P.: Extended formulations in mixed-integer convex programming. In: Louveaux, Q., Skutella, M. (eds.) IPCO 2016. LNCS, vol. 9682, pp. 102–113. Springer, Cham (2016). doi:10.1007/978-3-319-33461-5_9
9. Vielma, J.P.: Mixed integer linear programming formulation techniques. SIAM Rev. **57**(1), 3–57 (2015)

High Degree Sum of Squares Proofs, Bienstock-Zuckerberg Hierarchy and CG Cuts

Monaldo Mastrolilli[✉]

IDSIA, 6928 Manno, Switzerland
monaldo@idsia.ch

Abstract. Chvátal-Gomory (CG) cuts captures useful and efficient linear programs that the bounded degree Lasserre/Sum-of-Squares (SOS) hierarchy fails to capture. We present an augmented version of the SOS hierarchy for 0/1 integer problems that implies the Bienstock-Zuckerberg hierarchy by using high degree polynomials (when expressed in the standard monomial basis). It follows that for a class of polytopes (e.g. set covering and packing problems), the SOS approach can optimize, up to an arbitrarily small error, over the polytope resulting from any constant rounds of CG cuts in polynomial time.

1 Introduction

The Lasserre/Sum-of-Squares (SOS) hierarchy [13,15,18] is a systematic procedure for constructing a sequence of increasingly tight semidefinite relaxations. The SOS hierarchy is parameterized by its *level* d, such that the formulation gets tighter as d increases, and a solution of accuracy $\varepsilon > 0$ can be found in time $(mn \log(1/\varepsilon))^{O(d)}$ where n is the number of variables and m the number of constraints in the original problem. It is known that the hierarchy converges to the 0/1 polytope in n levels and captures the convex relaxations used in the best available approximation algorithms for a wide variety of optimization problems (see e.g. [2,5] and the references therein).

In a recent paper Kurpisz, Leppänen and the author [11] characterize the set of 0/1 integer linear problems that still have an (arbitrarily large) integrality gap at level $n - 1$. These problems are the "hardest" for the SOS hierarchy in this sense. In another paper, the same authors [12] consider a problem that is solvable in $O(n \log n)$ time and prove that the integrality gap of the SOS hierarchy is unbounded at level $\Omega(\sqrt{n})$ even after incorporating the objective function as a constraint (a classical trick that sometimes helps to improve the quality of the relaxation). All these "SOS-hard" instances have a "covering nature".

Chvátal-Gomory (CG) rounding is a popular cut generating procedure that is often used in practice (see e.g. [6]). There are several prominent examples of CG-cuts in polyhedral combinatorics, including the odd-cycle inequalities of the stable set polytope, the blossom inequalities of the matching polytope, the

---

Supported by the Swiss National Science Foundation project 200020-169022 "Lift and Project Methods for Machine Scheduling Through Theory and Experiments".

F. Eisenbrand and J. Koenemann (Eds.): IPCO 2017, LNCS 10328, pp. 405–416, 2017.
DOI: 10.1007/978-3-319-59250-3_33

simple Möbius ladder inequalities of the acyclic subdigraph polytope and the simple comb inequalities of the symmetric traveling salesman polytope, to name a few. Chvátal-Gomory cuts captures useful and efficient linear programs that the bounded degree SOS hierarchy fails to capture. Indeed, the "SOS-hard" instances studied in [11] are the "easiest" for CG cuts, in the sense that they are captured within the *first* CG closure. It is worth noting that it is NP-hard [14] to optimize a linear function over the first CG closure, an interesting contrast to lift-and-project hierarchies (like Sherali-Adams, Lovász-Schrijver, and SOS) where one can optimize in polynomial time for any constant number of levels.

Interestingly, Bienstock and Zuckerberg [4] prove that, in the case of set covering, one can separate over all CG-cuts to an arbitrary fixed precision in polynomial time. The result in [4] is based on another result [3] by the same authors, namely on a (positive semidefinite) lift-and-project operator (which we denote (BZ) herein) that is quite different from the previously proposed operators. This lift-and-project operator generates different variables for different relaxations. They showed that this flexibility can be very useful in attacking relaxations of some set covering problems.

These three methods, (SOS, CG, BZ), are to some extent incomparable, roughly meaning that there are instances where one succeeds while the other fails (see [1] for a comparison between SOS and BZ, the already cited [11] for "easy" cases for CG cuts that are "hard" for SOS, and finally note that clique constraints are "easy" for SOS but "hard" for CG cuts [16], to name a few).

One can think of the standard Lasserre/SOS hierarchy at level $O(d)$ as optimizing an objective function over linear functionals that sends n-variate polynomials of degree at most d to real numbers. The restriction to polynomials of degree d is the standard way (as suggested in [13,15] and used in most of the applications) to bound the complexity, implying a semidefinite program of size $n^{O(d)}$. However, this is not strictly necessary for getting a polynomial time algorithm and it can be easily extended by considering more general subspaces having a "small" (i.e. polynomially bounded) set of basis functions (see e.g. Chap. 3 in [5] and [7,8]). This is a less explored direction and it will play a key role in this paper. Indeed, the more general view of the SOS approach has been used so far to exploit very symmetric situations (see e.g. [7,8,17]). For symmetric cases the use of different basis functions has been proved to be very useful.

To the best of author's knowledge, in this paper we give the first example where a different SOS basis is proved to be useful in *asymmetric* situations. More precisely, we claim that is possible to reframe the Bienstock-Zuckerberg hierarchy [3] as an augmented version of the SOS hierarchy that uses high degree polynomials. Due to space limitations in Sect. 4 we consider the set cover problem, that is the main known application of the BZ approach; the general BZ framework is strongly based on the set cover case. More details will appear in the full version of the paper.

The resulting high degree SOS approach retains in one single unifying SOS framework the best from the standard bounded degree SOS hierarchy, incorporates the BZ approach and allows to get arbitrary good approximate fixed rank CG cuts for both, set covering and packing problems, in polynomial time (BZ guarantees this only for set covering problems). Moreover, the proposed framework is very simple and, assuming a basic knowledge in SOS machinery (see Sect. 2), it is fully defined by giving the supporting polynomials (see Definition 2). This is in contrast to the Bienstock-Zuckerberg's hierarchy that requires an elaborate description [3,19].

2   sos-Proofs over the Boolean Hypercube

Certifying that a polynomial $f(x)$ is non-negative over a semialgebraic set $\mathcal{F}$ is an important problem in optimization, as certificates of non-negativity can often be leveraged into optimization algorithms. In this paper we are interested in the case $\mathcal{F}$ is the set of feasible solutions of a 0/1 integer linear program:

$$\mathcal{F} := \{x \in \mathbb{R}^n : x_k^2 - x_k = 0 \quad \forall k \in [n], \ g_i(x) \geq 0 \quad \forall i \in [m]\} \tag{1}$$

where $x_k^2 - x_k = 0$ encodes $x_k \in \{0,1\}$ and each constraint $g_i(x) \geq 0$ is linear. It is known that the nonnegativity of a polynomial over $\mathcal{F}$ defined in (1) can be certified by showing a degree-n sum of squares (SOS) representation (see e.g. [13,15]). Computing degree-n SOS representation can be automatized by solving a semidefinite program (SDP) which is an optimization problem over positive semidefinite (PSD) matrices. However this may take in general exponential time. The "standard" (namely the "most used") way to bound the complexity is to consider the polynomials $q_i \in \mathbb{R}_n[\mathbf{x}]$ used in the degree-n SOS representation in the standard monomial basis and to restrict their degree to a constant d. If one restricts the degrees of the polynomials in the certificate to be at most some integer d, it turns out that the positivity certificate is given by a semidefinite program of size $n^{O(d)}$. Clearly this restriction imposes severe restrictions on the kind of proofs that can be obtained. This type of algorithm was proposed first by Shor [18] and the idea was taken further by Parrilo [15] and Lasserre [13]. However, this fact can be easily extended to other subspaces than the standard monomial basis of bounded degree, by considering subspaces having a "small", i.e. polynomially bounded set of basis functions (see e.g. [5]). This is a less explored direction and it will play a key role in this paper. The following introduces this point.

Definition 1. *For any fixed subspace $Q \subseteq \mathbb{R}[\mathbf{x}]/\mathbf{I}(\mathbb{Z}_2^n)$, we say that a polynomial $f \in \mathbb{R}[\mathbf{x}]$ that is non-negative over the semialgebraic set (1) admits a Q-SOS representation (or it is Q-SOS derivable and write $\mathcal{F} \vdash_Q f(x) \geq 0$) if*

$$f(x) \equiv s_0(x) + \sum_{i \in [m]} s_i(x) g_i(x) \pmod{\mathbf{I}(\mathbb{Z}_2^n)} \tag{2}$$

where $s_i \in \{s \in \mathbb{R}[\mathbf{x}] : s = \sum_{i=1}^r q_i(x)^2, \text{ for some } q_1, \ldots, q_r \in Q\}$. For a set $\mathcal{S} \subseteq \mathbb{R}_n[\mathbf{x}]$ let $\langle \mathcal{S} \rangle = span(\mathcal{S})$ denote the vector space spanned by $\mathcal{S}$. If $\langle \mathcal{S} \rangle = Q$ then $\mathcal{S}$ is called the Q-SOS spanning set.

The existence of a $\langle S \rangle$-SOS representation can be decided by solving a semidefinite programming feasibility problem whose matrix dimension is bounded by $O(|S|)$. We refer to [5] for details.

The dual point of view. Consider the minimization of a given polynomial $p(x)$ over the semialgebraic set (1). Let $\mathcal{G} = \{g_i(x), i \in [m]\}$. For any $S \subseteq \mathbb{R}_n[\mathbf{x}]$, a relaxation is given by the following conic program:

$$\max\{\gamma : p - \gamma \in \mathbf{cone}_{\langle S \rangle}(\mathcal{G})\} \tag{3}$$

where $\mathbf{cone}_{\langle S \rangle}(\mathcal{G}) = \{f(x) : f(x) = s_0(x) + \sum_{g \in \mathcal{G}} s_g(x)g(x)$ where $s_g(x) = \sum_i q_i(x)^2, q_i \in \langle S \rangle\}$ is the cone of nonnegative polynomials generated by $\langle S \rangle$. By definition, the dual of $\mathbf{cone}_{\langle S \rangle}(\mathcal{G})$ are the linear functionals[1] $\mathbf{cone}_{\langle S \rangle}^{dual}(\mathcal{G}) = \{l : \langle l, h \rangle \geq 0, \forall h \in \mathbf{cone}_{\langle S \rangle}(\mathcal{G})\}$ that take nonnegative values on it.

The dual program of (3) is a semidefinite program whose matrix dimension is bounded by $O(|S|)$. We will use SOS$_{\langle S \rangle}$ to denote this relaxation. When $\langle S \rangle = \mathbb{R}[\mathbf{x}]_d / \mathbf{I}(\mathbb{Z}_2^n)$, namely S is the standard monomial basis of degree $\leq d$, then SOS$_{\langle S \rangle}$ is the (standard) Lasserre/SOS-hierarchy parameterized by the degree d.

By duality, note that any SOS$_{\langle S \rangle}$ feasible solution satisfies all the linear inequalities that are $\langle S \rangle$-SOS derivable. We will use this fact in the remainder of the paper.

3   Preliminaries

In the following, whenever we use "$\equiv$" assume that the equivalence is modulo the vanishing ideal (mod $\mathbf{I}(\mathbb{Z}_2^n)$) (unless differently defined).

For any set $Z \subseteq [n]$ and given $I \subseteq Z$ define the *Kronecker delta* function $\delta_I^Z(x)$ by: $\delta_I^Z(x) := \prod_{i \in I} x_i \prod_{j \in Z \setminus I}(1 - x_j)$. Note that $\sum_{I \subseteq Z} \delta_I^Z(x) = 1$, $(\delta_I^Z(x))^2 \equiv \delta_I^Z(x)$ and $\delta_I^Z(x)\delta_J^Z(x) \equiv 0$ for any $I \neq J$ with $I, J \subseteq Z$. Therefore, it follows that $(\sum_I \delta_I^Z(x))^2 \equiv \sum_I \delta_I^Z(x)$. Finally observe that for any linear function $g(x) = \sum_{i \in S} g_i x_i - g_0$ with $S \subseteq [n]$ we have $\delta_I^Z(x)(\sum_{i \in S} g_i x_i - g_0) \equiv \delta_I^Z(x)(\sum_{i \in S \cap I} g_i - g_0 + \sum_{S \setminus Z} g_i x_i)$.

These basic facts will be used several times, in particular over the boolean hypercube we can restrict with no loss to polynomials from $\mathbb{R}[\mathbf{x}]/\mathbf{I}(\mathbb{Z}_2^n)$, i.e. n-degree multilinear polynomials (we use $\mathbb{R}[\mathbf{x}]$ to denote the polynomial ring over the reals in n variables $\mathbb{R}[x_1, \ldots, x_n]$ and $\mathbb{R}[\mathbf{x}]_d$ to denote the subspace of $\mathbb{R}[\mathbf{x}]$ of polynomials of degree at most $d \leq n$). So $\delta_I^Z(x)$ is the multilinear representation of $(\delta_I^Z(x))^2$ over the boolean hypercube and we will use them both interchangeably.

Consider any group G that is acting on monomials in $\mathbb{R}[\mathbf{x}]$ via $gx_i = x_{g(i)}$, for all $g \in G$ and $i \in [n]$. Let $f \in \mathbb{R}[n]$ be a real-valued G-invariant polynomial that is nonnegative over the boolean hypercube. By a simple interpolating argument, we have $f(x) \equiv (\sum_{I \in N^+} \delta_I^{[n]}(x)\sqrt{f(x_I)})^2$, and therefore $f(x)$ is congruent (mod $\mathbf{I}(\mathbb{Z}_2^n)$) to the square of a G-invariant polynomial.

---

[1] In some research communities such linear functional is called *pseudo-expectation*.

Lemma 1. *Consider any group G that is acting on monomials in $\mathbb{R}[\mathbf{x}]$ via $g x_i = x_{g(i)}$ for each $g \in G$ and $i \in [n]$. Any real-valued G-invariant polynomial $f \in \mathbb{R}[n]$ that is nonnegative over the boolean hypercube has a degree-n square representation $f(x) \equiv h(x)^2 \pmod{\mathbf{I}(\mathbb{Z}_2^n)}$, for some G-invariant polynomial $h \in \mathbb{R}[x]$.*

Let X be a nonempty set. A permutation σ of X is a bijection $\sigma : X \to X$. The set of all permutations of X is called the *symmetric group* of X and it is denoted by S_X. In the following for any $F \subseteq X$ we will consider the *stabilizer* of F in S_X, namely $stab_{S_X}(F)$ is the subgroup of S_X whose elements are permutations of set X that fix the elements from F. Note that $stab_{S_X}(F)$ is the symmetric subgroup $S_{X \setminus F}$ acting on X and leaving the points in F fixed. The set F is the $S_{X \setminus F}$ group's set of *fixed points* when acting on X.

For the main application of this paper (Sect. 4) the spanning set is given by products of $S_{X \setminus F}$-invariant polynomials (see Definition 2). Note that when $F = \emptyset$ an $S_{X \setminus F}$-invariant polynomial is standardly called a *symmetric* polynomial. Generalizing the latter terminology, we will also use $(X \setminus F)$-*symmetric* polynomial to denote a $S_{X \setminus F}$-invariant polynomial. Observe that *any* polynomial is $(X \setminus F)$-symmetric for some $F \subseteq X$.

From Lemma 1 any non-negative $(X \setminus F)$-symmetric polynomial is congruent $(\bmod \; \mathbf{I}(\mathbb{Z}_2^n))$ to the square of one $(X \setminus F)$-symmetric polynomial. This simple fact will play a central role in our derivations. In particular the following will be used several times in the following.

Corollary 1. *Consider any finite set of polynomials $\mathcal{S} \subseteq \mathbb{R}[\mathbf{x}]/\mathbf{I}(\mathbb{Z}_2^n)$ and let $Q = span(\mathcal{S})$. For any $F \subseteq X \subseteq [n]$, if the ring $(\mathbb{R}[\mathbf{x}]/\mathbf{I}(\mathbb{Z}_2^n))^{S_{X \setminus F}}$ of all $(X \setminus F)$-symmetric polynomials is a subspace of Q then any nonnegative $(X \setminus F)$-symmetric polynomial has a Q-SOS representation.*

A simple counting argument shows the following rough bound on the size of the spanning set of $(X \setminus F)$-symmetric polynomials.

Lemma 2. *For any $X \subseteq [n]$, let $Q_{X,t}$ denote the subspace of all $(X \setminus F)$-symmetric polynomials for all $F \subseteq X, |F| \leq t$. There is a spanning set $\mathcal{S}_{X,t}$ such that $Q_{X,t} \subseteq span(\mathcal{S}_{X,t})$ and $|\mathcal{S}_{X,t}| = n^{O(t)}$.*

4   Set Covering

Consider any $m \times n$ 0-1 matrix A, and let $\mathcal{F}_A$ be the feasible region for the 0-1 set covering problem defined by A:

$$\mathcal{F}_A = \{x \in \{0,1\}^n : Ax \geq e\} \tag{4}$$

where e is the vector of 1s. We denote by $A_i \subseteq \{1, \dots, n\}$ the set of indices of nonzeros in the i-th row of A (namely the *support* of the i-th constraint). By overloading notation, we also use A_i to denote the corresponding set of variables $\{x_j : j \in A_i\}$. We will assume that A is *minimal*, i.e. there is no $i \neq j$ such that $A_i \subseteq A_j$.

We will also use the following notation. For any $T, F \subseteq [n]$ with $T \cap F = \emptyset$, let $\mathcal{F}_{A_{(T,F)}}$ denote the subregion of $\mathcal{F}_A$ where $x_i = 1$, for $i \in T$, and $x_j = 0$, for $j \in F$. Let $A_{(T,F)}$ be the matrix that is obtained from A by removing all the rows where x_i appears for $i \in T$ (these constraints are satisfied when $x_i = 1$ for $i \in T$) and setting to zero the j-th column for $j \in F$. We will assume that $A_{(T,F)}$ is minimal by removing the dominated rows. Therefore, $\mathcal{F}_{A_{(T,F)}} = \{x \in \{0,1\}^n : A_{(T,F)}x \geq e, x_i = 1 \ \forall i \in T, x_j = 0 \ \forall j \in F\}$ and $\mathcal{F}_{A_{(T,F)}} \subseteq \mathcal{F}_A$.

4.1   The Spanning Polynomials for the Set Covering Problem

In this section we define the spanning polynomials for the set covering problem. For the sake of simplicity, we will assume that the collection of valid inequalities, i.e. $\{g_i(x) \geq 0, i \in [\ell]\}$, that are defined in the semialgebraic set (1) is given by $Ax \geq e$ and the nonnegative constraints $x \geq 0$. The latter is not strictly necessary, since $x_i = x_i^2$ and therefore $x_i \geq 0$, but this will simplify the exposition. We start with a simple structural observation regarding the set covering problem (see e.g. [19] for a proof).

Proposition 1. *Consider a set covering problem defined by any $m \times n$ 0-1 matrix B such that no two constraints overlap in any of the variables, namely for any $i, j \in [m]$ with $i \neq j$ we have $B_i \cap B_j = \emptyset$. When this holds then the linear constraints are convex hull defining: $conv(\mathcal{F}_B) = \{x \in [0,1]^n : Bx \geq e\}$.*

Remark 1. From Proposition 1 it follows that any valid inequality $a^\top x \geq a_0$ for $\mathcal{F}_B$ is valid also for the feasible region of the linear relaxation $\{x \in [0,1]^n : Bx \geq e\}$, i.e. $a^\top x \geq a_0$ can be derived as a nonnegative linear combination and right-hand-side weakening from $\{x \geq 0, Bx \geq e\}$: $a = \lambda^\top B + \gamma^\top I$ and $a_0 \leq \lambda^\top e$, for some $\lambda, \gamma \geq 0$ and where I denotes the $n \times n$ identity matrix.

By the previous observation the "interesting" variables are those that appear in more than one constraint. This gives the intuition why the $Q_A(t)$-sos polynomials that we are going to define are polynomials in these variables.

Definition 2. *For any $t \in [n]$ and $\mathcal{C}(t) = \{C : C \subseteq [m] \wedge |C| \leq t\}$, let $V_C = \bigcup_{i,j \in C, i \neq j} A_i \cap A_j$ be the set of variables occurring in more than one row with index from $C \in \mathcal{C}(t)$. The subspace of polynomials $Q_A(t)$ is (inductively) defined as the set of all polynomials $p(x) \in \mathbb{R}[\mathbf{x}]$ for which there exists a $C \in \mathcal{C}(t)$ and $I \subseteq C$ with $|I| \leq t$ such that $p(x)$ can be written as $p(x) = q(x)r(x)$, where $q(x)$ is $(V_C \backslash I)$-symmetric and, depending on $|I|$, $r(x)$ is either 1 (if $|I| \in \{0, t\}$) or $r(x) \in Q_{A_{(I, V_C \backslash I)}}(t - |I|)$ (else).*

By Lemma 2 a $Q_A(t)$-sos representation can be decided by solving a semidefinite programming feasibility problem of size $n^{O(t^2)}$.

For any given inequality $a^\top x - a_0 \geq 0$ with indices ordered so that $0 < a_1 \leq a_2 \leq \cdots \leq a_h$ and $a_j = 0$ for $j > h$, its *pitch* is the minimum integer $\pi = \pi(a, a_0)$ such that $\sum_{i=1}^{\pi} a_i - a_0 \geq 0$. The definition of pitch was introduced in [3, 19]. The main result of this section is the following.

Theorem 1. *Suppose $a^\top x - a_0 \geq 0$ is a valid inequality for $\mathcal{F}_A$ of pitch $\pi = \pi(a, a_0)$ with $a \geq 0$. Then $a^\top x - a_0$ admits a $Q_A(\pi)$-SOS representation.*

Corollary 2. *For any $k \geq 1$, any valid solution of the $\text{SOS}_{Q_A(k)}$ relaxation satisfies all the valid inequalities for $\mathcal{F}_A$ of pitch $\leq k$.*

Remark 2. Note that for the set-covering problem with a full-circulant constraint matrix (namely $\sum_{j \neq i} x_j \geq 1$ for each $i = 1, \ldots, n$) the pitch 2 valid inequality $\sum_{j=1}^n x_j \geq 2$ has rank at least $n - 3$ for a lifting operator stronger than the Sherali-Adams [3] and requires at least $\Omega(\log^{1-\varepsilon} n)$ levels [10] for the standard SOS hierarchy (conjectured to be n/4 in [3]). Viceversa, the augmented $\text{SOS}_{Q_A(k)}$ relaxation returns a solution that satisfies all the pitch 2 valid inequalities in polynomial time ($Q_A(2)$, see Definition 2, is sufficient for this purpose).

Proof of Theorem 1. The proof will be by induction on the pitch value. Consider *any* $m \times n$ 0-1 matrix A', and let $\mathcal{F}_{A'}$ be the feasible region for the 0-1 set covering problem defined by A': $\mathcal{F}_{A'} = \{x \in \{0, 1\}^n : A'x \geq e\}$. Assume that $a'^\top x - a'_0 \geq 0$ with $a' \geq 0$ is a valid inequality for $\mathcal{F}_{A'}$ of pitch π'. If $\pi' = 0$ we must have $a'_0 \leq 0$, so since $a' \geq 0$, $a'^\top x - a'_0$ has a trivial $Q_{A'}(\pi')$-SOS representation as conical combination of x_i, for $i \in [n]$. By induction hypothesis, from now on we will assume that for *any* $m \times n$ 0-1 matrix A' the claim holds for any constraint $a'^\top x - a'_0 \geq 0$ of pitch p, with $0 \leq p \leq \pi - 1$, that is valid for $\mathcal{F}_{A'}$.

We start describing a key structural property of valid inequalities for set covering that was proved in [3,19]. We observe that the statement of Lemma 3 below is slightly different from Proposition 4.22 in [19] (or Theorem 6.3 in [3]). The difference is given by Property (8) (see Lemma 3). This property is not explicitly given in [3,19], but it can be derived by their construction (further details will appear in the longer version of this paper).

Lemma 3. *[3, 19] Suppose $a^\top x - a_0 \geq 0$ is a valid inequality for $\mathcal{F}_A$ with $a \geq 0$. Let $supp(a)$ denote the support of a. Then there is a subset $C = C(a, a_0)$ of the rows of A with $|C| \leq \pi(a, a_0)$, such that*

$$A_i \subseteq supp(a) \quad \forall i \in C \tag{5}$$

$$B_i = A_i - \bigcup_{r \in C - \{i\}} A_r \neq \emptyset \quad \forall i \in C \tag{6}$$

$$(a^\top x - a_0)_{(\emptyset, V)} \geq 0 \, is \, valid \, for \, \mathcal{F}_B = \{x \in \{0, 1\}^n : \sum_{j \in B_i} x_j \geq 1, i \in C\} \tag{7}$$

$$\mathcal{F}_{A_{(\emptyset, V)}} \neq \emptyset \tag{8}$$

where $V := \bigcup_{\substack{i, j \in C \\ i \neq j}} A_i \cap A_j$ is the set of variables occurring in more than one row from C.

Consider any valid inequality $a^\top x - a_0 \geq 0$ for $\mathcal{F}_A$ of pitch $\pi = \pi(a, a_0)$ with $a \geq 0$. We show that $a^\top x - a_0$ admits a $Q_A(\pi)$-SOS representation. By Lemma 3 there is a subset $C = C(a, a_0)$ of the rows of A that satisfies (5)–(8) where V denotes the set of variables occurring in more than one row of C and $|C| \leq \pi$. The following polynomials have a $Q_A(\pi)$-SOS representation: δ_J^V, for $J \subseteq V$ with $|J| < \pi$, and $\delta_{\geq\pi}^V := \sum_{J \subseteq V, |J| \geq \pi} \delta_J^V$ (it is zero if $|V| < \pi$). Note that $\sum_{J \subseteq V, |J| < \pi} \delta_J^V + \delta_{\geq\pi}^V = 1$. It follows that

$$
a^\top x - a_0 = \underbrace{\left(\sum_{J \subseteq V, |J| < \pi} \delta_J^V + \delta_{\geq\pi}^V \right)}_{=1} (a^\top x - a_0)
$$

$$
= \underbrace{\delta_\emptyset^V (a^\top x - a_0)}_{\text{first}} + \underbrace{\left(\sum_{J \subseteq V, 0 < |J| < \pi} \delta_J^V \right)(a^\top x - a_0)}_{\text{second}} + \underbrace{(\delta_{\geq\pi}^V)(a^\top x - a_0)}_{\text{third}} \qquad (9)
$$

Therefore, showing that $a^\top x - a_0$ is $Q_A(\pi)$-SOS derivable boils down to prove that each of the summands in (9) is $Q_A(\pi)$-SOS derivable.

Let's start considering the first summand in (9), namely $\delta_\emptyset^V (a^\top x - a_0)$. By Lemma 3, first note that $(a^\top x - a_0)_{(\emptyset, V)} \geq 0$ is valid for $\mathcal{F}_B$ (see (7)). Moreover, no two constraints in $\mathcal{F}_B$ overlap in any of the variables and therefore, by Proposition 1, the linear relaxation is convex hull defining: $conv(\mathcal{F}_B) = \{x \in [0,1]^n : \sum_{j \in B_i} x_j \geq 1, i \in C\}$. This means (see Remark 1) that $(a^\top x - a_0)_{(\emptyset, V)}$ can be implied by a conical combination of the linear constraints in $conv(\mathcal{F}_B) = \{x \in [0,1]^n : \sum_{j \in B_i} x_j \geq 1, i \in C\}$. Note that these linear constraints are just a subset of the linear constraints from $\{x \in [0,1]^n : Ax \geq e\}$ after setting to zero all the variables from V. It follows that $(a^\top x - a_0)_{(\emptyset, V)} = \left(\lambda^\top (Ax - e) + \gamma^\top x + \mu\right)_{(\emptyset, V)}$ for some $\lambda, \gamma, \mu \geq 0$. For any $x_i \in V$ we have $\delta_\emptyset^V x_i \equiv 0$ (recall that whenever we use "$\equiv$" we assume that the equivalence is $(\bmod \; \mathbf{I}(\mathbb{Z}_2^n))$) and

$$
\delta_\emptyset^V (a^\top x - a_0) \equiv \delta_\emptyset^V \left(\lambda^\top (Ax - e) + \gamma^\top x + \mu\right)_{(\emptyset, V)} \equiv \delta_\emptyset^V \left(\lambda^\top (Ax - e) + \gamma^\top x + \mu\right)
$$

$$
\equiv \sum_{i \in C} \underbrace{(\delta_\emptyset^V \sqrt{\lambda_i})^2}_{s_i(x)} \underbrace{\left(\sum_{j \in A_i} x_j - 1\right)}_{g_i(x)} + \sum_{j \in supp(a)} \underbrace{(\delta_\emptyset^V \sqrt{\gamma_j})^2}_{s_j(x)} \underbrace{x_j}_{g_j(x)} + \underbrace{(\delta_\emptyset^V \sqrt{\mu})^2}_{s_0(x)}
$$

Note that the latter has exactly the form in (2), where $\delta_\emptyset^V \in Q_A(\pi)$, and therefore it shows that the first summand $\delta_\emptyset^V (a^\top x - a_0)$ is $Q_A(\pi)$-SOS derivable.

Consider a generic second type summand from (9), i.e. $\delta_J^V (a^\top x - a_0)$ with $J \subseteq V, 0 < |J| < \pi$. Note that $a^\top x - a_0 \geq 0$ is by assumption a valid inequality for any feasible integral solution. By Property (8) we know that by setting to zero all the variables from V we obtain a non-empty subset of feasible integral solutions. It follows that by setting $x_j = 1$, for $j \in J$, and $x_h = 0$, for $h \in V \setminus J$, we obtain a non-empty subset of feasible integral solutions, i.e. $\mathcal{F}_{A_{(J, V \setminus J)}} \neq \emptyset$ and

$(a^\top x - a_0)_{(J,V\setminus J)} \geq 0$ is a valid inequality for the solutions in $\mathcal{F}_{A_{(J,V\setminus J)}}$ (since $a^\top x - a_0 \geq 0$ is by assumption a valid inequality for any feasible integral solution). Moreover the pitch p of $(a^\top x - a_0)_{(J,V\setminus J)} \geq 0$ is strictly smaller than π, $0 \leq p \leq \pi - |J|$. It follows, by induction hypothesis that $(a^\top x - a_0)_{(J,V\setminus J)}$ has a $Q_{A_{(J,V\setminus J)}}(p)$-SOS representation, namely $(a^\top x - a_0)_{(J,V\setminus J)} \equiv s'_0(x) + \sum_i s'_i(x)g_i(x)_{(J,V\setminus J)}$ where $s'_i \in \{s' \in \mathbb{R}[\mathbf{x}] : s' = \sum_i q_i(x)^2, q_i \in Q_{A_{(J,V\setminus J)}}(p)\text{-SOS}\}$ and each $g_i(x)_{(J,V\setminus J)} \geq 0$ is a valid linear constraint for $\mathcal{F}_{A_{(J,V\setminus J)}}$, where $g_i(x)_{(J,V\setminus J)}$ is either $(\sum_{j\in A_h} x_j - 1)_{(J,V\setminus J)} \geq 0$ or $(x_j)_{(J,V\setminus J)} \geq 0$ for some $h \in [m], j \in [n]$. Then $\delta^V_J(a^\top x - a_0) \equiv \delta^V_J(a^\top x - a_0)_{(J,V\setminus J)} \equiv \delta^V_J\left(s'_0(x) + \sum_i s'_i(x)g_i(x)_{(J,V\setminus J)}\right) \equiv s'_0(x)\delta^V_J + \sum_i s'_i(x)\delta^V_J g_i(x)$. Recall that $0 \leq p \leq \pi - |J|$ and $s'_i(x) = \sum_j q_j(x)^2$ for $q_j \in Q_{A_{(J,V\setminus J)}}(p)$-SOS therefore $\delta^V_J q_j(x)^2 \equiv (\delta^V_J q_j(x))^2$ and $\delta^V_J q_j(x) \in Q_A(\pi)$-SOS (by Definition 2). It follows that the second summand is $Q_A(\pi)$-SOS derivable.

Finally, consider the third summand from (9), i.e. $\delta^V_{\geq\pi}(a^\top x - a_0)$. Recall that we are assuming that $0 < a_1 \leq a_2 \leq \cdots \leq a_h$ and $a_j = 0$ for $j > h$ for some $h \in [n]$, so the $supp(a) = \{1, \ldots, h\}$. Moreover the pitch $\pi \leq h$ is the minimum such that $\sum_{i=1}^\pi a_i - a_0 \geq 0$. Note that $V \subseteq supp(a)$ and therefore if $\delta^V_{\geq\pi}$ is a non-null polynomial then $|V| \geq \pi$ (we assume this in the following otherwise we are done for this case). Let $a'_i := a_i$ for $i = 1, \ldots, \pi$, $a'_i := a_\pi$ for $i = \pi+1, \ldots, a_h$ and $a'_i := 0$ for $i \in supp(a)\setminus V$. Note that for any $I \subseteq [\pi]$ and $J_1, J_2 \subseteq V\setminus[\pi]$ with $|J_1| = |J_2| \geq \pi - |I|$ we have for $\ell = 1, 2$ $\delta^V_{I\cup J_\ell}\left(\sum_{i\in V} a'_i x_i - a_0\right) \equiv$

$$\delta^V_{I\cup J_\ell}\left(\sum_{i\in I\cup J_\ell} a'_i - a_0\right) = \delta^V_{I\cup J_\ell}\left(\sum_{i\in I} a'_i + \overbrace{|J_\ell|a_\pi - a_0}^{\geq 0}\right).$$

Note that for any $k = \pi - |I|, \ldots, |V|$ the polynomial $\sum_{\substack{J\subseteq V\setminus[\pi] \\ |J|=k}} \delta^V_{I\cup J}$ is $(V\setminus[\pi])$-symmetric and therefore it belongs to $Q_A(\pi)$. It follows that

$$\delta^V_{\geq\pi}\left(\sum_{i=1}^h a_i x_i - a_0\right) = \delta^V_{\geq\pi}\left(\sum_{i\in V} a_i x_i - a_0\right) + \delta^V_{\geq\pi}\left(\sum_{i\in supp(a)\setminus V} a_i x_i\right)$$

$$= \delta^V_{\geq\pi}\left(\sum_{i\in V} a'_i x_i - a_0\right) + \delta^V_{\geq\pi}\left(\sum_{i\in supp(a)} (a_i - a'_i)x_i\right)$$

$$= \sum_{I\subseteq V\cap[\pi]} \overbrace{\sum_{k=\pi-|I|}^{|V|} \sum_{\substack{J\subseteq V\setminus[\pi] \\ |J|=k}} \delta^V_{I\cup J}}^{=\delta^V_{\geq\pi}}\left(\sum_{i\in V} a'_i x_i - a_0\right) + \sum_{i\in supp(a)} ((a_i - a'_i)\delta^V_{\geq\pi})x_i$$

$$\equiv \sum_{I\subseteq V\cap[\pi]} \sum_{k=\pi-|I|}^{|V|}\left(\sum_{\substack{J\subseteq V\setminus[\pi] \\ |J|=k}} \delta^V_{I\cup J}\left(\overbrace{\sum_{i\in I} a'_i + ka_\pi - a_0}^{\geq 0}\right)\right) + \sum_{i\in supp(a)} \overbrace{((a_i - a'_i)\delta^V_{\geq\pi})}^{\geq 0}x_i$$

$$\equiv \underbrace{\sum_{I\subseteq V\cap[\pi]} \sum_{k=\pi-|I|}^{|V|}\left(\overbrace{\sum_{\substack{J\subseteq V\setminus[\pi] \\ |J|=k}} \delta^V_{I\cup J}}^{(V\setminus[\pi])\text{-symmetric}} \sqrt{\sum_{i\in I} a'_i + ka_\pi - a_0}\right)^2}_{s_0(x)} + \sum_{i\in supp(a)} \underbrace{\left(\sqrt{a_i - a'_i}\, \overbrace{\delta^V_{\geq\pi}}^{V\text{-symm.}}\right)^2}_{s_i(x)} \underbrace{x_i}_{g_i(x)}$$

The latter has exactly the form in (2), and each polynomial under the square is from $Q_A(\pi)$, and therefore the third summand $\delta^V_{\geq\pi}(a^\top x - a_0)$ is $Q_A(\pi)$-SOS derivable and the claim follows.   □

5 Covering/Packing sos PTAS for Fixed Rank CG Closure

For an arbitrary fixed precision $\varepsilon > 0$ and fixed positive integer q, choose π such that $\left(\frac{\pi+1}{\pi}\right)^q \leq 1 + \varepsilon$. Bienstock and Zuckerberg (see Lemma 2.1 in [4]) prove that any solution that satisfies the set of valid inequalities of pitch π can be rounded to approximate all the CG-cuts constraint of rank q to precision $\varepsilon > 0$. It follows that the SOS approach with high degree polynomials described in this paper computes fixed rank CG $(1 + \varepsilon)$-approximate solutions for any fixed $\varepsilon > 0$ in polynomial time as well (PTAS).

Interestingly, in the following we observe that for the packing problem the standard SOS hierarchy with bounded degree polynomials is sufficient to obtain fixed rank CG $(1 - \varepsilon)$-approximate solutions. It follows that the SOS approach can be used for approximating CG cuts of any fixed rank and to any arbitrary precision for both, packing and set covering problems (BZ guarantees this only for set covering problems).

Approximate fixed-rank CG closure for packing problems. Consider any $m \times n$ nonnegative matrix A, and let $\mathcal{P}$ be the feasible region for the 0–1 set packing problem defined by A: $\mathcal{P} = \{x \in \{0,1\}^n : Ax \leq b\}$ where $b \in \mathbb{R}^m_+$. For an integer $t \geq 0$, denote by $P^{(t)}$ the t-th CG closure and let $cg^{(t)}(c) := \max\{c^\top x : x \in P^{(t)}\}$. Without loss of generality, we will assume that $c \in \mathbb{R}^n_+$ (otherwise it is always optimal to set $x_i = 0$ whenever $c_i \leq 0$).

We can extend the definition of pitch also for packing inequalities as follows. For any given packing inequality $a_0 - a^\top x \geq 0$ with $a_0, a \geq 0$ and indices ordered so that $0 < a_1 \leq a_2 \leq \cdots \leq a_h$ and $a_j = 0$ for $j > h$, its *pitch* is the *maximum* integer $\pi = \pi(a, a_0)$ such that $a_0 - \sum_{i=1}^{\pi} a_i \geq 0$. For example, classical clique inequality $\sum_{i\in clique} x_i \leq 1$ have pitch equal to one.

The following result for packing problems can be seen as the dual of Theorem 1 for set cover. It can be easily obtained by using the so called "Decomposition Theorem" due to Karlin, Mathieu, and Nguyen [9]. For completeness, here we sketch a direct simple proof that follows the approach used throughout this paper.

Lemma 4. *Suppose $a_0 - a^\top x \geq 0$ is a valid inequality for $\mathcal{P}$ of pitch $\pi = \pi(a, a_0)$ with $a_0, a \geq 0$. Then $a_0 - a^\top x$ admits a $\mathbb{R}_{\pi+1}[\mathbf{x}]$-SOS representation.*[2]

Proof (sketch). Let $S = supp(a)$ and $x^I = \prod_{i\in I} x_i$. Note that for any $I \subseteq S$ we have $x^I(a_0 - a^\top x) \equiv x^I(a_0 - \sum_{i\in I} a_i - \sum_{i\notin I} a_i x_i) \pmod{\mathbf{I}(\mathbb{Z}_2^n)}$. Let $F := \{I \subseteq S : (a_0 - \sum_{i\in I} a_i) < 0\}$ and $T := \{J \subseteq S : J \notin F\}$ (and therefore if we set to one

---

[2] This is the standard bounded degree SOS proof system.

all the variables x_i with $i \in I$ for any $I \in F$ then the assumed valid inequality $a_0 - a^\top x \geq 0$ is violated). Let $V := \{x \in \mathbb{R}^n : x^I = 0 \ \forall I \in F, \ x_k^2 - x_k = 0 \ \forall k \in [n]\}$ and note that any feasible integral solution belongs to V. Any δ_j^S is actually equivalent (mod $\mathbf{I}(V)$) to a polynomial $\bar{\delta}_j^S$ of degree at most π (obtained from δ_j^S by zeroing all the monomials x^I with $I \in F$ and therefore at least all the monomials of degree larger than π). Note that $\sum_{I \subseteq [n]} \bar{\delta}_I^S = \sum_{I \subseteq T} \bar{\delta}_I^S = 1$, $(\bar{\delta}_I^S)^2 \equiv \bar{\delta}_I^S$ (mod $\mathbf{I}(V)$) and $\bar{\delta}_I^S(a_0 - a^\top x) \equiv \bar{\delta}_I^S(a_0 - \sum_{i \in I} a_i)$ (mod $\mathbf{I}(V)$). Then

$$a_0 - a^\top x = \left(a_0 - a^\top x\right) \overbrace{\left(\sum_{I \subseteq T} \bar{\delta}_I^S\right)}^{=1} \equiv \sum_{I \in T} \underbrace{\overbrace{\left(a_0 - \sum_{i \in I} a_i\right)}^{\geq 0} (\bar{\delta}_I^S)^2}_{s_0(x)} \ (\text{mod } \mathbf{I}(V))$$

From the above equivalence we see that $a_0 - a^\top x$ can be written (mod $\mathbf{I}(V)$) as a conical combination of squares of polynomials of degree at most π.

By definition of the equivalences (mod $\mathbf{I}(V)$) and (mod $\mathbf{I}(\mathbb{Z}_2^n)$), we can now easily transform the equivalence (mod $\mathbf{I}(V)$) into the equivalence (mod $\mathbf{I}(\mathbb{Z}_2^n)$) as given by (2) by adding some polynomials from $\mathbf{I}(V) \setminus \mathbf{I}(\mathbb{Z}_2^n)$. It is easy to argue that these polynomials have degree $O(\pi)$. (More details will appear in the longer version of this paper.) ☐

Let $Pr(d)$ denote the set of feasible solutions for $\mathbb{R}_{d+1}[\mathbf{x}]$-SOS projected to the original variables. The following simple result shows that fixed rank CG closures of packing problems can be approximated to any arbitrarily precision in polynomial time by using the SOS hierarchy.

Theorem 2. *For each integer $t \geq 0$ and $\varepsilon > 0$ there are integers $d = d(t, \varepsilon)$ such that $\max\{c^\top x : x \in Pr(d)\} \leq (1 + \varepsilon)cg^{(t)}$, for any $c \in \mathbb{R}_+^n$.*

Proof. For any fixed $\varepsilon > 0$ and integer $t \geq 0$ choose $d > 0$ integral large enough that $((d+1)/d)^t \leq 1 + \varepsilon$. Consider the solution $x^{(\ell)}$ obtained by multiplying any given solution $x \in Pr(d)$ by a factor equal to $(\frac{d}{d+1})^\ell$. It follows that $\max\{c^\top x : x \in Pr(d)\}$ is not larger than a factor of $(\frac{d+1}{d})^\ell$ of the value of $x^{(\ell)}$. Now the claim follows by showing that $x^{(t)}$ is feasible for the rank-t CG closure.

The proof is by induction on t. As a base of induction note that when $t = 0$ then clearly $x^{(0)}$ satisfies all the original constraints. Assume now, by induction hypothesis, that the claim is true for any rank equal to $(t - 1)$ with $t \geq 1$ and we need to show that it is valid also for rank-t. If the pitch of a generic rank-t valid inequality for $P^{(t)}$ is at most d then by Lemma 4 it follows that any feasible solution $x \in Pr(d)$ (and therefore $x^{(\ell)}$) satisfies this inequality. Otherwise, consider a generic rank-t valid inequality $\lfloor a_0 \rfloor - a^\top x \geq 0$ of pitch larger than d, where $a_0 - a^\top x \geq 0$ is any valid inequality from the closure $P^{(t-1)}$. By induction hypothesis note that $a_0 - a^\top x^{(t-1)} \geq 0$. Since the pitch is higher than d then $a_0 > d$ (vector a can be assumed, w.l.o.g., to be nonnegative and integral) and therefore $\frac{a_0}{\lfloor a_0 \rfloor} \leq \frac{d+1}{d}$ and by multiplying the solution $x^{(t-1)} \in P^{(t-1)}$ by $d/(d+1)$ we obtain a feasible solution for the rank-t CG closure. ☐

Open Problems. It would be nice to understand if it is possible (i) to generalize Theorem 1 to work with general covering problems, (ii) to get a PTAS to approximate all the CG-cuts constraints for more general problems.

References

1. Au, Y.H., Tunçel, L.: Elementary polytopes with high lift-and-project ranks for strong positive semidefinite operators. arXiv preprint arXiv:1608.07647 (2016)
2. Bansal, N.: Hierarchies reading group. http://www.win.tue.nl/nikhil/hierarchies/
3. Bienstock, D., Zuckerberg, M.: Subset algebra lift operators for 0-1 integer programming. SIAM J. Optim. **15**(1), 63–95 (2004)
4. Bienstock, D., Zuckerberg, M.: Approximate fixed-rank closures of covering problems. Math. Program. **105**(1), 9–27 (2006)
5. Blekherman, G., Parrilo, P.A., Thomas, R.R.: Semidefinite Optimization and Convex Algebraic Geometry, vol. 13. Siam, Philadelphia (2013)
6. Conforti, M., Cornuejols, G., Zambelli, G.: Integer Programming. Springer Publishing Company, Incorporated (2014)
7. Fawzi, H., Saunderson, J., Parrilo, P.A.: Equivariant semidefinite lifts and sum-of-squares hierarchies. SIAM J. Optim. **25**(4), 2212–2243 (2015)
8. Gatermann, K., Parrilo, P.A.: Symmetry groups, semidefinite programs, and sums of squares. J. Pure Appl. Algebra **192**(1), 95–128 (2004)
9. Karlin, A.R., Mathieu, C., Nguyen, C.T.: Integrality gaps of linear and semidefinite programming relaxations for knapsack. In: Günlük, O., Woeginger, G.J. (eds.) IPCO 2011. LNCS, vol. 6655, pp. 301–314. Springer, Heidelberg (2011). doi:10.1007/978-3-642-20807-2_24
10. Kurpisz, A., Leppänen, S., Mastrolilli, M.: Sum-of-squares hierarchy lower bounds for symmetric formulations. In: Louveaux, Q., Skutella, M. (eds.) IPCO 2016. LNCS, vol. 9682, pp. 362–374. Springer, Cham (2016). doi:10.1007/978-3-319-33461-5_30
11. Kurpisz, A., Leppänen, S., Mastrolilli, M.: On the hardest problem formulations for the 0/1 lasserre hierarchy. Math. Oper. Res. **42**(1), 135–143 (2017)
12. Kurpisz, A., Leppänen, S., Mastrolilli, M.: An unbounded sum-of-squares hierarchy integrality gap for a polynomially solvable problem. Mathematical Programming (2017, to appear)
13. Lasserre, J.B.: Global optimization with polynomials and the problem of moments. SIAM J. Optim. **11**(3), 796–817 (2001)
14. Letchford, A.N., Pokutta, S., Schulz, A.S.: On the membership problem for the 0, 1/2-closure. Oper. Res. Lett. **39**(5), 301–304 (2011)
15. Parrilo, P.A.: Semidefinite programming relaxations for semialgebraic problems. Math. Program. **96**(2), 293–320 (2003)
16. Pudlák, P.: Lower bounds for resolution and cutting plane proofs and monotone computations. J. Symb. Log. **62**(3), 981–998 (1997)
17. Raymond, A., Saunderson, J., Singh, M., Thomas, R.R.: Symmetric sums of squares over k-subset hypercubes. arXiv preprint arXiv:1606.05639 (2016)
18. Shor, N.: Class of global minimum bounds of polynomial functions. Cybern. Syst. Anal. **23**(6), 731–734 (1987)
19. Zuckerberg, M.: A set theoretic approach to lifting procedures for 0, 1 integer programming. Ph.D. thesis, Columbia University (2004)

Enumeration of Integer Points in Projections of Unbounded Polyhedra

Danny Nguyen$^{(\boxtimes)}$ and Igor Pak

Department of Mathematics, UCLA, Los Angeles, CA 90095, USA
{ldnguyen,pak}@math.ucla.edu

Abstract. We extend the Barvinok–Woods algorithm for enumeration of integer points in projections of polytopes to unbounded polyhedra. To achieve this, we employ a new structural result on projections of semilinear subsets of the integer lattice.

1 Introduction

1.1 The Results

Integer linear programming in fixed dimension is a classical subject [Len83]. The pioneering result by Lenstra [Len83] shows that the *feasibility* of integer linear programming in a fixed dimension n can be decided in polynomial time:

$$(\circ) \qquad \exists \mathbf{x} \in \mathbb{Z}^n : A\mathbf{x} \leq \bar{b}.$$

This result was extended by Kannan [Kan90], who showed that *parametric integer linear programming* in fixed dimensions can be decided in polynomial time:

$$(\circ\circ) \qquad \forall \mathbf{y} \in P \cap \mathbb{Z}^n \ \exists \mathbf{x} \in \mathbb{Z}^m : A\mathbf{x} + B\mathbf{y} \leq \bar{b}.$$

Both results rely on difficult results in geometry of numbers and can be viewed geometrically: $(\circ)$ asks whether a polyhedron $Q = \{A\mathbf{x} \leq \bar{b}\} \subseteq \mathbb{R}^n$ has an integer point. Similarly, $(\circ\circ)$ asks whether every integer point in the polyhedron P is the projection of an integer point in the polyhedron $Q = \{A\mathbf{x} + B\mathbf{y} \leq \bar{b}\} \subseteq \mathbb{R}^{m+n}$.

Barvinok [Bar93] famously showed that the number of integer points in polytopes in a fixed dimension n can be computed in polynomial time. He used a technology of *short generating functions* (GF) to enumerate the integer points in general (possibly unbounded) rational polyhedra in $\mathbb{R}^d$ in the following form:

$$(*) \qquad f(\mathbf{t}) = \sum_{i=1}^{N} \frac{c_i \, \mathbf{t}^{\bar{a}_i}}{(1 - \mathbf{t}^{\bar{b}_{i1}}) \cdots (1 - \mathbf{t}^{\bar{b}_{i k_i}})},$$

where $c_i \in \mathbb{Q}$, $\bar{a}_i, \bar{b}_{ij} \in \mathbb{Z}^n$ and $\mathbf{t}^{\bar{a}} = t_1^{a_1} \cdots t_n^{a_n}$ if $\bar{a} = (a_1, \ldots, a_n) \in \mathbb{Z}^n$. Under the substitution $\mathbf{t} \leftarrow 1$ in $(*)$, one can count the number of integer points in a (bounded) polytope Q, and thus solves $(\circ)$ quantitatively for the bounded case. In general, one can also succinctly represent integer points in the

© Springer International Publishing AG 2017
F. Eisenbrand and J. Koenemann (Eds.): IPCO 2017, LNCS 10328, pp. 417–429, 2017.
DOI: 10.1007/978-3-319-59250-3_34

intersections, unions and complements of general (possibly unbounded) rational polyhedra [Bar08, BP99] in $\mathbb{R}^n$ using short GFs.

Barvinok's algorithm was extended to projections of polytopes by Barvinok and Woods [BW03], see Theorem 2. The result has a major technical drawback: while it does generalize Kannan's result for bounded P and Q as in (∞), it does not apply for unbounded polyhedra. The main result of this paper is an extension of Barvinok's algorithm to the unbounded case (Theorem 3).

Example 1. Consider $Q = \{(x, y, z) \in \mathbb{R}^3_+ : x = 2y + 5z\}$. Then $Q \cap \mathbb{Z}^3$ projected on $\mathbb{Z}$ has a short GF $\frac{1}{(1-t^2)(1-t^5)} - \frac{t^{10}}{(1-t^2)(1-t^5)} = 1 + t^2 + t^4 + t^5 + t^7 +$ etc.

Our main tool is a structural result describing projections of *semilinear sets*, which are defined as disjoint union of intersections of polyhedra and lattice cosets. More precisely, we prove that such projections are also semilinear and give bound on the (combinatorial) complexity of the projection (Theorem 1). In combination with the Barvinok–Woods theorem we obtain the extension to unbounded polyhedra.

1.2 Connections and Applications

After Lenstra's algorithm, many other methods for fast integer programming in fixed dimensions have been bound (see [Eis03, FT87]). Kannan's algorithm was strengthened in [ES08]. Barvinok's algorithm has been simplified and improved in [DK97, KV08]. Both Barvinok's and Barvinok–Woods' algorithms have been implemented and used for practical computation [D+04, Köp07, V+07].

Let us emphasize that in the context of parametric integer programming, there are two main reasons to study unbounded polyhedra:

(1) Working with short GFs of integer points in unbounded polyhedra allows to compute to various integral sums and valuations over convex polyhedra. We refer to [B+12, Bar08, BV07] for many examples and further references.

(2) For a fixed unbounded polyhedron Q and a varying polytope P in (∞), one can count the number of points in the projection of Q within P. This is done by intersecting Q with a box of growing size and then projecting it. The Barvinok–Woods algorithm is called multiple times for different boxes. Our approach allows one to call the Barvinok–Woods algorithm only once to project Q (unbounded), and then call a more economical Barvinok's algorithm to compute the intersection with P. We refer to [ADL16] for an explicit example related to the famous *Frobenius Problem* requiring such an application.

In conclusion, let us mention that semilinear sets are well studied subjects in both computer science and logic. The study of semilinear sets has numerous applications in computer science, such as analysis of *number decision diagrams* (see [Ler05]), and *context-free languages* (see [Par66]). The fact that semilinear sets are closed under taking projections is not new (see [GS64]). Woods [W15] also characterized semilinear sets as exactly those sets with rational generating functions, which also implies closedness under projections. In our paper, we prove the structural result on projections of semilinear sets by a direct argument, without using tools from logic. By doing so, we obtain effective polynomial bounds for the number of polyhedral pieces and the facet complexity of each piece in the projection. We refer to [Gin66] for background on semilinear sets, and to [CH16] for most recent developments.

2   Notations

We use $\mathbb{N} = \{0, 1, 2, \ldots\}$, $\mathbb{Z}_+ = \{1, 2, \ldots\}$ and $\mathbb{R}_+ = \{x \in \mathbb{R} : x \geq 0\}$.

All constant vectors are denoted $\overline{a}, \overline{c}, \overline{n}, \overline{v}, \overline{b}, \overline{d}$, etc.

For a number N, we also denote by N the vector with all entries equal to N.

Matrices are denoted A, B, C, etc.

Variables are denoted x, y, z, etc.; vectors of variables are denoted $\mathbf{x}, \mathbf{y}, \mathbf{z}$, etc.

If $x_j \leq y_j$ for every index j in vectors $\mathbf{x}$ and $\mathbf{y}$, we write $\mathbf{x} \leq \mathbf{y}$.

GF is an abbreviation for "*generating function*".

Multi-variable generating functions are denoted by $f(\mathbf{t}), g(\mathbf{t}), h(\mathbf{t})$, etc.

Polyhedron is an intersection of finitely many closed half-spaces in $\mathbb{R}^n$.

Polytope is a bounded polyhedron.

Polyhedra/polytopes are denoted by P, Q, R, etc.

The *affine dimension* of P is denoted by $\dim(P)$.

Integer lattices are denoted by $\mathcal{L}, \mathcal{T}, \mathcal{U}, \mathcal{W}$, etc.

Let $\mathrm{rank}(\mathcal{L})$ denotes the *rank* of $\mathcal{L}$.

Patterns are denoted by $\boldsymbol{L}, \boldsymbol{T}, \boldsymbol{S}, \boldsymbol{U}, \boldsymbol{W}$, etc.

The function $\phi(\cdot)$ denotes the binary length of a number, vector, matrix.

For a polyhedron Q described by a linear system $A\mathbf{x} \leq \overline{b}$, we denote by $\phi(Q)$ the total length $\phi(A) + \phi(\overline{b})$.

For a lattice $\mathcal{L}$ generated by a matrix A, $\phi(\mathcal{L})$ denotes $\phi(A)$.

3   Structure of a Projection

3.1   Semilinear Sets and Their Projections

In this section, we assume all dimensions m, n, etc., are fixed. We emphasize that all lattices mentioned are of full rank. All inputs are in binary.

Definition 1. Given a set $X \subseteq \mathbb{R}^{n+1}$, the *projection* of X, denoted by $\mathrm{proj}(X)$, is defined as

$$\mathrm{proj}(X) := \{(x_2, \ldots, x_n) : (x_1, x_2, \ldots, x_{n+1}) \in X\} \subseteq \mathbb{R}^n.$$

For any $\mathbf{y} \in \mathrm{proj}(Q)$, denote by $\mathrm{proj}^{-1}(\mathbf{y}) \subseteq X$ the preimage of $\mathbf{y}$ in X.

Definition 2. Let $\mathcal{L} \subseteq \mathbb{Z}^n$ be a full-rank lattice. A *pattern $\boldsymbol{L}$ with period $\mathcal{L}$* is a union of finitely many (integer) cosets of $\mathcal{L}$. For any other lattice $\mathcal{L}'$, if $\boldsymbol{L}$ can be expressed as a finite union of cosets of $\mathcal{L}'$, then we also call $\mathcal{L}'$ a period of $\boldsymbol{L}$.

Given a rational polyhedron Q and a pattern $\boldsymbol{L}$, the set $Q \cap \boldsymbol{L}$ is called a *patterned polyhedron*. When the pattern $\boldsymbol{L}$ is not emphasized, we simply call Q a *patterned polyhedron with period $\mathcal{L}$*.

Definition 3. A *semilinear* set X is a set of the form

$$X = \bigsqcup_{i=1}^{k} Q_i \cap \boldsymbol{L}_i, \tag{1}$$

where each $Q_i \cap \boldsymbol{L}_i$ is a patterned polyhedron with period $\mathcal{L}_i$, and the polyhedra Q_i are a pairwise disjoint.[1] The *period length* $\psi(X)$ of X is defined as

$$\psi(X) = \sum_{i=1}^{k} \phi(Q_i) + \phi(\mathcal{L}_i).$$

Note that $\psi(X)$ does not depend on the number of cosets in each $\boldsymbol{L}_i$. Define

$$\eta(X) := \sum_{i=1}^{k} \eta(Q_i),$$

where each $\eta(Q_i)$ is the number of facets of the polyhedron Q_i.

Our main structural result is the following theorem.

Theorem 1. *Let $m \in \mathbb{N}$ be fixed. Let $X \subseteq \mathbb{Z}^m$ be a semilinear set of the form* (1). *Let $T : \mathbb{R}^m \to \mathbb{R}^n$ be a linear map satisfying $T(\mathbb{Z}^m) \subseteq \mathbb{Z}^n$. Then $T(X)$ is also a semilinear set, and there exists a decomposition*

$$T(X) = \bigsqcup_{j=1}^{r} R_j \cap \boldsymbol{T}_j, \tag{2}$$

where each $R_j \cap \boldsymbol{T}_j$ is a patterned polyhedron in $\mathbb{R}^n$ with period $\mathcal{T}_j \subseteq \mathbb{Z}^n$. The polyhedra R_j and lattices $\mathcal{T}_j$ can be found in time $poly(\psi(X))$. Moreover,

$$r = \eta(X)^{O(m!)} \quad and \quad \eta(R_j) = \eta(X)^{O(m!)}, \ 1 \le j \le r.$$

Remark 1. The above result describes all pieces R_j and periods $\mathcal{T}_j$ in polynomial time. However, it does not explicitly describe the patterns $\boldsymbol{T}_j$. The latter is actually an NP-hard problem (see [W04, Proposition 5.3.2]).

---

[1] In Theoretical CS literature, semilinear sets are often explicitly presented by generators, which makes some operations like projections easy to compute, while structural properties harder to establish (see e.g. [CH16] and the references therein).

For the proof of Theorem 1, we need a technical lemma:

Lemma 1. Let $n \in \mathbb{N}$ be fixed. Consider a patterned polyhedron $(Q \cap L) \subseteq \mathbb{R}^{n+1}$ with period $\mathcal{L}$. There exists a decomposition

$$\text{proj}(Q \cap L) = \bigsqcup_{j=0}^{r} R_j \cap T_j, \tag{3}$$

where each $R_j \cap T_j$ is a patterned polyhedron in $\mathbb{R}^n$ with period $\mathcal{T}_j \subseteq \mathbb{Z}^n$. The polyhedra R_j and lattices $\mathcal{T}_j$ can be found in time $\text{poly}(\phi(Q) + \phi(\mathcal{L}))$. Moreover,

$$r = O\big(\eta(Q)^2\big) \quad and \quad \eta(R_j) = O\big(\eta(Q)^2\big), \quad for\, all \; 0 \le j \le r.$$

We postpone the proof of the lemma until Subsect. 3.3.

3.2   Proof of Theorem 1

Definition 4. A *copolyhedron* $P \subseteq \mathbb{R}^n$ is a polyhedron with possibly some open facets. If P is a rational copolyhedron, we denote by $\lfloor P \rfloor$ the (closed) polyhedron obtained from P by sharpening each open facet $(\overline{a}\mathbf{x} < b)$ of P to $(\overline{a}\mathbf{x} \le b - 1)$, after scaling $\overline{a}$ and b to integers. Clearly, we have $P \cap \mathbb{Z}^n = \lfloor P \rfloor \cap \mathbb{Z}^n$.

Recall that X has the form (1) with each $Q_i \cap L_i$ having period $\mathcal{L}_i$. Define:

$$\widehat{Q}_i := \big\{ (\mathbf{x}, \mathbf{y}) : \mathbf{y} = T(\mathbf{x}) \text{ and } \mathbf{x} \in Q_i \big\} \subseteq \mathbb{R}^{m+n}. \tag{4}$$

Consider a pattern $U_i = L_i \oplus \mathbb{Z}^n \subseteq \mathbb{Z}^{m+n}$ with period $\mathcal{U}_i = \mathcal{L}_i \oplus \mathbb{Z}^n$. Then $\widehat{Q}_i \cap U_i$ is a patterned polyhedron in $\mathbb{R}^{m+n}$. By (4), we have:

$$T(Q_i \cap L_i) = S(\widehat{Q}_i \cap U_i) \quad \text{and} \quad T(X) = \bigcup_{i=1}^{r} S(\widehat{Q}_i \cap U_i),$$

where S is a vertical projection mapping $(\mathbf{x}, \mathbf{y}) \in \mathbb{R}^{m+n}$ to $\mathbf{y} \in \mathbb{R}^n$. We can write $S = S_1 \circ \cdots \circ S_m$, where each $S_i : \mathbb{R}^{i+n} \to \mathbb{R}^{i+n-1}$ is a projection along the x_{i+n} coordinate. We repeatedly apply Lemma 1 on $S_m, \ldots, S_1$.

Start by applying Lemma 1 on S_m, we have:

$$S_m(\widehat{Q}_i \cap U_i) = \bigsqcup_{j=1}^{r_i} (R_{ij} \cap T_{ij}), \tag{5}$$

where each $R_{ij} \cap T_{ij}$ is a patterned polyhedron in $\mathbb{Z}^{m+n-1}$ with period $\mathcal{T}_{ij}$. Note that two pieces R_{ij} and $R_{i'j'}$ can be overlapping if $i \ne i'$. However, we can refine all R_{ij} into polynomially many disjoint copolyhedra P_d, so that

$$\bigcup_{i=1}^{k} \bigcup_{j=1}^{r_i} R_{ij} = \bigsqcup_{d=1}^{e} P_d. \tag{6}$$

For each P_d we can also find a pattern $\boldsymbol{W}_d$ with period $\mathcal{W}_d \in \mathbb{Z}^{m+n-1}$. The (full-rank) period $\mathcal{W}_d$ can be taken as the intersection of polynomially many (full-rank) periods $\mathcal{T}_{ij}$ for which $P_d \subseteq R_{ij}$. We round each P_d to $\lfloor P_d \rfloor$ (see Definition 4). From (5) and (6) we have:

$$\bigcup_{i=1}^{k} S_m(\widehat{Q}_i \cap U_i) = \bigsqcup_{d=1}^{e} (\lfloor P_d \rfloor + \boldsymbol{W}_d).$$

A similar argument applies to $S_1 \circ \cdots \circ S_1$. In the end, we have (2).

Using Lemma 1, we can bound the number of polyhedra r_i in (5), and also the number of facets $\eta(R_{ij})$ for each R_{ij}. This gives us a bound on e, the number of refined pieces in (6). By a careful analysis, after all m projections, the total number r of pieces in the final decomposition (2) is at most $\rho(X)^{O(m!)}$. Each piece R_j also has at most $\rho(X)^{O(m!)}$ facets. $\qquad \square$

3.3   Proof of Lemma 1

The proof is by induction on n. The case $n = 0$ is trivial. Assume $n \geq 1$.

Let $\boldsymbol{L} \subseteq \mathbb{Z}^{n+1}$ be a full-rank pattern with period $\mathcal{L}$ as in the lemma. Then, the projection of $\boldsymbol{L}$ onto $\mathbb{Z}^n$ is another pattern $\boldsymbol{L}'$ with full-rank period $\mathcal{L}' = \mathrm{proj}(\mathcal{L})$. Since $\mathcal{L}$ is of full rank, we can define

$$\ell = \min\{t \in \mathbb{Z}_+ : (t, 0, \ldots, 0) \in \mathcal{L}\}. \tag{7}$$

Let $R = \mathrm{proj}(Q)$. Assume Q is described by the system $A\mathbf{x} \leq \bar{b}$. Recall the *Fourier–Motzkin elimination method* (see [Sch86, Sect. 12.2]), which gives the facets of R from those of Q. First, rewrite and group the inequalities in $A\mathbf{x} \leq \bar{b}$ into

$$A_1\mathbf{y} + \bar{b}_1 \leq x_1, \quad x_1 \leq A_2\mathbf{y} + \bar{b}_2 \quad \text{and} \quad A_3\mathbf{y} \leq \bar{b}_3, \tag{8}$$

where $\mathbf{y} = (x_2, \ldots, x_{n+1}) \in \mathbb{R}^n$. Then R is described by a system $C\mathbf{y} \leq \bar{d}$, which consists of $(A_3\mathbf{y} \leq \bar{b}_3)$ and $(\bar{a}_1\mathbf{y} + b_1 \leq \bar{a}_2\mathbf{y} + b_2)$ for every possible pair of rows $\bar{a}_1\mathbf{y} + b_1$ and $\bar{a}_2\mathbf{y} + b_2$ from the first two systems in (8). Moreover, we can decompose

$$R = \bigsqcup_{j=1}^{r} P_j, \tag{9}$$

where each P_j is a copolyhedron, so that over each P_j, the largest row in $A_1\mathbf{y} + \bar{b}_1$ is $\bar{a}_{j1}\mathbf{y} + b_{j1}$ and the smallest row in $A_2\mathbf{y} + \bar{b}_2$ is $\bar{a}_{j2}\mathbf{y} + b_{j2}$. Thus, for every $\mathbf{y} \in P_j$, we have $\mathrm{proj}^{-1}(\mathbf{y}) = [\alpha_j(\mathbf{y}), \beta_j(\mathbf{y})]$, where $\alpha_j(\mathbf{y}) = \bar{a}_{j1}\mathbf{y} + b_{j1}$ and $\beta_j(\mathbf{y}) = \bar{a}_{j2}\mathbf{y} + b_{j2}$ are affine rational functions. Let $m = \eta(Q)$. Note that the system $C\mathbf{y} \leq \bar{d}$ contains at most $O(m^2)$ inequalities, i.e., $\eta(R) = O(m^2)$. Also, we have $r = O(m^2)$ and $\eta(P_j) = O(m)$ for $1 \leq j \leq r$.

For each $\mathbf{y} \in R$, the preimage $\text{proj}^{-1}(\mathbf{y}) \subseteq Q$ is a segment in the direction x_1. Denote by $|\text{proj}^{-1}(\mathbf{y})|$ the length of this segment. Now we refine the decomposition in (9) to

$$R = R_0 \sqcup R_1 \sqcup \cdots \sqcup R_r, \quad \text{where} \tag{10}$$

(a) Each R_j is a copolyhedron in $\mathbb{R}^n$, with $\eta(R_j) = O(m^2)$ and $r = O(m^2)$.
(b) For every $\mathbf{y} \in R_0$, we have the length $|\text{proj}^{-1}(\mathbf{y})| \geq \ell$.
(c) For every $\mathbf{y} \in R_j$ $(1 \leq j \leq r)$, we have the length $|\text{proj}^{-1}(\mathbf{y})| < \ell$. Furthermore, we have $\text{proj}^{-1}(\mathbf{y}) = [\alpha_j(\mathbf{y}), \beta_j(\mathbf{y})]$, where α_j and β_j are affine rational functions in $\mathbf{y}$.

This refinement can be obtained as follows. First, define

$$R_0 = \text{proj}[Q \cap (Q + \ell\overline{v}_1)] \subseteq R,$$

where $\overline{v}_1 = (1, 0, \ldots, 0)$. The facets of R_0 can be found from those of $Q \cap (Q + \ell\overline{v}_1)$ again by Fourier–Motzkin elimination, and also $\eta(R_0) = O(m^2)$. Observe that $|\text{proj}^{-1}(\mathbf{y})| \geq \ell$ if and only if $\mathbf{y} \in R_0$. Define $R_j := P_j \backslash R_0$ for $1 \leq j \leq r$. Recall that for every $\mathbf{y} \in P_j$, we have $\text{proj}^{-1}(\mathbf{y}) = [\alpha_j(\mathbf{y}), \beta_j(\mathbf{y})]$. Therefore,

$$R_j = P_j \backslash R_0 = \{\mathbf{y} \in P_j : |\text{proj}^{-1}(\mathbf{y})| < \ell\} = \{\mathbf{y} \in P_j : \alpha_j(\mathbf{y}) + \ell > \beta_j(\mathbf{y})\}.$$

It is clear that each R_j is a copolyhedron satisfying condition (c). Moreover, for each $1 \leq j \leq r$, we have $\eta(R_j) \leq \eta(P_j) + 1 = O(m)$. By (9), we can decompose:

$$R = R_0 \sqcup (R \backslash R_0) = R_0 \bigsqcup_{j=1}^{r}(P_j \backslash R_0) = \bigsqcup_{j=0}^{r} R_j.$$

This decomposition satisfies all conditions (a)–(c) and proves (10). Note also that by converting each R_j to $\lfloor R_j \rfloor$, we do not lose any integer points in R. Let us show that the part of $\text{proj}(Q \cap L)$ within R_0 has a simple pattern:

Lemma 2. $\text{proj}(Q \cap L) \cap R_0 = R_0 \cap L'$.

Proof. Recall that $\text{proj}(L) = L'$, which implies LHS $\subseteq$ RHS. On the other hand, for every $\mathbf{y} \in L'$, there exists $\mathbf{x} \in L$ such that $\mathbf{y} = \text{proj}(\mathbf{x})$. If $\mathbf{y} \in R_0 \cap L'$, we also have $|\text{proj}^{-1}(\mathbf{y})| \geq \ell$ by condition b), with ℓ defined in (7). The point $\mathbf{x}$ and the segment $\text{proj}^{-1}(\mathbf{y})$ lie on the same vertical line. Therefore, since $|\text{proj}^{-1}(\mathbf{y})| \geq \ell$, we can find another $\mathbf{x}'$ such that $\mathbf{x}' \in \text{proj}^{-1}(\mathbf{y}) \subseteq Q$ and also $\mathbf{x}' - \mathbf{x} \in \mathcal{L}$. Since L has period $\mathcal{L}$, we have $\mathbf{x}' \in L$. This implies $\mathbf{x}' \in Q \cap L$, and $\mathbf{y} \in \text{proj}(Q \cap L)$. Therefore we have RHS $\subseteq$ LHS, and the lemma holds.

It remains to show that $\text{proj}(Q \cap L) \cap R_j$ also has a pattern for every $j > 0$. By condition (c), every such R_j has a "thin" preimage. Let $Q_j = \text{proj}^{-1}(R_j) \subseteq Q$. If $\dim(R_j) < n$, we have $\dim(Q_j) < n + 1$. In this case we can apply the inductive hypothesis. Otherwise, assume $\dim(R_j) = n$. For convenience, we refer to R_j and Q_j as just R and Q. We can write $R = R' + D$, where $R' \subseteq R$ is a polytope and D is the recession cone of R.

Consider $\mathbf{y} \in R$, $\overline{v} \in D$ and $\lambda > 0$. Since $\mathbf{y} + \lambda \overline{v} \in R$, from (c) we have $\text{proj}^{-1}(\mathbf{y} + \lambda \overline{v}) = [\alpha(\mathbf{y} + \lambda \overline{v}), \beta(\mathbf{y} + \lambda \overline{v})]$. Denote by $\widetilde{\alpha}$ and $\widetilde{\beta}$ the linear parts of the affine maps α and β. By property of affine maps, we have:

$$\text{proj}^{-1}(\mathbf{y} + \lambda \overline{v}) = [\alpha(\mathbf{y} + \lambda \overline{v}), \beta(\mathbf{y} + \lambda \overline{v})] = [\alpha(\mathbf{y}) + \lambda \widetilde{\alpha}(\overline{v}), \ \beta(\mathbf{y}) + \lambda \widetilde{\beta}(\overline{v})]. \quad (11)$$

Therefore,

$$|\text{proj}^{-1}(\mathbf{y} + \lambda \overline{v})| = \beta(\mathbf{y}) - \alpha(\mathbf{y}) + \lambda [\widetilde{\beta} - \widetilde{\alpha}](\overline{v}).$$

Since $(\mathbf{y} + \lambda \overline{v}) \in R$, by c) we have:

$$0 \leq |\text{proj}^{-1}(\mathbf{y} + \lambda \overline{v})| = \beta(\mathbf{y}) - \alpha(\mathbf{y}) + \lambda [\widetilde{\beta} - \widetilde{\alpha}](\overline{v}) < \ell.$$

Because $\lambda > 0$ is arbitrary, we must have $[\widetilde{\beta} - \widetilde{\alpha}](\overline{v}) = 0$. This holds for all $\overline{v} \in D$. We conclude that $[\widetilde{\beta} - \widetilde{\alpha}]$ vanishes on the whole subspace $H := \text{span}(D)$, i.e., for any $\overline{v} \in H$ we have $\widetilde{\alpha}(\overline{v}) = \widetilde{\beta}(\overline{v})$. Thus, we can rewrite (11) as

$$\text{proj}^{-1}(\mathbf{y} + \lambda \overline{v}) = [\alpha(\mathbf{y}), \beta(\mathbf{y})] + \lambda \widetilde{\alpha}(\overline{v}) = \text{proj}^{-1}(\mathbf{y}) + \lambda \widetilde{\alpha}(\overline{v}). \quad (12)$$

Define $C := \widetilde{\alpha}(D)$ and $G := \widetilde{\alpha}(H)$. Note that $\text{span}(C) = G$, because $\text{span}(D) = H$. Recall that $R = R' + D$ with R' a polytope. In (12), we let $\mathbf{y}$ vary over R', λ vary over $\mathbb{R}_+$ and $\overline{v}$ vary over D. The LHS becomes $Q = \text{proj}^{-1}(R)$. The RHS becomes $\text{proj}^{-1}(R') + C$. Therefore, we have $Q = \text{proj}^{-1}(R') + C$. Since $\text{proj}^{-1}(R')$ is a polytope, we conclude that C is the recession cone for Q.

Because $\text{proj}^{-1}(\mathbf{y}) = [\alpha(\mathbf{y}), \beta(\mathbf{y})]$ for every $\mathbf{y} \in R$, the last n coordinates in $\alpha(\mathbf{y})$ and $\beta(\mathbf{y})$ are equal to $\mathbf{y}$. This also holds for $\widetilde{\alpha}(\mathbf{y})$ and $\widetilde{\beta}(\mathbf{y})$, i.e., $\text{proj}(\widetilde{\alpha}(\mathbf{y})) = \text{proj}(\widetilde{\alpha}(\mathbf{y})) = \mathbf{y}$. This implies $\text{proj}(G) = H$, because $G = \widetilde{\alpha}(H)$. In other words, $\widetilde{\alpha}$ is the inverse map for proj on G (see Fig. 1).

Recall that $Q \cap L$ is a patterned polyhedron with period $\mathcal{L}$, and $\text{proj}(Q) = R$. Define $\mathcal{S} := \mathcal{L} \cap G$ and $\mathcal{T} := \text{proj}(\mathcal{S}) \subset \text{proj}(G) = H$. Since $\mathcal{L}$ is full-rank, we have $\text{rank}(\mathcal{S}) = \dim(G)$. Since $\widetilde{\alpha}$ and proj are inverse maps, we have $\mathcal{S} = \widetilde{\alpha}(\mathcal{T})$.

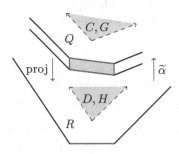

Fig. 1. R and $Q = \text{proj}^{-1}(R)$, with R' and $\text{proj}^{-1}(R')$ shown in blue. The cones C and D span G and H, respectively. (Color figure online)

We claim that $\text{proj}(Q \cap L) \subset R$ is a patterned polyhedron with period $\mathcal{T}$. Indeed, consider any two points $\mathbf{y}_1, \mathbf{y}_2 \in R$ with $\mathbf{y}_2 - \mathbf{y}_1 \in \mathcal{T}$. Assume that $\mathbf{y}_1 \in \text{proj}(Q \cap L)$, i.e., there exists $\mathbf{x}_1 \in Q \cap L$ with $\text{proj}(\mathbf{x}_1) = \mathbf{y}_1$. We show that $\mathbf{y}_2 \in \text{proj}(Q \cap L)$. First, we have $\text{proj}^{-1}(\mathbf{y}_1) = [\alpha(\mathbf{y}_1), \beta(\mathbf{y}_1)]$ and $\text{proj}^{-1}(\mathbf{y}_2) = [\alpha(\mathbf{y}_2), \beta(\mathbf{y}_2)]$. Let $\bar{v} = \mathbf{y}_2 - \mathbf{y}_1 \in \mathcal{T} \subset H$. By (12), we have:

$$[\alpha(\mathbf{y}_2), \beta(\mathbf{y}_2)] = \text{proj}^{-1}(\mathbf{y}_2) = \text{proj}^{-1}(\mathbf{y}_1 + \bar{v}) = [\alpha(\mathbf{y}_1), \beta(\mathbf{y}_1)] + \tilde{\alpha}(\bar{v}). \quad (13)$$

Thus, we have $\alpha(\mathbf{y}_1) - \beta(\mathbf{y}_1) = \alpha(\mathbf{y}_2) - \beta(\mathbf{y}_2)$. In other words, the points $\alpha(\mathbf{y}_1), \beta(\mathbf{y}_1), \alpha(\mathbf{y}_2)$ and $\beta(\mathbf{y}_2)$ form a parallelogram. By $\text{proj}(\mathbf{x}_1) = \mathbf{y}_1$, we have:

$$\mathbf{x}_1 \in \text{proj}^{-1}(\mathbf{y}_1) = [\alpha(\mathbf{y}_1), \beta(\mathbf{y}_1)] \subseteq Q.$$

So $\mathbf{x}_1$ lies on the edge $[\alpha(\mathbf{y}_1), \beta(\mathbf{y}_1)]$ of the parallelogram. Therefore, we can find another point $\mathbf{x}_2$ lying on the other edge $[\alpha(\mathbf{y}_2), \beta(\mathbf{y}_2)] = \text{proj}^{-1}(\mathbf{y}_2)$ with

$$\mathbf{x}_2 - \mathbf{x}_1 = \alpha(\mathbf{y}_2) - \alpha(\mathbf{y}_1) = \tilde{\alpha}(\mathbf{y}_2 - \mathbf{y}_1) = \tilde{\alpha}(\bar{v}) \in \tilde{\alpha}(\mathcal{T}) = \mathcal{S}.$$

This $\mathbf{x}_2$ satisfies $\text{proj}(\mathbf{x}_2) = \mathbf{y}_2$. Recall that $\mathbf{x}_1 \in L$, with L having period $\mathcal{L}$. Since $\mathbf{x}_2 - \mathbf{x}_1 \in \mathcal{S} \subset \mathcal{L}$, we have $\mathbf{x}_2 \in L$. This implies $\mathbf{x}_2 \in Q \cap L$ and $\mathbf{y}_2 \in \text{proj}(Q \cap L)$.

So we have established that $\text{proj}(Q \cap L) \subset R$ is a patterned polyhedron with period $\mathcal{T}$. Note that

$$\text{rank}(\mathcal{T}) = \text{rank}(\mathcal{S}) = \dim(G) = \dim(H) = \dim(D).$$

If $\dim(D) = n$ then $\mathcal{T}$ is full-rank. If $\dim(D) < n$, recall that $R = R' + D$ where R' is a polytope, and $\text{span}(D) = H$. Let $H^\perp$ be the complement subspace to H in $\mathbb{R}^n$, and $R^\perp$ be the projection of R' onto $H^\perp$. Since $R^\perp$ is bounded, we can take a large enough lattice $\mathcal{T}^\perp \subset H^\perp$ such that there are no two points $\mathbf{z}_1 \neq \mathbf{z}_2 \in R^\perp$ with $\mathbf{z}_1 - \mathbf{z}_2 \in \mathcal{T}^\perp$. Now the lattice $\mathcal{T}^\perp \oplus \mathcal{T}$ is full-rank, which can be taken as a period for $\text{proj}(Q \cap L)$.

To summarize, for every piece R_j and $Q_j = \text{proj}^{-1}(R_j)$, $1 \leq j \leq r$, the projection $\text{proj}(Q_j \cap L) \subset R_j$ has period $\mathcal{T}_j$. Thus $\text{proj}(Q_j \cap L)$ is a patterned polyhedron. This completes the proof. $\qquad\square$

4 Finding Short GF for Unbounded Projection

4.1 Barvinok–Woods Algorithm

In this section, we are again assuming that dimensions m and n are fixed. We recall the Barvinok–Woods algorithm, which finds in polynomial time a short GF for the projection of integer points in a polytope:

Theorem 2 ([BW03]). *Let $m, n \in \mathbb{N}$ be fixed. Given a rational (bounded) polytope $Q = \{\mathbf{x} \in \mathbb{R}^m : A\mathbf{x} \leq \bar{b}\}$, and a linear transformation $T : \mathbb{R}^m \to \mathbb{R}^n$*

represented as a matrix $T \in \mathbb{Z}^{n \times m}$, *there is a polynomial time algorithm to compute a short GF for* $T(Q \cap \mathbb{Z}^m)$ *as:*

$$g(\mathbf{t}) = \sum_{\mathbf{y} \in T(Q \cap \mathbb{Z}^m)} \mathbf{t}^{\mathbf{y}} = \sum_{i=1}^{M} \frac{c_i \, \mathbf{t}^{\overline{a}_i}}{(1 - \mathbf{t}^{\overline{b}_{i1}}) \dots (1 - \mathbf{t}^{\overline{b}_{is}})}, \tag{14}$$

where $c_i = p_i/q_i \in \mathbb{Q}$, $\overline{a}_i, \overline{b}_{ij} \in \mathbb{Z}^n$, $\overline{b}_{ij} \neq 0$ *for all* i, j, *and* s *is a constant depending only on* m. *Moreover,* g *has length* $\phi(g) = poly(\phi(Q) + \phi(T))$, *where*

$$\phi(g) = \sum_{i} \lceil \log_2 |p_i \, q_i| + 1 \rceil + \sum_{i,j} \lceil \log_2 a_{ij} + 1 \rceil + \sum_{i,j,k} \lceil \log_2 b_{ijk} + 1 \rceil. \tag{15}$$

Using Theorem 1, we extend Theorem 2 as follows:

Theorem 3. *Let* $m, n \in \mathbb{N}$ *be fixed. Given a possibly unbounded polyhedron* $Q \subseteq \mathbb{R}^m$, *and a linear transformation* $T : \mathbb{R}^m \to \mathbb{R}^n$ *which satisfies* $T(Q) \subseteq \mathbb{R}^n_+$, *there is a time algorithm to compute a short GF for* $T(Q \cap \mathbb{Z}^m)$ *as in* (14).

Remark 2. The extra condition $T(Q) \subseteq \mathbb{R}^n_+$ is to make that the power series $\sum \mathbf{t}^{\mathbf{y}}$ of $T(Q \cap \mathbb{Z}^m)$ converges on a non-empty open domain to the computed short GF. More generally, when $T(Q)$ has pointed recession cone, we can apply an appropriate unimodular transformation to make sure $T(Q) \subseteq \mathbb{R}^n_+$. For the most general case when $T(Q)$ could contain some infinite lines, we can resort to the theory of valuation (see [Bar08, BP99]) to make sense of the infinite GF.

We need a standard result on polyhedron triangulation to prove the theorem:

Proposition 1 (see e.g. [Mei93]). *Fix* $n \in \mathbb{N}$. *Let* $R = \{\mathbf{x} \in \mathbb{R}^n : C\mathbf{x} \leq \overline{d}\}$ *be a possibly unbounded polyhedron. There is a decomposition*

$$R = \bigsqcup_{k=1}^{t} R_k \oplus D_k, \tag{16}$$

where each R_k *is a copolytope, and each* D_k *is a simple cone. Each part* $R_k \oplus D_k$ *is a direct sum, with* R_k *and* D_k *affinely independent. All* R_k *and* D_k *can be found in time* $poly(\phi(R))$.

4.2   Proof of Theorem 3

WLOG, we can assume $\dim(Q) = m$ and $\dim(T(Q)) = n$. Clearly, the set $X = Q \cap \mathbb{Z}^m$ is a semilinear, and we want to find a short GF for $T(X)$.

First, we argue that for any bounded polytope $P \subset \mathbb{R}^n$, a short GF for $T(X) \cap P$ can be found in time $poly(\phi(Q) + \phi(P))$. Assume P is given by a system $C\mathbf{y} \leq \overline{d}$. For any $\overline{v} \in P$, we have $\overline{v} \in T(X)$ if and only if the following system has a solution $\mathbf{x} \in \mathbb{Z}^m$:

$$\begin{cases} A\mathbf{x} \leq \overline{b} \\ T(\mathbf{x}) = \overline{v} \end{cases}. \tag{17}$$

By bound on integer programming solutions (see [Sch86, Corollary 17.1b]), it is equivalent to find such an $\mathbf{x}$ with length bounded polynomially in the length of the system (17). The parameter $\overline{v}$ lies in P, which is bounded. Therefore, we can find $N \in \mathbb{N}$ of length $\phi(N) = \text{poly}(\phi(P) + \phi(Q))$, such that (17) is equivalent to:

$$\begin{cases} A\mathbf{x} \leq \overline{b} \\ CT(\mathbf{x}) \leq \overline{d} \\ -N \leq \mathbf{x} \leq N \end{cases}.$$

Applying Theorem 2 to $\widehat{Q}$, we get a short GF $g(\mathbf{t})$ for $T(\widehat{Q} \cap \mathbb{Z}^m) = T(X) \cap P$.

Now we are back to finding a short GF for the entire projection $T(X)$. Applying Theorem 1 to X, we have a decomposition:

$$T(X) = \bigsqcup_{j=1}^{r} R_j \cap \boldsymbol{T}_j. \tag{18}$$

We proceed to find a short GF g_j for each patterned polyhedron $R_j \cap \boldsymbol{T}_j$ with period $\mathcal{T}_j$. For convenience, we refer to R_j, $\boldsymbol{T}_j$, $\mathcal{T}_j$, g_j simply as R, $\boldsymbol{T}$, $\mathcal{T}$ and g. By Proposition 1, we can decompose

$$R = \bigsqcup_{i=1}^{t_j} R_i \oplus D_i \quad \text{and} \quad R \cap \boldsymbol{T} = \bigsqcup_{i=1}^{t_j} (R_i \oplus D_i) \cap \boldsymbol{T}. \tag{19}$$

Recall from Theorem 1 that $\mathcal{T}$ has full rank. Let $d_i = \dim(D_i)$ and $\overline{v}_i^1, \ldots, \overline{v}_i^{d_i}$ be the generating rays of the (simple) cone D_i. For each $\overline{v}_i^t$, we can find $n_t \in \mathbb{Z}_+$ such that $\overline{w}_i^t = n_t \overline{v}_i^t \in \mathcal{T}$. Let P_i and $\mathcal{T}_i$ be the parallelepiped and lattice spanned by $\overline{w}_i^1, \ldots, \overline{w}_i^{d_i}$, respectively. We have $D_i = P_i + \mathcal{T}_i$ and therefore

$$R_i \oplus D_i = R_i \oplus (P_i + \mathcal{T}_i) = (R_i \oplus P_i) + \mathcal{T}_i. \tag{20}$$

Each $R_i \oplus P_i$ is a copolytope. Note that Theorem 2 is stated for (closed) polytopes. We round each $R_i \oplus P_i$ to $\lfloor R_i \oplus P_i \rfloor$, where $\lfloor . \rfloor$ was described in Definition 4 (Sect. 3.2). By the earlier argument, we can find a short GF $h_i(\mathbf{t})$ for $T(X) \cap (R_i \oplus P_i) = (R_i \oplus P_i) \cap \boldsymbol{T}$. Since $\mathcal{T}_i \subseteq \mathcal{T}$, the pattern $\boldsymbol{T}$ also has period $\mathcal{T}_i$. By (20), we can get the short GF $f_i(\mathbf{t})$ for $(R_i \oplus D_i) \cap \boldsymbol{T}$ as

$$f_i(\mathbf{t}) = \sum_{\mathbf{y} \in (R_i \oplus D_i) \cap \boldsymbol{T}} \mathbf{t}^{\mathbf{y}} = \sum_{\mathbf{y} \in (R_i \oplus P_i) \cap \boldsymbol{T}} \mathbf{t}^{\mathbf{y}} \cdot \sum_{\mathbf{y} \in \mathcal{T}_i} \mathbf{t}^{\mathbf{y}} = h_i(\mathbf{t}) \prod_{t=1}^{d_i} \frac{1}{1 - \mathbf{t}^{\overline{w}_i^t}}. \tag{21}$$

By (19), we obtain

$$g(\mathbf{t}) = \sum_{\mathbf{y} \in R \cap \boldsymbol{T}} \mathbf{t}^{\mathbf{y}} = \sum_{1 \leq i \leq t_j} f_i(\mathbf{t}). \tag{22}$$

In summary, we obtained a short GF $g_j(\mathbf{t})$ for each piece $R_j \cap \boldsymbol{T}_j$ $(1 \leq j \leq r)$. Summing over all j in (18), we get a short GF for $T(X)$, as desired. □

Acknowledgements. We are greatly indebted to Sasha Barvinok and Sinai Robins for introducing us to the subject. We are thankful to Iskander Aliev, Matthias Aschenbrenner, Artëm Chernikov, Jesús De Loera, Lenny Fukshansky, Oleg Karpenkov and Kevin Woods for interesting conversations and helpful remarks. We also thank the anonymous referees for helpful references, which we cannot fully accommodate due to space limit. The second author was partially supported by the NSF.

References

[ADL16] Aliev, I., De Loera, J.A., Louveaux, Q.: Parametric polyhedra with at least k lattice points: their semigroup structure and the k-Frobenius problem. In: Beveridge, A., Griggs, J.R., Hogben, L., Musiker, G., Tetali, P. (eds.) Recent Trends in Combinatorics. Springer, Switzerland (2016)

[B+12] Baldoni, V., Berline, N., De Loera, J.A., Köppe, M., Vergne, M.: Computation of the highest coefficients of weighted Ehrhart quasi-polynomials of rational polyhedra. Found. Comput. Math. **12**, 435–469 (2012)

[Bar93] Barvinok, A.: A polynomial time algorithm for counting integral points in polyhedra when the fimension is fixed. In: Proceedings of the 34th FOCS, IEEE, Los Alamitos, CA, pp. 566–572 (1993)

[Bar08] Barvinok, A.: Integer Points in Polyhedra. EMS, Zürich (2008)

[BP99] Barvinok, A., Pommersheim, J.E.: An algorithmic theory of lattice points in polyhedra. In: New Perspectives in Algebraic Combinatorics, pp. 91–147. Cambridge University Press, Cambridge (1999)

[BW03] Barvinok, A., Woods, K.: Short rational generating functions for lattice point problems. J. Amer. Math. Soc. **16**, 957–979 (2003)

[BV07] Berline, N., Vergne, M.: Local Euler-Maclaurin formula for polytopes. Mosc. Math. J. **7**, 355–386 (2007)

[CH16] Chistikov, D., Haase, C.: The taming of semi-linear set. In: Proceedings of the ICALP 2016, pp. 127:1–127:13 (2016)

[D+04] De Loera, J.A., Haws, D., Hemmecke, R., Huggins, P., Sturmfels, B., Yoshida, R.: Short rational functions for toric algebra and applications. J. Symbolic Comput. **38**, 959–973 (2004)

[D+04] De Loera, J.A., Hemmecke, R., Tauzer, J., Yoshida, R.: Effective lattice point counting in rational convex polytopes. J. Symbolic Comput. **38**, 1273–1302 (2004)

[DK97] Dyer, M., Kannan, R.: On Barvinok's algorithm for counting lattice points in fixed dimension. Math. Oper. Res. **22**, 545–549 (1997)

[Eis03] Eisenbrand, F.: Fast integer programming in fixed dimension. In: Battista, G., Zwick, U. (eds.) ESA 2003. LNCS, vol. 2832, pp. 196–207. Springer, Heidelberg (2003). doi:10.1007/978-3-540-39658-1_20

[ES08] Eisenbrand, F., Shmonin, G.: Parametric integer programming in fixed dimension. Math. Oper. Res. **33**, 839–850 (2008)

[FT87] Frank, A., Tardos, É.: An application of simultaneous Diophantine approximation in combinatorial optimization. Combinatorica **7**, 49–65 (1987)

[Gin66] Ginsburg, S.: The Mathematical Theory of Context Free Languages. McGraw-Hill, New York (1966)

[GS64] Ginsburg, S., Spanier, E.: Bounded ALGOL-like languages. Trans. Amer. Math. Soc. **113**, 333–368 (1964)

[Kan90] Kannan, R.: Test sets for integer programs, ∀∃ sentences. In: Polyhedral Combinatorics, pp. 39–47. AMS, Providence (1990)

[Köp07] Köppe, M.: A primal Barvinok algorithm based on irrational decompositions. SIAM J. Discrete Math. **21**, 220–236 (2007)

[KV08] Köppe, M., Verdoolaege, S.: Computing parametric rational generating functions with a primal Barvinok algorithm. Electron. J. Combin. **15**(1), 19 (2008). RP 16

[Len83] Lenstra, H.: Integer programming with a fixed number of variables. Math. Oper. Res. **8**, 538–548 (1983)

[Ler05] Leroux, J.: A polynomial time presburger criterion and synthesis for number decision diagrams. In: Proceedings of the 20th LICS, IEEE, Chicago, IL, pp. 147–156 (2005)

[Mei93] Meiser, S.: Point location in arrangement of hyperplanes. Inform. Comput. **106**, 286–303 (1993)

[Par66] Parikh, R.: On context-free languages. J. Assoc. Comput. Mach. **13**, 570–581 (1966)

[Sch86] Schrijver, A.: Theory of Linear and Integer Programming. Wiley, Chichester (1986)

[V+07] Verdoolaege, S., Seghir, R., Beyls, K., Loechner, V., Bruynooghe, M.: Counting integer points in parametric polytopes using Barvinok's rational functions. Algorithmica **48**, 37–66 (2007)

[W04] Woods, K.: Rational Generating Functions and Lattice Point Sets, Ph.D. thesis, University of Michigan, 112 p. (2004)

[W15] Woods, K.: Presburger arithmetic, rational generating functions, and quasi-polynomials. J. Symb. Log. **80**, 433–449 (2015)

Excluded t-Factors in Bipartite Graphs: A Unified Framework for Nonbipartite Matchings and Restricted 2-Matchings

Kenjiro Takazawa$^{(\boxtimes)}$

Department of Industrial and Systems Engineering,
Faculty of Science and Engineering, Hosei University,
Tokyo 184-8584, Japan
takazawa@hosei.ac.jp

Abstract. We propose a new framework of optimal t-matchings excluding prescribed t-factors in bipartite graphs. It is a generalization of the nonbipartite matching problem and includes a number of generalizations such as the triangle-free 2-matching, square-free 2-matching, and even factor problems. We demonstrate a unified understanding of those generalizations by designing a combinatorial algorithm for our problem under a reasonable assumption, which is broad enough to include the specific problems listed above. We first present a min-max theorem and a combinatorial algorithm for the unweighted version. We further provide a linear programming formulation with dual integrality and a primal-dual algorithm for the weighted version. A key ingredient of our algorithm is a technique of shrinking forbidden structures, which commonly extends the techniques of shrinking odd cycles, triangles, and squares in Edmonds' blossom algorithm, in the triangle-free 2-matching algorithm, and in the square-free 2-matching algorithm, respectively.

1 Introduction

Since matching theory [16] was established, a number of generalizations of the matching problem have been proposed up to the present date. Examples include path-matchings [4], even factors [5,19], triangle-free 2-matchings [3,18], simple square-free 2-matchings [10,19], simple $K_{t,t}$-free t-matchings [8], simple K_{t+1}-free t-matchings [1], 2-matchings covering prescribed edge cuts [2,12], and $\mathcal{U}$-feasible 2-matchings [22]. For most of those generalizations, important results in matching theory, such as a min-max theorem, polynomial algorithms, and a linear programming formulation with dual integrality, are extended. However, while some similar structures are found, in most cases they are studied separately and little connection among them is discovered.

In the present paper, we propose a new framework of *optimal t-matchings excluding prescribed t-factors*, to demonstrate a unified understanding of those generalizations. Our framework includes all of the generalizations listed above, and the traveling salesman problem (TSP) as well. This implies some intractability of the framework, but we propose a tractable class which includes many of

© Springer International Publishing AG 2017
F. Eisenbrand and J. Koenemann (Eds.): IPCO 2017, LNCS 10328, pp. 430–441, 2017.
DOI: 10.1007/978-3-319-59250-3_35

the efficiently solvable classes of the above problems. Our main contribution is a min-max theorem and a combinatorial polynomial algorithm which commonly extend those for the matching and triangle-free 2-matching problems in nonbipartite graphs and the simple square-free 2-matching and $K_{t,t}$-free t-matching problems in bipartite graphs.

A key ingredient of our algorithm is a technique of shrinking excluded t-factors. This technique commonly extends the techniques of shrinking odd cycles, triangles, and squares in a matching algorithm [7], in a triangle-free 2-matching algorithm [3], and in square-free 2-matching algorithms in bipartite graphs [10, 19], respectively. We demonstrate that our framework is tractable in the class where this shrinking technique works.

1.1   Previous Work

The problems most relevant to our work are the *even factor, triangle-free 2-matching*, and *simple square-free* 2-matching problems.

The even factor problem [5] is a generalization of the nonbipartite matching problem, which admits a further generalization: the basic/independent even factor problem [5, 11] is its generalization including matroid intersection. Let $D = (V, A)$ be a digraph. A subset of arcs $F \subseteq A$ is a *path-cycle factor* if it is a vertex-disjoint collection of directed cycles (dicycles) and directed paths (dipaths). Equivalently, an arc subset F is a path-cycle factor if, in the subgraph (V, F), the indegree and outdegree of every vertex are at most one. An *even factor* is a path-cycle factor excluding dicycles of odd length (odd dicycles).

While the maximum even factor problem is NP-hard, in *odd-cycle symmetric* digraphs it enjoys min-max theorems, an Edmonds-Gallai decomposition, and polynomial-time algorithms. A digraph is called *odd-cycle symmetric* if every odd dicycle has its reverse dicycle. Moreover, a maximum-weight even factor can be found in polynomial time in odd-cycle symmetric weighted digraphs, which are odd-cycle symmetric digraphs with arc-weight such that the total weight of the arcs in an odd dicycle is equal to that of its reverse dicycle. The maximum-weight matching problem is straightforwardly reduced to the maximum-weight even factor problem in odd-cycle symmetric weighted digraphs. The assumption of odd-cycle symmetry of (weighted) digraphs is justified by its relation to discrete convexity. For more detail and references of the aforementioned results, the readers are referred to a survey paper [20].

The triangle-free 2-matching and simple square-free 2-matching problems are examples of the *restricted 2-matching problem*, a main objective of which is to provide a tight relaxation of the TSP. Let $G = (V, E)$ be an undirected graph which may have parallel edges but may not have loops. For a positive integer t, an edge set $F \subseteq E$ is called a *t-matching* (resp., *t-factor*) if every vertex in V has at most (resp., exactly) t incident edges in F. A 2-matching is called *triangle-free* if it excludes cycles of length three. Note that a triangle-free 2-matching may contain parallel edges. For the maximum-weight triangle-free 2-matching problem in which all parallel copies of each edge have the same weight, a combinatorial algorithm together with a totally dual integral formulation is presented in [3,18].

An edge set is called *simple* if it excludes parallel edges. If we restrict 2-matchings to be simple, the triangle-free 2-matching problem becomes much more complicated [9]. More generally, for a positive integer k, a simple 2-matching is called $C_{\leq k}$-*free* if it excludes cycles of length at most k. Finding a maximum simple $C_{\leq k}$-free 2-matching is NP-hard for $k \geq 5$, and is open for $k = 4$. In contrast, the simple $C_{\leq 4}$-free 2-matching problem becomes tractable in bipartite graphs. We often refer to a simple $C_{\leq 4}$-free 2-matching in a bipartite graph as a *square-free 2-matching*. Throughout this paper, a square-free 2-matching always means a simple $C_{\leq 4}$-free 2-matching in a bipartite graph, unless otherwise stated. For the square-free 2-matching problem in bipartite graphs, min-max theorems, combinatorial algorithms, and decomposition theorems are established. For the weighted case, while finding a maximum-weight square-free 2-matching in a bipartite graph is NP-hard, it is solvable in polynomial time if the weight is *vertex-induced* on each C_4 (see Sect. 2 for definition). This assumption on the weight is again justified by its relation to discrete convexity. See [21,22] for more detail on the aforementioned results.

It should be noted that Pap [19] presented combinatorial algorithms for the even factor and square-free 2-matching problems in the same paper. Indeed, these algorithms are based on similar techniques of shrinking odd cycles and C_4's, and may imply some similarity of these two problems. However, to the best of our knowledge, a comprehensive theory including both of these problems has not been proposed.

1.2   Our Contribution

In the present paper, we discuss $\mathcal{U}$-*feasible t-matchings*: for an undirected graph $G = (V, E)$ and $\mathcal{U} \subseteq 2^V$, a t-matching F is $\mathcal{U}$-*feasible* if it excludes a t-factor in U for each $U \in \mathcal{U}$ (see Definition 1 for a formal description). The optimal $\mathcal{U}$-feasible t-matching problem generalizes not only the $\mathcal{U}$-feasible 2-matching problem [22], but also all of the aforementioned generalizations of the matching problem. Thus, it could be recognized that $\mathcal{U}$-feasibility is a common generalization of the blossom constraint for the nonbipartite matching problem and the subtour elimination constraint for the TSP.

In this paper, we present a min-max theorem and an efficient combinatorial algorithm for the maximum $\mathcal{U}$-feasible t-matching problem in bipartite graphs under a plausible assumption on $(G, \mathcal{U}, t)$. These results commonly extend those for nonbipartite matchings, even factors, triangle-free 2-matchings, and square-free 2-matchings. We remark here that the $\mathcal{U}$-feasible t-matching problem in *bipartite* graphs is a generalization of the *nonbipartite* matching problem. Our algorithm runs in $O(t(|V|^3\alpha + |V|^2\beta))$ time, where α and β are the time for checking feasibility of an edge set and expanding the shrunk structures, respectively. The complexities α and β are typically small, i.e., constant or $O(|V|)$, in the above specific cases.

We further solve the maximum-weight $\mathcal{U}$-feasible t-matching problem in bipartite graphs, under the same assumption on $(G, \mathcal{U}, t)$ and a certain assumption on the edge weights. We establish a linear programming description with dual integrality and a primal-dual algorithm with running time $O(t(|V|^3(|E| + \alpha) + |V|^2 \beta))$.

Imposing some assumption on $(G, \mathcal{U}, t)$ would be reasonable in order to have $\mathcal{U}$-feasible t-matchings tractable. (Recall that it can describe Hamilton cycles.) Indeed, we assume for the excluded t-factors that the expanding technique is always valid (see Definition 3). This assumption is broad enough to include the instances reduced from nonbipartite matchings, even factors in odd-cycle symmetric digraphs, triangle-free 2-matchings in nonbipartite graphs, square-free 2-matchings, and simple $K_{t,t}$-free t-matchings in bipartite graphs.

In the weighted case, the assumption on the edge weights is that the weights are *vertex-induced* for each $U \in \mathcal{U}$ (see Definition 2). We note that this assumption exactly corresponds to the previous assumptions for the maximum-weight even factor, square-free 2-matching, and simple $K_{t,t}$-free t-matching problems. Those previous assumptions are plausible from the viewpoint of discrete convexity [14, 15]. This would be an example of a unified understanding of the previous results on even factors and square-free 2-matchings.

2   Our Framework

Let $G = (V, E)$ be an undirected graph which may have parallel edges. An edge e connecting $u, v \in V$ is denoted by $\{u, v\}$. If G is a digraph, then an arc from u to v is denoted by (u, v). For $X \subseteq V$, let $G[X] = (X, E[X])$ denote the subgraph of G induced by X, that is, $E[X] = \{\{u, v\} \mid u, v \in X, \{u, v\} \in E\}$. Similarly, for $F \subseteq E$, define $F[X] = \{\{u, v\} \mid u, v \in X, \{u, v\} \in F\}$. If $X, Y \subseteq V$ are disjoint, then $F[X, Y]$ denotes the set of edges in F connecting X and Y.

For $v \in V$, let $\delta(v) \subseteq E$ denote the set of edges incident to v. For $F \subseteq E$ and $v \in V$, let $\deg_F(v) = |F \cap \delta(v)|$. Recall that F is a t-*matching* if $\deg_F(v) \le t$ for each $v \in V$, and a t-*factor* if $\deg_F(v) = t$ for every $v \in V$.

Definition 1. *For a graph $G = (V, E)$ and $\mathcal{U} \subseteq 2^V$, a t-matching $F \subseteq E$ is called $\mathcal{U}$-feasible if $|F[U]| \le \lfloor (t|U| - 1)/2 \rfloor$ for each $U \in \mathcal{U}$.*

Equivalently, a t-matching F in G is not $\mathcal{U}$-feasible if $F[U]$ is a t-factor in $G[U]$ for some $U \in \mathcal{U}$. This concept is a generalization of that for $\mathcal{U}$-feasible 2-matchings introduced in [22].

In what follows, we consider the maximum $\mathcal{U}$-feasible t-matching problem, in which the goal is to find a $\mathcal{U}$-feasible t-matching F maximizing $|F|$. We further deal with the maximum-weight $\mathcal{U}$-feasible t-matching problem, in which the objective is to find a $\mathcal{U}$-feasible t-matching F maximizing $w(F) = \sum_{e \in F} w(e)$ for a given edge-weight vector $w \in \mathbf{R}_+^E$. For a vector $x \in \mathbf{R}^E$ and $F \subseteq E$, in general we denote $x(F) = \sum_{e \in F} x(e)$. In discussing the weighted version, we assume that w is *vertex-induced* on each $U \in \mathcal{U}$.

Definition 2. *For a graph $G = (V, E)$, a vertex subset $U \subseteq V$, and an edge-weight $w \in \mathbf{R}^E$, w is called* vertex-induced on U *if there exists a function $\pi_U : U \to \mathbf{R}$ on U such that $w(\{u, v\}) = \pi_U(u) + \pi_U(v)$ for each $\{u, v\} \in E[U]$.*

The reductions of the aforementioned generalizations of the matching problem to the $\mathcal{U}$-feasible t-matching problem appear in the full version of the paper.

3   Maximum $\mathcal{U}$-Feasible t-Matching

In this section, we exhibit a min-max theorem and a combinatorial algorithm for the maximum $\mathcal{U}$-feasible t-matching problem in bipartite graphs. Our algorithm commonly extends those for nonbipartite matchings [7], even factors [19], triangle-free 2-matchings [3], and square-free 2-matchings [10, 19].

As a preliminary, we present a weak duality relation. Let $G = (V, E)$ be an undirected graph and $\mathcal{U} \subseteq 2^V$. At this point, G do not need to be bipartite. For $X \subseteq V$, define $\mathcal{U}_X \subseteq \mathcal{U}$ and $C_X \subseteq X$ by $\mathcal{U}_X = \{U \in \mathcal{U} \mid U$ forms a component in $G[X]\}$, and $C_X = X \setminus \bigcup_{U \in \mathcal{U}_X} U$. Then the following inequality holds for an arbitrary $\mathcal{U}$-feasible t-matching F and $X \subseteq V$.

Lemma 1. *For an arbitrary $\mathcal{U}$-feasible t-matching $F \subseteq E$ and $X \subseteq V$, it holds that*

$$|F| \leq t|X| + |E[C_{V \setminus X}]| + \sum_{U \in \mathcal{U}_{V \setminus X}} \left\lfloor \frac{t|U| - 1}{2} \right\rfloor. \tag{1}$$

Proof. The lemma follows from

$$2|F[X]| + |F[X, V \setminus X]| \leq t|X|, \tag{2}$$

$$|F[V \setminus X]| \leq |E[C_{V \setminus X}]| + \sum_{U \in \mathcal{U}_{V \setminus X}} \left\lfloor \frac{t|U| - 1}{2} \right\rfloor. \tag{3}$$

3.1   Algorithm

From now on, we assume bipartiteness of the graph. Let $G = (V, E)$ be an undirected bipartite graph. Denote the two color classes of V by V^+ and V^-. For $X \subseteq V$, denote $X^+ = X \cap V^+$ and $X^- = X \cap V^-$. The endvertices of an edge $e \in E$ in V^+ and V^- are denoted by $\partial^+ e$ and $\partial^- e$, respectively.

We begin with the description of shrinking a forbidden structure $U \in \mathcal{U}$. For concise notation, we denote the input graph by $\hat{G} = (\hat{V}, \hat{E})$ and the graph in hand, i.e., the graph resulted from possibly repeated shrinkings, by $G = (V, E)$. Consequently, we have that $\mathcal{U} \in 2^{\hat{V}}$. Denote the solution in hand by $F \subseteq E$. Intuitively, shrinking of U consists of identifying all vertices in U^+ and in U^- to obtain new vertices $u_U{}^+$ and $v_U{}^-$, respectively, and deleting all the edges in $E[U]$. A formal description is as follows.

Procedure SHRINK(U). Let $u_U{}^+$ and $v_U{}^-$ be new vertices, and reset the end-vertices of an edge $e \in \hat{E} \setminus \hat{E}[U]$ with $\partial^+ e = u$ and $\partial^- e = v$ as $\partial^+ e := u_U{}^+$ if $u \in U^+$ and $\partial^- e := v_U{}^-$ if $v \in U^-$. Update G by $V^+ := (V^+ \setminus U^+) \cup \{u_U{}^+\}$, $V^- := (V^- \setminus U^-) \cup \{v_U{}^-\}$, and $E := E \setminus \hat{E}[U]$. Finally, $F := F \cap E$ and return (G, F).

We refer to a vertex $v \in V$ as a *natural vertex* if v is a vertex in the original graph $\hat{G}$, and as a *pseudovertex* if it is a newly added vertex in shrinking some $U \in \mathcal{U}$. We denote the set of the natural vertices by V_n, and that of the pseudovertices by V_p. For $X \subseteq \hat{V}$, define $X_n = X \cap V_n$ and $X_p = \bigcup\{u_U{}^+, v_U{}^- \mid U \subseteq X,\ u_U{}^+, v_U{}^- \in V_p\}$. For $X \subseteq V$, define $\hat{X} \subseteq \hat{V}$ by $\hat{X} = X_n \cup \bigcup\{U^+ \mid u_U{}^+ \in X \cap V_p\} \cup \bigcup\{U^- \mid v_U{}^- \in X \cap V_p\}$.

Procedure EXPAND(G, F) is to execute the reverse operation of SHRINK(U) for all shrunk $U \in \mathcal{U}$. A key point is that $\lfloor (t|U| - 1)/2 \rfloor$ edges are added to F from $\hat{E}[U]$ for each $U \in \mathcal{U}$.

Procedure EXPAND(G, F). Let $G := \hat{G}$. For each inclusionwise maximal $U \in \mathcal{U}$ which is shrunk, add $F_U \subseteq \hat{E}[U]$ of $\lfloor (t|U| - 1)/2 \rfloor$ edges to F, so that F is a $\mathcal{U}$-feasible t-matching in $\hat{G}$. Now return (G, F).

In Procedure EXPAND(G, F), the existence of F_U is not trivial. In order to attain that $\hat{F} = F \cup \bigcup\{F_U \mid U \in \mathcal{U}$ is a maximal shrunk set$\}$ is a t-matching in $\hat{G}$, $F \subseteq E$ and $F_U \subseteq \hat{E}[U]$ should satisfy

$$\deg_F(u) \leq \begin{cases} t & (u \in V_n), \\ 1 & (u \in V_p) \end{cases} \tag{4}$$

$$\deg_{F_U}(u) \begin{cases} = t - 1 & (u \in U \text{ is incident to an edge in } F[U, V \setminus U]), \\ \leq t & (u \in U \text{ is not incident to an edge in } F[U, V \setminus U]). \end{cases} \tag{5}$$

To achieve this, we maintain that F satisfies the degree constraint (4). Moreover, we assume that there exists F_U satisfying $|F_U| = \lfloor (t|U| - 1)/2 \rfloor$ and (5) for an arbitrary F with (4) and every maximal shrunk set $U \in \mathcal{U}$. This assumption is formally defined in the following way.

Definition 3. *Let $\hat{G} = (\hat{V}, \hat{E})$ be a bipartite graph, $\mathcal{U} \subseteq 2^{\hat{V}}$, and t be a positive integer. For arbitrary $U_1, \dots, U_l \in \mathcal{U}$ that are pairwise disjoint, let $G = (V, E)$ denote the graph obtained from $\hat{G}$ by executing SHRINK(U_1), ..., SHRINK(U_l), and let $F \subseteq E$ be an arbitrary edge set satisfying (4). We say that $(\hat{G}, \mathcal{U}, t)$ admits expansion if there exists $F_{U_i} \subseteq \hat{E}[U_i]$ satisfying $|F_{U_i}| = \lfloor (t|U_i| - 1)/2 \rfloor$ and (5) for each $i = 1, \dots, l$.*

In what follows we assume that $(\hat{G}, \mathcal{U}, t)$ admits expansion. This is exactly the class of $(G, \mathcal{U}, t)$ to which our algorithm is applicable.

Furthermore, we should take $\mathcal{U}$-feasibility of $\hat{F}$ into account. We refer to F in G as *feasible* if $\hat{F}$ is $\mathcal{U}$-feasible. If there are several possibilities of F_U, we say that F is $\mathcal{U}$-feasible if there is at least one $\mathcal{U}$-feasible $\hat{F}$. In other words, if F satisfying (4) is not feasible, then there exists $U \in \mathcal{U}$ such that

$$\deg_F(v) = \begin{cases} t & (v \in U_\mathrm{n}), \\ 1 & (v \in U_\mathrm{p}), \end{cases} \tag{6}$$

and $\hat{F}$ shall have a t-factor in $\hat{G}[U]$.

We are now ready for the entire description of our algorithm. The algorithm begins with $G = \hat{G}$ and an arbitrary $\mathcal{U}$-feasible t-matching $F \subseteq \hat{E}$, typically $F = \emptyset$. We first construct an auxiliary digraph.

Procedure AUXILIARYDIGRAPH(G, F). Construct a digraph (V, A) defined by $A = \{(u, v) | u \in V^+, v \in V^-, \{u, v\} \in E \backslash F\} \cup \{(v, u) | u \in V^+, v \in V^-, \{u, v\} \in F\}$. Define the sets of source vertices $S \subseteq V^+$ and sink vertices $T \subseteq V^-$ by $S = \{u \in V_\mathrm{n}^+ \mid \deg_F(u) \leq t - 1\} \cup \{u_U^+ \in V_\mathrm{p}^+ \mid \deg_F(u_U^+) = 0\}$ and $T = \{v \in V_\mathrm{n}^- \mid \deg_F(v) \leq t - 1\} \cup \{v_U^- \in V_\mathrm{p}^- \mid \deg_F(v_U^-) = 0\}$. Now return $D = (V, A; S, T)$.

Suppose that there exists a directed path $P = (e_1, f_1, \ldots, e_l, f_l, e_{l+1})$ in D from S to T. Note that $e_i \in E \backslash F$ $(i = 1, \ldots, l+1)$ and $f_i \in F$ $(i = 1, \ldots, l)$. Denote the symmetric difference $(F \backslash P) \cup (P \backslash F)$ of F and P by $F \triangle P$. If $F \triangle P$ is feasible, we execute AUGMENT(G, F, P) below, and then EXPAND(G, F).

Procedure AUGMENT(G, F, P). Let $F := F \triangle P$ and return F.

If $F \triangle P$ is not feasible, we apply SHRINK(U) after determining a set $U \in \mathcal{U}$ to be shrunk by the following procedure.

Procedure VIOLATINGSET(G, F, P). For $j = 1, \ldots, l$, define $F_j = (F \backslash \{f_1, \ldots f_j\}) \cup \{e_1, \ldots, e_j\}$. Also define $F_0 = F$ and $F_{l+1} = F \triangle P$. Let j^* be the minimum index j such that F_j is not feasible, let $U \in \mathcal{U}$ be an arbitrary set satisfying (6) for $F = F_{j^*}$. Now let $F := F_{j^*-1}$, and return (F, U).

Finally, if D does not have a directed path from S to T, we determine $X \subseteq \hat{V}$ minimizing the right-hand side of (1) as follows.

Procedure MINIMIZER(G, F). Let $R \subseteq V$ be the set of vertices reachable from S, and $X := (V^+ \backslash R^+) \cup R^-$. If a natural vertex $v \in V^- \backslash X$ has t edges in F connecting R^+ and v, then $X := X \cup \{v\}$. If a pseudovertex $v^-_U \in V^- \backslash X$ has one edge in F connecting R^+ and v^-_U, then $X := X \cup \{v^-_U\}$. Return $X := \hat{X}$.

We then apply EXPAND(G, F) and the algorithm terminates by returning $F \subseteq \hat{E}$ and $X \subseteq \hat{V}$.

3.2 Min-Max Theorem: Strong Duality

In this section, we strengthen Lemma 1 to be a min-max relation and prove the validity of our algorithm in Sect. 3.1. That is, we show that the output (F, X) of the algorithm satisfies (1) with equality. This constructively proves the following min-max relation for the class of $(G, \mathcal{U}, t)$ admitting expansion.

Theorem 1. *Let $G = (V, E)$ be a bipartite graph, $\mathcal{U} \subseteq 2^V$, and t be a positive integer such that $(G, \mathcal{U}, t)$ admits expansion. Then, the maximum size of a $\mathcal{U}$-feasible t-matching is equal to the minimum of*

$$\min_{X \subseteq V} \left\{ t|X| + |E[C_{V \setminus X}]| + \sum_{U \in \mathcal{U}_{V \setminus X}} \left\lfloor \frac{t|U| - 1}{2} \right\rfloor \right\}.$$

Proof. It suffices to prove equality in (2) and (3) for the algorithm output $(\hat{F}, \hat{X})$.

First, since X is defined based on reachability in the auxiliary digraph D, $F[X] = \emptyset$ holds when no directed path from S to T is found. Moreover, it is not difficult to see that $v_U^+ \in R$ holds for every pseudovertex v_U^+. Hence it follows that $\hat{F}[\hat{X}] = \emptyset$.

Second, for every $v \in \hat{X}$, $\deg_{\hat{F}}(v) = t$ holds, and thus (2) holds by equality.

Finally, edges in $\hat{G}[\hat{V} \setminus \hat{X}]$ are in F before the last EXPAND(G, F) or obtained by expanding pseudovertices $u_U{}^+$ and $v_U{}^-$, which are isolated vertices in $G[V \setminus X]$. This means that U forms a component in $\hat{G}[\hat{V} \setminus \hat{X}]$, and thus the equality in (3) follows.

4 Weighted $\mathcal{U}$-Feasible t-Matching

In this section, we extend the min-max theorem and the algorithm presented in Sect. 3 to the maximum-weight $\mathcal{U}$-feasible t-matching problem. Recall that G is a bipartite graph in which every edge may have parallel copies with the same weight, and $(G, \mathcal{U}, t)$ admits expansion. We assume that w is vertex-induced on each $U \in \mathcal{U}$, which commonly extends the assumptions for the maximum-weight square-free and even factor problems.

4.1 Linear Program

Described below is a linear programming relaxation of the maximum-weight $\mathcal{U}$-feasible t-matching problem, where the variable is $x \in \mathbf{R}^E$:

$$
\begin{aligned}
\text{(P)} \quad \text{maximize} \quad & \sum_{e \in E} w(e)x(e) \\
\text{subject to} \quad & x(\delta(v)) \le t && (v \in V), \\
& x(E[U]) \le \left\lfloor \frac{t|U| - 1}{2} \right\rfloor && (U \in \mathcal{U}), \\
& 0 \le x(e) \le 1 && (e \in E).
\end{aligned}
$$

We shall remark that the second constraint, describing $\mathcal{U}$-feasibility, is a common extension of the blossom constraint for the nonbipartite matching problem (put $t = 1$), and the subtour elimination constraints for the TSP (put $t = 2$).

Its dual program, where the variables are $p \in \mathbf{R}^V$, $q \in \mathbf{R}^E$, and $r \in \mathbf{R}^{\mathcal{U}}$, is given as follows:

(D) minimize $t \sum_{v \in V} p(v) + \sum_{e \in E} q(e) + \sum_{U \in \mathcal{U}} \left\lfloor \frac{t|U|-1}{2} \right\rfloor r(U)$

subject to $p(u) + p(v) + q(e) + \sum_{U \in \mathcal{U}:\, e \in E[U]} r(U) \geq w(e) \qquad (e = \{u, v\} \in E),$

$\qquad\qquad p(v) \geq 0 \qquad\qquad\qquad\qquad\qquad\qquad (v \in V),$
$\qquad\qquad q(e) \geq 0 \qquad\qquad\qquad\qquad\qquad\qquad (e \in E),$
$\qquad\qquad r(U) \geq 0 \qquad\qquad\qquad\qquad\qquad\qquad (U \in \mathcal{U}).$

Define $w' \in \mathbf{R}^E$ by $w'(e) = p(u) + p(v) + q(e) + \sum_{U \in \mathcal{U}:\, e \in E[U]} r(U) - w(e)$ for $e = \{u, v\} \in E$. The complementary slackness conditions for (P) and (D) are as follows.

$$x(e) > 0 \Longrightarrow w'(e) = 0 \qquad\qquad\qquad\qquad (e \in E), \qquad (7)$$

$$p(v) > 0 \Longrightarrow x(\delta(v)) = t \qquad\qquad\qquad\qquad (v \in V), \qquad (8)$$

$$q(e) > 0 \Longrightarrow x(e) = 1 \qquad\qquad\qquad\qquad (e \in E), \qquad (9)$$

$$r(U) > 0 \Longrightarrow x(E[U]) = \left\lfloor \frac{t|U| - 1}{2} \right\rfloor \qquad (U \in \mathcal{U}). \qquad (10)$$

4.2 Primal-Dual Algorithm

In this section, we exhibit a combinatorial primal-dual algorithm for the maximum-weight $\mathcal{U}$-feasible t-matching problem in bipartite graphs, where $(G, \mathcal{U}, t)$ admits expansion and w is vertex-induced for each $U \in \mathcal{U}$.

We maintain primal and dual feasible solutions satisfying (7), (9), (10), and (8) for every $v \in V^-$. The algorithm terminates when (8) is attained for every $v \in V^+$. Again denote the input graph by $\hat{G} = (\hat{V}, \hat{E})$, and the graph in hand, i.e., the graph resulted from possibly repeated shrinkings, by $G = (V, E)$. The variables in the algorithm are $F \subseteq E$, $p \in \mathbf{R}^{\hat{V}}$, $q \in \mathbf{R}^{\hat{E}}$, and $r \in \mathbf{R}^{\mathcal{U}}$. Note that p and q are always defined on the original vertex and edge sets, respectively.

In the beginning, we set

$$F = \emptyset, \qquad p(v) = \begin{cases} \max\{w(e) \mid e \in \delta(v)\} & (v \in V^+), \\ 0 & (v \in V^-), \end{cases} \qquad (11)$$

$$q(e) = 0 \quad (e \in E), \qquad r(U) = 0 \quad (U \in \mathcal{U}).$$

The auxiliary digraph D is constructed as follows. Major differences from Sect. 3.1 are that we only use an edge e with $w'(e) = 0$, and a vertex in V^+ can become a sink vertex.

Procedure AUXILIARYDIGRAPH(G, F, p, q, r). Define a digraph (V, A) by $A = \{(\partial^+ e, \partial^- e) \mid e \in E \setminus F, w'(e) = 0\} \cup \{(\partial^- e, \partial^+ e) \mid e \in F\}$. The sets of source vertices $S \subseteq V^+$ and sink vertices $T \subseteq V^+ \cup V^-$ are defined by

$$S = \{u \in V_n^+ \mid \deg_F(v) \le t - 1, \, p(u) > 0\}$$
$$\cup \{u_U^+ \in V_p^+ \mid \deg_F(u_U^+) = 0, \, p(u) > 0 \text{ for some } u \in U\}$$
$$T = \{v \in V_n^- \mid \deg_F(v) \le t - 1\} \cup \{v_U^- \in V_p^- \mid \deg_F(v_U^-) = 0\}$$
$$\cup \{u \in V_n^+ \mid \deg_F(u) = t, \, p(u) = 0\}$$
$$\cup \{u_U^+ \in V_p^+ \mid \deg_F(u_U^+) = 1, \, p(u) = 0 \text{ for some } u \in U\}.$$

Return $D = (V, A; S, T)$,

Suppose that D has a directed path P from S to T, and let $F' := \cdot F \triangle P$.

If F' is feasible, we execute AUGMENT(G, F, P), which is the same as in Sect. 3.1. Note that, if P ends in a vertex in $T \cap V^+$, then $|F|$ does not increase. However, in this case the number of vertices satisfying (8) increases by one, and we get closer to the termination condition ((8) for every vertex).

If F' is not feasible, apply VIOLATINGSET(G, F, P) as in Sect. 3.1. For the output U of VIOLATINGSET(G, F, P), execute MODIFY(G, F, U) below if $p(u) = 0$ for some $u \in U^+$. Otherwise apply SHRINK(U).

Procedure MODIFY(G, F, U). Let $u^* \in U^+$ satisfy $p(u^*) = 0$. Then find $K \subseteq E[U]$ such that

$$\deg_K(u) = \begin{cases} t & (u \in U_n^+ \setminus \{u^*\}), \\ t - 1 & (u = u^*), \\ 0 & (u = u_{U'}^+ \in U_p^+, u^* \in U'), \\ \deg_{F[U]}(u) & (u \in U_n^- \cup U_p^-). \end{cases}$$

Now return $F := (F \setminus F[U]) \cup K$.

If D does not have a directed path from S to T, then update the dual variables p, q, and r by procedure DUALUPDATE(G, F, p, q, r) described below.

Procedure DUALUPDATE(G, F, p, q, r). Let $R \subseteq V$ be the set of vertices reachable from S in the auxiliary digraph D. Then,

$$p(v) := \begin{cases} p(v) - \epsilon & (v \in \hat{R}^+), \\ p(v) + \epsilon & (v \in \hat{R}^-), \\ p(v) & (v \in \hat{V} \setminus \hat{R}), \end{cases}$$

$$q(e) := \begin{cases} q(e) + \epsilon & (\partial^+ e \in \hat{R}^+, \partial^- e \in \hat{V}^- \setminus \hat{R}^-), \\ q(e) & (v \in \hat{V}^- \setminus \hat{R}^-), \end{cases}$$

$$r(U) := \begin{cases} r(U) + \epsilon & (u_U^+ \in R^+, v_U^- \in V^- \setminus R^-), \\ r(U) - \epsilon & (u_U^+ \in V^+ \setminus R^+, v_U^- \in R^+), \\ r(U) & (\text{otherwise}), \end{cases}$$

where

$$\epsilon = \min\{\epsilon_1, \epsilon_2, \epsilon_3\}, \quad \epsilon_1 = \min\{w'(\{u, v\}) \mid u \in \hat{R}^+, v \in \hat{V}^- \setminus \hat{R}^-\},$$

$$\epsilon_2 = \min\{p(u) \mid u \in \hat{R}^+\}, \quad \epsilon_3 = \min\{r(U) \mid u_U^+ \in \hat{V}^+ \setminus \hat{R}^+, v_U^- \in \hat{R}^-\}.$$

Then return (p, q, r).

Finally, we expand every U satisfying $r(U) = 0$ after AUGMENT(G, F, P), MODIFY(G, F, U), and DUALUPDATE(G, F, p, q, r). We note that, if any $U' \subsetneq U$ satisfies $r_{U'} > 0$, which implies that U' had been shrunk before U was shrunk, then U' is maintained to be shrunk.

Procedure EXPAND(G, F, r). For each shrunk $U \in \mathcal{U}$ with $r(U) = 0$, execute the following procedures. Update G by replacing u_U^+ and v_U^- by the graph induced by $U_n \cup U_p$ just before SHRINK(U) is applied. Determine $F_U \subseteq E[U_n \cup U_p]$ of $(t|U_n| + |U_p|)/2 - 1$ edges so that $F' = F \cup F_U$ can be extended to a $\mathcal{U}$-feasible t-matching in $\hat{G}$. Then return $F := F'$.

The algorithm constructively proves the following theorem for the integrality of (P) and (D). This is a common extension of dual integrality theorems for nonbipartite matchings [6], even factors [13], triangle-free 2-matchings [3], and square-free 2-matchings [17].

Theorem 2. *If $(G, \mathcal{U}, t)$ admits expansion and w is vertex-induced on each $U \in \mathcal{U}$, then the linear program (P) has an integer optimal solution. Moreover, the dual program (D) also has an integer optimal solution such that $\{U \in \mathcal{U} \mid r(U) > 0\}$ is a laminar family.*

5   Conclusion

We have presented a new framework of the optimal $\mathcal{U}$-feasible t-matching problem. Then we have established a min-max theorem and a combinatorial algorithm under the reasonable assumption that G is bipartite, $(G, \mathcal{U}, t)$ admits expansion, and w is vertex-induced on each $U \in \mathcal{U}$. Our problem under these assumptions can describe a number of generalizations of the matching problem, such as the matching and triangle-free 2-matching problem in nonbipartite graphs, and the square-free 2-matching problem in bipartite graphs. We have also seen that $\mathcal{U}$-feasibility is a common generalization of the blossom constraints for the nonbipartite matching problem and the subtour elimination constraints for the TSP. We anticipate that this unified perspective provides a new approach to the TSP utilizing matching theory.

Acknowledgements. The author is obliged to Yutaro Yamaguchi for the helpful comments on the draft of the paper. The author is also thankful to anonymous referees for their careful reading and comments. This research is partially supported by JSPS KAKENHI Grant Number 16K16012.

References

1. Bérczi, K., Végh, L.A.: Restricted b-matchings in degree-bounded graphs. In: Eisenbrand, F., Shepherd, F.B. (eds.) IPCO 2010. LNCS, vol. 6080, pp. 43–56. Springer, Heidelberg (2010). doi:10.1007/978-3-642-13036-6_4
2. Boyd, S., Iwata, S., Takazawa, K.: Finding 2-factors closer to TSP tours in cubic graphs. SIAM J. Discrete Math. **27**, 918–939 (2013)
3. Cornuéjols, G., Pulleyblank, W.: A matching problem with side conditions. Discrete Math. **29**, 135–159 (1980)
4. Cunningham, W.H., Geelen, J.F.: The optimal path-matching problem. Combinatorica **17**, 315–337 (1997)
5. Cunningham, W.H., Geelen, J.F.: Vertex-disjoint dipaths and even dicircuits, unpublished (2001)
6. Cunningham, W.H., Marsh III, A.B.: A primal algorithm for optimum matching. Math. Program. Study **8**, 50–72 (1978)
7. Edmonds, J.: Paths, trees, and flowers. Can. J. Math. **17**, 449–467 (1965)
8. Frank, A.: Restricted t-matchings in bipartite graphs. Discrete Appl. Math. **131**, 337–346 (2003)
9. Hartvigsen, D.: Extensions of matching theory. Ph.D. thesis, Carnegie Mellon University (1984)
10. Hartvigsen, D.: Finding maximum square-free 2-matchings in bipartite graphs. J. Comb. Theor. Ser. B **96**, 693–705 (2006)
11. Iwata, S., Takazawa, K.: The independent even factor problem. SIAM J. Discrete Math. **22**, 1411–1427 (2008)
12. Kaiser, T., Škrekovski, R.: Cycles intersecting edge-cuts of prescribed sizes. SIAM J. Discrete Math. **22**, 861–874 (2008)
13. Király, T., Makai, M.: On polyhedra related to even factors. In: Bienstock, D., Nemhauser, G. (eds.) IPCO 2004. LNCS, vol. 3064, pp. 416–430. Springer, Heidelberg (2004). doi:10.1007/978-3-540-25960-2_31
14. Kobayashi, Y., Szabó, J., Takazawa, K.: A proof of Cunningham's conjecture on restricted subgraphs and jump systems. J. Comb. Theor. Ser. B **102**, 948–966 (2012)
15. Kobayashi, Y., Takazawa, K.: Even factors, jump systems, and discrete convexity. J. Comb. Theor. Ser. B **99**, 139–161 (2009)
16. Lovász, L., Plummer, M.D.: Matching Theory. AMS Chelsea Publishing, Providence (2009)
17. Makai, M.: On maximum cost $K_{t,t}$-free t-matchings of bipartite graphs. SIAM J. Discrete Math. **21**, 349–360 (2007)
18. Pap, G.: A TDI description of restricted 2-matching polytopes. In: Bienstock, D., Nemhauser, G. (eds.) IPCO 2004. LNCS, vol. 3064, pp. 139–151. Springer, Heidelberg (2004). doi:10.1007/978-3-540-25960-2_11
19. Pap, G.: Combinatorial algorithms for matchings, even factors and square-free 2-factors. Math. Program. **110**, 57–69 (2007)
20. Takazawa, K.: Even factors: algorithms and structure. RIMS Kôkyûroku Bessatsu, B23, pp. 233–252. Kyoto University, Research Institute for Mathematical Sciences (2010)
21. Takazawa, K.: Decomposition theorems for square-free 2-matchings in bipartite graphs. In: Mayr, E.W. (ed.) WG 2015. LNCS, vol. 9224, pp. 373–387. Springer, Heidelberg (2016). doi:10.1007/978-3-662-53174-7_27
22. Takazawa, K.: Finding a maximum 2-matching excluding prescribed cycles in bipartite graphs. Discrete Optim. (to appear)

Equilibrium Computation in Atomic Splittable Singleton Congestion Games

Tobias Harks[1] and Veerle Timmermans[2(✉)]

[1] Institute of Mathematics, University of Augsburg, 86135 Augsburg, Germany
`tobias.harks@math.uni-augsburg.de`
[2] Department of Quantitative Economics, Maastricht University,
6200 MD Maastricht, The Netherlands
`v.timmermans@maastrichtuniversity.nl`

Abstract. We devise the first polynomial time algorithm computing a pure Nash equilibrium for atomic splittable congestion games with singleton strategies and player-specific affine cost functions. Our algorithm is purely combinatorial and computes the *exact* equilibrium assuming rational input. The idea is to compute a pure Nash equilibrium for an associated *integrally-splittable* singleton congestion game in which the players can only split their demands in integral multiples of a common packet size. While integral games have been considered in the literature before, no polynomial time algorithm computing an equilibrium was known. Also for this class, we devise the first polynomial time algorithm and use it as a building block for our main algorithm.

1 Introduction

One of the core topics in algorithmic game theory is the complexity of computing equilibria. As pointed out by several researchers (e.g., [6,9]), the computational tractability of a solution concept contributes to its credibility as a plausible prediction of the outcome of competitive environments in practice. The most accepted solution concept in non-cooperative game theory is the Nash equilibrium – a strategy profile, from which no player wants to unilaterally deviate. While a Nash equilibrium generally exists only in mixed strategies, the practically important class of congestion games admits pure Nash equilibria, see Rosenthal [28]. In the classical model of Rosenthal, a pure strategy of a player consists of a subset of resources, and the congestion cost of a resource depends only on the number of players choosing the same resource. Over the last decade, the algorithmic game theory community has intensively studied the complexity of computing equilibria for congestion games. As the first seminal work in this area, Fabrikant et al. [11] showed that the problem of computing a pure Nash equilibrium is PLS-complete for network congestion games. Ackermann et al. [1] strengthened this result to hold even for network congestion games with linear cost functions. On the other hand, there are polynomial algorithms for symmetric network congestion games [11], for matroid congestion games with player-specific cost functions [1,2], for polymatroid congestion games with player-specific cost

© Springer International Publishing AG 2017
F. Eisenbrand and J. Koenemann (Eds.): IPCO 2017, LNCS 10328, pp. 442–454, 2017.
DOI: 10.1007/978-3-319-59250-3_36

functions and polynomially bounded demands [15,16] and for so-called total uni-modular congestion games [26]. For further results regarding the computation of approximate equilibria in congestion games see [4,5,7,29].

For *atomic splittable congestion games*, the problem of computing an equilibrium is much less explored in the literature. In such a game, there is a finite set of resources and a finite set of players. In addition, each player is associated with a positive demand and a collection of allowable subsets of resources. A strategy for a player is a (possibly fractional) distribution of the player-specific demand over the allowable subsets. This quite basic model has several applications, e.g., packet-routing in communication networks (see [20,21,25]), traffic networks [18] and logistics networks [8]. We are only aware of two works that derive a polynomial time algorithm for equilibrium computation: (1) For affine player-independent cost functions, there exists a convex potential whose global minima are pure Nash equilibria, see Cominetti et al. [8]. Thus, for any $\epsilon > 0$ one can compute an ϵ-approximate equilibrium in polynomial time by convex programming methods. (2) Huang [19] also considered affine player-independent cost functions, and he devised a combinatorial algorithm computing an exact equilibrium for routing games on symmetric s-t graphs that are so-called *well-designed*. This condition is met for instance by series-parallel graphs. His proof technique also uses the convex potential.

Our Results. We study atomic splittable *singleton* congestion games with player-specific affine cost functions and develop the first polynomial time algorithm computing a pure Nash equilibrium. From now on we use equilibrium as shortcut for pure Nash equilibrium. Our algorithm is purely combinatorial and computes an *exact* equilibrium. The main ideas and constructions are as follows. By analyzing the first order necessary optimality conditions of an equilibrium, it can be shown that any equilibrium is *rational* as it is a solution to a system of linear equations with rational coefficients (assuming rational input). Using that equilibria are unique for singleton games (see [3,27]), we further derive that the constraint matrix of the equation system is non-singular, allowing for an explicit representation of the equilibrium by Cramer's rule (using determinants of the constraint- and their sub-matrices). This way, we obtain an explicit lower bound on the minimum demand value for any used resource in the equilibrium. We further show that the unique equilibrium is also the unique equilibrium for an associated *integrally-splittable* game in which the players may only distribute the demands in *integer multiples* of a common *packet size* of some value $k^* \in \mathbb{Q}_{>0}$ over the resources. While we are not able to compute k^* exactly, we can efficiently compute some sufficiently small $k_0 \leq k^*$ with the property that an equilibrium for the k_0-integrally-splittable game allows us to determine the set of resources on which a player will put a positive amount of load in the atomic splittable equilibrium. Once these *support sets* are known, an atomic splittable equilibrium can be computed in polynomial time by solving a system of linear equations. This way, we can reduce the problem of computing the exact equilibrium for an atomic splittable game to computing an equilibrium for an associated k_0-integrally-splittable game. Integrally-splittable congestion games have been studied before by Tran-Tanh et al. [30] for the case of player-independent convex cost functions and later by Harks et al. [15,16] for polymatroid strategy spaces and player-specific convex cost functions. In particular, Harks et al. devised an algorithm with running time $n^2 m(\delta/k_0)^3$, where n is the number of players, m the number of resources, and δ is an upper bound on the demand of the players (cf. Corollary 5.2 [16]). As δ is encoded in binary, however, the algorithm is only pseudo-polynomial even for player-specific affine cost functions.

We devise a polynomial time algorithm for integrally-splittable singleton congestion games with player-specific affine cost functions. Our algorithm works as follows. For a game with initial packet size k_0, we start by finding an equilibrium for packet size $k = k_0 \cdot 2^q$ for some q of order $O(\log(\delta/k_0))$, satisfying only a part of the player-specific demands. Then we repeat the following two actions:

1. We half the packet size from k to $k/2$ and construct a $k/2$-equilibrium using the k-equilibrium. Here, a k-equilibrium denotes an equilibrium for an integrally-splittable game with common packet size k. We show that this can be done in polynomial time by repeatedly performing the following operations given a k-equilibrium: (a) Among players who can improve, we find the player that benefits most by moving one packet of size $k/2$; (b) If necessary, we perform a sequence of backward-shuffles of packets to correct the *load decrease* caused by the first packet movement (this is called a *backward path*); (c) If necessary, we perform a sequence of forward-shuffles of packets to correct the load increase caused by the first packet movement (this is called a *forward path*). (a)–(c) is iterated until a $k/2$-equilibrium for the currently scheduled demand is reached. For strategy profile x we define $\Delta(x)$ to be a vector that contains the cost for moving one packet to the currently cheapest resource, for each combination of a player and resource. We show that after each iteration $\Delta(x)$ lexicographically increases, which implies that we converge to a $k/2$-equilibrium.

2. For each player i we repeat the following step: if the current packet size k is smaller than the currently unscheduled demand of player i, we add one more packet for this particular player to the game and recompute the equilibrium. This part of the algorithm has also been used in the algorithm in [15,30].

After q iterations, we have scheduled all demands and obtain an equilibrium for the desired packet size k_0. Polynomial running time of the algorithm is shown by several structural results on the sensitivity of equilibria with respect to packet sizes $2k$ and k. Specifically, we derive bounds on the difference of the resulting global load vectors as well as the individual load vectors of players. We use these insights to show that $\Delta(x)$ reaches a lexicographical maximum in a polynomial number of steps. Overall, compared to the existing algorithms of [15,30], our algorithm has two main innovations: packet sizes are decreased exponentially (yielding polynomial running time in δ) and k-equilibrium computation for an intermediate packet size k is achieved via a careful construction of a sequence of single packet movements (backward- and forward paths) from a given $2k$-equilibrium (ensuring its polynomial length).

Related Work. Atomic splittable network congestion games with player-independent cost functions have been studied (seemingly independent) by Orda et al. [25] and Haurie and Marcotte [18] and Marcotte [23]. Both lines of research mentioned that Rosens' existence result for concave games on compact strategy spaces implies the existence of pure Nash equilibria. Marcotte [23] proposed four numerical algorithms computing a pure Nash equilibrium and he shows local convergence results. Meunier and Pradeau [24] developed a pivoting-algorithm (similar to Lemke's algorithm) for nonatomic network congestion games with affine player-specific cost functions. Gairing et al. [12] considered nonatomic routing games on parallel links with affine player-specific cost functions. They developed a convex potential function that can be minimized within arbitrary precision in polynomial time. Deligkas et al. [10] considered general concave games with compact action spaces and investigated algorithms computing an approximate equilibrium. They discretize the compact strategy space and use the Lipschitz constants of utility functions to show that only a finite number of representative strategy profiles need to be considered for obtaining an approximate equilibrium (see also Lipton et al. [22] for a similar approach). The running time of the algorithm, however, depends on an upper bound of the norm of strategy vectors, thus, implying only a pseudo-polynomial algorithm for our setting.

2   Preliminaries

Atomic Splittable Singleton Games. An atomic splittable singleton congestion game is given by the tuple: $\mathcal{G} := (N, E, (d_i)_{i \in N}, (E_i)_{i \in N}, (c_{i,e})_{i \in N, e \in E_i})$, where $E = \{e_1, \ldots, e_m\}$ is a finite set of resources and $N = \{1, \ldots, n\}$ is a finite set of players. Each player $i \in N$ is associated with a demand $d_i \in \mathbb{Q}_{\geq 0}$ and a set of allowable resources $E_i \subseteq E$. A strategy for player $i \in N$ is a (possibly fractional) distribution of the demand d_i over the singletons in E_i. Thus, one can represent the strategy space of every player $i \in N$ by the polytope: $\mathcal{S}_i(d_i) := \{x_i \in \mathbb{R}_{\geq 0}^{|E_i|} \mid \sum_{e \in E_i} x_{i,e} = d_i\}$. The combined strategy space is denoted by $\mathcal{S} := \prod_{i \in N} \mathcal{S}_i(d_i)$ and we denote by $x = (x_i)_{i \in N}$ the overall strategy profile. We define $x_{i,e} := (x_i)_e$ as the load of player i on $e \in E_i$ and $x_{i,e} = 0$ when $e \in E \setminus E_i$. The total load on e is given as $x_e := \sum_{i \in N} x_{i,e}$. Resources have player-specific affine cost functions $c_{i,e}(x_e) = a_{i,e} x_e + b_{i,e}$ with $a_{i,e} \in \mathbb{Q}_{>0}$ and $b_{i,e} \in \mathbb{Q}_{\geq 0}$ for all $i \in N$ and $e \in E_i$. The total cost of player i in strategy distribution x is defined as: $\pi_i(x) = \sum_{e \in E} c_{i,e}(x_e) x_{i,e}$. For $i \in N$, we write $\mathcal{S}_{-i}(d_{-i}) = \prod_{j \neq i} \mathcal{S}_j(d_j)$ and $x = (x_i, x_{-i})$ meaning that $x_i \in \mathcal{S}_i(d_i)$ and $x_{-i} \in \mathcal{S}_{-i}(d_{-i})$. A strategy profile x is an *equilibrium* if $\pi_i(x) \leq \pi_i(y_i, x_{-i})$ for all $i \in N$ and $y_i \in \mathcal{S}_i(d_i)$. A pair $(x, (y_i, x_{-i})) \in \mathcal{S} \times \mathcal{S}$ is called an *improving move* of player i, if $\pi_i(x_i, x_{-i}) > \pi_i(y_i, x_{-i})$. The marginal cost for player i on resource e is defined as: $\mu_{i,e}(x) = c_{i,e}(x_e) + x_{i,e} c'_{i,e}(x_e) = a_{i,e}(x_e + x_{i,e}) + b_{i,e}$.

Lemma 1 (cf. Harks [14]). *Strategy profile x is an equilibrium if and only if the following holds: when $x_{i,e} > 0$, then $\mu_{i,e}(x) \leq \mu_{i,e'}(x)$ for all $e' \in E_i$.*

Using that the strategy space is compact and cost functions are convex, Kakutanis' fixed point theorem implies the existence of an equilibrium. Uniqueness is proven by Richmann and Shimkin [27] and Bhaskar et al. [3].

Game $\mathcal{G}$ is called symmetric whenever $E_i = E$ for all $i \in N$. We can project any asymmetric game $\mathcal{G}$ on a symmetric game $\mathcal{G}'$ by setting $b_{i,e}$ sufficiently high whenever $e \notin E_i$. Therefore, in the rest of this paper only symmetric games are considered. Details can be found in the full version of this paper [17].

Integral Singleton Games. A k-integral game is given by the tuple $\mathcal{G}_k := (N, E, (d_i)_{i \in N}, (c_{i,e})_{i \in N, e \in E})$ with $k \in \mathbb{Q}_{>0}$. Here, players cannot split their load fractionally, but only in multiples of k. Assume d_i is a multiple of k, then the strategy space for player i is the following set: $S_i(d_i, k) := \{x_i \in \mathbb{Q}_{\geq 0}^E \mid x_{i,e} = kq, q \in \mathbb{N}_{\geq 0}, \sum_{e \in E} x_{i,e} = d_i\}$. In this game, k is also called the *packet size*. When E, N and $(c_{i,e})_{i \in N, e \in E}$ are clear from the context, we write $\mathcal{G}_k((d_i)_{i \in N})$ instead.

For player-specific affine cost functions the (discrete) marginal increase and decrease are defined as follows: $\mu_{i,e}^{+k}(x) = (x_{i,e} + k)c_{i,e}(x_e + k) - x_{i,e}c_{i,e}(x_e) = ka_{i,e}(x_e + x_{i,e} + k)$ and $\mu_{i,e}^{-k}(x) = x_{i,e}c_{i,e}(x_e) - (x_{i,e} - k)c_{i,e}(x_e - k) = ka_{i,e}(x_e + x_{i,e} - k)$ if $x_{i,e} > 0$ and $-\infty$ otherwise.

Lemma 2 (cf. Groenevelt [13]). *Strategy profile x is an equilibrium in a k-integral congestion game if and only if: when $x_{i,e} > 0$, then $\mu_{i,e}^{-k}(x) \leq \mu_{i,e'}^{+k}(x)$ for all $e' \in E$.*

Define $\mu_{i,\min}^{+k}(x) := \min_{e \in E_i}\{\mu_{i,e}^{+k}(x)\}$ and $\mu_{i,\max}^{-k}(x) := \max_{e \in E_i}\{\mu_{i,e}^{-k}(x)\}$. Then strategy profile x is an equilibrium in a k-integral congestion game if and only if $\mu_{i,\max}^{-k}(x) \leq \mu_{i,\min}^{+k}(x)$ for all $i \in N$.

3   Reduction to Integrally-Splittable Games

We show that the problem of finding an atomic splittable equilibrium reduces to the problem of finding a k_0-splittable equilibrium for some $k_0 \in \mathbb{Q}_{>0}$.

Theorem 1. *Let x be the unique equilibrium of an atomic splittable singleton game $\mathcal{G}$. Then, there exist $k^* \in \mathbb{Q}_{>0}$ such that x is also the unique equilibrium for the k^*-integral splittable game $\mathcal{G}_{k^*}$.*

Proof. We define the support set I_i for each player as $I_i := \{e \in E \mid x_{i,e} > 0\}$. Lemma 1 implies that if x is an equilibrium, and $x_{i,e} > 0, x_{i,e'} > 0$, then $\mu_{i,e}(x) = \mu_{i,e'}(x)$. Define $p := \sum_{i \in N} |I_i| \leq nm$. Then, if the correct support set I_i of each player is known, the equilibrium can be computed by solving the following set of p linear equations on p variables: (a) Every player $i \in N$ should satisfy its demand-constraint: $\sum_{e \in I_i} x_{i,e} = d_i$. (b) For every player $i \in N$, there are $|I_i| - 1$ equations of type $\mu_{i,e}(x) = \mu_{i,e'}(x)$ for $e, e' \in I_i$, which we write as $a_{i,e}(x_e + x_{i,e}) - a_{i,e'}(x_{e'} + x_{i,e'}) = b_{i,e} - b_{i,e'}$. Note that x_e is not an extra variable, but an abbreviation for $\sum_{i \in N} x_{i,e}$. From now on we refer to this set of equalities as the system $Ax = b$. Note that as the equilibrium exists and is unique, A is non-singular. Using Cramer's Rule, the unique solution is given by:

$x_{i,e} = |\det(A_{i,e})|/|\det(A)|$, where $A_{i,e}$ is the matrix formed by replacing the column that corresponds to value $x_{i,e}$ in A by b.

We define $\mathcal{I} := \{\{a_{i,e}, b_{i,e} \mid i \in N, e \in E_i\} \cup \{d_i \mid i \in N\} \cup \{1\}\}$, $a_{max} :=$ $\max(\mathcal{I})$ and $a_{gcd} := \max\{a \in \mathbb{Q}_{>0} \mid \forall b \in \mathcal{I}, \exists \ell \in \mathbb{N} \text{ such that } b = a \cdot \ell)\}$. Then, as all values in A and b depend on adding and subtracting values in $\mathcal{I}$, $|\det(A_{i,e})|$ is an integer multiple of $(a_{gcd})^p$ and, hence, an integer multiple of $(a_{gcd})^{nm}$. Thus, all player-specific loads are an integer multiple of $(a_{gcd})^{nm}/|\det(A)|$ and, if we define $k^* = (a_{gcd})^{nm}/|\det(A)|$, x is an equilibrium for the k^*-integral splittable game. The proof that k^*-splittable equilibria are unique can be found in [17]. $\square$

Note that we do not know matrix A beforehand, but we do know that $|A_{i,j}| \leq 2a_{max}$. Using Hadamard's inequality we find that $|\det(A)| \leq (2a_{max})^{nm}(nm)^{nm/2}$. Hence, $k^* \geq (a_{gcd}^{nm})/((2a_{max})^{nm}(nm)^{nm/2})$. For the atomic splittable equilibrium x and any k-integral-splittable equilibrium x_k, we first prove that $|x_e - (x_k)_e| < mk$ and $|x_{i,e} - (x_k)_{i,e}| < m^2 k$ (Lemma 3.2 and 3.2 in the full version of this paper [17]). Then, given the equilibrium for some sufficiently small k_0, we are able compute the correct supports set of each player in the atomic splittable equilibrium, and compute the exact atomic splittable equilibrium.

Theorem 2. *Given an atomic splittable congestion game $\mathcal{G}$ and an equilibrium x_{k_0} for k_0-splittable game $\mathcal{G}_{k_0}$, where: $k_0 = (a_{gcd}^{nm})/(2m^2\lceil(2a_{max})^{nm}(nm)^{nm/2}\rceil)$, we can compute in $O((nm)^3)$ the exact atomic splittable equilibrium x for $\mathcal{G}$.*

Proof. First note that all demands d_i are integer multiples of k_0, as d_i is an integer multiple of a_{gcd}, and both $2m^2$ and $\lceil(2a_{max})^{nm}(nm)^{nm/2}\rceil$ are integer. Theorem 1 implies that there exists a k^* such that the atomic splittable equilibrium is also an equilibrium for the k^*-integral splittable game, and that $k_0 \leq k^*/(2m^2)$. We check for each $i \in N$ and $e \in E$ if $(x_{k_0})_{i,e} \geq m^2 k_0$. If this is the case, we prove that $x_{i,e} > 0$. On the contrary, if we assume that $x_{i,e} = 0$, then $(x_{k_0})_{i,e} - x_{i,e} \geq m^2 k_0$, which contradicts the fact that $|x_{i,e} - (x_k)_{i,e}| < m^2 k$. Thus, $x_{i,e} > 0$. On the other hand, if $(x_{k_0})_{i,e} < m^2 k_0$, we can use a similar argument to conclude that $x_{i,e} = 0$. Hence, given an equilibrium (x_{k_0}) for k_0-splittable game $\mathcal{G}_{k_0}$, we can compute the correct support sets I_i for all $i \in N$, where $I_i := \{e \in E \mid (x_{k_0})_{i,e} \geq m^2 k_0\}$. Given the correct support sets, we can easily compute the correct, exact equilibrium by solving the system $Ax = b$ of at most nm linear equations in running time $O((nm)^3)$. $\square$

It is left to compute an equilibrium x_{k_0} for integral game $\mathcal{G}_{k_0}$ with packet size $k_0 = a_{gcd}^{nm}/(2m^2\lceil(2a_{max})^{nm}(nm)^{nm/2}\rceil)$. Such integral games have been studied in the literature before, see Harks et al. [15,16]. In particular, [16, Algorithm 1] has running time $O(nm(\delta/k_0)^3)$. Here δ is an upper bound on the player specific demands. In general, δ is not bounded in k_0, thus, the running time is not polynomially bounded in the size of the input.

4   A Polynomial Algorithm for Integral Games

We develop a *polynomial time* algorithm that computes an equilibrium for any k-integral splittable singleton game with player-specific affine cost functions. We use elements of [15, Algorithm 1] to construct a new algorithm with running time $O(n^2 m^{14} \log(\delta/k))$. We first introduce some new notation. For two strategy profiles x and y we denote their Hamming distance by $H(x, y) := \sum_{e \in E} |x_e - y_e|$ and for two vectors $x_i, y_i \in \mathbb{R}^{|E|}$ we denote their Hamming distance by $H(x_i, y_i) := \sum_{e \in E} |x_{i,e} - y_{i,e}|$. For two resources $e^-, e^+ \in E$ with $y_{i,e^-} = x_{i,e^-} - k$, $y_{i,e^+} = x_{i,e^+} + k$ and $y_{i,e} = x_{i,e}$ for all $e \in E \backslash \{e^-, e^+\}$, we denote $(x_i)_{e^- \to e^+} := y_i$. If x is a strategy profile for some game $\mathcal{G}_k$ and $y_i = (x_i)_{e^- \to e^+}$, we denote $x_{i:e^- \to e^+} := (y_i, x_{-i})$. We define a *restricted best response*:

Definition 1. *Let x be a strategy profile for game $\mathcal{G}_k((d_i)_{i \in N})$. Assume there exists $e^-, e^+ \in E$ such that $\mu_{i,\max}^{-k}(x) > \mu_{i,\min}^{+k}(x)$, $e^- \in \arg\max\{\mu_{i,e}^{-k}(x)\}$ and $e^+ \in \arg\min\{\mu_{i,e}^{+k}(x)\}$. Then, we term strategy $y_i = (x_i)_{e^- \to e^+}$ a restricted best response to x for player i.*

Note that when y_i is a restricted best response to x_i, $H(x_i, y_i) = 2k$. We first describe two subroutines, termed ADD and RESTORE. The first subroutine, ADD, is described in [17, Algorithm 1] and consists of lines 4–10 of [15, Algorithm 1]. Given an equilibrium x_k for game $\mathcal{G}_k((d_i)_{i \in N})$, it computes an equilibrium for the game, where the demand for player j is increased by k. This new packet is placed on a resource $e' \in \min_{e \in E}\{\mu_{j,e}^{+k}(x_k)\}$. In effect, the load on resource e' increases and only those players using $x_{i,e'} > 0$ can potentially decrease their cost by a deviation. In this case, Harks et al. proved in [15, Theorem 3.2] that a best response can be obtained by a restricted best response moving a packet away from e', decreasing the marginal cost for all other players on this resource to their original level. Thus, only one packet is moved throughout, preserving the invariant that only players using a resource to which the packet is moved may have an incentive to profitably deviate.

The second subroutine, RESTORE, takes as input an equilibrium x_{2k} for packet size $2k$ and game $\mathcal{G}_k((d_i)_{i \in N})$, and constructs an equilibrium for packet size k. This algorithm makes use of two sub-algorithms: [17, Algorithms 2 and 5], which determine a backward path- and a forward path of restricted best responses respectively. In a backward path we are given a resource e_1^- and a strategy profile x_1^b. In iteration q, we decide if there exists a player that has a restricted best response from some e_{q+1}^- to e_q^- where we obtain x_{q+1}^b from x_q^b. If no player has a restricted best response to resource e_q^-, we check if $(x_q^b)_{e_q^-} > (x_{2k})_{e_q^-} - 2mk$. If so, we end our backward path. Else, we look for a player that has an improving move in which she shifts one packet from some e_{q+1}^- to e_q^-, and then continue the backward path. Note that in each step we preserve the invariant that $H(x_1^b, x_q^b) \in \{0, 2k\}$. A forward path is very similar to a backward path, but we change the perspective. Thus, given a resource e_q^+ and a strategy profile x_q^f, we check in iteration q if there exists a player that has a restricted best response from e_q^+ to some e_{q+1}^+. Both algorithms (back- and forward path) can be seen as a special instantiation of a general restricted best response dynamic (cf. [15,30]).

We are now ready to define subroutine RESTORE. Initialize x by equilibrium x_{2k}. While x is not an equilibrium for $\mathcal{G}_k$, we iterate the following. Among players who can improve, we find the player j that benefits most from a restricted best response. We carry out a restricted best response for player j and move a packet from some resource e_1^- to some e_1^+. Then we compute a backward path, starting in resource e_1^-. If the resulting strategy profile has Hamming distance zero with x, we stop this iteration and overwrite x by the resulting strategy profile. Else, we compute a forward path, starting in e_1^+ and overwrite x by the resulting strategy profile. The pseudo-code of subroutine RESTORE can be found in Algorithm 1.

Input: equilibrium x_{2k} for $\mathcal{G}_{2k}((d_i')_{i \in N})$
1 $x \leftarrow x_{2k}$;
2 **while** x *not an equilibrium for* $\mathcal{G}_k((d_i')_{i \in N})$ **do**
3 $\quad j \leftarrow \arg \max_{i \in N} \{ \mu_{i,\min}^{+k}(x) - \mu_{i,\max}^{-k}(x) \}$;
4 $\quad$ Choose $e_1^- \in \arg \max \{ \mu_{j,e}^{-k}(x) \}$ and $e_1^+ \in \arg \min \{ \mu_{j,e}^{+k}(x) \}$;
5 $\quad x_1^b \leftarrow x_{j:e_1^- \to e_1^+}$;
6 $\quad (x_{q_b}^b, e_{q_b}^-) \leftarrow \mathrm{BP}(x_k, x_1^b, e_1^-, \mathcal{G}_k)$;
7 $\quad$ **if** $e_1^+ \neq e_{q_b}^-$ **then**
8 $\quad \quad x_1^f \leftarrow x_{q_b}^b$;
9 $\quad \quad (x_{q_f}^f, e_{q_f}^+) \leftarrow \mathrm{FP}(x_k, x_1^f, e_1^+, \mathcal{G}_k)$;
10 $\quad \quad x' \leftarrow x_{q_f}^f$;
11 $\quad$ **else**
12 $\quad \quad x' \leftarrow x_{q_b}^b$;
13 $\quad$ **end**
14 $\quad x \leftarrow x'$;
15 **end**
16 **return** x;

Algorithm 1. Subroutine RESTORE$(x, k, (d_i')_{i \in N}, \mathcal{G})$

Using the subroutines ADD and RESTORE we develop PACKETHALVER, which computes an equilibrium x_{k_0} for the k_0-splittable game $\mathcal{G}_{k_0}((d_i)_{i \in N})$. In this algorithm we start with an equilibrium x_k for $\mathcal{G}_k((d_i')_{i \in N})$, where $d_i' = 0$ for all $i \in N$, $k = 2^{q_1} k_0$ and $q_1 = \arg \min_{q \in \mathbb{N}} \{ 2^q k_0 > \max_{i \in N} d_i \}$. Note that this game has a trivial equilibrium, where $(x_k)_{i,e} = 0$ for all $i \in N$ and $e \in E$. We repeat the following two steps: (a) Given an equilibrium x_k for $\mathcal{G}_k((d_i')_{i \in N})$, we construct an equilibrium for $\mathcal{G}_{k/2}((d_i')_{i \in N})$ using subroutine RESTORE and set k to $k/2$. (b) For each player $i \in N$ we check if $d_i - d_i' \geq k$. If so, we increase d_i' by k and recompute equilibrium x_k using subroutine ADD. After q_1 iterations PACKETHALVER returns an equilibrium x_{k_0} for $\mathcal{G}_{k_0}((d_i)_{i \in N})$. The pseudo-code of PACKETHALVER can be found in Algorithm 2.

Input: Integral splittable congestion game $\mathcal{G}_{k_0}((d_i)_{i\in N})$.

```
1  q_1 ← arg min_{q∈ℕ}{2^q k_0 > max_{i∈N} d_i}; k ← 2^{q_1} k_0; d'_i ← 0; x_k ← (0)_{e∈E,i∈N};
2  for 1,...,q_1 − 1 do
3  |    k ← k/2;
4  |    x_k ← RESTORE(x_{2k}, 𝒢_k((d'_i)_{i∈N}));
5  |    for i ∈ N do
6  |    |    if d_i − d'_i > k then
7  |    |    |    x_k ← ADD(x_k, i, 𝒢_k((d'_i)_{i∈N}));
8  |    |    |    d'_i ← d'_i + k;
9  |    |    end
10 |    end
11 end
12 return x_k;
```

Algorithm 2. Algorithm PACKETHALVER($\mathcal{G}_{k_0}((d_i)_{i\in N})$)

5   Correctness

We prove that PACKETHALVER returns an equilibrium for game $\mathcal{G}_{k_0}((d_i)_{i\in N})$. In order to do so, we first need to verify that the two subroutines ADD and RESTORE are correct. Subroutine ADD is proven to be correct by Harks, Peis, and Klimm [15], thus, it is left to verify correctness of RESTORE and PACKETHALVER. To verify the correctness of subroutine RESTORE($x_{2k}, \mathcal{G}_k((d_i)_{i\in N})$), we need to prove that RESTORE terminates. We define: $\Delta(x) := (\mu_{i,\min}^{+k}(x) - \mu_{i,e}^{-k}(x))_{i\in N; e\in E}$. Let $\Delta_{\min}(x)$ be the minimum value in $\Delta(x)$. Note that when all elements in $\Delta(x)$ are non-negative, or, equivalently, when $\Delta_{\min}(x)$ is non-negative, x is an equilibrium. Our goal is to show that after each iteration in the while-loop (lines 2–15 of RESTORE) $\Delta(x)$ increases according to a certain lexicographical order defined as follows. Given two vectors $u, v \in \mathbb{R}^n$, we say that v is *sorted lexicographically larger* than u, if there is an index $k \in \{1,...,n\}$ such that $u_{\phi(i)} = v_{\psi(i)}$ for all $i < k$ and $u_{\phi(k)} < v_{\psi(k)}$, where ϕ and ψ are permutations that sort u and v non-decreasingly. We write $u <_{lex} v$. If $u_{\phi(i)} = v_{\psi(i)}$ for all $i \in \{1,...,n\}$, we write $u =_{lex} v$.

Proving that $\Delta(x)$ sorted lexicographically increases implies that RESTORE does not cycle, and thus, as the strategy space is finite, terminates. In general, under the hypothesis that $\Delta(x)$ lexicographically increases, we obtain the following strategy profiles within a while-loop (lines 2–15 of RESTORE): $x \to x_1^b \to x_2^b \to \cdots \to x_{q_b}^b = x_1^f \to x_2^f \to \cdots \to x_{q_f}^f = x'$. We introduce two types of vectors that help us prove that $\Delta(x) <_{lex} \Delta(x')$. Assume that in iteration q of the backward path a player moves a packet from e_{q+1}^- to e_q^-. We define:

$$B_{i,e}^{q,-k}(x) = \begin{cases} \mu_{i,e}^{-k}(x) + k^2 a_{i,e} & \text{if } e_q^- \neq e_1^+ \text{ and } e = e_q^-, \\ \mu_{i,e}^{-k}(x) - k^2 a_{i,e} & \text{if } e_q^- \neq e_1^+ \text{ and } e = e_1^+, \\ \mu_{i,e}^{-k}(x) & \text{otherwise.} \end{cases}$$

$$B_{i,\min}^{q,+k}(x) = \begin{cases} \mu_{i,\min}^{+k}(x) + k^2 a_{i,e} & \text{if } e_q^- \neq e_1^+ \text{ and } e_q^- = \arg\min \mu_{i,e}^{+k}(x), \\ \mu_{i,\min}^{+k}(x) - k^2 a_{i,e} & \text{if } e_q^- \neq e_1^+ \text{ and } e_1^+ = \arg\min \mu_{i,e}^{+k}(x), \\ \mu_{i,\min}^{+k}(x) & \text{otherwise.} \end{cases}$$

As k is fixed within RESTORE, we write $B_{i,e}^{q,-}(x)$ and $B_{i,\min}^{q,+}(x)$ instead. We define: $B^q(x) = (B_{i,\min}^{q,+}(x) - B_{i,e}^{q,-}(x))_{i\in N; e\in E}$. Similarly, assume that in iteration q of the forward path a packet is moved from e_q^+ to e_{q+1}^+. We define:

$$F_{i,e}^{q,-}(x) = \begin{cases} \mu_{i,e}^{-k}(x) - k^2 a_{i,e} & \text{if } e = e_q^+, \\ \mu_{i,e}^{-k}(x) & \text{otherwise.} \end{cases}$$

$$F_{i,\min}^{q,+}(x) = \begin{cases} \mu_{i,\min}^{+k}(x) - k^2 a_{i,e_q^f} & \text{if } e_q^+ = \arg\min \mu_{i,e}^{+k}(x), \\ \mu_{i,\min}^{+k}(x) & \text{otherwise.} \end{cases}$$

We define: $F^q(x) = (F_{i,\min}^{q,+}(x) - F_{i,e}^{q,-}(x))_{i\in N; e\in E}$.

In order to show $\Delta(x) <_{lex} \Delta(x')$, we first prove that $\Delta(x) <_{lex} B^1(x_1^b) <_{lex} \cdots <_{lex} B^{q_b}(x_1^b)$ and $F^1(x_1^f) <_{lex} \cdots <_{lex} F^{q_f}(x_{q_f}^f)$. Hence, the backward path and the forward path end after a finite number of steps. Then we connect $\Delta(x), B^q(x)$ and $F^q(x)$ by proving that if $\Delta(x) <_{lex} B^{q_b}(x_{q_b}^b)$, then $\Delta(x) <_{lex} F^1(x_{q_b}^b)$, and if $\Delta(x) <_{lex} F^{q_f}(x_{q_f}^f)$, then $\Delta(x) <_{lex} \Delta(x_{q_f}^f)$. The formal statements and their proofs can be found the full version of this paper [17].

Lemma 3. *Let x and x' be defined as in the while-loop (lines 2–15) of RESTORE. Then $\Delta(x) <_{lex} \Delta(x')$, and $\Delta_{\min}(x)$ occurs less in $\Delta(x')$ than in $\Delta(x)$.*

The proof of Lemma 3 can be found in the full version of this paper [17]. As $\Delta(x)$ lexicographically increases after each loop, RESTORE terminates. It is left to prove that PACKETHALVER returns an equilibrium for game $\mathcal{G}_{k_0}((d_i)_{i\in N})$.

Theorem 3. *Given a k_0-integral splittable singleton game with affine player-specific cost functions $\mathcal{G}_{k_0}$, PACKETHALVER returns an equilibrium for $\mathcal{G}_{k_0}$.*

Proof. Strategy profile x' is initialized as the all-zero strategy profile, hence, an equilibrium for the game $\mathcal{G}_{2^{q_1}k_0}(\mathbf{0})$. Assume that in iteration q we enter the for-loop in PACKETHALVER with an equilibrium x for game $\mathcal{G}_{2^{q_1-q+1}k_0}$ with demands $d_i' = d_i - (d_i \mod 2^{q_1-q+1}k_0)$. Algorithm RESTORE computes an equilibrium for packet size $2^{q_1-q}k_0$ and demands $d_i' = d_i - (d_i \mod 2^{q_1-q}k_0)$. In lines 5–10 of PACKETHALVER we check for each player $i \in N$ if her unscheduled load satisfies $d_i - d_i' \geq 2^{q_1-q}k_0$. If so, we schedule one extra packet for player i using subroutine ADD. Thus, after the q'th iteration in the for-loop, we obtain an equilibrium for packet size $2^{q_1-q}k_0$ and demands $d_i' = d_i - (d_i \mod 2^{q_1-q}k_0)$. Hence, after the q_1'th iteration we obtain an equilibrium for packet size $2^0 k = k$ and demands $d_i' = d_i - (d_i \mod k) = d_i$, which is an equilibrium for game $\mathcal{G}_{k_0}((d_i)_{i\in N})$. $\square$

6   Running Time

We prove that the running time of PACKETHALVER is polynomially bounded in n, m, $\log k$ and $\log \delta$, where δ is the upper bound on player specific demands d_i. For this, we first need to analyze the running time of the two subroutines ADD and RESTORE. In [16] Harks et al. proved that it takes time $nm(\delta/k)^2$ to execute ADD. If their algorithm is applied to games with singleton strategy spaces and player-specific affine cost functions, the running time reduces to $O(nm^4)$. Proofs for this statement can be found in the full version of this paper [17].

We analyze the running time of RESTORE. The crucial idea is that for each strategy profile y (for a game with packet size k) obtained during the execution of RESTORE, we have both $|(y_e - (x_{2k})_e| \leq 2mk$ and $|y_{i,e} - (x_{2k})_{i,e}| < 2m^2k$ for all $i \in N$ and $e \in E$. This enable us to prove that a backward and forward path of restricted best responses are found within in polynomial time, and that at most $O(nm^6)$ iterations are needed for RESTORE to terminate. This results in the following lemma:

Lemma 4. RESTORE *has running time* $O(n^2m^{14})$.

All lemma's and proofs needed to prove this statement can be found in the full version of this paper [17]. Finally, we prove the following theorem.

Theorem 4. PACKETHALVER *runs in time* $O(n^2m^{14}\log(\delta/k_0))$.

Proof. Note that we picked $q_1 \in \mathbb{N}$ to be the smallest number such that $2^{q_1}k_0$ exceeds d_i for all player-specific demands d_i. This implies that q_1 is bounded in $O(\log(\delta/k_0))$, where δ is an upper bound on the player-specific demands. Thus, we execute lines 3–10 $O(\log(\delta/k_0))$ times. In line 4 we call RESTORE, which runs in $O(n^2m^{14})$. In line 5–9 we execute ADD (which runs in $O(nm^6)$) at most n times. Thus, the computation time of lines 5–10 is $O(n^2m^6)$. This implies that it takes time $O(n^2m^{14})$ to go through a complete iteration in the for loop. Thus, PACKETHALVER runs in time $O(n^2m^{14}\log(\delta/k_0))$.   □

In the full version of this paper [17] we show that $\log(1/k_0)$ is polynomially bounded in the input. Thus, we can compute an atomic splittable equilibrium by first computing the k_0 splittable equilibrium using the algorithm above. Then, we compute the exact equilibrium in time $O((nm)^3)$. Thus, we can compute an atomic splittable equilibrium in running time: $O\left((nm)^3 + n^2m^{14}\log\left(\frac{\delta}{k_0}\right)\right)$.

References

1. Ackermann, H., Röglin, H., Vöcking, B.: On the impact of combinatorial structure on congestion games. J. ACM **55**(6), 1–22 (2008)
2. Ackermann, H., Röglin, H., Vöcking, B.: Pure Nash equilibria in player-specific and weighted congestion games. Theoret. Comput. Sci. **410**(17), 1552–1563 (2009)
3. Bhaskar, U., Fleischer, L., Hoy, D., Huang, C.-C.: Equilibria of atomic flow games are not unique. Math. Oper. Res. **40**(3), 634–654 (2015)

4. Caragiannis, I., Fanelli, A., Gravin, N., Skopalik, A.: Efficient computation of approximate pure Nash equilibria in congestion games. In: FOCS, Palm Springs, CA, USA, pp. 532–541 (2011)
5. Caragiannis, I., Fanelli, A., Gravin, N., Skopalik, A.: Approximate pure Nash equilibria in weighted congestion games: existence, efficient computation, and structure. ACM Trans. Econ. Comput. **3**(1), 2 (2015)
6. Chen, X., Deng, X., Teng, S.-H.: Settling the complexity of computing two-player Nash equilibria. J. ACM **56**(3), 14:1–14:55 (2009)
7. Chien, S., Sinclair, A.: Convergence to approximate Nash equilibria in congestion games. Games Econom. Behav. **71**(2), 315–327 (2011)
8. Cominetti, R., Correa, J.R., Stier-Moses, N.E.: The impact of oligopolistic competition in networks. Oper. Res. **57**(6), 1421–1437 (2009)
9. Daskalakis, C., Goldberg, P.W., Papadimitriou, C.H.: The complexity of computing a Nash equilibrium. SIAM J. Comput. **39**(1), 195–259 (2009)
10. Deligkas, A., Fearnley, J., Spirakis, P.: Lipschitz continuity and approximate equilibria. In: Gairing, M., Savani, R. (eds.) SAGT 2016. LNCS, vol. 9928, pp. 15–26. Springer, Heidelberg (2016). doi:10.1007/978-3-662-53354-3_2
11. Fabrikant, A., Papadimitriou, C., Talwar, K.: The complexity of pure Nash equilibria. In: Babai, L. (ed.) STOC, pp. 604–612 (2004)
12. Gairing, M., Monien, B., Tiemann, K.: Routing (un-)splittable flow in games with player-specific linear latency functions. ACM Trans. Algorithms **7**(3), 1–31 (2011)
13. Groenevelt, H.: Two algorithms for maximizing a separable concave function over a polymatroid feasible region. Eur. J. Oper. Res. **54**(2), 227–236 (1991)
14. Harks, T.: Stackelberg strategies and collusion in network games with splittable flow. Theory Comput. Syst. **48**, 781–802 (2011)
15. Harks, T., Klimm, M., Peis, B.: Resource competition on integral polymatroids. In: Liu, T.-Y., Qi, Q., Ye, Y. (eds.) WINE 2014. LNCS, vol. 8877, pp. 189–202. Springer, Cham (2014). doi:10.1007/978-3-319-13129-0_14
16. Harks, T., Klimm, M., Peis, B.: Sensitivity analysis for convex separable optimization over integral polymatroids (2016). https://arxiv.org/pdf/1611.05372.pdf
17. Harks, T., Timmermans, V.: Equilibrium computation in atomic splittable singleton congestion games (2016). https://arxiv.org/pdf/1612.00190.pdf
18. Haurie, A., Marcotte, P.: On the relationship between Nash-Cournot and Wardrop equilibria. Networks **15**, 295–308 (1985)
19. Huang, C.-C.: Collusion in atomic splittable routing games. Theory Comput. Syst. **52**(4), 763–801 (2013)
20. Korilis, Y., Lazar, A., Orda, A.: Capacity allocation under noncooperative routing. IEEE Trans. Aut. Contr. **42**(3), 309–325 (1997)
21. Korilis, Y.A., Lazar, A.A., Orda, A.: Architecting noncooperative networks. IEEE J. Sel. Areas Commun. **13**(7), 1241–1251 (1995)
22. Lipton, R.J., Markakis, E., Mehta, A.: Playing large games using simple strategies. In: Proceedings 4th ACM Conference on Electronic Commerce (EC-2003), San Diego, California, USA, June 9–12, pp. 36–41 (2003)
23. Marcotte, P.: Algorithms for the network oligopoly problem. J. Oper. Res. Soc. **38**(11), 1051–1065 (1987)
24. Meunier, F., Pradeau, T.: A lemke-like algorithm for the multiclass network equilibrium problem. In: Chen, Y., Immorlica, N. (eds.) WINE 2013. LNCS, vol. 8289, pp. 363–376. Springer, Heidelberg (2013). doi:10.1007/978-3-642-45046-4_30
25. Orda, A., Rom, R., Shimkin, N.: Competitive routing in multi-user communication networks. IEEE/ACM Trans. Network. **1**, 510–521 (1993)

26. A. D. Pia, M. Ferris, and C. Michini. Totally unimodular congestion games. In: Proceedings of the 28th Annual ACM-SIAM Symposium on Discrete Algorithms (to appear, 2017)
27. Richman, O., Shimkin, N.: Topological uniqueness of the Nash equilibrium for selfish routing with atomic users. Math. Oper. Res. **32**(1), 215–232 (2007)
28. Rosenthal, R.: A class of games possessing pure-strategy Nash equilibria. Internat. J. Game Theor. **2**(1), 65–67 (1973)
29. Skopalik, A., Vöcking, B.: Inapproximability of pure Nash equilibria. In: Proceedings of the 40th Annual ACM Syposium Theory Computing, pp. 355–364 (2008)
30. Tran-Thanh, L., Polukarov, M., Chapman, A., Rogers, A., Jennings, N.: On the existence of pure strategy Nash equilibria in integer-splittable weighted congestion games. In: Persiano, G. (ed.) SAGT, pp. 236–253 (2011)

Author Index

Printed in the United States
By Bookmasters